# LINEAR ALGEBRA

## A Concrete Introduction

SECOND EDITION

# LINEAR ALGEBRA

## A Concrete Introduction

**Dennis M. Schneider**
Knox College, Galesburg, IL

with the assistance of
**Manfred Steeg**
**Frank H. Young**

**Macmillan Publishing Company**
New York
**Collier Macmillan Publishers**
London

Macmillan Publishing Company
866 Third Avenue, New York, New York 10022

Collier Macmillan Canada, Inc.

**Library of Congress Cataloging-in-Publication Data**

Schneider, Dennis M.
Linear algebra, a concrete introduction.

Includes index.
1. Algebras, Linear. I. Steeg, Manfred.
II. Young, Frank H. (Frank Hood),
III. Title.
QA184.S37 1987 512'.5 85-15327
ISBN 0-02-406910-8

Printing: 12345678 Year: 789012345

ISBN 0-02-406910-8

To Jean, Bob, and Mark

# Preface

Linear algebra is deeply rooted in analytic geometry and systems of linear equations. Until the last twenty-five or thirty years, students were first introduced to this material in courses on analytic geometry and the theory of equations. When they subsequently studied an abstract treatment of linear algebra, they were equipped to deal with it because they were already familiar with some of the concrete problems that gave birth to the subject.

Recently, all of this has changed. Courses on analytic geometry and the theory of equations have long since vanished from the curriculum. As a result, most students enter a course on linear algebra with a very modest understanding of vectors and analytic geometry in $R^2$ and $R^3$ and perhaps some computational skills with matrices and linear systems. This is not a sufficient background for studying an abstract presentation of linear algebra. *Before students can understand and appreciate an abstraction, they must have a clear understanding of what is being abstracted and appreciate the need for the abstraction.*

My aim in writing this book is to present the abstract concepts of linear algebra as growing naturally out of concrete problems. The development of the material moves in a pedagogically sensitive way from the concrete to the abstract. Abstract concepts are introduced after the student has a clear understanding of what is being abstracted and appreciates the need for the abstraction. Thus students are introduced to the process of abstraction as a way of making it easier to conceptualize material and cope with problems that have already been encountered. They are also introduced to the process of understanding and constructing mathematical arguments.

In view of these remarks, we chose as our point of departure systems of linear equations. In Chapter 1 we introduce vectors, matrices, and develop the Gaussian elimination process. Gaussian elimination is first developed for the simplest and most important matrices, the nonsingular matrices. We define a matrix to be nonsingular if it can be reduced by forward elimination to a triangular matrix having no zeros on its diagonal. Once in this form, back substitution easily produces a solution. Echelon matrices are then introduced as the next simplest class of matrices that can be handled by back substitution, and then we show that forward elimination will reduce any matrix to an echelon matrix. The concepts of existence and uniqueness of solutions emerge as a major theme of the chapter.

Chapter 2 begins by introducing subspaces of $R^n$ as a natural generalization to higher dimensions of lines and planes in $R^3$. The question of existence of solutions of a linear system motivates the concept of a spanning set. Similarly, the question of uniqueness of solutions of a linear system motivates the concept of linear independence. Thus existence and uniqueness correspond to spanning and independence (i.e., a basis). The dimension of a subspace is introduced as a measure of its size and leads naturally to the concept of coordinates.

In Chapter 3 we use inconsistent systems and the geometrical construction of dropping a perpendicular from a point to a line or plane to motivate the need to extend the concepts of length and orthogonality to higher-dimensional spaces.

Chapter 4 (which requires calculus and may be omitted) extends the concepts of Chapters 2 and 3 to the function space setting.

In Chapter 5 we again return to the question of existence and uniqueness. Existence is used to motivate the concepts of the image of a linear transformation and uniqueness to motivate the definition of a one-to-one linear transformation. Sections 5.3 and 5.4 (which may be omitted) deal with the way a matrix represents a linear transformation. Section 5.5 (which may also be omitted) introduces eigenvalues and eigenvectors of a linear transformation and discusses the possibility of representing a linear transformation by a diagonal matrix.

It is not until Chapter 6 (which may be omitted or covered simultaneously with Chapters 4 and 5) that we finally give an abstract definition of a vector space, an inner product space, and a linear transformation between two abstract vector spaces. At this point the reader is prepared to appreciate how these definitions provide a single conceptual framework for dealing with problems in linear algebra.

Chapter 7 covers determinants. Sections 7.1 and 7.2 contain the material needed for Chapter 8.

Eigenvalues and eigenvectors are discussed in Chapter 8. Our discussion focuses on the problem of diagonalizing a matrix and the spectral theorem for symmetric matrices. This material leads naturally to a discussion of quadratic forms. In Section 8.8 (which requires Section 5.5) we bring together our results on eigenvalues and eigenvectors of linear transformations and matrices. In the last four sections of this chapter, we apply our results to systems of linear differential equations.

Chapter 9 provides an introduction to some numerical techniques that are useful for solving problems in linear algebra with a computer. The material in this chapter may be integrated with the rest of the text if desired.

All matters concerning complex numbers are left to Chapter 10. There we extend the material in the text to $C^m$ and prove the spectral theorem for Hermitian matrices. The material in this chapter may be integrated with the rest of the text if desired.

The concepts of linear algebra permeate all areas of pure and applied mathematics. The subject provides an excellent introduction to abstract mathematics and the process of understanding and constructing mathematical arguments. Its connection with other areas of science, social science, and technology provide a wealth of applications. The applications that are presented in the text should provide the reader with a sense of the vast scope and rich nature of the subject. We believe that the theory and the applications illuminate one another. Not only does a knowledge of the theory help one to understand the applications, but a knowledge of the applications helps one to understand the theory. The applications that we have chosen are real, not artificial. Each application is carefully motivated, explained, and developed.

There are a wealth of problems in the book. The problem sets begin with computational problems and are followed by more substantial problems. Answers to problems marked with an asterisk are provided at the end of the book.

Since we have found that at this stage most students find set theoretical notation and the sigma notation more of a hindrance than a help, we have avoided their use in the text. We have also avoided using mathematical induction.

The book is organized to provide a great deal of flexibility. There is more material in this book than can be covered in a one-semester or one-quarter course. The material in Chapters 1, 2, and 3, Sections 5.1 and 5.2, Sections 7.1 and 7.2, and Sections 8.1 through 8.6 form the core of any course in linear algebra. Time available and other considerations will determine what, if any, additional material will be covered. We have included a dependency diagram to help in the selection of additional material.

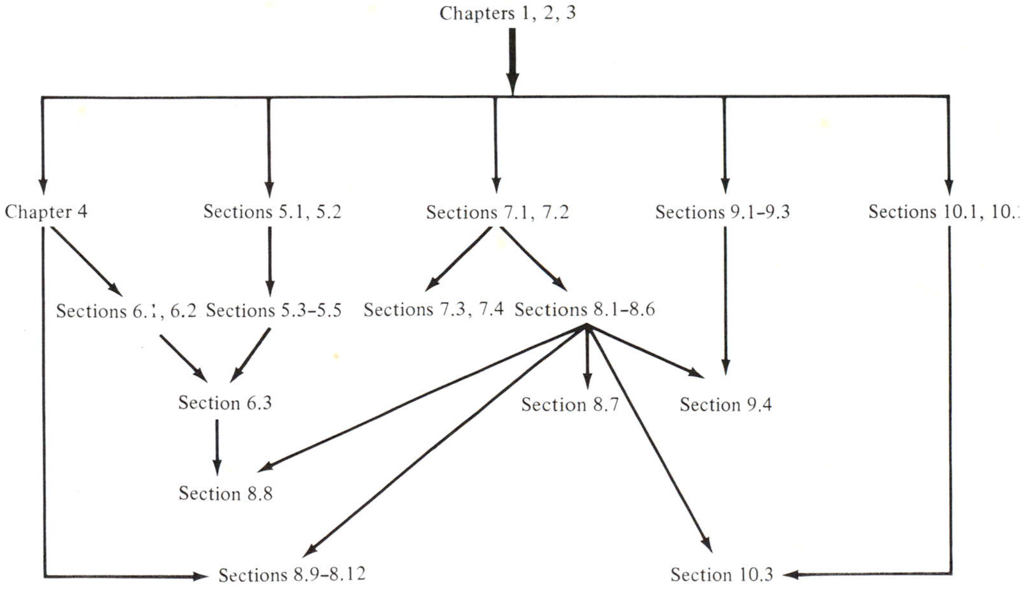


Instructors who wish to teach a course that stresses the computational aspects and/or the applications of linear algebra will want to omit many of the proofs given in the text. There is plenty of material to support this type of course. Finally, the book can be used for both a calculus-based and a non-calculus-based course.

Probably no book is ever written without a lot of help and encouragement from others. Certainly, this book was not. I am grateful to Manfred Steeg and Frank Young for their contributions to the first edition. To the students at Knox College who have suffered through using photocopies of this book in the various stages of its development, I wish to express my sincere thanks. Thanks also to the many readers (students, colleagues, and reviewers) whose thoughtful comments and criticisms have greatly improved the text. Special thanks to Tony Horowitz and Kevin MacNeil. They read the entire manuscript, and their input improved significantly the quality of the book. Thanks are also due to Kevin MacNeil, Linda Brands, Becky Halm, Miltiades Georgiou, and Rou Su Hsing for assistance in solving the problems; to Ruby James, Mavis Meadows, and Jon Ripperger for their excellent typing of the manuscript; and to Karen Fealey, Miltiades Georgiou, Joanne Lynch, and Mary Vlastnik for their careful proofreading of the manuscript. I am indebted to the entire staff of Macmillan, especially Gary Ostedt, Mathematics Editor, and J. Edward Neve and Mariann Cortissoz, Production Supervisors, for their professional work and total cooperation. Finally, I thank my wife, Jean, for her assistance, encouragement, and support with all aspects of this work. To her, and my two sons, Bob and Mark, I dedicate this book.

D.M.S.

# Contents

## 3
## Inconsistent Systems, Inner Products, and Projections 162



## 4
## Function Spaces 216



## 5
## Linear Transformations 238



## 6
## Abstract Vector Spaces 301



## 7
## Determinants 323

# 8
## Eigenvalues and Eigenvectors 347



# 9
## Numerical Methods 419



# 10
## The Vector Space $C^m$ 437

# LINEAR ALGEBRA

## A Concrete Introduction

# 1

# Systems of Linear Equations

Linear algebra is deeply rooted in the theory of systems of linear equations. The formulation of many problems in mathematics and its applications lead to a system of simultaneous linear equations. Gaussian elimination provides a systematic and computationally efficient method for solving linear systems. We first develop this procedure for the most important and simplest systems. These are systems of $n$ equations in $n$ unknowns that have a unique solution. Gaussian elimination is then extended to systems of $m$ equations in $n$ unknowns. The procedure produces all solutions of a system that has solutions and it identifies those systems that do not have solutions.

## 1.1 Systems of Linear Equations

Since we are going to be talking a great deal about systems of linear equations, we had best make sure that we know what a single linear equation is. Our old friend, the general equation of a line in the plane,

$$ax + by = c,$$

is an example of what is called a linear equation in the variables $x$ and $y$. (Every line in the plane can be written in this form; in contrast, the more familiar equation $y = mx + b$ excludes vertical lines.) A linear equation in the variables $x_1$ and $x_2$ is an equation that can be expressed in the form

$$a_1x_1 + a_2x_2 = b.$$

This is the equation of a line in the $x_1x_2$-plane. In general, a **linear equation in the $n$ variables** (or **unknowns**) $x_1, x_2, \ldots, x_n$ is an equation that can be expressed in the form

$$a_1x_1 + a_2x_2 + \cdots + a_nx_n = b,$$

where the **coefficients** $a_1, a_2, \ldots, a_n$ and the **constant term** $b$ are numbers. (*Warning:* When $n > 2$ a linear equation no longer represents a line.) We adopt the usual notation of using subscripts because this makes it easier to understand and manipulate equations involving several variables and we do not have to worry about running out of letters. Some examples of linear equations are:

$$x_1 + 7x_2 = 3 \qquad x_1 - 3x_2 + x_4 = \frac{5}{2}$$
$$0.5x_1 = 3x_2 - 7 \qquad x_1 + \phantom{3x_2+} x_3 + x_5 = 1$$

Some examples of equations that are not linear are:

$$x_1 + x_1x_3 = 5 \qquad x_1^2 + 2x_2 + x_3 = 4$$
$$e^{x_1} + x_2 = 2 \qquad \frac{1}{x_1} + x_2 = 0$$

Any equation that contains a power of a variable ($x_i^r$, where $r \neq 1$), a product of two or more variables (e.g., $x_ix_j$), or one or more variables as arguments for trigonometric, logarithmic, or exponential functions is not a linear equation.

One natural problem that arises about a linear equation is to find all solutions of the equation. By a **solution** of a linear equation we mean a collection of values of the variables which when substituted into the equation makes the equation true. For example, $x_1 = 2$, $x_2 = -1$ is a solution of the equation $x_1 + x_2 = 1$. Another solution is $x_1 = 0, x_2 = 1$. Any two numbers that add up to 1 will work. However, if we specify one of the variables, the other variable is uniquely determined. For example, if we choose $x_2 = 5$, then $x_1$ must be $-4$; and if we choose $x_1 = 3$, $x_2$ must equal $-2$. The situation is that one variable is free to take on any value and the other variable is determined by the value of this variable. The variable that is allowed to take on any value is called the **free variable** and the variable that is determined in terms of the free variable is called the **determined variable**. The choice of the determined variable is not unique . To standardize terminology, we will always choose the first or leading variable in an equation to be the determined variable and the remaining variables to be free variables. Any solution to our equation can be described by

$$x_1 = 1 - x_2, \qquad x_2 = x_2.$$

The first equation says that $x_1$ is determined by $x_2$ and the second equation $x_2 = x_2$ indicates that $x_2$ is free to assume any value. Since any solution can be found from these equations, we say that they give the **general solution** of our equation.

Let us pursue for a moment an important relationship between an equation like

$$x_1 + 5x_2 = 7$$

and its general solution

$$x_1 = -5x_2 + 7, \qquad x_2 = x_2.$$

This equation and its general solution both describe the same set of points in the $x_1x_2$-plane, but in essentially different ways. The equation $x_1 + 5x_2 = 7$ describes a set $S$ of points $(x_1, x_2)$ in the plane by describing the relationship which $x_1$ and $x_2$ must satisfy to belong to the set. In other words, it constrains membership in the set $S$ by requiring the coordinates of a point $(x_1, x_2)$ to satisfy the equation $x_1 + 5x_2 = 7$. For example, (2, 1) belongs to the set $S$ because $2 + 5 \cdot 1 = 7$, while (2, 2) does not belong to the set $S$ because $2 + 5 \cdot 2 \neq 7$. For this reason, a linear equation $a_1x_1 + a_2x_2 = b$ is often thought of as a **constraint equation**.

On the other hand, the general solution

$$x_1 = -5x_2 + 7, \qquad x_2 = x_2$$

describes the same set $S$, but in a totally different way. With these equations we generate the solutions of the equation $x_1 + 5x_2 = 7$ by assigning an arbitrary value to $x_2$ and then computing $x_1$ from the equation $x_1 = -5x_2 + 7$. The free variable $x_2$ is called a **parameter** and the equations

$$x_1 = -5x_2 + 7, \qquad x_2 = x_2$$

are called **parametric equations**. (The parameter $x_2$ can actually be assigned another name, say $t$, and the equations written in the form

$$x_1 = -5t + 7, \qquad x_2 = t.)$$

One more example. Given the equation $x_1 - x_2 + 4x_3 = 1$, we choose the first variable $x_1$ to be the determined variable and the remaining variables ($x_2$ and $x_3$) to be the free variables. The general solution of the equation is given by

$$x_1 = x_2 - 4x_3 + 1, \qquad x_2 = x_2, \qquad x_3 = x_3.$$

Returning to the familiar setting of the plane, suppose that we are given two lines $l_1$ and $l_2$ and asked to determine whether the lines intersect. After a moment's reflection about the geometry of a plane we conclude that there are three possibilities (see Figure 1.1):

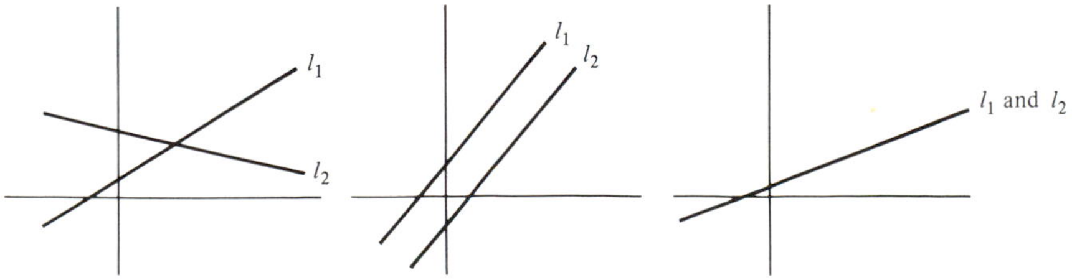


Figure 1.1

1. The lines are not parallel and hence intersect in a unique point.
2. The lines are parallel and hence do not intersect.
3. The lines coincide and hence intersect everywhere.

Now let's consider the same problem from an algebraic point of view. Suppose that we are given two linear equations (i.e., equations for our lines $l_1$ and $l_2$)

$$\begin{aligned} a_{11}x_1 + a_{12}x_2 &= b_1 \\ a_{21}x_1 + a_{22}x_2 &= b_2 \end{aligned} \tag{1}$$

and asked to determine whether there are numbers $\alpha_1$ and $\alpha_2$ such that $x_1 = \alpha_1$ and $x_2 = \alpha_2$ satisfy *both* of these equations. Since such a pair of numbers $\alpha_1$ and $\alpha_2$ exists if and only if the point $(\alpha_1, \alpha_2)$ in the $x_1x_2$-plane lies on both lines, our geometric reasoning leads us to conclude that there are three possibilities:

1. There is a unique solution of the system.
2. There is no solution of the system.
3. There are infinitely many solutions of the system.

An example of two equations that have a unique solution is

$$\begin{aligned} x_1 + \phantom{3}x_2 &= 4 \\ 2x_1 + 3x_2 &= 9. \end{aligned}$$

Geometrically, the first equation is the equation of a line with slope $-1$ and $x_2$-intercept 4, and the second equation is the equation of a line with slope $-\frac{2}{3}$ and $x_2$-intercept 3. (Here we have assumed that the horizontal axis is the $x_1$-axis and the vertical axis is the $x_2$-axis.) It is easy to verify that the lines intersect at $(3, 1)$ and hence $x_1 = 3$, $x_2 = 1$ is a solution of the system of equations. In fact, it is the only solution. For if we multiply the first equation by $-2$ and add it to the second equation, the second equation becomes $x_2 = 1$. Substituting this value of $x_2$ into the first equation gives $x_1 = 3$.

An example of two parallel lines is

$$\begin{aligned} x_1 + 3x_2 &= 1 \\ x_1 + 3x_2 &= 2. \end{aligned}$$

Both equations represent lines with slope $-\frac{1}{3}$, but the $x_2$-intercept of the first line is $\frac{1}{3}$ while that of the second line is $\frac{2}{3}$. Hence the lines do not intersect. It is also very easy to verify algebraically that the two equations do not have a solution. For if $x_1 = \alpha_1$, $x_2 = \alpha_2$ were a solution to both equations, then 1 would be equal to 2, since both are equal to $\alpha_1 + 3\alpha_2$. This is clearly impossible.

Finally, an example of two lines that coincide is

$$\begin{aligned} x_1 + \phantom{2}x_2 &= 1 \\ 2x_1 + 2x_2 &= 2. \end{aligned}$$

Both lines have slope $-1$ and $x_2$-intercept 1. Algebraically, the second equation is simply 2 times the first equation. Thus if $x_1 = \alpha_1$, $x_2 = \alpha_2$ is a solution of the first equation, then $\alpha_1 + \alpha_2 = 1$. Hence $2\alpha_1 + 2\alpha_2 = 2(\alpha_1 + \alpha_2) = 2$ and $x_1 = \alpha_1$,

$x_2 = \alpha_2$ is also a solution of the second equation. Conversely, if $x_1 = \beta_1$, $x_2 = \beta_2$ is a solution of the second equation, then $2\beta_1 + 2\beta_2 = 2$. Hence $\beta_1 + \beta_2 = 1$ and $x_1 = \beta_1$, $x_2 = \beta_2$ is also a solution of the first equation.

The system (1) is an example of two linear equations in the two variables (or unknowns) $x_1$ and $x_2$. The general form of a **system of $m$ linear equations in the $n$ variables** (or **unknowns**) $x_1, x_2, \ldots, x_n$ is

$$\begin{array}{ccccccccc} a_{11}x_1 & + & a_{12}x_2 & + & \cdots & + & a_{1n}x_n & = & b_1 \\ a_{21}x_1 & + & a_{22}x_2 & + & \cdots & + & a_{2n}x_n & = & b_2 \\ \vdots & & \vdots & & & & \vdots & & \vdots \\ a_{m1}x_1 & + & a_{m2}x_2 & + & \cdots & + & a_{mn}x_n & = & b_m. \end{array} \tag{2}$$

We will call such a system an $m \times n$ (read $m$ by $n$) linear system. The double subscripting of the coefficients has been arranged so that the first subscript refers to the equation and the second subscript refers to the variable. In other words, $a_{ij}$ is the coefficient of $x_j$ in the $i$th equation. Similarly, $b_i$ is the constant term in the $i$th equation.

For example, the following is a linear system with two equations and three unknowns:

$$\begin{aligned} x_1 + \phantom{2}x_2 + \phantom{2}x_3 &= 4 \\ 3x_1 - 2x_2 + 2x_3 &= 6. \end{aligned} \tag{3}$$

Here $a_{11} = 1$, $a_{12} = 1$, $a_{13} = 1$, $a_{21} = 3$, $a_{22} = -2$, $a_{23} = 2$, $b_1 = 4$, and $b_2 = 6$.

In view of our definition of a solution of a linear equation, it should come as no surprise that we define a **solution of a system of linear equations** (2) to be a collection of values of the variables which is simultaneously a solution to *every* linear equation in the system. For example, $x_1 = 2$, $x_2 = 1$, $x_3 = 1$ is a solution of the system (3), as is $x_1 = -2$, $x_2 = 0$, $x_3 = 6$. In fact, you can easily verify by direct substitution that the system has infinitely many solutions which are given by

$$x_1 = -\frac{4}{5}x_3 + \frac{14}{5}, \qquad x_2 = -\frac{1}{5}x_3 + \frac{6}{5}, \qquad x_3 = x_3. \tag{4}$$

Since $x_1$ and $x_2$ are determined in terms of $x_3$, they are called the determined variables and $x_3$ is called the free variable. You probably know that each of the equations in (3) is the equation of a plane in 3-space (if you don't, accept it for now; we will develop this fact shortly). From geometry we know that given two planes in 3-space, one of three things happens (see Figure 1.2):

1. The planes are not parallel and hence intersect in a line. The corresponding system has infinitely many solutions.
2. The planes are parallel and hence do not intersect. The corresponding system has no solution.
3. The planes coincide. The corresponding system has infinitely many solutions.

The equations in system (3) represent two planes that are not parallel and (4) gives parametric equations of the line of intersection of the planes with $x_3$ considered as

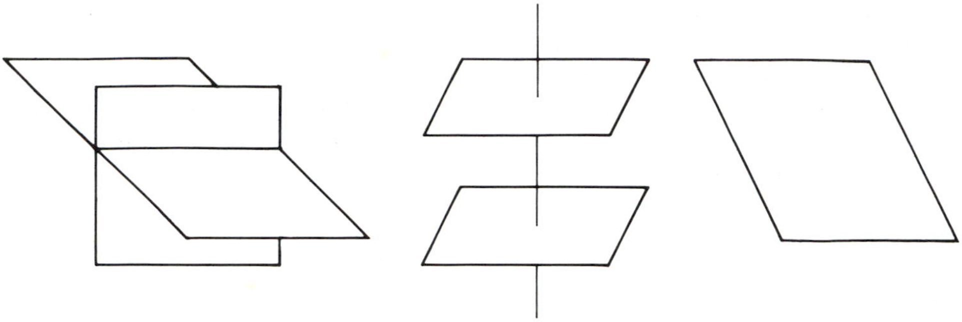

Figure 1.2

the parameter. (Again, if you are not familiar with parametric equations of a line, don't panic. They will be developed shortly.) Examples of systems that represent parallel planes and planes which coincide are:

$$\begin{aligned} x_1 - x_2 + 7x_3 &= 1 \\ x_1 - x_2 + 7x_3 &= 5 \end{aligned} \quad \text{and} \quad \begin{aligned} 2x_1 + 5x_2 - x_3 &= -1 \\ 10x_1 + 25x_2 - 5x_3 &= -5. \end{aligned}$$

Finally, consider a system of three equations in three unknowns:

$$\begin{aligned} a_{11}x_1 + a_{12}x_2 + a_{13}x_3 &= b_1 \\ a_{21}x_1 + a_{22}x_2 + a_{23}x_3 &= b_2 \\ a_{31}x_1 + a_{32}x_2 + a_{33}x_3 &= b_3. \end{aligned}$$

Each of these equations represents a plane in 3-space. From geometry we know that the four possibilities are:

1. The planes intersect in a unique point (see Figure 1.3). The corresponding system has a unique solution.
2. The planes intersect in a line (see Figure 1.4). The corresponding system has infinitely many solutions.

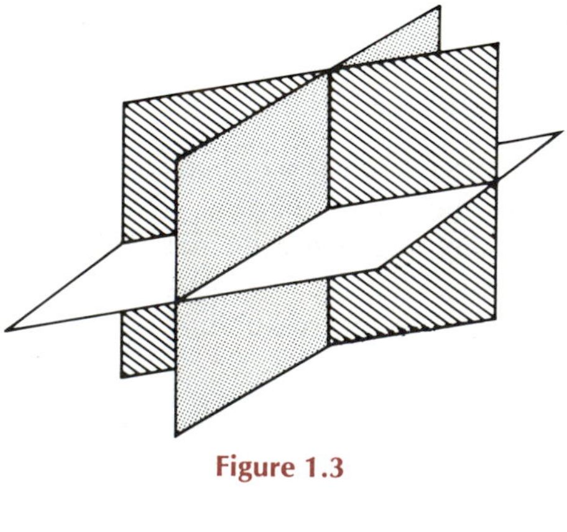

Figure 1.3

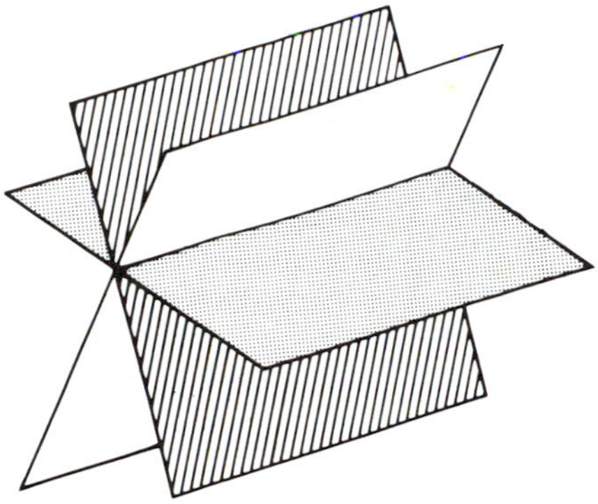

Figure 1.4

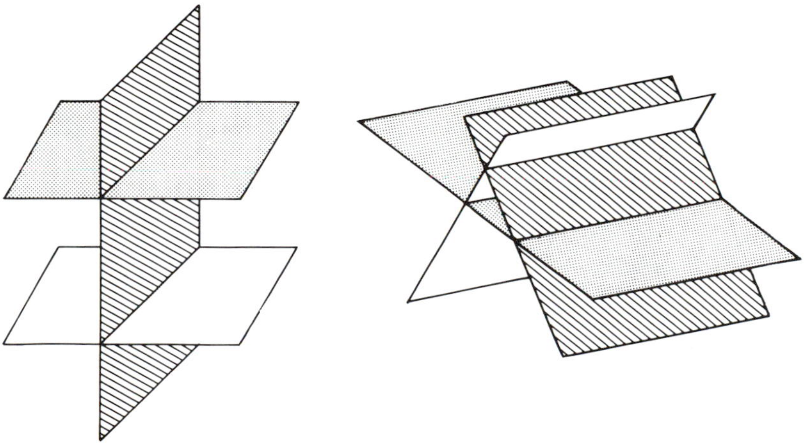

(a) (b)

Figure 1.5

3. The planes do not intersect. This happens when at least two of the planes are parallel and distinct or when each plane is parallel to the line of intersection of the other two planes (see Figure 1.5). The corresponding solutions.
4. The three planes coincide. The corresponding system has infinitely many solutions.

Our examples so far show that a linear system may or may not have solutions. Systems that have at least one solution are called **consistent**, and systems that have no solutions are called **inconsistent**. Remember these and all definitions; they will be used frequently.

# PROBLEMS 1.1

**1.** Complete the following definitions.
(a) A linear equation in the variables $x_1$, $x_2, x_3$ is....
(b) A system of three linear equations in the four unknowns $x_1, x_2, x_3, x_4$ is....
(c) A solution of a system of linear equation is....
(d) A system of linear equations is inconsistent if....

**2.** What is our agreement about how to choose the free and determined variables in a linear equation?

***3.** Which of the following equations are linear? If the equation is linear, identify the free and determined variables.
(a) $3x_1 - 2x_2 + x_3 = 0$
(b) $2x_1 + 3x_2 + x_4 = x_3 - 8$
(c) $x_1x_2 + x_3 = 2$
(d) $2x_1 + x_2^2 - x_3 = x_4$
(e) $\dfrac{x_1 - x_2}{x_3} = x_4$
(f) $x_1 + x_2 - x_5 = 4$

**4.** Identify the linear equations.
(a) $3x = 5y - z$
(b) $5x_1 = 3x_2 - 1$
(c) $x_1 + x_7 = 0$
(d) $\log(x_1) = 4$
(e) $\dfrac{1}{x_1} + x_2 = -52$
(f) $\dfrac{x_2}{x_4} = \dfrac{x_3}{x_1}$

***5.** Which of the following are solutions of the equation $2x_1 - 5x_2 + x_3 = 3$?
(a) $x_1 = 1, x_2 = -1, x_3 = -3$
(b) $x_1 = 10, x_2 = 2.5, x_3 = -4.5$
(c) $x_1 = 1, x_2 = -1, x_3 = -4$
(d) $x_1 = \frac{2}{9}, x_2 = -\frac{5}{9}, x_3 = -\frac{2}{9}$

**6.** Which of the following are solutions of the equation $3x_1 - 2x_2 + 4x_3 + x_4 = 0$?
(a) $x_1 = 1, x_2 = 1, x_3 = 1, x_4 = 4$
(b) $x_1 = 1, x_2 = 2, x_3 = 1, x_4 = 3$
(c) $x_1 = 2, x_2 = 1, x_3 = -1, x_4 = 0$
(d) $x_1 = 2, x_2 = 1, x_3 = -1, x_4 = 1$

***7.** Determine which of the following are solutions of the system

$$\begin{aligned} x_1 + 2x_2 - 2x_3 &= 3 \\ -x_1 + x_2 - 5x_3 &= 0 \\ 3x_1 + 3x_2 + x_3 &= 6. \end{aligned}$$

(a) $x_1 = 1, x_2 = 1, x_3 = 0$
(b) $x_1 = -7, x_2 = 8, x_3 = 3$
(c) $x_1 = 9, x_2 = 0, x_3 = 3$
(d) $x_1 = 3, x_2 = 0, x_3 = 0$
(e) $x_1 = 9, x_2 = -6, x_3 = -3$
(f) $x_1 = 1, x_2 = 2, x_3 = 2$

**8.** Determine which of the following are solutions of the system

$$\begin{aligned} 3x_1 - 2x_2 - 3x_3 - 2x_4 &= -1 \\ x_1 + x_2 - x_3 + x_4 &= 8 \\ 2x_1 + 3x_2 + x_3 \qquad &= 21. \end{aligned}$$

(a) $x_1 = 1, x_2 = 1, x_3 = 1, x_4 = -1$
(b) $x_1 = 5, x_2 = 3, x_3 = 2, x_4 = 2$
(c) $x_1 = 2, x_2 = 6, x_3 = -1, x_4 = -1$
(d) $x_1 = 0, x_2 = -2, x_3 = 1, x_4 = 1$

**9.** Explain why the system

$$\begin{aligned} x_1 + x_2 + 2x_3 &= 1 \\ 3x_1 + 3x_2 + 6x_3 &= 2 \end{aligned}$$

cannot have any solutions.

***10.** Find three different values for $b$ that will make the following system inconsistent.

$$\begin{aligned} x_1 + x_2 + 2x_3 &= 1 \\ 3x_1 + 3x_2 + 6x_3 &= b \end{aligned}$$

**11.** For which values of $b$ will the following system have solutions?

$$\begin{aligned} x_1 + 2x_2 - x_3 &= 4 \\ 2x_1 + 4x_2 - 2x_3 &= b \end{aligned}$$

***12.** Given the system

$$\begin{aligned} 2x_1 + x_2 - x_3 &= 4 \\ x_1 - x_2 + 3x_3 &= 2, \end{aligned}$$

* Asterisks indicate problems which have answers in the answer section.

find an equation with the property that when it is included in the system the resulting system of three equations is inconsistent.

**13.** Verify that an infinite number of solutions of the system

$$\begin{aligned} 3x_1 - 2x_2 - 3x_3 - 2x_4 &= -1 \\ x_1 + x_2 - x_3 + x_4 &= 8 \\ 2x_1 + 3x_2 + x_3 \qquad &= 21 \end{aligned}$$

is given by $x_1 = 3 + t$, $x_2 = 5 - t$, $x_3 = x_4 = t$, where $t$ is any number.

**14.** Verify that an infinite number of solutions of the system

$$\begin{aligned} x_1 \qquad + x_3 + x_4 &= 7 \\ x_1 + x_2 \qquad - x_4 &= 4 \\ x_2 - x_3 - 2x_4 &= -3 \end{aligned}$$

is given by $x_1 = 2 + t - s$, $x_2 = 3 - t + 2s$, $x_3 = 4 - t$, $x_4 = 1 + s$, where $s$ and $t$ are arbitrary numbers.

**15.** Verify that a solution of

$$\begin{aligned} a_{11}x_1 + a_{12}x_2 &= b_1 \\ a_{21}x_1 + a_{22}x_2 &= b_2 \end{aligned}$$

is given by

$$x_1 = \frac{b_1a_{22} - b_2a_{12}}{a_{11}a_{22} - a_{12}a_{21}}$$
$$x_2 = \frac{a_{11}b_2 - a_{21}b_1}{a_{11}a_{22} - a_{12}a_{21}}$$

provided that $a_{11}a_{22} - a_{12}a_{21} \neq 0$.

***16.** Given a geometric argument to prove that if there are two distinct solutions of the system

$$\begin{aligned} a_{11}x_1 + a_{12}x_2 &= b_1 \\ a_{21}x_1 + a_{22}x_2 &= b_2, \end{aligned}$$

then there are infinitely many solutions of the system.

**17.** Give a geometric description of the possible ways three lines in the plane can intersect. Then interpret your answer in terms of a system of three equations in two unknowns.

**18.** Give an algebraic and geometric argument to prove that each of the following systems is consistent.

*(a) $a_{11}x_1 + a_{12}x_2 = 0$
$a_{21}x_1 + a_{22}x_2 = 0$

*(b) $a_{11}x_1 + a_{12}x_2 = 0$
$a_{21}x_1 + a_{22}x_2 = 0$
$a_{31}x_1 + a_{32}x_2 = 0$

*(c) $a_{11}x_1 + a_{12}x_2 + a_{13}x_3 = 0$
$a_{21}x_1 + a_{22}x_2 + a_{23}x_3 = 0$

*(d) $a_{11}x_1 + a_{12}x_2 + a_{13}x_3 = 0$
$a_{21}x_1 + a_{22}x_2 + a_{23}x_3 = 0$
$a_{31}x_1 + a_{32}x_2 + a_{33}x_3 = 0$

**19.** Find a generalization of Problem 18.

# 1.2
# Some Examples of Systems of Linear Equations

Most students of high school algebra have been subjected to problems such as this:

> Find three numbers whose sum is 20 and such that (1) the first plus twice the second plus three times the third equals 44 and (2) twice the sum of the first and second minus four times the third equals −14. [This problem is found (together with 19 others involving digits, water tanks, boats, work, and freight trains) in H. B. Fine, *A College Algebra*, Ginn & Co., 1904, pp. 150–152.]

This problem is equivalent to finding the solution of the following system of linear equations (where $x_1$, $x_2$, and $x_3$ are the three numbers we are trying to find).

$$\begin{aligned} x_1 + x_2 + x_3 &= 20 \\ x_1 + 2x_2 + 3x_3 &= 44 \\ 2x_1 + 2x_2 - 4x_3 &= -14 \end{aligned}$$

Problems such as this, although of interest to professional (or habitual) problem solvers, are not important applications of linear equations. They give practice in translating English into the language of mathematics as well as practice in computation but do not give the student adequate motivation for studying the mathematics that is being used. In this section we give several practical and important examples where systems of linear equations arise naturally.

## Electric Circuits

Most people think of electricity as something that "flows" through wires. Indeed, it is usually convenient to think of electricity as electrons flowing through wires. When we think of something flowing we naturally think of the "pressure" behind the flow and the "quantity" of substance flowing. For electrical circuits, the "pressure" behind the electrons is measured in volts and the "quantity" of electrons flowing, called the current, is measured in amperes or amps. For the sake of simplicity we will consider only direct-current (dc) circuits, circuits in which the electricity travels in one direction in each wire.

Let us consider the electric circuit given by Figure 1.6. This circuit has four junctions, places where many wires come together ($J_1$, $J_2$, $J_3$, and $J_4$). There are also six branches with currents $I_1, I_2, \ldots, I_6$ and one source of electricity in the branch from $J_4$ to $J_1$. Each branch of this circuit has been (arbitrarily) assigned an arrow indicating a direction of flow. The actual direction of flow will be given by the sign of the current in that branch. A positive current will mean a flow in the direction of the arrow; a negative current will flow in the opposite direction.

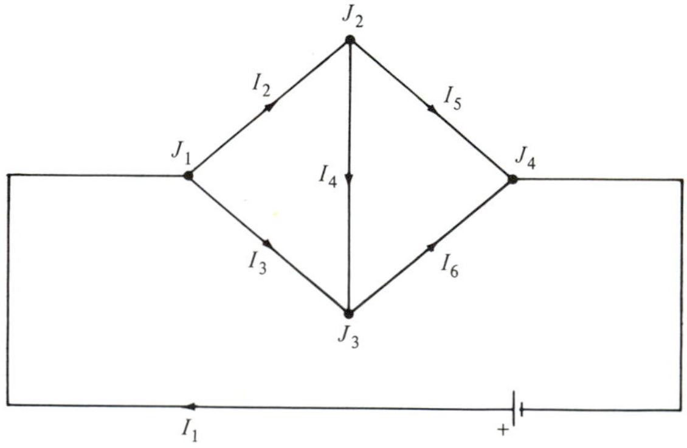


Figure 1.6

We can use an ammeter to measure both the direction and the amount of current flowing in each wire of this circuit. If the currents in all the wires that come together at a junction are added, it is found that the sum is zero. This is not too surprising—it expresses the fact that the substance (the electrons) which is flowing is not being created or destroyed at the junction. In brief, what goes in must equal what comes out. This is one of two basic laws regarding electric circuits which were first formulated by G. R. Kirchhoff in 1845.

At $J_1$ we have $I_1$ amps flowing in and $I_2 + I_3$ amps flowing out. By Kirchhoff's law, $I_1 - I_2 - I_3 = 0$. This is a linear equation with currents as the variables. There will be one equation for each junction. Looking at all four junctions we get the following four equations.

$$\begin{array}{rcrcrcrcrcrcl} I_1 & - & I_2 & - & I_3 & & & & & & & = 0 & \text{(junction } J_1\text{)} \\ & & I_2 & & & - & I_4 & - & I_5 & & & = 0 & \text{(junction } J_2\text{)} \\ & & & & I_3 & + & I_4 & & & - & I_6 & = 0 & \text{(junction } J_3\text{)} \\ -I_1 & & & & & & & + & I_5 & + & I_6 & = 0 & \text{(junction } J_4\text{)} \end{array}$$

We conclude that the values of the currents flowing in the six branches of the circuit of Figure 1.6 must satisfy (be a solution of) this system of linear equations.

It is clear that this system of equations does not have a unique solution. We could double the current in each branch and still satisfy Kirchhoff's law at each junction. Thus a knowledge of the physical problem (the electric circuit) has given us information about a mathematical problem (the system of equations). The reverse is also possible. In particular, the theory developed in Chapter 2 will enable us to determine the minimum number of currents that must be measured before the equations above give us complete knowledge of all currents in the circuit.

## Pricing Goods in a Closed Economy

Suppose that there are three people in a closed economy, for example, three astronauts orbiting the earth in a space station. No goods enter or leave the station. Suppose that each of these three astronauts is producing one product and that each product is divided up among the three astronauts. Let us be specific about how the goods produced in our simple economy are distributed. Suppose that astronaut 1 produces a certain quantity of item $A$ each day, which is divided up so that $\frac{2}{5}$ goes to astronaut 1, $\frac{1}{5}$ to astronaut 2, and $\frac{2}{5}$ to astronaut 3. Let astronaut 2 produce item $B$ and let $\frac{1}{4}$ of the amount produced be given to astronaut 1 and an equal amount to astronaut 3. Finally, let astronaut 3 produce item $C$, with $\frac{1}{2}$ of the production going to astronaut 1 and $\frac{1}{4}$ to astronaut 2. We now have complete information about the way the goods in this economy are distributed.

This small community does not need money to accomplish the exchange of goods. But we usually have a closed economy with a much larger number of producers. For example, the earth is a closed economy which consists of a very large number of producing units. Such communities require some common medium of exchange (i.e., money). We then have a way to measure the relative value of each of the goods produced and we can assign prices to the products.

Suppose that the value of the goods consumed by one of the astronauts is greater than the value of the goods produced by that astronaut. This astronaut's supply of money will decrease every day. This cannot continue indefinitely. If our economy is to be stable, if it is to be in a state of equilibrium, we must reject this possibility. Similarly, to have a stable economy, we must reject the possibility of an astronaut consuming goods less valuable than those produced. Thus, in a stable economy, the value of the goods produced by each astronaut must equal the value of the goods consumed by that astronaut.

Assuming that our mini-economy is stable, let us attempt to assign prices to the goods being produced. Let $p_i$ be the value (price) of the goods produced by astronaut $i$ in one day. At the end of each day the first astronaut has $\frac{2}{5}$ of his own production, $\frac{1}{4}$ of the second astronaut's production, and $\frac{1}{2}$ of the third astronaut's production. The value of these goods is $\frac{2}{5}p_1 + \frac{1}{4}p_2 + \frac{1}{2}p_3$. But the value of the goods produced by the first astronaut is $p_1$. Thus the prices must satisfy the equation

$$\frac{2}{5}p_1 + \frac{1}{4}p_2 + \frac{1}{2}p_3 = p_1.$$

A similar equation can be found for each astronaut and we have the following system of equations in the variables $p_1, p_2, p_3$.

$$\begin{aligned}
\frac{2}{5}p_1 + \frac{1}{4}p_2 + \frac{1}{2}p_3 &= p_1 \qquad \text{(astronaut 1)}\\
\frac{1}{5}p_1 + \frac{1}{2}p_2 + \frac{1}{4}p_3 &= p_2 \qquad \text{(astronaut 2)}\\
\frac{2}{5}p_1 + \frac{1}{4}p_2 + \frac{1}{4}p_3 &= p_3 \qquad \text{(astronaut 3)}
\end{aligned} \tag{1}$$

One solution to this system is $p_1 = 5$, $p_2 = 4$, $p_3 = 4$. Another solution is $p_1 = 10$, $p_2 = 8$, $p_3 = 8$. But from the point of view of economics these two solutions are essentially the same. Merely doubling all prices does not change the relative values of the goods. Another solution of the system is $p_1 = p_2 = p_3 = 0$. Such a solution has no economic meaning. It does not measure the relative values of the goods in a meaningful way.

Two important questions must now be asked. First, given a closed economy, does there always exist a meaningful set of prices for the goods which will yield a stable economy, that is, an economy in which the value of the goods consumed by each individual equals the value of the goods produced by that individual? Second, if such a set of prices exists, is any other set of prices satisfying the same conditions in some sense the same? These questions are of critical importance to the economist. In Chapter 2 we use linear algebra to show that the answer to each of these questions is indeed affirmative.

## Curve Fitting in Analytic Geometry

In elementary analytic geometry it is shown that the equation of a nonvertical straight line in the $xy$-plane is of the form $y = mx + b$. The constants $m$ and $b$ are the slope and $y$-intercept of the line. There are occasions when the equation of the line is unknown (we do not know $m$ and $b$) but some points on the line are known. If the points $(x_i, y_i)$, $i = 1, 2, \ldots, n$, are all on the line, then for each $i = 1, 2, \ldots, n$, we have

$$y_i = mx_i + b.$$

This is a system of $n$ equation in the two unknowns $m$ and $b$. (The reader will note our sudden change of notation. The standard notation of analytic geometry is so familiar that it is pedagogically unsound to alter it.) Note that $n$ can be any positive integer, including 1.

To make our example more concrete, let us take three points in the plane: $(0, 0)$, $(1, 1)$, and $(2, 1)$. Then our system of equations is

$$\begin{aligned} 0 &= m \cdot 0 + b \\ 1 &= m \cdot 1 + b \\ 1 &= m \cdot 2 + b. \end{aligned}$$

Since the first equation says that $b = 0$, we may substitute that value for $b$ in the last two equations. They then become $m = 1$ and $2m = 1$. This is clearly impossible. We conclude that no straight line can pass through these three points.

If, on the other hand, we take the two points $(2, 3)$ and $(4, 7)$, we get the equations

$$\begin{aligned} 3 &= m \cdot 2 + b \\ 7 &= m \cdot 4 + b. \end{aligned}$$

The unique solutions is $m = 2$, $b = -1$. These results agree with out geometric intuition. Three points do not necessarily lie on a straight line, whereas two points *always* determine a unique line.

Suppose that we ask another question about the three points $(0, 0)$, $(1, 1)$, and $(2, 1)$. Can a quadratic curve, one of the form $y = ax^2 + bx + c$, be fitted to these points? If so, then the following equations must be satisfied:

$$\begin{aligned} 0 &= a \cdot 0 + b \cdot 0 + c \\ 1 &= a \cdot 1 + b \cdot 1 + c \\ 1 &= a \cdot 4 + b \cdot 2 + c. \end{aligned}$$

The first equation says that $c = 0$, so we have

$$\begin{aligned} c &= 0 \\ a + \phantom{2}b &= 1 \\ 4a + 2b &= 1. \end{aligned}$$

There is a unique solution; $a = -\frac{1}{2}$, $b = \frac{3}{2}$, $c = 0$. So we can fit a unique quadratic polynomial to these three points, namely $y = -\frac{1}{2}x^2 + \frac{3}{2}x$.

In this example we have seen some interesting connections between geometric ideas (lines, curves) and algebraic ideas (systems of equations). In subsequent chapters we convert the geometry of Euclidean space into the algebra of systems of equations. We also convert the algebra of systems of equations into the geometry of Euclidean space. The net result will be a more thorough understanding of both.

One further note. We saw above that a system of equations with no solution could arise naturally when studying a certain problem. In Chapter 3 we discuss how such inconsistent systems arise in other situations. We also consider the question of what can be done with such systems.

## PROBLEMS 1.2

**1.** (These problems are for students who enjoy word problems. They are taken from Fine's book.) In each of the following problems certain unknowns satisfy a system of linear equations. Identify the unknowns and exhibit the correct system. Do not solve the system.

*(a) Find a number of two digits from the following data: (1) twice the first digit plus three times the second equals 37; (2) if the order of the digits be reversed, the number is diminished by 9.

*(b) The sum of three numbers is 51. If the first number be divided by the second, the quotient is 2 and the remainder is 5; but if the second number be divided by the third, the quotient is 3 and the remainder 2. What are the numbers?

(c) In a certain number of three digits, the second digit is equal to the sum of the first and the third, the sum of the second and the third digits is 8, and if the first and third digits are interchanged, the number is increased by 99. Find the number.

*(d) $A$ gave $B$ as much money as $B$ had; then $B$ gave $A$ as much money as $A$ had left. Finally, $A$ gave $B$ as much money as $B$ then had left. $A$ then had \$16 and $B$ \$24. How much had each originally?

(e) Find the fortunes of three men, $A$, $B$, and $C$, from the following data: $A$ and $B$ together have $p$ dollars; $B$ and $C$, $q$ dollars; $C$ and $A$, $r$ dollars.

*(f) Given three alloys of the following composition: $A$, 5 parts (by weight) gold, 2 silver, 1 lead; $B$, 2 parts gold, 5 silver, 1 lead; $C$, 3 parts gold, 1 silver, 4 lead. To obtain 9 ounces of an alloy containing equal quantities (by weight) of gold, silver, and lead, how many ounces of $A$, $B$, and $C$ must be taken together?

(g) Two vessels, $A$ and $B$ contain mixtures of alcohol and water. A mixture of 3 parts from $A$ and 2 parts from $B$ will contain 40% of alcohol; and a mixture of 1 part from $A$ and 2 parts from $B$ will contain 32% of alcohol. What are the percentages of alcohol in $A$ and $B$, respectively?

**2.** What system of equations must the currents in each circuit satisfy?

*(a)

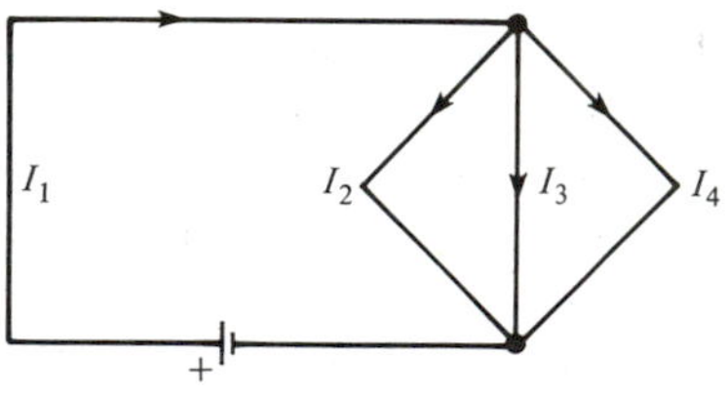


(b)

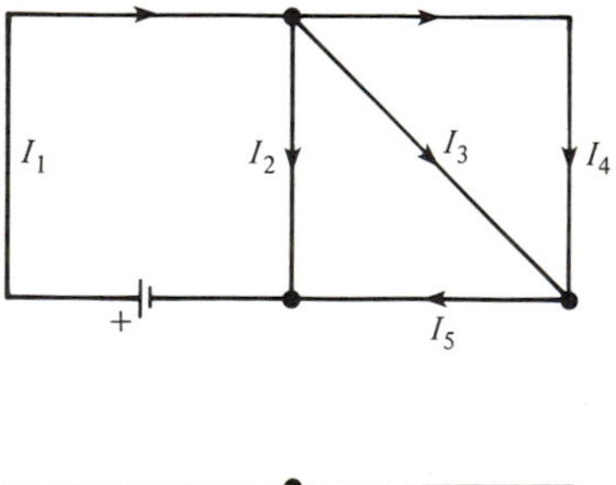


*(c)

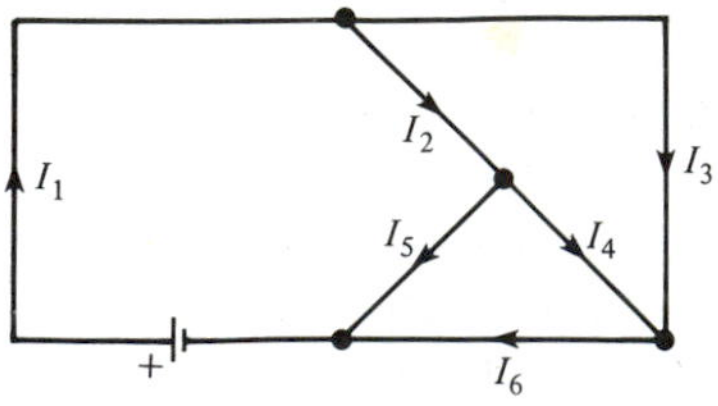

(d) 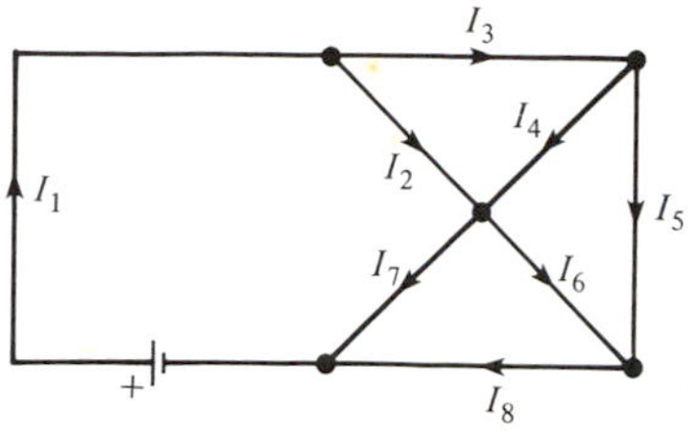


*(e) 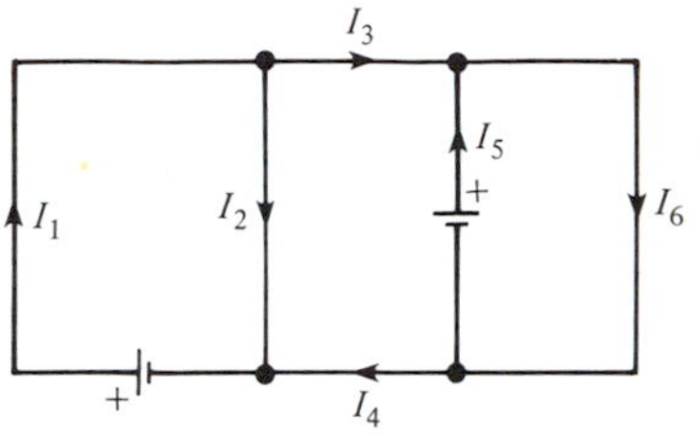


(f) 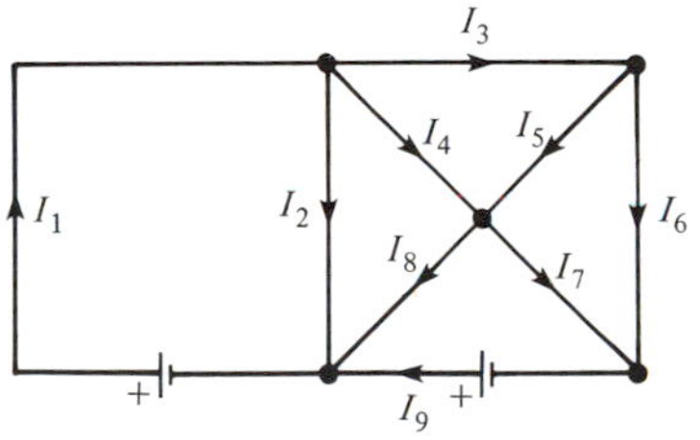


3. Kirchhoff's second law states that the sum of the voltage drops around any closed circuit must be zero. In our more intuitive language this says that the sum of the "pressures" around a circular path must be zero. In essence this law prohibits perpetual flow of electricity unless there is a source for the electricity (e.g., a battery).

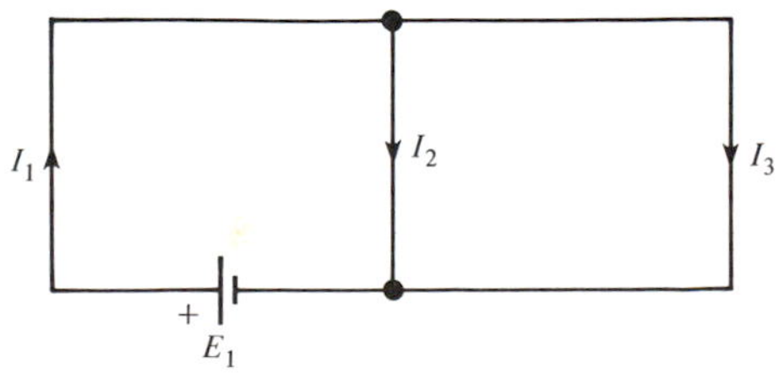


The voltage $E$ and the current $I$ in any branch of a circuit satisfy the relation $E = IR$ (Ohm's law), where $R$ is the resistance (in ohms) in the branch. For example, consider the circuit pictured. This circuit has three closed loops; the loop consisting of the branches with currents $I_1$ and $I_2$, the loop consisting of the branches with currents $I_2$ and $I_3$, and the loop consisting of the branches with currents $I_1$ and $I_3$. If we let $R_i$ be the resistance in the branch with current $I_i$, we have the equations

$$\begin{aligned} R_1I_1 + R_2I_2 - E_1 &= 0 \\ R_2I_2 - R_3I_3 \qquad &= 0 \\ R_1I_1 + R_3I_3 - E_1 &= 0, \end{aligned}$$

each of which corresponds to one of the three closed loops in the circuit. Notice that the voltage from the battery ($E_1$) has been subtracted because the voltage increase caused by the battery is a negative voltage drop.

For each of the circuits in Problem 2, find the additional equations satisfied by the currents if:

*(a) All batteries produce 1 volt and all branches have a resistance of 1 ohm.
*(b) All batteries produce 1 volt and the branch with current $I_j$ has a resistance of $j$ ohms.

4. The prices in a closed economy satisfy the following system of equations:

$$\begin{aligned} \frac{1}{2}p_1 + \frac{1}{3}p_2 + \frac{1}{4}p_3 &= p_1 \\ \frac{1}{3}p_1 + \frac{1}{3}p_2 + \frac{1}{2}p_3 &= p_2 \\ \frac{1}{6}p_1 + \frac{1}{3}p_2 + \frac{1}{4}p_3 &= p_3. \end{aligned}$$

*(a) What fraction of production does producer 1 retain? sell to producer 2? sell to producer 3?
(b) What fraction of production does producer 2 retain? sell to producer 1? sell to producer 3?
(c) What fraction of production does producer 3 retain? sell to producer 2? sell to producer 1?

5. Consider a closed economy consisting of five producers. The number in the $i$th row and $j$th column of the following table represents the fraction of the $j$th producer's daily production that is consumed by the $i$th producer.

| | 1 | 2 | 3 | 4 | 5 |
|---|---|---|---|---|---|
| **1** | $\frac{1}{2}$ | 0 | $\frac{1}{5}$ | 0 | $\frac{1}{12}$ |
| **2** | $\frac{1}{8}$ | $\frac{1}{3}$ | $\frac{1}{5}$ | $\frac{2}{3}$ | $\frac{1}{3}$ |
| **3** | $\frac{1}{8}$ | $\frac{4}{9}$ | 0 | 0 | $\frac{1}{4}$ |
| **4** | $\frac{1}{8}$ | $\frac{1}{9}$ | $\frac{2}{5}$ | 0 | $\frac{1}{4}$ |
| **5** | $\frac{1}{8}$ | $\frac{1}{9}$ | $\frac{1}{5}$ | $\frac{1}{3}$ | $\frac{1}{12}$ |

Let $p_i$, $i = 1, 2, 3, 4, 5$, be the price for 1 unit of item $i$. What system of linear equations must these prices satisfy in order to have a stable economy?

*6. Four people stranded on a desert island form a closed economy. The number in the $i$th row and the $j$th column of the following table represents the fraction of the $i$th person's daily production that is consumed by the $j$th person. What system of equations do the prices satisfy?

| | 1 | 2 | 3 | 4 |
|---|---|---|---|---|
| **1** | $\frac{1}{3}$ | $\frac{1}{6}$ | $\frac{1}{3}$ | $\frac{1}{6}$ |
| **2** | $\frac{1}{5}$ | $\frac{2}{5}$ | $\frac{1}{5}$ | $\frac{1}{5}$ |
| **3** | $\frac{1}{6}$ | $\frac{1}{6}$ | $\frac{1}{2}$ | $\frac{1}{6}$ |
| **4** | $\frac{1}{8}$ | $\frac{1}{4}$ | $\frac{1}{8}$ | $\frac{1}{2}$ |

7. Explain why in a stable economy we must reject the possibility of a person consuming goods less valuable than those produced by that person.

8. Verify that $p_1 = -10$, $p_2 = -8$, $p_3 = -8$ is a solution of the linear system (1) in the text. Discuss the economic meaning of these prices.

9. The following system of equations cannot be the equations satisfied by the prices in a stable mini-economy. Why?

$$\begin{aligned}
\frac{1}{2}p_1 + \frac{1}{2}p_2 + \frac{1}{2}p_3 &= p_1 \\
\frac{1}{2}p_1 + \frac{1}{3}p_2 + \frac{1}{4}p_3 &= p_2 \\
\frac{1}{2}p_1 + \frac{1}{4}p_2 + \frac{1}{4}p_3 &= p_3
\end{aligned}$$

*10. Suppose that we have an $n$-person stable economy. Then the prices for the goods satisfy the following linear system.

$$\begin{aligned}
a_{11}p_1 + a_{12}p_2 + \cdots + a_{1n}p_n &= p_1 \\
a_{21}p_1 + a_{22}p_2 + \cdots + a_{2n}p_n &= p_2 \\
\vdots \qquad\qquad \vdots \qquad\qquad\quad \vdots \quad &\quad \vdots \\
a_{n1}p_1 + a_{n2}p_2 + \cdots + a_{nn}p_n &= p_n
\end{aligned}$$

Show that the value of

$$a_{1j} + a_{2j} + a_{3j} + \cdots + a_{nj}$$

does not depend on $j$. What is its value?

11. What equations must $m$ and $b$ satisfy if the points $(1, 5)$ and $(-1, 1)$ are to lie on the straight line $y = mx + b$?

*12. What equations must $m$ and $b$ satisfy if the points $(1, -1)$, $(5, 7)$, $(-2, -7)$, and $(0, -3)$ are to lie on the straight line $y = mx + b$?

*13. What equations must $a$, $b$, and $c$ satisfy if the quadratic polynomial $y = ax^2 + bx + c$ is to pass through the points $(0, -2)$, $(1, 6)$, $(-2, 0)$, $(4, 66)$?

14. Consider a cubic curve $y = ax^3 + bx^2 + cx + d$. What system of linear equations must $a$, $b$, $c$, and $d$ satisfy so that the curve passes through the points $(0, 1)$, $(1, 2)$, $(3, -1)$, $(-1, -1)$?

15. What system of linear equations must $a$, $b$, $c$, $d$, and $e$ satisfy so that the curve $y = ax^4 + bx^3 + cx^2 + dx + e$ passes through the points $(1, 1)$, $(-1, 1)$, $(2, 2)$, $(-2, 3)$, $(3, 10)$?

# 1.3 Vectors

So far we have defined a system of linear equations and have seen many examples of problems whose mathematical formulation gives rise to a system of linear equations. Of course, to solve these problems we must solve the systems. Before turning to a systematic procedure for solving systems of linear equations, we introduce matrix and vector notation to describe these systems. Armed with this notation we will be able to describe a linear system in a single equation, and we will be able to efficiently implement our solution procedure by performing operations on the matrix of the system rather than on the system itself. Finally, as with all good mathematical concepts, vectors and matrices take on a life of their own. They will be with us throughout the rest of the book.

So much for our preview of coming attractions. Now let's get down to the business at hand. The general form of a system of linear equations is

$$\begin{aligned} a_{11}x_1 + a_{12}x_2 + \cdots + a_{1n}x_n &= b_1 \\ a_{21}x_1 + a_{22}x_2 + \cdots + a_{2n}x_n &= b_2 \\ \vdots \qquad \vdots \qquad\qquad \vdots \quad &\quad \vdots \\ a_{m1}x_1 + a_{m2}x_2 + \cdots + a_{mn}x_n &= b_m. \end{aligned} \tag{1}$$

The system is specified by the coefficients of the variables (the $a_{ij}$'s), the constant terms (the $b_i$'s), and the variables themselves (the $x_j$'s). We are going to treat each of these collections as a single object. We begin by looking at the column of numbers on the right. We let

$$\mathbf{b} = \begin{bmatrix} b_1 \\ b_2 \\ \vdots \\ b_m \end{bmatrix} \tag{2}$$

and we call it an $m$-**dimensional column vector**. The number $b_i$ is called the $i$**th coordinate** or the $i$**th component** of **b**. Of course, since we called (2) a *column* vector, there must be *row* vectors. Indeed there are, as well as an operation that turns column vectors into row vectors, and vice versa. An $m$-**dimensional row vector v** is a row of $m$ real numbers

$$\mathbf{v} = [v_1 \quad v_2 \quad \cdots \quad v_m].$$

As with column vectors, $v_i$ is called the $i$**th coordinate** or the $i$**th component** of **v**. The operation that converts a row vector into a column vector, and vice versa, is called the transpose and is defined as follows. The **transpose** of a row vector

$\mathbf{v} = [v_1 \quad v_2 \quad \cdots \quad v_m]$ is the column vector

$$\mathbf{v}^T = \begin{bmatrix} v_1 \\ v_2 \\ \vdots \\ v_m \end{bmatrix},$$

and the transpose of the column vector

$$\mathbf{w} = \begin{bmatrix} w_1 \\ w_2 \\ \vdots \\ w_m \end{bmatrix}$$

is the row vector $\mathbf{w}^T = [w_1 \quad w_2 \quad \cdots \quad w_m]$. Thus the transpose lays down a column vector and stands up a row vector.

The variables in the equation will also be represented by an $n$-dimensional column vector

$$\mathbf{x} = \begin{bmatrix} x_1 \\ x_2 \\ \vdots \\ x_n \end{bmatrix}.$$

Now consider the collection of coefficients. Looking at the system (1) and ignoring everything except the coefficients, we see a two-dimensional array of numbers. Such an array is called a **matrix** and is denoted by

$$A = \begin{bmatrix} a_{11} & a_{12} & a_{13} & \cdots & a_{1n} \\ a_{21} & a_{22} & a_{23} & \cdots & a_{2n} \\ a_{31} & a_{32} & a_{33} & \cdots & a_{3n} \\ \vdots & \vdots & \vdots & & \vdots \\ a_{m1} & a_{m2} & a_{m3} & \cdots & a_{mn} \end{bmatrix}. \tag{3}$$

The matrix has $m$ rows and $n$ columns and is called an $m \times n$ matrix (read $m$ by $n$ matrix). Each column of the matrix is an $m$-dimensional column vector and each row is an $n$-dimensional row vector. Thus a matrix can be thought of as a collection of $m$ row vectors or as a collection of $n$ column vectors. The **entry** $a_{ij}$ is in the $i$th row and the $j$th column of the matrix. That is, the first subscript denotes the row and the second subscript denotes the column. We will often denote the matrix (3) by $A = [a_{ij}]$. An $n \times n$ matrix is called a **square** matrix.

The careful reader will have noticed that an $n$-dimensional column vector is an $n \times 1$ matrix and an $n$-dimensional row vector is a $1 \times n$ matrix.

Now that we have names and notations for the three collections of numbers (coefficients, constant terms, variables) in the linear system (1), we are going to define operations on row vectors, column vectors, and matrices so that the system (1) will be represented by the single equation

$$A\mathbf{x} = \mathbf{b}.$$

First we introduce a new concept.

An ordered $n$-tuple of real numbers

$$\mathbf{x} = (x_1, x_2, \ldots, x_n)$$

is called a **vector** (or an $n$-**vector**) and $x_i$ is called the $i$th **component** of $\mathbf{x}$. The vector $\mathbf{x} = (x_1, x_2, \ldots, x_n)$ is not the $n$-dimensional row vector $[x_1 \quad x_2 \quad \cdots \quad x_n]$ and it is not the $n$-dimensional column vector

$$\begin{bmatrix} x_1 \\ x_2 \\ \vdots \\ x_n \end{bmatrix}.$$

But almost! There is a correspondence between vectors and $n$-dimensional row vectors. The vector

$$\mathbf{x} = (x_1, x_2, \ldots, x_n)$$

corresponds to the $n$-dimensional row vector

$$[x_1 \quad x_2 \quad \cdots \quad x_n].$$

Similarly, there is a correspondence between vectors and $n$-dimensional column vectors. The vector

$$\mathbf{x} = (x_1, x_2, \ldots, x_n)$$

corresponds to the $n$-dimensional column vector

$$\begin{bmatrix} x_1 \\ x_2 \\ \vdots \\ x_n \end{bmatrix}.$$

Thus an $n$-vector can be used to represent either an $n$-dimensional row vector or an $n$-dimensional column vector.

Let $R^n$ denote the collection of all $n$-tuples of real numbers. Said otherwise, $R^n$ denotes the collection of all $n$-vectors each of whose components is a real number. Thus vectors $\mathbf{v} = (v_1, v_2, \ldots, v_n)$ in $R^n$ can be used to represent *both* $n$-dimensional row vectors

$$[v_1 \quad v_2 \quad \cdots \quad v_n] \tag{4}$$

and $n$-dimensional column vectors

$$\begin{bmatrix} v_1 \\ v_2 \\ \vdots \\ v_n \end{bmatrix}. \tag{5}$$

We will abuse the notation somewhat and continue to use $\mathbf{v}$ to denote both the $n$-dimensional row vector (4) and the $n$-dimensional column vector (5).

Since a vector $\mathbf{v} = (v_1, v_2)$ in $R^2$ is just an ordered pair of real numbers, it can be thought of as the coordinates of a point in the plane. Similarly, a vector $\mathbf{v} = (v_1, v_2, v_3)$ in $R^3$ can be viewed as the coordinates of a point in 3-space. Although we cannot visualize higher-dimensional spaces, we continue to think of a vector $\mathbf{v} = (v_1, v_2, \ldots, v_n)$ in $R^n$ as the coordinates of a point in $n$-space (see Figure 1.7).

There are two basic algebraic operations on vectors, addition of vectors and multiplication of vectors by a number. Before we define these operations we need to define equality of vectors. Two $n$-vectors $\mathbf{u} = (u_1, u_2, \ldots, u_n)$ and $\mathbf{v} = (v_1, v_2, \ldots, v_n)$ are **equal** if $u_1 = v_1$, $u_2 = v_2, \ldots, u_n = v_n$. Thus two vectors are equal if and only if they are equal componentwise. Now we can define the **sum** of two vectors $\mathbf{u} = (u_1, u_2, \ldots, u_n)$ and $\mathbf{v} = (v_1, v_2, \ldots, v_n)$ in $R^n$ to be the $n$-vector

$$\mathbf{u} + \mathbf{v} = (u_1 + v_1, u_2 + v_2, \ldots, u_n + v_n).$$

Since adding two vectors merely amounts to adding corresponding components, we say that addition is defined componentwise. Of course, addition of two vectors is defined if and only if the two vectors have the same size; that is, both vectors are in $R^n$. If $\alpha$ is a number, we define the product of $\alpha$ and $\mathbf{v}$ to be

$$\alpha\mathbf{v} = (\alpha v_1, \alpha v_2, \ldots, \alpha v_n).$$

Thus to multiply a vector by a number we merely multiply each of its components by the number. We will often use lowercase Greek letters to denote numbers used to multiply vectors. Whatever the type of letters used, when numbers multiply vectors they are called **scalars**. The multiplication of a vector by a scalar is called **scalar multiplication**.

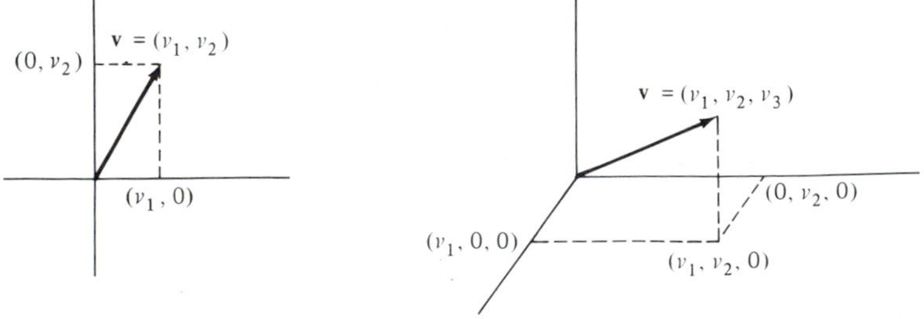


Figure 1.7

As with numbers, we define $-\mathbf{v}$ by the equation

$$-\mathbf{v} = (-1)\mathbf{v}$$

and subtraction by the equation

$$\mathbf{u} - \mathbf{v} = \mathbf{u} + (-\mathbf{v}).$$

The vector each of whose components is 0 is called the **zero vector** and is denoted by $\mathbf{0}$.

### EXAMPLE 1

If $\mathbf{u} = (1, 2, 3, 4)$ and $\mathbf{v} = (1, -1, 1, -1)$, then

$$\begin{aligned} \mathbf{u} + \mathbf{v} &= (1 + 1, 2 - 1, 3 + 1, 4 - 1) = (2, 1, 4, 3) \\ 2\mathbf{u} &= (2, 4, 6, 8) \\ 3\mathbf{u} + 5\mathbf{v} &= (3, 6, 9, 12) + (5, -5, 5, -5) = (8, 1, 14, 7) \\ -\mathbf{v} &= (-1, 1, -1, 1) \\ 2\mathbf{u} - \mathbf{v} &= (1, 5, 5, 9) \end{aligned}$$

The following algebraic identities are a direct consequence of the definitions of addition and scalar multiplication

Let $\mathbf{u}$, $\mathbf{v}$, and $\mathbf{w}$ be vectors in $R^n$ and let $\alpha$ and $\beta$ be scalars.

I. Properties of addition

(1) $\mathbf{u} + \mathbf{v} = \mathbf{v} + \mathbf{u}$ (commutativity)

(2) $(\mathbf{u} + \mathbf{v}) + \mathbf{w} = \mathbf{u} + (\mathbf{v} + \mathbf{w})$ (associativity)

(3) $\mathbf{u} + \mathbf{0} = \mathbf{u}$ (zero vector)

(4) $\mathbf{u} - \mathbf{u} = \mathbf{0}$ (additive inverse)

II. Properties of scalar multiplication

(5) $(\alpha\beta)\mathbf{u} = \alpha(\beta\mathbf{u})$ (associativity)

(6) $1 \cdot \mathbf{u} = \mathbf{u}$ (scalar identity)

(7) $\alpha(\mathbf{u} + \mathbf{v}) = \alpha\mathbf{u} + \alpha\mathbf{v}$ (distributivity)

(8) $(\alpha + \beta)\mathbf{u} = \alpha\mathbf{u} + \beta\mathbf{u}$ (distributivity)

Each of these identities is proved by observing that the operations are defined componentwise and the components are numbers that satisfy the same identities. For example, to prove that $\mathbf{u} + \mathbf{v} = \mathbf{v} + \mathbf{u}$, we must prove that for each $i = 1, 2, \ldots, n$, the $i$th component of $\mathbf{u} + \mathbf{v}$ is equal to the $i$th component of $\mathbf{v} + \mathbf{u}$; after all, two vectors are equal if and only if they are equal componentwise. Since addition is defined componentwise, the $i$th component of $\mathbf{u} + \mathbf{v}$ is $u_i + v_i$ and $i$th component of $\mathbf{v} + \mathbf{u}$ is $v_i + u_i$. Of course, $u_i + v_i = v_i + u_i$ because $u_i$ and $v_i$ are real numbers and addition of real numbers is commutative. The remaining identities are proved similarly.

**Proof of 2** $(\mathbf{u} + \mathbf{v}) + \mathbf{w} = \mathbf{u} + (\mathbf{v} + \mathbf{w})$ since

$$(u_i + v_i) + w_i = u_i + (v_i + w_i) \qquad \text{for} \quad i = 1, 2, \ldots, n.$$

**Proof of 3** $\mathbf{u} + \mathbf{0} = \mathbf{u}$ since

$$u_i + 0 = u_i \qquad \text{for} \quad i = 1, 2, \ldots, n.$$

**Proof of 4** $\mathbf{u} - \mathbf{u} = \mathbf{0}$ since

$$u_i - u_i = 0 \qquad \text{for} \quad i = 1, 2, \ldots, n.$$

**Proof of 5** $(\alpha\beta)\mathbf{u} = \alpha(\beta\mathbf{u})$ since

$$(\alpha\beta)u_i = \alpha(\beta u_i) \qquad \text{for} \quad i = 1, 2, \ldots, n.$$

**Proof of 6** $1 \cdot \mathbf{u} = \mathbf{u}$ since

$$1 \cdot u_i = u_i \qquad \text{for} \quad i = 1, 2, \ldots, n.$$

**Proof of 7** $\alpha(\mathbf{u} + \mathbf{v}) = \alpha\mathbf{u} + \alpha\mathbf{v}$ since

$$\alpha(u_i + v_i) = \alpha u_i + \alpha v_i \qquad \text{for} \quad i = 1, 2, \ldots, n.$$

**Proof of 8** $(\alpha + \beta)\mathbf{u} = \alpha\mathbf{u} + \beta\mathbf{u}$ since

$$(\alpha + \beta)u_i = \alpha u_i + \beta u_i \qquad \text{for} \quad i = 1, 2, \ldots, n.$$

Having defined addition and scalar multiplication for vectors in $R^n$, it is clear how we should define these operations for row vectors and column vectors. Two $n$-dimensional row vectors

$$\mathbf{u} = [u_1 \quad u_2 \quad \cdots \quad u_n] \qquad \text{and} \qquad \mathbf{v} = [v_1 \quad v_2 \quad \cdots \quad v_n]$$

are represented by the two vectors

$$\mathbf{u} = (u_1, u_2, \ldots, u_n) \qquad \text{and} \qquad \mathbf{v} = (v_1, v_2, \ldots, v_n)$$

in $R^n$. The sum of these two vectors is

$$\mathbf{u} + \mathbf{v} = (u_1 + v_1, u_2 + v_2, \ldots, u_n + v_n).$$

We define the sum of the row vectors $\mathbf{u}$ and $\mathbf{v}$ to be the row vector specified by this vector. Thus

$$\mathbf{u} + \mathbf{v} = [u_1 + v_1 \quad u_2 + v_2 \quad \cdots \quad u_n + v_n].$$

For exactly the same reasons, we define a scalar times an $n$-dimensional row vector $\mathbf{u}$ by the equation

$$\alpha\mathbf{u} = [\alpha u_1 \quad \alpha u_2 \quad \cdots \quad \alpha u_n].$$

Addition and scalar multiplication of column vectors is handled in the same way for the same reason. If

$$\mathbf{u} = \begin{bmatrix} u_1 \\ u_2 \\ \vdots \\ u_n \end{bmatrix} \qquad \text{and} \qquad \mathbf{v} = \begin{bmatrix} v_1 \\ v_2 \\ \vdots \\ v_n \end{bmatrix},$$

then

$$\mathbf{u} + \mathbf{v} = \begin{bmatrix} u_1 + v_1 \\ u_2 + v_2 \\ \vdots \\ u_n + v_n \end{bmatrix}$$

and

$$\alpha\mathbf{u} = \begin{bmatrix} \alpha u_1 \\ \alpha u_2 \\ \vdots \\ \alpha u_n \end{bmatrix}.$$

It follows from the way we have defined things that the addition and scalar multiplication of row or column vectors corresponds to the addition and scalar multiplication of the vectors that represent them.

## PROBLEMS 1.3

**1.** Complete the following definitions.
(a) If $\mathbf{u}$ and $\mathbf{v}$ are $n$-vectors, then their sum is....
(b) $R^n$ is....
(c) The transpose of the row vector $[u_1 \quad u_2 \quad \cdots \quad u_n]$ is....
(d) An $n \times m$ matrix is....
(e) The vector $-\mathbf{v}$ is....

**2.** Let $\mathbf{u} = (4, -3, 2, 1)$ and $\mathbf{v} = (4, 3, 2)$.
(a) What is the third coordinate of $\mathbf{u}$? the third coordinate of $\mathbf{v}$?
(b) Explain why we cannot compute $\mathbf{u} + \mathbf{v}$.

***3.** Let $\mathbf{u} = (3, -1)$, $\mathbf{v} = (2, 5)$, and $\mathbf{w} = (0, 4)$. Compute $\mathbf{u} + \mathbf{v}$, $8\mathbf{u}$, $-3\mathbf{w}$, and $6\mathbf{u} + 2\mathbf{v} - \mathbf{w}$.

**4.** Let $\mathbf{u} = (1, -2, 4)$, $\mathbf{v} = (-3, 0, 9)$, and $\mathbf{w} = (0, 0, 5)$. Compute $\mathbf{u} - \mathbf{v}$, $-3\mathbf{u} + 5\mathbf{v}$, $\mathbf{u} + \mathbf{v} - 6\mathbf{w}$.

**5.** If $\mathbf{u} = (1, 3, 0, 6)$ and $\mathbf{v} = (2, 1, -1, 4)$, find a 4-vector $\mathbf{x}$ such that:
*(a) $3\mathbf{x} = 2\mathbf{u} + \mathbf{v}$
(b) $3\mathbf{u} + 2\mathbf{x} = \mathbf{u} - 5\mathbf{v}$

**6.** Show that $0 \cdot \mathbf{u} = \mathbf{0}$ and $\alpha \cdot \mathbf{0} = \mathbf{0}$

***7.** Show that if $\mathbf{u} + \mathbf{v} = \mathbf{0}$, then $\mathbf{u} = -\mathbf{v}$.

**8.** Show that $(-1)(-\mathbf{u}) = \mathbf{u}$.

**9.** Show that if $\mathbf{u} + \mathbf{v} = \mathbf{v}$, then $\mathbf{u} = \mathbf{0}$.

**10.** Show that if $\mathbf{u} + \mathbf{v} = \mathbf{u} + \mathbf{w}$, then $\mathbf{v} = \mathbf{w}$.

**11.** Show that if $\mathbf{x} + \mathbf{u} = \mathbf{v}$, then $\mathbf{x} = \mathbf{v} - \mathbf{u}$.

***12.** Show that if $\alpha\mathbf{u} = \mathbf{0}$, then $\alpha = 0$ or $\mathbf{u} = \mathbf{0}$.

**13.** Show that if $(\alpha + \beta)\mathbf{u} = (\alpha + \gamma)\mathbf{u}$ and $\mathbf{u} \neq 0$, then $\beta = \gamma$.

**14.** Show that $(\mathbf{v}^T)^T = \mathbf{v}$.

**15.** Show that $(\mathbf{u} + \mathbf{v})^T = \mathbf{u}^T + \mathbf{v}^T$.

**16.** Show that $(\alpha\mathbf{u})^T = \alpha\mathbf{u}^T$.

# 1.4
# Geometry of Vectors

Now for some geometry. For many purposes it is convenient to view a vector **v** as an arrow whose tail is at the origin and whose head is at the point **v**. This interpretation of a vector allows us to draw nice geometric pictures and is used to represent such physical concepts as force, velocity, and acceleration. For example, both addition and subtraction have nice geometric interpretations when we represent the vectors as arrows. In $R^2$, if $\mathbf{u} = (u_1, u_2)$ and $\mathbf{v} = (v_1, v_2)$, then $\mathbf{u} + \mathbf{v} = (u_1 + v_1, u_2 + v_2)$ is the diagonal of the parallelogram determined by **u** and **v** (see Figure 1.8). Subtraction can be interpreted similarly. Since $\mathbf{u} - \mathbf{v}$ is the vector that when added to **v** gives **u**, we have the picture shown in Figure 1.9.

Multiplication of the vector $\mathbf{u} = (u_1, u_2)$ by a positive scalar $\alpha$ stretches ($\alpha > 1$) or shrinks ($0 < \alpha < 1$) the vector. If $\alpha < 0$, then $\alpha\mathbf{u}$ points in the opposite direction from **u** and is stretched ($|\alpha| > 1$) or shrunk ($0 < |\alpha| < 1$) (see Figure 1.10). In any case, $\alpha\mathbf{u}$ lies on the same line through the origin as **u**. Two vectors that lie on the same line through the origin are called **collinear**.

With slightly more difficulty we could draw these pictures in $R^3$. In higher dimensions we will continue to represent vectors by arrows, but of course we will not draw coordinate axes. We merely specify the origin **0** of our higher-dimensional space and draw our vectors as arrows whose tails are at **0** (see Figure 1.11).

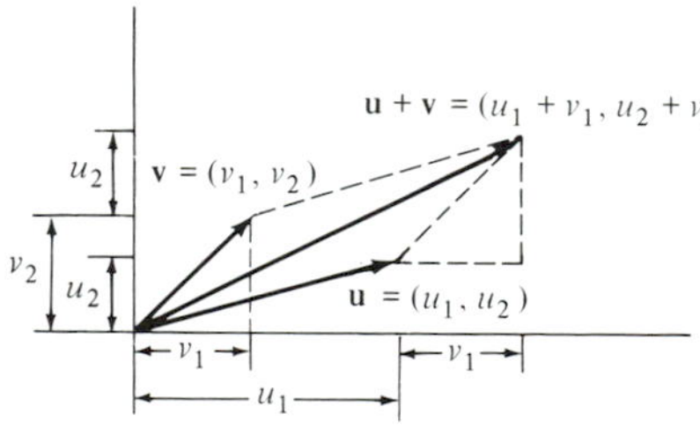


Figure 1.8

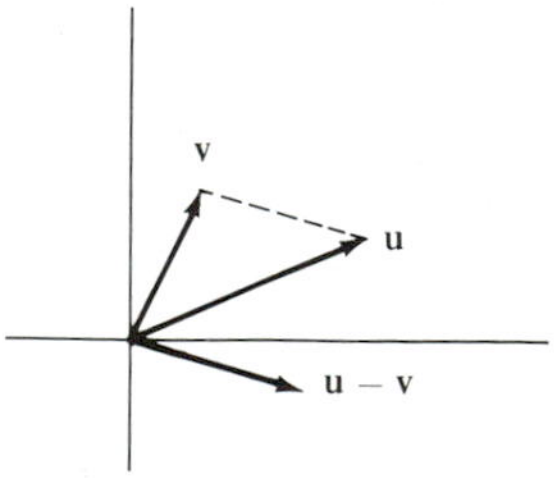


Figure 1.9

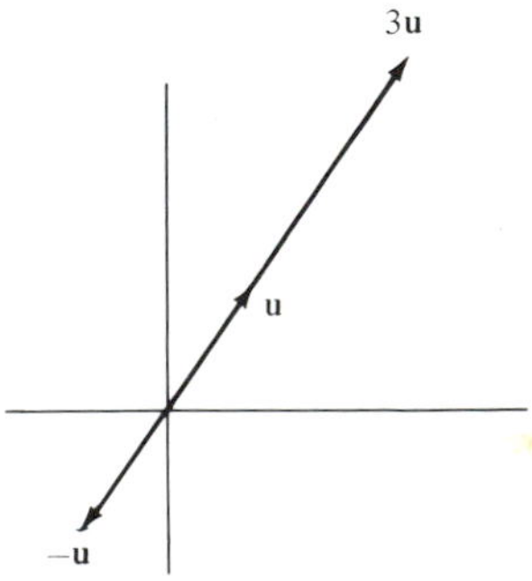


Figure 1.10

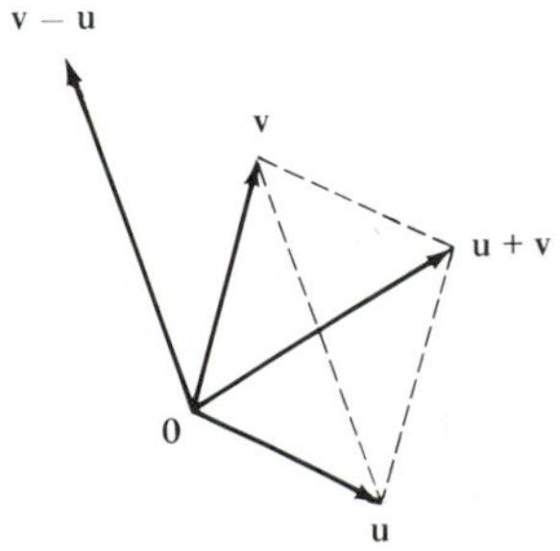


Figure 1.11

We are now in a position to determine vector equations for lines and planes. We begin with lines in $R^2$. If $\mathbf{u} = (u_1, u_2) \neq (0,0)$ and $\alpha$ is a scalar, then $\alpha\mathbf{u}$ lies on the same line through the origin as $\mathbf{u}$. Conversely, any point $\mathbf{x} = (x_1, x_2)$ on this line is a scalar multiple of $\mathbf{u}$. For if $x_1 \neq 0$, there is a scalar $\alpha$ such that $x_1 = \alpha u_1$. From similar triangles it then follows that $x_2 = \alpha u_2$, so that $\mathbf{x} = \alpha\mathbf{u}$. If $u_1 = 0$, then $x_1 = 0$ and there is a scalar $\alpha$ such that $x_2 = \alpha u_2$. Hence $\mathbf{x} = \alpha\mathbf{u}$ (see Figure 1.12). Thus $\mathbf{x}$ is on the line if and only if $\mathbf{x}$ is a scalar multiple of $\mathbf{u}$. This allows us to define

$$\mathbf{x} = \alpha\mathbf{u}, \qquad \alpha \text{ a scalar},$$

to be a vector equation of the line through the origin in the direction of $\mathbf{u}$. The beauty of this equation is that it does not care about the fact that $\mathbf{x}$ and $\mathbf{u}$ are 2-vectors. *Given a nonzero vector* $\mathbf{u}$ *in* $R^n$, *we define*

$$\mathbf{x} = \alpha\mathbf{u}, \qquad \alpha \textit{ a scalar},$$

*to be a vector equation of the line in* $R^n$ *through the origin in the direction of* $\mathbf{u}$.

For example, if $\mathbf{u} = (1, 2)$, then a vector equation of the line in $R^2$ through the origin in the direction of $\mathbf{u}$ is

$$\mathbf{x} = \alpha\mathbf{u} \qquad \text{or} \qquad (x_1, x_2) = \alpha(1, 2).$$

Equating components we obtain $x_1 = \alpha, x_2 = 2\alpha$. These are parametric equations of the line. If we eliminate the parameter, we obtain a Cartesian equation of the line $2x_1 - x_2 = 0$.

If $\mathbf{u} = (3, 2, 1)$, then a vector equation of the line in $R^3$ through the origin in the direction of $\mathbf{u}$ is $\mathbf{x} = \alpha\mathbf{u}$, or

$$(x_1, x_2, x_3) = \alpha(3, 2, 1).$$

Equating components, we obtain parametric equations of the line:

$$x_1 = 3\alpha, \qquad x_2 = 2\alpha, \qquad x_3 = \alpha.$$

Note that if we eliminate the parameter $\alpha$, we need a system of two Cartesian equations

$$x_1 - 3x_3 = 0, \qquad x_2 - 2x_3 = 0$$

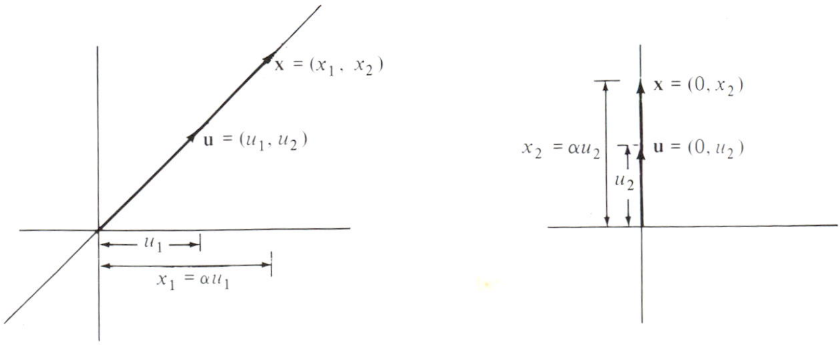


Figure 1.12

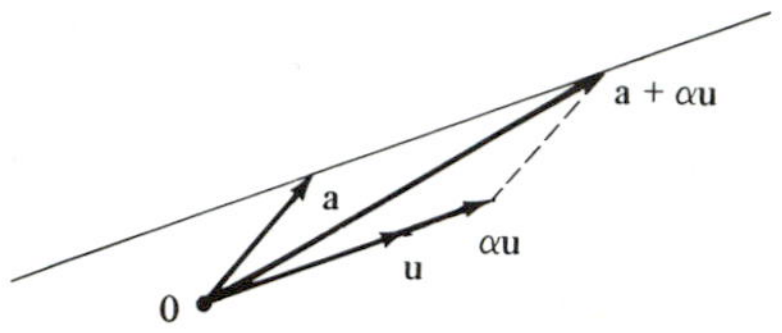


Figure 1.13

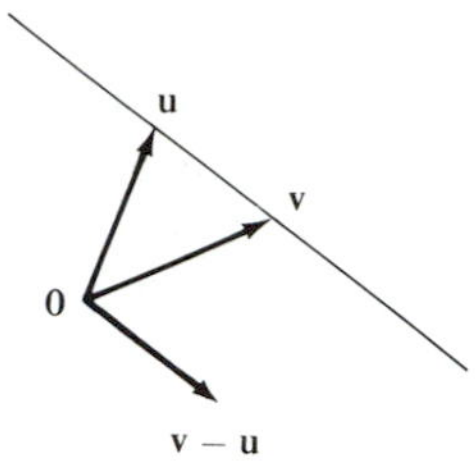


Figure 1.14

to specify the line. It is impossible to specify a line in $R^3$ with a single linear Cartesian equation.

If $\mathbf{u} = (1, 0, 2, 5, 7)$, then a vector equation of the line in $R^5$ through the origin in the direction of $\mathbf{u}$ is $\mathbf{x} = \alpha\mathbf{u}$, or $(x_1, x_2, x_3, x_4, x_5) = \alpha(1, 0, 2, 5, 7)$. Again equating components we obtain $x_1 = \alpha$, $x_2 = 0$, $x_3 = 2\alpha$, $x_4 = 5\alpha$, $x_5 = 7\alpha$, which are parameteric equations of the line.

It should be clear from the parallelogram law for addition (see Figure 1.13) that a *vector equation of the line through a vector* **a** *parallel to a nonzero vector* **u** *is given by*

$$\mathbf{x} = \mathbf{a} + \alpha\mathbf{u}, \qquad \alpha \textit{ a scalar}.$$

For example, a vector equation of the line in $R^3$ through $\mathbf{a} = (1, 1, 1)$ and parallel to $\mathbf{u} = (1, 2, -1)$ is

$$(x_1, x_2, x_3) = (1, 1, 1) + \alpha(1, 2, -1).$$

Equating components, we obtain parametric equations of the line:

$$x_1 = 1 + \alpha, \qquad x_2 = 1 + 2\alpha, \qquad x_3 = 1 - \alpha.$$

If **u** and **v** are distinct points in $R^n$, we see from our geometric interpretation of subtraction that the line through **u** and **v** is parallel to the vector $\mathbf{v} - \mathbf{u}$ (see Figure 1.14). *Thus a vector equation of the line through* **u** *and* **v** *is*

$$\mathbf{x} = \mathbf{u} + \alpha(\mathbf{v} - \mathbf{u}), \qquad \alpha \textit{ a scalar}.$$

It is sometimes useful to write this equation in the form

$$\mathbf{x} = (1 - \alpha)\mathbf{u} + \alpha\mathbf{v}.$$

For example, a vector equation of the line through $\mathbf{u} = (1, 1, 1)$ and $\mathbf{v} = (2, -1, 2)$ is

$$(x_1, x_2, x_3) = (1, 1, 1) + \alpha(1, -2, 1).$$

## EXAMPLE 1

To determine whether the vector $\mathbf{w} = (4, 3, -2, -1)$ lies on the line through $\mathbf{u} = (1, 0, 1, 2)$ and $\mathbf{v} = (2, 1, 0, 1)$, we must determine whether **w** is a solution of the equation

$$\mathbf{x} = (1, 0, 1, 2) + \alpha(1, 1, -1, -1).$$

That is, we must determine if there is a scalar $\alpha$ such that

$$4 = 1 + \alpha, \qquad 3 = \alpha, \qquad -2 = 1 - \alpha, \qquad -1 = 2 - \alpha.$$

Clearly, $\alpha = 3$ is a solution and therefore $\mathbf{w}$ is on the line.

Now what about planes in $R^3$? Let $\mathbf{u}$ and $\mathbf{v}$ be noncollinear vectors in $R^3$. Consider a vector of the form $\alpha\mathbf{u} + \beta\mathbf{v}$, where $\alpha$ and $\beta$ are scalars. It follows from the parallelogram law for addition that $\alpha\mathbf{u} + \beta\mathbf{v}$ lies in the plane through the origin, $\mathbf{u}$, and $\mathbf{v}$. Conversely, given any vector $\mathbf{x}$ in this plane, it is geometrically obvious that $\mathbf{x}$ can be expressed as a sum of scalar multiples of $\mathbf{u}$ and $\mathbf{v}$. To see this, draw a line through $\mathbf{x}$ parallel to $\mathbf{u}$ (see Figure 1.15). Let $\mathbf{v}'$ be the point where this line intersects the line through the origin and $\mathbf{v}$. As we have just seen, $\mathbf{v}' = \beta\mathbf{v}$ for some scalar $\beta$. Similarly, the line through $\mathbf{x}$ parallel to $\mathbf{v}$ intersects the line through the origin and $\mathbf{u}$ at the point $\mathbf{u}' = \alpha\mathbf{u}$. From our construction, $\mathbf{x} = \mathbf{u}' + \mathbf{v}' = \alpha\mathbf{u} + \beta\mathbf{v}$.

Thus if $\mathbf{u}$ and $\mathbf{v}$ are noncollinear vectors in $\mathbf{R}^3$, we define

$$\mathbf{x} = \alpha\mathbf{u} + \beta\mathbf{v}, \qquad \text{where } \alpha \text{ and } \beta \text{ are scalars,}$$

to be a vector equation of the plane in $R^3$ through the origin and the vectors $\mathbf{u}$ and $\mathbf{v}$. Again, the beauty of this equation is that it does not care about the fact that $\mathbf{x}$, $\mathbf{u}$, and $\mathbf{v}$ are vectors in $R^3$. Consequently, *given two noncollinear vectors* $\mathbf{u}$ *and* $\mathbf{v}$ *in* $R^n$, *we define*

$$\mathbf{x} = \alpha\mathbf{u} + \beta\mathbf{v}, \qquad \textit{where } \alpha \textit{ and } \beta \textit{ are scalars,}$$

*to be a vector equation of the plane in* $R^n$ *passing through the origin,* $\mathbf{u}$, *and* $\mathbf{v}$.

For example, a vector equation of the plane in $R^3$ through the origin, $\mathbf{u} = (1, 2, 1)$, and $\mathbf{v} = (-1, 1, 0)$ is

$$(x_1, x_2, x_3) = \alpha(1, 2, 1) + \beta(-1, 1, 0).$$

Similarly, a vector equation of the plane in $R^5$ through the origin, $\mathbf{u} = (1, -1, 2, 1, 1)$,

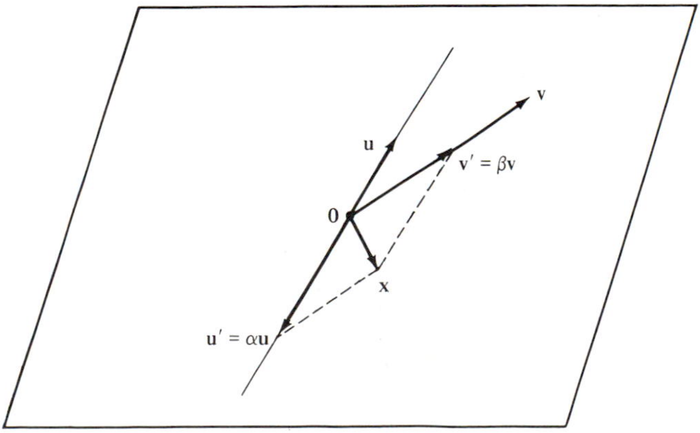


Figure 1.15

and $\mathbf{v} = (1, 1, 2, 2, -1)$ is

$$\mathbf{x} = \alpha(1, -1, 2, 1, 1) + \beta(1, 1, 2, 2, -1).$$

Arguments similar to those given for lines should convince you that (see Figure 1.16):

1. *If* $\mathbf{u}$ *and* $\mathbf{v}$ *are noncollinear vectors in* $R^n$*, an equation of the plane through a point* $\mathbf{a}$ *in* $R^n$ *and parallel to the plane* $\mathbf{x} = \alpha\mathbf{u} + \beta\mathbf{v}$ *is*

$$\mathbf{x} = \mathbf{a} + \alpha\mathbf{u} + \beta\mathbf{v}.$$

2. *An equation of the plane through three noncollinear vectors* $\mathbf{u}$, $\mathbf{v}$, *and* $\mathbf{w}$ *is*

$$\mathbf{x} = \mathbf{u} + \alpha(\mathbf{v} - \mathbf{u}) + \beta(\mathbf{w} - \mathbf{u}).$$

*Note:* When a plane passes through the origin, then the plane literally contains any vector in the plane. On the other hand, when a plane does not pass through the origin, it does not literally contain any vector. It is the tip of the vector, i.e., the point determined by the vector, that is on the plane.

For example, an equation of the plane through (1, 1, 0), (1, 0, 1), and (0, 1, 1) is

$$(x_1, x_2, x_3) = (1, 1, 0) + \alpha(0, -1, 1) + \beta(-1, 0, 1).$$

An equation of the plane through the origin parallel to this plane is

$$(x_1, x_2, x_3) = \alpha(0, -1, 1) + \beta(-1, 0, 1).$$

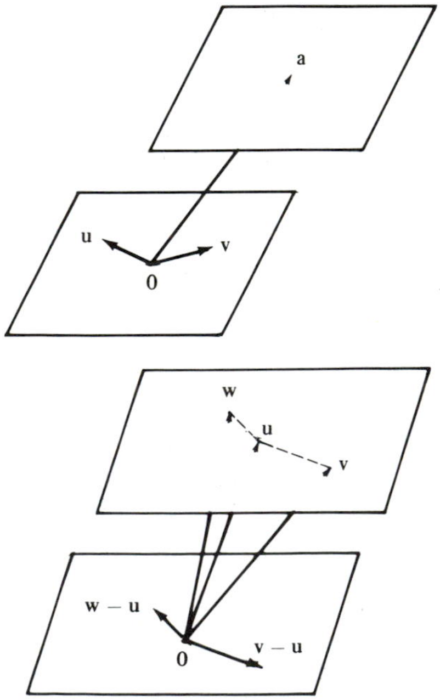


Figure 1.16

### EXAMPLE 2

Consider the problem of determining whether the line through (10, 1, 4) parallel to (−2, −3, −5) intersects the plane through the origin and the vectors (1, 2, 4) and (−2, 2, 3). An equation of the line is

$$\mathbf{x} = (10, 1, 4) + \alpha(-2, -3, -5)$$

and an equation of the plane is

$$\mathbf{x} = \beta(1, 2, 4) + \gamma(-2, 2, 3).$$

The line and the plane intersect if and only if there are scalars $\alpha_0, \beta_0, \gamma_0$ such that

$$(10, 1, 4) + \alpha_0(-2, -3, -5) = \beta_0(1, 2, 4) + \gamma_0(-2, 2, 3).$$

Equating components, we are led to the system of equations

$$\begin{aligned} 2\alpha_0 + \beta_0 - 2\gamma_0 &= 10 \\ 3\alpha_0 + 2\beta_0 + 2\gamma_0 &= 1 \\ 5\alpha_0 + 4\beta_0 + 3\gamma_0 &= 4. \end{aligned}$$

Here is another problem whose solution is determined by a system of linear equations. It turns out that $\alpha_0 = 1, \beta_0 = 2, \gamma_0 = -3$ is a solution of the system. As a check we determine the vector on the line corresponding to $\alpha = 1$ and the vector in the plane corresponding to $\beta = 2$ and $\gamma = -3$:

$$\begin{aligned} (10, 1, 4) + (-2, -3, -5) &= (8, -2, -1) \\ 2(1, 2, 4) - 3(-2, 2, 3) &= (8, -2, -1). \end{aligned}$$

## PROBLEMS 1.4

**1.** Complete the following definitions.
 (a) A vector equation of the line through **u** and **v** is....
 (b) A vector equation of a plane through three noncollinear vectors **u**, **v**, and **w** is....

**2.** In each part, find an equation of the line.
 *(a) The line through the origin and (1, 2, 3, 1).
 *(b) The line through (2, 1, 0, 5) and parallel to (1, 2, 3, 1).
 (c) The line through (3, 0, 3) and (5, 1, 1).
 (d) The line through (1, 1, 1) and parallel to the line in part (c).

***3.** Determine which of the following vectors lie on the line through (1, 2, 1) and (−2, 3, −1).
 (a) (7, 0, 5) (b) (−1, 0, 2)
 (c) $\left(\frac{5}{2}, \frac{3}{2}, 2\right)$

**4.** Determine which of the following vectors lie on the line through (3, 1, 1) and parallel to (5, 4, −1).
 (a) (−7, −7, 3) (b) (8, 5, 1)
 (c) (2, −19, 6)

**5.** In each part, determine whether the three vectors lie on a line.
 *(a) (1, 7), (−1, 3), (0, 5)
 (b) (−3, 1, 1), (−2, −1, 4), (−4, 3, −2)
 *(c) (3, −4, 1), (−1, 5, 2), (2, 1, 3)
 (d) (3, 1, 3, 1), (1, −1, 1, −1), (5, −1, 5, 1)
 (e) (1, 2, 1, −1), (3, 1, 0, 4), (−5, 5, 4, −16)

*6. Show that the vectors (1, 0, 1), (2, 1, 2), (4, 3, 3), (3, 2, 2) form the vertices of a parallelogram.

7. Let **u**, **v**, and **w** be vectors. Show that **w**, **u** + **w**, **v** + **w**, **u** + **v** + **w** form the vertices of a parallelogram.

*8. For what value of $x_3$ will the line through (2, 4, 4) and (1, 2, 1) be parallel to the line through (2, 5, 8) and $(0, 1, x_3)$?

In Problems 9 through 18, find a vector equation of the plane.

*9. The plane through the origin and (1, 2, 3), (−1, 1, −1).

10. The plane through (1, 1, 1) and parallel to the plane in Problem 9.

*11. The plane through (1, 0, 0), (0, 1, 0), (0, 0, 1).

12. The plane through (1, 1, 1) and parallel to the plane $\mathbf{x} = (2, -1, 3) + \alpha(5, 5, -1) + \beta(3, 3, 3)$.

*13. The plane through (5, 5, 0) and parallel to the plane in Problem 11.

14. The plane through (−2, 5, 1) and parallel to the plane $\mathbf{x} = \alpha(-2, 1, 1) + \beta(0, 0, 1)$.

*15. The plane parallel to the $x_1x_2$-plane and containing the vector (1, 5, 4).

16. The plane containing the vector (−1, 0, 1) and the line $\mathbf{x} = (1, 1, 1) + \alpha(1, 7, -1)$.

*17. The plane parallel to the $x_3$-axis and containing the vectors (1, 0, 0) and (0, 1, 0).

18. The plane parallel to (1, 3, 1) and containing the vectors (1, 1, 1) and (−2, 1, 2).

*19. Describe geometrically the possible ways a line and a plane in $R^3$ can intersect.

20. Show that the line $\mathbf{x} = \alpha\mathbf{u}$ and the plane $\mathbf{x} = \alpha'\mathbf{v} + \beta'\mathbf{w}$ intersect.

*21. Show that the two planes

$$\mathbf{x} = \alpha\mathbf{u} + \beta\mathbf{v} \quad \text{and} \quad \mathbf{x} = \alpha'\mathbf{u}' + \beta'\mathbf{v}'$$

intersect.

22. Show that the equations

$$\begin{aligned} x &= 2 + \alpha - 2\beta \\ y &= -1 + \beta \\ z &= \alpha + 2\beta \end{aligned}$$

define a plane in $R^3$.

*23. Consider the line through (1, 1, 1) and parallel to (1, 2, 3). Determine an equation of a plane that contains this line. How many such planes are there?

24. Consider the line through (1, 0, 1) and parallel to (1, 2, 3).
(a) Show that the line does not contain the origin.
(b) Find an equation of the plane containing this line and the origin.

*25. Show that there is one and only one plane that contains a given line and a given vector not on the line.

26. Suppose that the vector **b** lies in the plane

$$\mathbf{x} = \mathbf{a} + \alpha\mathbf{u} + \beta\mathbf{v}.$$

Prove that the line determined by **a** and **b** lies in the plane.

*27. Suppose that the vector **a** does not lie on the plane

$$\mathbf{x} = \alpha\mathbf{u} + \beta\mathbf{v}.$$

Show that the line

$$\mathbf{x} = \mathbf{a} + \gamma\mathbf{u}$$

does not intersect the plane.

28. Suppose that the vector **b** lies in the plane

$$\mathbf{x} = \alpha\mathbf{u} + \beta\mathbf{v}$$

and that the vector **a** does not lie in the plane. Show that the line

$$\mathbf{x} = \mathbf{a} + \gamma\mathbf{b}$$

does not intersect the plane.

29. Suppose that **a** and **u** are not collinear and **b** is on the line $\mathbf{x} = \mathbf{a} + \alpha\mathbf{u}$. Show that $2\mathbf{b}$ is not on the line.

30. Consider the line through two vectors **u** and **v** in $R^2$.
   *(a) Show that for any $\alpha$, $(1 - \alpha)\mathbf{u} + \alpha\mathbf{v}$ lies on this line.
   *(b) Show that if $0 \le \alpha \le 1$, then the vector $(1 - \alpha)\mathbf{u} + \alpha\mathbf{v}$ lies on the line segment joining **u** and **v**.
   (c) Give a geometric argument based on the parallelogram law to show that if $\alpha = 1/2$, then $(1 - \alpha)\mathbf{u} + \alpha\mathbf{v}$ is the midpoint of the line segment joining **u** and **v**.
   (d) If $\alpha = 1/3$, where is $(1 - \alpha)\mathbf{u} + \alpha\mathbf{v}$ on this line segment?
   (e) Generalize parts (c) and (d).

31. Let **u**, **v**, **w** be noncollinear vectors.
   *(a) Show that if $\alpha + \beta + \gamma = 1$, then $\alpha\mathbf{u} + \beta\mathbf{v} + \gamma\mathbf{w}$ lies on the plane determined by **u**, **v**, and **w**.
   (b) Show that every vector in this plane can be written as $\alpha\mathbf{u} + \beta\mathbf{v} + \gamma\mathbf{w}$ when $\alpha + \beta + \gamma = 1$. Hence an equation of a plane through **u**, **v**, and **w** is $\mathbf{x} = \alpha\mathbf{u} + \beta\mathbf{v} + \gamma\mathbf{w}$, where $\alpha + \beta + \gamma = 1$.

32. Let **u** and **v** be noncollinear vectors in $R^2$. Show that the parallelogram determined by **u** and **v** is the collection of all vectors of the form

$$\alpha\mathbf{u} + \beta\mathbf{v}, \qquad 0 \le \alpha \le 1, 0 \le \beta \le 1.$$

33. Show that the diagonals of a parallelogram bisect each other.

*34. Find necessary and sufficient conditions on **a** and **u** that the line $\mathbf{x} = \mathbf{a} + \alpha\mathbf{u}$ passes through the origin.

*35. Let **u** and **v** be noncollinear vectors. If $\mathbf{a} \neq 0$, can the plane $\mathbf{x} = \mathbf{a} + \alpha\mathbf{u} + \beta\mathbf{v}$ pass through the origin?

36. Suppose **u**, **v** are noncollinear and $\mathbf{u}'$, $\mathbf{v}'$ are noncollinear.
   *(a) When will the equations

$$\mathbf{x} = \alpha\mathbf{u} + \beta\mathbf{v}$$

   and

$$\mathbf{x} = \alpha'\mathbf{u}' + \beta'\mathbf{v}'$$

   describe the same plane?
   (b) When will the equations

$$\mathbf{x} = \mathbf{a} + \alpha\mathbf{u} + \beta\mathbf{v}$$

   and

$$\mathbf{x} = \mathbf{a}' + \alpha'\mathbf{u}' + \beta'\mathbf{v}'$$

   describe the same plane?

# 1.5 Matrices

We turn our attention now to matrices. Vectors in $R^n$ specify both $n$-dimensional row vectors and $n$-dimensional column vectors. An $n$-dimensional row vector can be viewed as a $1 \times n$ matrix and an $n$-dimensional column vector can be viewed as an $n \times 1$ matrix. The relation of equality and the operations of addition and scalar multiplication on row and column vectors are consistent with these operations on vectors. Thus we certainly want the relation of equality and the operations of addition and scalar multiplication of matrices to be consistent with these operations on vectors. In light of this remark, we frame our definitions componentwise, beginning with equality. Two $m \times n$ matrices $A$ and $B$ are **equal** if and only if $a_{ij} = b_{ij}$ for $i = 1, 2, \ldots, m$ and $j = 1, 2, \ldots, n$. Hence equality makes sense only for matrices of the same size; the same is true for addition. If $A = [a_{ij}]$ and $B = [b_{ij}]$ are $m \times n$

matrices and if $\alpha$ is a scalar, then:

1. The **sum** $A + B$ is defined to be the $m \times n$ matrix whose $ij$th entry is $a_{ij} + b_{ij}$:

$$A + B = [a_{ij}] + [b_{ij}] = [a_{ij} + b_{ij}].$$

2. The product $\alpha A$ (called **scalar multiplication**) is defined to be the $m \times n$ matrix whose $ij$th entry is $\alpha a_{ij}$:

$$\alpha A = \alpha[a_{ij}] = [\alpha a_{ij}].$$

Writing this out in more detail, let

$$A = \begin{bmatrix} a_{11} & a_{12} & \cdots & a_{1n} \\ a_{21} & a_{22} & \cdots & a_{2n} \\ \vdots & \vdots & & \vdots \\ a_{m1} & a_{m2} & \cdots & a_{mn} \end{bmatrix} \quad \text{and} \quad B = \begin{bmatrix} b_{11} & b_{12} & \cdots & b_{1n} \\ b_{21} & b_{22} & \cdots & b_{2n} \\ \vdots & \vdots & & \vdots \\ b_{m1} & b_{m2} & \cdots & b_{mn} \end{bmatrix}.$$

Then

$$A + B = \begin{bmatrix} a_{11} + b_{11} & a_{12} + b_{12} & \cdots & a_{1n} + b_{1n} \\ a_{21} + b_{21} & a_{22} + b_{22} & \cdots & a_{2n} + b_{2n} \\ \vdots & \vdots & & \vdots \\ a_{m1} + b_{m1} & a_{m2} + b_{m2} & \cdots & a_{mn} + b_{mn} \end{bmatrix}$$

and

$$\alpha A = \begin{bmatrix} \alpha a_{11} & \alpha a_{12} & \cdots & \alpha a_{1n} \\ \alpha a_{21} & \alpha a_{22} & \cdots & \alpha a_{2n} \\ \vdots & \vdots & & \vdots \\ \alpha a_{m1} & \alpha a_{m2} & \cdots & \alpha a_{mn} \end{bmatrix}.$$

We adopt the same notation for subtraction of matrices as we do for numbers and vectors:

$$-A = (-1)A \quad \text{and} \quad A - B = A + (-1)B.$$

The $m \times n$ matrix each of whose components is zero is called the **zero matrix** and is denoted by 0. The zero matrix comes in all sizes; for every $m$ and $n$ there is an $m \times n$ zero matrix.

Addition and scalar multiplication of matrices satisfy the same algebraic identities as vectors satisfy.

Let $A$, $B$, and $C$ be $m \times n$ matrices and let $\alpha$ and $\beta$ be scalars.

I. Properties of addition

| | |
|---|---|
| (1) $A + B = B + A$ | (commutativity) |
| (2) $(A + B) + C = A + (B + C)$ | (associativity) |
| (3) $A + 0 = A$ | (zero matrix) |
| (4) $A - A = 0$ | (additive inverse) |

II. Properties of scalar multiplication

(5) $(\alpha\beta)A = \alpha(\beta A)$ (associativity)

(6) $1 \cdot A = A$ (scalar identity)

(7) $\alpha(A + B) = \alpha A + \alpha B$ (distributivity)

(8) $(\alpha + \beta)A = \alpha A + \beta A$ (distributivity)

As with vectors, each identity follows from the fact that the operations are defined componentwise and the components of the matrix are numbers that satisfy these identities. For example, to prove (1) we let $A = [a_{ij}]$ and $B = [b_{ij}]$ be two $m \times n$ matrices. Then $A + B$ is the $m \times n$ matrix whose $ij$th component is $a_{ij} + b_{ij}$ and $B + A$ is the $m \times n$ matrix whose $ij$th component is $b_{ij} + a_{ij}$. Since addition of numbers is commutative, $a_{ij} + b_{ij} = b_{ij} + a_{ij}$. Therefore, $A + B = B + A$.

This proof can be written very compactly as follows.

$$A + B = [a_{ij}] + [b_{ij}] = [a_{ij} + b_{ij}] = [b_{ij} + a_{ij}] = [b_{ij}] + [a_{ij}] = B + A.$$

Without this notation it becomes

$$\begin{aligned} A + B &= \begin{bmatrix} a_{11} & a_{12} & \cdots & a_{1n} \\ a_{21} & a_{22} & \cdots & a_{2n} \\ \vdots & \vdots & & \vdots \\ a_{m1} & a_{m2} & \cdots & a_{mn} \end{bmatrix} + \begin{bmatrix} b_{11} & b_{12} & \cdots & b_{1n} \\ b_{21} & b_{22} & \cdots & b_{2n} \\ \vdots & \vdots & & \vdots \\ b_{m1} & b_{m2} & \cdots & b_{mn} \end{bmatrix} \\ &= \begin{bmatrix} a_{11} + b_{11} & a_{12} + b_{12} & \cdots & a_{1n} + b_{1n} \\ a_{21} + b_{21} & a_{22} + b_{22} & \cdots & a_{2n} + b_{2n} \\ \vdots & \vdots & & \vdots \\ a_{m1} + b_{m1} & a_{m2} + b_{m2} & \cdots & a_{mn} + b_{mn} \end{bmatrix} \\ &= \begin{bmatrix} b_{11} + a_{11} & b_{12} + a_{12} & \cdots & b_{1n} + a_{1n} \\ b_{21} + a_{21} & b_{22} + a_{22} & \cdots & b_{2n} + a_{2n} \\ \vdots & \vdots & & \vdots \\ b_{m1} + a_{m1} & b_{m2} + a_{m2} & \cdots & b_{mn} + a_{mn} \end{bmatrix} \\ &= B + A. \end{aligned}$$

Good notation is worth a lot. But to be useful it must be understood. Write out the proofs of the remaining properties using our compact notation. Then write out a couple of them without this notation to make sure that you understand what is being said.

We have one more operation to define to complete our program of representing the linear system

$$\begin{aligned} a_{11}x_1 + a_{12}x_2 + \cdots + a_{1n}x_n &= b_1 \\ a_{21}x_1 + a_{22}x_2 + \cdots + a_{2n}x_n &= b_2 \\ \vdots \qquad \vdots \qquad\quad \vdots \quad &\quad \vdots \\ a_{m1}x_1 + a_{m2}x_2 + \cdots + a_{mn}x_n &= b_m \end{aligned} \tag{1}$$

by the single matrix equation $A\mathbf{x} = \mathbf{b}$. You guessed it! We must define the product of a matrix with a vector. We can view the system as the vector equation since two vectors are equal if and only if they are equal componentwise.

$$\begin{bmatrix} a_{11}x_1 + a_{12}x_2 + \cdots + a_{1n}x_n \\ a_{21}x_1 + a_{22}x_2 + \cdots + a_{2n}x_n \\ \vdots \quad\quad \vdots \quad\quad\quad \vdots \\ a_{m1}x_1 + a_{m2}x_2 + \cdots + a_{mn}x_n \end{bmatrix} = \begin{bmatrix} b_1 \\ b_2 \\ \vdots \\ b_m \end{bmatrix} \tag{2}$$

It is apparent now how to define $A\mathbf{x}$ so that $A\mathbf{x} = \mathbf{b}$:

$$A\mathbf{x} = \begin{bmatrix} a_{11} & a_{12} & \cdots & a_{1n} \\ a_{21} & a_{22} & \cdots & a_{2n} \\ \vdots & \vdots & & \vdots \\ a_{m1} & a_{m2} & \cdots & a_{mn} \end{bmatrix} \begin{bmatrix} x_1 \\ x_2 \\ \vdots \\ x_n \end{bmatrix} = \begin{bmatrix} a_{11}x_1 + a_{12}x_2 + \cdots + a_{1n}x_n \\ a_{21}x_1 + a_{22}x_2 + \cdots + a_{2n}x_n \\ \vdots \quad\quad \vdots \quad\quad\quad \vdots \\ a_{m1}x_1 + a_{m2}x_2 + \cdots + a_{mn}x_n \end{bmatrix}.$$

There is our definition of the **product** of a matrix and a column vector. Clearly, the product is defined only if the number of columns of the matrix is equal to the number of rows of the column vector. An $m \times n$ matrix $A$ times an $n$-dimensional column vector $\mathbf{x}$ is the $m$-dimensional column vector whose $i$th component is obtained from the $i$th row of $A$ and the column vector $\mathbf{x}$ by multiplying corresponding components and adding the result. Most people do this by running their left index finger along the $i$th row of $A$ and their right index finger down the column $\mathbf{x}$, multiplying corresponding entries and adding the result. For example, the second component on the right side is the number

$$[a_{21} \quad a_{22} \quad \cdots \quad a_{2n}] \begin{bmatrix} x_1 \\ x_2 \\ \vdots \\ x_n \end{bmatrix} = a_{21}x_1 + a_{22}x_2 + \cdots + a_{2n}x_n.$$

Forming sums of products of corresponding components in this way is an extremely useful vector operation.

***Definition*** If $\mathbf{u}$ and $\mathbf{v}$ are vectors in $R^n$, we define their **inner product** (or **dot product**) by the equation

$$\langle \mathbf{u}, \mathbf{v} \rangle = u_1v_1 + u_2v_2 + \cdots + u_nv_n.$$

Notice that $\langle \mathbf{u}, \mathbf{v} \rangle$ can be expressed as a matrix product:

$$\langle \mathbf{u}, \mathbf{v} \rangle = u_1v_1 + u_2v_2 + \cdots + u_nv_n = [u_1 \quad u_2 \quad \cdots \quad u_n] \begin{bmatrix} v_1 \\ v_2 \\ \vdots \\ v_n \end{bmatrix}.$$

If both $\mathbf{u}$ and $\mathbf{v}$ represent column vectors, then we can express this equation as

$$\langle \mathbf{u}, \mathbf{v} \rangle = \mathbf{u}^T\mathbf{v}.$$

### EXAMPLE 1

(a) $$\begin{bmatrix} 2 & 4 & -2 & 3 \\ -1 & 0 & 2 & 3 \end{bmatrix}\begin{bmatrix} 4 \\ 1 \\ -1 \\ 3 \end{bmatrix}$$

$$= \begin{bmatrix} 2\cdot 4 + 4\cdot 1 + (-2)\cdot(-1) + 3\cdot 3 \\ (-1)\cdot 4 + 0\cdot 1 + 2\cdot(-1) + 3\cdot 3 \end{bmatrix} = \begin{bmatrix} 23 \\ 3 \end{bmatrix}$$

(b) $$\begin{bmatrix} 1 & -3 \\ 2 & 5 \\ 1 & 4 \end{bmatrix}\begin{bmatrix} -1 \\ 4 \end{bmatrix} = \begin{bmatrix} 1\cdot(-1) + (-3)\cdot 4 \\ 2\cdot(-1) + 5\cdot 4 \\ 1\cdot(-1) + 4\cdot 4 \end{bmatrix} = \begin{bmatrix} -13 \\ 18 \\ 15 \end{bmatrix}$$

(c) $$\begin{bmatrix} 2 & 3 & -1 \\ 1 & 0 & 1 \\ 4 & -3 & 1 \end{bmatrix}\begin{bmatrix} 1 \\ 2 \\ 3 \end{bmatrix} = \begin{bmatrix} 2\cdot 1 + 3\cdot 2 + (-1)\cdot 3 \\ 1\cdot 1 + 0\cdot 2 + 1\cdot 3 \\ 4\cdot 1 + (-3)\cdot 2 + 1\cdot 3 \end{bmatrix} = \begin{bmatrix} 5 \\ 4 \\ 1 \end{bmatrix}$$

(d) If $\mathbf{u} = (1, 2, -1)$ and $\mathbf{v} = (-1, 5, 1)$, then

$$\langle \mathbf{u}, \mathbf{v} \rangle = 1\cdot(-1) + 2\cdot 5 + (-1)\cdot 1 = 8.$$

We motivated the definition of the product of a matrix and a column vector by viewing the system of equations (1) a single vector equation (2). The left side of (2) can be further decomposed as follows:

$$\begin{bmatrix} a_{11}x_1 + a_{12}x_2 + \cdots + a_{1n}x_n \\ a_{21}x_1 + a_{22}x_2 + \cdots + a_{2n}x_n \\ \vdots \qquad \vdots \qquad \vdots \\ a_{m1}x_1 + a_{m2}x_2 + \cdots + a_{mn}x_n \end{bmatrix} = x_1\begin{bmatrix} a_{11} \\ a_{21} \\ \vdots \\ a_{m1} \end{bmatrix} + x_2\begin{bmatrix} a_{12} \\ a_{22} \\ \vdots \\ a_{m2} \end{bmatrix} + \cdots + x_n\begin{bmatrix} a_{1n} \\ a_{2n} \\ \vdots \\ a_{mn} \end{bmatrix}.$$

Thus we can interpret the product $A\mathbf{x}$ as $x_1$ times the first column of $A$ plus $x_2$ times the second column of $A$ plus and so on plus $x_n$ times the last column of $A$. That is, $A\mathbf{x}$ is a sum of scalar multiples of the columns of $A$. Remember that! It is an extremely important and useful way to view the product $A\mathbf{x}$.

### EXAMPLE 2

Let $A$ be the matrix

$$A = \begin{bmatrix} 3 & 2 & 3 \\ 2 & -1 & 1 \\ 1 & 1 & -1 \end{bmatrix}.$$

Then

$$A\mathbf{x} = \begin{bmatrix} 3 & 2 & 3 \\ 2 & -1 & 1 \\ 1 & 1 & -1 \end{bmatrix} \begin{bmatrix} x_1 \\ x_2 \\ x_3 \end{bmatrix} = \begin{bmatrix} 3x_1 + 2x_2 + 3x_3 \\ 2x_1 - x_2 + x_3 \\ x_1 + x_2 - x_3 \end{bmatrix}$$

$$= x_1 \begin{bmatrix} 3 \\ 2 \\ 1 \end{bmatrix} + x_2 \begin{bmatrix} 2 \\ -1 \\ 1 \end{bmatrix} + x_3 \begin{bmatrix} 3 \\ 1 \\ -1 \end{bmatrix}.$$

In particular, if

$$\mathbf{x} = \begin{bmatrix} 2 \\ -1 \\ 3 \end{bmatrix},$$

then $A\mathbf{x}$ is twice the first column of $A$ minus the second column of $A$ plus three times the third column of $A$.

Since sums of scalar multiples of vectors occur so frequently in linear algebra, we make the following definition.

***Definition*** A **linear combination** of vectors $\mathbf{v}_1, \mathbf{v}_2, \ldots, \mathbf{v}_n$ is a sum of scalar multiples of $\mathbf{v}_1, \mathbf{v}_2, \ldots, \mathbf{v}_n$. That is, it is a sum of the form

$$\alpha_1\mathbf{v}_1 + \alpha_2\mathbf{v}_2 + \cdots + \alpha_n\mathbf{v}_n,$$

where $\alpha_1, \alpha_2, \ldots, \alpha_n$ are scalars, called the **coefficients**.

The concept of a linear combination of vectors is central to all of linear algebra. As we have just seen, $A\mathbf{x}$ is a linear combination of the columns of $A$ whose coefficients are the components of $\mathbf{x}$. In particular, if we let

$$\mathbf{e}_1 = \begin{bmatrix} 1 \\ 0 \\ 0 \\ \vdots \\ 0 \end{bmatrix}, \quad \mathbf{e}_2 = \begin{bmatrix} 0 \\ 1 \\ 0 \\ \vdots \\ 0 \end{bmatrix}, \quad \ldots, \quad \mathbf{e}_n = \begin{bmatrix} 0 \\ 0 \\ 0 \\ \vdots \\ 1 \end{bmatrix},$$

then if $A$ is an $m \times n$ matrix,

$$A\mathbf{e}_i = i\text{th column of } A.$$

**Result 1** Let $A$ and $B$ be $m \times n$ matrices. If $A\mathbf{x} = B\mathbf{x}$ for all $\mathbf{x}$ in $R^n$, then $A = B$.

**Proof** Since $A\mathbf{x} = B\mathbf{x}$ for all $\mathbf{x}$ in $R^n$, $A\mathbf{e}_i = B\mathbf{e}_i$ for $i = 1, 2, \ldots, n$. Thus $A$ and $B$ have the same columns and hence are equal.

A **solution** of the system $A\mathbf{x} = \mathbf{b}$ where $A$ is an $m \times n$ matrix and $\mathbf{b}$ is an $m$-dimensional column vector is an $n$-dimensional column vector $\mathbf{x}$ such that $A\mathbf{x}$ is equal to $\mathbf{b}$. We call the vector $\mathbf{x}$ in $R^n$ that represents the column vector $\mathbf{x}$ a **solution vector**. Note that a solution of the system $A\mathbf{x} = \mathbf{b}$ expresses $\mathbf{b}$ as a linear combination of the columns of $A$.

# PROBLEMS 1.5

**1.** Complete the following definitions.
   (a) Two $m \times n$ matrices $A$ and $B$ are equal if....
   (b) The sum of two $m \times n$ matrices $A$ and $B$ is....
   (c) If $\alpha$ is a scalar and $A$ is an $m \times n$ matrix, then $\alpha A$ is....
   (d) The zero matrix is....
   (e) If $A$ is an $m \times n$ matrix and $\mathbf{x}$ is an $n$-dimensional column vector, then $A\mathbf{x}$ is....
   (f) The inner product of two vectors $\mathbf{u}$ and $\mathbf{v}$ in $R^n$ is....
   (g) A linear combination of the vectors $\mathbf{v}_1, \mathbf{v}_2, \ldots, \mathbf{v}_n$ is....

Problems 2 through 7 refer to the following matrices.

$$A = \begin{bmatrix} -3 & 1 & 2 & 0 \\ 9 & 6 & -4 & 2 \end{bmatrix}$$

$$B = \begin{bmatrix} 1 & -1 & 1 & 5 \\ 0 & 3 & 0 & 2 \end{bmatrix}$$

$$C = \begin{bmatrix} 0 & 0 & 0 \\ -1 & 2 & 3 \\ 0 & 1 & 6 \end{bmatrix}$$

$$D = \begin{bmatrix} 5 & 1 & 0 \\ 1 & 1 & 1 \\ -1 & 1 & 4 \end{bmatrix}$$

$$E = \begin{bmatrix} 2 & 2 & 1 \\ -5 & 1 & 2 \\ 6 & 0 & 6 \end{bmatrix}$$

$$F = \begin{bmatrix} 1 & 2 & -1 \\ 2 & 1 & 2 \\ -1 & 4 & 3 \\ 2 & 0 & 7 \end{bmatrix}$$

$$G = \begin{bmatrix} 3 & 0 & 0 & 0 \\ 0 & 2 & 0 & 0 \\ 0 & 0 & -7 & 0 \\ 0 & 0 & 0 & 1 \end{bmatrix}$$

***2.** Compute $A + B$, $-2A$, $3A - B$, and $2A + 3B$.

**3.** Compute $C + D - E$, $C - (D + E)$, and $2C - 2D - 2E$.

***4.** Let $\mathbf{u}$ and $\mathbf{v}$ be the four-dimensional column vectors determined by the vectors $(0, 2, -1, 1)$ and $(3, 2, 1, -2)$, respectively. Compute $A\mathbf{u}$, $A\mathbf{v}$, $A(\mathbf{u} + \mathbf{v})$, $B\mathbf{u}$, $B(\mathbf{u} + \mathbf{v})$, $A(3\mathbf{v})$, and $G\mathbf{v}$.

**5.** Let $\mathbf{u}$ and $\mathbf{v}$ be the three-dimensional column vectors determined by $(4, -1, 2)$ and $(-1, 2, 1)$, respectively. Compute $(C + D)\mathbf{u}$, $(C + D + E)\mathbf{u}$, and $F\mathbf{v}$.

**6.** For a column vector $\mathbf{x}$ of the correct size, express each of the following as a linear combination of the columns of the matrix.
   (a) $A\mathbf{x}$ (b) $D\mathbf{x}$ (c) $F\mathbf{x}$ (d) $G\mathbf{x}$

**7.** Let

$$\mathbf{b} = \begin{bmatrix} 1 \\ 2 \end{bmatrix}, \quad \mathbf{c} = \begin{bmatrix} 0 \\ 0 \\ 0 \end{bmatrix},$$

$$\mathbf{d} = \begin{bmatrix} -1 \\ 0 \\ 0 \end{bmatrix}, \quad \mathbf{e} = \begin{bmatrix} 2 \\ 3 \\ 4 \\ 5 \end{bmatrix}.$$

Write each of the following as a system of equations.
   (a) $A\mathbf{x} = \mathbf{b}$ (b) $D\mathbf{x} = \mathbf{c}$
   (c) $C\mathbf{x} = \mathbf{d}$ (d) $G\mathbf{x} = \mathbf{e}$

**8.** Rewrite each of the following systems of equations as a matrix equation.

*(a)
$$\begin{aligned} -x_1 + x_2 + 2x_3 &= 1 \\ 2x_1 - 3x_2 - x_3 &= 5 \\ -2x_1 + x_2 + 7x_3 &= 9 \end{aligned}$$

(b)
$$\begin{aligned} x_1 - 3x_2 + x_3 + x_4 &= 0 \\ 2x_1 - 6x_2 - x_3 - x_4 &= 0 \\ -x_1 + 3x_2 + x_3 + 2x_4 &= 0 \end{aligned}$$

*(c)
$$\begin{aligned} 3x_1 - x_2 + x_3 &= 1 \\ 3x_1 + x_2 \qquad &= 0 \\ x_2 - 3x_3 &= -1 \\ 6x_1 + x_2 - x_3 &= 2 \end{aligned}$$

(d)
$$\begin{aligned} x_1 + 2x_2 + 3x_3 &= 1 \\ 2x_1 + x_2 - x_3 &= 0 \\ 2x_1 - x_2 \qquad &= 0 \end{aligned}$$

*(e)
$$\begin{aligned} x_1 + x_2 + x_3 + x_4 + x_5 &= 1 \\ 2x_3 \qquad - x_5 &= 1 \\ 2x_1 + 2x_2 + x_3 - x_4 - x_5 &= 1 \\ x_1 + x_2 - x_3 - x_4 + x_5 &= 1 \end{aligned}$$

(f)
$$\begin{aligned} 3x_1 - x_2 &= 0 \\ 2x_1 + x_2 &= -2 \\ 7x_1 + x_2 &= -4 \end{aligned}$$

*(g) $2x_1 + 3x_2 = 1$

(h) $-x_1 + x_3 - 2x_4 = 5$

**9.** Compute the inner product of each of the following pairs of vectors.

*(a) $(1, 2), (-2, 1)$

(b) $(1, 1), (-4, 3)$

*(c) $(1, 2, 3), (3, 2, 1)$

(d) $(0, 0, 0), (1, 2, 7)$

*(e) $(2, 1, -3, 5), (4, 3, -2, 1)$

(f) $(1, -1, 1, -1), (2, -5, 1, 1)$

***10.** If $A$ is the $m \times n$ zero matrix, then $A\mathbf{x} = \mathbf{0}$ for all $\mathbf{x}$ in $R^n$. Is the converse true?

**11.** If $A$ is an $m \times n$ matrix and $\mathbf{0}$ is the $n$-dimensional column vector each of whose components is zero, show that

$$A\mathbf{0} = \mathbf{0}.$$

Problems 12 and 18 refer to properties (2) through (8) of addition and scalar multiplication of matrices.

***12.** Prove property (2).

**13.** Prove property (3).

**14.** Prove property (4).

***15.** Prove property (5).

**16.** Prove property (6).

**17.** Prove property (7).

**18.** Prove property (8).

# 1.6
# Matrix Multiplication

There is a natural way to define the product of two matrices in terms of the machinery we have developed. Let $A$ and $B$ be two matrices. Consider $B$ as a collection of column vectors and define the product $AB$ to be the matrix whose $j$th column is $A$ times the $j$th column of $B$. Of course this means that if $A$ is $m \times n$, then each column of $B$ must be $n$-vector; however, $B$ can have any number of columns.

***Definition*** Let $A$ be an $m \times n$ matrix, let $B$ be an $n \times p$ matrix, and let $\mathbf{b}_j$ denote the $j$th column of $B$. Then we define the product $AB$ to be the $m \times p$ matrix

$$AB = A[\mathbf{b}_1 \quad \mathbf{b}_2 \quad \cdots \quad \mathbf{b}_p] = [A\mathbf{b}_1 \quad A\mathbf{b}_2 \quad \cdots \quad A\mathbf{b}_p].$$

The product of two matrices is defined as long as the number of columns in the first matrix is equal to the number of rows in the second matrix.

For example, let

$$A = \begin{bmatrix} 2 & 3 \\ -1 & 4 \end{bmatrix} \quad \text{and} \quad B = \begin{bmatrix} 1 & 3 & -1 \\ 2 & 2 & 0 \end{bmatrix}.$$

Since

$$A\mathbf{b}_1 = \begin{bmatrix} 8 \\ 7 \end{bmatrix}, \quad A\mathbf{b}_2 = \begin{bmatrix} 12 \\ 5 \end{bmatrix}, \quad A\mathbf{b}_3 = \begin{bmatrix} -2 \\ 1 \end{bmatrix},$$

we have

$$AB = \begin{bmatrix} 8 & 12 & -2 \\ 7 & 5 & 1 \end{bmatrix}.$$

Since the product $A\mathbf{b}_j$ is computed one entry at a time, the product $AB$ can also be computed one entry at a time (i.e., componentwise). If we call the product $C = AB$, then $c_{ij}$ is the $i$th coordinate of the $j$th column of $C$. That is, $c_{ij}$ is the $i$th coordinate of $A\mathbf{b}_j$, and this is the inner product of the $i$th row of $A$ with $\mathbf{b}_j$. Thus

$$c_{ij} = [a_{i1} \quad a_{i2} \quad \cdots \quad a_{in}] \begin{bmatrix} b_{1j} \\ b_{2j} \\ \vdots \\ b_{nj} \end{bmatrix} = a_{i1}b_{1j} + a_{i2}b_{2j} + \cdots + a_{in}b_{nj}.$$

Thus $c_{ij}$ can be computed by running the left index finger along the $i$th row of $A$ while running the right index finger down the $j$th column of $B$, forming the sum of the products of corresponding components.

$$\begin{bmatrix} a_{11} & \cdots & a_{1n} \\ \vdots & & \vdots \\ a_{i1} & \cdots & a_{in} \\ \vdots & & \vdots \\ a_{m1} & \cdots & a_{mn} \end{bmatrix} \begin{bmatrix} b_{11} & \cdots & b_{1j} & \cdots & b_{1p} \\ \vdots & & \vdots & & \vdots \\ b_{n1} & \cdots & b_{nj} & \cdots & b_{np} \end{bmatrix} = \begin{bmatrix} c_{11} & \cdots & c_{1j} & \cdots & c_{1p} \\ \vdots & & \vdots & & \vdots \\ c_{i1} & \cdots & c_{ij} & \cdots & c_{ip} \\ \vdots & & \vdots & & \vdots \\ c_{m1} & \cdots & c_{mj} & \cdots & c_{mp} \end{bmatrix}$$

Besides the 0 matrix, there is another special matrix that deserves a special name. Just as the 0 matrix plays the role in matrix addition that the number zero plays in ordinary addition of numbers, this matrix plays the role in matrix multiplication that the number 1 plays in ordinary multiplication of numbers. It is called the $n \times n$ **identity matrix**. It is defined to be the $n \times n$ matrix $[a_{ij}]$ such that

1. $a_{ii} = 1$ for $i = 1, 2, \ldots, n$.
2. $a_{ij} = 0$ for $i \neq j$.

We denote it by $I$ (for identity); it looks like

$$I = \begin{bmatrix} 1 & 0 & \cdots & 0 & 0 \\ 0 & 1 & \cdots & 0 & 0 \\ \vdots & \vdots & & \vdots & \vdots \\ 0 & 0 & \cdots & 1 & 0 \\ 0 & 0 & \cdots & 0 & 1 \end{bmatrix}.$$

In any matrix the entries $a_{ii}$ are called the **diagonal entries**. A square matrix whose only nonzero entries are its diagonal entries is called a **diagonal matrix**. A typical diagonal matrix looks like

$$\begin{bmatrix} a_{11} & 0 & \cdots & 0 & 0 \\ 0 & a_{22} & \cdots & 0 & 0 \\ \vdots & \vdots & & \vdots & \vdots \\ 0 & 0 & \cdots & a_{n-1,n-1} & 0 \\ 0 & 0 & \cdots & 0 & a_{nn} \end{bmatrix}.$$

As a nice shorthand notation for a diagonal matrix, we suppress writing all the zeros:

$$\begin{bmatrix} a_{11} & & & \\ & a_{22} & & \\ & & \ddots & \\ & & & a_{nn} \end{bmatrix}.$$

The identity matrix is a diagonal matrix with each of its diagonal entries 1.

$$I = \begin{bmatrix} 1 & & & \\ & 1 & & \\ & & \ddots & \\ & & & 1 \end{bmatrix}$$

Here are the most important identities satisfied by matrix multiplication.

## Properties of Matrix Multiplication

Let $A$, $B$, and $C$ be matrices and $\alpha$ a scalar.

(1) $(AB)C = A(BC)$.
(2) $A(B + C) = AB + AC$.
(3) $(A + B)C = AC + BC$.
(4) $A(\alpha B) = (\alpha A)B = \alpha(AB)$.
(5) $AI = A$ and $IA = A$.
(6) $A0 = 0$ and $0A = 0$.

Some comments are in order. First, all matrices must have the correct size so that all the indicated operations are defined. For example, in (1) the number of columns in $A$ must equal the number of rows in $B$ and the number of columns in $B$ must be equal to the number of rows in $C$. Similarly, in (2) $B$ and $C$ must have the same size and the number of columns in $A$ must equal the number of rows in $B$ (or $C$).

Second, one very familiar identity is missing, the commutative law $AB = BA$. Of course, it is missing because it is not true. Usually, $AB \neq BA$. Almost any example you write down will illustrate this. In fact, try to find two matrices at least one of which is not diagonal for which $AB = BA$. Even for both products to be defined both $A$ and $B$ must be $n \times n$ matrices. In Problem 10 we list several other familiar identities that are not satisfied by matrix multiplication.

Now for some proofs. The first identity is the hardest. We will compute the $ij$th entry on both sides and show that they are equal. The $ij$th entry of $(AB)C$ is the inner product of the $i$th row of $AB$ and the $j$th column of $C$. The entries in the $i$th row of $AB$ result from the inner product of the $i$th row of $A$ with the columns of $B$: the $k$th entry in the $i$th row of $AB$ is the inner product of the $i$th row of $A$ with the $k$th column of $B$:

$$\begin{bmatrix} a_{i1} & a_{i2} & \cdots & a_{in} \end{bmatrix} \begin{bmatrix} b_{1k} \\ b_{2k} \\ \vdots \\ b_{nk} \end{bmatrix} = a_{i1}b_{1k} + a_{i2}b_{2k} + \cdots + a_{in}b_{nk}.$$

Thus the $i$th row of $AB$ is (letting $p$ denote the number of columns in $B$)

$$[a_{i1}b_{11} + \cdots + a_{in}b_{n1} \quad a_{i1}b_{12} + \cdots + a_{in}b_{n2} \quad \cdots \quad a_{i1}b_{1p} + \cdots + a_{in}b_{np}]$$

and the $ij$th entry of $(AB)C$ is (letting $q$ denote the number of columns of $C$)

$$\begin{aligned} &c_{1j}(a_{i1}b_{11} + \cdots + a_{in}b_{n1}) + c_{2j}(a_{i1}b_{12} + \cdots + a_{in}b_{n2}) \\ &\qquad + \cdots + c_{pj}(a_{i1}b_{1p} + \cdots + a_{in}b_{np}). \end{aligned} \tag{1}$$

Now we compute the $ij$th entry of $A(BC)$. It is the inner product of the $i$th row of $A$ with the $j$th column of $BC$. The $j$th column of $BC$ is the inner product of the rows of $B$ with the $j$th column of $C$:

$$\begin{bmatrix} b_{11}c_{1j} + \cdots + b_{1p}c_{pj} \\ b_{21}c_{1j} + \cdots + b_{2p}c_{pj} \\ \vdots \qquad\qquad \vdots \\ b_{n1}c_{1j} + \cdots + b_{np}c_{pj} \end{bmatrix}$$

Thus the $ij$th entry of $A(BC)$ is

$$\begin{aligned} &a_{i1}(b_{11}c_{1j} + \cdots + b_{1p}c_{pj}) + a_{i2}(b_{21}c_{1j} + \cdots + b_{2p}c_{pj}) \\ &\qquad + \cdots + a_{in}(b_{n1}c_{1j} + \cdots + b_{np}c_{pj}). \end{aligned} \tag{2}$$

Is (1) equal to (2)? Sure! The terms are just collected differently in the two expressions. Thus $(AB)C$ and $A(BC)$ have the same entries and hence are equal.

We will also prove (2). The $ij$th entry in $A(B + C)$ is the inner product of the $i$th row of $A$ with the $j$th column of $B + C$:

$$a_{i1}(b_{1j} + c_{1j}) + a_{i2}(b_{2j} + c_{2j}) + \cdots + a_{in}(b_{nj} + c_{nj}).$$

The $ij$th entry of $AB + AC$ is

$$a_{i1}b_{1j} + a_{i1}c_{1j} + a_{i2}b_{2j} + a_{i2}c_{2j} + \cdots + a_{in}b_{nj} + a_{in}c_{nj}.$$

These two expressions are clearly equal.

The proofs of the remaining properties are left as exercises for the reader.

In Section 1.3 we defined the transpose of a $1 \times n$ matrix and an $n \times 1$ matrix. There is a natural extension of this definition to any matrix. If $A$ is an $m \times n$ matrix, the **transpose** of $A$, denoted by $A^T$, is the $n \times m$ matrix obtained from $A$ by interchanging its rows and columns. The $ij$th entry of $A^T$ is $a_{ji}$, that is, the $ji$th entry of $A$. For instance, if

$$A = \begin{bmatrix} 1 & 3 & -4 \\ 2 & 1 & 0 \\ 3 & 1 & 4 \end{bmatrix}, \qquad \text{then} \quad A^T = \begin{bmatrix} 1 & 2 & 3 \\ 3 & 1 & 1 \\ -4 & 0 & 4 \end{bmatrix}.$$

The most important identities involving transposes are the following.

## Properties of the Transpose

Let $A$ and $B$ be matrices and $\alpha$ a scalar.

(1) $(A^T)^T = A$.
(2) $(\alpha A)^T = \alpha A^T$.
(3) $(A + B)^T = A^T + B^T$.
(4) $(AB)^T = B^T A^T$.

As usual we assume that all matrices have the correct size so that the indicated operations are defined.

The most interesting identity and the most difficult to prove is (4). Let $A$ be an $m \times n$ matrix and $B$ an $n \times p$ matrix. Then $AB$ is $m \times p$ and $(AB)^T$ is $p \times m$. Also, $B^T$ is $p \times n$, $A^T$ is $n \times m$, and $B^T A^T$ is $p \times m$. Thus the two sides of (4) are matrices of the same size. To prove that they are equal, we show that they have the same $ij$th element. The $ij$th element of $(AB)^T$ is the $ji$th element of $AB$, which in turn is the inner product of the $j$th row of $A$ with the $i$th column of $B$:

$$ij\text{th element of } (AB)^T = a_{j1}b_{1i} + a_{j2}b_{2i} + \cdots + a_{jn}b_{ni}.$$

Now the $ij$th element of $B^T A^T$ is the inner product of the $i$th row of $B^T$ with the $j$th column of $A^T$. But the $i$th row of $B^T$ is the $i$th column of $B$ and the $j$th column of $A^T$ is the $j$th row of $A$. Therefore, the $ij$th element of $B^T A^T$ is the inner product of the $i$th column of $B$ with the $j$th row of $A$:

$$ij\text{th element of } B^T A^T = b_{1i}a_{j1} + b_{2i}a_{j2} + \cdots + b_{ni}a_{jn}.$$

Thus $(AB)^T$ and $B^T A^T$ have the same entry in every position. This completes the proof of (4). The proofs of (1), (2), and (3) are left for your enjoyment.

We conclude this section with some comments about notation. Let $A$ be an $m \times n$ matrix.

1. If $\mathbf{x}$ is a vector in $R^n$ (i.e., an $n$-tuple of real numbers), then by $A\mathbf{x}$ we mean $A$ times the $n$-dimensional column vector determined by $\mathbf{x}$.

2. If **b** is a vector in $R^m$, then in the equation

$$A\mathbf{x} = \mathbf{b},$$

**b** denotes the $m$-dimensional column vector determined by the vector **b**.
3. If **x** is a vector in $R^m$, then in the expression

$$\mathbf{x}^T A,$$

**x** denotes the $m$-dimensional column vector determined by the vector **x**.
4. If **x** is in $R^n$ and **y** is in $R^m$, then in the expression

$$\langle A\mathbf{x}, \mathbf{y}\rangle,$$

$A\mathbf{x}$ denotes the vector in $R^m$ determined by the $m$-dimensional column vector $A\mathbf{x}$.
5. In general, whenever a vector multiplies a matrix or is multiplied by a matrix, the vector represents a column vector in the equation.

With these conventions in mind, we can express the inner product of two vectors **x** and **y** as a matrix product:

$$\langle \mathbf{x}, \mathbf{y}\rangle = \mathbf{x}^T\mathbf{y}.$$

## PROBLEMS 1.6

**1.** Complete each of the following definitions.
(a) The product of an $m \times n$ matrix $A$ and an $n \times k$ matrix $B$ is....
(b) The $n \times n$ identity matrix is....
(c) An $n \times n$ diagonal matrix is....
(d) The transpose of an $n \times m$ matrix is....

***2.** Let

$$A = \begin{bmatrix} 2 & 1 & 3 \\ 4 & -1 & 2 \end{bmatrix},$$

$$B = \begin{bmatrix} 2 & 3 \\ 0 & -1 \\ 1 & -2 \end{bmatrix}, \quad \text{and} \quad \mathbf{v} = \begin{bmatrix} 1 \\ 2 \end{bmatrix}.$$

Compute $AB$, $(AB)\mathbf{v}$, $B\mathbf{v}$, and $A(B\mathbf{v})$.

**3.** Let $A$ and $B$ be as in Problem 2 and let

$$\mathbf{v} = \begin{bmatrix} 2 \\ 0 \\ 1 \end{bmatrix}.$$

Compute $BA$, $(BA)\mathbf{v}$, $A\mathbf{v}$, and $B(A\mathbf{v})$.

***4.** Let $A = [1 \quad 2 \quad -1]$ and

$$B = \begin{bmatrix} 2 \\ 0 \\ 1 \end{bmatrix}.$$

Compute $AB$ and $BA$.

**5.** Let

$$A = \begin{bmatrix} 5 & 2 \\ -3 & 1 \end{bmatrix},$$

$$B = \begin{bmatrix} 2 & -1 & 3 \\ 0 & 4 & 1 \end{bmatrix},$$

$$C = \begin{bmatrix} 1 & 3 \\ 2 & -4 \end{bmatrix}, \quad D = \begin{bmatrix} -2 & 1 & 3 \\ -1 & 0 & 2 \\ 4 & 1 & 4 \end{bmatrix},$$

$$E = [2 \quad 1 \quad 3], \quad F = \begin{bmatrix} -1 \\ 3 \\ 1 \end{bmatrix}.$$

Compute the following matrices.
*(a) $AB$, $AC$, $BD$, $BF$, $EF$, and $FE$.
(b) $EE^T$, $F^TD$, $(A + C)B$, $D(E^T + F)$, and $3C^TB$.

**6.** Show that $A^3[=AAA]=0$, where

$$A=\begin{bmatrix}1&1&3\\5&2&6\\-2&-1&-3\end{bmatrix}.$$

**7.** Show that $A^2=A$, where

$$A=\begin{bmatrix}2&-2&-4\\-1&3&4\\1&-2&-3\end{bmatrix}.$$

**8.** Explain why the product of an $m\times r$ matrix and an $r\times n$ matrix is an $m\times n$ matrix.

**9.** Let

$$A=\begin{bmatrix}1&1&-1\\2&0&3\\4&3&-2\end{bmatrix}\quad\text{and}$$

$$B=\begin{bmatrix}-9&1&3\\16&2&-5\\6&1&-2\end{bmatrix}.$$

Compute $AB$ and $BA$.

**10.** In (a) through (e) we have listed several familiar identities which are not satisfied by matrix multiplication. For each of these identities, give examples of matrices for which the identity fails.

*(a) If $AB=0$, then $A=0$ or $B=0$.
(b) If $AB=AC$, and $A\neq 0$, then $B=C$.
(c) $(A+B)^2=A^2+2AB+B^2$.
(d) $A^2-B^2=(A-B)(A+B)$.
(e) If $A^2=I$, then $A=I$.

**11.** Let

$$w_1=a_{11}y_1+a_{12}y_2$$
$$w_2=a_{21}y_1+a_{22}y_2$$

and

$$y_1=b_{11}x_1+b_{12}x_2$$
$$y_2=b_{21}x_1+b_{22}x_2.$$

(a) Find $c_{11}, c_{12}, c_{21}$, and $c_{22}$ such that

$$w_1=c_{11}x_1+c_{12}x_2$$
$$w_2=c_{21}x_1+c_{22}x_2.$$

(b) Interpret all of these systems as matrix equations. What is the relationship between the three coefficient matrices?
(c) Investigate this problem when there are three $y$'s ($y_1, y_2$, and $y_3$) and four $x$'s ($x_1, x_2, x_3$, and $x_4$).

**12.** Show that if one column of a matrix $B$ is a multiple of another column of $B$, the same is true for the corresponding columns of $AB$.

**13.** Let $A$ an $m\times n$ matrix and $\mathbf{y}$ be a vector in $R^m$.

(a) Show that

$$\mathbf{y}^TA=[y_1\ \cdots\ y_m]\begin{bmatrix}a_{11}&\cdots&a_{1n}\\\vdots&&\vdots\\a_{m1}&\cdots&a_{mn}\end{bmatrix}$$
$$=y_1[a_{11}\ \cdots\ a_{1n}]+\cdots+y_m[a_{m1}\ \cdots\ a_{mn}].$$

Thus $\mathbf{y}^TA$ is a linear combination of the rows of $A$ whose coefficients are the components of $\mathbf{y}$.

(b) Explain why the $i$th row of a product $AB$ is a linear combination of the rows of $A$ whose coefficients are the numbers in the $i$th row of $A$.

**14.** Let $D$ be the diagonal matrix

$$\begin{bmatrix}3&&\\&1&\\&&-1\end{bmatrix}.$$

*(a) Describe how a vector is changed when it is multiplied by $D$.
(b) Compute $D^2$, $D^3$, and $D^4$.
(c) On the basis of the answers to part (b), describe $D^k$ for $k>0$.
(d) Describe the relationship between $D^n$ and $D$ for an arbitrary diagonal matrix $D$.
(e) If $D$ is a diagonal matrix, describe $AD$ and $DA$.

**15.** Find all $2\times 2$ matrices $A$ that satisfy:
*(a) $A^2=I$ (b) $A^2=0$

**16.** Define the **trace** of an $n \times n$ matrix $A$ to be the sum of the diagonal entries. If the trace of $A$ is denoted tr $A$ and $A = [a_{ij}]$, then

$$\operatorname{tr} A = a_{11} + a_{22} + \cdots + a_{nn}.$$

Let $A$ and $B$ be $n \times n$ matrices.

(a) Show that $\operatorname{tr}(\alpha A) = \alpha \cdot \operatorname{tr} A$.

(b) Show that $\operatorname{tr}(A + B) = \operatorname{tr} A + \operatorname{tr} B$.

*(c) Show that $\operatorname{tr}(AB) = \operatorname{tr}(BA)$.

*(d) Do there exist square matrices $A$ and $B$ such that

$$AB - BA = I?$$

**17.** A square matrix $A$ is called **symmetric** if $A^T = A$ and **skew-symmetric** if $A^T = -A$.

(a) Prove that every diagonal element of a skew-symmetric matrix is zero.

(b) Describe what a symmetric matrix looks like. Do the same for a skew-symmetric matrix.

(c) Find all $n \times n$ matrices that are both symmetric and skew-symmetric.

*(d) For any square matrix $A$, show that $A^TA$ is symmetric.

(e) For any square matrix $A$, prove that $\frac{1}{2}(A + A^T)$ is a symmetric matrix. Why does $A$ have to be square?

(f) For any square matrix $A$, prove that $\frac{1}{2}(A - A^T)$ is a skew-symmetric matrix.

(g) Show that any square matrix $A$ can be expressed as a sum $A = S + K$, where $S$ is a symmetric matrix and $K$ is a skew-symmetric matrix.

(h) Show that the representation above is unique; that is, show that if $A = S + K = S' + K'$, where $S$, $S'$ are symmetric and $K$, $K'$ are skew-symmetric, then $S = S'$ and $K = K'$.

**18.** Prove that $(A + B)C = AC + BC$.

**19.** Prove that $A(\alpha B) = (\alpha A)B = \alpha(AB)$.

**20.** Prove that $AI = A$ and $IA = A$.

**21.** Prove that $A0 = 0$ and $0A = 0$.

**22.** Prove that $(A^T)^T = A$.

**23.** Prove that $(\alpha A)^T = \alpha A^T$.

**24.** Prove that $(A + B)^T = A^T + B^T$.

**25.** Let $P$ be a matrix with exactly one 1 in each row and column, and with all other entries 0.

(a) Find all $3 \times 3$ matrices with this property.

(b) If $A$ is any $3 \times 3$ matrix, for each of the matrices in part (a) describe $AP$ in terms of the columns of $A$.

(c) For each of the matrices in part (a), describe $PA$ in terms of the rows of $A$.

(d) If $P$ is $n \times n$, explain why the columns of $P$ are the columns $\mathbf{e}_1, \mathbf{e}_2, \ldots, \mathbf{e}_n$ of the $n \times n$ identity matrix in some order.

(e) If $A$ and $P$ are $n \times n$, describe $AP$ in terms of the columns of $A$.

(f) If $A$ and $P$ are $n \times n$, describe $PA$ in terms of the rows of $A$.

(g) Is it clear why a matrix such as $P$ is called a permutation matrix?

**26.** In each of the following, determine the possible sizes of $A$, $B$, $C$, $D$, so that the given equation makes sense.

*(a) $AB = BA$

(b) $AB = AC$

*(c) $A(BC) = A(B + C)$

(d) $AB = CA$

(e) $A + B = B(C + D)$

(f) $AB + BC + CD = 0$

**27.** A square matrix $A = [a_{ij}]$ is called **upper triangular** if $a_{ij} = 0$ whenever $i > j$. Let $A$ and $B$ be $n \times n$ upper triangular matrices.

(a) Show that $AB$ is upper triangular.

(b) Use part (a) to show that for any positive integer $n$, $A^n$ is upper triangular.

(c) State and prove analogous statements for lower triangular matrices.

***28.** Let $A$ be an $m \times n$ matrix. Show that for any vector $\mathbf{x}$ in $R^n$ and $\mathbf{y}$ in $R^m$,

$$\langle A\mathbf{x}, \mathbf{y} \rangle = \langle \mathbf{x}, A^T\mathbf{y} \rangle.$$

[*Hint:* Express the inner product $\langle A\mathbf{x}, \mathbf{y} \rangle$ as a matrix product and use the fact that $(AB)^T = B^T A^T$.]

**29.** Let $A = [a_{ij}]$ be an $n \times n$ matrix.
(a) Show that

$$\langle A\mathbf{e}_i, \mathbf{e}_j \rangle = a_{ji}$$

and

$$\langle \mathbf{e}_i, A\mathbf{e}_j \rangle = a_{ij}$$

where $\mathbf{e}_i$ is the vector in $R^n$ whose $i$th component is 1 and whose other components are 0.

(b) Show that if $\langle A\mathbf{x}, \mathbf{y} \rangle = 0$ for all $\mathbf{x}$ and $\mathbf{y}$ in $R^n$, then $A = 0$.
(c) Are parts (a) and (b) true when $A$ is an $m \times n$ matrix?

**30.** Let $A$ be an $n \times n$ matrix. Show that if

$$\langle A\mathbf{x}, \mathbf{y} \rangle = \langle \mathbf{x}, A\mathbf{y} \rangle$$

for all $\mathbf{x}$ and $\mathbf{y}$ in $R^n$, then $A$ is symmetric (i.e., $A = A^T$).

**31.** Let $A$ be a symmetric matrix (i.e., $A = A^T$). Show that if $A^2 = 0$, then $A = 0$.

# 1.7 Gaussian Elimination: Nonsingular Systems

Now it is time to learn how to solve a system of equations. We begin with the simplest situation, a square system.

Some square systems are easy to solve. For example, to solve

$$\begin{aligned} 2x_1 + x_2 - \phantom{3}x_3 &= 5 \\ x_2 + \phantom{3}x_3 &= 3 \\ 3x_3 &= 6 \end{aligned} \tag{1}$$

we should obviously work from the bottom up as follows. The last equation gives $x_3 = 2$. Substituting into the second equation gives $x_2 = 1$, and then substituting into the first equation gives $x_1 = 3$. This procedure is called **back substitution**.

The matrix equation of this system is

$$A\mathbf{x} = \mathbf{b} \qquad \text{or} \qquad \begin{bmatrix} 2 & 1 & -1 \\ 0 & 1 & 1 \\ 0 & 0 & 3 \end{bmatrix} \begin{bmatrix} x_1 \\ x_2 \\ x_3 \end{bmatrix} = \begin{bmatrix} 5 \\ 3 \\ 6 \end{bmatrix}.$$

This system is completely determined by the coefficient matrix $A$ and the vector $\mathbf{b}$. We can capture this information in a single matrix called the **augmented matrix**. The augmented matrix of our system is

$$[A : \mathbf{b}] \qquad \text{or} \qquad \left[\begin{array}{ccc:c} 2 & 1 & -1 & 5 \\ 0 & 1 & 1 & 3 \\ 0 & 0 & 3 & 6 \end{array}\right].$$

The dotted line is there as a reminder that the matrix $A$ has been augmented by the column vector $\mathbf{b}$. Each row of the augmented matrix $[A : \mathbf{b}]$ represents an equation in a system of equations, and hence performing any operation on the equations of the system is the same as performing that operation on the rows of the augmented matrix.

The process of back substitution can be accomplished by performing operations on the equations of the system or on the rows of the augmented matrix $[A : \mathbf{b}]$. For example, from the third equation $3x_3 = 6$ in our system we obtain $x_3 = 2$ by multiplying both sides of the equation by $\frac{1}{3}$. We accomplish the same thing by multiplying the third row of the augmented matrix by $\frac{1}{3}$.

$$\begin{aligned} 2x_1 + x_2 - x_3 &= 5 \\ x_2 + x_3 &= 3 \\ x_3 &= 2 \end{aligned} \qquad \begin{bmatrix} 2 & 1 & -1 & : & 5 \\ 0 & 1 & 1 & : & 3 \\ 0 & 0 & 1 & : & 2 \end{bmatrix}$$

Substituting $x_3 = 2$ into the second equation can be accomplished by multiplying the third equation by $-1$ and adding it to the second equation or by multiplying the third row of the augmented matrix by $-1$ and adding it to the second row.

$$\begin{aligned} 2x_1 + x_2 - x_3 &= 5 \\ x_2 \qquad &= 1 \\ x_3 &= 2 \end{aligned} \qquad \begin{bmatrix} 2 & 1 & -1 & : & 5 \\ 0 & 1 & 0 & : & 1 \\ 0 & 0 & 1 & : & 2 \end{bmatrix}$$

Next, substituting $x_2 = 1$ and $x_3 = 2$ amounts to multiplying the third equation (row of the augmented matrix) by 1 and adding it to the first equation (row) and then multiplying the second equation (row) by $-1$ and adding it to the first equation (row).

$$\begin{aligned} 2x_1 \qquad\qquad &= 6 \\ x_2 \qquad &= 1 \\ x_3 &= 2 \end{aligned} \qquad \begin{bmatrix} 2 & 0 & 0 & : & 6 \\ 0 & 1 & 0 & : & 1 \\ 0 & 0 & 1 & : & 2 \end{bmatrix}$$

Finally, we obtain $x_1 = 3$ by multiplying the first equation (row) by $\frac{1}{2}$.

$$\begin{aligned} x_1 \qquad\qquad &= 3 \\ x_2 \qquad &= 1 \\ x_3 &= 2 \end{aligned} \qquad \begin{bmatrix} 1 & 0 & 0 & : & 3 \\ 0 & 1 & 0 & : & 1 \\ 0 & 0 & 1 & : & 2 \end{bmatrix}$$

Thus we have transformed the original system

$$\begin{aligned} 2x_1 + x_2 - x_3 &= 5 \\ x_2 + x_3 &= 3 \\ 3x_3 &= 6 \end{aligned} \quad \text{to the system} \quad \begin{aligned} x_1 \qquad\qquad &= 3 \\ x_2 \qquad &= 1 \\ x_3 &= 2 \end{aligned}$$

and we have transformed the augmented matrix

$$\begin{bmatrix} 2 & 1 & -1 & : & 5 \\ 0 & 1 & 1 & : & 3 \\ 0 & 0 & 3 & : & 6 \end{bmatrix} \quad \text{to the augmented matrix} \quad \begin{bmatrix} 1 & 0 & 0 & : & 3 \\ 0 & 1 & 0 & : & 1 \\ 0 & 0 & 1 & : & 2 \end{bmatrix}.$$

From either the equations or the augmented matrix we see that the solution of the system is

$$x_1 = 3, \qquad x_2 = 1, \qquad x_3 = 2.$$

You are probably thinking that we have introduced a lot of machinery just to solve a pretty simple system of equations. If this were the end of the story, we would accept the criticism. Keep the faith; there is more to come. But first we summarize what we have done.

The system (1) was easy to solve because of the shape of its coefficient matrix. An $n \times n$ matrix $A$ is called **upper triangular** if $a_{ij} = 0$ whenever $i > j$. In Theorem 1 we will prove that *any system whose coefficient matrix $A$ is an upper triangular matrix with $a_{ii} \neq 0$ for all $i$ can be solved by back substitution.* The procedure transforms the augmented matrix $[A : \mathbf{b}]$ to an augmented matrix $[I : \mathbf{d}]$, where $I$ is the $n \times n$ identity matrix. Thus $A\mathbf{x} = \mathbf{b}$ is transformed to the system $I\mathbf{x} = \mathbf{d}$. Since $I\mathbf{x} = \mathbf{x}$, the solution of the system is $\mathbf{x} = \mathbf{d}$. The two row operations used to transform $[A : \mathbf{b}]$ to $[I : \mathbf{d}]$ are:

1. Add a multiple of one equation (row) to another equation (row).
2. Multiply an equation (row) by a nonzero constant.

***Theorem 1*** *Let $A$ be an $n \times n$ upper triangular matrix. If all the diagonal entries of $A$ are nonzero (that is, $a_{ii} \neq 0$ for all $i$), then $A$ can be transformed to the identity matrix using back substitution.*

**Proof** Suppose that $a_{ii} \neq 0$ for all $i$. Multiplying the last row of $A$ by $1/a_{nn}$, we transform $A$ into a matrix whose $n, n$ entry is 1.

$$\begin{bmatrix} a_{11} & a_{12} & \cdots & a_{1,n-1} & a_{1n} \\ 0 & a_{22} & \cdots & a_{2,n-1} & a_{2n} \\ \vdots & \vdots & & \vdots & \vdots \\ 0 & 0 & \cdots & a_{n-1,n-1} & a_{n-1,n} \\ 0 & 0 & \cdots & 0 & 1 \end{bmatrix}$$

We now zero out every other entry in the last column by multiplying the last row by $-a_{n-1,n}$ and adding it to the $(n-1)$st row, then multiplying the last row by $-a_{n-2,n}$ and adding it to the $(n-2)$nd row, and so on. In general, we zero out the $i, n$ entry by multiplying the last row by $-a_{in}$ and adding it to the $i$th row. When we complete this process we have transformed $A$ into

$$\begin{bmatrix} a_{11} & a_{12} & \cdots & a_{1,n-1} & 0 \\ 0 & a_{22} & \cdots & a_{2,n-1} & 0 \\ \vdots & \vdots & & \vdots & \vdots \\ 0 & 0 & \cdots & a_{n-1,n-1} & 0 \\ 0 & 0 & \cdots & 0 & 1 \end{bmatrix}.$$

Because the matrix is upper triangular the first $n - 1$ rows and columns will not be changed. Thus the diagonal entries of this submatrix are still nonzero and we may repeat the process on the first $n-1$ rows and columns to change $a_{n-1,n-1}$ to 1 and change every preceding entry in the $(n-1)$st column to a zero. Continuing in this way we will clearly arrive at the identity matrix.

What do we do with a system that does not have a triangular shape? Obviously, if we can convert it to one that has a triangular shape, we can solve it. For example, suppose that we are given the square system and we form its augmented matrix

$$\begin{aligned} x_1 + 2x_2 + 3x_3 &= 6 \\ 2x_1 + 7x_2 + x_3 &= 13 \\ x_1 + 4x_2 - 6x_3 &= 1 \end{aligned} \qquad \begin{bmatrix} 1 & 2 & 3 & : & 6 \\ 2 & 7 & 1 & : & 13 \\ 1 & 4 & -6 & : & 1 \end{bmatrix}.$$

We can eliminate $x_1$ from the second and third equations or make the 2,1 and 3,1 entry of the augmented matrix zero as follows. Multiply the first equation (row) by $-2$ and add it to the second equation (row), and then multiply the first equation (row) by $-1$ and add it to the third equation (row).

$$\begin{aligned} x_1 + 2x_2 + 3x_3 &= 6 \\ 3x_2 - 5x_3 &= 1 \\ 2x_2 - 9x_3 &= -5 \end{aligned} \qquad \begin{bmatrix} 1 & 2 & 3 & : & 6 \\ 0 & 3 & -5 & : & 1 \\ 0 & 2 & -9 & : & -5 \end{bmatrix}$$

Now we eliminate $x_2$ from the third equation or make the 3,2 entry of the augmented matrix zero by multiplying the second equation (row) by $-\frac{2}{3}$ and adding it to the third equation (row).

$$\begin{aligned} x_1 + 2x_2 + 3x_3 &= 6 \\ 3x_2 - 5x_3 &= 1 \\ -\frac{17}{3}x_3 &= -\frac{17}{3} \end{aligned} \qquad \begin{bmatrix} 1 & 2 & 3 & : & 6 \\ 0 & 3 & -5 & : & 1 \\ 0 & 0 & -\frac{17}{3} & : & -\frac{17}{3} \end{bmatrix}$$

We can now use back substitution to transform the system to

$$\begin{aligned} x_1 \qquad\qquad &= -1 \\ x_2 \qquad &= 2 \\ x_3 &= 1 \end{aligned} \qquad \begin{bmatrix} 1 & 0 & 0 & : & -1 \\ 0 & 1 & 0 & : & 2 \\ 0 & 0 & 1 & : & 1 \end{bmatrix}.$$

The solution vector is $\mathbf{x} = (-1, 2, 1)$.

This procedure for converting a system to a triangular form is called **forward elimination**. It involves the systematic use of the same two operations used for back substitution: adding a multiple of one equation (row) to another equation (row) and multiplying an equation (row) by a nonzero constant. The first equation is used to eliminate $x_1$ from all subsequent equations. Then the new second equation is used to eliminate $x_2$ from the third and subsequent equations. The procedure continues in the obvious way. After each series of eliminations is completed, we move down one equation and over to the next variable and begin a new series of eliminations.

Let's try another example.

$$
\begin{aligned}
x_1 + x_2 + 3x_3 + x_4 &= -1 \\
2x_1 + 2x_2 - x_3 + 2x_4 &= 5 \\
2x_1 + 3x_2 + 4x_3 + 2x_4 &= 3 \\
2x_1 + 3x_2 + 4x_3 + 4x_4 &= 3
\end{aligned}
\qquad
\begin{bmatrix}
1 & 1 & 3 & 1 & : & -1 \\
2 & 2 & -1 & 2 & : & 5 \\
2 & 3 & 4 & 2 & : & 3 \\
2 & 3 & 4 & 4 & : & 3
\end{bmatrix}
$$

Eliminating $x_1$ from the second, third, and fourth equations, we obtain

$$
\begin{aligned}
x_1 + x_2 + 3x_3 + x_4 &= -1 \\
-7x_3 &= 7 \\
x_2 - 2x_3 &= 5 \\
x_2 - 2x_3 + 2x_4 &= 5
\end{aligned}
\qquad
\begin{bmatrix}
1 & 1 & 3 & 1 & : & -1 \\
0 & 0 & -7 & 0 & : & 7 \\
0 & 1 & -2 & 0 & : & 5 \\
0 & 1 & -2 & 2 & : & 5
\end{bmatrix}.
$$

The procedure we just described requires us to use the second equation to eliminate $x_2$ from the third and fourth equations. Since this is clearly impossible, we have a problem. For this example the solution is clear. Since the order of the equations is immaterial, we merely interchange the second and third equations:

$$
\begin{aligned}
x_1 + x_2 + 3x_3 + x_4 &= -1 \\
x_2 - 2x_3 &= 5 \\
-7x_3 &= 7 \\
x_2 - 2x_3 + 2x_4 &= 5
\end{aligned}
\qquad
\begin{bmatrix}
1 & 1 & 3 & 1 & : & -1 \\
0 & 1 & -2 & 0 & : & 5 \\
0 & 0 & -7 & 0 & : & 7 \\
0 & 1 & -2 & 2 & : & 5
\end{bmatrix}.
$$

We can now use the second equation to eliminate $x_2$ from all subsequent equations.

$$
\begin{aligned}
x_1 + x_2 + 3x_3 + x_4 &= -1 \\
x_2 - 2x_3 &= 5 \\
-7x_3 &= 7 \\
2x_4 &= 0
\end{aligned}
\qquad
\begin{bmatrix}
1 & 1 & 3 & 1 & : & -1 \\
0 & 1 & -2 & 0 & : & 5 \\
0 & 0 & -7 & 0 & : & 7 \\
0 & 0 & 0 & 2 & : & 0
\end{bmatrix}
$$

The system is now triangular and we may apply back substitution to transform it to

$$
\begin{aligned}
x_1 &= -1 \\
x_2 &= 3 \\
x_3 &= -1 \\
x_4 &= 0
\end{aligned}
\qquad
\begin{bmatrix}
1 & 0 & 0 & 0 & : & -1 \\
0 & 1 & 0 & 0 & : & 3 \\
0 & 0 & 1 & 0 & : & -1 \\
0 & 0 & 0 & 1 & : & 0
\end{bmatrix}.
$$

Thus the solution vector is $\mathbf{x} = (-1, 3, -1, 0)$.

We have added one more operation to our list; we will not need any more. The **three basic row operations** are:

1. Add a multiple of one equation (row) to another equation (row).
2. Multiply an equation (row) by a nonzero constant.
3. Interchange two equations (rows).

The use of these three basic row operations to reduce a matrix to triangular form is called the **forward elimination procedure**.

Interchanging two equations or two rows of a matrix is called **pivoting**. Pivoting will allow forward elimination to continue—sometimes. Clearly, there must be something to pivot with! For example, pivoting will not cure the system

$$\begin{aligned} x_1 + x_2 + x_3 + x_4 &= 2 \\ x_3 + x_4 &= 1 \\ x_3 - x_4 &= -3 \\ x_3 + 2x_4 &= 3 \end{aligned} \qquad \begin{bmatrix} 1 & 1 & 1 & 1 & : & 2 \\ 0 & 0 & 1 & 1 & : & 1 \\ 0 & 0 & 1 & -1 & : & -3 \\ 0 & 0 & 1 & 2 & : & 3 \end{bmatrix}. \tag{2}$$

The problem is with the coefficient matrix.

***Definition*** A square matrix that cannot be transformed to a triangular matrix with $a_{ii} \neq 0$ for all $i$ using the three basic row operations is called **singular**. A matrix that can be so transformed is called **nonsingular**.

In particular, an upper triangular matrix with $a_{ii} \neq 0$ for all $i$ is nonsingular. On the other hand, if we reduce a matrix to triangular form and find a zero on the diagonal, can we conclude from this definition that the matrix is singular? No! To conclude that a matrix is singular from this definition we must show that *all* reductions of the matrix to triangular form result in a matrix having a zero on its diagonal. It turns out that if one reduction of a square matrix to a triangular matrix results in the triangular matrix with a zero on its diagonal, then *any* reduction of the matrix to a triangular matrix results in a triangular matrix having a zero on its diagonal (in fact, in the same position). But this is a theorem; it is not the definition (see Section 1.9).

By Theorem 1, back substitution will transform an upper triangular matrix whose diagonal entries are not zero to the identity matrix. We have proved half of the following theorem.

***Theorem 2*** *An $n \times n$ matrix $A$ is nonsingular if and only if it can be transformed into the identity matrix using the three basic row operations.*

**Proof** It remains to prove that if $A$ is an $n \times n$ matrix which can be transformed to the identity matrix using the three basic row operations, then $A$ is nonsingular. This follows immediately from the definition of nonsingular, since the identity matrix is a special case of a triangular matrix all of whose diagonal entries are nonzero.

**EXAMPLE 1**

The following matrices are singular:

$$\begin{bmatrix} 2 & 0 \\ 0 & 0 \end{bmatrix}, \quad \begin{bmatrix} 1 & 2 & 3 \\ 0 & 1 & 2 \\ 0 & 0 & 0 \end{bmatrix}, \quad \begin{bmatrix} 1 & 2 & -1 \\ 2 & -2 & 3 \\ 7 & 2 & 3 \end{bmatrix}, \quad \begin{bmatrix} 0 & 1 & 1 \\ 2 & 1 & 4 \\ 2 & 2 & 5 \end{bmatrix}.$$

As we remarked above, it is difficult to verify using the definition that these matrices are singular. Nevertheless, try your hand at reducing them to triangular matrices with no zeros on the diagonal. The following matrices are nonsingular:

$$\begin{bmatrix} 1 & 2 \\ 0 & 5 \end{bmatrix}, \quad \begin{bmatrix} 1 & 2 \\ -1 & 2 \end{bmatrix}, \quad \begin{bmatrix} 1 & 2 & 3 \\ 0 & 1 & 2 \\ 0 & 0 & 1 \end{bmatrix}, \quad \begin{bmatrix} 2 & 1 & -2 \\ 6 & 4 & 4 \\ 10 & 8 & 6 \end{bmatrix}.$$

To verify that the second and the fourth matrices are nonsingular, you will have to do a few computations.

We now have a complete solution procedure that will solve any square system $A\mathbf{x} = \mathbf{b}$ with a nonsingular coefficient matrix $A$. We call the procedure **Gaussian elimination** (after K. F. Gauss (1777–1855), who was probably the greatest mathematician of all time). The procedure transform $[A : \mathbf{b}]$ to $[I : \mathbf{d}]$ using forward elimination followed by back substitution. Forward elimination reduces the system $A\mathbf{x} = \mathbf{b}$ to a triangular system $E\mathbf{x} = \mathbf{c}$ whose coefficient matrix has no zeros on its diagonal. Back substitution further reduces $E\mathbf{x} = \mathbf{c}$ to $I\mathbf{x} = \mathbf{d}$. The solution of the system is $\mathbf{x} = \mathbf{d}$.

***Theorem 3*** *Let $A$ be an $n \times n$ nonsingular matrix. For any vector $\mathbf{b}$ in $R^n$, the system $A\mathbf{x} = \mathbf{b}$ has one and only one solution.*

**Proof** Since $A$ is nonsingular, given any vector $\mathbf{b}$ in $R^n$, the system $[A : \mathbf{b}]$ can be transformed by Gaussian elimination into the system $[I : \mathbf{d}]$. Thus $\mathbf{x} = \mathbf{d}$ is the unique solution.

Theorem 3 is an example of an **existence and uniqueness theorem**. The existence part is the fact that for every $\mathbf{b}$ in $R^n$, the system $A\mathbf{x} = \mathbf{b}$ has a solution. The uniqueness part is the fact that it has only one solution.

A system of the form $A\mathbf{x} = \mathbf{0}$ is called a **homogeneous system**. These systems always have at least one solution, namely $\mathbf{x} = \mathbf{0}$; it is called the **trivial solution**. Any other solution is called a **nontrivial solution**. Theorem 3 gives:

***Theorem 4*** *Let $A$ be an $n \times n$ nonsingular matrix. Then the homogeneous system $A\mathbf{x} = \mathbf{0}$ has one and only one solution, namely the trivial solution $\mathbf{x} = \mathbf{0}$.*

There is one last piece of business we need to take care of. I think it is pretty clear to you that although the three basic row operations do change the matrix of a system, they do not change the solutions of a system. Two linear systems are called **equivalent** if they have the same solutions. The result we must prove is:

***Theorem 5*** *Each of the three basic row operations transforms a system of equations into an equivalent system of equations.*

**Proof** Obviously, interchanging a pair of equations and multiplying an equation by a nonzero constant do not change the solutions of a system. Let $E_1, E_2, \ldots, E_i, \ldots, E_j, \ldots, E_m$ represent the equations of a system. Multiplying one of these equations (say $E_j$) by a constant $\alpha$ and adding the result to another equation (say $E_i$) transforms the system to one whose equations are $E_1, E_2, \ldots, E_i + \alpha E_j, \ldots, E_j, \ldots, E_m$. The two systems differ only in the $i$th equation. Hence to prove that the two systems are equivalent, we must show that any solution of the first system satisfies the $i$th equation of the second system and any solution of the second system satisfies the $i$th equation of the first. Suppose that we have a solution of the first system. Then it satisfies $E_i, E_j, \alpha E_j$ and hence it satisfies $E_i + \alpha E_j$. Conversely, if we have a solution of the second system, then it satisfies $E_i + \alpha E_j$, $E_j$, $(-\alpha)E_j$ and hence it satisfies $E_i + \alpha E_j + (-\alpha)E_j = E_i$. This completes the proof.

### EXAMPLE 2

A mistake that is often made is to replace two linear equations, say

$$\begin{aligned} x_1 - \phantom{3}x_2 &= 5 \\ 2x_1 + 3x_2 &= 2 \end{aligned} \tag{*}$$

by their sum,

$$3x_1 + 2x_2 = 7. \tag{**}$$

These system are *not* equivalent! It is true that any solution of (*) is a solution of (**); the sum of two identities is an identity. But (**) has many solutions that are not solutions of (*). For example, $\mathbf{x} = (1, 2)$, $\mathbf{x} = (5, -4)$, $\mathbf{x} = (3, -1)$. Geometrically, the solution set of (*) is the vector $(\frac{17}{5}, -\frac{8}{5})$, whereas that of (**) is a line.

## PROBLEMS 1.7

**1.** Complete each of the following definitions.

(a) A square matrix is upper triangular if....

(b) A square matrix is singular if....

(c) If square matrix is nonsingular if....

(d) Two systems of equations are equivalent if....

(e) The three basic row operations are....

(f) Forward elimination consists of....

**2.** In each part, write the system of equations as a matrix equation. Then form the augmented matrix and solve the system.

*(a) $\begin{aligned} x_1 + x_2 &= 2 \\ x_1 - x_2 &= 4 \end{aligned}$

(b) $\begin{aligned} x_1 - \phantom{2}x_2 &= -1 \\ -x_1 + 2x_2 &= 4 \end{aligned}$

*(c) $\begin{aligned} x_1 + 4x_2 - x_3 &= 1 \\ x_1 + x_2 + x_3 &= 0 \\ 2x_1 \phantom{+ 4x_2} + 3x_3 &= 0 \end{aligned}$

(d) $\begin{aligned} 2x_1 + 4x_2 - 2x_3 \phantom{- x_4} &= 0 \\ 3x_1 + 6x_2 + 5x_3 - x_4 &= -4 \\ -x_1 \phantom{+ 6x_2} + 3x_3 - 2x_4 &= 2 \\ 2x_1 + 3x_2 - x_3 + x_4 &= -1 \end{aligned}$

**3.** In each part, you are given a matrix $A$

and a vector **b**. Solve the system $A\mathbf{x} = \mathbf{b}$.

*(a) $\begin{bmatrix} 2 & 1 \\ 4 & -1 \end{bmatrix}$, $\begin{bmatrix} 7 \\ 17 \end{bmatrix}$

(b) $\begin{bmatrix} 3 & 7 \\ 2 & 4 \end{bmatrix}$, $\begin{bmatrix} 4 \\ 7 \end{bmatrix}$

*(c) $\begin{bmatrix} 1 & 1 & 1 \\ -1 & 1 & 0 \\ 0 & 1 & 1 \end{bmatrix}$, $\begin{bmatrix} -2 \\ 0 \\ -1 \end{bmatrix}$

(d) $\begin{bmatrix} 1 & -1 & 1 \\ 4 & -3 & -1 \\ 3 & 1 & 2 \end{bmatrix}$, $\begin{bmatrix} 3 \\ 6 \\ 23 \end{bmatrix}$

*(e) $\begin{bmatrix} -2 & -2 & -2 \\ 0 & -2 & 6 \\ 2 & 3 & 1 \end{bmatrix}$, $\begin{bmatrix} 0 \\ 0 \\ 0 \end{bmatrix}$

(f) $\begin{bmatrix} 1 & 2 & 3 \\ 2 & 1 & -1 \\ 2 & -1 & 0 \end{bmatrix}$, $\begin{bmatrix} 1 \\ 0 \\ 0 \end{bmatrix}$

*(g) $\begin{bmatrix} 1 & 1 & 1 & 1 \\ 2 & 1 & -1 & 2 \\ 1 & 2 & 1 & -1 \\ 1 & 1 & -2 & 1 \end{bmatrix}$, $\begin{bmatrix} 1 \\ 9 \\ -6 \\ 7 \end{bmatrix}$

(h) $\begin{bmatrix} 1 & 1 & 0 & 0 & 0 & 0 \\ 1 & 2 & 1 & 0 & 0 & 0 \\ 0 & 1 & 2 & 1 & 0 & 0 \\ 0 & 0 & 1 & 2 & 1 & 0 \\ 0 & 0 & 0 & 1 & 2 & 1 \\ 0 & 0 & 0 & 0 & 1 & 2 \end{bmatrix}$, $\begin{bmatrix} 1 \\ 3 \\ 3 \\ 1 \\ -3 \\ -5 \end{bmatrix}$

**4.** Show that each of the following matrices is nonsingular.

(a) $\begin{bmatrix} 1 & 1 & 1 \\ 0 & 2 & 3 \\ 5 & 5 & 1 \end{bmatrix}$ (b) $\begin{bmatrix} 1 & -1 & 1 \\ 2 & 0 & 3 \\ 3 & -1 & 2 \end{bmatrix}$

(c) $\begin{bmatrix} 3 & 2 & 0 \\ 2 & 3 & 2 \\ 0 & 2 & 3 \end{bmatrix}$ (d) $\begin{bmatrix} 1 & 2 & 3 \\ 1 & 1 & 2 \\ 0 & 1 & 2 \end{bmatrix}$

(e) $\begin{bmatrix} 2 & 0 & 0 & 1 \\ 1 & 2 & 0 & -1 \\ 0 & 1 & 3 & 5 \\ 0 & 0 & 2 & 4 \end{bmatrix}$

(f) $\begin{bmatrix} 1 & 1 & 1 & 1 \\ 1 & 1 & 1 & -1 \\ 1 & 1 & -1 & -1 \\ 1 & -1 & -1 & -1 \end{bmatrix}$

***5.** Let $\mathbf{v} = (3, 1, 2)$, $\mathbf{w} = (0, 1, -1)$, and

$$A = \begin{bmatrix} 1 & 0 & 1 \\ 0 & 1 & 1 \\ 1 & 1 & 0 \end{bmatrix}.$$

Find a vector $\mathbf{x} = (x_1, x_2, x_3)$ such that $A\mathbf{x} + 2\mathbf{w} = \mathbf{v} - 3\mathbf{x}$.

**6.** (a) What system of linear equations must $a$, $b$, and $c$ satisfy so that the parabola $y = ax^2 + bx + c$ passes through the points $(-1, 8)$, $(1, -2)$, and $(3, 12)$.

(b) Find a solution to this system using the Gaussian elimination procedure.

***7.** Find numbers $a$, $b$, $c$ so that the parabola $y = ax^2 + bx + c$ passes through the points $(1, 0)$, $(2, 3)$, $(-1, 6)$.

**8.** Find a polynomial $p(x) = a + bx + cx^2 + dx^3$ such that $p(1) = 2$, $p(2) = 3$, $p(-3) = 18$, $p(5) = 210$.

**9.** In each part show that the lines intersect.

*(a) $\mathbf{x} = (1, 1, 1) + \alpha(1, 2, 3)$ and $\mathbf{x} = (2, 1, 2) + \beta(1, 4, 5)$

(b) $\mathbf{x} = (4, 4, 1) + \alpha(3, 2, 7)$ and $\mathbf{x} = (4, 2, -3) + \beta(1, -1, -1)$

**10.** In each part, find the point of intersection of the line and the plane.

*(a) $\mathbf{x} = (1, 1, 1) + \alpha(-1, 0, -5)$
$\mathbf{x} = \beta(1, 2, 5) + \gamma(1, 3, 1)$

(b) $\mathbf{x} = \alpha(1, 2, 3)$
$\mathbf{x} = (1, 0, 1) + \beta(1, 0, 1) + \gamma(-1, -3, -2)$

**11.** Let $A$ and $B$ be square matrices. Show that if $B$ is nonsingular and $A$ can be transformed to $B$ using the three row operations, then $A$ is also nonsingular.

**12.** Show that if $A$ is an $n \times n$ matrix and there is a vector $\mathbf{b}$ in $R^n$ for which the system $A\mathbf{x} = \mathbf{b}$ is inconsistent, then $A$ is singular.

**13.** Show that if $A\mathbf{x} = \mathbf{0}$ has a nontrivial solution, then $A$ is singular.

# 1.8 Echelon Systems

The Gaussian elimination algorithm produces a solution to any square system $A\mathbf{x} = \mathbf{b}$ whose coefficient matrix $A$ is nonsingular. The algorithm has two parts. Forward elimination reduces the system $A\mathbf{x} = \mathbf{b}$ to a triangular system $E\mathbf{x} = \mathbf{c}$ whose coefficient matrix $E$ has no zeros on its diagonal; back substitution then reduces $E\mathbf{x} = \mathbf{c}$ to $I\mathbf{x} = \mathbf{d}$.

Our purpose now is to generalize this algorithm to singular and rectangular systems. We began the preceding section by observing that back substitution solves any system whose coefficient matrix is an upper triangular matrix with no zeros on its diagonal. In trying to generalize the Gaussian algorithm, it is quite natural then to begin by trying to generalize the class of matrices that can be solved by back substitution.

The crucial property of a nonsingular triangular matrix which allowed back substitution to proceed is that *its first nonzero entry in each row is farther to the right than the first nonzero entry in all preceding rows.* Let's see what back substitution does to some systems whose coefficient matrices have this property.

### EXAMPLE 1

Consider the system

$$\begin{aligned} 2x_1 + x_2 - \phantom{3}x_3 + \phantom{6}x_4 &= 5 \\ x_2 + \phantom{3}x_3 - \phantom{6}x_4 &= 3 \\ 3x_3 + 6x_4 &= 6. \end{aligned} \tag{1}$$

It would be difficult not to notice that by moving the terms involving $x_4$ to the right side of these equations, we transform these equations into a triangular system in the variables $x_1$, $x_2$, and $x_3$.

$$\begin{aligned} 2x_1 + x_2 - \phantom{3}x_3 &= 5 - \phantom{6}x_4 \\ x_2 + \phantom{3}x_3 &= 3 + \phantom{6}x_4 \\ 3x_3 &= 6 - 6x_4 \end{aligned}$$

We can now use back substitution to solve for $x_1$, $x_2$, and $x_3$ in terms of $x_4$.

$$\begin{aligned} x_1 \qquad\qquad &= 3 - 3x_4 \\ x_2 \qquad &= 1 + 3x_4 \\ x_3 &= 2 - 2x_4 \end{aligned}$$

The variable $x_4$ is free to take on any value. No matter what value is chosen for $x_4$, it, together with the values for $x_1$, $x_2$, and $x_3$ determined from these equations, will be a solution to the system. For example, when $x_4 = 7$, then $x_1 = -18$, $x_2 = 22$, $x_3 = -12$, $x_4 = 7$ is a particular solution of the system. If we use the equation $x_4 = x_4$ to denote the fact that $x_4$ is free, every solution of the system is given by

$$x_1 = 3 - 3x_4, \qquad x_2 = 1 + 3x_4, \qquad x_3 = 2 - 2x_4, \qquad x_4 = x_4.$$

Since every solution can be found from these equations, they are said to give the **general solution** of the system. We can write the general solution in vector form as

$$(x_1, x_2, x_3, x_4) = (3, 1, 2, 0) + x_4(-3, 3, -2, 1).$$

Notice that this is an equation of a line in $R^4$.

### EXAMPLE 2

The system

$$\begin{aligned} x_1 + 2x_2 + 3x_3 + x_4 - \phantom{2}x_5 &= \phantom{-}2 \\ x_3 + x_4 + \phantom{2}x_5 &= -1 \\ x_4 - 2x_5 &= \phantom{-}4 \end{aligned} \tag{2}$$

is a triangular system in the variables $x_1$, $x_3$, and $x_4$.

$$\begin{aligned} x_1 + 3x_3 + x_4 &= \phantom{-}2 - 2x_2 + \phantom{2}x_5 \\ x_3 + x_4 &= -1 \phantom{- 2x_2} - \phantom{2}x_5 \\ x_4 &= \phantom{-}4 \phantom{- 2x_2} + 2x_5 \end{aligned}$$

Using back substitution, we can solve for $x_1$, $x_3$, and $x_4$ in terms of $x_2$ and $x_5$.

$$\begin{aligned} x_1 &= 13 - 2x_2 + 8x_5 \\ x_3 &= -5 \phantom{- 2x_2} - 3x_5 \\ x_4 &= \phantom{-}4 \phantom{- 2x_2} + 2x_5 \end{aligned}$$

Whatever values are chosen for $x_2$ and $x_5$, they, together with the values of $x_1$, $x_3$, and $x_4$ determined by these equations, will be a solution of the system. The general solution is

$$x_1 = 13 - 2x_2 + 8x_5, \qquad x_2 = x_2, \qquad x_3 = -5 - 3x_5,$$
$$x_4 = 4 + 2x_5, \qquad x_5 = x_5.$$

In vector form it is

$$(x_1, x_2, x_3, x_4, x_5) = (13, 0, -5, 4, 0) + x_2(-2, 1, 0, 0, 0) + x_5(8, 0, -3, 2, 1).$$

Notice that this is an equation of a plane in $R^5$.

As these examples indicate, any system whose coefficient matrix has the property that the first nonzero entry in any row is farther to the right than the first nonzero entry in all preceding rows can be converted to triangular form by moving some of the variables to the right side. Back substitution then solves for the remaining variables in terms of these variables. Since the variables moved to the right side are free to take on any value, they are called the **free variables**. The remaining variables are determined in terms of these variables and hence are called the **determined variables**.

Note that system (2) can also be considered to be a triangular system in the variables $x_1$, $x_2$, and $x_4$. Similarly, system (2) can also be considered triangular in three different sets of variables: $x_1$, $x_3$, and $x_5$; $x_2$, $x_3$, and $x_4$; $x_2$, $x_3$, and $x_5$. The choice of free and determined variables is not unique. To standardize our procedure, we will always make the same choice as we made in the examples. That is, when our system is in the desired form *we will always choose the variables that appear as the first variables of the equations of the system to be the determined variables.*

There is no need to actually move the free variables to the right side of the equation at the start. Beginning with the last equation, eliminate the last determined variable from all preceding equations. Then move to the next-to-last equation and eliminate the next-to-last determined variable from all preceding equations. Continue in the prescribed manner. Applying this procedure to systems (1) and (2), we obtain

$$\begin{aligned} x_1 \qquad\qquad + 3x_4 &= 3 \\ x_2 \qquad - 3x_4 &= 1 \\ x_3 + 2x_4 &= 2 \end{aligned} \qquad\qquad \begin{aligned} x_1 + 2x_2 \qquad\qquad - 8x_5 &= 13 \\ x_3 \qquad + 3x_5 &= -5 \\ x_4 - 2x_5 &= 4. \end{aligned}$$

This observation allows us to work directly with the augmented matrix of the system. Consider the augmented matrices of systems (1) and (2).

$$\left[\begin{array}{cccccc} 2 & 1 & -1 & 1 & : & 5 \\ 0 & 1 & 1 & -1 & : & 3 \\ 0 & 0 & 3 & 6 & : & 6 \end{array}\right] \quad \left[\begin{array}{ccccccc} 1 & 2 & 3 & 1 & -1 & : & 2 \\ 0 & 0 & 1 & 1 & 1 & : & -1 \\ 0 & 0 & 0 & 1 & -2 & : & 4 \end{array}\right] \tag{3}$$

The determined variables correspond to the first nonzero entry in each row. Working backward, we use the first nonzero entry in each row to zero out the preceding entries in that column. Carry out the procedure and verify that it transforms the augmented matrices (3) into the following augmented matrices:

$$\left[\begin{array}{cccccc} 1 & 0 & 0 & 3 & : & 3 \\ 0 & 1 & 0 & -3 & : & 1 \\ 0 & 0 & 1 & 2 & : & 2 \end{array}\right] \quad \left[\begin{array}{ccccccc} 1 & 2 & 0 & 0 & -8 & : & 13 \\ 0 & 0 & 1 & 0 & 3 & : & -5 \\ 0 & 0 & 0 & 1 & -2 & : & 4 \end{array}\right]. \tag{4}$$

From these matrices it is very easy to write down the general solution:

$$\begin{aligned} x_1 &= 3 - 3x_4 \\ x_2 &= 1 + 3x_4 \\ x_3 &= 2 - 2x_4 \\ x_4 &= \qquad x_4 \end{aligned} \qquad \begin{aligned} x_1 &= 13 - 2x_2 + 8x_5 \\ x_2 &= \qquad\quad x_2 \\ x_3 &= -5 \qquad\quad - 3x_5 \\ x_4 &= 4 \qquad\quad + 2x_5 \\ x_5 &= \qquad\qquad\quad x_5. \end{aligned}$$

We have supplemented the equations with the equations that indicate the free variables. In vector form these equations become:

$$(x_1, x_2, x_3, x_4) = (3, 1, 2, 0) + x_4(-3, 3, -2, 1),$$
$$(x_1, x_2, x_3, x_4, x_5) = (13, 0, -5, 4, 0) + x_2(-2, 1, 0, 0, 0) + x_5(8, 0, -3, 2, 1).$$

As these examples indicate, there is a general type of system that will yield to back substitution. They are the systems whose coefficient matrices have a "generalized triangular" shape. That is, the first nonzero term in each row is farther to the right than the first nonzero term in all previous rows. We place an additional normalizing condition that all zero rows occur below the nonzero rows and make the following definition.

***Definition*** A matrix is in **echelon form** (or is an **echelon matrix**) if:

(a) The first nonzero entry in each row of the matrix is farther to the right than the first nonzero entry in all preceding rows.
(b) All zero rows are below the nonzero rows.

The coefficient matrices of systems (1) and (2) are in echelon form. Any triangular matrix whose zero rows are below its nonzero rows is in echelon form. Other examples are:

$$\begin{bmatrix} 0 & 1 & 0 & 2 \\ 0 & 0 & 0 & 5 \end{bmatrix}, \quad \begin{bmatrix} 2 & 0 \\ 0 & 1 \\ 0 & 0 \\ 0 & 0 \end{bmatrix}, \quad \begin{bmatrix} 4 \\ 0 \\ 0 \end{bmatrix}, \quad \begin{bmatrix} 0 & -1 & 2 \end{bmatrix}.$$

However, the following matrices are not in echelon form.

$$\begin{bmatrix} 4 \\ 1 \\ 0 \end{bmatrix}, \quad \begin{bmatrix} 1 & 2 & 3 \\ 0 & 0 & 0 \\ 0 & 0 & 2 \end{bmatrix}, \quad \begin{bmatrix} 0 & 1 & 2 \\ 2 & 1 & 0 \\ 0 & 0 & 1 \end{bmatrix}$$

The first nonzero entry in each row of an echelon matrix has a special significance, and hence a special name. It corresponds to a determined variable and is called a **pivot**. Some important facts about pivots are the following:

1. Each nonzero row of an echelon matrix contains exactly one pivot.
2. Each column of an echelon matrix contains at most one pivot. A column of an echelon matrix that contains a pivot is called a **pivot column**. A column that does not contain a pivot is called a **nonpivot column**.
3. *The number of nonzero rows of an echelon matrix is equal to the number of pivot columns*, for the number of nonzero rows is equal to the number of pivots (by 1) and the number of pivots is equal to the number of pivot columns (by 2).

When an echelon matrix $E$ is the coefficient matrix of a linear system $E\mathbf{x} = \mathbf{c}$, then:

1. The determined variables of the system correspond to the pivot columns of $E$.
2. The free variables of the system correspond to the nonpivot columns of $E$.

What does back substitution do to an echelon matrix? It replaces every pivot with a 1 and replaces every entry above every pivot with a zero.

***Definition*** An echelon matrix is called a **reduced echelon matrix** if

(a) Each pivot is a 1.
(b) Each pivot column contains exactly one nonzero entry (namely, the pivot).

### EXAMPLE 3

The following matrices are in reduced echelon form.

$$[1 \quad 2 \quad 3], \quad \begin{bmatrix} 0 & 0 \\ 0 & 0 \end{bmatrix}, \quad \begin{bmatrix} 0 & 1 \\ 0 & 0 \end{bmatrix}, \quad \begin{bmatrix} 1 & 0 & 0 & 1 \\ 0 & 1 & 0 & 2 \\ 0 & 0 & 1 & 3 \end{bmatrix}$$

The following matrices are not in reduced echelon form.

$$\begin{bmatrix} 1 & 2 \\ 0 & 1 \end{bmatrix}, \quad \begin{bmatrix} 1 & 0 & 0 & 1 \\ 0 & 1 & 1 & 2 \\ 1 & 0 & 0 & 3 \end{bmatrix}$$

Thus back substitution transforms an echelon matrix to a reduced echelon matrix. For example, the coefficient matrices of systems (1) and (2),

$$\begin{bmatrix} 2 & 1 & -1 & 1 \\ 0 & 1 & 1 & -1 \\ 0 & 0 & 3 & 6 \end{bmatrix}, \quad \begin{bmatrix} 1 & 2 & 3 & 1 & -1 \\ 0 & 0 & 1 & 1 & 1 \\ 0 & 0 & 0 & 1 & -2 \end{bmatrix},$$

are in echelon form and these matrices were reduced by back substitution to the reduced echelon matrices

$$\begin{bmatrix} 1 & 0 & 0 & 3 \\ 0 & 1 & 0 & -3 \\ 0 & 0 & 1 & 2 \end{bmatrix}, \quad \begin{bmatrix} 1 & 2 & 0 & 0 & -8 \\ 0 & 0 & 1 & 0 & 3 \\ 0 & 0 & 0 & 1 & -2 \end{bmatrix}.$$

## PROBLEMS 1.8

**1.** Complete each of the following definitions.
   (a) A matrix is in echelon form if....
   (b) A pivot in an echelon matrix is....
   (c) A reduced echelon matrix is....

***2.** Which of the following matrices is in echelon form? If a matrix is in echelon form, identify the pivots.

(a) $\begin{bmatrix} 2 & -1 & 1 \\ 0 & 3 & 1 \end{bmatrix}$

(b) $\begin{bmatrix} 1 & 0 & 2 \\ 0 & 3 & 1 \\ 0 & 5 & 0 \end{bmatrix}$

(c) $\begin{bmatrix} 1 & -1 & 0 & 1 & 0 \\ 0 & 0 & 1 & 2 & 0 \\ 0 & 0 & 0 & 0 & 1 \end{bmatrix}$

(d) $\begin{bmatrix} 1 & 0 & 3 & 0 \\ 0 & 2 & 1 & 0 \\ 0 & 0 & 0 & 0 \\ 0 & 0 & 0 & 0 \end{bmatrix}$

(e) $\begin{bmatrix} 1 & 3 & 0 & 1 \\ 0 & 2 & 0 & 2 \\ 0 & 0 & 0 & 1 \\ 0 & 0 & 2 & 0 \end{bmatrix}$

(f) $\begin{bmatrix} 1 & 0 & 1 & 0 & 2 \\ 0 & 1 & 1 & 1 & 3 \\ 0 & 0 & 0 & 0 & 0 \end{bmatrix}$

(g) $\begin{bmatrix} 0 & 0 & 0 \\ 0 & 1 & 0 \\ 0 & 0 & 1 \end{bmatrix}$

(h) $\begin{bmatrix} 0 & 1 & 0 & 1 \\ 0 & 0 & 1 & 0 \\ 0 & 0 & 0 & 0 \end{bmatrix}$

(i) $\begin{bmatrix} -1 & 2 & -3 & 4 & -5 & 6 \end{bmatrix}$

***3.** Which of the matrices in Problem 2 are in reduced echelon form?

**4.** For each of the echelon matrices $A$ given in Problem 2, find all solutions of the linear system $A\mathbf{x} = \mathbf{0}$.

**5.** Write each of the following systems in matrix form and solve the system.

*(a) $\begin{aligned} x_1 + 2x_2 &= 6 \\ 3x_2 &= 6 \end{aligned}$

(b) $\begin{aligned} 2x_1 + x_2 &= 5 \\ -x_2 &= 4 \end{aligned}$

(c) $\begin{aligned} 3x_1 + 4x_2 &= 7 \\ 3x_2 &= 4 \end{aligned}$

*(d) $\begin{aligned} x_1 + 2x_2 + 3x_3 &= 9 \\ x_3 &= 1 \end{aligned}$

(e) $\begin{aligned} 2x_1 + x_2 + x_3 &= 5 \\ 3x_2 + 2x_3 &= 1 \end{aligned}$

*(f) $\begin{aligned} x_1 + 2x_2 + 3x_3 &= 9 \\ 2x_2 + x_3 &= 7 \\ x_3 &= 1 \end{aligned}$

(g) $\begin{aligned} 2x_1 + x_2 + x_3 + x_4 &= 6 \\ 2x_3 + x_4 &= 8 \end{aligned}$

*(h) $\begin{aligned} 2x_1 + 3x_2 + x_3 - x_4 + x_5 &= 6 \\ x_3 + x_5 &= 4 \\ x_4 - x_5 &= 2 \end{aligned}$

(i) $\begin{aligned} 2x_1 + x_2 + x_3 &= 5 \\ 3x_2 + 2x_3 &= 1 \\ x_3 &= 2 \end{aligned}$

(j) $\begin{aligned} x_1 + 2x_2 &= 3 \\ x_2 + 2x_3 &= -2 \\ x_3 + 2x_4 &= 4 \end{aligned}$

(k) $\begin{aligned} x_1 + 2x_2 - 3x_3 + x_4 - x_5 &= -2 \\ x_2 - 2x_3 + x_4 - 2x_5 &= -1 \\ 2x_3 - x_4 + x_5 &= -3 \\ x_4 - 2x_5 &= 1 \end{aligned}$

*(l) $\begin{aligned} 2x_1 + x_2 + x_3 + x_4 &= 2 \\ x_2 + x_3 + x_6 &= 3 \\ x_3 &= -1 \\ x_5 + x_6 &= -4 \end{aligned}$

(m) $x_1 + 2x_2 - x_3 = 0$

(n) $x_1 + 2x_2 - x_3 = 5$

(o) $-x_2 + x_3 = 0$

*(p) $x_1 + 2x_2 + 3x_3 + 4x_4 + 5x_5 + 6x_6 = 17$

**6.** In each part you are given the general solution in vector form of a linear system. Find an augmented reduced echelon matrix that led to this solution.

*(a) $\begin{bmatrix} x_1 \\ x_2 \\ x_3 \end{bmatrix} = \begin{bmatrix} -1 \\ 3 \\ 0 \end{bmatrix}$

(b) $\begin{bmatrix} x_1 \\ x_2 \\ x_3 \end{bmatrix} = \begin{bmatrix} 2 \\ 0 \\ 4 \end{bmatrix} + x_2 \begin{bmatrix} -2 \\ 1 \\ 0 \end{bmatrix}$

*(c) $\begin{bmatrix} x_1 \\ x_2 \\ x_3 \end{bmatrix} = x_2 \begin{bmatrix} 3 \\ 1 \\ 0 \end{bmatrix} + x_3 \begin{bmatrix} 1 \\ 0 \\ 1 \end{bmatrix}$

(d) $\begin{bmatrix} x_1 \\ x_2 \\ x_3 \\ x_4 \end{bmatrix} = \begin{bmatrix} 1 \\ 0 \\ 3 \\ 0 \end{bmatrix} + x_2 \begin{bmatrix} -1 \\ 1 \\ 0 \\ 0 \end{bmatrix} + x_4 \begin{bmatrix} 3 \\ 0 \\ -1 \\ 1 \end{bmatrix}$

(e) $\begin{bmatrix} x_1 \\ x_2 \\ x_3 \\ x_4 \end{bmatrix} = \begin{bmatrix} 0 \\ 2 \\ 1 \\ 1 \end{bmatrix} + x_1 \begin{bmatrix} 1 \\ 0 \\ 0 \\ 0 \end{bmatrix}$

(f) $$\begin{bmatrix} x_1 \\ x_2 \\ x_3 \\ x_4 \\ x_5 \end{bmatrix} = \begin{bmatrix} 1 \\ 0 \\ 1 \\ 0 \\ 0 \end{bmatrix} + x_2 \begin{bmatrix} 2 \\ 1 \\ 0 \\ 0 \\ 0 \end{bmatrix} + x_4 \begin{bmatrix} -1 \\ 0 \\ 0 \\ 1 \\ 0 \end{bmatrix} + x_5 \begin{bmatrix} 0 \\ 0 \\ 0 \\ 0 \\ 1 \end{bmatrix}$$

(g) $$\begin{bmatrix} x_1 \\ x_2 \\ x_3 \\ x_4 \\ x_5 \\ x_6 \end{bmatrix} = \begin{bmatrix} -\frac{1}{2} \\ 4 \\ -1 \\ 0 \\ 0 \\ 0 \end{bmatrix} + x_4 \begin{bmatrix} -\frac{1}{2} \\ 0 \\ 0 \\ 1 \\ 0 \\ 0 \end{bmatrix} + x_6 \begin{bmatrix} \frac{1}{2} \\ 1 \\ 0 \\ 0 \\ -1 \\ 1 \end{bmatrix}$$

**7.** Let $R$ be a reduced echelon matrix.

*(a) If $R\mathbf{x} = \mathbf{d}$ is inconsistent, what can you say about the augmented matrix $[R : \mathbf{d}]$?

(b) If $R\mathbf{x} = \mathbf{d}$ has a unique solution, what can you say about the augmented matrix $[R : \mathbf{d}]$?

(c) If $R\mathbf{x} = \mathbf{d}$ has infinitely many solutions, what can you say about the augmented matrix $[R : \mathbf{d}]$?

**8.** Let $E$ be an $m \times n$ echelon matrix.

*(a) Show that $E\mathbf{x} = \mathbf{c}$ is consistent for every $\mathbf{c}$ in $R^m$ if and only if $E$ has no zero rows.

(b) Show that every consistent system $E\mathbf{x} = \mathbf{c}$ has a unique solution if and only if every column of $E$ is a pivot column.

(c) Show that the homogeneous system $E\mathbf{x} = \mathbf{0}$ has a unique solution if and only if every consistent system $E\mathbf{x} = \mathbf{c}$ has a unique solution.

(d) Show that if $n > m$, then any consistent system $E\mathbf{x} = \mathbf{c}$ has infinitely many solutions.

(e) Show that if $n < m$, then there are vectors $\mathbf{c}$ in $R^m$ for which the system is inconsistent.

**9.** Let $E$ be an $n \times n$ echelon matrix.

*(a) Show that $E$ has a zero on its diagonal if and only if $E$ has a zero row.

(b) Show that $E$ is nonsingular if and only if $E$ has no zero rows.

(c) Show that $E$ is nonsingular if and only if the only solution of the homogeneous system $E\mathbf{x} = \mathbf{0}$ is the trivial solution.

(d) Show that $E$ is nonsingular if and only if $E\mathbf{x} = \mathbf{c}$ has a unique solution for every $\mathbf{c}$ in $R^n$.

**10.** Problems 9(b), (c), (d) state three equivalent conditions for a square echelon matrix to be nonsingular. State three parallel conditions for a square echelon matrix to be singular.

# 1.9 Gaussian Elimination

To complete our program we turn to the task of showing that forward elimination will transform any matrix into echelon form.

In Section 1.7 forward elimination was stopped dead in its tracks by the system

$$\begin{bmatrix} 1 & 1 & 1 & 1 \\ 0 & 0 & 1 & 1 \\ 0 & 0 & 1 & -1 \\ 0 & 0 & 1 & 2 \end{bmatrix} \begin{bmatrix} x_1 \\ x_2 \\ x_3 \\ x_4 \end{bmatrix} = \begin{bmatrix} 2 \\ 1 \\ -3 \\ 3 \end{bmatrix}.$$

The 2,2 entry in the matrix is zero and there is no subsequent row to pivot with. Since we are trying to convert this matrix to an echelon matrix, it is clear what we should do at this point; move to the 2,3 entry and make each subsequent entry in the third column zero.

$$\begin{bmatrix} 1 & 1 & 1 & 1 & : & 2 \\ 0 & 0 & 1 & 1 & : & 1 \\ 0 & 0 & 0 & -2 & : & -4 \\ 0 & 0 & 0 & 1 & : & 2 \end{bmatrix}$$

Our next move is equally clear: move to the 3,4 entry and make each subsequent entry in the fourth column zero.

$$\begin{bmatrix} 1 & 1 & 1 & 1 & : & 2 \\ 0 & 0 & 1 & 1 & : & 1 \\ 0 & 0 & 0 & -2 & : & -4 \\ 0 & 0 & 0 & 0 & : & 0 \end{bmatrix}$$

The matrix is now in echelon form and back substitution will transform it to the reduced echelon matrix

$$\begin{bmatrix} 1 & 1 & 0 & 0 & : & 1 \\ 0 & 0 & 1 & 0 & : & -1 \\ 0 & 0 & 0 & 1 & : & 2 \\ 0 & 0 & 0 & 0 & : & 0 \end{bmatrix}.$$

The first, third, and fourth columns are the pivot columns and the second column is the nonpivot column. Thus $x_1$, $x_3$, and $x_4$ are the determined variables, and $x_2$ is the free variable. The general solution in vector form is the line

$$(x_1, x_2, x_3, x_4) = (1, 0, -1, 2) + x_2(-1, 1, 0, 0).$$

This example should make it clear that by using forward elimination we can transform any matrix to an echelon matrix. Begin by finding the first column that does not consist entirely of zeros. Pivot a nonzero entry into the first entry of that column and use forward elimination to change each subsequent entry in the column to a zero. The first row and the columns up to and including this column remain unchanged during the rest of the procedure. Now repeat the process on the submatrix consisting of the remaining rows and columns. That is, find the first column of the submatrix that does not consist entirely of zeros. Pivot a nonzero entry into the first row of this submatrix and use forward elimination to replace each subsequent entry in the column with a zero. The first two rows and the columns up to and including this column remain unchanged during the rest of the procedure. Now repeat the process on the submatrix consisting of the remaining rows and columns. When the process terminates, the matrix is in echelon form.

Our program is now complete. We have extended the Gaussian algorithm for square systems developed in Section 1.7 to arbitrary systems. It consists of forward elimination followed by back substitution. Forward elimination reduces a matrix to an echelon matrix and back substitution reduces the echelon matrix to a reduced echelon matrix. Thus given a consistent system $A\mathbf{x} = \mathbf{b}$, forward elimination transforms it to an equivalent system $E\mathbf{x} = \mathbf{c}$ with $E$ an echelon matrix. Then back substitution transforms $E\mathbf{x} = \mathbf{c}$ to $R\mathbf{x} = \mathbf{d}$ with $R$ a reduced echelon matrix.

$$[A : \mathbf{b}] \xrightarrow[\text{elimination}]{\text{forward}} [E : \mathbf{c}] \xrightarrow[\text{substitution}]{\text{back}} [R : \mathbf{d}]$$

## EXAMPLE 1

Consider the system $A\mathbf{x} = \mathbf{b}$ whose augmented matrix is

$$[A : \mathbf{b}] = \begin{bmatrix} 1 & -3 & 3 & 1 & 1 & : & 34 \\ 2 & -6 & -1 & -2 & -5 & : & -8 \\ 3 & -9 & -5 & 1 & -11 & : & -20 \end{bmatrix}.$$

Forward elimination reduces this matrix to

$$[E : \mathbf{c}] = \begin{bmatrix} 1 & -3 & 3 & 1 & 1 & : & 34 \\ 0 & 0 & -7 & -4 & -7 & : & -76 \\ 0 & 0 & 0 & 6 & 0 & : & 30 \end{bmatrix}.$$

Back substitution further reduces this matrix to

$$[R : \mathbf{d}] = \begin{bmatrix} 1 & -3 & 0 & 0 & -2 & : & 5 \\ 0 & 0 & 1 & 0 & 1 & : & 8 \\ 0 & 0 & 0 & 1 & 0 & : & 5 \end{bmatrix}.$$

The free variables are $x_2$ and $x_5$. The general solution is the plane

$$(x_1, x_2, x_3, x_4, x_5) = (5, 0, 8, 5, 0) + x_2(3, 1, 0, 0, 0) + x_5(2, 0, -1, 0, 1).$$

## EXAMPLE 2

Consider the system $A\mathbf{x} = \mathbf{b}$ whose augmented matrix is

$$[A : \mathbf{b}] = \begin{bmatrix} 1 & 2 & : & 5 \\ -1 & 1 & : & 1 \\ 1 & 1 & : & 6 \end{bmatrix}.$$

Forward elimination reduces $[A : \mathbf{b}]$ to

$$[E : \mathbf{c}] = \begin{bmatrix} 1 & 2 & : & 5 \\ 0 & 3 & : & 6 \\ 0 & 0 & : & 3 \end{bmatrix}.$$

The last row of this matrix represents the equation

$$0x_1 + 0x_2 = 3.$$

Since this equation is inconsistent, the system $E\mathbf{x} = \mathbf{c}$ is inconsistent (any solution of this system must be a solution of each equation of the system). Finally, since the system $E\mathbf{x} = \mathbf{c}$ is equivalent to the system $A\mathbf{x} = \mathbf{b}$, the system $A\mathbf{x} = \mathbf{b}$ is inconsistent.

Consider a homogeneous system $A\mathbf{x} = \mathbf{0}$. Row operations performed on the augmented matrix $[A : \mathbf{0}]$ will not change the augmented column of zeros. Therefore, forward elimination reduces $[A : \mathbf{0}]$ to $[E : \mathbf{0}]$ and back substitution reduces $[E : \mathbf{0}]$ to $[R : \mathbf{0}]$. Clearly, there is no need to augment the matrix $A$ with a column of zeros. To solve a homogeneous system $A\mathbf{x} = \mathbf{0}$ we simply reduce $A$ to $E$ to $R$.

### EXAMPLE 3

Let us find the general solution to the homogeneous system $A\mathbf{x} = \mathbf{0}$, where

$$A = \begin{bmatrix} 1 & 2 & 3 & 1 \\ 2 & 4 & 5 & 1 \\ 1 & 2 & 2 & 0 \end{bmatrix}.$$

Forward elimination reduces $A$ to

$$E = \begin{bmatrix} 1 & 2 & 3 & 1 \\ 0 & 0 & -1 & -1 \\ 0 & 0 & 0 & 0 \end{bmatrix}.$$

Back substitution reduces $E$ to

$$R = \begin{bmatrix} 1 & 2 & 0 & -2 \\ 0 & 0 & 1 & 1 \\ 0 & 0 & 0 & 0 \end{bmatrix}.$$

The free variables are $x_2$ and $x_4$. The general solution is the plane

$$(x_1, x_2, x_3, x_4) = x_2(-2, 1, 0, 0) + x_4(2, 0, -1, 1).$$

In certain applications it is necessary to solve several systems

$$A\mathbf{x} = \mathbf{b}_1, \quad A\mathbf{x} = \mathbf{b}_2, \quad \ldots, \quad A\mathbf{x} = \mathbf{b}_k,$$

which all have the same coefficient matrix $A$. To solve each of these systems we would reduce the augmented matrices

$$[A : \mathbf{b}_1], \quad [A : \mathbf{b}_2], \quad \ldots, \quad [A : \mathbf{b}_k]$$

to reduced echelon systems

$$[R : \mathbf{d}_1], \quad [R : \mathbf{d}_2], \quad \ldots, \quad [R : \mathbf{d}_k].$$

But the reduction of $A$ to $R$ is identical for each of these $k$ augmented matrices. We can reduce the work significantly by forming the augmented matrix

$$[A : \mathbf{b}_1 \quad \mathbf{b}_2 \quad \cdots \quad \mathbf{b}_k]$$

and reducing it to

$$[R : \mathbf{d}_1 \quad \mathbf{d}_2 \quad \cdots \quad \mathbf{d}_k],$$

thereby solving the $k$ different systems simultaneously with only *one* reduction of $A$ to $R$.

### EXAMPLE 4

Let

$$A = \begin{bmatrix} 2 & 3 \\ 3 & 4 \end{bmatrix}, \qquad \mathbf{b}_1 = \begin{bmatrix} 1 \\ 2 \end{bmatrix}, \qquad \mathbf{b}_2 = \begin{bmatrix} 0 \\ 1 \end{bmatrix}, \qquad \mathbf{b}_3 = \begin{bmatrix} 1 \\ 0 \end{bmatrix}.$$

To solve the three systems $A\mathbf{x} = \mathbf{b}_1$, $A\mathbf{x} = \mathbf{b}_2$, $A\mathbf{x} = \mathbf{b}_3$, first form the augmented matrix

$$[A : \mathbf{b}_1 \quad \mathbf{b}_2 \quad \mathbf{b}_3] = \begin{bmatrix} 2 & 3 & : & 1 & 0 & 1 \\ 3 & 4 & : & 2 & 1 & 0 \end{bmatrix}.$$

Forward elimination reduces this augmented matrix to

$$[E : \mathbf{c}_1 \quad \mathbf{c}_2 \quad \mathbf{c}_3] = \begin{bmatrix} 2 & 3 & : & 1 & 0 & 1 \\ 0 & -\frac{1}{2} & : & \frac{1}{2} & 1 & -\frac{3}{2} \end{bmatrix},$$

and back substitution reduces it further to

$$[R : \mathbf{d}_1 \quad \mathbf{d}_2 \quad \mathbf{d}_3] = \begin{bmatrix} 1 & 0 & : & 2 & 3 & -4 \\ 0 & 1 & : & -1 & -2 & 3 \end{bmatrix}.$$

The three solutions can now be read from the appropriate columns. The vector solution of

$$\begin{aligned} A\mathbf{x} &= \mathbf{b}_1 \quad \text{is} \quad \mathbf{x} = (2, -1), \\ A\mathbf{x} &= \mathbf{b}_2 \quad \text{is} \quad \mathbf{x} = (3, -2), \\ A\mathbf{x} &= \mathbf{b}_3 \quad \text{is} \quad \mathbf{x} = (-4, 3). \end{aligned}$$

### EXAMPLE 5

Consider the plane through the origin and the vectors $\mathbf{u} = (1, 1, 2)$ and $\mathbf{v} = (1, -1, 1)$. We will use Gaussian elimination to derive a Cartesian equation for this plane. Its vector equation is

$$\mathbf{x} = \alpha\mathbf{u} + \beta\mathbf{v}$$

or

$$(x_1, x_2, x_3) = \alpha(1, 1, 2) + \beta(1, -1, 1),$$

where $\alpha$ and $\beta$ are scalars. Equating components, we obtain

$$\begin{aligned} x_1 &= \alpha + \beta \\ x_2 &= \alpha - \beta \\ x_3 &= 2\alpha + \beta. \end{aligned}$$

The matrix form of these equation is

$$\begin{bmatrix} x_1 \\ x_2 \\ x_3 \end{bmatrix} = \begin{bmatrix} 1 & 1 \\ 1 & -1 \\ 2 & 1 \end{bmatrix} \begin{bmatrix} \alpha \\ \beta \end{bmatrix}.$$

Thus a vector $\mathbf{x}$ is in the plane if and only if this system of equations has a solution. Form the augmented matrix of the system

$$\begin{bmatrix} 1 & 1 & : & x_1 \\ 1 & -1 & : & x_2 \\ 2 & 1 & : & x_3 \end{bmatrix}$$

and reduce it to echelon form,

$$\begin{bmatrix} 1 & 1 & : & x_1 \\ 0 & -2 & : & -x_1 + x_2 \\ 0 & 0 & : & -\frac{3}{2}x_1 - \frac{1}{2}x_2 + x_3 \end{bmatrix}.$$

Since $\mathbf{x} = (x_1, x_2, x_3)$ is in the plane if and only if this system has a solution, $(x_1, x_2, x_3)$ is in the plane if and only if

$$-\frac{3}{2}x_1 - \frac{1}{2}x_2 + x_3 = 0.$$

This is a Cartesian equation of the plane.

Converting a Cartesian equation of a plane to a vector equation of a plane is simple. For example, to find a vector equation of the plane whose Cartesian equation is

$$2x_1 - x_2 + 5x_3 = 0,$$

we simply write down the general solution of the equation. The free variables are $x_2$ and $x_3$ and the general solution is

$$\mathbf{x} = (x_1, x_2, x_3) = x_2(1/2, 1, 0) + x_3(-5/2, 0, 1).$$

## Gauss–Jordan Elimination

There is an alternative solution procedure, called **Gauss–Jordan elimination**, which is frequently used. In this procedure the pivot equation is first divided by the coefficient of the pivot variable. Then the pivot variable is eliminated from the prior equations as well as the subsequent ones. It can be shown that the Gauss–Jordan procedure requires 50% more multiplications than the Gaussian elimination procedure (see Chapter 9). If the only time-consuming operation were multiplication (as is indeed the case when computations are performed on a computer), then the Gauss–Jordan procedure would clearly be less efficient. But a significant portion of the work involved when doing calculations by hand is the writing of all the intermediate systems. Since Gauss–Jordan requires much less writing, it can be done at least as quickly as Gaussian elimination.

For most of the remainder of this book we will use the phrase "Gaussian elimination" as an abbreviation for either of these two complete solution procedures. We suggest that the student understand both procedures. Routine computations can be done by either method.

## EXAMPLE 6

Let us solve the system $A\mathbf{x} = \mathbf{b}$, where

$$A = \begin{bmatrix} 1 & 3 & 4 \\ 2 & 9 & 6 \\ 1 & 5 & 6 \end{bmatrix}, \qquad \mathbf{b} = \begin{bmatrix} 8 \\ 27 \\ 15 \end{bmatrix},$$

using Gauss–Jordan elimination. The first step is to use the 1,1 entry to replace the 2,1 and 3,1 entry of $[A : \mathbf{b}]$ with a zero.

$$\begin{bmatrix} 1 & 3 & 4 & : & 8 \\ 0 & 3 & -2 & : & 11 \\ 0 & 2 & 2 & : & 7 \end{bmatrix}$$

Now make the second pivot 1 and use the 2,2 entry to replace all other entries in the second column with a zero.

$$\begin{bmatrix} 1 & 0 & 6 & : & -3 \\ 0 & 1 & -\frac{2}{3} & : & \frac{11}{3} \\ 0 & 0 & \frac{10}{3} & : & -\frac{1}{3} \end{bmatrix}$$

Finally, make the third pivot 1 and replace all other entries in the third column with a zero.

$$\begin{bmatrix} 1 & 0 & 0 & : & -\frac{12}{5} \\ 0 & 1 & 0 & : & \frac{18}{5} \\ 0 & 0 & 1 & : & -\frac{1}{10} \end{bmatrix}.$$

The solution is $\mathbf{x} = (-\frac{12}{5}, \frac{18}{5}, -\frac{1}{10})$

We conclude this section by developing some results about the possible echelon and reduced echelon form of a matrix. These results will enable us to define an important number associated with a matrix, namely its rank. We begin our discussion with the special case of a nonsingular matrix.

By Theorem 2 of Section 1.7, a matrix is nonsingular if and only if it can be reduced to the identity matrix by Gaussian elimination. Thus:

**Result 1** The reduced echelon form of any nonsingular matrix is the identity matrix.

### EXAMPLE 7

Forward elimination reduces

$$A = \begin{bmatrix} 1 & 2 & 3 \\ 2 & 5 & 1 \\ 1 & 4 & -6 \end{bmatrix} \quad \text{to} \quad E = \begin{bmatrix} 1 & 2 & 3 \\ 0 & 1 & -5 \\ 0 & 0 & 1 \end{bmatrix}.$$

Since $E$ is triangular with no zeros on its diagonal, $A$ is nonsingular. Back substitution further reduces $E$ to $I$.

This result says that all nonsingular matrices have the same reduced echelon form, the identity matrix. Therefore, the reduced echelon form of a nonsingular matrix $A$ is unique. No matter how the reduction is carried out, the result is the same. This same fact is true of any matrix. The proof is, however, rather involved and we shall omit it.

***Theorem 1*** *The reduced echelon form of a matrix is unique.*

The corresponding theorem for the echelon form of a matrix is not true. It is quite possible to reduce a matrix to different echelon matrices. For example, the matrix

$$A = \begin{bmatrix} 2 & 2 & 1 & 1 \\ 2 & 3 & 4 & 1 \\ 2 & 1 & -2 & 1 \end{bmatrix}$$

can be reduced to the echelon matrix

$$E_1 = \begin{bmatrix} 2 & 2 & 1 & 1 \\ 0 & 1 & 3 & 0 \\ 0 & 0 & 0 & 0 \end{bmatrix}$$

or by first interchanging the first and second rows it can be reduced to the echelon matrix

$$E_2 = \begin{bmatrix} 2 & 3 & 4 & 1 \\ 0 & -1 & -3 & 0 \\ 0 & 0 & 0 & 0 \end{bmatrix}.$$

However, the reduced echelon form of both of these matrices is the same, namely

$$R = \begin{bmatrix} 1 & 0 & -\frac{5}{2} & \frac{1}{2} \\ 0 & 1 & 3 & 0 \\ 0 & 0 & 0 & 0 \end{bmatrix}.$$

Although a matrix $A$ can be reduced to different echelon matrices, there are two important properties shared by all possible echelon forms of $A$.

> ***Theorem 2*** *Let $E_1$ and $E_2$ be two echelon forms of a matrix $A$.*
> (*a*) *$E_1$ and $E_2$ have the same pivot columns and hence the same number of pivot columns.*
> (*b*) *$E_1$ and $E_2$ have the same number of nonzero rows, which is the same as the number of pivot columns.*

For example, the pivot columns of the echelon forms $E_1$ and $E_2$ in our example are the first and second columns. Hence both echelon forms have two pivot columns. Both also have two nonzero rows.

To prove the first statement in the theorem, suppose that $E_1$ and $E_2$ are both echelon forms of $A$. Back substitution reduces these matrices to reduced echelon matrices $R_1$ and $R_2$ by replacing each pivot with a 1 and replacing each entry above each pivot with a 0. Therefore, $E_1$ and $R_1$ have the same pivot columns and $E_2$ and $R_2$ have the same pivot columns. But by Theorem 1, $R_1 = R_2$. Therefore, $E_1$ and $E_2$ have the same pivot columns. This proves the first statement. The second statement follows immediately from the first since the number of nonzero rows in any echelon matrix is equal to the number of pivot columns.

Because any two echelon forms of a matrix $A$ have the same pivot columns, we define the **pivot columns** of $A$ to be the columns of $A$ that correspond to the pivot columns of any echelon form of $A$.

***Definition*** The **rank** of a matrix $A$ is the number of pivot columns of $A$.

By Theorem 2 the rank of $A$ is also equal to the number of nonzero rows in any echelon form of $A$.

> ***Theorem 3*** *Let $A$ be an $n \times n$ matrix. $A$ is nonsingular if and only if the rank of $A$ is $n$.*

> **Proof** This follows immediately from the fact that (see Theorem 2 of Section 1.7) $A$ is nonsingular if and only if the reduced echelon form of $A$ is the $n \times n$ identity matrix $I$.

In Section 1.7 we defined a square matrix to be singular if it cannot be reduced to a triangular matrix having no zeros on its diagonal. To verify that a matrix is singular using this definition, we must try all possible ways of reducing the matrix to a triangular matrix using the three basic row operations to see that we cannot obtain a triangular matrix having no zeros on its diagonal. However, by Theorem 2(a), if a square matrix can be transformed to an echelon matrix having a zero on its diagonal, then any other reduction of $A$ to an echelon matrix will result in an echelon matrix having a zero in the same diagonal entry. Thus a matrix is singular if and only if anytime it is reduced to an echelon matrix using forward elimination, there is at least one zero on the diagonal.

# PROBLEMS 1.9

**1.** Complete each of the following definitions.
(a) The Gaussian elimination procedure consists of ....
(b) The Gauss–Jordan elimination procedure consists of ....
(c) The rank of a matrix is ....

**2.** In each part, do the following for the given matrix $A$ and vector $\mathbf{b}$.
(i) Find the rank of $A$.
(ii) Find the general solution of $A\mathbf{x} = \mathbf{0}$.
(iii) Find the general solution of $A\mathbf{x} = \mathbf{b}$.
(iv) Express $\mathbf{b}$ as a linear combination of columns of $A$.

*(a) $\begin{bmatrix} 2 & 3 \\ 2 & 4 \end{bmatrix}, \begin{bmatrix} 6 \\ 8 \end{bmatrix}$

(b) $\begin{bmatrix} 1 & 6 \\ 3 & 1 \end{bmatrix}, \begin{bmatrix} 7 \\ 4 \end{bmatrix}$

*(c) $\begin{bmatrix} 1 & 3 \\ 2 & 6 \end{bmatrix}, \begin{bmatrix} 2 \\ 7 \end{bmatrix}$

(d) $\begin{bmatrix} 2 & 1 & 2 \\ 4 & 5 & -3 \end{bmatrix}, \begin{bmatrix} 5 \\ 3 \end{bmatrix}$

*(e) $\begin{bmatrix} 4 & 3 & 2 \\ 1 & -4 & 1 \end{bmatrix}, \begin{bmatrix} 0 \\ 1 \end{bmatrix}$

(f) $\begin{bmatrix} 0 & 1 & 2 & 3 \\ 0 & 2 & 2 & -2 \end{bmatrix}, \begin{bmatrix} 1 \\ 0 \end{bmatrix}$

*(g) $\begin{bmatrix} 1 & 1 & 0 \\ 1 & 2 & 1 \\ 0 & 1 & 2 \end{bmatrix}, \begin{bmatrix} 1 \\ 1 \\ 1 \end{bmatrix}$

(h) $\begin{bmatrix} 1 & 0 & 3 \\ 2 & -3 & 4 \\ 1 & -7 & -2 \end{bmatrix}, \begin{bmatrix} 2 \\ 6 \\ 6 \end{bmatrix}$

*(i) $\begin{bmatrix} 1 & 1 & 1 \\ 6 & 6 & -2 \\ 0 & 3 & -1 \end{bmatrix}, \begin{bmatrix} 0 \\ -8 \\ -3 \end{bmatrix}$

(j) $\begin{bmatrix} 1 & 1 & -1 & 1 \\ 1 & 2 & 0 & 2 \\ 1 & 1 & 0 & 0 \end{bmatrix}, \begin{bmatrix} 7 \\ 15 \\ 17 \end{bmatrix}$

*(k) $\begin{bmatrix} 0 & 1 & -1 \\ 1 & -1 & 1 \\ 2 & 1 & 2 \end{bmatrix}, \begin{bmatrix} 1 \\ 0 \\ 2 \end{bmatrix}$

(l) $\begin{bmatrix} 1 & -1 & -3 \\ 1 & -3 & -13 \\ -3 & -4 & 4 \end{bmatrix}, \begin{bmatrix} 2 \\ 14 \\ 0 \end{bmatrix}$

*(m) $\begin{bmatrix} 2 & 3 & 1 & -1 \\ 2 & -1 & 1 & -1 \\ 4 & 6 & 2 & -2 \end{bmatrix}, \begin{bmatrix} 2 \\ 4 \\ 6 \end{bmatrix}$

(n) $\begin{bmatrix} 2 & 3 & 1 & -1 \\ 2 & -1 & 1 & -1 \\ 4 & 2 & 2 & -2 \end{bmatrix}, \begin{bmatrix} 2 \\ 4 \\ 6 \end{bmatrix}$

*(o) $\begin{bmatrix} 1 & 2 & -3 \\ 1 & -4 & -13 \\ -3 & -6 & 4 \end{bmatrix}, \begin{bmatrix} 4 \\ 14 \\ 2 \end{bmatrix}$

(p) $\begin{bmatrix} 1 & -2 & 7 \\ 2 & -4 & 1 \\ 1 & 1 & 1 \end{bmatrix}, \begin{bmatrix} 1 \\ 0 \\ 1 \end{bmatrix}$

*(q) $\begin{bmatrix} 1 & 1 & 1 & 1 & 1 \\ 1 & 0 & 0 & 0 & 1 \\ 1 & 1 & 0 & 0 & 0 \end{bmatrix}, \begin{bmatrix} -2 \\ 4 \\ 5 \end{bmatrix}$

(r) $\begin{bmatrix} 1 & 2 & 1 & 3 \\ 1 & 2 & 3 & -1 \\ 2 & 4 & 1 & 1 \end{bmatrix}, \begin{bmatrix} 1 \\ 5 \\ 7 \end{bmatrix}$

*(s) $\begin{bmatrix} 1 & -6 & 0 & 2 \\ 2 & -3 & 0 & -4 \\ 0 & 1 & 2 & 3 \\ 3 & -12 & 1 & 1 \end{bmatrix}, \begin{bmatrix} 1 \\ 1 \\ 1 \\ 1 \end{bmatrix}$

(t) $\begin{bmatrix} 1 & 2 & 1 & 1 & 1 \\ -1 & -2 & 0 & -1 & -3 \\ 2 & 4 & 3 & 2 & 0 \\ 0 & 0 & -1 & 0 & 2 \end{bmatrix}, \begin{bmatrix} 5 \\ 2 \\ 17 \\ -7 \end{bmatrix}$

*(u) $[1 \quad 2 \quad -1 \quad -2 \quad 3]$, $[1]$

(v) $[3 \quad 1 \quad -2 \quad 3]$, $[5]$

**3.** In each part, find the general solution in vector form of the linear systems $A\mathbf{x} = \mathbf{b}_k$ for each $k$.

*(a) $A = \begin{bmatrix} 2 & -4 & 0 \\ -1 & 2 & 3 \\ 1 & 1 & 2 \end{bmatrix}$, $\mathbf{b}_1 = \begin{bmatrix} 2 \\ 0 \\ 1 \end{bmatrix}$,

$\mathbf{b}_2 = \begin{bmatrix} 4 \\ 2 \\ 1 \end{bmatrix}$, $\mathbf{b}_3 = \begin{bmatrix} 6 \\ 0 \\ 3 \end{bmatrix}$, $\mathbf{b}_4 = \begin{bmatrix} 6 \\ 0 \\ 0 \end{bmatrix}$,

$\mathbf{b}_5 = \begin{bmatrix} -2 \\ 5 \\ 4 \end{bmatrix}$

(b) $A = \begin{bmatrix} 1 & 0 & -1 & 2 \\ 0 & 0 & 1 & 2 \\ 2 & 1 & -2 & 5 \\ 3 & 1 & -2 & 9 \end{bmatrix}$,

$\mathbf{b}_1 = \begin{bmatrix} 0 \\ 0 \\ 0 \\ 0 \end{bmatrix}$, $\mathbf{b}_2 = \begin{bmatrix} 0 \\ 1 \\ 1 \\ 1 \end{bmatrix}$, $\mathbf{b}_3 = \begin{bmatrix} 2 \\ 1 \\ 2 \\ 5 \end{bmatrix}$

(c) $A = \begin{bmatrix} 1 & 4 & 3 \\ 2 & 5 & 4 \\ 1 & -3 & -2 \end{bmatrix}$, $\mathbf{b}_1 = \begin{bmatrix} 1 \\ 0 \\ 0 \end{bmatrix}$,

$\mathbf{b}_2 = \begin{bmatrix} 0 \\ 1 \\ 0 \end{bmatrix}$, $\mathbf{b}_3 = \begin{bmatrix} 0 \\ 0 \\ 1 \end{bmatrix}$

**4.** For each of the following, find the value of $\alpha$ that makes the system $A\mathbf{x} = \mathbf{b}$ consistent.

*(a) $A = \begin{bmatrix} 1 & 1 & 2 \\ 2 & 3 & -1 \\ 3 & 4 & 1 \end{bmatrix}$, $\mathbf{b} = \begin{bmatrix} 2 \\ 5 \\ \alpha \end{bmatrix}$

(b) $A = \begin{bmatrix} 3 & -5 \\ 6 & 7 \\ -3 & 39 \end{bmatrix}$, $\mathbf{b} = \begin{bmatrix} 1 \\ 2 \\ \alpha \end{bmatrix}$

**5.** For each of the following systems $A\mathbf{x} = \mathbf{b}$, for what values of $\alpha$ (if any) does the system have no solutions? Exactly one solution? Infinitely many solutions?

*(a) $A = \begin{bmatrix} 1 & \alpha \\ \alpha & 1 \end{bmatrix}$, $\mathbf{b} = \begin{bmatrix} 1 \\ 4 \end{bmatrix}$

(b) $A = \begin{bmatrix} 1 & 2 \\ \alpha & 2 \end{bmatrix}$, $\mathbf{b} = \begin{bmatrix} 1 \\ 5 \end{bmatrix}$

*(c) $A = \begin{bmatrix} 1 & 1 & 1 \\ \alpha & 0 & 1 \\ 0 & \alpha & 1 \end{bmatrix}$, $\mathbf{b} = \begin{bmatrix} 1 \\ 2 \\ 2 \end{bmatrix}$

(d) $A = \begin{bmatrix} 1 & 2 & \alpha \\ \alpha & \alpha & 0 \\ \alpha & 3\alpha & 8 \end{bmatrix}$, $\mathbf{b} = \begin{bmatrix} 1 \\ 1 \\ 3 \end{bmatrix}$

(e) $A = \begin{bmatrix} 1 & -1 & -1 \\ -1 & 1 & \alpha \\ \alpha & \alpha & 1 \end{bmatrix}$, $\mathbf{b} = \mathbf{0}$

**6.** For each of the following, find a Cartesian equation of the plane whose vector equation is given.

*(a) $\mathbf{x} = \alpha(1,1,0) + \beta(1,0,1)$
(b) $\mathbf{x} = (1,1,1) + \alpha(1,1,0) + \beta(1,0,1)$
*(c) $\mathbf{x} = (1,0,1) + \alpha(1,1,1) + \beta(2,-1,-1)$
(d) $\mathbf{x} = (1,5,6) + \alpha(0,0,1) + \beta(0,1,1)$
*(e) $\mathbf{x} = (0,1,0) + \alpha(5,0,7) + \beta(2,-7,0)$
(f) $\mathbf{x} = (2,3,4) + \alpha(1,0,1) + \beta(0,1,0)$

**7.** In each part, determine whether the two given planes intersect. If they intersect, determine whether they define the same plane or intersect in a line. Finally, if they intersect in a line, determine an equation of that line.

*(a) $\mathbf{x} = (1,0,0) + \alpha(-1,1,0) + \beta(-1,0,1)$
$\mathbf{x} = \gamma(0,1,0) + \delta(1,0,1)$

*(b) $\mathbf{x} = (1,2,5) + \alpha(1,0,1) + \beta(1,2,1)$
$\mathbf{x} = \gamma(4,2,4) + \delta(0,1,0)$

*(c) $\mathbf{x} = (1,2,1) + \alpha(1,1,0) + \beta(0,1,1)$
$\mathbf{x} = (-1,4,5) + \gamma(1,-2,-3) + \delta(2,3,1)$

(d) $\mathbf{x} = (3,5,1) + \alpha(5,0,6) + \beta(1,0,0)$
$\mathbf{x} = (7,9,-2) + \gamma(3,1,4) + \delta(1,-1,0)$

(e) $\mathbf{x} = (-2,3,5) + \alpha(2,1,1) + \beta(2,2,2)$
$\mathbf{x} = (7,-1,1) + \gamma(1,1,1) + \delta(0,1,1)$

(f) $\mathbf{x} = (1,1,1) + \alpha(2,1,-5) + \beta(-3,1,5)$
$\mathbf{x} = (0,0,4) + \gamma(0,1,-1) + \delta(1,0,-2)$

***8.** Is there a parabola of the form $y = ax^2 + bx + c$ that passes through the points $(0, 5)$, $(-1, 4)$, $(1, 8)$, $(-2, 5)$? Is there a cubic equation of the form $y = ax^3 + bx^2 + c$ that passes through these points?

**9.** Consider a stable economy consisting of three persons. The number in the $i$th row and $j$th column of the following table represents the fraction of the $i$th person's daily production which is consumed by the $j$th person.

| | **1** | **2** | **3** |
|---|---|---|---|
| **1** | $\frac{1}{4}$ | $\frac{1}{2}$ | $\frac{1}{4}$ |
| **2** | $\frac{1}{3}$ | $\frac{1}{6}$ | $\frac{1}{2}$ |
| **3** | $\frac{1}{5}$ | $\frac{2}{5}$ | $\frac{2}{5}$ |

(a) What system of linear equations must the prices $p_1, p_2$, and $p_3$ satisfy?

*(b) Find the general solution of this system.

(c) Explain why all price vectors for this economy are essentially the same.

***10.** Find a $3 \times 3$ matrix $X$ (if possible) such that

$$\begin{bmatrix} 1 & 2 & 3 \\ 2 & 5 & 1 \\ 4 & 9 & 7 \end{bmatrix} X = \begin{bmatrix} 6 & 0 & 8 \\ 8 & -1 & 7 \\ 20 & -1 & 23 \end{bmatrix}.$$

**11.** Let $A$ be an $n \times n$ matrix. Show that $A$ is singular if and only if rank $(A) < n$.

**12.** Prove that an $n \times n$ matrix $A$ is singular if and only if when $A$ is reduced to an echelon matrix $E$, $E$ has at least one zero row.

**13.** Show that an $n \times n$ matrix $A$ is singular if and only if the homogeneous system $A\mathbf{x} = \mathbf{0}$ has a nontrivial solution.

***14.** Let

$$A = \begin{bmatrix} 1 & 2 & 3 & 4 \\ 0 & 5 & 0 & 0 \\ 6 & 7 & 8 & 0 \\ 0 & 9 & 10 & 0 \end{bmatrix}.$$

Show that $A$ is nonsingular. (*Hint:* This can be done without any computations whatsoever.)

# 1.10
## Theorems on Linear Systems

Up until this point in our development we have focused our attention on finding the solutions of a fixed system of equations

$$A\mathbf{x} = \mathbf{b}.$$

That is, given an $m \times n$ matrix $A$ and a vector $\mathbf{b}$ in $R^m$, we have been interested in finding all solution vectors $\mathbf{x}$ in $R^n$. We proved that Gaussian elimination will solve any consistent system. The procedure decomposes into two parts. Forward elimination reduces the system $A\mathbf{x} = \mathbf{b}$ to an echelon system $E\mathbf{x} = \mathbf{c}$, and back substitution further reduces the echelon system to a reduced echelon system $R\mathbf{x} = \mathbf{d}$.

$$[A : \mathbf{b}] \xrightarrow[\text{elimination}]{\text{forward}} [E : \mathbf{c}] \xrightarrow[\text{substitution}]{\text{back}} [R : \mathbf{d}]$$

Since all three systems

$$A\mathbf{x} = \mathbf{b}, \qquad E\mathbf{x} = \mathbf{c}, \qquad R\mathbf{x} = \mathbf{d}$$

have the same solutions, all questions about the solutions to the system $A\mathbf{x} = \mathbf{b}$ can be answered from the system $E\mathbf{x} = \mathbf{c}$ or $R\mathbf{x} = \mathbf{d}$.

We turn our attention in this section to some more general and subtle questions about linear systems. The first question has to do with existence of solutions. Instead of beginning with *both* a matrix $A$ *and* a vector $\mathbf{b}$ and then asking for all *solution* (if there are any) of the system, we begin with only a matrix $A$ and then ask: For what vectors $\mathbf{b}$ does the system have a solution? The change is from finding the solutions of a particular system $A\mathbf{x} = \mathbf{b}$ to finding those vectors $\mathbf{b}$ for which the system has a solution. Our question decomposes naturally into the following two questions.

**Question 1** Let $A$ be an $m \times n$ matrix. Is there a necessary and sufficient condition on the matrix $A$ that will guarantee that for every $\mathbf{b}$ in $R^m$ the system $A\mathbf{x} = \mathbf{b}$ is consistent?

**Question 2** Let $A$ be an $m \times n$ matrix. If it is not true that the system $A\mathbf{x} = \mathbf{b}$ is consistent for every $\mathbf{b}$ in $R^m$, can we determine all vectors $\mathbf{b}$ in $R^m$ for which the system $A\mathbf{x} = \mathbf{b}$ *is* consistent?

We emphasize the fact that these are questions about existence. First we are asking for a condition on an $m \times n$ matrix $A$ which will guarantee that for any vector $\mathbf{b}$ in $R^m$ there exists a solution vector $\mathbf{x}$ in $R^n$ of the system $A\mathbf{x} = \mathbf{b}$. And if there is such a condition on the matrix $A$ and the condition is not satisfied (so that it is not true that the system $A\mathbf{x} = \mathbf{b}$ is consistent for every $\mathbf{b}$ in $R^m$), we are asking if we can nevertheless determine all those vectors $\mathbf{b}$ in $R^m$ for which the system is consistent.

In the special case when $A$ is a square $n \times n$ matrix, Theorem 3 of Section 1.7 gives a sufficient condition for $A\mathbf{x} = \mathbf{b}$ to be consistent for all $\mathbf{b}$ in $R^n$. If an $n \times n$ matrix $A$ is nonsingular, then $A\mathbf{x} = \mathbf{b}$ is consistent for every $\mathbf{b}$ in $R^n$. Our proof of that theorem was based on the Gaussian elimination procedure. If $A$ is nonsingular, then it can be transformed by Gaussian elimination to the identity matrix, and hence for any $\mathbf{b}$ in $R^n$ the system $A\mathbf{x} = \mathbf{b}$ can be transformed to the equivalent consistent system $I\mathbf{x} = \mathbf{d}$.

We will now show that the Gaussian elimination procedure enables us to answer both of our questions for an *arbitrary* $m \times n$ matrix.

***Theorem 1*** *(Existence of Solutions) Let $A$ be an $m \times n$ matrix and suppose that $A$ has been reduced to an echelon matrix $E$ by forward elimination.*

*(a) The system $A\mathbf{x} = \mathbf{b}$ is consistent for every $\mathbf{b}$ in $R^m$ if and only if $E$ has no zero rows.*

*(b) If $E$ has one or more zero rows, then the system $A\mathbf{x} = \mathbf{b}$ is consistent if and only if when $[A : \mathbf{b}]$ is reduced to $[E : \mathbf{c}]$ each zero row of $E$ is matched by a corresponding zero entry in $\mathbf{c}$.*

**Proof of (a)** Suppose that for every vector $\mathbf{b}$ in $R^m$ the system $A\mathbf{x} = \mathbf{b}$ is consistent and that $E$ has one or more zero rows. Then certainly the last row of

$E$ is zero. If $\mathbf{c}$ is any vector in $R^m$ whose $m$th component is not zero, the system $E\mathbf{x} = \mathbf{c}$ is inconsistent. The reason is clear: the $m$th equation of the system is the inconsistent equation

$$0 \cdot x_1 + 0 \cdot x_2 + \cdots + 0 \cdot x_n = c_m, \qquad c_m \neq 0.$$

Since $E$ can be transformed back to $A$ using the three basic row operations, the same sequence of operations will transform the augmented matrix $[E : \mathbf{c}]$ of the inconsistent system $E\mathbf{x} = \mathbf{c}$ to the augmented matrix $[A : \mathbf{b}]$. Since the system $A\mathbf{x} = \mathbf{b}$ is equivalent to the inconsistent system $E\mathbf{x} = \mathbf{c}$, it is also inconsistent. Thus the assumption that $E$ has one or more zero rows leads to the conclusion that there is at least one vector $\mathbf{b}$ such that $A\mathbf{x} = \mathbf{b}$ is inconsistent, a contradiction.

The proof of the converse is almost identical. Suppose that $E$ has no zero rows and that for some vector $\mathbf{b}$ in $R^m$ the system $A\mathbf{x} = \mathbf{b}$ is inconsistent. Forward elimination reduces $[A : \mathbf{b}]$ to $[E : \mathbf{c}]$ and since $A\mathbf{x} = \mathbf{b}$ and $E\mathbf{x} = \mathbf{c}$ are equivalent systems, $E\mathbf{x} = \mathbf{c}$ must also be inconsistent. Impossible! $E$ has no zero rows; therefore, for any vector $\mathbf{c}$ in $R^m$ the system $E\mathbf{x} = \mathbf{c}$ is consistent (and back substitution will find all solutions). Thus the assumption that $A\mathbf{x} = \mathbf{b}$ is inconsistent for some vector $\mathbf{b}$ leads to a contradiction. The proof of part (a) is now complete. The proof of part (b) is similar and is left for your enjoyment (see Problem 18).

Recall that the rank of a matrix $A$ is the number of nonzero rows in $E$ where $E$ is an echelon form of $A$. If $A$ is $m \times n$, then $A$ has rank $m$ if and only if $E$ has no zero rows. Therefore we can state Theorem 1(a) in terms of rank as follows.

***Theorem 2*** *Let $A$ be an $m \times n$ matrix. Then $A\mathbf{x} = \mathbf{b}$ has a solution for every vector $\mathbf{b}$ in $R^m$ if and only if $A$ has rank $m$.*

To appreciate the significance of part (b) of Theorem 1, we must understand the relationship between $\mathbf{b}$ and $\mathbf{c}$. When $[A : \mathbf{b}]$ is reduced to $[E : \mathbf{c}]$, what are the components of $\mathbf{c}$? A moment's thought about how forward elimination proceeds reveals that the components of $\mathbf{c}$ are sums of multiples of the components of $\mathbf{b}$. For example, forward elimination reduces

$$\begin{bmatrix} 1 & 2 & 3 & 1 & : & b_1 \\ 3 & 2 & 5 & 1 & : & b_2 \\ 4 & 4 & 8 & 2 & : & b_3 \end{bmatrix} \quad \text{to} \quad \begin{bmatrix} 1 & 2 & 3 & 1 & : & b_1 \\ 0 & -4 & -4 & -2 & : & b_2 - 3b_1 \\ 0 & 0 & 0 & 0 & : & b_3 - b_2 - b_1 \end{bmatrix}.$$

Thus

$$\mathbf{c} = \begin{bmatrix} b_1 \\ b_2 - 3b_1 \\ b_3 - b_2 - b_1 \end{bmatrix}.$$

Each component of $\mathbf{c}$ is a sum of multiples of the components of $\mathbf{b}$. The system

$A\mathbf{x} = \mathbf{b}$ is consistent if and only if the zero row in $E$ is matched by a corresponding zero entry in $\mathbf{c}$. Thus the system $A\mathbf{x} = \mathbf{b}$ has a solution if and only if the components of $\mathbf{b}$ satisfy the constraint $b_3 - b_2 - b_1 = 0$.

### EXAMPLE 1

Forward elimination reduces the augmented matrix

$$\begin{bmatrix} 1 & 2 & -1 & : & b_1 \\ 2 & 4 & -2 & : & b_2 \\ -4 & -8 & 4 & : & b_3 \end{bmatrix} \quad \text{to} \quad \begin{bmatrix} 1 & 2 & -1 & : & b_1 \\ 0 & 0 & 0 & : & b_2 - 2b_1 \\ 0 & 0 & 0 & : & b_3 + 4b_1 \end{bmatrix}.$$

Thus

$$\mathbf{c} = \begin{bmatrix} b_1 \\ b_2 - 2b_1 \\ b_3 + 4b_1 \end{bmatrix}.$$

Each component of $\mathbf{c}$ is a sum of multiples of the components of $\mathbf{b}$. The system $A\mathbf{x} = \mathbf{b}$ is consistent if and only if the two zero rows in $E$ are matched by corresponding zero entries in $\mathbf{c}$. Thus the system $A\mathbf{x} = \mathbf{b}$ has a solution if and only if the components of $\mathbf{b}$ satisfy the two equations

$$\begin{aligned} b_2 - 2b_1 &= 0 \\ b_3 + 4b_1 &= 0. \end{aligned}$$

As an application of Theorem 1(b), let us show that any plane in $R^3$ through the origin has an equation of the form

$$a_1x_1 + a_2x_2 + a_3x_3 = 0,$$

where $a_1$, $a_2$, and $a_3$ are appropriate scalars (see Example 5 of Section 1.9). Suppose that $\mathbf{u}$ and $\mathbf{v}$ are two *fixed noncollinear* vectors in the plane. (That is, suppose that neither $\mathbf{u}$ nor $\mathbf{v}$ is a scalar multiple of the other. Geometrically, this means that the vectors do not point in the same or opposite directions.) A vector equation of the plane is

$$\mathbf{x} = \alpha\mathbf{u} + \beta\mathbf{v}.$$

Equating components, we obtain the system of equations

$$\begin{aligned} x_1 &= \alpha u_1 + \beta v_1 \\ x_2 &= \alpha u_2 + \beta v_2 \\ x_3 &= \alpha u_3 + \beta v_3, \end{aligned}$$

or in the matrix form

$$\begin{bmatrix} x_1 \\ x_2 \\ x_3 \end{bmatrix} = \begin{bmatrix} u_1 & v_1 \\ u_2 & v_2 \\ u_3 & v_3 \end{bmatrix} \begin{bmatrix} \alpha \\ \beta \end{bmatrix}.$$

(Note that $u_1, u_2, u_3, v_1, v_2, v_3$ are constants throughout this discussion.) Thus a vector $\mathbf{x}$ is in the plane if and only if this system of equations has a solution. Form the augmented matrix of the system,

$$\begin{bmatrix} u_1 & v_1 & : & x_1 \\ u_2 & v_2 & : & x_2 \\ u_3 & v_3 & : & x_3 \end{bmatrix}.$$

If $u_1 \neq 0$, then forward elimination reduces this matrix to

$$\begin{bmatrix} u_1 & v_1 & : & x_1 \\ 0 & u_1 v_2 - u_2 v_1 & : & u_1 x_2 - u_2 x_1 \\ 0 & u_1 v_3 - u_3 v_1 & : & u_1 x_3 - u_3 x_1 \end{bmatrix}.$$

At least one of $u_1 v_2 - u_2 v_1$ and $u_1 v_3 - u_3 v_1$ is nonzero. For if both were zero, then

$$(v_1, v_2, v_3) = \frac{v_1}{u_1}(u_1, u_2, u_3),$$

contrary to the assumption that $\mathbf{u}$ and $\mathbf{v}$ were noncollinear. Thus with possibly a row interchange, we can further reduce the matrix to

$$\begin{bmatrix} u_1 & v_1 & : & x_1 \\ 0 & * & : & * \\ 0 & 0 & : & * \end{bmatrix},$$

where the 2,2 entry of the matrix is not zero. The 3,3 entry of this matrix is

$$u_1 x_3 - u_3 x_1, \qquad \text{or} \qquad u_1 x_2 - u_2 x_1, \qquad \text{or}$$
$$u_1 x_3 - u_3 x_1 - (u_1 x_2 - u_2 x_1)\frac{u_1 v_3 - u_3 v_1}{u_1 v_2 - u_2 v_1}.$$

Since $(x_1, x_2, x_3)$ lies on the plane if and only if this system is consistent, the equation of the plane is

$$u_1 x_3 - u_3 x_1 = 0, \qquad \text{or} \qquad u_1 x_2 - u_2 x_1 = 0, \qquad \text{or}$$
$$(u_1 x_3 - u_3 x_1)(u_1 v_2 - u_2 v_1) - (u_1 v_3 - u_3 v_1)(u_1 x_2 - u_2 x_1) = 0.$$

Each of these equations can be brought into the form $a_1 x_1 + a_2 x_2 + a_3 x_3 = 0$ with at most a little algebra.

If $u_1 = 0$, we interchange the first row with the second or third row and proceed as above. Thus we have proved the first part of the following result.

**Result 1** Equations of planes in $\mathbf{R}^3$.

(a) A plane in $R^3$ through the origin has a Cartesian equation of the form

$$a_1 x_1 + a_2 x_2 + a_3 x_3 = 0$$

with $(a_1, a_2, a_3) \neq \mathbf{0}$.

(b) Any equation of this form with $(a_1, a_2, a_3) \neq \mathbf{0}$ is the equation of a plane in $R^3$ through the origin.

**Proof of (b)** The vector form of the general solution of the equation $a_1x_1 + a_2x_2 + a_3x_3 = 0$ is:

$$(x_1, x_2, x_3) = x_2\left(-\frac{a_2}{a_1}, 1, 0\right) + x_3\left(-\frac{a_3}{a_1}, 0, 1\right) \qquad \text{if } a_1 \neq 0,$$

$$(x_1, x_2, x_3) = x_1(1, 0, 0) + x_3\left(0, -\frac{a_3}{a_2}, 1\right) \qquad \text{if } a_1 = 0 \text{ and } a_2 \neq 0,$$

$$(x_1, x_2, x_3) = x_1(1, 0, 0) + x_2(0, 1, 0) \qquad \text{if } a_1 = 0 \text{ and } a_2 = 0.$$

Each of these equations is a vector equation of a plane.

These examples illustrate a general method for determining the collection of all vectors **b** such that $A\mathbf{x} = \mathbf{b}$ is consistent. Form the augmented matrix $[A : \mathbf{b}]$ and reduce it to $[E : \mathbf{c}]$. The components of **c** are sums of multiples of the components of **b**. Each zero row in $E$ generates an equation that the components of **b** must satisfy for the system $A\mathbf{x} = \mathbf{b}$ to be consistent. For obvious reasons these equations are called **constraint equations**.

Now that we have investigated the question of existence, we turn to the question of uniqueness. That is, when does a consistent system $A\mathbf{x} = \mathbf{b}$ have a unique solution? The answer this time is all tied up with free variables. *Free variables are responsible for nonuniqueness; therefore, uniqueness corresponds to the absence of free variables.* These statements follow immediately from the structure of the general solution of a consistent system, which we will investigate now. Let us begin with the simplest case, a homogeneous system. Homogeneous systems are always consistent.

Consider the homogeneous system $A\mathbf{x} = \mathbf{0}$, where

$$A = \begin{bmatrix} 1 & 5 & 1 & 1 & 0 \\ 1 & 5 & 2 & 4 & 1 \\ 2 & 10 & 0 & -4 & 0 \\ 0 & 0 & 1 & 3 & 1 \end{bmatrix}.$$

The reduced echelon form of $A$ is

$$R = \begin{bmatrix} 1 & 5 & 0 & -2 & 0 \\ 0 & 0 & 1 & 3 & 0 \\ 0 & 0 & 0 & 0 & 1 \\ 0 & 0 & 0 & 0 & 0 \end{bmatrix}.$$

The nonpivot columns are the second and the fourth columns, so the free variables are $x_2$ and $x_4$. The vector form of the general solution is easy to write down.

$$\mathbf{x} = (x_1, x_2, x_3, x_4, x_5) = x_2(-5, 1, 0, 0, 0) + x_4(2, 0, -3, 1, 0)$$

Contemplate the relationship between the form of this solution and the form of the reduced echelon matrix $R$. The general solution is a linear combination of the vectors *whose coefficients are the free variables.* The vector that $x_2$ multiplies is the solution vector corresponding to $x_2 = 1$ and $x_4 = 0$. It has a 1 as its second entry and a 0 as its fourth entry. Similarly, the vector that $x_4$ multiplies is the solution vector corresponding to $x_2 = 0$ and $x_4 = 1$. It has a 0 as its second entry and a 1 as its fourth entry.

This example illustrates a procedure for finding the general solution to a homogeneous system $A\mathbf{x} = \mathbf{0}$. First reduce $A$ to its reduced echelon form, $R$. If there are no free variables, then $\mathbf{x} = \mathbf{0}$ is the only solution. If there are free variables, then there are *infinitely many* nontrivial solutions. In fact, to each free variable $x_i$ there corresponds a nonzero solution vector $\mathbf{v}_i$ obtained by setting $x_i = 1$ and setting all other free variables equal to zero. Every solution of $A\mathbf{x} = \mathbf{0}$ is a linear combination of these nonzero vectors $\mathbf{v}_i$.

***Theorem 3*** *The general solution of the homogeneous system $A\mathbf{x} = \mathbf{0}$ is a linear combination of the solution vectors obtained by in turn setting each free variable to* 1 *and all other free variables to* 0. *In particular, the system $A\mathbf{x} = \mathbf{0}$ has infinitely many solutions if and only if there are free variables; it has a unique solution if and only if there are no free variables.*

We have settled the question of uniqueness for homogeneous systems. A homogeneous system $A\mathbf{x} = \mathbf{0}$ has a unique solution if and only if there are no free variables. Since free variables correspond to nonpivot columns, the responsibility for uniqueness shifts to the columns of $A$. The homogeneous system $A\mathbf{x} = \mathbf{0}$ has a unique solution if and only if every column of $A$ is a pivot column. Since the rank of $A$ is the number of pivot columns of $A$, we have

***Theorem 4*** *Let $A$ be an $m \times n$ matrix. The homogeneous system $A\mathbf{x} = \mathbf{0}$ has a unique solution if and only if the rank of $A$ is $n$.*

Now let us consider a nonhomogeneous system with the same coefficient matrix.

$$\begin{bmatrix} 1 & 5 & 1 & 1 & 0 \\ 1 & 5 & 2 & 4 & 1 \\ 2 & 10 & 0 & -4 & 0 \\ 0 & 0 & 1 & 3 & 1 \end{bmatrix} \begin{bmatrix} x_1 \\ x_2 \\ x_3 \\ x_4 \\ x_5 \end{bmatrix} = \begin{bmatrix} 4 \\ 12 \\ -6 \\ 8 \end{bmatrix}.$$

Reducing $[A : \mathbf{b}]$ to $[R : \mathbf{d}]$ gives

$$\begin{bmatrix} 1 & 5 & 0 & -2 & 0 & : & -3 \\ 0 & 0 & 1 & 3 & 0 & : & 7 \\ 0 & 0 & 0 & 0 & 1 & : & 1 \\ 0 & 0 & 0 & 0 & 0 & : & 0 \end{bmatrix}.$$

The vector form of the general solution is

$$(x_1, x_2, x_3, x_4, x_5) = (-3, 0, 7, 0, 1) + x_2(-5, 1, 0, 0, 0) + x_4(2, 0, -3, 1, 0).$$

Look at what is revealed about the structure of the solution. The vector form of the general solution of this nonhomogeneous system is a vector plus the vector form of the general solution of the associated homogeneous system. This vector is the solution of the nonhomogeneous system corresponding to both free variables set to zero. A solution that corresponds to a particular assignment of values to the free variables is called a **particular solution**. Thus we call $\mathbf{x} = (-3, 0, 7, 0, 1)$ a particular solution. We have proved (at least for this example) that the general solution of the nonhomogeneous system consists of a particular solution of the nonhomogeneous system plus the general solution of the associated homogeneous system.

***Theorem 5*** *Suppose that $A\mathbf{x} = \mathbf{b}$ is consistent. The general solution to the nonhomogeneous system $A\mathbf{x} = \mathbf{b}$ is a particular solution of the nonhomogeneous system plus the general solution to the associated homogeneous system $A\mathbf{x} = \mathbf{0}$.*

**Proof** Let $\mathbf{v}_p$ be a solution to $A\mathbf{x} = \mathbf{b}$. We must show two things: (i) If $\mathbf{u}$ is a solution of the homogeneous system, then $\mathbf{v}_p + \mathbf{u}$ is a solution of $A\mathbf{x} = \mathbf{b}$; (ii) if $\mathbf{v}$ is any solution to $A\mathbf{x} = \mathbf{b}$, then there is a solution $\mathbf{u}$ of the homogeneous system such that $\mathbf{v} = \mathbf{v}_p + \mathbf{u}$. Well, (i) is easy:

$$A(\mathbf{v}_p + \mathbf{u}) = A\mathbf{v}_p + A\mathbf{u} = \mathbf{b} + \mathbf{0} = \mathbf{b}.$$

To prove (ii), we make the following observation. If there is such a vector $\mathbf{u}$ such that $\mathbf{v} = \mathbf{v}_p + \mathbf{u}$, then it must be the vector $\mathbf{v} - \mathbf{v}_p$. Thus all we have to do then is to set $\mathbf{u} = \mathbf{v} - \mathbf{v}_p$ and show that it is a solution of the homogeneous system:

$$A\mathbf{u} = A(\mathbf{v} - \mathbf{v}_p) = A\mathbf{v} - A\mathbf{v}_p = \mathbf{b} - \mathbf{b} = \mathbf{0}.$$

The proof is complete.

This theorem settles the question of uniqueness for nonhomogeneous systems. For it follows from this theorem that a consistent system $A\mathbf{x} = \mathbf{b}$ has a unique solution if and only if the associated homogeneous system $A\mathbf{x} = \mathbf{0}$ has a unique solution. Since $A\mathbf{x} = \mathbf{0}$ has a unique solution if and only if the rank of $A$ is $n$ (Theorem 4), we have proved:

***Theorem 6*** *Let $A$ be an $m \times n$ matrix. The following are equivalent.*

(*a*) *Any consistent system $A\mathbf{x} = \mathbf{b}$ has a unique solution.*
(*b*) *The homogeneous system $A\mathbf{x} = \mathbf{0}$ has a unique solution, namely the trivial solution.*
(*c*) *The rank of $A$ is $n$.*

## PROBLEMS 1.10

**1.** In each part, find the constraint equations that a vector $\mathbf{b}$ must satisfy for the system $A\mathbf{x} = \mathbf{b}$ to be consistent.

*(a) $A = \begin{bmatrix} 1 & 1 & 1 \\ 1 & -1 & 1 \\ 1 & 0 & 1 \end{bmatrix}$

(b) $A = \begin{bmatrix} 1 & -1 & 2 \\ 1 & 1 & 0 \\ 1 & 0 & 1 \end{bmatrix}$

*(c) $A = \begin{bmatrix} 1 & 2 & -1 & 3 \\ 2 & 5 & -4 & 8 \\ -1 & 1 & -5 & 3 \end{bmatrix}$

(d) $A = \begin{bmatrix} 1 & 0 & -1 \\ 0 & 1 & 1 \\ 1 & 2 & 0 \\ 0 & 0 & 1 \end{bmatrix}$

*(e) $A = \begin{bmatrix} 1 & 2 & 3 \\ 2 & 3 & 4 \\ 3 & 4 & 5 \\ 4 & 5 & 6 \end{bmatrix}$

(f) $A = \begin{bmatrix} 1 & 1 & 1 \\ 0 & 1 & 0 \\ 0 & 0 & 1 \\ 1 & 2 & 2 \end{bmatrix}$

*(g) $A = \begin{bmatrix} 1 & 0 & 1 & 2 \\ 0 & 1 & 1 & 2 \\ 1 & 1 & 1 & 3 \end{bmatrix}$

(h) $A = \begin{bmatrix} 1 & -1 & 1 & 0 & 0 \\ 0 & 0 & 1 & 1 & 1 \\ 1 & -1 & 2 & 1 & 1 \\ -1 & 1 & -3 & -2 & -2 \end{bmatrix}$

*(i) $A = \begin{bmatrix} 1 & 1 & 1 \\ 1 & -1 & 2 \\ 2 & 4 & 1 \\ 2 & -2 & 4 \\ 0 & 2 & -1 \end{bmatrix}$

(j) $A = \begin{bmatrix} 0 & 1 & 1 \\ 0 & 2 & -1 \\ 3 & 2 & 5 \\ 2 & 1 & 3 \end{bmatrix}$

**2.** For each of the following systems, describe geometrically how the planes are situated with respect to each other and say whether the solution set is empty, a point, a line, or a plane (see Section 1.1).

*(a)
$$\begin{aligned} x + y + z &= 5 \\ 3x - y + 2z &= 1 \\ x + 5y + 2z &= 8 \end{aligned}$$

(b)
$$\begin{aligned} x + 2y + 3z &= 4 \\ 5x + 6y + 7z &= 8 \\ 9x + 10y + 11z &= 12 \end{aligned}$$

*(c)
$$\begin{aligned} x + 2y + 3z &= 4 \\ 2x + 3y + 4z &= 5 \\ x - y - z &= 0 \end{aligned}$$

(d)
$$\begin{aligned} x + 2y + 3z &= 1 \\ x + 4y + 5z &= 1 \\ 2x + 4y + 6z &= 3 \end{aligned}$$

*(e)
$$\begin{aligned} x + y + 2z &= 1 \\ 2x + 2y + 4z &= 3 \\ 3x + 3y + 6z &= 3 \end{aligned}$$

**3.** For each of the following, find a vector equation of the plane whose Cartesian equation is given.

*(a) $2x + y - z = 0$

(b) $2x + y - z + 1 = 0$

*(c) $2x + 3y + z = 0$

(d) $2x + 3y + z - 5 = 0$

*(e) $x - y + 2z = 0$

(f) $x - y + 2z = 3$

**4.** In each of the following, find a system of equations whose solution set is the given plane.

*(a) $\mathbf{x} = (1, 2, 3, 2, 1) + \alpha(1, 2, 1, 1, 0) + \beta(0, 1, 3, 1, 4)$

(b) $\mathbf{x} = (5, 0, 1, 2) + \alpha(1, 2, 3, 1) + \beta(1, 1, 0, 3)$

(c) $\mathbf{x} = (-2, 1, 2, -1) + \alpha(1, 4, 3, 5) + \beta(-1, -3, 2, 6)$

***5.** Explain why a homogeneous system of three equations in four unknowns must have a nontrivial solution.

**6.** Find all $m \times n$ matrices whose rank is zero.

**7.** What can you say about the reduced echelon form of the augmented matrix of an inconsistent linear system? In other words, how can you identify an inconsistent system after reducing its augmented matrix?

**8.** What can you say about the reduced echelon form of the augmented matrix of a linear system with a unique solution? In other words, how can you identify a system with a unique solution after reducing its augmented matrix?

**9.** What can you say about the reduced echelon form of the augmented matrix of a linear system with infinitely many solutions? In other words, how can you identify a system with an infinite number of solutions after reducing its augmented matrix?

*10. Suppose that $A$ is a $4 \times 3$ matrix. Which of the following statements could be true about the system

$$A\begin{bmatrix} x_1 \\ x_2 \\ x_3 \end{bmatrix} = \begin{bmatrix} 1 \\ 2 \\ 3 \\ 4 \end{bmatrix}?$$

Give an example of $A$ for the ones you say could happen, and a reason why the others could not happen.
(a) There are no solutions.
(b) There is exactly one solution.
(c) There are exactly two solutions.
(d) There are infinitely many solutions.

11. Show that any plane in $R^3$ has a Cartesian equation of the form $a_1x_1 + a_2x_2 + a_3x_3 = d$. Then show that the converse is also true.

*12. Let $A$ be an $m \times n$ matrix, where $m < n$, and let $\mathbf{b}$ be a vector in $R^m$. Explain why the system $A\mathbf{x} = \mathbf{b}$ cannot have just one solution.

13. Let $A$ be an $m \times n$ matrix, where $m < n$. Explain why the system $A\mathbf{x} = \mathbf{0}$ must have a solution $\mathbf{x} \neq \mathbf{0}$.

14. Let $A$ be an $n \times k$ matrix and $B$ a $k \times n$ matrix. Show that $AB$ is singular if $k < n$.

15. Suppose that $A$ is an $n \times n$ matrix with the property that for every vector $\mathbf{b}$ in $R^n$, $A\mathbf{x} = \mathbf{b}$ has a solution. Explain why $A$ is nonsingular.

16. Let $A$ be an $m \times n$ matrix. Show that the following are equivalent.
(a) For every consistent system $A\mathbf{x} = \mathbf{b}$ the solution is unique.
(b) The homogeneous system $A\mathbf{x} = \mathbf{0}$ has a unique solution.

17. Let $A$ be an $n \times n$ singular matrix. Show that for every $\mathbf{b}$ in $R^m$, the system $A\mathbf{x} = \mathbf{b}$ has no solution or infinitely many solutions.

*18. Let $A$ be an $m \times n$ matrix of rank $m$. Show that if $C$ is any $m \times p$ matrix, there is an $n \times p$ matrix $B$ such that $AB = C$.

19. Let $A$ be an $m \times n$ matrix with rank $n$, and let $B$ be an $n \times p$ matrix with rank $p$. Show that $AB$ has rank $p$.

20. Let $A$ be an $m \times n$ matrix.
*(a) Prove that if the vectors that form the columns of $A$ sum to 0, then rank $(A) < n$.
(b) Prove that if some linear combination of the columns of $A$ is $\mathbf{0}$, then rank $(A) < n$.

21. Let $A$ be an $m \times n$ matrix. Prove that if the vectors that form the rows of $A$ sum to $\mathbf{0}$, then rank $(A) < m$. (*Hint:* Apply Problem 20 to $A^T$.)

*22. Suppose that the set of vectors $\mathbf{b}$ such that $A\mathbf{x} = \mathbf{b}$ is consistent is equal to the set of vectors $\mathbf{c}$ such that $B\mathbf{x} = \mathbf{c}$ is consistent. Does it follow that the homogeneous systems

$$A\mathbf{x} = \mathbf{0} \qquad \text{and} \qquad B\mathbf{x} = \mathbf{0}$$

have the same solutions? Either prove this statement or give an example to show that it is false.

23. Suppose that the homogeneous systems

$$A\mathbf{x} = \mathbf{0} \qquad \text{and} \qquad B\mathbf{x} = \mathbf{0}$$

have the same solutions . Does it follow that the set of vectors $\mathbf{b}$ for which $A\mathbf{x} = \mathbf{b}$ is consistent is the same as the set of vectors $\mathbf{c}$ such that $B\mathbf{x} = \mathbf{c}$ is consistent? Either prove the statement or give an example to show that it is false.

24. Prove part (b) of Theorem 1.

# 1.11 Nonsingular Matrices Revisited

In general, existence and uniqueness are separate properties of a system. An $m \times n$ system can have one without the other. Existence corresponds to no zero rows, uniqueness to no nonpivot columns. For example, if

$$A = \begin{bmatrix} 1 & 2 & 3 & 4 \\ 0 & 1 & 2 & 3 \\ 0 & 0 & 1 & 2 \end{bmatrix},$$

then for every $\mathbf{b}$ in $R^3$ the corresponding system $A\mathbf{x} = \mathbf{b}$ always has a solution (there are no zero rows), but for any $\mathbf{b}$ in $R^3$ there are infinitely many solutions to $A\mathbf{x} = \mathbf{b}$ (there are free variables). On the other hand, if

$$A = \begin{bmatrix} 1 & 2 & 3 \\ 0 & 1 & 2 \\ 0 & 0 & 1 \\ 0 & 0 & 0 \end{bmatrix},$$

the corresponding system $A\mathbf{x} = \mathbf{b}$ is inconsistent whenever $b_4 \neq 0$. However, when the system has a solution (when $b_4 = 0$), then it has only one solution (there are no free variables).

For square systems, however, existence and uniqueness always go together. For if $A$ is $n \times n$, then:

1. $A\mathbf{x} = \mathbf{b}$ is consistent for every $\mathbf{b}$ in $R^n$ if and only if the rank of $A$ is $n$ (see Theorem 2 of Section 1.10).
2. Any consistent system $A\mathbf{x} = \mathbf{b}$ has a unique solution if and only if the rank of $A$ is $n$ (see Theorem 6 of Section 1.10).

Thus for square matrices existence corresponds to rank $n$, which in turn corresponds to uniqueness. By Theorem 3 of Section 1.9, the $n \times n$ matrices with rank $n$ are the nonsingular matrices.

***Theorem 1*** *Let $A$ be an $n \times n$ matrix. The following are equivalent.*

(*a*) *$A$ is nonsingular.*
(*b*) *The rank of $A$ is $n$.*
(*c*) *For every $\mathbf{b}$ in $R^n$, $A\mathbf{x} = \mathbf{b}$ has a unique solution.*
(*d*) *The homogeneous equation $A\mathbf{x} = \mathbf{0}$ has a unique solution* (*namely, the trivial solution $\mathbf{x} = \mathbf{0}$*).

**Proof** We prove that (a) is equivalent to (b) in Theorem 3 of Section 1.9, and we just observed that (b) is equivalent to each of (c) and (d).

By Theorem 3 of Section 1.9, an $n \times n$ matrix $A$ is singular if and only if the rank of $A$ is less than $n$. Thus for singular matrices we lose both existence and uniqueness:

we lose existence because any echelon form of $A$ has at least one zero row; we lose uniqueness because there is at least one nonpivot column.

We now develop a very interesting and important consequence of Theorem 1 for nonsingular matrices. Let $A$ be an $n \times n$ nonsingular matrix. Let $\mathbf{e}_i$ be the vector in $R^n$ whose $i$th component is 1 and whose other components are zero:

$$\begin{aligned} \mathbf{e}_1 &= (1, 0, 0, \ldots, 0) \\ \mathbf{e}_2 &= (0, 1, 0, \ldots, 0) \\ &\vdots \\ \mathbf{e}_n &= (0, 0, 0, \ldots, 1). \end{aligned}$$

Of course, if we form the matrix having these vectors as its columns, we obtain the $n \times n$ identity matrix $I$:

$$I = [\mathbf{e}_1 \quad \mathbf{e}_2 \quad \cdots \quad \mathbf{e}_n].$$

From Theorem 1, each of the systems

$$A\mathbf{x} = \mathbf{e}_1, \quad A\mathbf{x} = \mathbf{e}_2, \quad \ldots, \quad A\mathbf{x} = \mathbf{e}_n$$

has a unique solution. Let $\mathbf{b}_i$ denote the solution to the $i$th equation. Therefore,

$$A\mathbf{b}_1 = \mathbf{e}_1, \quad A\mathbf{b}_2 = \mathbf{e}_2, \quad \ldots, \quad A\mathbf{b}_n = \mathbf{e}_n.$$

Interpreting these equations in terms of matrix multiplication, we obtain

$$A[\mathbf{b}_1 \quad \mathbf{b}_2 \quad \cdots \quad \mathbf{b}_n] = [\mathbf{e}_1 \quad \mathbf{e}_2 \quad \cdots \quad \mathbf{e}_n]$$

or

$$AB = I,$$

where $B = [\mathbf{b}_1 \quad \mathbf{b}_2 \quad \cdots \quad \mathbf{b}_n]$. Thus we have shown:

**Result 1** If $A$ is nonsingular, there is a unique matrix $B$ such that $AB = I$.

## EXAMPLE 1

Let

$$A = \begin{bmatrix} 1 & 1 & 1 \\ 2 & 3 & 3 \\ 3 & 4 & 5 \end{bmatrix}.$$

To find a matrix $B$ such that $AB = I$, where $I$ is the $3 \times 3$ identity matrix, we must solve the three systems

$$A\mathbf{x} = \mathbf{e}_1, \qquad A\mathbf{x} = \mathbf{e}_2, \qquad A\mathbf{x} = \mathbf{e}_3.$$

Form the augmented matrix

$$[A : I] = \left[\begin{array}{ccc:ccc} 1 & 1 & 1 & 1 & 0 & 0 \\ 2 & 3 & 3 & 0 & 1 & 0 \\ 3 & 4 & 5 & 0 & 0 & 1 \end{array}\right]$$

and use Gaussian elimination to reduce it to

$$[I : B] = \begin{bmatrix} 1 & 0 & 0 & : & 3 & -1 & 0 \\ 0 & 1 & 0 & : & -1 & 2 & -1 \\ 0 & 0 & 1 & : & -1 & -1 & 1 \end{bmatrix}.$$

Thus the matrix $B$ is

$$B = \begin{bmatrix} 3 & -1 & 0 \\ -1 & 2 & -1 \\ -1 & -1 & 1 \end{bmatrix}.$$

Armed with this result, we can prove the following theorem.

***Theorem 2*** *The transpose of a nonsingular matrix is nonsingular.*

**Proof** Let $A$ be an $n \times n$ nonsingular matrix. We are to show that $A^T$ is also nonsingular. By Theorem 1 it is sufficient to show that the only solution of the homogeneous equation $A^T\mathbf{x} = \mathbf{0}$ is the trivial solution $\mathbf{x} = \mathbf{0}$. Suppose that $A^T\mathbf{x} = \mathbf{0}$. By properties 1 and 4 of transposes, $\mathbf{x}^T A = (A^T\mathbf{x})^T = \mathbf{0}^T$. Since $A$ is nonsingular, the previous result guarantees that there is an $n \times n$ matrix $B$ such that $AB = I$. Then

$$\mathbf{x}^T = \mathbf{x}^T I = \mathbf{x}^T(AB) = (\mathbf{x}^T A)B = \mathbf{0}^T B = \mathbf{0}^T.$$

Therefore, $\mathbf{x} = \mathbf{0}$ and we have shown that the only solution to the homogeneous equation $A^T\mathbf{x} = \mathbf{0}$ is the trivial solution. Thus $A^T$ is nonsingular.

This is a remarkable result. It says that if a square matrix can be reduced to the identity and if we interchange the rows and columns of the matrix, then the new matrix can also be reduced to the identity.

In light of the fact that matrix multiplication is not commutative, the following result is also remarkable.

**Result 2** If $A$ is an $n \times n$ nonsingular matrix and $B$ is an $n \times n$ matrix such that $AB = I$, then $BA = I$ also.

**Proof** Let $A$ be a nonsingular matrix, and suppose that $AB = I$. Our objective is to prove that $BA = I$. Taking transposes, this is equivalent to proving that $A^T B^T = I^T = I$. We have just proved that $A^T$ is nonsingular (since $A$ is nonsingular) and hence by Result 1 there is an $n \times n$ matrix $C$ such that $A^T C = I$. But $C = B^T$ since

$$\begin{aligned} C &= IC = I^T C = (AB)^T C = (B^T A^T)C = B^T(A^T C) \\ &= B^T I = B^T. \end{aligned}$$

Thus $A^T B^T = I$, so that $BA = I$, as asserted.

Combining Results 1 and 2, we have

***Theorem 3*** *Let $A$ be an $n \times n$ matrix. If $A$ is nonsingular, then there is a unique $n \times n$ matrix $B$ such that*

$$AB = I = BA.$$

The fact that the matrix $B$ is unique came from Result 1. We can also prove this fact directly.

**Result 3** Suppose that $A$ is an $n \times n$ matrix and that there is an $n \times n$ matrix $B$ such that

$$AB = I = BA.$$

Then $B$ is unique.

**Proof** Suppose to the contrary that there is another matrix $C$ such that $AC = I = CA$. Then

$$C = IC = (BA)C = B(AC) = BI = B.$$

Now for the converse of Theorem 3.

***Theorem 4*** *Let $A$ be an $n \times n$ matrix. If there is an $n \times n$ matrix $B$ such that $AB = I = BA$, then $A$ is nonsingular. Moreover, the matrix $B$ is unique.*

**Proof** To prove that $A$ is nonsingular it is sufficient to prove that the homogeneous system $A\mathbf{x} = \mathbf{0}$ has only the trivial solution. If $A\mathbf{x} = \mathbf{0}$, then

$$\mathbf{x} = I\mathbf{x} = (BA)\mathbf{x} = B(A\mathbf{x}) = B\mathbf{0} = \mathbf{0}.$$

This completes the proof.

When $a$ is a nonzero number there is a unique number $b$ such that $ab = 1 = ba$. The number $b$ is called the inverse of $a$ and is denoted by $a^{-1}$. A nonsingular matrix is similar to a nonzero number. If $A$ is nonsingular, there is a unique matrix $B$ such that $AB = I = BA$. Thus we call $B$ the **inverse** of $A$ and denote it by $A^{-1}$. (*Warning:* With numbers, $b^{-1} = 1/b$. There is no parallel statement for matrices; that is, there is no operation of division for matrices.) Since nonsingular matrices are those that have inverses, they are also called **invertible matrices**. Thus an $n \times n$ square matrix $A$ is nonsingular (or invertible) if and only if there is an $n \times n$ matrix $A^{-1}$ such that

$$AA^{-1} = I = A^{-1}A.$$

It follows from Result 2 that the matrix $B$ computed in Example 1 is the inverse of $A$. The example also illustrates a general procedure for computing the inverse of a nonsingular matrix. Since $AA^{-1} = I$, the columns of $A^{-1}$ are the solutions to the systems $A\mathbf{x} = \mathbf{e}_i$. Therefore, when the augmented matrix $[A : I]$ is reduced to reduced echelon form, the result is $[I : A^{-1}]$.

**EXAMPLE 2**

To compute the inverse of

$$A = \begin{bmatrix} 1 & 2 \\ -1 & 3 \end{bmatrix},$$

form the augmented matrix

$$[A \,:\, I] = \begin{bmatrix} 1 & 2 & : & 1 & 0 \\ -1 & 3 & : & 0 & 1 \end{bmatrix}$$

and reduce it to

$$[I \,:\, A^{-1}] = \begin{bmatrix} 1 & 0 & : & \frac{3}{5} & -\frac{2}{5} \\ 0 & 1 & : & \frac{1}{5} & \frac{1}{5} \end{bmatrix}.$$

Thus

$$A^{-1} = \begin{bmatrix} \frac{3}{5} & -\frac{2}{5} \\ \frac{1}{5} & \frac{1}{5} \end{bmatrix} = \frac{1}{5}\begin{bmatrix} 3 & -2 \\ 1 & 1 \end{bmatrix}.$$

**EXAMPLE 3**

Let

$$A = \begin{bmatrix} 1 & 1 & 1 \\ 1 & 2 & 1 \\ 2 & 3 & 2 \end{bmatrix}$$

and suppose that we are asked to determine whether $A$ is nonsingular and if it is to determine its inverse. Applying forward elimination to the augmented matrix

$$\begin{bmatrix} 1 & 1 & 1 & : & 1 & 0 & 0 \\ 1 & 2 & 1 & : & 0 & 1 & 0 \\ 2 & 3 & 2 & : & 0 & 0 & 1 \end{bmatrix},$$

we obtain

$$\begin{bmatrix} 1 & 1 & 1 & : & 1 & 0 & 0 \\ 0 & 1 & 0 & : & -1 & 1 & 0 \\ 0 & 0 & 0 & : & -1 & -1 & 1 \end{bmatrix}.$$

Since the rank of $A$ is 2, $A$ is singular and hence it does not have an inverse.

We can give a simple formula for the inverse of a nonsingular $2 \times 2$ matrix. By Theorem 1 the matrix

$$A = \begin{bmatrix} a & b \\ c & d \end{bmatrix}$$

is nonsingular if and only if $A$ has rank 2. Suppose that $a \neq 0$. Subtracting $c/a$ times the first row from the second reduces $A$ to the echelon matrix

$$\begin{bmatrix} a & b \\ 0 & d - \dfrac{bc}{a} \end{bmatrix}.$$

Thus $A$ has rank 2 if and only if

$$0 \neq d - \frac{bc}{a} = \frac{ad - bc}{a},$$

that is, if and only if $ad - bc \neq 0$. This is also true when $a = 0$. In that case, by merely interchanging the rows of $A$, we reduce $A$ to the echelon matrix

$$\begin{bmatrix} c & d \\ 0 & b \end{bmatrix}.$$

Hence $A$ has rank 2 if and only if both $c \neq 0$ and $b \neq 0$, that is, $0 \neq -bc = ad - bc$. Thus the $2 \times 2$ matrix $A$ is nonsingular if and only if $ad - bc \neq 0$. The number $ad - bc$ is called the determinant of $A$. (Determinants are discussed in Chapter 7.) If $A$ is nonsingular, an easy calculation shows that

$$A^{-1} = \frac{1}{ad - bc}\begin{bmatrix} d & -b \\ -c & a \end{bmatrix}.$$

For example, the matrix $\begin{bmatrix} 2 & 3 \\ 4 & 6 \end{bmatrix}$ is singular since $2 \cdot 6 - 3 \cdot 4 = 0$. The matrix $\begin{bmatrix} 3 & 2 \\ 5 & 1 \end{bmatrix}$ is nonsingular since $3 \cdot 1 - 2 \cdot 5 = -7 \neq 0$. Its inverse is

$$\begin{bmatrix} 3 & 2 \\ 5 & 1 \end{bmatrix}^{-1} = \frac{1}{-7}\begin{bmatrix} 1 & -2 \\ -5 & 3 \end{bmatrix} = \begin{bmatrix} -\frac{1}{7} & \frac{2}{7} \\ \frac{5}{7} & -\frac{3}{7} \end{bmatrix}.$$

Let $A$ be nonsingular and consider the role of $A^{-1}$ in the equation $AA^{-1} = I = A^{-1}A$. It is a square matrix for which there is another square matrix, namely $A$, such that

$$A^{-1}A = I = AA^{-1}.$$

Thus $A^{-1}$ is nonsingular and its inverse is $A$. Since the inverse of $A^{-1}$ is denoted by $(A^{-1})^{-1}$, we have $(A^{-1})^{-1} = A$. This is part (a) of the following theorem.

***Theorem 5*** *Let $A$ be an $n \times n$ nonsingular matrix.*

(*a*) *$A^{-1}$ is also nonsingular and $(A^{-1})^{-1} = A$.*
(*b*) *$A^T$ is nonsingular and $(A^T)^{-1} = (A^{-1})^T$.*
(*c*) *If $B$ is also an $n \times n$ nonsingular matrix, then $AB$ is nonsingular and*

$$(AB)^{-1} = B^{-1}A^{-1}.$$

**Proof of (b)** We have already proved that $A^T$ is nonsingular. Since

$$A^T(A^{-1})^T = (A^{-1}A)^T = I^T = I$$

and

$$(A^{-1})^T A^T = (AA^{-1})^T = I^T = I,$$

it follows from Theorem 4 that the inverse of $A^T$ is $(A^{-1})^T$: $(A^T)^{-1} = (A^{-1})^T$.

**Proof of (c)** Since

$$(B^{-1}A^{-1})(AB) = B^{-1}(A^{-1}A)B = B^{-1}IB = B^{-1}B = I$$

and

$$(AB)(B^{-1}A^{-1}) = A(BB^{-1})A^{-1} = AIA^{-1} = AA^{-1} = I,$$

it follows from Theorem 4 that $AB$ is nonsingular and its inverse is $B^{-1}A^{-1}$: $(AB)^{-1} = B^{-1}A^{-1}$.

The following theorem should come as no surprise.

***Theorem 6*** *Let $A$ be an $n \times n$ nonsingular matrix. For every $\mathbf{b}$ in $R^n$, the unique solution of the system $A\mathbf{x} = \mathbf{b}$ is*

$$\mathbf{x} = A^{-1}\mathbf{b}.$$

**Proof** Since $A$ is nonsingular, $A$ has an inverse $A^{-1}$ and for every $\mathbf{b}$ in $R^n$ the system $A\mathbf{x} = \mathbf{b}$ has a unique solution. Thus to prove the theorem we must show that if $\mathbf{x}$ is the solution of $A\mathbf{x} = \mathbf{b}$, then $\mathbf{x} = A^{-1}\mathbf{b}$ and conversely if $\mathbf{x} = A^{-1}\mathbf{b}$, then $\mathbf{x}$ is a solution of the system $A\mathbf{x} = \mathbf{b}$. If $\mathbf{x}$ is a solution of $A\mathbf{x} = \mathbf{b}$, then multiplying both sides of this equation on the left by $A^{-1}$, we obtain

$$\mathbf{x} = I\mathbf{x} = (A^{-1}A)\mathbf{x} = A^{-1}(A\mathbf{x}) = A^{-1}\mathbf{b}.$$

Conversely, if $\mathbf{x} = A^{-1}\mathbf{b}$, then multiplying both sides of the equation on the left by $A$, we obtain

$$A\mathbf{x} = A(A^{-1}\mathbf{b}) = (AA^{-1})\mathbf{b} = I\mathbf{b} = \mathbf{b}.$$

The proof is complete.

## EXAMPLE 4

Consider the system $A\mathbf{x} = \mathbf{b}$, where

$$A = \begin{bmatrix} 1 & -2 & 3 \\ 3 & 0 & 1 \\ 2 & 2 & 1 \end{bmatrix}, \qquad \mathbf{b} = \begin{bmatrix} 1 \\ 2 \\ 3 \end{bmatrix}.$$

By Theorem 6 the solution of the system is

$$\mathbf{x} = A^{-1}\mathbf{b}.$$

A straightforward application of this formula requires us to compute $A^{-1}$ and then form the product $A^{-1}\mathbf{b}$.

To compute $A^{-1}$ we form the augmented matrix

$$[A : I] = \begin{bmatrix} 1 & -2 & 3 & : & 1 & 0 & 0 \\ 3 & 0 & 1 & : & 0 & 1 & 0 \\ 2 & 2 & 1 & : & 0 & 0 & 1 \end{bmatrix}.$$

Forward elimination reduces this matrix to

$$\begin{bmatrix} 1 & -2 & 3 & : & 1 & 0 & 0 \\ 0 & 6 & -8 & : & -3 & 1 & 0 \\ 0 & 0 & 3 & : & 1 & -1 & 1 \end{bmatrix}$$

and back substitution reduces it further to

$$\begin{bmatrix} 1 & 0 & 0 & : & -\frac{2}{18} & \frac{8}{18} & -\frac{2}{18} \\ 0 & 1 & 0 & : & -\frac{1}{18} & -\frac{5}{18} & \frac{8}{18} \\ 0 & 0 & 1 & : & \frac{6}{18} & -\frac{6}{18} & \frac{6}{18} \end{bmatrix}.$$

Thus

$$A^{-1} = \frac{1}{18}\begin{bmatrix} -2 & 8 & -2 \\ -1 & -5 & 8 \\ 6 & -6 & 6 \end{bmatrix}.$$

Thus the solution of $A\mathbf{x} = \mathbf{b}$ is

$$\mathbf{x} = A^{-1}\mathbf{b} = \frac{1}{18}\begin{bmatrix} -2 & 8 & -2 \\ -1 & -5 & 8 \\ 6 & -6 & 6 \end{bmatrix}\begin{bmatrix} 1 \\ 2 \\ 3 \end{bmatrix} = \frac{1}{18}\begin{bmatrix} 8 \\ 13 \\ 12 \end{bmatrix}.$$

In the previous example we computed the solution of a system $A\mathbf{x} = \mathbf{b}$ with a nonsingular coefficient matrix $A$ by computing $A^{-1}$ and then forming the product $A^{-1}\mathbf{b}$. Let us compare this procedure with our standard procedure for solving *any* system. When $A$ is an $n \times n$ nonsingular matrix, its reduced echelon form is the $n \times n$ identity matrix I. Thus

$$[A : \mathbf{b}] \to [I : \mathbf{d}]$$

and $\mathbf{x} = \mathbf{d}$ is the unique solution of the system. By Theorem 6 the unique solution is $\mathbf{x} = A^{-1}\mathbf{b}$. Thus

$$\mathbf{d} = A^{-1}\mathbf{b},$$

and when $A$ is reduced to $I$,

$$[A : \mathbf{b}] \to [I : A^{-1}\mathbf{b}].$$

For instance, if $A$ and $\mathbf{b}$ are as in Example 4, forward elimination reduces

$$[A : \mathbf{b}] = \begin{bmatrix} 1 & -2 & 3 & : & 1 \\ 3 & 0 & 1 & : & 2 \\ 2 & 2 & 1 & : & 3 \end{bmatrix} \quad \text{to} \quad \begin{bmatrix} 1 & -2 & 3 & : & 1 \\ 0 & 6 & -8 & : & -1 \\ 0 & 0 & 3 & : & 2 \end{bmatrix},$$

and back substitution further reduces this matrix to

$$\begin{bmatrix} 1 & 0 & 0 & : & \frac{8}{18} \\ 0 & 1 & 0 & : & \frac{13}{18} \\ 0 & 0 & 1 & : & \frac{12}{18} \end{bmatrix}.$$

The solution is

$$\mathbf{x} = \frac{1}{18}\begin{bmatrix} 8 \\ 13 \\ 12 \end{bmatrix} = A^{-1}\mathbf{b}.$$

Clearly, solving a system $A\mathbf{x} = \mathbf{b}$ with a nonsingular coefficient matrix by finding $A^{-1}$ and then forming the product $A^{-1}\mathbf{b}$ is much more work than solving the system using Gaussian elimination. The reason is that computing the inverse of an $n \times n$ nonsingular matrix $A$ involves solving a system with $n$ different right-hand sides $\mathbf{e}_1, \mathbf{e}_2, \ldots, \mathbf{e}_n$. That is, it requires us to reduce

$$[A \; : \; I] \rightarrow [I \; : \; A^{-1}].$$

Thus solving a system $A\mathbf{x} = \mathbf{b}$ where $A$ is nonsingular by finding $A^{-1}$ and then computing $A^{-1}\mathbf{b}$ involves solving a system with $n$ *different* right-hand sides and then performing a final multiplication. On the other hand, Gaussian elimination only involves solving a system with *one* right-hand side.

In spite of the fact that computing the inverse of a nonsingular matrix is a computationally difficult task, it is an extremely useful concept in both pure and applied mathematics.

We conclude this section with a discussion of an application that involves finding the inverse of a certain nonsingular matrix. In Section 1.2 we introduced a model of a closed economy. In that model we assumed that each producing unit has an unlimited supply of raw materials. In a normal economy the output of each producing unit depends upon the availability of certain goods which are produced by others. If we wish to analyze the interrelationships between various producing units, we must use a model of the economy that focuses on production and consumption rather than on pricing. Such a model (called the input–output model) has been developed by W. W. Leontief (b. 1906), winner of the Nobel Prize in Economic Science in 1973.

To describe the Leontief input–output model, we first divide the economy into certain numbered sectors where all production takes place, together with an "open" sector where the consumption unrelated to production takes place. This open sector accounts for all consumption by individuals in the economy as contrasted with the consumption necessary for production. Suppose that the production sectors are (1) agriculture, (2) manufacturing, and (3) service. We let $d_i$ denote the quantity of sector $i$ production that is demanded by the open sector and $x_i$ denote the total production of sector $i$. The model describes the production and consumption of all goods by means of a system of equations. Each one of these equations expresses the rather obvious fact that the total production of any sector is the sum of that part of the sector's output used in production and that part of the sector's output consumed (demanded) by the individuals.

$$\begin{aligned} x_1 &= a_{11}x_1 + a_{12}x_2 + a_{13}x_3 + d_1 \\ x_2 &= a_{21}x_1 + a_{22}x_2 + a_{23}x_3 + d_2 \\ x_3 &= a_{31}x_1 + a_{32}x_2 + a_{33}x_3 + d_3 \end{aligned} \tag{1}$$

The coefficients $a_{ij}$, the input coefficients, measure the production that is used as input for the production of goods. Specifically, $a_{ij}$ is the number of units of sector $i$ production consumed in producing a unit of sector $j$ production. All of the input coefficients are nonnegative. The matrix $A$ is called a **consumption matrix**. Furthermore, the $x_i$'s and the $d_i$'s are all nonnegative. If $\mathbf{x} = (x_1, x_2, x_3)$ is the production vector, $\mathbf{d} = (d_1, d_2, d_3)$ is the demand vector, and $A = [a_{ij}]$ is the matrix of input coefficients, then the Leontief input–output model states that

$$\mathbf{x} = A\mathbf{x} + \mathbf{d}$$

or, equivalently,

$$(I - A)\mathbf{x} = \mathbf{d}. \tag{2}$$

The model is used to compute the production $\mathbf{x}$ necessary to satisfy a given demand $\mathbf{d}$. We hope that the economy as described by our model will have a production vector $\mathbf{x}$ for every possible demand vector $\mathbf{d}$. This happens if and only if $I - A$ is a nonsingular matrix and in that case equation (2) has a unique solution given by

$$\mathbf{x} = (I - A)^{-1}\mathbf{d}. \tag{3}$$

Since for each demand vector $\mathbf{d}$ the production vector $\mathbf{x}$ obtained from (3) must have nonnegative components, the entries of $(I - A)^{-1}$ must be all nonnegative. For if $(I - A)^{-1}$ had a negative entry in the $ij$th position, then the production vector $\mathbf{x} = (I - A)^{-1}\mathbf{e}_j$ would not be nonnegative.

We suspect that if the economy described by the equations (1) can actually have all its sectors producing (all $x_i$'s positive) and have something left over for the open sector (at least one $d_i$ positive), then it could somehow manage to satisfy any demand from the open sector. This leads us to conjecture that the following theorem is true.

***Theorem 7*** *Given a matrix $A$ with nonnegative entries. If there exists one nonzero vector* $\mathbf{d}$ *(with nonnegative entries) such that the equation* $(I - A)\mathbf{x} = \mathbf{d}$ *has a solution* $\mathbf{x}$ *with all entires positive, then the matrix $I - A$ has an inverse and every entry in the inverse is nonnegative.*

This theorem is indeed true, but its proof requires advanced techniques.

## PROBLEMS 1.11

**1.** Complete the following definitions.
 (a) A matrix is called invertible if ....
 (b) The inverse of a nonsingular matrix is ....

**2.** In each part, find (if possible) the inverse of the matrix $A$.

*(a) $A = \begin{bmatrix} 1 & 3 \\ -2 & 6 \end{bmatrix}$

(b) $A = \begin{bmatrix} 2 & -6 \\ -4 & 12 \end{bmatrix}$

*(c) $A = \begin{bmatrix} 4 & 10 \\ 2 & 5 \end{bmatrix}$

(d) $A = \begin{bmatrix} 1 & -1 & 1 \\ 4 & 5 & 6 \\ 5 & 7 & 9 \end{bmatrix}$

*(e) $A = \begin{bmatrix} 1 & -1 & 1 \\ 2 & 0 & 3 \\ 3 & -1 & 2 \end{bmatrix}$

(f) $A = \begin{bmatrix} 1 & 1 & 1 \\ 0 & 2 & 3 \\ 5 & 5 & 1 \end{bmatrix}$

*(g) $A = \begin{bmatrix} 3 & 2 & 0 \\ 2 & 3 & 2 \\ 0 & 2 & 3 \end{bmatrix}$

(h) $A = \begin{bmatrix} 1 & 2 & -3 \\ 1 & -2 & 1 \\ 5 & -2 & -3 \end{bmatrix}$

*(i) $A = \begin{bmatrix} 1 & 2 & 3 \\ 1 & 1 & 2 \\ 0 & 1 & 2 \end{bmatrix}$

(j) $A = \begin{bmatrix} 2 & 0 & 0 & 1 \\ 1 & 2 & 0 & -1 \\ 0 & 1 & 3 & 5 \\ 0 & 0 & 2 & 4 \end{bmatrix}$

*(k) $A = \begin{bmatrix} 1 & 1 & 1 & 1 \\ 1 & 1 & 1 & -1 \\ 1 & 1 & -1 & -1 \\ 1 & -1 & -1 & -1 \end{bmatrix}$

(l) $A = \begin{bmatrix} 1 & -1 & 2 & 3 \\ 0 & 2 & 1 & 2 \\ 0 & 0 & -1 & 3 \\ 0 & 0 & 0 & 1 \end{bmatrix}$

*(m) $A = \begin{bmatrix} 1 & 2 & 3 & 1 \\ 1 & 3 & 3 & 2 \\ 2 & 4 & 3 & 3 \\ 1 & 1 & 1 & 1 \end{bmatrix}$

**3.** In each part, do the following.

(i) Find $A^{-1}$.

(ii) Use $A^{-1}$ to solve $A\mathbf{x} = \mathbf{b}$.

(iii) Write **b** as a linear combination of the columns of $A$.

*(a) $A = \begin{bmatrix} 1 & 1 & 1 \\ 0 & 2 & 3 \\ 3 & 5 & 2 \end{bmatrix}, \quad b = \begin{bmatrix} 1 \\ -1 \\ 3 \end{bmatrix}$

(b) $A = \begin{bmatrix} 1 & 1 & 1 \\ 0 & 1 & 1 \\ 1 & 0 & 1 \end{bmatrix}, \quad b = \begin{bmatrix} 2 \\ 1 \\ 1 \end{bmatrix}$

*(c) $A = \begin{bmatrix} 1 & 1 & 1 \\ 0 & 2 & 3 \\ 5 & 5 & 1 \end{bmatrix}, \quad b = \begin{bmatrix} 8 \\ 2 \\ 0 \end{bmatrix}$

(d) $A = \begin{bmatrix} 1 & 1 & 0 \\ 2 & 1 & 0 \\ 0 & 0 & 1 \end{bmatrix}, \quad b = \begin{bmatrix} 1 \\ 1 \\ 1 \end{bmatrix}$

**4.** For each of the following matrices $A$, find $A^{-1}$ and check your answer by matrix multiplication.

*(a) $A = \begin{bmatrix} 1 & a & b \\ 0 & 1 & c \\ 0 & 0 & 1 \end{bmatrix}$

(b) $A = \begin{bmatrix} 1 & a & 0 & 0 \\ 0 & 1 & b & 0 \\ 0 & 0 & 1 & c \\ 0 & 0 & 0 & 1 \end{bmatrix}$

**5.** Find (if possible) a matrix $B$ such that

*(a) $AB = I$, where $A = \begin{bmatrix} 1 & 2 & 3 \\ 2 & 5 & 7 \end{bmatrix}$.

*(b) $AB = I$, where $A = \begin{bmatrix} 2 & 0 \\ 3 & 1 \\ 5 & 2 \end{bmatrix}$.

(c) $AB = I$, where $A = \begin{bmatrix} 1 & 1 & 1 & 1 \\ 0 & 1 & 2 & 1 \\ 1 & 1 & 2 & 1 \end{bmatrix}$.

***6.** Let $B = \begin{bmatrix} 5 & -4 \\ 2 & 2 \end{bmatrix}$,

$$C = \begin{bmatrix} 21 & -18 \\ -17 & 16 \end{bmatrix}, \quad \text{and}$$

$$D = \begin{bmatrix} 1 & 2 \\ 3 & 4 \end{bmatrix}.$$

Find a $2 \times 2$ matrix $A$ such that $2AB = C + 3D$.

*7. Let $A$ be an $m \times n$ matrix and let $I$ be the $m \times m$ identity matrix. Show that if the rank of $A$ is equal to $m$, then there is an $n \times m$ matrix $B$ such that $AB = I$.

8. Let $A$ be an $n \times n$ nonsingular matrix.
   (a) Show that if $B$ is an $n \times p$ matrix, there is a unique $n \times p$ matrix $D$ such that $AD = B$.
   (b) Show that if $B$ is an $m \times n$ matrix, there is a unique $m \times n$ matrix $C$ such that $CA = B$.

*9. Let $A$ be a symmetric nonsingular matrix. Show that $A^{-1}$ is also symmetric. Is the corresponding statement true for skew-symmetric matrices?

10. Suppose that $d_1, d_2, \ldots, d_n$ are nonzero numbers. Show that

$$\begin{bmatrix} 0 & 0 & \cdots & 0 & d_1 \\ 0 & 0 & \cdots & d_2 & 0 \\ \vdots & \vdots & & \vdots & \vdots \\ \vdots & \vdots & & \vdots & \vdots \\ 0 & d_{n-1} & \cdots & 0 & 0 \\ d_n & 0 & \cdots & 0 & 0 \end{bmatrix}$$

is nonsingular and find its inverse.

*11. Is the sum of two nonsingular matrices a nonsingular matrix?

*12. Suppose that $A$ is an $n \times n$ matrix such that $(I - A)(I + A) = 0$.
   (a) Show that $A$ is nonsingular and that $A^{-1} = A$.
   (b) Does it follow from

$$(I - A)(I + \text{A}) = 0$$

   that $I - A = 0$ or $I + A = 0$?

13. Let $A$ be an $n \times n$ matrix such that $2A^2 - 6A + 2I = 0$. Show that $A$ is nonsingular and that $A^{-1} = 3I - A$.

14. Suppose that $A$ is a square matrix and $A^2 - 8A + 2I = 0$. Show that $A$ is nonsingular and that $A^{-1} = 4I - \frac{1}{2}A$.

15. Suppose that $A$, $B$, and $C$ are $n \times n$ nonsingular matrices. Prove that $ABC$ is nonsingular and find $(ABC)^{-1}$

16. Let $A$ and $B$ be $n \times n$ matrices. Show that if $AB$ is nonsingular, then both $A$ and $B$ are nonsingular. (*Hint:* First show that $B$ is nonsingular.)

17. Let $A$ be an upper triangular nonsingular matrix.
   *(a) Show that the solution of $A\mathbf{x} = \mathbf{e}_1$ is a multiple of $\mathbf{e}_1$. (*Hint:* Solve the system by back substitution.)
   (b) Show that the solution of $A\mathbf{x} = \mathbf{e}_2$ is a linear combination of $\mathbf{e}_1$ and $\mathbf{e}_2$.
   (c) Show that the solution of $A\mathbf{x} = \mathbf{e}_i$ is a linear combination of $\mathbf{e}_1, \mathbf{e}_2, \ldots, \mathbf{e}_i$.
   (d) Show that the inverse of $A$ is also an upper triangular matrix.
   (e) Is the same true for lower triangular matrices?

18. Let $A$ be a square matrix.
   *(a) Show that $I - A^n = (I - A)(A^{n-1} + A^{n-2} + \cdots + A + I)$.
   (b) Show that if $A^n = 0$, then $I - A$ is nonsingular.

19. If all row sums of a consumption matrix are less than 1, prove that there is a production vector $\mathbf{x}$ such that $(I - A)\mathbf{x}$ has all positive entries.

20. If all column sums of a consumption matrix are less than 1, use Theorem 7 to prove that $I - A$ has an inverse and every entry of $(I - A)^{-1}$ is nonnegative.

21. Prove the converse of Theorem 7.

22. A square matrix $A$ is called **nilpotent** if there is a positive integer $n$ such that $A^n = 0$. Show that any nilpotent matrix is singular. (*Hint:* Suppose $A$ is nonsingular.)

*23. Let $A$ be an $m \times m$ nonsingular matrix and let $B$ be an $m \times n$ matrix. Explain why it is that when $[A : B]$ is reduced to $[I : C]$, $C = A^{-1}B$.

**24.** Let $A = \begin{bmatrix} a & b \\ c & d \end{bmatrix}$, where $a$, $b$, $c$, and $d$ are all nonnegative, be the matrix of input coefficients for a two-production-sector economy.

(a) Show that the matrix $I - A$ will have an inverse with nonnegative entries if and only if the following two conditions hold:
(1) $a < 1, d < 1$.
(2) $(1 - a)(1 - d) > bc$.

(b) Discuss the economic meaning of each condition above.

# SUPPLEMENTARY PROBLEMS

**1.** In each part, let $A$ be an $m \times n$ matrix and let $r$ be its rank. Mark each of the following statements true if the statement is true for all $m \times n$ matrices and mark it false if there is at least one $m \times n$ matrix for which the statement is false. Give a brief explanation.

*(a) If $m < n$, then $A\mathbf{x} = \mathbf{0}$ has a nontrivial solution.

(b) If $r < m$, then $A\mathbf{x} = \mathbf{0}$ has a nontrivial solution.

*(c) If $r < n$, then for every $\mathbf{b}$ in $R^m$ the equation has infinitely many solutions.

(d) If $r = n = m$, then $A$ is nonsingular.

*(e) If $n < m$, then every consistent system $A\mathbf{x} = \mathbf{b}$ has a unique solution.

(f) If $r = m$, then $A\mathbf{x} = \mathbf{b}$ has at least one solution for every $\mathbf{b}$ in $R^m$.

*(g) If $r = n$, then $A\mathbf{x} = \mathbf{b}$ has a unique solution for every $\mathbf{b}$ in $R^m$.

(h) If for every $\mathbf{b}$ in $R^m$ the system $A\mathbf{x} = \mathbf{b}$ has at most one solution, then $r = n$.

*(i) If for every $\mathbf{b}$ in $R^m$ the system $A\mathbf{x} = \mathbf{b}$ has exactly one solution, then $r = m = n$.

(j) $A\mathbf{x} = \mathbf{0}$ has a unique solution, then $r = n$.

*(k) If $A\mathbf{x} = \mathbf{b}$ has at least one solution for every $\mathbf{b}$ in $R^m$, then $r = m$.

(l) If $A\mathbf{x} = \mathbf{0}$ has a nontrivial solution, then every consistent system $A\mathbf{x} = \mathbf{b}$ has infinitely many solutions.

*(m) If $m < n$, then $A\mathbf{x} = \mathbf{b}$ is consistent for every $\mathbf{b}$ in $R^m$.

(n) If $A$ is triangular, then $r$ is the number of nonzero diagonal entries of $A$.

*(o) When $A$ is reduced to an echelon matrix $E$, $E$ has $r$ zero rows.

(p) $r$ is the number of pivot columns in any echelon form of $A$.

*(q) $r$ is the number of nonzero rows of $A$.

(r) $r$ is the number of determined variables.

*(s) $r$ might be larger than $m$ or $n$ but never larger than both $m$ and $n$.

(t) If $A$ is singular, then $r < n$.

*(u) If $m = n$ and $A$ is diagonal, then $A$ is nonsingular if and only if the diagonal entries of $A$ are nonzero.

(v) If two rows of $A$ are equal, then $r < m$.

**2.** For each part, find:
(i) The rank of $A$.
(ii) The general solution of $A\mathbf{x} = \mathbf{0}$.
(iii) The general solution of $A\mathbf{x} = \mathbf{b}$.

*(a) $A = \begin{bmatrix} 1 & 2 & 1 & -2 \\ -2 & -4 & -1 & 0 \\ 4 & 8 & 6 & -16 \end{bmatrix}$,

$\mathbf{b} = \begin{bmatrix} 1 \\ -3 \\ 2 \end{bmatrix}$

(b) $A = \begin{bmatrix} 1 & 2 & 3 & 4 \\ 1 & 3 & 5 & 7 \\ 1 & 0 & -1 & -2 \end{bmatrix}$,

$\mathbf{b} = \begin{bmatrix} 5 \\ 11 \\ -6 \end{bmatrix}$

*(c) $A = \begin{bmatrix} 1 & 2 & 3 & 4 & 5 \\ 1 & 3 & 5 & 7 & 4 \\ 1 & 4 & 7 & 10 & 3 \\ 1 & 2 & 4 & 6 & 8 \end{bmatrix}$,

$\mathbf{b} = \begin{bmatrix} 6 \\ 13 \\ 20 \\ 13 \end{bmatrix}$

(d) $A = \begin{bmatrix} 1 & -2 & 2 & -1 & 0 \\ 2 & 3 & 4 & 5 & 0 \\ 1 & 1 & 1 & 1 & 3 \\ 2 & 1 & 4 & 7 & 0 \end{bmatrix}$,

$\mathbf{b} = \begin{bmatrix} 3 \\ -1 \\ -2 \\ -7 \end{bmatrix}$

**3.** For each part, find $A^{-1}$ and solve $A\mathbf{x} = \mathbf{b}$.

*(a) $A = \begin{bmatrix} 3 & -1 & 1 \\ 1 & -1 & 1 \\ 2 & 1 & 3 \end{bmatrix}$, $\mathbf{b} = \begin{bmatrix} 1 \\ 2 \\ 0 \end{bmatrix}$

(b) $A = \begin{bmatrix} 1 & 1 & -1 \\ 2 & 5 & -1 \\ 2 & 0 & -3 \end{bmatrix}$, $\mathbf{b} = \begin{bmatrix} 1 \\ 0 \\ 0 \end{bmatrix}$

*(c) $A = \begin{bmatrix} 1 & 0 & 1 \\ 0 & 1 & 1 \\ 0 & 0 & 1 \end{bmatrix}$, $\mathbf{b} = \begin{bmatrix} 1 \\ 1 \\ 2 \end{bmatrix}$

(d) $A = \begin{bmatrix} 1 & 0 & 3 \\ 1 & 2 & 5 \\ 1 & 3 & 2 \end{bmatrix}$, $\mathbf{b} = \begin{bmatrix} 0 \\ 1 \\ 0 \end{bmatrix}$

**4.** Find the inverses of $A$, $B$, $A^T$, and $AB$, where

$$A = \begin{bmatrix} 1 & -1 & 1 \\ 0 & 1 & 2 \\ 1 & 0 & 2 \end{bmatrix} \quad \text{and}$$

$$B = \begin{bmatrix} 0 & 2 & 0 \\ 1 & 1 & 2 \\ 2 & 0 & 1 \end{bmatrix}.$$

***5.** Determine which of the following matrices are nonsingular.

(a) $\begin{bmatrix} -3 & 5 \\ 0 & 2 \end{bmatrix}$

(b) $\begin{bmatrix} 6 & -18 \\ -5 & 15 \end{bmatrix}$

(c) $\begin{bmatrix} 1 & -1 & 0 \\ 0 & 0 & 2 \\ 3 & 2 & 1 \end{bmatrix}$

(d) $\begin{bmatrix} -1 & 0 & 5 \\ 2 & 0 & 1 \\ 3 & 0 & -4 \end{bmatrix}$

(e) $\begin{bmatrix} 0 & 0 & 2 \\ 0 & 1 & 2 \\ 3 & 4 & 5 \end{bmatrix}$

(f) $\begin{bmatrix} 1 & 2 & 0 & 1 \\ 1 & 3 & 0 & 0 \\ 1 & 4 & 1 & 0 \\ 1 & 5 & 0 & 1 \end{bmatrix}$

(g) $\begin{bmatrix} 1 & 0 & 1 & 0 & 0 \\ -1 & 0 & -1 & 1 & 0 \\ 2 & 1 & 1 & 1 & 1 \\ 0 & 1 & -1 & -1 & 1 \\ 0 & 2 & 1 & -2 & 2 \end{bmatrix}$

**6.** What system of equations must the currents in the following circuit satisfy? Solve the system.

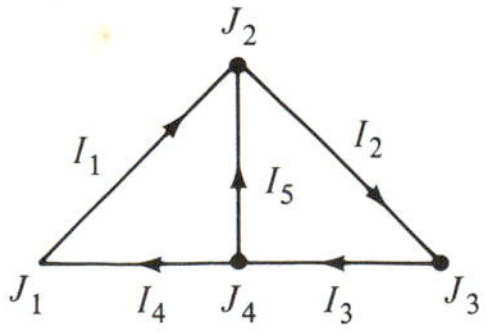

***7.** Give a geometric description of the solution sets of the following linear systems.

(a) $\begin{bmatrix} 2 & 1 \\ 4 & 2 \end{bmatrix}\begin{bmatrix} x_1 \\ x_2 \end{bmatrix} = \begin{bmatrix} 3 \\ 6 \end{bmatrix}$

(b) $\begin{bmatrix} 2 & 3 & 1 \\ 0 & 2 & -2 \\ -1 & 1 & -3 \end{bmatrix}\begin{bmatrix} x_1 \\ x_2 \\ x_3 \end{bmatrix} = \begin{bmatrix} 4 \\ 0 \\ -2 \end{bmatrix}$

(c) $\begin{bmatrix} 1 & -2 & 4 \end{bmatrix}\begin{bmatrix} x_1 \\ x_2 \\ x_3 \end{bmatrix} = [9]$

**8.** Find two $2 \times 2$ matrices $A$ and $B$ for which $AB = 0$ and $BA \neq 0$.

***9.** Two people attempt to find the general solution to $A\mathbf{x} = \mathbf{b}$. Person one's answer is

$$\mathbf{x} = \begin{bmatrix} 4 \\ 0 \\ 1 \\ 0 \end{bmatrix} + \alpha_1 \begin{bmatrix} -2 \\ 1 \\ 0 \\ 0 \end{bmatrix} + \alpha_2 \begin{bmatrix} 1 \\ 0 \\ 2 \\ 1 \end{bmatrix}.$$

Person two's answer is

$$\mathbf{x} = \begin{bmatrix} 4 \\ 1 \\ 5 \\ 2 \end{bmatrix} + \beta_1 \begin{bmatrix} 0 \\ 1 \\ 4 \\ 2 \end{bmatrix} + \beta_2 \begin{bmatrix} -3 \\ 1 \\ -2 \\ -1 \end{bmatrix}.$$

Did they arrive at the same answer?

**10.** (a) Given an $m \times n$ matrix $A$ and a vector $\mathbf{b}$ in $R^n$, how would you solve the equation $\mathbf{x}^T\mathbf{A} = \mathbf{b}^T$?

(b) Explain why it has a solution if and only if $\mathbf{b}$ is a linear combination of the rows of $A$.

**11.** (a) Show that if $A$, $B$, and $C$ are matrices and $A$ is nonsingular, then $AB = AC$ implies that $B = C$.

(b) Find matrices $A$, $B$, and $C$ such that $A \neq 0$, $AB = AC$, and $B \neq C$.

**12.** Suppose that $A$ is a $2 \times 3$ matrix such that

$$A\begin{bmatrix} 1 \\ 2 \\ 3 \end{bmatrix} = \begin{bmatrix} 4 \\ 5 \end{bmatrix} \quad \text{and} \quad A\begin{bmatrix} 6 \\ 7 \\ 8 \end{bmatrix} = \begin{bmatrix} 9 \\ 10 \end{bmatrix}.$$

(a) Write (3, 5) as a linear combination of (4, 5) and (9, 10).

*(b) Find a vector $\mathbf{x}$ such that $A\mathbf{x} = \begin{bmatrix} 3 \\ 5 \end{bmatrix}$.

(c) Write (3, 2, 1) as a linear combination of (1, 2, 3) and (6, 7, 8).

(d) What is

$$A\begin{bmatrix} 3 \\ 2 \\ 1 \end{bmatrix}?$$

**13.** An $n \times n$ square matrix $A$ is called **idempotent** if $A^2 = A$.

*(a) Show that if $A$ is idempotent and nonsingular, then $A = I$.

(b) Describe all idempotent, diagonal $n \times n$ matrices.

(c) Suppose that $A = \begin{bmatrix} a & b \\ 0 & d \end{bmatrix}$, $A \neq 0$, $A \neq I$, and $A$ is idempotent. Show that either $a = 1, d = 0$ or $a = 0, d = 1$.

(d) Find two different, nondiagonal, nontriangular idempotent $2 \times 2$ matrices.

***14.** If $A$ is an $n \times n$ diagonal matrix that is nonsingular, find $A^{-1}$.

# 2

# Fundamental Concepts of Linear Algebra

Chapter 1 dealt with systems of linear equations, vectors, and matrices. Vectors and matrices were introduced to package a system of linear equations into a single matrix equation. Of course, this required us to define operations on vectors and matrices. Having defined addition and scalar multiplication, vectors and matrices began to take on a life of their own. So far we have concentrated almost exclusively on their algebraic properties. But we had a glimpse at least of the geometric life of vectors. All scalar multiples of a vector in $R^n$ determine a line through the origin in $R^n$. All linear combinations of any two noncollinear vectors in $R^3$ determine a plane through the origin in $R^3$. We are going to define a new concept, that of a subspace of $R^n$, which will generalize the geometric notion of a plane (or line) in $R^3$ to $R^n$. The concept of a subspace will provide us with a geometric understanding of the theorems we have proved about linear systems which is not available from a purely algebraic point of view. Common properties of various collections of vectors associated with linear systems that might have gone unnoticed will be revealed. The result will be a deeper understanding of systems of linear equations. Finally, as with vectors and matrices, the concept of a subspace of $R^n$ will take on a life of its own. It will give birth to further concepts all of which (as we will see) apply to much more general objects than vectors in $R^n$.

# 2.1
# $R^n$ and Its Subspaces

Recall that $R^n$ is the collection of all ordered $n$-tuples of real numbers. The elements of $R^n$ are called vectors (or $n$-vectors). The relation of equality and the operations of addition and scalar multiplication on vectors in $R^n$ are defined componentwise. Two vectors in $R^n$ are equal if their corresponding components are equal; two vectors in $R^n$ are added by adding their corresponding components; a scalar and a vector in $R^n$ are multiplied by multiplying each component of the vector by the scalar. The identities satisfied by these operations on $R^n$ are:

I. Properties of addition
   (1) $\mathbf{x} + \mathbf{y} = \mathbf{y} + \mathbf{x}$
   (2) $\mathbf{x} + (\mathbf{y} + \mathbf{z}) = (\mathbf{x} + \mathbf{y}) + \mathbf{z}$
   (3) $\mathbf{x} + \mathbf{0} = \mathbf{x}$
   (4) $\mathbf{x} - \mathbf{x} = \mathbf{0}$
II. Properties of scalar multiplication
   (5) $\alpha(\beta\mathbf{x}) = (\alpha\beta)\mathbf{x}$
   (6) $1 \cdot \mathbf{x} = \mathbf{x}$
   (7) $\alpha(\mathbf{x} + \mathbf{y}) = \alpha\mathbf{x} + \alpha\mathbf{y}$
   (8) $(\alpha + \beta)\mathbf{x} = \alpha\mathbf{x} + \beta\mathbf{x}$

From now on, $R^n$ will denote the collection of all $n$-tuples of real numbers together with the two operations of addition and scalar multiplication that satisfy the identities (1) through (8). The two operations impose an algebraic structure on $R^n$. In mathematics, when a collection of objects has a structure imposed on it, it is often called a **space**. An adjective is always added to specify the nature or structure of the objects in the space. Since our objects are vectors, we call $R^n$ a **vector space**.

The vector space $R^n$ contains many subsets with algebraic properties analogous to those of $R^n$ itself. These special subsets are called subspaces and they provide a generalization to higher dimensions of lines and planes in $R^3$ through the origin. To motivate the definition of a subspace, we present an example.

Consider a plane $V$ through the origin in $R^3$. It is geometrically obvious from the parallelogram law of addition that the sum of two vectors in the plane is a vector in the plane. Similarly, the plane contains any scalar multiple of a nonzero vector $\mathbf{v}$ in the plane because it contains the entire line through the origin and the given vector $\mathbf{v}$ (see Figure 2.1).

Thus a plane $V$ is a subset of $R^3$ with the following properties:

1. $V$ contains the zero vector.
2. $V$ contains the sum of any two vectors in $V$.
3. $V$ contains any scalar multiple of a vector in $V$.

The last two properties are called **closure** properties.

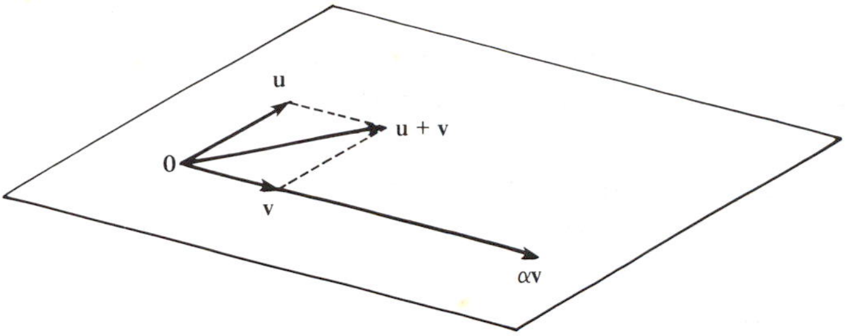


Figure 2.1

***Definition*** A collection $V$ of vectors in $R^n$ is said to be **closed under addition** if $\mathbf{u} + \mathbf{v}$ is in $V$ whenever $\mathbf{u}$ and $\mathbf{v}$ are in $V$. Similarly, $V$ is said to be **closed under scalar multiplication** of $\alpha\mathbf{u}$ is in $V$ whenever $\mathbf{u}$ is in $V$ and $\alpha$ is a scalar. Note that closure under scalar multiplication requires that if $\mathbf{u}$ is in $V$, then $\alpha\mathbf{u}$ is in $V$ *for every real number* $\alpha$.

Consequently, a plane $V$ in $R^3$ through the origin is a subset of $R^3$ that contains the origin and is closed under addition and scalar multiplication. Thus $V$ shares three important properties with $R^3$. Both contain the zero vector. Adding two vectors in $R^3$ results in a vector in $R^3$. Adding two vectors in $R^3$ that also belong to $V$ results in a vector in $R^3$ that also belongs to $V$. Multiplying a vector in $R^3$ by a scalar results in a vector in $R^3$. Multiplying a vector in $R^3$ that belongs to $V$ by a scalar results in a vector in $R^3$ that belongs to $V$. Thus we can consider vector addition and scalar multiplication as being defined just for vectors in $V$! That is, if we begin with $V$ and ignore the rest of $R^3$ and define addition and scalar multiplication componentwise, then the sum of two vectors in $V$ will be in $V$ and any scalar multiple of a vector in $V$ will be in $V$. Moreover, properties (1) through (8) of addition and scalar multiplication continue to hold when all vectors $\mathbf{u}$, $\mathbf{v}$, and $\mathbf{w}$ are restricted to lie in $V$. Why? Because vectors in $V$ are vectors in $R^3$ and these properties hold for all 3-vectors. We say that $V$ "inherits" these properties from $R^n$. Therefore, if we consider vector addition and scalar multiplication as being defined just for vectors in $V$, then these operations on $V$ continue to satisfy properties (1) through (8). So $V$ has an algebraic structure analogous to that of $R^3$ itself. Any result about $R^3$ which is a consequence of the fact that $R^3$ contains the zero vector and that vector addition and scalar multiplication satisfy properties (1) through (8) will also be a result about $V$.

At this point we ask you to verify that a line in $R^3$ through the origin is closed under vector addition and scalar multiplication. Also, convince yourself that the algebraic structure of such a line is analogous to the algebraic structure of $R^3$ itself. That is, we can consider vector addition and scalar multiplication as being defined just for vectors on such a line, and when this is done, properties (1) through (8) of addition and scalar multiplication continue to hold.

Thus lines and planes in $R^3$ that contain the origin share these crucial properties: they contain the origin; they are closed under vector addition; and they are closed under scalar multiplication. From these three facts it follows that lines and planes in $R^3$ through the origin have an algebraic structure like that of $R^3$ itself. The stage is now set for the following definition and theorem.

***Definition*** A subset $V$ of $R^n$ is called a **subspace** of $R^n$ if:

(a) $V$ contains the zero vector.

(b) $V$ is closed under vector addition. That is, $\mathbf{u} + \mathbf{v}$ is in $V$ whenever $\mathbf{u}$ and $\mathbf{v}$ are in $V$.

(c) $V$ is closed under scalar multiplication. That is, $\alpha\mathbf{u}$ is in $V$ whenever $\mathbf{u}$ is in $V$ and $\alpha$ is a scalar.

***Theorem 1*** *Let $V$ be a subspace of $R^n$. Then vector addition and scalar multiplication can be considered as being defined just for vectors in $V$. Moreover, when this is done, properties* (1) *through* (8) *of addition and scalar multiplication continue to hold.*

**Proof** The fact that vector addition and scalar multiplication can be considered as being defined just for vectors in $V$ is because $V$ is closed under vector addition and scalar multiplication. That properties (1) through (8) continue to hold in this context follows because all vectors in $V$ are vectors in $R^n$ and these properties hold for vectors in $R^n$. For example, consider property (1). Let $\mathbf{u}$ and $\mathbf{v}$ be vectors in $V$. Then $\mathbf{u}$ and $\mathbf{v}$ are also vectors in $R^n$. Thus $\mathbf{u} + \mathbf{v} = \mathbf{v} + \mathbf{u}$ because this holds for every pair of vectors in $R^n$.

We called $R^n$ a vector space because it is a set of vectors with two operations defined on it that satisfy properties (1) through (8). But by Theorem 1 a subspace of $R^n$ is itself a set of vectors with two operations defined on it that satisfy properties (1) through (8). The "sub" in the definition of a subspace $V$ of $R^n$ is there to call attention to the fact that $V$ is contained in $R^n$. When it is not important that $V$ is contained in $R^n$, we will refer to $V$ itself as a vector space.

### EXAMPLE 1

Both lines and planes in $R^3$ passing through the origin are subspaces of $R^3$. On the other hand, lines or planes in $R^3$ that do not pass through the origin are not subspaces. They do not contain the origin; they are not closed under addition; they are not closed under scalar multiplication [see Figure 2.2(a), (b)].

### EXAMPLE 2

Certainly, $R^n$ is a subspace of itself. Every other subspace of $R^n$ is called a **proper subspace**. Since every subspace of $R^n$ contains the zero vector, the smallest possible subspace of $R^n$ is the subset consisting only of the zero vector. We denote this subset by $\{\mathbf{0}\}$. It is closed under addition and scalar multiplication because $\mathbf{0} + \mathbf{0} = \mathbf{0}$ and

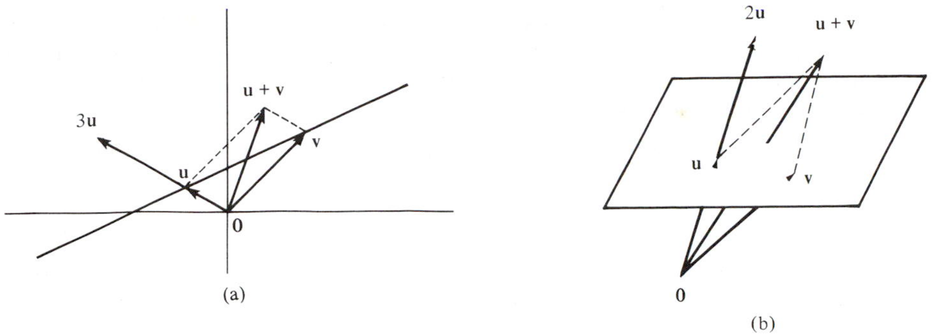


Figure 2.2

$\alpha \cdot \mathbf{0} = \mathbf{0}$ for any scalar $\alpha$. Therefore, $\{\mathbf{0}\}$ is a subspace, called the **trivial subspace**. Any other subspace of $R^n$ is called **nontrivial**.

A subspace of $R^n$ is, first of all, a subset of $R^n$. Any subset $V$ of $R^n$ can be described as the collection of vectors $\mathbf{v}$ in $R^n$ that satisfy some condition. For example, if $\mathbf{w}$ is a nonzero vector in $R^3$, then the line in $R^3$ through $\mathbf{0}$ and containing $\mathbf{w}$ is the collection $V$ of all vectors $\mathbf{v}$ in $R^3$ such that for some scalar $\alpha$, $\mathbf{v} = \alpha\mathbf{w}$. The condition for membership in $V$ is being a scalar multiple of $\mathbf{w}$. A plane in $R^3$ containing $\mathbf{0}$ is the collection $V$ of all possible linear combinations of two noncollinear vectors $\mathbf{w}_1$ and $\mathbf{w}_2$. In this case, the condition for membership in $V$ is being a linear combination of $\mathbf{w}_1$ and $\mathbf{w}_2$. We saw in Result 1 of Section 1.10 that a plane in $R^3$ containing $\mathbf{0}$ can also be described as the set of all 3-tuples $(x_1, x_2, x_3)$ satisfying an equation of the form $ax_1 + bx_2 + cx_3 = 0$. The condition for membership here is being a solution of a linear equation.

Since a subset $V$ of $R^n$ is described as the collection of all vectors $\mathbf{v}$ in $R^n$ that satisfy some stated condition, to verify that $V$ is a subspace we must verify that:

1. The zero vector satisfies the condition.
2. The sum of two vectors that satisfy the condition also satisfies the condition.
3. Any scalar multiple of a vector that satisfies the condition also satisfies the condition.

## EXAMPLE 3

Let $V$ be the subset of $R^3$ consisting of all vectors $\mathbf{v} = (v_1, v_2, v_3)$ that satisfy the condition that $v_3 = 2v_1$. Certainly, the zero vector $\mathbf{0} = (0, 0, 0)$ satisfies this condition. Suppose that $\mathbf{v}$ and $\mathbf{w}$ are two vectors that satisfy the condition (that is, suppose that $\mathbf{v}$ and $\mathbf{w}$ are in $V$). Thus $v_3 = 2v_1$ and $w_3 = 2w_1$. To determine whether the sum $\mathbf{v} + \mathbf{w}$ is in $V$, we must determine whether the sum $\mathbf{v} + \mathbf{w} = (v_1 + w_1, v_2 + w_2, v_3 + w_3)$ satisfies the condition for membership in $V$. So the question is: Is $v_3 + w_3 = 2(v_1 + w_1)$? Sure! $v_3 = 2v_1$ and $w_3 = 2w_1$, so $v_3 + w_3 = 2v_1 + 2w_1 =$

$2(v_1 + w_1)$. Thus $\mathbf{v} + \mathbf{w}$ is in $V$ and $V$ is closed under addition. Now we check scalar multiplication. Let $\alpha$ be a scalar and $\mathbf{v}$ a vector in $V$. To determine whether $\alpha\mathbf{v}$ is in $V$, we must determine whether $\alpha\mathbf{v} = (\alpha v_1, \alpha v_2, \alpha v_3)$ satisfies the condition for membership in $V$. Since $v_3 = 2v_1$ ($\mathbf{v}$ belongs to $V$), $\alpha v_3 = \alpha(2v_1) = 2(\alpha v_1)$. Therefore, $\alpha\mathbf{v}$ is in $V$ and $V$ is closed under scalar multiplication. Thus $V$ is a subspace of $R^3$.

## EXAMPLE 4

Let us consider an example of a subset $V$ of $R^3$ where membership in $V$ requires that two conditions be satisfied. Let $V$ consist of all vectors $\mathbf{v} = (v_1, v_2, v_3)$ satisfying the conditions that $v_1 + v_2 = 0$ and $v_2 + v_3 = 0$. Clearly, $\mathbf{0}$ is in $V$. Suppose that $\mathbf{v} = (v_1, v_2, v_3)$ and $\mathbf{w} = (w_1, w_2, w_3)$ are in $V$. Then

$$\begin{aligned} v_1 + v_2 &= 0 \qquad \text{and} \qquad v_2 + v_3 = 0 \\ w_1 + w_2 &= 0 \qquad \text{and} \qquad w_2 + w_3 = 0. \end{aligned}$$

Does the sum $\mathbf{v} + \mathbf{w} = (v_1 + w_1, v_2 + w_2, v_3 + w_3)$ satisfy the conditions for membership in $V$? Sure!

$$\begin{aligned} (v_1 + w_1) + (v_2 + w_2) &= (v_1 + v_2) + (w_1 + w_2) = 0 + 0 = 0 \\ (v_2 + w_2) + (v_3 + w_3) &= (v_2 + v_3) + (w_2 + w_3) = 0 + 0 = 0 \end{aligned}$$

Therefore, $\mathbf{v} + \mathbf{w}$ belongs to $V$ and $V$ is closed under addition. $V$ is also closed under scalar multiplication. For if $\alpha$ is a scalar and $\mathbf{v}$ is in $V$, then

$$\alpha\mathbf{v} = (\alpha v_1, \alpha v_2, \alpha v_3)$$

and since

$$v_1 + v_2 = 0 \qquad \text{and} \qquad v_2 + v_3 = 0,$$

it follows that

$$\begin{aligned} \alpha v_1 + \alpha v_2 &= \alpha(v_1 + v_2) = \alpha 0 = 0 \\ \alpha v_2 + \alpha v_3 &= \alpha(v_2 + v_3) = \alpha 0 = 0. \end{aligned}$$

Thus $\alpha\mathbf{v}$ is in $V$, as asserted. Therefore, $V$ is indeed a subspace.

## EXAMPLE 5

We gave a geometric argument to verify that a plane in $R^3$ passing through the origin is a subspace of $R^3$. We can also verify this fact algebraically. A plane through the origin in $R^3$ consists of all possible linear combinations of two noncollinear vectors $\mathbf{w}_1$ and $\mathbf{w}_2$. Thus the condition a vector $\mathbf{v}$ in $R^3$ must satisfy to be in the plane is that there are scalars $\beta_1$ and $\beta_2$ such that $\mathbf{v} = \beta_1\mathbf{w}_1 + \beta_2\mathbf{w}_2$. Certainly, the zero vector satisfies this condition: $\mathbf{0} = 0\mathbf{w}_1 + 0\mathbf{w}_2$. Let $\mathbf{u}$ and $\mathbf{v}$ be vectors in the plane. Then there are scalars $\beta_1, \beta_2$ and $\gamma_1, \gamma_2$ such that

$$\mathbf{u} = \beta_1\mathbf{w}_1 + \beta_2\mathbf{w}_2 \qquad \text{and} \qquad \mathbf{v} = \gamma_1\mathbf{w}_1 + \gamma_2\mathbf{w}_2.$$

Since

$$\mathbf{u} + \mathbf{v} = (\beta_1 + \gamma_1)\mathbf{w}_1 + (\beta_2 + \gamma_2)\mathbf{w}_2,$$

it follows that $\mathbf{u} + \mathbf{v}$ is in the plane because it is a linear combination of $\mathbf{w}_1$ and $\mathbf{w}_2$. Similarly, if $\alpha$ is a scalar, then since

$$\alpha\mathbf{u} = (\alpha\beta_1)\mathbf{w}_1 + (\alpha\beta_2)\mathbf{w}_2,$$

it follows that $\alpha\mathbf{u}$ is in the plane because it, too, is a linear combination of $\mathbf{w}_1$ and $\mathbf{w}_2$.

## EXAMPLE 6

Consider the subset of $V$ of $R^2$ consisting of all vectors $\mathbf{v} = (v_1, v_2)$ that satisfy the condition that $v_1$ and $v_2$ are integers. Certainly, $\mathbf{0} = (0, 0)$ meets the condition for membership in $V$. Suppose that $\mathbf{v} = (v_1, v_2)$ and $\mathbf{w} = (w_1, w_2)$ are two vectors in $V$. Membership in $V$ requires that $v_1$, $v_2$, $w_1$, and $w_2$ be integers. The sum $\mathbf{v} + \mathbf{w} = (v_1 + w_1, v_2 + w_2)$ is therefore in $V$ since it meets the condition for membership in $V$; $v_1 + w_1$ and $v_2 + w_2$ are integers because the sum of integers is an integer. Now let $\alpha$ be a scalar and $\mathbf{v} = (v_1, v_2)$ be a vector in $V$. Is $\alpha\mathbf{v}$ in $V$? Not always! If $\alpha$ is an integer, then $\alpha v_1$ and $\alpha v_2$ are both integers and therefore $\alpha\mathbf{v}$ is in $V$. But if $\alpha = \frac{1}{2}$ and $\mathbf{v} = (1, 2)$, then $\alpha\mathbf{v} = (\frac{1}{2}, 1)$ is not in $V$. For $V$ to be closed under scalar multiplication, given *any* scalar $\alpha$ and *any* vector $\mathbf{v}$ in $V$ we must have $\alpha\mathbf{v}$ in $V$. Since this is not the case in our example, $V$ is not a subspace of $R^2$.

## EXAMPLE 7

Let $V$ be the subset of $R^2$ consisting of all vectors $\mathbf{v} = (v_1, v_2)$ that satisfy the condition that $v_1 v_2 \geq 0$. Geometrically, $V$ is simply the set of all vectors in $R^2$ that lie in the first or third quadrant of the plane. Clearly, the zero vector $\mathbf{0} = (0, 0)$ satisfies $0 \cdot 0 \geq 0$, so it is in $V$. Let $\mathbf{u} = (u_1, u_2)$ and $\mathbf{v} = (v_1, v_2)$ be vectors in $V$. From the membership condition, $u_1 u_2 \geq 0$ and $v_1 v_2 \geq 0$. Does $\mathbf{u} + \mathbf{v} = (u_1 + v_1, u_2 + v_2)$ meet the membership condition? That is, does it follow that $(u_1 + v_1)(u_2 + v_2) \geq 0$? Since $(u_1 + v_1)(u_2 + v_2) = u_1 u_2 + u_1 v_2 + v_1 u_2 + v_1 v_2$ and since $u_1 u_2 \geq 0$ and $v_1 v_2 \geq 0$,

$$u_1 u_2 + u_1 v_2 + v_1 u_2 + v_1 v_2 \geq 0$$

if and only if

$$u_1 v_2 + v_1 u_2 \geq -(u_1 u_2 + v_1 v_2).$$

There is absolutely no reason to believe that this last condition is satisfied. All we know is that $u_1$, $u_2$ have the same sign and $v_1$, $v_2$ have the same sign. That is not sufficient to guarantee that the mixed product satisfies

$$u_1 v_2 + v_1 u_2 \geq -(u_1 u_2 + v_1 v_2).$$

For example, if $\mathbf{u} = (1, 2)$ and $\mathbf{v} = (-2, -1)$, then

$$u_1 v_2 + v_1 u_2 = (1)(-1) + (-2)(2) = -5$$

and

$$-[u_1 u_2 + v_1 v_2] = -[(1)(2) + (-2)(-1)] = -4$$

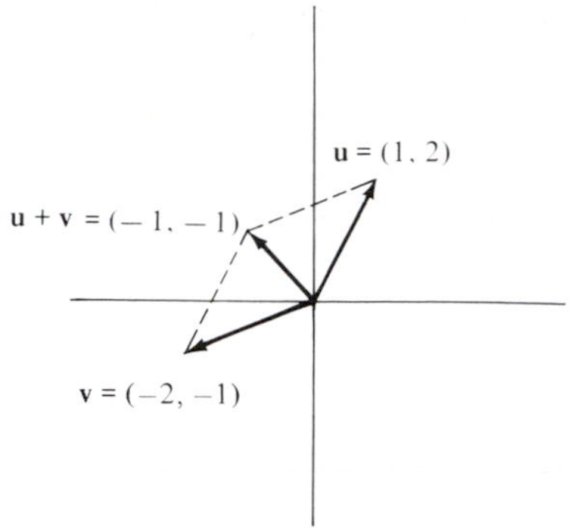


Figure 2.3

(see Figure 2.3). So $V$ is not closed under addition and hence is not a subspace of $R^2$. It is interesting to note, however, that $V$ is closed under scalar multiplication (see Problem 3).

As these last two examples indicate, a subset of $R^n$ can be closed under one of the operations of addition and scalar multiplication without being closed under the other. Thus the two closure properties are independent; neither one can be deduced from the other.

## EXAMPLE 8

We have found that each of the following are subspaces of $R^3$:

1. $\{\mathbf{0}\}$.
2. Lines through the origin.
3. Planes through the origin.
4. $R^3$ itself.

**Question** Are there other subspaces of $R^3$? No! These are, in fact, the only subspaces of $R^3$. For suppose that $V$ is a subspace of $R^3$. Then $V$ contains $\mathbf{0}$. If $V$ contains no nonzero vectors, then $V = \{\mathbf{0}\}$. Otherwise, there is a $\mathbf{u} \neq \mathbf{0}$ in $V$. Since $V$ is closed under scalar multiplication, $V$ contains the entire line $\mathbf{x} = t\mathbf{u}$ through the origin in the direction of $\mathbf{u}$. Either $V$ is this line or it contains a vector $\mathbf{v}$ not on the line. Since $V$ is closed under addition and scalar multiplication, $V$ must then contain all possible linear combinations of $\mathbf{u}$ and $\mathbf{v}$. That is, $V$ must contain the plane through the origin determined by $\mathbf{u}$ and $\mathbf{v}$. Now either $V$ is this plane, or it contains a vector $\mathbf{w}$ not on this plane. In the latter case, $V = R^3$. The proof is from geometry. We must show that if $\mathbf{x}$ is any vector in $R^3$, then $\mathbf{x}$ is in $V$. We will show this by showing that $\mathbf{x}$ is the sum of two vectors, one from the plane determined by $\mathbf{u}$ and $\mathbf{v}$, and the other a scalar multiple of $\mathbf{w}$. Let $\mathbf{x}$ be in $R^3$. Draw a line from $\mathbf{x}$ to the plane determined by $\mathbf{u}$ and $\mathbf{v}$ *which is parallel to* $\mathbf{w}$, and let $\mathbf{y}$ be the point where the line intersects the plane. From our construction the vector $\mathbf{x} - \mathbf{y}$ is a multiple of $\mathbf{w}$, say

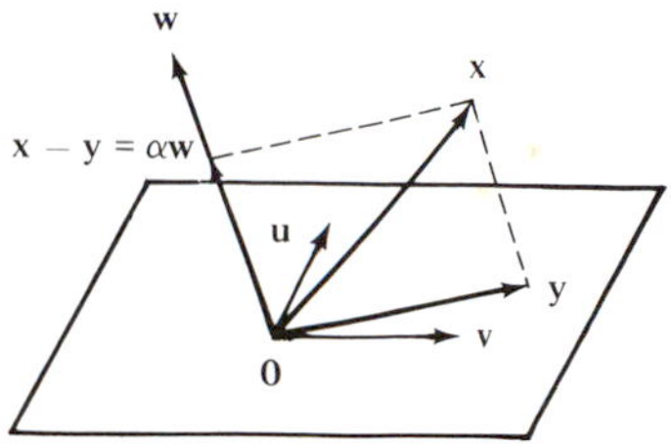


Figure 2.4

$\mathbf{x} - \mathbf{y} = \alpha\mathbf{w}$. The parallelogram law for addition then yields that $\mathbf{x} = \mathbf{y} + \alpha\mathbf{w}$. Thus $\mathbf{x}$ is in $V$, being a sum of two vectors in $V$ (see Figure 2.4).

Our classification of the subspaces of $R^3$ is based on the size of the subspace. In Problem 13 we invite you to find a similar classification of the subspaces of $R^2$. To extend our classification to $R^4$ and beyond, we need a generalization of the notion of the "size" of a subspace. We will accomplish this generalization in Section 2.5.

## PROBLEMS 2.1

**1.** Complete each of the following definitions.

(a) A collection $V$ of vectors in $R^n$ is closed under addition if....

(b) A collection $V$ of vectors in $R^n$ is closed under scalar multiplication if....

(c) A subset $V$ of $R^n$ is called a subspace if....

(d) $V$ is a proper subspace of $R^n$ if....

(e) The trivial subspace is....

(f) A subspace is nontrivial if....

**2.** In each part, determine whether the given collection of vectors is a subspace of $R^m$ for an appropriate $m$. If the collection forms a subspace, show that it does. If the collection fails to be a subspace, list the conditions that fail.

*(a) The collection of all 3-vectors $(x_1, x_2, x_3)$ such that $x_1 + x_2 - x_3 = 0$.

(b) The collection of all 3-vectors $(x_1, x_2, x_3)$ such that $x_1 = 5x_3$ and $x_1 + x_2 - x_3 = 0$.

*(c) The collection of all 5-vectors whose second and fourth components are equal.

(d) The collection of all 3-vectors whose first component is negative.

*(e) The collection of all 4-vectors whose components satisfy $x_1 = x_2 = x_3 = x_4$.

(f) The collection of all 5-vectors whose components satisfy $x_1 + x_2 + x_3 \leq x_4 + x_5$.

*(g) The collection of all 3-vectors whose components satisfy $x_1 = 0$ or $x_3 = 0$.

(h) The collection of all 3-vectors whose components satisfy $x_1 - x_2 = 0$ and $x_2 x_3 = 0$.

*(i) The collection of all 3-vectors whose components satisfy $x_1 + x_2 = x_3$ and $x_2 = 4x_3$.

(j) The collection of all 3-vectors whose components satisfy $x_1 x_2 = x_3$.

*(k) The collection of all 3-vectors whose components satisfy $x_1^2 + x_2^2 + x_3^2 = 1$.

(l) The collection of all 3-vectors whose components satisfy $x_1 + x_2 = x_3$ and $2x_2 - x_3 = x_1$.

*(m) The collection of all 4-vectors whose

components satisfy $x_1 \geq 0$, $x_2 \geq 0$, $x_3 \leq 0$, $x_4 = 0$.

(n) The collection of all 2-vectors whose components satisfy $2x_1 - 3x_2 = 1$.

*(o) The collection of all 4-vectors whose components satisfy $x_1x_2x_3x_4 = 0$.

(p) The collection of all 2-vectors whose components satisfy $x_2$ is an integer.

*(q) The collection of all 3-vectors whose components satisfy $x_1 = 1$.

(r) The collection of all 3-vectors whose components satisfy $x_1x_2 \neq 0$.

*(s) The collection of all vectors **w** such that $\mathbf{w} = \mathbf{u} + \mathbf{v}$ where the components of **u** satisfy $u_1 + u_2 - u_3 = 0$ and the components of **v** satisfy $v_1 = v_2$.

**3.** Show that the set $V$ in Example 7 is closed under scalar multiplication. Give both an algebraic and a geometric argument.

***4.** Let $A$ and $B$ be $m \times n$ matrices. Show that the collection of all vectors **x** in $R^n$ such that $A\mathbf{x} = B\mathbf{x}$ is a subspace of $R^n$.

***5.** Show that the set of all vectors of the form

$$(1, 2, 1) + \alpha(1, 1, 0) + \beta(0, 1, 1)$$

is a subspace of $R^3$.

**6.** Let $A$ be an $n \times n$ matrix.

(a) Show that the subset of $R^n$ consisting of all vectors **x** such that $A\mathbf{x} = \mathbf{x}$ is a subspace of $R^n$.

(b) Show that the subset of $R^n$ consisting of all vectors **x** such that $A\mathbf{x} = -2\mathbf{x}$ is a subspace of $R^n$.

(c) State and prove a generalization.

**7.***(a) Does the plane $x - 2y + z = 1$ represent a subspace of $R^3$? Why?

(b) Does the line $x + 2y = 2$ represent a subspace of $R^2$? Why?

**8.** We saw in Result 1 of Section 1.10 that $a_1x_1 + a_2x_2 + a_3x_3 = 0$ is a Cartesian equation of a plane in $R^3$. Thus a plane can be described as the set $V$ of vectors $\mathbf{x} = (x_1, x_2, x_3)$ in $R^3$ which satisfy the condition that

$$a_1x_1 + a_2x_2 + a_3x_3 = 0.$$

Use this condition to verify that $V$ is a subspace of $R^3$.

**9.** Let **w** be a nonzero vector in $R^3$ and let $V$ be the line through the origin and **w**. Thus a vector **v** in $R^3$ belongs to $V$ if and only if there is a scalar $\beta$ such that $\mathbf{v} = \beta\mathbf{w}$. Use this condition to verify that $V$ is a subspace.

**10.** Show that the subset $V$ of $R^4$ consisting of all possible linear combinations of $(1, 0, 1, 0)$ and $(1, 1, 1, 1)$ is a vector space.

**11.** (a) Show that the subset $V$ of $R^2$ consisting of all possible linear combinations of $(1, 1)$ and $(1, 2)$ is a subspace.

(b) Show that $V = R^2$. Give both an algebraic and a geometric argument.

**12.** Find all subspaces of $R^1$.

***13.** Find all subspaces of $R^2$.

**14.** Show that if a subspace $V$ of $R^3$ is a line, then $V$ is the intersection of two planes.

**15.** Suppose that **v** and **w** are solutions of a nonhomogeneous system $A\mathbf{x} = \mathbf{b}$, $\mathbf{b} \neq \mathbf{0}$. Show that $\mathbf{v} + \mathbf{w}$ and $\alpha\mathbf{v}$, where $\alpha \neq 1$, are not solutions of $A\mathbf{x} = \mathbf{b}$. Thus the set of solutions of a nonhomogeneous system does not form a subspace.

***16.** Let $V$ be a nonempty subset of $R^n$ (that is, let $V$ be a subset of $R^n$ which contains something). Show that if $V$ is closed under addition and scalar multiplication, then $V$ is a subspace. (*Hint:* All you must do is show that **0** is in $V$.)

**17.** Let $V$ be a subset of $R^n$ such that:

1. **0** is in $V$.
2. If **u** and **v** are in $V$ and $\alpha$ and $\beta$ are scalars, then $\alpha\mathbf{u} + \beta\mathbf{v}$ is in $V$.

(a) Show that $V$ is a subspace.

(b) Show that the condition that **0** is in $V$ may be replaced by the condition that $V$ is nonempty.

***18.** Let $U$ and $V$ be subspaces of $R^n$. The set of all vectors that are in $U$ or $V$ is called the **union** of $U$ and $V$ and is denoted by $U \cup V$. Show that $U \cup V$ is a subspace of $R^n$ if and only if one of $U$ or $V$ is a subspace of the other.

**19.** Suppose that $U$ and $V$ are subspaces of $R^m$. The **sum** of $U$ and $V$, denoted by $U + V$, is defined as the set of all vectors that can be expressed as a sum of a vector in $U$ and a vector in $V$. Thus a vector $\mathbf{w}$ belongs to $U + V$ if and only if there are vectors $\mathbf{u}$ in $U$ and $\mathbf{v}$ in $V$ such that $\mathbf{w} = \mathbf{u} + \mathbf{v}$.

*(a) Show that $U + V$ is a subspace of $R^m$.

(b) Show that both $U$ and $V$ are subspaces of $U + V$.

(c) Describe $U + V$ if $U$ and $V$ are two distinct straight lines in $R^3$ passing through the origin.

(d) Describe $U + V$ if $U$ is a plane and $V$ is a straight line in $R^3$ both of which pass through the origin.

**20.** Suppose that $U$ and $V$ are subspaces of $R^n$. Then the set of all vectors that are in both $U$ and $V$ is called the **intersection** of $U$ and $V$. We denote the intersection of $U$ and $V$ by $U \cap V$.

(a) Show that $U \cap V$ is a subspace of $R^n$.

(b) Suppose that $U$ and $V$ are two distinct planes in $R^3$ passing through the origin. Describe their intersection.

(c) Describe $U \cap V$ if $U$ and $V$ are two distinct straight lines passing through the origin.

**21.** A subset $U$ of $R^n$ is called an **affine subspace** if there is a subspace $V$ of $R^n$ and a vector $\mathbf{a}$ in $R^n$ such that

$$U = \mathbf{a} + V.$$

We say that $U$ is a translation of the subspace $V$ (see Figure).

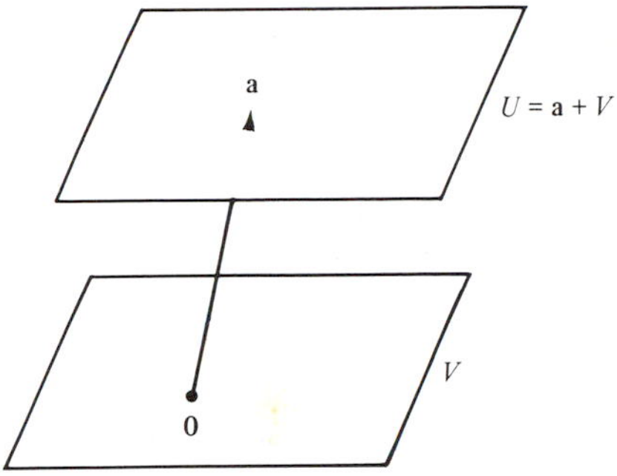


*(a) Show that a line in $R^2$ not passing through the origin is an affine subspace of $R^2$.

(b) Show that a plane in $R^3$ not passing through the origin is an affine subspace of $R^3$.

(c) Show that an affine subspace $U$ is a subspace if and only if $\mathbf{0}$ is in $U$.

(d) Show that any line or plane in $R^3$ is an affine subspace.

(e) Show that if $U$ is an affine subspace and $\mathbf{x}$ is in $U$, then the collection $\mathbf{x} - U$ of all vectors of the form $\mathbf{x} - \mathbf{u}$ for $\mathbf{u}$ in $U$ is a subspace.

(f) If $U$ is an affine subspace and $\mathbf{u}_1, \mathbf{u}_2, \ldots, \mathbf{u}_k$ are in $U$, then $\alpha_1\mathbf{u}_1 + \alpha_2\mathbf{u}_2 + \cdots + \alpha_k\mathbf{u}_k$ is in $U$ if $\alpha_1 + \alpha_2 + \cdots + \alpha_k = 1$.

## 2.2 Spanning

Existence and uniqueness of solutions of a linear system $A\mathbf{x} = \mathbf{b}$ was one of the major themes in Chapter 1. We will see that both of these notions are associated with subspaces of $R^n$. Existence of solutions is determined by the membership of $\mathbf{b}$ in a subspace (called the column space of $A$) and uniqueness of solutions is determined by the size of another subspace (called the nullspace of $A$). Finally, the fact that row operations transform a system $A\mathbf{x} = \mathbf{b}$ into an equivalent echelon system $E\mathbf{x} = \mathbf{c}$ is reflected in the identity of two subspaces (called the row space of $A$ and the row space of $E$).

We begin with the most fundamental question: existence. Given a matrix $A$, find all vectors $\mathbf{b}$ such that the system $A\mathbf{x} = \mathbf{b}$ has at least one solution. In Chapter 1 we found an algebraic description of these vectors. The system $A\mathbf{x} = \mathbf{b}$ is consistent if and only if $\mathbf{b}$ satisfies the constraint equations obtained from the zero rows of the echelon form $[E : \mathbf{c}]$ of the augmented matrix $[A : \mathbf{b}]$.

There is also a simple geometric description of these vectors; they form a subspace.

***Theorem 1*** *Let $A$ be an $m \times n$ matrix. The subset $V$ of $R^m$ consisting of all vectors $\mathbf{b}$ such that $A\mathbf{x} = \mathbf{b}$ is consistent is a subspace of $R^m$.*

**Proof** Since the homogeneous system $A\mathbf{x} = \mathbf{0}$ is consistent, $V$ contains the zero vector. Let $\mathbf{b}_1$ and $\mathbf{b}_2$ be two vectors in $V$. Then $A\mathbf{x} = \mathbf{b}_1$ and $A\mathbf{x} = \mathbf{b}_2$ have solutions, say $\mathbf{x} = \mathbf{v}_1$ and $\mathbf{x} = \mathbf{v}_2$, respectively. Since

$$A(\mathbf{v}_1 + \mathbf{v}_2) = A\mathbf{v}_1 + A\mathbf{v}_2 = \mathbf{b}_1 + \mathbf{b}_2,$$

$\mathbf{x} = \mathbf{v}_1 + \mathbf{v}_2$ is a solution of $A\mathbf{x} = \mathbf{b}_1 + \mathbf{b}_2$. Hence $\mathbf{b}_1 + \mathbf{b}_2$ is in $V$ and we have proved that $V$ is closed under addition. $V$ is also closed under scalar multiplication. For if $\alpha$ is a scalar, then $\mathbf{x} = \alpha\mathbf{v}_1$ is a solution of $A\mathbf{x} = \alpha\mathbf{b}_1$ because

$$A(\alpha\mathbf{v}_1) = \alpha A\mathbf{v}_1 = \alpha\mathbf{b}_1.$$

Thus $V$ is indeed a subspace.

There is yet another description of this subspace which is based on the interpretation of $A\mathbf{x}$ as a linear combination of the columns of $A$ whose coefficients are the components of $\mathbf{x}$. Let $A$ be an $m \times n$ matrix and let $\mathbf{v}_1, \mathbf{v}_2, \ldots, \mathbf{v}_n$ denote the columns of $A$:

$$A = [\mathbf{v}_1 \quad \mathbf{v}_2 \quad \cdots \quad \mathbf{v}_n].$$

Then

$$A\mathbf{x} = x_1\mathbf{v}_1 + x_2\mathbf{v}_2 + \cdots + x_n\mathbf{v}_n.$$

It follows, then, that the system $A\mathbf{x} = \mathbf{b}$ is consistent if and only if $\mathbf{b}$ is a linear combination of the columns $A$. This means that the collection of all vectors $\mathbf{b}$ in $R^m$ such that $A\mathbf{x} = \mathbf{b}$ is consistent consists precisely of all possible linear combinations of the columns of $A$. This explains why we call this subspace the "column space" of $A$.

***Definition*** Let $A$ be an $m \times n$ matrix. The subspace of $R^m$ consisting of all vectors $\mathbf{b}$ for which the system $A\mathbf{x} = \mathbf{b}$ is consistent is called the **column space** of the matrix $A$. We will denote the column space of $A$ by $C(A)$.

Thus the vectors in the column space of a matrix $A$ can be described in two ways.

1. A vector $\mathbf{b}$ is in the column space of $A$ if and only if $A\mathbf{x} = \mathbf{b}$ is consistent, and this holds if and only if $\mathbf{b}$ satisfies the constraint equations obtained from the zero rows of the echelon form $[E : \mathbf{c}]$ of the augmented matrix $[A : \mathbf{b}]$.
2. A vector $\mathbf{b}$ is in the column space of $A$ if and only if $\mathbf{b}$ is a linear combination of the columns of $A$.

These two descriptions of $C(A)$ illustrate two general methods for describing a subspace $V$ of $R^m$. The first is to impose constraints that the components of a vector must satisfy in order to be in the subspace $V$. The second is to specify a set of vectors in $V$ such that every vector in $V$ is a linear combination of the given vectors. When a subspace can be generated from a set of vectors by forming linear combinations, we say that the set of vectors span the subspace.

***Definition*** Let $V$ be a subspace of $R^m$. A set of vectors $\mathbf{v}_1, \mathbf{v}_2, \ldots, \mathbf{v}_k$ in $V$ is said to **span** $V$ if every vector in $V$ can be expressed as a linear combination of the vectors $\mathbf{v}_1, \mathbf{v}_2, \ldots, \mathbf{v}_k$. We also say that the subspace $V$ is **spanned by** the vectors $\mathbf{v}_1, \mathbf{v}_2, \ldots, \mathbf{v}_k$ or that the span of these vectors is $V$.

In the terminology of spanning, the columns of $A$ span the column space of $A$.

**EXAMPLE 1**

Let

$$A = \begin{bmatrix} 2 & 1 \\ 0 & 1 \\ 2 & 1 \end{bmatrix}, \qquad \mathbf{d}_1 = \begin{bmatrix} 3 \\ -1 \\ 3 \end{bmatrix}, \qquad \text{and} \qquad \mathbf{d}_2 = \begin{bmatrix} 3 \\ 1 \\ 8 \end{bmatrix}.$$

The vector $\mathbf{d}_1$ lies in $C(A)$ since $\mathbf{x} = (2, -1)$ is a solution vector of the system $A\mathbf{x} = \mathbf{d}_1$. Thus

$$\mathbf{d}_1 = \begin{bmatrix} 3 \\ -1 \\ 3 \end{bmatrix} = 2\begin{bmatrix} 2 \\ 0 \\ 2 \end{bmatrix} + (-1)\begin{bmatrix} 1 \\ 1 \\ 1 \end{bmatrix}.$$

Reducing

$$[A : \mathbf{b}] = \begin{bmatrix} 2 & 1 & : & b_1 \\ 0 & 1 & : & b_2 \\ 2 & 1 & : & b_3 \end{bmatrix} \qquad \text{to} \qquad [E : \mathbf{c}] = \begin{bmatrix} 2 & 1 & : & b_1 \\ 0 & 1 & : & b_2 \\ 0 & 0 & : & b_3 - b_1 \end{bmatrix},$$

we see that $C(A)$ consists of all vectors $\mathbf{b} = (b_1, b_2, b_3)$ such that $b_3 - b_1 = 0$. Thus the vector $\mathbf{d}_2$ does not lie in $C(A)$ since $d_3 - d_1 = 8 - 3 = 5 \neq 0$. The system $A\mathbf{x} = \mathbf{d}_2$ is inconsistent. Now consider the goemetry of the situation. The column space of $A$ is the subspace of $R^3$ spanned by the columns of $A$. This subspace is

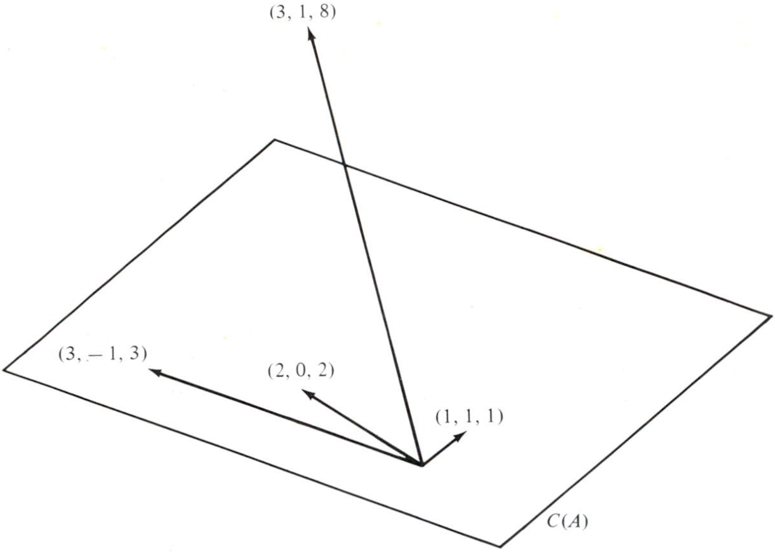


**Figure 2.5**

a plane through the origin because the columns of $A$ are not collinear. Thus the system $A\mathbf{x} = \mathbf{b}$ has a solution if and only if $\mathbf{b}$ lies in this plane. The vector $\mathbf{d}_1$ is in the plane and hence the system $A\mathbf{x} = \mathbf{d}_1$ is consistent. On the other hand, the vector $\mathbf{d}_2$ is not in the plane and consequently the system $A\mathbf{x} = \mathbf{d}_2$ is inconsistent (see Figure 2.5).

There is another way of looking at spanning which does not involve linear systems. The column space of an $m \times n$ matrix $A$ is the set of all possible linear combinations of the columns of $A$. Each column of $A$ is a vector in $R^m$, and we have shown that the collection of all possible linear combinations of these vectors form a subspace of $R^m$. Forgetting altogether about the matrix $A$, we have proved the following result.

***Theorem 2*** *Let $\mathbf{v}_1, \mathbf{v}_2, \ldots, \mathbf{v}_n$ be vectors in $R^m$. The subset $V$ of $R^m$ consisting of all possible linear combinations of $\mathbf{v}_1, \mathbf{v}_2, \ldots, \mathbf{v}_n$ is a subspace of $R^m$. We call this subspace the* **subspace spanned by** $\mathbf{v}_1, \mathbf{v}_2, \ldots, \mathbf{v}_n$.

We asserted that we have proved this result. How? We simply form the matrix $A$ whose columns are $\mathbf{v}_1, \mathbf{v}_2, \ldots, \mathbf{v}_n$ and observe that $V = C(A)$. In other words, *the subspace spanned by a list of vectors is just the column space of the matrix having the vectors in the list as its columns.*

Notice that if $V$ is the subspace spanned by $\mathbf{v}_1, \mathbf{v}_2, \ldots, \mathbf{v}_n$, then these vectors span $V$. Similarly, if $\mathbf{v}_1, \mathbf{v}_2, \ldots, \mathbf{v}_n$ span $V$, then $V$ is the subspace spanned by the vectors. The difference is simply a matter of emphasis. If we begin with a subspace, then we are interested in a set of vectors in $V$ that span $V$. If we begin with a set of vectors in $R^m$, we are interested in the subspace that these vectors span.

### EXAMPLE 2

Let $\mathbf{v}$ be a nonzero vector in $R^3$. Then the subspace of $R^3$ spanned by $\mathbf{v}$ is the line through the origin containing $\mathbf{v}$. If $\mathbf{v}_1$ and $\mathbf{v}_2$ are vectors in $R^3$ that are not collinear, the subspace of $R^3$ spanned by $\mathbf{v}_1$ and $\mathbf{v}_2$ is the plane through the origin containing $\mathbf{v}_1$ and $\mathbf{v}_2$.

Given a list of vectors $\mathbf{v}_1, \mathbf{v}_2, \ldots, \mathbf{v}_n$ in $R^m$ and another vector $\mathbf{b}$ in $R^m$, we pose the question: Is $\mathbf{b}$ a linear combination of $\mathbf{v}_1, \mathbf{v}_2, \ldots, \mathbf{v}_n$? Letting $A$ be the matrix whose columns are these vectors,

$$A = [\mathbf{v}_1 \quad \mathbf{v}_2 \quad \cdots \quad \mathbf{v}_n],$$

the question is whether $\mathbf{b}$ is in the column space of $A$. Hence $\mathbf{b}$ is a linear combination of the $\mathbf{v}_i$'s if and only if the system $A\mathbf{x} = \mathbf{b}$ has a solution.

### EXAMPLE 3

Let

$$\mathbf{v}_1 = (-1, 3, 1), \qquad \mathbf{v}_2 = (0, 2, 4), \qquad \mathbf{v}_3 = (1, 0, 2), \qquad \text{and} \qquad \mathbf{b} = (-3, 12, 12).$$

Is $\mathbf{b}$ a linear combination of $\mathbf{v}_1, \mathbf{v}_2, \mathbf{v}_3$? From the preceding discussion we answer the question by forming the matrix $A$ whose columns are $\mathbf{v}_1, \mathbf{v}_2$, and $\mathbf{v}_3$ and determining whether the system $A\mathbf{x} = \mathbf{b}$ is consistent. Gaussian elimination reduces the augmented matrix

$$\begin{bmatrix} -1 & 0 & 1 & : & -3 \\ 3 & 2 & 0 & : & 12 \\ 1 & 4 & 2 & : & 12 \end{bmatrix} \quad \text{to} \quad \begin{bmatrix} 1 & 0 & 0 & : & 2 \\ 0 & 1 & 0 & : & 3 \\ 0 & 0 & 1 & : & -1 \end{bmatrix}.$$

The system is indeed consistent and the solution is

$$\mathbf{x} = (2, 3, -1).$$

Thus $\mathbf{b}$ is a linear combination of $\mathbf{v}_1, \mathbf{v}_2$, and $\mathbf{v}_3$:

$$\mathbf{b} = 2\mathbf{v}_1 + 3\mathbf{v}_2 - \mathbf{v}_3.$$

### EXAMPLE 4

Determine whether the vector $\mathbf{b}$ is a linear combination of the vectors $\mathbf{v}_1$ and $\mathbf{v}_2$, where

$$\mathbf{v}_1 = (1, 3, -5), \qquad \mathbf{v}_2 = (2, 9, 2) \qquad \text{and} \qquad \mathbf{b} = (1, 1, 4).$$

Form the matrix $A$ whose columns are $\mathbf{v}_1$ and $\mathbf{v}_2$. The vector $\mathbf{b}$ is a linear combination of $\mathbf{v}_1$ and $\mathbf{v}_2$ if and only if $\mathbf{b}$ is in $C(A)$. Gaussian elimination reduces

$$\begin{bmatrix} 1 & 2 & : & 1 \\ 3 & 9 & : & 1 \\ -5 & 2 & : & 4 \end{bmatrix} \quad \text{to} \quad \begin{bmatrix} 1 & 2 & : & 1 \\ 0 & 3 & : & -2 \\ 0 & 0 & : & 17 \end{bmatrix}.$$

Thus the system $A\mathbf{x} = \mathbf{b}$ is inconsistent and therefore $\mathbf{b}$ is not a linear combination of $\mathbf{v}_1$ and $\mathbf{v}_2$.

Given an $m \times n$ matrix $A$ and a vector $\mathbf{b}$ in $R^m$, the system $A\mathbf{x} = \mathbf{b}$ has a solution if and only if $\mathbf{b}$ lies in $C(A)$. Consequently, the system $A\mathbf{x} = \mathbf{b}$ has a solution *for every vector* $\mathbf{b}$ in $R^m$ if and only if *every vector* $\mathbf{b}$ in $R^m$ lies in $C(A)$, that is, $C(A) = R^m$. In other words, the system $A\mathbf{x} = \mathbf{b}$ has a solution for every $\mathbf{b}$ in $R^m$ if and only if the columns of $A$ span $R^m$. Moreover, Theorem 2 of Section 1.10 states that $A\mathbf{x} = \mathbf{b}$ is consistent for every $\mathbf{b}$ in $R^m$ if and only if the rank of $A$ is $m$. Thus we have proved the following theorem.

***Theorem 3*** *Let $A$ be an $m \times n$ matrix. The following are equivalent.*

(*a*) *The system $A\mathbf{x} = \mathbf{b}$ is consistent for every $\mathbf{b}$ in $R^m$.*
(*b*) *The columns of $A$ span $R^m$, that is, $C(A) = R^m$.*
(*c*) *The rank of $A$ is $m$.*

We proved this theorem by proving that both (b) and (c) are equivalent to (a). It is a good exercise to prove directly that (b) is equivalent to (c) (see Problem 15).

### EXAMPLE 5

An obvious but important example of a set of vectors that span all of $R^m$ is the set

$$\begin{aligned} \mathbf{e}_1 &= (1, 0, 0, \ldots, 0) \\ \mathbf{e}_2 &= (0, 1, 0, \ldots, 0) \\ &\vdots \\ \mathbf{e}_m &= (0, 0, 0, \ldots, 1). \end{aligned}$$

Every vector $\mathbf{b}$ in $R^m$ is a linear combination of these vectors because the corresponding system

$$I\mathbf{x} = \mathbf{b}$$

has a solution, namely $\mathbf{x} = \mathbf{b}$. Thus

$$\mathbf{b} = b_1\mathbf{e}_1 + b_2\mathbf{e}_2 + \cdots + b_m\mathbf{e}_m.$$

### EXAMPLE 6

To determine whether the vectors

$$(1, 0, 0), \quad (2, 2, 4), \quad (-1, 0, 1)$$

span $R^3$, use forward elimination to reduce the matrix

$$A = \begin{bmatrix} 1 & 2 & -1 \\ 0 & 2 & 0 \\ 0 & 4 & 1 \end{bmatrix} \quad \text{to} \quad E = \begin{bmatrix} 1 & 2 & -1 \\ 0 & 2 & 0 \\ 0 & 0 & 1 \end{bmatrix}.$$

Since $E$ has no zero rows, the rank of $A$ is 3 and the vectors span $R^3$.

### EXAMPLE 7

Let $A$ be an $m \times n$ matrix and suppose that the rank of $A$ is $r$.

1. If $r = m$, then $A\mathbf{x} = \mathbf{b}$ has a solution for all $\mathbf{b}$ in $R^m$. That is, $C(A) = R^m$.
2. If $r < m$, then $C(A)$ is a proper subspace of $R^m$; that is, the columns of $A$ do not span $R^m$. Thus there is a vector $\mathbf{b}$ in $R^m$ such that $A\mathbf{x} = \mathbf{b}$ has no solution.
3. If $r < n$, then for any $\mathbf{b}$ in $C(A)$ the system $A\mathbf{x} = \mathbf{b}$ has infinitely many solutions. For if $r$ (the number of pivot columns) is less than $n$ (the number of columns), the system $A\mathbf{x} = \mathbf{b}$ has free variables. From Theorem 6 of Section 1.10 it follows that the system $A\mathbf{x} = \mathbf{b}$ has infinitely many solutions.
4. If $m < n$, then $r \leq m$ since the number of nonzero rows is at most equal to the number of rows. If $r = m$, then by (1) the system $A\mathbf{x} = \mathbf{b}$ is consistent for every $\mathbf{b}$ in $R^m$. And since $m < n$, it follows that $r < n$, and hence by (3) the system has infinitely many solutions for every $\mathbf{b}$ in $R^m$. If $r < m$, then a similar analysis shows that for any vector $\mathbf{b}$ in $R^m$, the system $A\mathbf{x} = \mathbf{b}$ has no solutions or infinitely many solutions.

***Theorem 4*** *$R^m$ cannot be spanned by fewer than $m$ vectors.*

**Proof** Let $\mathbf{v}_1, \mathbf{v}_2, \ldots, \mathbf{v}_n$ be vectors in $R^m$ and form the matrix $A$ having these vectors as its columns. $A$ is an $m \times n$ matrix. If $r$ is the rank of $A$, then obviously $r \leq m$ and $r \leq n$. Therefore, if $n < m$, then $r < m$. Thus by Theorem 3 the columns of $A$ do not span $R^m$.

### EXAMPLE 8

Given a list of five or fewer vectors in $R^6$, Theorem 4 allows us to conclude without any computations at all that the vectors cannot span $R^6$. On the other hand, a list of six or more vectors in $R^6$ will require some further analysis to determine whether the vectors span $R^6$.

We conclude this section by introducing two more subspaces associated with a matrix $A$. The first of these subspaces is related to the question of uniqueness of solutions of a system of equations.

***Theorem 5*** *Let $A$ be an $m \times n$ matrix. The subset of $R^n$ consisting of all solution vectors of the homogeneous system $A\mathbf{x} = \mathbf{0}$ is a subspace of $R^n$. It is called the* **nullspace** *of $A$ and is denoted by $N(A)$.*

**Proof** The zero vector $\mathbf{0}$ is in $N(A)$ since $A\mathbf{0} = \mathbf{0}$. If $\mathbf{u}$ and $\mathbf{v}$ are two vectors in $N(A)$ and $\alpha$ is a scalar, then $\mathbf{u} + \mathbf{v}$ and $\alpha\mathbf{u}$ are also in $N(A)$ because

$$A(\mathbf{u} + \mathbf{v}) = A\mathbf{u} + A\mathbf{v} = \mathbf{0} + \mathbf{0} = \mathbf{0}$$

and

$$A(\alpha\mathbf{u}) = \alpha A\mathbf{u} = \alpha \cdot \mathbf{0} = \mathbf{0}.$$

Thus $N(A)$ is a subspace.

***Theorem 6*** *Let $A$ be a matrix. The vectors obtained as solution vectors of $A\mathbf{x} = \mathbf{0}$ by in turn setting each free variable to* 1 *and all other free variables to* 0 *span the nullspace of $A$.*

**Proof** This is simply a restatement in the language of spanning of Theorem 3 of Section 1.10.

## EXAMPLE 9

Let $A = [1 \quad 2 \quad 1]$. The nullspace of $A$ consists of all solution vectors of $A\mathbf{x} = \mathbf{0}$, that is, $N(A)$ consists of all vectors $\mathbf{x} = (x_1, x_2, x_3)$ in $R^3$ that satisfy

$$x_1 + 2x_2 + x_3 = 0.$$

Since $x_2$ and $x_3$ are free variables, the general solution of this equation is

$$\mathbf{x} = x_2(-2, 1, 0) + x_3(-1, 0, 1).$$

Hence $N(A)$ is the subspace of $R^3$ spanned by $(-2, 1, 0)$ and $(-1, 0, 1)$. Geometrically, it is a plane through the origin in $R^3$.

## EXAMPLE 10

To find the nullspace of

$$A = \begin{bmatrix} 1 & 2 & 1 \\ 1 & 3 & 2 \end{bmatrix},$$

we use Gaussian elimination to reduce $A$ to

$$E = \begin{bmatrix} 1 & 0 & -1 \\ 0 & 1 & 1 \end{bmatrix}.$$

Hence $x_3$ is the free variable and the general solution of $A\mathbf{x} = \mathbf{0}$ is

$$\mathbf{x} = x_3(1, -1, 1).$$

Thus $N(A)$ is the subspace spanned by $(1, -1, 1)$. Geometrically, it is a line in $R^3$ passing through the origin.

## EXAMPLE 11

Earlier in this section we pointed out that there are two general methods for describing a subspace of $R^m$. The first is to impose constraint equations and the second is to specify a spanning set. The first method expresses the subspace as the nullspace of a matrix and the second expresses the subspace as the column space of another matrix. For instance, suppose that $V$ is the subspace of $R^4$ consisting of all vectors in $R^4$ that satisfy the constraints

$$v_1 + v_2 + 2v_3 = 0 \qquad \text{and} \qquad v_1 + v_3 + v_4 = 0.$$

Then clearly $V$ is simply the nullspace of the matrix

$$\begin{bmatrix} 1 & 1 & 2 & 0 \\ 1 & 0 & 1 & 1 \end{bmatrix}.$$

By reducing the matrix to the reduced echelon matrix

$$\begin{bmatrix} 1 & 0 & 1 & 1 \\ 0 & 1 & 1 & -1 \end{bmatrix},$$

we find that the vectors $(-1,-1,1,0)$ and$(-1,1,0,1)$ span the nullspace of the matrix. Thus the subspace can also be expressed as the column space of the matrix

$$\begin{bmatrix} -1 & -1 \\ -1 & 1 \\ 1 & 0 \\ 0 & 1 \end{bmatrix}.$$

On the other hand, suppose that $V$ is the subspace of $R^4$ spanned by the vectors $\mathbf{v}_1 = (-2,3,1,0)$ and $\mathbf{v}_2 = (-1,3,0,1)$. If $A$ is the matrix having these vectors as its columns, then $V = C(A)$. A vector $\mathbf{v}$ is in $V$ if and only if the system $A\mathbf{x} = \mathbf{v}$ is consistent. Gaussian elimination reduces the augmented matrix

$$\begin{bmatrix} -2 & -1 & : & v_1 \\ 3 & 3 & : & v_2 \\ 1 & 0 & : & v_3 \\ 0 & 1 & : & v_4 \end{bmatrix} \quad \text{to} \quad \begin{bmatrix} 1 & 0 & : & v_3 \\ 0 & 1 & : & v_4 \\ 0 & 0 & : & v_2 - 3v_3 - 3v_4 \\ 0 & 0 & : & v_1 + 2v_3 + v_4 \end{bmatrix}.$$

Thus the constraint equations are

$$\begin{aligned} v_1 + 2v_3 + \phantom{3}v_4 &= 0 \\ v_2 - 3v_3 - 3v_4 &= 0, \end{aligned}$$

and $V$ is simply the nullspace of the matrix

$$\begin{bmatrix} 1 & 0 & 2 & 1 \\ 0 & 1 & -3 & -3 \end{bmatrix}.$$

Besides the column space and the nullspace of an $m \times n$ matrix $A$, there is a third subspace associated with $A$. It is the subspace of $R^n$ consisting of all linear combinations of the rows of $A$.

***Definition*** Let $A$ be an $m \times n$ matrix. The subspace of $R^n$ spanned by the rows of $A$ is called the **row space** of $A$. We denote the row space of $A$ by $R(A)$.

The fact that Gaussian elimination transforms a system of equations into an equivalent system of equations is reflected in the fact that row operations do not change the row space of $A$.

**Result 1** If a matrix $B$ is obtained from a matrix $A$ from a sequence of row operations, then $A$ and $B$ have the same row space.

**Proof** Let $A$ be a matrix and let us denote the rows of $A$ by $\mathbf{r}_1, \mathbf{r}_2, \ldots, \mathbf{r}_m$:

$$A = \begin{bmatrix} \text{———} & \mathbf{r}_1 & \text{———} \\ \text{———} & \mathbf{r}_2 & \text{———} \\ & \vdots & \\ \text{———} & \mathbf{r}_m & \text{———} \end{bmatrix}.$$

It should be clear that interchanging two rows of $A$ does not change the row space of $A$. If we multiply the $i$th row of $A$ by a nonzero constant $\beta$, we obtain a matrix $B$ whose rows are

$$\mathbf{r}_1, \mathbf{r}_2, \ldots, \mathbf{r}_{i-1}, \beta\mathbf{r}_i, \mathbf{r}_{i+1}, \ldots, \mathbf{r}_m.$$

Any linear combination of the rows of $A$ is a linear combination of the rows of $B$ because

$$\alpha_1\mathbf{r}_1 + \cdots + \alpha_i\mathbf{r}_i + \cdots + \alpha_m\mathbf{r}_m = \alpha_1\mathbf{r}_1 + \cdots + \left(\frac{\alpha_i}{\beta}\right)(\beta\mathbf{r}_i) + \cdots + \alpha_m\mathbf{r}_m.$$

Similarly, any linear combination of the rows of $B$ is a linear combination of the rows of $A$ because

$$\alpha_1\mathbf{r}_1 + \cdots + \alpha_i(\beta\mathbf{r}_i) + \cdots + \alpha_m\mathbf{r}_m = \alpha_1\mathbf{r}_1 + \cdots + (\alpha_i\beta)\mathbf{r}_i + \cdots + \alpha_m\mathbf{r}_m.$$

Therefore, $A$ and $B$ have the same row space. Finally, suppose that $B$ is obtained from $A$ by adding a multiple of one row of $A$ (say $\beta\mathbf{r}_j$) to another row of $A$ (say $\mathbf{r}_i$). Then the rows of $B$ are

$$\mathbf{r}_1, \ldots, \mathbf{r}_i + \beta\mathbf{r}_j, \ldots, \mathbf{r}_j, \ldots, \mathbf{r}_m.$$

The equations

$$\begin{aligned}\alpha_1\mathbf{r}_1 &+ \cdots + \alpha_i\mathbf{r}_i + \cdots + \alpha_j\mathbf{r}_j + \cdots + \alpha_m\mathbf{r}_m \\ &= \alpha_1\mathbf{r}_1 + \cdots + \alpha_i(\mathbf{r}_i + \beta\mathbf{r}_j) + \cdots + (\alpha_j - \alpha_i\beta)\mathbf{r}_j + \cdots + \alpha_m\mathbf{r}_m\end{aligned}$$

and

$$\begin{aligned}\alpha_1\mathbf{r}_1 &+ \cdots + \alpha_i(\mathbf{r}_i + \beta\mathbf{r}_j) + \cdots + \alpha_j\mathbf{r}_j + \cdots + \alpha_m\mathbf{r}_m \\ &= \alpha_1\mathbf{r}_1 + \cdots + \alpha_i\mathbf{r}_i + \cdots + (\alpha_i\beta + \alpha_j)\mathbf{r}_j + \cdots + \alpha_m\mathbf{r}_m\end{aligned}$$

show that any vector that is a linear combination of the rows of $A$ is also a linear combination of the rows of $B$, and vice versa. Thus $R(A) = R(B)$. Since performing any single row operation on a matrix does not change its row space, performing any sequence of row operations does not change the row space. The result is therefore proved.

In particular, if we reduce $A$ to an echelon matrix $E$, $R(A) = R(E)$. Since the nonzero rows of $E$ clearly span $R(E)$, they also span $R(A)$. This proves the following theorem.

***Theorem 7*** *If $A$ is reduced to an echelon matrix $E$ by row operations, then $R(A) = R(E)$ and the nonzero rows of $E$ span $R(A)$.*

For example, reducing

$$A = \begin{bmatrix} 1 & 2 & -1 \\ 2 & -2 & 3 \\ 7 & 2 & 3 \end{bmatrix} \quad \text{to} \quad E = \begin{bmatrix} 1 & 2 & -1 \\ 0 & -6 & 5 \\ 0 & 0 & 0 \end{bmatrix},$$

we find that the row space of $A$ is spanned by the vectors $(1, 2, -1)$ and $(0, -6, 5)$.

## PROBLEMS 2.2

**1.** Complete each of the following definitions.

(a) The column space of a matrix $A$ is....

(b) The nullspace of a matrix $A$ is....

(c) The row space of a matrix $A$ is....

(d) Vectors $\mathbf{v}_1, \mathbf{v}_2, \ldots, \mathbf{v}_k$ span a subspace $V$ if....

(e) The subspace spanned by $\mathbf{v}_1, \mathbf{v}_2, \ldots, \mathbf{v}_k$ is....

**2.** In each part, determine whether the vector $\mathbf{b}$ is a linear combination of the $\mathbf{v}_i$'s.

*(a) $\mathbf{b} = (3, -5)$, $\mathbf{v}_1 = (1, 2)$, $\mathbf{v}_2 = (-2, 6)$

(b) $\mathbf{b} = (1, 1, 1)$, $\mathbf{v}_1 = (1, 0, 1)$, $\mathbf{v}_2 = (0, 3, 5)$

*(c) $\mathbf{b} = (2, 1, 3)$, $\mathbf{v}_1 = (2, 0, 1)$, $\mathbf{v}_2 = (1, 3, -1)$

(d) $\mathbf{b} = (2, 4, 6)$, $\mathbf{v}_1 = (1, 1, 0)$, $\mathbf{v}_2 = (2, 2, 1)$, $\mathbf{v}_3 = (0, 0, 2)$, $\mathbf{v}_4 = (1, 2, 1)$

*(e) $\mathbf{b} = (2, -2, \frac{1}{6}, \frac{1}{6})$, $\mathbf{v}_1 = (1, -1, 0, 0)$, $\mathbf{v}_2 = (2, 0, 1, 1)$, $\mathbf{v}_3 = (0, 3, 1, 1)$

(f) $\mathbf{b} = (0, 1, 0, 1, 0)$, $\mathbf{v}_1 = (1, 2, 2, 1, 1)$, $\mathbf{v}_2 = (\frac{2}{3}, 1, \frac{4}{3}, 1, \frac{2}{3})$

**3.** Which of the following sets of vectors span $R^3$? If they span, show that they do. If not, find an explicit vector in $R^3$ that is not a linear combination of the given vectors.

*(a) $(2, 0, 0), (1, 1, 0), (-1, 3, 4)$

(b) $(-3, 0, 0), (1, 0, 1), (2, 0, -3)$

*(c) $(5, 0, 0), (-1, 2, 0)$

(d) $(1, 0, 2), (1, 1, 1), (0, 2, 1), (1, 1, 2)$

*(e) $(1, 2, 3), (1, 3, 5), (2, -2, 1)$

(f) $(2, 4, 8), (-2, 1, 1), (10, 5, 13), (-2, 1, 3)$

**4.** Which of the following sets of vectors span $R^4$? If they span, show that they do. If they do not span, find an explicit vector in $R^4$ that is not a linear combination of the given vectors.

*(a) $(1, 0, 0, 0), (0, 3, 0, 0), (2, 1, 2, 0), (1, 1, 1, 0)$

(b) $(-2, 0, 0, 0), (1, 1, 0, 0), (-1, 2, 1, 1)$

*(c) $(1, 0, 0, 0), (1, 1, 0, 0), (1, 1, 1, 0), (1, 1, 1, 1)$

(d) $(0, 1, 1, 0), (1, 1, 1, 1), (-1, 1, -1, 1), (1, 2, 3, 4)$

*(e) $(1, 0, 3, -1), (0, 1, 1, 1), (3, -2, 5, -2), (1, -1, 0, 1)$

(f) $(1, 0, -7, 2), (2, 0, 0, 0), (3, 3, 5, -1), (4, 0, 6, 0)$

**5.** For each of the following, describe the subspace spanned by the given vectors by finding constraints that a vector must satisfy to be in the subspace.

*(a) $(1, 0, 0), (1, 0, 1), (1, 1, 1), (0, 2, 4)$

(b) $(1, 0, 0, 1), (0, 0, 1, 0)$

*(c) $(-1, 2, 0, 2), (2, -4, 1, 4), (-1, 3, 1, 1)$

(d) $(1, 2, 2, 2), (0, 1, 3, 4), (1, 2, 3, 2)$

*(e) $(1, 4, -5, 2, 1), (2, -3, 1, 4, -15)$

(f) $(1, 2, 3, 4, 5), (2, 4, 6, 8, 9), (2, 3, 4, 5, 6)$

**6.** For each of the following matrices, describe its column space by finding the constraints a vector $\mathbf{b}$ must satisfy in order to be in the column space.

*(a) $\begin{bmatrix} 3 & -1 \\ 9 & -3 \\ 6 & -2 \end{bmatrix}$

(b) $\begin{bmatrix} 1 & 2 & 0 \\ 3 & 1 & 1 \\ 5 & 5 & 1 \end{bmatrix}$

*(c) $\begin{bmatrix} 1 & 2 & 0 \\ 1 & 1 & 2 \\ 0 & 2 & 3 \end{bmatrix}$

(d) $\begin{bmatrix} 0 & 0 & 0 \\ 0 & 0 & 0 \\ 0 & 0 & 0 \end{bmatrix}$

*(e) $\begin{bmatrix} 1 & 1 & 1 \\ -1 & 1 & 2 \\ 1 & 3 & -5 \end{bmatrix}$

(f) $\begin{bmatrix} 1 & 2 & -1 & 2 \\ 2 & 4 & 1 & 7 \\ 3 & 6 & -2 & 7 \end{bmatrix}$

*(g) $\begin{bmatrix} 1 & -1 & 2 & 1 \\ -1 & 2 & -3 & 1 \\ -1 & 3 & -3 & 1 \\ 2 & 1 & 2 & 6 \end{bmatrix}$

(h) $\begin{bmatrix} 1 & 1 & 1 & 1 \\ 0 & 1 & 0 & 1 \\ 1 & 2 & 1 & 2 \\ 1 & 0 & 1 & 0 \end{bmatrix}$

*(i) $\begin{bmatrix} 1 \\ 2 \\ 3 \\ -5 \\ -9 \end{bmatrix}$

**7.** For each of the following matrices, find a set of vectors that span the nullspace of the matrix.

*(a) $\begin{bmatrix} 2 & -3 \\ 4 & -6 \end{bmatrix}$

(b) $\begin{bmatrix} 1 & 1 & -1 & 3 \\ 2 & 2 & 1 & 0 \end{bmatrix}$

*(c) $\begin{bmatrix} 1 & 2 & 0 & 1 & 1 \\ 3 & 6 & 1 & 2 & 1 \end{bmatrix}$

(d) $\begin{bmatrix} 0 & 1 & 0 & 1 & 0 \\ 1 & 2 & 0 & 1 & 0 \\ 0 & 1 & 0 & 1 & 2 \end{bmatrix}$

*(e) $\begin{bmatrix} 0 & 0 & 0 & 0 \\ 0 & 0 & 0 & 0 \\ 0 & 0 & 0 & 0 \end{bmatrix}$

(f) $\begin{bmatrix} 0 & 1 & 4 & 0 & 0 \\ 0 & 3 & 12 & 0 & 1 \\ 0 & 1 & 0 & 0 & 1 \end{bmatrix}$

*(g) $\begin{bmatrix} 0 & 1 & 2 & 3 & 4 \\ 1 & 2 & 3 & 4 & 5 \\ 2 & 3 & 4 & 5 & 6 \end{bmatrix}$

(h) $\begin{bmatrix} 1 & 2 & 3 & 4 & 5 \end{bmatrix}$

*(i) $\begin{bmatrix} 1 \\ 2 \\ 3 \\ 4 \end{bmatrix}$

***8.** For each matrix $A$ in Problem 6, find a matrix $B$ such that $N(B) = C(A)$.

***9.** For each matrix $A$ in Problem 7, find a matrix $B$ such that $C(B) = N(A)$.

**10.** For each of the following, determine whether the two lists of vectors span the same subspace.

*(a) $(1, 0, -1), (0, 1, 1)$ and $(1, 1, 0)$, $(4, 1, -3), (-2, 3, 5)$

(b) $(0, 1, 1), (-1, -3, 4)$ and $(1, -1, 2)$, $(2, 0, 1), (4, -2, 5)$

*(c) $(-3, 1, 1, 1), (-4, 1, 1, 1)$, $(-3, 0, -1, 1)$ and $(-3, 1, 0, 0)$, $(-3, 0, 1, 2), (1, 0, 1, 0)$

(d) $(0, 1, 1, 1), (-1, -1, 0, 7)$, $(1, 28, 21, 14)$ and $(1, 0, 0, 7)$, $(2, 1, 0, 14), (3, 1, 1, 21)$

**11.** Show that the following subspaces are equal.

*(a) $R(A)$ and $C(A^T)$

(b) $C(A)$ and $R(A^T)$

**12.** Let $A$ be an $m \times n$ matrix. Show that if

$m < n$, then the nullspace of $A$ is a nontrivial subspace of $R^n$.

*13. Find all $m \times n$ matrices whose nullspace is $R^n$.

14. Suppose that an $m \times n$ matrix $A$ is reduced to an echelon matrix $E$.
(a) Explain why $N(A) = N(E)$.
(b) Explain why $C(A) \neq C(E)$.

15. Prove directly that conditions (b) and (c) of Theorem 3 are equivalent.

16. Let $A$ be an $n \times n$ upper triangular matrix. Explain why $R(A) = C(A) = R^n$ if the diagonal entries of $A$ are nonzero. What happens if some diagonal entry is zero?

*17. Let $\mathbf{u}_1$, $\mathbf{u}_2$, and $\mathbf{u}_3$ be vectors in $R^m$. Show that if $\mathbf{u}_3$ is a linear combination of $\mathbf{u}_1$ and $\mathbf{u}_2$, then $\mathbf{u}_1$, $\mathbf{u}_2$ span the same subspace as $\mathbf{u}_1$, $\mathbf{u}_2$, and $\mathbf{u}_3$.

18. Suppose that the vector $\mathbf{v}$ is a linear combination of the vectors $\mathbf{v}_1, \mathbf{v}_2, \ldots, \mathbf{v}_n$. Explain why the subspaces spanned by $\mathbf{v}_1, \mathbf{v}_2, \ldots, \mathbf{v}_n$ and $\mathbf{v}_1, \mathbf{v}_2, \ldots, \mathbf{v}_n, \mathbf{v}$ are identical.

19. Let $\mathbf{u}$, $\mathbf{v}$, and $\mathbf{w}$ be vectors in $R^m$ and suppose that there are scalars $\alpha$, $\beta$, and $\gamma$ such that

$$\alpha\mathbf{u} + \beta\mathbf{v} + \gamma\mathbf{w} = \mathbf{0} \qquad \text{and} \qquad \alpha, \beta \neq 0.$$

Show that $\mathbf{u}$ and $\mathbf{w}$ span the same subspace as $\mathbf{v}$ and $\mathbf{w}$.

*20. Let $V$ be the subspace of $R^m$ spanned by the vectors $\mathbf{v}_1$, $\mathbf{v}_2$, and $\mathbf{v}_3$. Show that the vectors $\mathbf{u}_1 = \mathbf{v}_1$, $\mathbf{u}_2 = \mathbf{v}_1 + \mathbf{v}_2$, and $\mathbf{u}_3 = \mathbf{v}_1 + \mathbf{v}_2 + \mathbf{v}_3$ also span $V$.

21. Under what conditions do the planes

$$\mathbf{x} = \alpha\mathbf{u} + \beta\mathbf{v} \quad \text{and} \quad \mathbf{x} = \alpha'\mathbf{u}' + \beta'\mathbf{v}'$$

define the same subspace?

*22. Let $\mathbf{v}_1, \mathbf{v}_2, \ldots, \mathbf{v}_n$ be vectors in $R^m$, and let $V$ be the subspace spanned by these vectors. Show that if $W$ is any subspace that contains $\mathbf{v}_1, \mathbf{v}_2, \ldots, \mathbf{v}_n$, then $W$ contains the whole subspace $V$. Thus the subspace spanned by a set of vectors $\mathbf{v}_1, \mathbf{v}_2, \ldots, \mathbf{v}_n$ is the smallest subspace containing $\mathbf{v}_1, \mathbf{v}_2, \ldots, \mathbf{v}_n$.

23. Suppose that the vectors $\mathbf{u}_1, \mathbf{u}_2, \ldots, \mathbf{u}_m$ are linear combinations of the vectors $\mathbf{v}_1, \mathbf{v}_2, \ldots, \mathbf{v}_n$. Explain why the subspace $U$ spanned by the $\mathbf{u}_i$'s is a subspace of the subspace $V$ spanned by the $\mathbf{v}_i$'s.

24. Let $A$ be an $m \times n$ matrix and $B$ be an $n \times p$ matrix.
*(a) Show that the nullspace of $B$ is a subspace of the nullspace of $AB$.
*(b) Show that if $m = n$ and $A$ is nonsingular, then $N(B) = N(AB)$.
(c) Show that $C(AB)$ is a subspace of $C(A)$.
(d) Show that if $n = p$ and $B$ is nonsingular, then $C(AB) = C(A)$.

25. Let $A$ be an $m \times n$ matrix.
(a) Show that the set of all vectors $\mathbf{x}$ in $R^m$ such that

$$\mathbf{x}^T A = \mathbf{0}$$

form a subspace of $R^m$. For obvious reasons it is called the **left-nullspace** of $A$.
(b) Show that the left-nullspace of $A$ is equal to the nullspace of $A^T$.

26. (a) Show that a vector $\mathbf{b}$ is in $R(A)$ if and only if there is a vector $\mathbf{x}$ such that $\mathbf{x}^T A = \mathbf{b}^T$.
(b) Use part (a) to show that $R(BC)$ is a subspace of $R(C)$, where $B$ is an $m \times n$ matrix and $C$ is an $n \times p$ matrix.
(c) If $m = n$ and $B$ is nonsingular, prove that $R(BC) = R(C)$.

27. Let $A$ and $B$ be two $m \times n$ matrices. For each of the following statements, either prove that the statement is true or give an example to show that it is false.
*(a) If $R(A) = R(B)$, then $N(A) = N(B)$.
(b) If $R(A) = R(B)$, then $C(A) = C(B)$.
*(c) If $N(A) = N(B)$, then $R(A) = R(B)$.
(d) If $N(A) = N(B)$, then $C(A) = C(B)$.
*(e) If $C(A) = C(B)$, then $R(A) = R(B)$.
(f) If $C(A) = C(B)$, then $N(A) = N(B)$.

**28.** (a) If $U$ is the subspace of $R^5$ consisting of all vectors whose components satisfy $x_1 + x_2 + x_4 = 0$ and $x_2 + x_3 - x_4 = 0$ and $V$ is the subspace of $R^5$ consisting of all vectors whose components satisfy $x_1 + x_2 + x_3 + x_4 + x_5 = 0$, show that $U \cap V = N(A)$, where

$$A = \begin{bmatrix} 1 & 1 & 0 & 1 & 0 \\ 0 & 1 & 1 & -1 & 0 \\ 1 & 1 & 1 & 1 & 1 \end{bmatrix}.$$

(b) State and prove a generalization of part (a).

**29.** If $U$ is the subspace of $R^m$ spanned by $\mathbf{u}_1, \mathbf{u}_2, \ldots \mathbf{u}_r$ and $V$ is the subspace of $R^m$ spanned by $\mathbf{v}_1, \mathbf{v}_2, \ldots, \mathbf{v}_s$, show that $U + V$ is spanned by $\mathbf{u}_1, \mathbf{u}_2, \ldots, \mathbf{u}_r, \mathbf{v}_1, \mathbf{v}_2, \ldots, \mathbf{v}_s$.

**30.** (a) Show that if $x_i$ is a free variable, then the solution to the homogeneous equation $A\mathbf{x} = \mathbf{0}$ having $x_i = 1$ and all other free variables equal to 0 enables us to express the $i$th column of $A$ as a linear combination of the pivot columns of $A$.

(b) Show that each nonpivot column of $A$ is a linear combination of the previous pivot columns of $A$.

(c) Use part (b) to give another proof that the pivot columns of $A$ span $C(A)$.

***31.** For each of the matrices in Problem 6, express each nonpivot column as a linear combination of the previous pivot columns using the method of Problem 30.

**32.** Show that the set of solutions of a nonhomogeneous system is an affine subspace.

**33.** Let $A$ be an $m \times m$ idempotent matrix (i.e., $A^2 = A$).

*(a) Show that $C(A)$ consists of all vectors $\mathbf{x}$ such that $A\mathbf{x} = \mathbf{x}$.

(b) Show that $N(A)$ consists of all vectors of the form $\mathbf{x} - A\mathbf{x}$.

(c) Show that $C(A) \cap N(A) = \{\mathbf{0}\}$.

(d) Show that $R^m = C(A) + N(A)$.

# 2.3 Linear Independence

The column space provides us with a geometric characterization of the algebraic question of existence of a solution to a linear system. Given an $m \times n$ matrix $A$, the linear system $A\mathbf{x} = \mathbf{b}$ is consistent if and only if $\mathbf{b}$ lies in the column space of $A$, that is, if and only if $\mathbf{b}$ lies in the subspace spanned by the columns of $A$. The linear system $A\mathbf{x} = \mathbf{b}$ is consistent for every $\mathbf{b}$ in $R^m$ if and only if $C(A) = R^m$, that is, if and only if the columns of $A$ span $R^m$. *Thus existence of solutions corresponds to spanning.*

We turn now to the question of uniqueness. We assume that we are given a consistent system $A\mathbf{x} = \mathbf{b}$ and ask for a geometric condition that the solution be unique. By Theorem 6 of Section Section 1.10 we know that for a given matrix $A$, uniqueness of the solution to every consistent system $A\mathbf{x} = \mathbf{b}$ is equivalent to uniqueness of the solution to the single homogeneous system $A\mathbf{x} = \mathbf{0}$. That is, $A\mathbf{x} = \mathbf{b}$ has a unique solution for every vector $\mathbf{b}$ in $C(A)$ if and only if $A\mathbf{x} = \mathbf{0}$ has no nontrivial solutions. Thus a geometric condition for uniqueness of solutions is that $N(A) = \{\mathbf{0}\}$.

***Theorem 1*** *Let $A$ be an $m \times n$ matrix. The system $A\mathbf{x} = \mathbf{b}$ has a unique solution for every $\mathbf{b}$ in $C(A)$ if and only if the nullspace of $A$ is the trivial subspace*: $N(A) = \{\mathbf{0}\}$.

Consider now the homogeneous system $A\mathbf{x} = \mathbf{0}$ from the point of view that $A\mathbf{x}$ is a linear combination of the columns of $A$ whose coefficients are the components of $\mathbf{x}$. Let $A$ be an $m \times n$ matrix and let $\mathbf{v}_1, \mathbf{v}_2, \ldots, \mathbf{v}_n$ denote the columns of $A$:

$$A = [\mathbf{v}_1 \quad \mathbf{v}_2 \quad \cdots \quad \mathbf{v}_n].$$

The homogeneous system $A\mathbf{x} = \mathbf{0}$ becomes

$$x_1\mathbf{v}_1 + x_2\mathbf{v}_2 + \cdots + x_n\mathbf{v}_n = \mathbf{0}.$$

Thus $N(A) = \{\mathbf{0}\}$ if and only if the only way to express $\mathbf{0}$ as a linear combination of the columns of $A$ is with all coefficients equal to zero. Vectors that have this property are called independent.

***Definition*** Let $\mathbf{v}_1, \mathbf{v}_2, \ldots, \mathbf{v}_n$ be vectors in $R^m$.

(a) These vectors are called **linearly independent** (or simply **independent**) if the *only* solution of the equation

$$\alpha_1\mathbf{v}_1 + \alpha_2\mathbf{v}_2 + \cdots + \alpha_n\mathbf{v}_n = \mathbf{0}$$

is the trivial solution.

(b) These vectors are called **linearly dependent** (or simply **dependent**) if they are not independent. That is, $\mathbf{v}_1, \mathbf{v}_2, \ldots, \mathbf{v}_n$ are dependent if there are scalars $\alpha_1, \alpha_2, \ldots, \alpha_n$ *not all zero* such that

$$\alpha_1\mathbf{v}_1 + \alpha_2\mathbf{v}_2 + \cdots + \alpha_n\mathbf{v}_n = \mathbf{0}.$$

In this case, we say that $\mathbf{0}$ is a **nontrivial linear combination** of the vectors $\mathbf{v}_1, \mathbf{v}_2, \ldots, \mathbf{v}_n$.

From our previous discussion we have the following result.

***Theorem 2*** *Let $A$ be an $m \times n$ matrix.*

*(a) The columns of $A$ are independent if and only if* $N(A) = \{\mathbf{0}\}$.
*(b) The columns of $A$ are dependent if and only if* $N(A) \neq \{\mathbf{0}\}$.

## EXAMPLE 1

Let $A$ be the matrix

$$A = \begin{bmatrix} 1 & -1 & 1 \\ 3 & -2 & -1 \\ 2 & 1 & -2 \\ 4 & 1 & 3 \end{bmatrix}$$

To determine whether the columns of $A$ are independent, reduce $A$ to the echelon matrix

$$E = \begin{bmatrix} 1 & -1 & 1 \\ 0 & 1 & -4 \\ 0 & 0 & 8 \\ 0 & 0 & 0 \end{bmatrix}$$

Since every column of $E$ is a pivot column, $N(A) = \{\mathbf{0}\}$. By Theorem 2 the columns of $A$ are independent.

On the other hand, reducing the matrix

$$A = \begin{bmatrix} 1 & 1 & -1 & 1 \\ 1 & 0 & 1 & 1 \\ 0 & 1 & -2 & -1 \end{bmatrix}$$

to the echelon matrix

$$E = \begin{bmatrix} 1 & 1 & -1 & 1 \\ 0 & -1 & 2 & 0 \\ 0 & 0 & 0 & -1 \end{bmatrix}$$

we find that the third column of $E$ is not a pivot column. Thus $N(A)$ is not the trivial subspace and we conclude from Theorem 2 that the columns of $A$ are dependent. Any nonzero vector in $N(A)$ allows us to express $\mathbf{0}$ as a nontrivial linear combination of the columns of $A$. For example, $(-1, 2, 1, 0)$ is a nonzero vector in $N(A)$ and

$$(-1)(1,1,0) + 2\cdot(1,0,1) + 1\cdot(-1,1,-2) + 0\cdot(1,1,-1) = (0,0,0).$$

Given a list of vectors $\mathbf{v}_1, \mathbf{v}_2, \ldots, \mathbf{v}_n$ in $R^m$, how do we determine whether they are independent or dependent? Obviously, we form the matrix $A$ having these vectors as its columns and find the nullspace of $A$. If it is the trivial subspace, the vectors are independent; if it is a nontrivial subspace, the vectors are dependent.

## EXAMPLE 2

Are the vectors $\mathbf{v}_1 = (1, -2)$ and $\mathbf{v}_2 = (3, 1)$ independent or dependent? To answer the question, we form the matrix $A$ having these vectors as its columns and reduce $A$ to an echelon matrix $E$. We obtain

$$A = \begin{bmatrix} 1 & 3 \\ -2 & 1 \end{bmatrix}, \qquad E = \begin{bmatrix} 1 & 3 \\ 0 & 7 \end{bmatrix}.$$

Since every column of $E$ is a pivot column, $N(A) = \{\mathbf{0}\}$. Thus the vectors are independent.

### EXAMPLE 3

To determine whether the vectors $\mathbf{v}_1 = (2, -2, 4)$, $\mathbf{v}_2 = (3, 0, 1)$, and $\mathbf{v}_3 = (-1, 1, -2)$ are independent or dependent, reduce the matrix

$$A = \begin{bmatrix} 2 & 3 & -1 \\ -2 & 0 & 1 \\ 4 & 1 & -2 \end{bmatrix} \quad \text{to} \quad E = \begin{bmatrix} 2 & 3 & -1 \\ 0 & 3 & 0 \\ 0 & 0 & 0 \end{bmatrix}.$$

Since the third column of $E$ is not a pivot column, $N(A) \neq \{\mathbf{0}\}$ and the vectors are dependent.

### EXAMPLE 4

The vectors $\mathbf{e}_1, \mathbf{e}_2, \ldots, \mathbf{e}_n$ are independent in $R^m$ *for* $n \leq m$, for if $A$ is the matrix having these vectors as its columns, $A$ is an $m \times n$ reduced echelon matrix with every column a pivot column. Therefore, $N(A) = \{\mathbf{0}\}$ and the vectors are independent.

A system $A\mathbf{x} = \mathbf{b}$ is consistent if and only if $\mathbf{b}$ is in the subspace spanned by the columns of $A$. In this sense existence corresponds to spanning. In the following theorem we prove that a consistent system $A\mathbf{x} = \mathbf{b}$ has a unique solution if and only if the columns of $A$ are independent. *Hence uniqueness corresponds to independence.*

***Theorem 3*** *Let $A$ be an $m \times n$ matrix. The following are equivalent.*

(*a*) *The system $A\mathbf{x} = \mathbf{b}$ has a unique solution for every $\mathbf{b}$ in $C(A)$.*
(*b*) *The columns of $A$ are independent; that is, $N(A) = \{\mathbf{0}\}$.*
(*c*) *The rank of $A$ is $n$.*

**Proof** The proof follows immediately from Theorem 6 of Section 1.10.

Let $\mathbf{v}_1, \mathbf{v}_2, \ldots, \mathbf{v}_n$ be independent vectors in $R^m$ and suppose that a vector $\mathbf{v}$ is a linear combination of these vectors. Then $\mathbf{v}$ is in the column space of the $m \times n$ matrix having $\mathbf{v}_1, \mathbf{v}_2, \ldots, \mathbf{v}_n$ as its columns. By Theorem 3, $\mathbf{v}$ has a unique expression as a linear combination of $\mathbf{v}_1, \mathbf{v}_2, \ldots, \mathbf{v}_n$. Thus any vector that can be expressed as a linear combination of an independent set of vectors can be so expressed in *one and only one way*. In other words, every vector in the subspace spanned by independent set of vectors can be expressed as a linear combination of these vectors in one and only one way. *Independence corresponds to uniqueness.*

Since independence corresponds to uniqueness, dependence corresponds to nonuniqueness. From this point of view Theorem 3 can be stated as follows.

***Theorem 3′*** *Let $A$ be an $m \times n$ matrix. The following are equivalent.*

(*a*) *For every $\mathbf{b}$ in $C(A)$, the system $A\mathbf{x} = \mathbf{b}$ has infinitely many solutions.*
(*b*) *The columns of $A$ are dependent.*
(*c*) *The rank of $A$ is less than $n$.*

Suppose that $\mathbf{v}_1, \mathbf{v}_2, \ldots, \mathbf{v}_n$ are dependent vectors in $R^m$ and that a vector $\mathbf{v}$ in $R^m$ is a linear combination of $\mathbf{v}_1, \mathbf{v}_2, \ldots, \mathbf{v}_n$. Form the matrix $A$ having $\mathbf{v}_1, \mathbf{v}_2, \ldots, \mathbf{v}_n$ as its columns: $A = [\mathbf{v}_1\ \mathbf{v}_2 \ldots \mathbf{v}_n]$. Then $\mathbf{v}$ is in $C(A)$ and the columns of $A$ are dependent. By Theorem 3′ the system $A\mathbf{x} = \mathbf{v}$ has infinitely many solutions. Thus any vector that can be expressed as a linear combination of a dependent set of vectors can be expressed in an infinite number of ways as a linear combination of the dependent vectors. In other words, every vector in the subspace spanned by a dependent set of vectors $\mathbf{v}_1, \mathbf{v}_2, \ldots, \mathbf{v}_n$ can be expressed as a linear combination of these dependent vectors in an infinite number of ways.

### EXAMPLE 5

Let $A$ be an $m \times n$ matrix and suppose that the rank of $A$ is $r$.

1. If $r = n$, then for every vector $\mathbf{b}$ in $R^m$ the system $A\mathbf{x} = \mathbf{b}$ has *at most* one solution. If $\mathbf{b}$ is in $C(A)$, then it has exactly one solution; if $\mathbf{b}$ is not in $C(A)$, then it has no solution.
2. If $r < n$, then for every vector $\mathbf{b}$ in $R^m$ the system $A\mathbf{x} = \mathbf{b}$ has no solution or it has infinitely many solutions. If $\mathbf{b}$ is in $C(A)$, then it has infinitely many solutions; if $\mathbf{b}$ is not in $C(A)$, then of course it has no solution.
3. If $m < n$, then since $r \le m$ it follows that $r \le m < n$. Thus by (2) given a vector $\mathbf{b}$ in $R^m$, the system $A\mathbf{x} = \mathbf{b}$ has no solution or infinitely many solutions.

***Theorem 4*** *$R^m$ cannot contain more than m independent vectors.*

**Proof** Let $\mathbf{v}_1, \mathbf{v}_2, \ldots, \mathbf{v}_n$ be vectors in $R^m$ and form the matrix $A$ having these vectors as its columns. $A$ is an $m \times n$ matrix. If $r$ is the rank of $A$, then obviously $r \le m$ and $r \le n$. Therefore, if $m < n$, it follows that $r < n$. Thus by Theorem 3′ the columns of $A$ are dependent.

### EXAMPLE 6

Given any list of five or more vectors in $R^4$, we know without any computations at all that the vectors are dependent. On the other hand, a list of at most four vectors in $R^4$ will require further analysis to determine whether the vectors in the list are independent.

By definition, the vectors $\mathbf{v}_1, \mathbf{v}_2, \ldots, \mathbf{v}_n$ in $R^m$ are independent if the *only* solution of the equation

$$\alpha_1\mathbf{v}_1 + \alpha_2\mathbf{v}_2 + \cdots + \alpha_n\mathbf{v}_n = \mathbf{0}$$

is the trivial solution $\alpha_1 = \alpha_2 = \cdots = \alpha_n = 0$. Consequently, the vectors $\mathbf{v}_1, \mathbf{v}_2, \ldots, \mathbf{v}_n$ in $R^m$ are independent if whenever $\alpha_1, \alpha_2, \ldots, \alpha_n$ are scalars such that

$$\alpha_1\mathbf{v}_1 + \alpha_2\mathbf{v}_2 + \cdots + \alpha_n\mathbf{v}_n = \mathbf{0},$$

it follows that $\alpha_1 = \alpha_2 = \cdots = \alpha_n = 0$. Thus one way to show that a collection of vectors $\mathbf{v}_1, \mathbf{v}_2, \ldots, \mathbf{v}_n$ is independent is to proceed as follows:

1. Suppose $\alpha_1, \alpha_2, \ldots, \alpha_n$ are scalars such that
$$\alpha_1\mathbf{v}_1 + \alpha_2\mathbf{v}_2 + \cdots + \alpha_n\mathbf{v}_n = \mathbf{0}.$$
2. Use whatever information is known about the $\mathbf{v}_i$'s to show that the $\alpha_i$'s must be zero.

For example, let us prove that a single nonzero vector is independent. Let $\mathbf{v}$ be a nonzero vector in $R^m$ and suppose that $\alpha$ is a scalar such that

$$\alpha\mathbf{v} = \mathbf{0}.$$

Since $\mathbf{v} \neq \mathbf{0}$ we conclude that $\alpha = 0$. Thus $\mathbf{v}$ is independent.

As another example, let us prove that two noncollinear vectors in $R^m$ are independent. Let $\mathbf{v}_1$ and $\mathbf{v}_2$ be noncollinear vectors in $R^m$. Suppose that there are scalars $\alpha_1$ and $\alpha_2$ such that

$$\alpha_1\mathbf{v}_1 + \alpha_1\mathbf{v}_2 = \mathbf{0}.$$

If $\alpha_1 \neq 0$, then we can solve this equation for $\mathbf{v}_1$ to obtain

$$\mathbf{v}_1 = -\frac{\alpha_2}{\alpha_1}\mathbf{v}_2.$$

This is impossible since $\mathbf{v}_1$ and $\mathbf{v}_2$ are not collinear. Therefore, $\alpha_1 = 0$. Thus the equation $\alpha_1\mathbf{v}_1 + \alpha_2\mathbf{v}_2 = \mathbf{0}$ reduces to

$$\alpha_2\mathbf{v}_2 = \mathbf{0}.$$

Since $\mathbf{v}_1$ and $\mathbf{v}_2$ are not collinear, $\mathbf{v}_2 \neq 0$. Thus $\alpha_2 = 0$ also. We conclude that $\mathbf{v}_1$ and $\mathbf{v}_2$ are independent.

By definition, vectors which are not independent are dependent. Given a collection $\mathbf{v}_1, \mathbf{v}_2, \ldots, \mathbf{v}_n$ of vectors in $R^m$, it follows that the vectors are dependent if there are scalars $\alpha_1, \alpha_2, \ldots, \alpha_n$ *not all zero* such that

$$\alpha_1\mathbf{v}_1 + \alpha_2\mathbf{v}_2 + \cdots + \alpha_n\mathbf{v}_n = \mathbf{0}.$$

Thus to prove that a collection of vectors $\mathbf{v}_1, \mathbf{v}_2, \ldots, \mathbf{v}_n$ is dependent, we must prove that $\mathbf{0}$ can be expressed as a nontrivial linear combination of these vectors.

For example, we prove that the zero vector is dependent. Since

$$1 \cdot \mathbf{0} = \mathbf{0}$$

and $1 \neq 0$, the zero vector is indeed dependent.

More generally, *any set of vectors that contains the zero vector is dependent*. To prove this, suppose that $\mathbf{v}_1, \mathbf{v}_2, \ldots, \mathbf{v}_n$ are vectors in $R^m$ and that one of these vectors, say $\mathbf{v}_i$, is the zero vector. Then

$$0\mathbf{v}_1 + \cdots + 0\mathbf{v}_{i-1} + 1\mathbf{v}_i + 0\mathbf{v}_{i+1} + \cdots + 0\mathbf{v}_n = \mathbf{0}.$$

Thus $\mathbf{0}$ is a nontrivial linear combination of the vectors $\mathbf{v}_1, \mathbf{v}_2, \ldots, \mathbf{v}_n$ and so they are dependent.

As another example, let us prove that two collinear vectors in $R^m$ are dependent. Suppose that $\mathbf{v}_1$ and $\mathbf{v}_2$ are collinear vectors in $R^m$. By definition, this means that $\mathbf{v}_1$ and $\mathbf{v}_2$ lie on the same line through the origin, and this means that one of these vectors is a scalar multiple of the other. Thus

$$\mathbf{v}_1 = \alpha\mathbf{v}_2 \qquad \text{or} \qquad \mathbf{v}_2 = \beta\mathbf{v}_1.$$

Hence

$$\mathbf{v}_1 - \alpha\mathbf{v}_2 = \mathbf{0} \qquad \text{or} \qquad -\beta\mathbf{v}_1 + \mathbf{v}_2 = \mathbf{0}.$$

In either case we have expressed $\mathbf{0}$ as a nontrivial linear combination of $\mathbf{v}_1$ and $\mathbf{v}_2$. It follows then that $\mathbf{v}_1$ and $\mathbf{v}_2$ are dependent.

We can now describe completely the geometric meaning of independence and dependence for two vectors. We proved that two noncollinear vectors are independent and two collinear vectors are dependent. It follows immediately that two independent vectors are noncollinear (for if they were not noncollinear they would be collinear and hence dependent) and two dependent vectors are collinear (if they were not collinear, then they would be noncollinear and hence independent). Hence (see Figure 2.6):

1. Two vectors are independent if and only if they are noncollinear.
2. Two vectors are dependent if and only if they are collinear.

It follows immediately from (2) that if two vectors $\mathbf{v}_1$ and $\mathbf{v}_2$ in $R^m$ are dependent, then at least one of them is a scalar multiple of the other. Conversely, if one of the vectors is a scalar multiple of the other, then the two vectors are dependent. Now suppose that $\mathbf{v}_1, \mathbf{v}_2$, and $\mathbf{v}_3$ are three dependent vectors in $R^m$. Then there are scalars $\alpha_1, \alpha_2$, and $\alpha_3$ not all zero such that

$$\alpha_1\mathbf{v}_1 + \alpha_2\mathbf{v}_2 + \alpha_3\mathbf{v}_3 = \mathbf{0}.$$

At least one of the scalars $\alpha_1, \alpha_2$, and $\alpha_3$ is not zero. If, for example, $\alpha_2$ is not zero we can solve this equation for $\mathbf{v}_2$ and obtain

$$\mathbf{v}_2 = -(\alpha_1/\alpha_2)\mathbf{v}_1 - (\alpha_3/\alpha_2)\mathbf{v}_3.$$

Similarly, if $\alpha_1$ or $\alpha_3$ is not zero we can solve for $\mathbf{v}_1$ or $\mathbf{v}_3$, respectively. Thus at least one of the vectors is a linear combination of the remaining vectors. Conversely, if

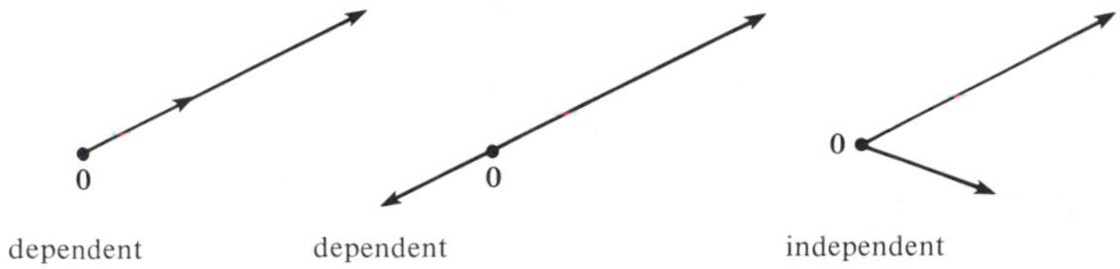


**Figure 2.6**

one of these vectors, say $\mathbf{v}_2$, is a linear combination of the other two vectors,

$$\mathbf{v}_2 = \alpha_1\mathbf{v}_1 + \alpha_3\mathbf{v}_3,$$

then

$$\mathbf{0} = \alpha_1\mathbf{v}_1 + (-1)\mathbf{v}_2 + \alpha_3\mathbf{v}_3,$$

and it follows that the three vectors are dependent.

***Theorem 5*** *Let $\mathbf{v}_1, \mathbf{v}_2, \ldots, \mathbf{v}_n$ be vectors in $R^m$. The following are equivalent.*

(*a*) *The vectors $\mathbf{v}_1, \mathbf{v}_2, \ldots, \mathbf{v}_n$ are dependent.*
(*b*) *It is possible to express one of the vectors, say $\mathbf{v}_i$, as a linear combination of the remaining vectors:*

$$\mathbf{v}_i = \alpha_1\mathbf{v}_1 + \cdots + \alpha_{i-1}\mathbf{v}_{i-1} + \alpha_{i+1}\mathbf{v}_{i+1} + \cdots + \alpha_n\mathbf{v}_n.$$

**Proof** First suppose that the vectors are dependent. Then there are scalars $\alpha_1, \alpha_2, \ldots, \alpha_n$ not all zero such that

$$\alpha_1\mathbf{v}_1 + \alpha_2\mathbf{v}_2 + \cdots + \alpha_n\mathbf{v}_n = \mathbf{0}.$$

At least one of the scalars, say $\alpha_i$, is not zero. Thus we can solve for the vector $\mathbf{v}_i$:

$$\mathbf{v}_i = \beta_1\mathbf{v}_1 + \cdots + \beta_{i-1}\mathbf{v}_{i-1} + \beta_{i+1}\mathbf{v}_{i+1} + \cdots + \beta_n\mathbf{v}_n,$$

where $\beta_j = -\alpha_j/\alpha_i$ for $j \neq i$. This shows that $\mathbf{v}_i$ is a linear combination of the remaining vectors.

Conversely, suppose that one of the vectors, say $\mathbf{v}_i$, is a linear combination of the remaining vectors:

$$\mathbf{v}_i = \gamma_1\mathbf{v}_1 + \cdots + \gamma_{i-1}\mathbf{v}_{i-1} + \gamma_{i+1}\mathbf{v}_{i+1} + \cdots + \gamma_n\mathbf{v}_n.$$

Subtracting $\mathbf{v}_i$ from both sides of this equation gives

$$\mathbf{0} = \gamma_1\mathbf{v}_1 + \cdots + \gamma_{i-1}\mathbf{v}_{i-1} - \mathbf{v}_i + \gamma_{i+1}\mathbf{v}_{i+1} + \cdots + \gamma_n\mathbf{v}_n,$$

which expresses $\mathbf{0}$ as a linear combination of the vectors with not all coefficients equal to 0 (the coefficient of $\mathbf{v}_i$ is $-1$). Thus the vectors are dependent.

## EXAMPLE 7

Let us prove that if three vectors are dependent, they lie in the same plane through the origin. First suppose that no two of the vectors are collinear. Then the vectors are dependent if and only if one of the vectors is a linear combination of the remaining two vectors and this holds if and only if one of the vectors lies in the plane spanned by the other two. Hence all three vectors lie in the same plane. Now suppose that at least two of the vectors are collinear. The two collinear vectors span a line through the origin. If the third vector is not on this line, then all three vectors lie in

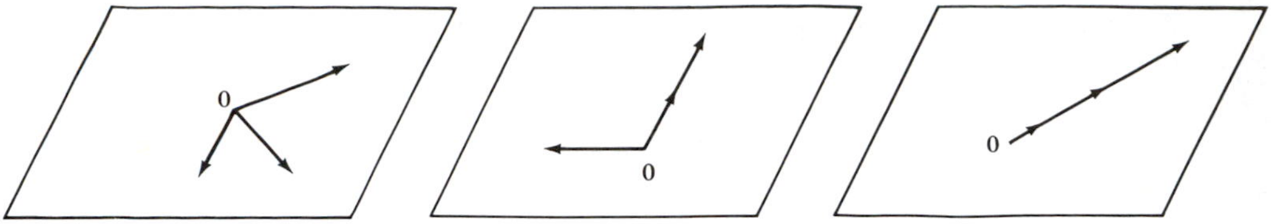


Figure 2.7

the plane determined by the third vector and any vector on the line. If the third vector is also on the line, then all three of the vectors lie in any plane containing the line (see Figure 2.7).

We conclude this section with some important results about the three subspaces associated with a matrix.

Theorem 6 of Section 2.2 states that the nullspace of a matrix $A$ is spanned by the set of solutions of $A\mathbf{x} = \mathbf{0}$ obtained by in turn setting each free variable to 1 and all other free variables to 0. These vectors are also independent. Before giving the proof in general, we illustrate it with an example. Let $A$ be the matrix

$$A = \begin{bmatrix} 2 & -4 & 2 \\ -1 & 2 & -1 \\ 3 & -6 & 3 \end{bmatrix}.$$

Forward elimination reduces $A$ to

$$E = \begin{bmatrix} 1 & -2 & 1 \\ 0 & 0 & 0 \\ 0 & 0 & 0 \end{bmatrix}.$$

The free variables are $x_2$ and $x_3$. Setting in turn each free variable to 1 and the remaining free variables to 0 gives the two vectors

$$\begin{array}{lll} (2, 1, 0) & \text{from} & x_2 = 1, x_3 = 0 \\ (-1, 0, 1) & \text{from} & x_2 = 0, x_3 = 1. \end{array}$$

These are the vectors that span the nullspace. Why are they independent? Because there is only one way to write $\mathbf{0}$ (the trivial solution) as a linear combination of these vectors; both free variables must be zero. For if

$$(0, 0, 0) = x_2(2, 1, 0) + x_3(-1, 0, 1) = (2x_2 - x_3, x_2, x_3),$$

then $x_2 = x_3 = 0$ and the vectors are indeed independent.

The situation in general is exactly the same. If the system has free variables, then any linear combination of the vectors generated in this way results in a vector having $x_i$ in its $i$th component if and only if $x_i$ is a free variable. Therefore, expressing the trivial solution $\mathbf{0}$ as a linear combination of these vectors forces all free variables to be zero. That is, the only way to write the trivial solution $\mathbf{x} = \mathbf{0}$ as a linear combination of these solutions is by having all free variables equal to 0. In the language of independence, the vectors are therefore independent.

***Theorem 6*** *The vectors obtained as solutions of the homogeneous system* $A\mathbf{x} = \mathbf{0}$ *by in turn setting each free variable to* 1 *and the remaining free variables to* 0 *are independent and span the nullspace of* $A$.

## EXAMPLE 8

Let us find an independent set of vectors that span the nullspace of

$$A = \begin{bmatrix} -1 & 2 & 1 & 1 & 3 \\ 1 & -2 & 1 & 7 & -13 \\ 0 & 0 & -1 & -4 & 5 \end{bmatrix}.$$

Since the reduced echelon form of $A$ is

$$R = \begin{bmatrix} 1 & -2 & 0 & 3 & -8 \\ 0 & 0 & 1 & 4 & -5 \\ 0 & 0 & 0 & 0 & 0 \end{bmatrix},$$

the free variables are $x_2$, $x_4$, and $x_5$. The solutions corresponding to

$$\begin{matrix} x_2 = 1 & x_4 = 0 & x_5 = 0, \\ x_2 = 0 & x_4 = 1 & x_5 = 0, \\ x_2 = 0 & x_4 = 0 & x_5 = 1, \end{matrix}$$

are, respectively,

$$\begin{aligned} \mathbf{u} &= (2, 1, 0, 0, 0) \\ \mathbf{v} &= (-3, 0, -4, 1, 0) \\ \mathbf{w} &= (8, 0, 5, 0, 1). \end{aligned}$$

These vectors are independent and span $N(A)$.

Now let's try to solve the analogous problem for the column space $C(A)$ of a matrix $A$. By Theorem 5 of Section 1.10 the general solution to the nonhomogeneous system expresses any solution to $A\mathbf{x} = \mathbf{b}$ as a particular solution plus the general solution of the associated homogeneous system $A\mathbf{x} = \mathbf{0}$. As we just observed above, the only solution of the homogeneous system having all free variables equal to 0 is the trivial solution $\mathbf{x} = \mathbf{0}$. Therefore, there is one and only one solution of the nonhomogeneous system $A\mathbf{x} = \mathbf{b}$ having all free variables equal to zero. To fully appreciate the significance of this statement, we must translate it to a statement about the columns of $A$. Let $\mathbf{v}_1, \mathbf{v}_2, \ldots, \mathbf{v}_n$ denote the columns of $A$: $A = [\mathbf{v}_1 \quad \mathbf{v}_2 \quad \cdots \quad \mathbf{v}_n]$. A solution of the equation $A\mathbf{x} = \mathbf{b}$ having the $i$th variable $x_i = 0$ corresponds to expressing $\mathbf{b}$ as a linear combination of the columns of $A$ without using the $i$th column:

$$\begin{aligned} \mathbf{b} &= x_1\mathbf{v}_1 + \cdots + x_{i-1}\mathbf{v}_{i-1} + 0 \cdot \mathbf{v}_i + x_{i+1}\mathbf{v}_{i+1} + \cdots + x_n\mathbf{v}_n \\ &= x_1\mathbf{v}_1 + \cdots + x_{i-1}\mathbf{v}_{i-1} + x_{i+1}\mathbf{v}_{i+1} + \cdots + x_n\mathbf{v}_n. \end{aligned}$$

Setting the $i$th variable to zero corresponds to striking out the $i$th column of $A$. It follows that setting all free variables to zero corresponds to striking out the nonpivot columns. Thus the solution of the nonhomogeneous system $A\mathbf{x} = \mathbf{b}$ having all free variables equal to zero expresses **b** as a linear combination of the pivot columns of $A$. We are now ready to state and prove our next theorem.

***Theorem 7*** *The pivot columns of a matrix $A$ are independent and span the column space of $A$.*

**Proof** As we have just observed, if **b** is in $C(A)$, there is a solution to $A\mathbf{x} = \mathbf{b}$ having all free variables equal to zero and this solution expresses **b** as a linear combination of the pivot columns of $A$. Thus the pivot columns span $C(A)$. It remains to prove that the pivot columns are independent. Writing **0** as a linear combination of the pivot columns of $A$ corresponds to a solution of $A\mathbf{x} = \mathbf{0}$ having all free variables 0. There is only one such solution, the trivial solution. Thus the only way to express **0** as a linear combination of the pivot columns is with all coefficients 0. The pivot columns are therefore independent.

## EXAMPLE 9

Forward elimination reduces

$$A = \begin{bmatrix} 1 & 3 & -2 & 1 \\ 2 & 6 & -2 & 8 \\ -1 & -3 & 8 & 17 \end{bmatrix} \quad \text{to} \quad E = \begin{bmatrix} 1 & 3 & -2 & 1 \\ 0 & 0 & 2 & 6 \\ 0 & 0 & 0 & 0 \end{bmatrix}.$$

Thus the first and third columns of $A$ are the pivot columns of $A$. They form an independent set of vectors that span $C(A)$.

Finally, we consider the row space of a matrix. We saw in Section 2.2 that when a matrix $A$ is reduced to an echelon matrix $E$, the nonzero rows of $E$ span the row space of $A$. The nonzero rows of $E$ are also independent. This follows from the form of the echelon matrix $E$. The form that we are referring to is the progression of the pivots from left to right as we move down the rows. For instance, let

$$E = \begin{bmatrix} 2 & 1 & 3 & 4 \\ 0 & 0 & 5 & 2 \\ 0 & 0 & 0 & 4 \\ 0 & 0 & 0 & 0 \end{bmatrix}.$$

If

$$\alpha_1(2, 1, 3, 4) + \alpha_2(0, 0, 5, 2) + \alpha_3(0, 0, 0, 4) = (0, 0, 0, 0),$$

then equating the first, third, and fourth components, we obtain

$$2\alpha_1 = 0, \qquad 3\alpha_1 + 5\alpha_2 = 0, \qquad 4\alpha_1 + 2\alpha_2 + 4\alpha_3 = 0.$$

Thus $\alpha_1 = \alpha_2 = \alpha_3 = 0$. Thus the nonzero rows of $E$ are indeed independent. Clearly, this reasoning can be extended to any echelon matrix.

***Theorem 8*** *Suppose that a matrix $A$ has been reduced to an echelon matrix $E$. Then the nonzero rows of $E$ are independent vectors that span $R(A)$.*

### EXAMPLE 10

Using Gaussian elimination, we can reduce the matrix

$$A = \begin{bmatrix} 1 & 2 & 0 & 1 \\ 2 & 5 & 1 & 4 \\ 3 & 8 & 2 & 7 \end{bmatrix} \quad \text{to} \quad E = \begin{bmatrix} 1 & 2 & 0 & 1 \\ 0 & 1 & 1 & 2 \\ 0 & 0 & 0 & 0 \end{bmatrix}.$$

Hence the vectors $(1, 2, 0, 1)$ and $(0, 1, 1, 2)$ are independent vectors that span $R(A)$.

Theorem 5 enables us to explain the significance of a zero row in the echelon form of a matrix. Suppose that $A$ is an $m \times n$ matrix and that when $A$ is reduced to an echelon matrix $E$, $E$ has one or more zero rows. For simplicity we will carry out the discussion when no pivoting has occurred in reducing $A$ to $E$. In this case, a row of $A$ that corresponds to a zero row of $E$ is a linear combination of the previous rows of $A$. Why? Because each step of the elimination process adds a multiple of a row to a subsequent row. By Theorem 5, the rows of $A$ are dependent. What is the significance of this fact in terms of a system of equations $A\mathbf{x} = \mathbf{b}$? Suppose that the system $A\mathbf{x} = \mathbf{b}$ is consistent. Since the systems

$$A\mathbf{x} = \mathbf{b} \qquad \text{and} \qquad E\mathbf{x} = \mathbf{c}$$

are equivalent, a zero row of $E$ is matched by a zero entry in $\mathbf{c}$ and corresponds to the equation

$$0x_1 + 0x_2 + \cdots + 0x_n = 0.$$

Thus the corresponding equation in the system $A\mathbf{x} = \mathbf{b}$ is a linear combination of the previous equations of the system. Hence it is *redundant*. That is, the information in this equation is already contained in the previous equations of the system. On the other hand, if the system $A\mathbf{x} = \mathbf{b}$ is inconsistent, then at least one zero row of $E$ is not matched by a corresponding zero entry in $\mathbf{c}$. Thus the corresponding equation of $A\mathbf{x} = \mathbf{b}$ is inconsistent with the previous equations of $A$.

## PROBLEMS 2.3

**1.** Complete the following definitions.
   (a) The vectors $\mathbf{v}_1, \mathbf{v}_2, \ldots, \mathbf{v}_n$ in $R^m$ are linearly independent if ....
   (b) The vectors $\mathbf{v}_1, \mathbf{v}_2, \ldots, \mathbf{v}_n$ in $R^m$ are linearly dependent if ....
   (c) A nontrivial linear combination of $\mathbf{v}_1, \mathbf{v}_2, \ldots, \mathbf{v}_n$ is ....

**2.** Which of the following sets of vectors are linearly independent?
   *(a) $(-1, 3), (5, 9)$
   (b) $(6, 8, 1), (0, 0, 0), (-1, 1, -1)$
   *(c) $(2, 1), (3, 1), (4, 1)$
   (d) $(4, 3, 0), (6, 1, 1), (0, 0, 1), (1, 1, 1), (1, 2, 3)$
   *(e) $(1, 2, 0), (2, 1, 1), (3, 1, 0)$
   (f) $(1, 1, 1, 1, 1), (0, 1, 0, 1, 0), (-1, 2, 1, 3, 1)$
   *(g) $(1, 5, 4, 6), (-3, 5, -7, 11), (0, 0, 0, 0), (5, 1, 1, 1)$
   (h) $(1, 0, 0, 0), (-1, 3, 1, 1), (1, 2, 3, 4), (7, 0, 7, 10)$
   *(i) $(1, 2, 3, 4), (2, 3, 4, 5), (3, 4, 5, 6), (4, 5, 6, 7)$

**3.** Show that the vectors $(1, 0, 3, 1)$, $(-1, 1, 0, 1)$, $(2, 3, 0, 0)$, and $(1, 1, 6, 3)$ are linearly dependent. Can each one of these vectors be expressed as a linear combination of the others?

4. Repeat Problem 3 for the vectors $(1, 1, 1, 1)$, $(0, 1, 1, 0)$, $(-1, 1, 0, 1)$, and $(2, 1, 1, 2)$.

5. In each part, do the following.
   (i) Find an independent set of vectors that span $C(A)$.
   (ii) Find an independent set of vectors that span $N(A)$.
   (iii) Find an independent set of vectors that span $R(A)$.
   (iv) Find the rank of $A$.

*(a) $\begin{bmatrix} 1 & 0 & -1 & -2 \\ 0 & 1 & 3 & 4 \end{bmatrix}$

(b) $\begin{bmatrix} 1 & 2 & 3 \\ 1 & 2 & 5 \end{bmatrix}$

*(c) $\begin{bmatrix} 1 & 1 & 0 & 0 & 1 \\ 0 & 0 & 1 & 1 & 0 \end{bmatrix}$

(d) $\begin{bmatrix} 0 & 1 & 4 \\ 0 & 2 & 5 \\ 0 & 3 & 6 \end{bmatrix}$

*(e) $\begin{bmatrix} 1 & 2 & -2 & 3 \\ 2 & 4 & -3 & 4 \\ 3 & 6 & -4 & 5 \end{bmatrix}$

(f) $\begin{bmatrix} 0 & 1 & 3 & 3 \\ 0 & 1 & 3 & 1 \\ 0 & 0 & 0 & 1 \end{bmatrix}$

*(g) $\begin{bmatrix} 1 & 1 & 1 & 0 & 0 \\ 0 & 0 & 1 & 1 & 1 \\ 1 & -1 & 2 & 1 & 1 \\ -1 & 1 & -3 & -2 & -2 \end{bmatrix}$

(h) $\begin{bmatrix} 1 & 1 & 1 & 1 & 1 \\ 1 & -1 & 3 & -1 & 5 \\ 1 & 5 & -3 & 5 & -7 \\ 0 & 4 & -4 & 4 & -8 \end{bmatrix}$

*(i) $\begin{bmatrix} 1 & 3 & 0 & 3 & 1 \\ 2 & 6 & 1 & 4 & 0 \\ -1 & -3 & 2 & -7 & 2 \\ 3 & 9 & 4 & 1 & 1 \end{bmatrix}$

6. If two of the vectors $\mathbf{v}_1, \mathbf{v}_2, \ldots, \mathbf{v}_n$ are equal, show that the vectors are linearly dependent.

*7. Show that if the vectors $\mathbf{v}_1, \mathbf{v}_2$, and $\mathbf{v}_3$ are independent, then the vectors $\mathbf{u}_1 = 2\mathbf{v}_1$, $\mathbf{u}_2 = \mathbf{v}_1 + \mathbf{v}_2$, and $\mathbf{u}_3 = -\mathbf{v}_1 + \mathbf{v}_3$ are also independent.

8.*(a) Show that the vectors $(1, 0, 0)$, $(y_1, 1, 0)$, $(z_1, z_2, 1)$ are independent for any scalars $\alpha$, $\beta$, and $\gamma$.
   (b) Show that the vectors

$$(x_1, 0, 0, 0), (y_1, y_2, 0, 0),$$
$$(z_1, z_2, z_3, 0), (w_1, w_2, w_3, w_4)$$

   are independent if and only if $x_1, y_2, z_3$, and $w_4$ are not zero.
   (c) Find and prove a generalization.

9. Let $(x_1, x_2)$ and $(y_1, y_2)$ be two vectors in $R^2$. Show that they are independent if and only if $x_1 y_2 - x_2 y_1 \neq 0$.

*10. For what values of $\lambda$ are the three vectors $(1, 1, 2)$, $(2, 3, 5)$, $(1, 3, \lambda)$ dependent?

11. Show that if the vectors $\mathbf{v}_1, \mathbf{v}_2, \ldots, \mathbf{v}_m, \mathbf{v}_{m+1}, \ldots, \mathbf{v}_n$ are independent, then the vectors $\mathbf{v}_1, \mathbf{v}_2, \ldots, \mathbf{v}_m$ are independent. More generally, explain why any subset of a linearly independent set of vectors must be linearly independent.

12. Let $\mathbf{u}$, $\mathbf{v}$, and $\mathbf{w}$ be three vectors in $R^m$. Prove that the vectors $\alpha\mathbf{u} - \beta\mathbf{v}$, $\gamma\mathbf{v} - \alpha\mathbf{w}$, $\beta\mathbf{w} - \gamma\mathbf{u}$ are dependent for any choice of the scalars $\alpha$, $\beta$, and $\gamma$.

13. Let $\mathbf{v}_1, \mathbf{v}_2, \ldots, \mathbf{v}_n$ be a set of vectors in $R^m$. Prove that they are independent if and only if none of the vectors is a linear combination of the remaining vectors. (*Hint:* Use Theorem 5.)

*14. Let $\mathbf{v}_1, \mathbf{v}_2, \ldots, \mathbf{v}_n$ be independent vectors in $R^m$. Prove that if $\mathbf{v}$ is in $R^m$ and $\mathbf{v}_1, \mathbf{v}_2, \ldots, \mathbf{v}_n, \mathbf{v}$ are independent, $\mathbf{v}$ is not in the subspace spanned by $\mathbf{v}_1, \mathbf{v}_2, \ldots, \mathbf{v}_n$.

15. Let $\mathbf{v}_1, \mathbf{v}_2, \ldots, \mathbf{v}_n$ be vectors in $R^m$. If one of the vectors, say $\mathbf{v}_i$, is a linear combination of the remaining vectors, prove that the $n - 1$ vectors $\mathbf{v}_1, \ldots, \mathbf{v}_{i-1}, \mathbf{v}_{i+1}, \ldots, \mathbf{v}_n$ span the same subspace as the $n$ vectors $\mathbf{v}_1, \mathbf{v}_2, \ldots, \mathbf{v}_n$.

**16.** Suppose that $\mathbf{v}_1, \mathbf{v}_2, \ldots, \mathbf{v}_n$ span a subspace $V$ of $R^m$. Prove that if one of the vectors can be removed from the list and the remaining $n - 1$ vectors still span $V$, then $\mathbf{v}_1, \mathbf{v}_2, \ldots, \mathbf{v}_n$ are dependent.

**17.** Suppose that $\mathbf{u}_1$, $\mathbf{u}_2$, and $\mathbf{u}_3$ are nonzero dependent vectors in a subspace $V$ of $R^m$.

*(a) Show that $\mathbf{u}_2$ is a linear combination of $\mathbf{u}_1$ or $\mathbf{u}_3$ is a linear combination of $\mathbf{u}_1$ and $\mathbf{u}_2$.

(b) State and prove a generalization of this result for a set of $n$ vectors.

**18.** Give a proof that the pivot columns of an echelon matrix are independent that is based on the form of the echelon matrix.

***19.** Let $A$ be an $n \times n$ matrix. Suppose that for every vector $\mathbf{b}$ in $R^n$, $A\mathbf{x} = \mathbf{b}$ has a solution. Show that the columns of $A$ are independent.

**20.** Suppose that $A$ is reduced to an echelon matrix $E$ with no pivoting. Explain why the rows of $A$ that correspond to the nonzero rows of $E$ span $R(A)$. Are they also independent? What happens if we allow pivoting?

**21.** Let $A$ be an upper triangular matrix. Are the nonzero rows of $A$ independent? If true, prove it. If false, give an example.

**22.** Let $A$ be an $m \times m$ matrix.

*(a) Show that the columns of $A$ are independent if and only if the rows of $A$ are independent.

(b) Show that the columns of $A$ span $R^m$ if and only if the rows of $A$ span $R^m$.

***23.** Suppose that the columns of the matrix $A$ are independent and the vectors $\mathbf{v}_1, \mathbf{v}_2, \ldots, \mathbf{v}_n$ are independent. Show that the vectors $A\mathbf{v}_1, A\mathbf{v}_2, \ldots, A\mathbf{v}_n$ are independent.

**24.** Suppose that $A$ is an $m \times n$ matrix and $\mathbf{b}_1, \mathbf{b}_2, \ldots, \mathbf{b}_k$ is an independent set of vectors in $R^m$. Suppose that $\mathbf{v}_i$ is a solution of $A\mathbf{x} = \mathbf{b}_i$, $i = 1, \ldots, k$. Show that $\mathbf{v}_1, \mathbf{v}_2, \ldots, \mathbf{v}_k$ is an independent set of vectors in $R^n$.

**25.** Let $A$ be an $n \times n$ matrix and let $\mathbf{v}_1, \mathbf{v}_2, \ldots, \mathbf{v}_n$ be $n$ vectors in $R^n$. Suppose that $A\mathbf{v}_1, A\mathbf{v}_2, \ldots, A\mathbf{v}_n$ are independent. Show that $A$ is nonsingular. (*Hint:* Use Problem 24.)

# 2.4
# Bases

As we have seen, existence corresponds to spanning and uniqueness corresponds to independence. A set of vectors spans a subspace if and only if the set of vectors contains enough information to describe the subspace completely. Such a set may, however, contain redundant information. This happens when the spanning set is dependent. If the spanning set is dependent, then every vector in the subspace can be expressed in an infinite number of ways as a linear combination of the spanning vectors. On the other hand, an independent set of vectors contains no redundant information. Whatever vectors can be described as a linear combination of an independent set of vectors can be described in only one way because independence corresponds to uniqueness. Thus when a subspace is spanned by an independent set of vectors, we have enough, and just enough, information to describe the subspace.

***Definition*** Let $V$ be a subspace of $R^m$. A set of vectors $\mathbf{v}_1, \mathbf{v}_2, \ldots, \mathbf{v}_n$ in $V$ which spans $V$ and are independent is called a **basis** for $V$.

## EXAMPLE 1

For each $m = 1, 2, \ldots$, we have found a basis for $R^m$, namely the vectors $\mathbf{e}_1, \mathbf{e}_2, \ldots, \mathbf{e}_m$. They form a basis for $R^m$ because they span $R^m$ (Example 5 of Section 2.2) and they are independent (Example 4 of Section 2.3). This basis is called the **standard basis** for $R^m$.

## EXAMPLE 2

Since the vectors $(1, 2)$ and $(-2, -3)$ are not collinear, it is clear geometrically that they form a basis for $R^2$. To prove this fact algebraically, form the matrix

$$A = \begin{bmatrix} 1 & -2 \\ 2 & -3 \end{bmatrix}$$

and reduce it to an echelon matrix,

$$E = \begin{bmatrix} 1 & -2 \\ 0 & 1 \end{bmatrix}.$$

Thus $A$ is a $2 \times 2$ matrix with rank 2. Therefore, the columns of $A$ span $R^2$ by Theorem 3 of Section 2.2 and they are independent by Theorem 3 of Section 2.3.

Our next theorem summarizes our findings about the column space, the nullspace, and the row space of a matrix in terms of bases.

***Theorem 1*** *Let $A$ be an $m \times n$ matrix.*

(*a*) *The pivot columns of $A$ form a basis for the column space of $A$.*
(*b*) *A basis for the nullspace of $A$ is given by the vectors that correspond to the solutions obtained by in turn setting each free variable to* 1 *and all other free variables to* 0.
(*c*) *When $A$ is reduced to an echelon matrix $E$, the nonzero rows of $E$ form a basis for the row space of $A$.*

In our next two theorems we formalize our remark that existence and uniqueness correspond to spanning and independence.

***Theorem 2*** *Let $\mathbf{v}_1, \mathbf{v}_2, \ldots, \mathbf{v}_n$ be a basis for a subspace $V$ of $R^m$. Then every vector in $V$ can be written uniquely as a linear combination of $\mathbf{v}_1, \mathbf{v}_2, \ldots, \mathbf{v}_n$.*

**Proof** Form the matrix $A$ having $\mathbf{v}_1, \mathbf{v}_2, \ldots, \mathbf{v}_n$ as its columns. Then $V = C(A)$ and $A$ has rank $n$. Therefore, for every $\mathbf{v}$ in $C(A)$, the system $A\mathbf{x} = \mathbf{v}$ has a unique solution. Thus every $\mathbf{v}$ in $V$ can be expressed as a linear combination of $\mathbf{v}_1, \mathbf{v}_2, \ldots, \mathbf{v}_n$ in one and only one way. The proof is complete.

***Theorem 3*** *Let $A$ be an $m \times m$ matrix. The following are equivalent.*

(*a*) *For every* $\mathbf{b}$ *in* $R^m$ *the system* $A\mathbf{x} = \mathbf{b}$ *has a unique solution.*
(*b*) *The columns of* $A$ *form a basis for* $R^m$.
(*c*) *The rank of* $A$ *is* $m$.
(*d*) *The matrix* $A$ *is nonsingular.*

**Proof** The theorem is an immediate consequence of Theorems 3 of Section 2.2, 3 of Section 2.3, and 3 of Section 1.9, and its proof is left for your enjoyment (see Problem 15).

## EXAMPLE 3

The columns of any $m \times m$ nonsingular matrix form a basis for $R^m$. Given a list of $m$ vectors in $R^m$, a general method for determining whether they form a basis for $R^m$ is to compute the rank of the $m \times m$ matrix having these vectors as its columns.

$R^m$ has a basis, namely $\mathbf{e}_1, \mathbf{e}_2, \ldots, \mathbf{e}_m$. In fact, it has many different bases; the columns (or rows) of any $m \times m$ nonsingular matrix form a basis for $R^m$. We turn now to the corresponding question for subspaces. Does every subspace $V$ of $R^m$ have a basis? The trivial subspace $\{\mathbf{0}\}$ does not contain a single independent vector. Therefore, there is no independent set of vectors that spans the trivial subspace. Said otherwise, the only possible basis for the trivial subspace is the empty set. We will find it a useful convention to define the empty set to be a basis for the trivial subspace.

To prove that every nontrivial subspace of $R^m$ has a basis, we need the following theorem.

***Theorem 4*** *Let* $\mathbf{v}_1, \mathbf{v}_2, \ldots, \mathbf{v}_n$ *be independent vectors in* $R^m$. *If a vector* $\mathbf{v}$ *in* $R^m$ *is not a linear combination of* $\mathbf{v}_1, \mathbf{v}_2, \ldots, \mathbf{v}_n$, *then the vectors* $\mathbf{v}_1, \mathbf{v}_2, \ldots, \mathbf{v}_n, \mathbf{v}$ *are independent.*

**Proof** Suppose that there are scalars $\alpha_1, \ldots, \alpha_n, \beta$ such that

$$\alpha_1\mathbf{v}_1 + \cdots + \alpha_n\mathbf{v}_n + \beta\mathbf{v} = \mathbf{0}. \tag{1}$$

If $\beta \neq 0$, then

$$\mathbf{v} = -\frac{1}{\beta}(\alpha_1\mathbf{v}_1 + \cdots + \alpha_n\mathbf{v}_n).$$

Thus if $\beta \neq 0$, then $\mathbf{v}$ is a linear combination of $\mathbf{v}_1, \ldots, \mathbf{v}_n$, contrary to our assumption. Thus $\beta$ must be zero and equation (1) becomes

$$\alpha_1\mathbf{v}_1 + \cdots + \alpha_n\mathbf{v}_n = \mathbf{0}.$$

Since we have assumed that $\mathbf{v}_1, \ldots, \mathbf{v}_n$ are independent, we conclude that $\alpha_1 = \cdots = \alpha_n = 0$. Hence $\alpha_1 = \cdots = \alpha_n = \beta = 0$, proving that $\mathbf{v}_1, \ldots, \mathbf{v}_n, \mathbf{v}$ are independent.

Now suppose that $V$ is a nontrivial subspace of $R^m$, and let $\mathbf{v}_1$ be a nonzero vector in $V$. If $V$ is the subspace spanned by $\mathbf{v}_1$, then $\mathbf{v}_1$ is a basis for $V$. Otherwise, there is a vector $\mathbf{v}_2$ in $V$ which is not in the subspace spanned by $\mathbf{v}_1$. By Theorem 4, $\mathbf{v}_1$ and $\mathbf{v}_2$ are independent. If they also span $V$, then they form a basis for $V$. If they do not span $V$, there must be a vector $\mathbf{v}_3$ in $V$ that is not a linear combination of $\mathbf{v}_1$ and $\mathbf{v}_2$. Again by Theorem 4, the vectors $\mathbf{v}_1$, $\mathbf{v}_2$, and $\mathbf{v}_3$ are independent. As long as the vectors do not span $V$, we can continue to enlarge our set of independent vectors in this way. The process must stop because $V$ is a subspace of $R^m$ and $R^m$ contains at most $m$ linearly independent vectors. Thus we must reach a point when our set of independent vectors span $V$. Thus $V$ has a basis.

***Theorem 5*** *Let $V$ be a subspace of $R^m$. Then $V$ has a basis.*

Compare the argument we gave to prove this theorem with the argument we gave to find all possible subspaces of $R^3$ in Example 8 of Section 2.1.

It is clear that our method of proof in Theorem 5 will also prove the following theorem (see Problem 15).

***Theorem 6*** *Let $V$ be a subspace of $R^m$. If $\mathbf{v}_1, \mathbf{v}_2, \ldots, \mathbf{v}_k$ are independent vectors in $V$, then there are vectors $\mathbf{v}_{k+1}, \ldots, \mathbf{v}_n$ in $V$ such that $\mathbf{v}_1, \ldots, \mathbf{v}_k, \mathbf{v}_{k+1}, \ldots, \mathbf{v}_n$ form a basis for $V$. Thus any independent set of vectors in a subspace $V$ can be enlarged to a basis for $V$.*

Expanding an independent set of vectors contained in a subspace $V$ of $R^m$ to a basis for $V$ can be accomplished as follows. Let $V$ be a subspace of $R^m$, and suppose that $\mathbf{v}_1, \mathbf{v}_2, \ldots, \mathbf{v}_k$ are independent vectors in $V$. Let $\mathbf{w}_1, \mathbf{w}_2, \ldots, \mathbf{w}_n$ be any set of vectors that span $V$. Form the matrix $A$ whose columns are vectors $\mathbf{v}_1, \mathbf{v}_2, \ldots, \mathbf{v}_k$ followed by $\mathbf{w}_1, \mathbf{w}_2, \ldots, \mathbf{w}_n$:

$$A = [\mathbf{v}_1 \quad \mathbf{v}_2 \quad \cdots \quad \mathbf{v}_k \quad \mathbf{w}_1 \quad \mathbf{w}_2 \quad \cdots \quad \mathbf{w}_n].$$

Then $V = C(A)$. Since $\mathbf{v}_1, \mathbf{v}_2, \ldots, \mathbf{v}_k$ are independent, when $A$ is reduced to an echelon matrix $E$, the first $k$ columns of $E$ will be pivot columns. Thus the pivot columns of $A$ will include the first $k$ columns of $A$. Consequently, the pivot columns of $A$ form a basis for $V = C(A)$ which contains the vectors $\mathbf{v}_1, \mathbf{v}_2, \ldots, \mathbf{v}_k$. Note that this procedure also provides an alternative proof of Theorem 6.

## EXAMPLE 4

Let $V$ be the subspace of $R^4$ consisting of all vectors $\mathbf{x} = (x_1, x_2, x_3, x_4)$ such that

$$x_1 + x_2 - x_3 + x_4 = 0.$$

Let us find a basis for $V$ that contains the vector $(0, 0, 1, 1)$. $V$ can be viewed as the nullspace of the matrix

$$A = [1 \quad 1 \quad -1 \quad 1].$$

A basis for $N(A)$ is found in the standard way (see Theorem 1) to be

$$(-1,1,0,0), \quad (1,0,1,0), \quad (-1,0,0,1).$$

Our goal is to find a basis for $N(A) = V$ that contains the given vector $(0,0,1,1)$. Form the matrix

$$B = \begin{bmatrix} 0 & -1 & 1 & -1 \\ 0 & 1 & 0 & 0 \\ 1 & 0 & 1 & 0 \\ 1 & 0 & 0 & 1 \end{bmatrix}$$

and reduce it to the echelon matrix

$$E = \begin{bmatrix} 1 & 0 & 1 & 0 \\ 0 & -1 & 1 & -1 \\ 0 & 0 & 1 & -1 \\ 0 & 0 & 0 & 0 \end{bmatrix}.$$

Thus the first three columns of $B$ are its pivot columns, and they form a basis for $C(B)$. Since $V = C(B)$, the vectors

$$(0,0,1,1), \quad (-1,1,0,0), \quad (1,0,1,0)$$

form a basis for $V$ that contains the given vector.

## EXAMPLE 5

To find a basis for $R^5$ that contains the vectors

$$\mathbf{v}_1 = (1,1,0,0,0) \qquad \text{and} \qquad \mathbf{v}_2 = (1,1,1,1,0),$$

form the matrix having these two vectors as its first two columns and having $\mathbf{e}_1, \mathbf{e}_2, \mathbf{e}_3, \mathbf{e}_4, \mathbf{e}_5$ as its remaining columns:

$$A = \begin{bmatrix} 1 & 1 & 1 & 0 & 0 & 0 & 0 \\ 1 & 1 & 0 & 1 & 0 & 0 & 0 \\ 0 & 1 & 0 & 0 & 1 & 0 & 0 \\ 0 & 1 & 0 & 0 & 0 & 1 & 0 \\ 0 & 0 & 0 & 0 & 0 & 0 & 1 \end{bmatrix}.$$

Reduce $A$ to

$$E = \begin{bmatrix} 1 & 1 & 1 & 0 & 0 & 0 & 0 \\ 0 & 1 & 0 & 0 & 1 & 0 & 0 \\ 0 & 0 & -1 & 1 & 0 & 0 & 0 \\ 0 & 0 & 0 & 0 & -1 & 1 & 0 \\ 0 & 0 & 0 & 0 & 0 & 0 & 1 \end{bmatrix}.$$

Thus columns 1, 2, 3, 5, 7 of $A$ are its pivot columns and hence the vectors

(1, 1, 0, 0, 0), (1, 1, 1, 1, 0), (1, 0, 0, 0, 0), (0, 0, 1, 0, 0), and (0, 0, 0, 0, 1) form a basis for $R^5$ that contains the given vectors $\mathbf{v}_1$ and $\mathbf{v}_2$.

Theorem 6 says that any independent set of vectors in a subspace can be expanded to a basis for the subspace. Our proof was based on the fact that (Theorem 4) beginning with an independent set of vectors, we can add to the set any vector that is not a linear combination of these vectors and the enlarged set will still be independent. The theorem that is in some sense dual to Theorem 4 is Theorem 7. It states that beginning with a spanning set for a subspace $V$, we can delete any vector from the set which is a linear combination of the remaining vectors and the reduced set of vectors will still span $V$. Using this theorem, we can then prove Theorem 8 below, which is dual to Theorem 6: Every spanning set for a subspace can be reduced to a basis for the subspace.

***Theorem 7*** *Suppose that the vectors* $\mathbf{v}_1, \mathbf{v}_2, \ldots, \mathbf{v}_n$ *span a subspace* $V$ *of* $R^m$. *The following are equivalent.*

(*a*) *The vectors* $\mathbf{v}_1, \mathbf{v}_2, \ldots, \mathbf{v}_n$ *are dependent.*
(*b*) *It is possible to remove at least one vector from the list* $\mathbf{v}_1, \mathbf{v}_2, \ldots, \mathbf{v}_n$ *in such a way that the remaining* $n - 1$ *vectors still span* $V$.

**Proof** First suppose that the vectors $\mathbf{v}_1, \mathbf{v}_2, \ldots, \mathbf{v}_n$ which span $V$ are dependent. By Theorem 5 of Section 2.3, we know that at least one of the vectors, say $\mathbf{v}_i$, is a linear combination of the remaining vectors $\mathbf{v}_1, \ldots, \mathbf{v}_{i-1}, \mathbf{v}_{i+1}, \ldots, \mathbf{v}_n$. We claim that these $n - 1$ vectors $\mathbf{v}_1, \ldots, \mathbf{v}_{i-1}, \mathbf{v}_{i+1}, \ldots, \mathbf{v}_n$ span $V$. Let $\mathbf{v}$ be in $V$. Since $\mathbf{v}_1, \mathbf{v}_2, \ldots, \mathbf{v}_n$ span $V$, $\mathbf{v}$ can be written as a linear combination of $\mathbf{v}_1, \mathbf{v}_2, \ldots, \mathbf{v}_n$,

$$\mathbf{v} = \beta_1 \mathbf{v}_1 + \cdots + \beta_{i-1}\mathbf{v}_{i-1} + \beta_i \mathbf{v}_i + \beta_{i+1}\mathbf{v}_{i+1} + \cdots + \beta_n \mathbf{v}_n.$$

Since $\mathbf{v}_i$ is a linear combination of the remaining $\mathbf{v}_j$'s,

$$\mathbf{v}_i = \alpha_1 \mathbf{v}_1 + \cdots + \alpha_{i-1}\mathbf{v}_{i-1} + \alpha_{i+1}\mathbf{v}_{i+1} + \cdots + \alpha_n \mathbf{v}_n.$$

Substituting the second equation into the first equation and collecting coefficients, we obtain

$$\mathbf{v} = (\alpha_1 \beta_i + \beta_1)\mathbf{v}_1 + \cdots + (\alpha_{i-1}\beta_i + \beta_{i-1})\mathbf{v}_{i-1} + (\alpha_{i+1}\beta_i + \beta_{i+1})\mathbf{v}_{i+1} + \cdots + (\alpha_n \beta_i + \beta_n)\mathbf{v}_n.$$

Thus $\mathbf{v}$ is also a linear combination of the $n - 1$ vectors $\mathbf{v}_1, \ldots, \mathbf{v}_{i-1}, \mathbf{v}_{i+1}, \ldots, \mathbf{v}_n$. This completes the proof that (a) implies (b). It remains to prove that (b) implies (a). Suppose that when the vector $\mathbf{v}_i$ is removed from the list $\mathbf{v}_1, \mathbf{v}_2, \ldots, \mathbf{v}_n$, the remaining vectors $\mathbf{v}_1, \ldots, \mathbf{v}_{i-1}, \mathbf{v}_{i+1}, \ldots, \mathbf{v}_n$ still span $V$. Since $\mathbf{v}_i$ is in $V$, it follows that $\mathbf{v}_i$ is a linear combination of the remaining vectors $\mathbf{v}_1, \ldots, \mathbf{v}_{i-1}, \mathbf{v}_{i+1}, \ldots, \mathbf{v}_n$. Theorem 5 of Section 2.3 then allows us to conclude that the vectors $\mathbf{v}_1, \mathbf{v}_2, \ldots, \mathbf{v}_n$ are dependent.

Thus when a set of vectors that spans a subspace $V$ is dependent, at least one of the vectors can be removed from the set in such a way that the remaining vectors still span $V$. The vector that is removed is redundant. The remaining vectors contain all the information needed to generate $V$. If these remaining vectors are themselves dependent, then yet another vector may be cast out in such a way that the vectors which remain still span $V$. When we can no longer cast out a vector in such a way that the remaining vectors still span $V$, what do we have? We have an independent set of vectors that spans $V$. Why? If the remaining vectors were not independent, then they would be dependent and we could cast out yet another vector and still span $V$. We summarize our discussion in the following theorem.

***Theorem 8*** *Suppose that a subspace $V$ is spanned by a dependent set of vectors. Then by deleting some of the vectors from the set we can obtain an independent set of vectors that spans $V$.*

*Thus a spanning set can always be reduced to a basis.*

We can give a different proof of this theorem which has the advantage that it gives an effective procedure for reducing a spanning set to a basis. Suppose that $\mathbf{v}_1, \mathbf{v}_2, \ldots, \mathbf{v}_k$ span $V$. Form the matrix $A$ whose columns are these vectors. Then $V = C(A)$. The pivot columns of $A$ form a basis for $C(A)$, and hence they are a subset of the spanning vectors that form a basis for $V$.

## EXAMPLE 6

Let $V$ be the subspace spanned by

$$\mathbf{v}_1 = (1, 2, 7), \qquad \mathbf{v}_2 = (2, 4, 14), \qquad \mathbf{v}_3 = (-1, 3, 3).$$

Let us find a basis for $V$ that consists of some subset of these vectors. Form the matrix

$$A = \begin{bmatrix} 1 & 2 & -1 \\ 2 & 4 & 3 \\ 7 & 14 & 3 \end{bmatrix}$$

and reduce $A$ to

$$E = \begin{bmatrix} 1 & 2 & -1 \\ 0 & 0 & 5 \\ 0 & 0 & 0 \end{bmatrix}.$$

The first and third columns of $A$ are its pivot columns. Thus $\mathbf{v}_1$ and $\mathbf{v}_3$ form a basis for $V$.

## EXAMPLE 7

Let $V$ be the subspace of $R^4$ spanned by the vectors $\mathbf{v}_1 = (1, 1, 2, 2)$, $\mathbf{v}_2 = (2, 3, 7, 6)$, $\mathbf{v}_3 = (1, 2, 5, 4)$, $\mathbf{v}_4 = (1, 0, 1, 2)$, and $\mathbf{v}_5 = (2, 1, 0, 1)$. Let $A$ be the matrix having these vectors as its *rows*. We reduce the matrix

$$A = \begin{bmatrix} 1 & 1 & 2 & 2 \\ 2 & 3 & 7 & 6 \\ 1 & 2 & 5 & 4 \\ 1 & 0 & 1 & 2 \\ 2 & 1 & 0 & 1 \end{bmatrix} \quad \text{to} \quad E = \begin{bmatrix} 1 & 1 & 2 & 2 \\ 0 & 1 & 3 & 2 \\ 0 & 0 & 2 & 2 \\ 0 & 0 & 0 & 0 \\ 0 & 0 & 0 & 0 \end{bmatrix}.$$

Since $V = R(A)$, the nonzero rows of the echelon matrix $E$, that is, the vectors $(1, 1, 2, 2)$, $(0, 1, 3, 2)$, and $(0, 0, 2, 2)$, form a basis for $V$. Note that only the first of the three basis vectors is one of the original five spanning vectors. If we want a basis for $V$ consisting of three of the original spanning vectors, we let $A$ be the matrix with columns $\mathbf{v}_1, \mathbf{v}_2, \mathbf{v}_3, \mathbf{v}_4, \mathbf{v}_5$ and reduce

$$A = \begin{bmatrix} 1 & 2 & 1 & 1 & 2 \\ 1 & 3 & 2 & 0 & 1 \\ 2 & 7 & 5 & 1 & 0 \\ 2 & 6 & 4 & 2 & 1 \end{bmatrix} \quad \text{to} \quad E = \begin{bmatrix} 1 & 2 & 1 & 1 & 2 \\ 0 & 1 & 1 & -1 & -1 \\ 0 & 0 & 0 & 2 & -1 \\ 0 & 0 & 0 & 0 & 0 \end{bmatrix}.$$

Then $V = C(A)$ and the columns of $A$ corresponding to the pivot columns of the echelon matrix $E$ form a basis for $V$. Thus $\mathbf{v}_1$, $\mathbf{v}_2$, and $\mathbf{v}_4$ are a basis for $V$.

## PROBLEMS 2.4

**1.** Complete each of the following definitions.

(a) A basis for a subspace $V$ of $R^n$ is....

(b) The standard basis for $R^n$ is....

**2.** In each part, determine whether the given vectors form a basis for $R^2$.

*(a) $(1, -1)$, $(3, 0)$

(b) $(2, -3)$, $(-6, 9)$

*(c) $(1, 1)$, $(2, 3)$, $(0, 2)$

(d) $(1, 1)$, $(0, 8)$

*(e) $(1, 0)$, $(0, 1)$, $(0, 0)$

(f) $(1, 1)$

**3.** In each part, determine whether the given vectors form a basis for $R^3$.

*(a) $(1, 1, 1)$, $(0, 2, 3)$, $(1, 0, 2)$

(b) $(1, 0, 1)$, $(2, 4, 8)$

*(c) $(3, 0, 1)$, $(1, 1, 1)$, $(4, 1, 2)$

(d) $(0, 0, 1)$, $(0, 1, 1)$, $(-1, 2, 2)$

*(e) $(1, -1, 3)$, $(2, 1, 2)$, $(3, -4, 2)$

(f) $(1, -2, 4)$, $(3, 2, 0)$, $(-1, -6, 8)$

**4.** In each part, determine whether the vectors form a basis for $R^4$.

*(a) $(1, 0, 1, 0)$, $(0, 1, -1, 2)$, $(0, 2, 2, 1)$, $(-2, 7, 1, 8)$

(b) $(1, 1, 1, 1)$, $(-1, 1, -1, 1)$, $(0, 1, 0, -1)$, $(1, 0, -1, 0)$

*(c) $(1, 0, 0, 0)$, $(1, 1, -1, -1)$, $(2, 5, 3, 7)$

(d) $(2, 1, 3, 1)$, $(-3, -1, 2, 1)$, $(2, 1, 2, -3)$, $(5, 2, 1, -1)$

*(e) $(5, 2, 1, 1)$, $(0, 0, 0, 0)$, $(-7, -2, 8, -3)$, $(2, 1, 3, -4)$

**5.** For each of the following collections $V$ of vectors, state why they form a subspace and find a basis for $V$.

*(a) $V$ is the set of all vectors $\mathbf{x}$ in $R^3$ such that $x_1 - 2x_2 + x_3 = 0$.

(b) $V$ is the set of all vectors $\mathbf{x}$ in $R^4$ such that $x_1 + x_2 + x_3 + x_4 = 0$ and $x_1 + x_2 - x_3 - x_4 = 0$.

*(c) $V$ is the set of all vectors $\mathbf{x}$ in $R^4$ such that $x_1 + 2x_2 - x_3 = 0$.

(d) $V$ is the set of all possible linear combinations of the vectors $(1,2,1,3)$, $(1,3,-2,-1)$, $(3,7,0,5)$, and $(0,-2,6,8)$.

*(e) $V$ is the set of all possible linear combinations of the vectors $(1,0,1,2,1)$, $(1,0,1,2,2)$, $(2,1,0,1,2)$, and $(1,1,-1,-1,0)$.

(f) $V$ is the set of all vectors in $R^3$ that lie on the plane through the origin and $(1,1,1)$, $(1,2,1)$.

*(g) $V$ is the set of all vectors in $R^3$ that lie on the plane through $(0,1,2)$, $(-2,5,5)$, $(-4,2,-6)$.

(h) $V$ is the set of all vectors in $R^3$ that lie on the line through $(2,-10,4)$, $(-3,15,-6)$.

*(i) $V$ is the set of all vectors that lie on the plane $\mathbf{x} = \alpha(1,2,1) + \beta(3,3,2)$ and the plane $2x_1 + 3x_2 - x_3 = 0$.

(j) $V$ is the set of all vectors that lie on the line $\mathbf{x} = (4,-2,6) + \alpha(2,-1,3)$ and the plane $2x_1 - x_2 + 2x_3 = 0$.

*(k) $V$ is the set of all vectors $\mathbf{x}$ in $R^5$ such that $A\mathbf{x} = B\mathbf{x}$ where

$$A = \begin{bmatrix} 2 & 2 & 0 & 1 & -3 \\ 3 & 3 & 3 & 0 & 0 \\ 0 & 6 & 3 & 1 & -1 \\ -1 & -2 & -2 & 3 & -3 \end{bmatrix}$$

and

$$B = \begin{bmatrix} 1 & 0 & -1 & -1 & 0 \\ 0 & -3 & -1 & 1 & -2 \\ -4 & -2 & -2 & 0 & 0 \\ 1 & 2 & 1 & 0 & 2 \end{bmatrix}.$$

**6.** For each of the following, find a basis for $R^n$ (for an appropriate $n$) which includes the given vectors.

*(a) $(1,2,3)$

(b) $(1,0,2)$, $(0,1,3)$

*(c) $(1,1,2)$

(d) $(1,2,2,4)$, $(1,1,1,2)$

*(e) $(1,1,0,0)$, $(1,2,3,1)$

(f) $(1,0,1,0)$, $(1,1,0,1)$

**7.** Find a basis for $R^1$.

**8.** Give a geometric construction which shows that $\mathbf{v}_1 = (1,2)$ and $\mathbf{v}_2 = (-2,3)$ form a basis for $R^2$.

**9.** For each of the following matrices $A$ and vectors $\mathbf{b}$, do the following.

(i) Find a basis for $C(A)$.
(ii) Find a basis for $N(A)$.
(iii) Find a basis for $R(A)$.
(iv) Find the rank of $A$.
(v) Find the general solution of $A\mathbf{x} = \mathbf{b}$.

*(a) $\begin{bmatrix} 1 & 2 & 4 & 5 \\ 2 & 3 & 5 & 6 \\ 3 & 4 & 6 & 7 \end{bmatrix}, \begin{bmatrix} 15 \\ 20 \\ 25 \end{bmatrix}$

(b) $\begin{bmatrix} 1 & -1 & 3 \\ 3 & -2 & 6 \\ -2 & 3 & 9 \end{bmatrix}, \begin{bmatrix} 0 \\ 1 \\ 1 \end{bmatrix}$

*(c) $\begin{bmatrix} 3 & -1 \\ -6 & 2 \\ 2 & 5 \end{bmatrix}, \begin{bmatrix} 1 \\ -2 \\ 12 \end{bmatrix}$

(d) $\begin{bmatrix} 1 & 2 & -1 \\ 2 & 4 & -2 \\ 1 & 2 & 2 \end{bmatrix}, \begin{bmatrix} 0 \\ 0 \\ 3 \end{bmatrix}$

*(e) $\begin{bmatrix} 1 & -1 & 0 & 2 \\ -2 & 2 & 0 & -4 \\ 3 & -3 & 0 & 6 \end{bmatrix}, \begin{bmatrix} 1 \\ 0 \\ 0 \end{bmatrix}$

(f) $\begin{bmatrix} 1 & 0 & 1 & 0 \\ -2 & 4 & 1 & -2 \\ -3 & 8 & 3 & -4 \end{bmatrix}, \begin{bmatrix} -1 \\ -4 \\ -9 \end{bmatrix}$

*(g) $\begin{bmatrix} 1 & 6 & 0 & 7 \\ 0 & -2 & 1 & -3 \\ 0 & 3 & 4 & -1 \\ 0 & 2 & 5 & -3 \end{bmatrix}, \begin{bmatrix} 21 \\ -8 \\ 1 \\ -4 \end{bmatrix}$

(h) $\begin{bmatrix} 1 & 6 & 4 \\ 0 & 2 & 2 \\ 3 & 5 & 1 \\ -1 & 1 & 3 \end{bmatrix}, \begin{bmatrix} 12 \\ 4 \\ 10 \\ 2 \end{bmatrix}$

*(i) $\begin{bmatrix} 1 & 2 & -1 & 1 & 2 \\ 1 & 2 & -1 & 1 & 2 \\ 2 & 4 & -2 & 2 & 5 \\ 2 & 4 & 0 & 1 & -1 \end{bmatrix}, \begin{bmatrix} 2 \\ 2 \\ 5 \\ 1 \end{bmatrix}$

(j) $\begin{bmatrix} 0 & -1 & 1 & 3 & 1 \\ 0 & 1 & 0 & -1 & 0 \\ 0 & 0 & 1 & 2 & 2 \\ 0 & -1 & 1 & 3 & 2 \end{bmatrix}, \begin{bmatrix} 0 \\ 1 \\ 1 \\ 0 \end{bmatrix}$

*(k) $\begin{bmatrix} 1 & -2 & 0 & 1 & 2 \\ 2 & 0 & 2 & 3 & 4 \\ -1 & 0 & 2 & 1 & 1 \\ 2 & 0 & -2 & 3 & 8 \\ 3 & 0 & 2 & 2 & 1 \end{bmatrix}, \begin{bmatrix} 2 \\ 7 \\ 2 \\ 3 \\ 7 \end{bmatrix}$

(l) $[3 \quad 4 \quad -7]$, $[3]$

*(m) $\begin{bmatrix} 0 \\ 9 \\ 3 \end{bmatrix}, \begin{bmatrix} 0 \\ -3 \\ -1 \end{bmatrix}$

(n) $\begin{bmatrix} 0 & 0 & 0 \\ 0 & 0 & 0 \\ 0 & 0 & 1 \end{bmatrix}, \begin{bmatrix} 0 \\ 0 \\ 5 \end{bmatrix}$

**10.** Show that the vectors $\mathbf{v}_1, \mathbf{v}_2, \mathbf{v}_3$ in $R^3$ form a basis for $R^3$ if and only if they do not lie on the same plane through the origin.

**11.** Let $\mathbf{v}_1, \mathbf{v}_2, \mathbf{v}_3, \mathbf{v}_4$ be a basis for $R^4$ and let $V$ be the subspace of $R^4$ spanned by $\mathbf{v}_3$ and $\mathbf{v}_4$. Show that if $\mathbf{v} = \alpha_1\mathbf{v}_1 + \alpha_2\mathbf{v}_2$ is in $V$, then $\mathbf{v} = \mathbf{0}$.

***12.** Given that $\mathbf{v}_1, \mathbf{v}_2$, and $\mathbf{v}_3$ form a basis for $V$. If

$$\mathbf{w}_1 = \mathbf{v}_1 + 2\mathbf{v}_2,$$
$$\mathbf{w}_2 = 3\mathbf{v}_1 + 2\mathbf{v}_2 + \mathbf{v}_3,$$
$$\mathbf{w}_3 = 4\mathbf{v}_1,$$

show that $\mathbf{w}_1, \mathbf{w}_2$, and $\mathbf{w}_3$ form a basis for $V$.

**13.** Do $\mathbf{e}_1 + \mathbf{e}_2$, $\mathbf{e}_2 + \mathbf{e}_3$, $\mathbf{e}_3 + \mathbf{e}_1$ form a basis for $R^3$? How about $\mathbf{e}_1 + \mathbf{e}_2$, $\mathbf{e}_2 + \mathbf{e}_3$, $\mathbf{e}_3 + \mathbf{e}_4$, $\mathbf{e}_4 + \mathbf{e}_1$ for $R^4$?

**14.** Suppose that the vectors $\mathbf{v}_1, \mathbf{v}_2, \mathbf{v}_3$ are a basis for a vector space $V$. Show that the vectors $\mathbf{u}_1 = \mathbf{v}_1 + \mathbf{v}_2$, $\mathbf{u}_2 = \mathbf{v}_2 + \mathbf{v}_3$, and $\mathbf{u}_3 = 2\mathbf{v}_3$ form a basis for $V$.

**15.** (a) Prove Theorem 3.
(b) Prove Theorem 6.

**16.** Let $\mathbf{v}_1, \mathbf{v}_2, \ldots, \mathbf{v}_n$ be a basis for a subspace $V$ of $R^m$. If $\mathbf{w}$ is in $V$ and $\mathbf{w}$ is not a linear combination of $\mathbf{v}_1, \mathbf{v}_2, \ldots, \mathbf{v}_{n-1}$, show that $\mathbf{v}_1, \mathbf{v}_2, \ldots, \mathbf{v}_{n-1}, \mathbf{w}$ form a basis for $V$.

***17.** Prove that if $\mathbf{v}_1, \mathbf{v}_2, \ldots, \mathbf{v}_n$ are vectors in a subspace $V$ of $R^m$ such that every vector $\mathbf{v}$ in $V$ has a unique expression as a linear combination of these vectors, then these vectors form a basis for $V$.

**18.** Let $\mathbf{v}_1, \mathbf{v}_2, \ldots, \mathbf{v}_n$ be a basis for a subspace $V$ of $R^m$.
(a) Show that if one of the vectors is deleted from the list, we no longer have a basis; the reduced list no longer spans $V$.
(b) Show that if a vector in $V$ is added to the list, we no longer have a basis; the expanded list is dependent.

**19.** Let $\mathbf{v}_1, \mathbf{v}_2, \ldots, \mathbf{v}_m$ be $m$ vectors in $R^m$. Prove the following.
*(a) If $\mathbf{v}_1, \mathbf{v}_2, \ldots, \mathbf{v}_m$ are independent, then they form a basis.
(b) If $\mathbf{v}_1, \mathbf{v}_2, \ldots, \mathbf{v}_m$ span $R^m$, they form a basis.

**20.** Suppose that $\mathbf{v}_1, \mathbf{v}_2, \ldots, \mathbf{v}_n$ form a basis for $V$. If $1 \leq k \leq n$, $\alpha_1, \ldots, \alpha_k$ are scalars with $\alpha_k \neq 0$, and $\mathbf{v} = \alpha_1\mathbf{v}_1 + \cdots + \alpha_k\mathbf{v}_k$, then

$$\mathbf{v}_1, \ldots, \mathbf{v}_{k-1}, \mathbf{v}, \mathbf{v}_{k+1}, \ldots, \mathbf{v}_n$$

form a basis for $V$.

**21.** Suppose that $\mathbf{v}_1, \mathbf{v}_2, \ldots, \mathbf{v}_k$ are independent vectors in a subspace $V$. Show that if these vectors do not form a basis for $V$, then there must exist a vector $\mathbf{v}_{k+1}$ in $V$ such that the set $\mathbf{v}_1, \mathbf{v}_2, \ldots, \mathbf{v}_k, \mathbf{v}_{k+1}$ is independent.

***22.** Is it true that if $\mathbf{v}_1, \mathbf{v}_2, \mathbf{v}_3$ are independent, then $\alpha_{11}\mathbf{v}_1, \alpha_{12}\mathbf{v}_1 + \alpha_{22}\mathbf{v}_2$, $\alpha_{13}\mathbf{v}_1 + \alpha_{23}\mathbf{v}_2 + \alpha_{33}\mathbf{v}_3$ are independent where the $\alpha_{ij}$ are scalars? Where the $\alpha_{ij}$ are nonzero scalars?

**23.** Show that $\mathbf{v}_1, \ldots, \mathbf{v}_4$ form a basis for $R^4$ if and only if the equation

$$\alpha_{1j}\mathbf{v}_1 + \alpha_{2j}\mathbf{v}_2 + \alpha_{3j}\mathbf{v}_3 + \alpha_{4j}\mathbf{v}_4 = \mathbf{e}_j$$

has a solution for $j = 1, 2, 3, 4$.

**24.** Let $\mathbf{v}_1, \mathbf{v}_2, \ldots, \mathbf{v}_n$ be vectors in $R^m$. Show that $k$ linear combinations of these vectors are independent if and only if the $k \times k$ matrix whose entries are the coefficients in these linear combinations is nonsingular.

**25.** Show that if $\mathbf{v}_1, \ldots, \mathbf{v}_n$ are independent and $\mathbf{v} = \alpha_1 \mathbf{v}_1 + \cdots + \alpha_n \mathbf{v}_n$, then $\mathbf{v} - \mathbf{v}_1, \mathbf{v} - \mathbf{v}_2, \ldots, \mathbf{v} - \mathbf{v}_n$ are independent if $\alpha_1 + \cdots + \alpha_n = 1$.

**26.** Let $A$ be an $n \times n$ matrix.
(a) Show that if all columns of $A$ sum to five, then rank $A - 5I < n$.
(b) Show that if all rows of $A$ sum to five, then rank $A - 5I < n$.
(c) Find and prove a generalization of parts (a) and (b).

**27.** Determine the intersection $U \cap V$ of the following subspaces $U$ and $V$.
(a) $U$ is the plane $x_1 + x_2 + x_3 = 0$ and $V$ is the plane $x_1 + 2x_2 + x_3 = 0$.
(b) $U$ is the subspace spanned by $(1, 2, 3, 4)$ and $V$ is the subspace spanned by $(1, 0, 1, 0)$, $(2, 3, 5, 0)$, $(1, 1, 1, 0)$.
(c) $U$ is the subspace spanned by $(3, -6, 1, 3)$, $(0, -2, 1, 1)$ and $V$ is the subspace spanned by $(1, -2, 1, 2)$, $(-1, 2, 1, 0)$, $(2, -4, 0, 1)$.

**28.** Let $U$ and $V$ be subspaces of $R^m$ with $U \cap V = \{\mathbf{0}\}$. If $\mathbf{u}_1, \mathbf{u}_2, \ldots, \mathbf{u}_r$ are a basis for $U$ and $\mathbf{v}_1, \mathbf{v}_2, \ldots, \mathbf{v}_s$ are a basis for $V$, show that $\mathbf{u}_1, \ldots, \mathbf{u}_r, \mathbf{v}_1, \ldots, \mathbf{v}_s$ are a basis for $U + V$.

**29.** Prove Theorem 4 by forming the augmented matrix $[A : \mathbf{v}]$ where $A$ is the matrix whose columns are $\mathbf{v}_1, \mathbf{v}_2, \ldots, \mathbf{v}_n$.

# 2.5
# Dimension

Consider a line and a plane in $R^3$. Why is the plane a bigger subspace than the line? Well, if the plane contains the line the answer is obvious [see Figure 2.8(a)]. But even if the plane does not contain the line [see Figure 2.8(b)] it is still geometrically obvious that the plane is bigger than the line. But why? Because a line consists of all possible multiples of one fixed vector and a plane consists of all possible linear combinations of two fixed noncollinear vectors. On a line we can move in only one fixed direction. But in a plane we can move in two independent directions. This corresponds to the fact that a line has a basis consisting of a single nonzero vector,

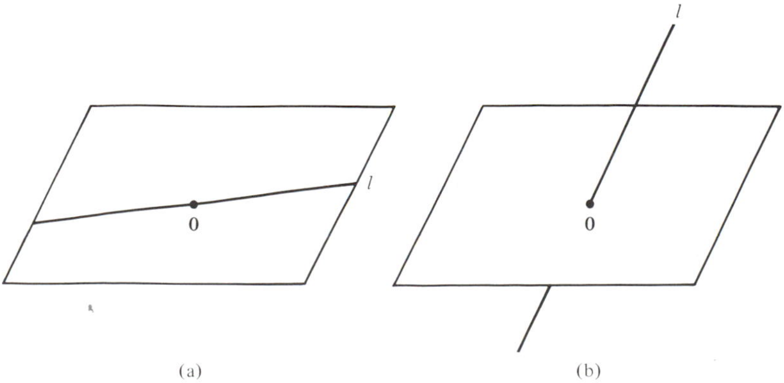


Figure 2.8

whereas a plane has a basis consisting of two noncollinear vectors. Thus a possible answer to our question is to say that a plane is a bigger subspace than a line because a plane has a basis consisting of two vectors and a line has a basis consisting of one vector. But this answer immediately raises another question. If we are going to define the size of a subspace to be the number of vectors in a basis for the subspace, we must know that any two bases for a subspace have the same number of vectors. Is this true? After all, any nontrivial subspace of $R^m$ has infinitely many bases. In particular, $R^m$ itself has infinitely many bases. Does every basis for $R^m$ have the same number of vectors?

***Theorem 1*** *Every basis for $R^m$ contains m vectors.*

**Proof** By Theorem 4 of Section 2.2, $R^m$ cannot be spanned by fewer than $m$ vectors. Therefore, a basis for $R^m$ must contain at least $m$ vectors. Theorem 4 of Section 2.3 states that $R^m$ cannot contain more than $m$ independent vectors. Therefore, a basis for $R^m$ must contain at most $m$ vectors. Combining these two facts completes the proof.

Thus we have an affirmative answer to our question for $R^m$. Is it also true that any two bases for a subspace $V$ of $R^m$ have the same number of vectors? The answer is again affirmative and constitutes one of the most important results of linear algebra. We begin its proof with the following theorem.

***Theorem 2*** *Let $V$ be a subspace of $R^m$. If $V$ is spanned by $n$ vectors, then no set of more than $n$ vectors in $V$ can be independent.*

**Proof** Suppose that $\mathbf{w}_1, \mathbf{w}_2, \ldots, \mathbf{w}_n$ span $V$ and that $\mathbf{v}_1, \mathbf{v}_2, \ldots, \mathbf{v}_k$ are vectors in $V$ with $k > n$. Our job is to show that the vectors $\mathbf{v}_1, \mathbf{v}_2, \ldots, \mathbf{v}_k$ are dependent. Let $A$ be the $m \times n$ matrix whose columns are $\mathbf{w}_1, \mathbf{w}_2, \ldots, \mathbf{w}_n$; thus $V = C(A)$. Since the $\mathbf{v}_j$'s are in $V$, it follows that each of the $k$ systems

$$A\mathbf{x} = \mathbf{v}_j, \qquad j = 1, 2, \ldots, k$$

is consistent. Let $\mathbf{b}_j, j = 1, \ldots, k$, be vectors in $R^n$ such that $A\mathbf{b}_j = \mathbf{v}_j$ and let $B$ be the $n \times k$ matrix whose columns are the vectors $\mathbf{b}_1, \mathbf{b}_2, \ldots, \mathbf{b}_k$. The product $C = AB$ is an $m \times k$ matrix whose columns are the vectors $\mathbf{v}_1, \mathbf{v}_2, \ldots, \mathbf{v}_k$. We will show that these vectors are dependent by showing that the nullspace of $C$ is not the trivial subspace. Since $B$ is an $n \times k$ matrix, its rank is at most $n$. Since $k > n$, the nullspace of $B$ is nontrivial. If $\mathbf{u}$ is a nonzero vector in $N(B)$, then

$$C\mathbf{u} = (AB)\mathbf{u} = A(B\mathbf{u}) = A\mathbf{0} = \mathbf{0}.$$

Thus $N(C) \neq \{\mathbf{0}\}$, which proves that its columns are dependent.

Now for the big theorem.

***Theorem 3*** *Let $V$ be a subspace of $R^m$. Any two bases for $V$ have the same number of vectors.*

**Proof** First suppose that $V$ is the trivial subspace. Then any basis for $V$ is empty and hence any two bases for $V$ have the same number of elements, namely 0. Now suppose that $V$ is a nontrivial subspace and that $\mathbf{v}_1, \mathbf{v}_2, \ldots, \mathbf{v}_n$ and $\mathbf{w}_1, \mathbf{w}_2, \ldots, \mathbf{w}_k$ are two bases for $V$. Since the $\mathbf{v}_i$'s span $V$ and the $\mathbf{w}_i$'s are independent, $k \leq n$. Similarly, since the $\mathbf{w}_i$'s span $V$ and the $\mathbf{v}_i$'s are independent, $n \leq k$. Therefore, $n = k$ and the proof is complete.

***Definition*** Let $V$ be a subspace of $R^m$. The number of elements in any basis for $V$ is called the **dimension** of $V$ and is denoted by dim $V$.

Thus the dimension of a subspace is a measure of its "size." The only zero-dimensional subspace is the trivial subspace. One-dimensional subspaces are the smallest nontrivial subspaces of $R^m$. Geometrically, they are lines through the origin. Two-dimensional subspaces are bigger. They are the subspaces that have bases consisting of two vectors. Any one-dimensional subspace is contained in a two-dimensional subspace of $R^m$ as long as $m \geq 2$. For if $V$ is a one-dimensional subspace and $\mathbf{v}$ is a basis for $V$, then $\mathbf{v}$ together with any nonzero vector $\mathbf{w}$ not in $V$ form a basis for a two-dimensional subspace that contains $V$. Geometrically, two-dimensional subspaces are planes through the origin. Similarly, three-dimensional subspaces are bigger than two-dimensional subspaces because any two-dimensional subspace is contained in a three-dimensional subspace of $R^m$ as long as $m \geq 3$. And so it goes.

Consequently, the concept of dimension provides a method for classifying all subspaces of $R^m$ according to size. The only zero-dimensional subspace of $R^m$ is the trivial subspace. The one-dimensional subspaces are lines through the origin. The two-dimensional subspaces are planes through the origin. The higher-dimensional subspaces of $R^m$ are higher-dimensional generalizations of planes. The biggest proper subspaces of $R^m$ are those subspaces with dimension $m - 1$. They are called **hyperplanes**. For instance, a line is a hyperplane in $R^2$ and a plane is a hyperplane in $R^3$.

To determine the dimension of a subspace $V$ of $R^m$, we have to find a basis for $V$. We proved that every subspace has a basis (Theorem 5 of Section 2.4), so our search will not be in vain. Once we find one basis for $V$, then the number $n$ of vectors in this basis is the dimension of $V$ and all bases for $V$ have $n$ vectors. For instance, $R^m$ has dimension $m$ because the standard basis for $R^m$ has $m$ vectors. Our next theorem gives the dimension of the three subspaces associated with an $m \times n$ matrix.

***Theorem 4*** *Let $A$ be an $m \times n$ matrix with rank $r$. Then*:

(a) $\dim C(A) = r$.
(b) $\dim N(A) = n - r$.
(c) $\dim R(A) = r$.

**Proof** Since $A$ has rank $r$, $A$ has $r$ pivot columns. By Theorem 1 of Section 2.4, these columns form a basis for $C(A)$ and hence $\dim C(A) = r$. Now for $N(A)$. By Theorem 1 of Section 2.4, $N(A)$ has a basis vector for every free variable. The free variables correspond to the nonpivot columns and since there are $n - r$ nonpivot columns, there are $n - r$ free variables. Therefore, $\dim N(A) = n - r$. Finally, since the rank of $A$ is the number of nonzero rows in any echelon form of $A$, and since these nonzero rows form a basis for $R(A)$ (Theorem 1 of Section 2.4), $\dim R(A) = r$.

From (a) and (c) we have the remarkable fact that the number of independent columns in a matrix is equal to the number of independent rows in the matrix.

The dimension of the nullspace of a matrix $A$ is called the **nullity** of $A$. Using this terminology, Theorem 5(b) can be written in the form

$$\text{rank } A + \text{nullity } A = n \qquad (A \text{ is } m \times n).$$

This is called the **rank plus nullity theorem**.

***Theorem 5*** *The column space of a matrix $A$ is equal to the row space of $A^T$: $C(A) = R(A^T)$. Consequently,*

$$\text{rank } A = \text{rank } A^T.$$

**Proof** The rows of $A^T$ are the columns of $A$. Therefore, the collection of all linear combinations of the columns of $A$ is equal to the collection of all linear combinations of the rows of $A^T$; that is, $C(A) = R(A^T)$. Finally,

$$\text{rank } A = \dim C(A) = \dim R(A^T) = \text{rank } A^T.$$

## EXAMPLE 1

In Section 1.2 we asked whether there always exists a meaningful set of prices for the goods in a closed economy. The example we considered consisted of three producers and we saw that in order to have a state of equilibrium, the price vector $\mathbf{p} = (p_1, p_2, p_3)$ must be a solution of the system

$$\begin{bmatrix} \frac{2}{5} & \frac{1}{4} & \frac{1}{2} \\ \frac{1}{5} & \frac{1}{2} & \frac{1}{4} \\ \frac{2}{5} & \frac{1}{4} & \frac{1}{4} \end{bmatrix} \begin{bmatrix} p_1 \\ p_2 \\ p_3 \end{bmatrix} = \begin{bmatrix} p_1 \\ p_2 \\ p_3 \end{bmatrix}.$$

As explained in Section 1.2, the entries in the $i$th column of the matrix describe how the goods produced by astronaut $i$ are divided among the three astronauts. Therefore, the sum of the entries in any column is 1. If we call the coefficient matrix $A$, we can rewrite the equation as $A\mathbf{p} = I\mathbf{p}$ or $(A - I)\mathbf{p} = \mathbf{0}$.

$$\begin{bmatrix} -\frac{3}{5} & \frac{1}{4} & \frac{1}{2} \\ \frac{1}{5} & -\frac{1}{2} & \frac{1}{4} \\ \frac{2}{5} & \frac{1}{4} & -\frac{3}{4} \end{bmatrix} \begin{bmatrix} p_1 \\ p_2 \\ p_3 \end{bmatrix} = \begin{bmatrix} 0 \\ 0 \\ 0 \end{bmatrix}.$$

The entries in each column of $A - I$ sum to zero. Therefore, the rows of $A - I$ are linearly dependent and hence the rank of $A - I$ is 2 or less. It follows that the nullity of $A$ is at least 1. This means that there is a nontrivial solution $\mathbf{p}$. The general solution of $(A - I)\mathbf{p} = \mathbf{0}$ is $\mathbf{p} = p_3(\frac{5}{4}, 1, 1)$ and the nullspace of $A - I$ is one-dimensional. We conclude that in this economy all sets of prices are essentially the same. All solutions except the trivial one assign the same relative values to the goods. Doubling $p_3$ is pure inflation. The only thing that changes is the "value" of the monetary unit.

## EXAMPLE 2

In Section 1.2 we saw that an electric circuit gives rise to a system of linear equations. The particular circuit that was discussed had four junctions and six branches. We deduced from Kirchhoff's s law that the current vector $\mathbf{I} = (I_1, I_2, I_3, I_4, I_5, I_6)$ has to be the solution of the homogeneous linear system $G\mathbf{I} = \mathbf{0}$, where

$$G = \begin{bmatrix} 1 & -1 & -1 & 0 & 0 & 0 \\ 0 & 1 & 0 & -1 & -1 & 0 \\ 0 & 0 & 1 & 1 & 0 & -1 \\ -1 & 0 & 0 & 0 & 1 & 1 \end{bmatrix}.$$

This $4 \times 6$ matrix (called the *incidence matrix* for the circuit) contains information about the design of the circuit. Each row of the matrix corresponds to a junction. For instance, the third row tells us that currents $I_3$ and $I_4$ flow into the third junction $J_3$ and that $I_6$ flows out of the junction. Each column of the matrix corresponds to a current. For instance, the fifth column indicates that $I_5$ flows from junction $J_2$ to junction $J_4$. In particular, each column contains besides zero entries exactly one $+1$ entry and one $-1$ entry. Therefore, the sum of the rows is a zero row and hence the rows of the matrix $G$ are linearly dependent. Since the first three rows of $G$ are linearly independent, the rank of $G$ is 3. Thus the nullspace of $G$ has dimension $6 - 3 = 3$. In fact, solving the system $G\mathbf{I} = \mathbf{0}$ leads to the general solution

$$\mathbf{I} = I_4(0, 1, -1, 1, 0, 0) + I_5(1, 1, 0, 0, 1, 0) + I_6(1, 0, 1, 0, 0, 1).$$

We now see that in this particular circuit we have to measure at least three currents before we can calculate all currents.

The concept of dimension also allows us to give a geometric description of the set of all solutions of a nonhomogeneous system $A\mathbf{x} = \mathbf{b}$. Since $\mathbf{b} \neq \mathbf{0}$, the set of solutions is not a subspace because it does not contain the zero vector ($A \cdot \mathbf{0} = \mathbf{0} \neq \mathbf{b}$). However, by Theorem 5 of Section 1.10, the set of solutions of $A\mathbf{x} = \mathbf{b}$ is a close relative of a subspace, namely, the nullspace of $A$.

***Theorem 6*** *Suppose that $A\mathbf{x} = \mathbf{b}$ is consistent and that $\mathbf{v}_p$ is a solution of the system. Then any solution of the system can be expressed in the form*

$$\mathbf{v} = \mathbf{v}_p + \mathbf{u},$$

*where $\mathbf{u}$ is a vector in the nullspace of $A$. Moreover, every vector of this form is a solution of the system.*

**Proof** This is simply a restatement of Theorem 5 of Section 1.10 in the language of nullspaces.

Let $A$ be an $m \times n$ matrix. The system $A\mathbf{x} = \mathbf{b}$ is consistent if and only if $\mathbf{b}$ lies in the column space of $A$. From this theorem it follows that all solutions of $A\mathbf{x} = \mathbf{b}$ are obtained by displacing the nullspace of $A$ by a single solution $\mathbf{v}_p$ of $A\mathbf{x} = \mathbf{b}$. The nullspace of $A$ is a subspace of $R^n$ of dimension $n - r$, where $r$ is the rank of $A$. Hence we might be tempted to speak of the set of solutions of $A\mathbf{x} = \mathbf{b}$ as having dimension $n - r$. Of course, strictly speaking, we defined the concept of dimension only for subspaces, and the set of solutions of $A\mathbf{x} = \mathbf{b}$ do not form a subspace unless $\mathbf{b} = \mathbf{0}$. The important point is that the "size" of the solution set of $A\mathbf{x} = \mathbf{b}$ depends only on the matrix $A$ and not on the particular vector $\mathbf{b}$ in the column space of $A$. If $\mathbf{c}$ is also in $C(A)$, the solution sets of $A\mathbf{x} = \mathbf{b}$ and $A\mathbf{x} = \mathbf{c}$ differ only in their displacement from the nullspace of $A$ and not in their "size." In particular, we see that $A\mathbf{x} = \mathbf{b}$ has a unique solution for every vector $\mathbf{b}$ in the column space of $A$ if and only if the nullspace of $A$ is trivial (i.e., $r = n$).

### EXAMPLE 3

Consider the systems $A\mathbf{x} = \mathbf{b}$ and $A\mathbf{x} = \mathbf{c}$, where

$$A = \begin{bmatrix} 1 & 2 \\ 2 & 4 \end{bmatrix}, \qquad \mathbf{b} = \begin{bmatrix} 3 \\ 6 \end{bmatrix}, \qquad \mathbf{c} = \begin{bmatrix} -1 \\ -2 \end{bmatrix}.$$

The nullspace of $A$ is the line in $R^2$ spanned by $\mathbf{u} = (-2, 1)$. The vector $\mathbf{v}_p = (3, 0)$ is a particular solution of $A\mathbf{x} = \mathbf{b}$. Thus the solutions of $A\mathbf{x} = \mathbf{b}$ are precisely the vectors

$$\mathbf{v} = \mathbf{v}_p + \alpha\mathbf{u},$$

where $\alpha$ is a scalar. These vectors form a line through $\mathbf{v}_p$ and parallel to the nullspace of $A$. Similarly, the vector $\mathbf{w}_p = (-1, 0)$ is a solution of $A\mathbf{x} = \mathbf{c}$, and hence the solutions of $A\mathbf{x} = \mathbf{c}$ are precisely the vectors

$$\mathbf{w} = \mathbf{w}_p + \alpha\mathbf{u},$$

where $\alpha$ is a scalar. These vectors form a line through $\mathbf{w}_p$ parallel to the nullspace of $A$ and of course also parallel to the line $\mathbf{v} = \mathbf{v}_p + \alpha\mathbf{u}$ (see Figure 2.9).

### EXAMPLE 4

Consider the systems $A\mathbf{x} = \mathbf{b}$ and $A\mathbf{x} = \mathbf{c}$, where

$$A = \begin{bmatrix} 1 & 0 & -1 \\ 2 & 0 & -2 \end{bmatrix}, \qquad \mathbf{b} = \begin{bmatrix} 2 \\ 4 \end{bmatrix}, \qquad \mathbf{c} = \begin{bmatrix} 3 \\ 6 \end{bmatrix}.$$

The nullspace of $A$ is the plane

$$\mathbf{x} = \alpha_1\mathbf{u}_1 + \alpha_2\mathbf{u}_2$$

spanned by

$$\mathbf{u}_1 = (1, 0, 1) \qquad \text{and} \qquad \mathbf{u}_2 = (0, 1, 0).$$

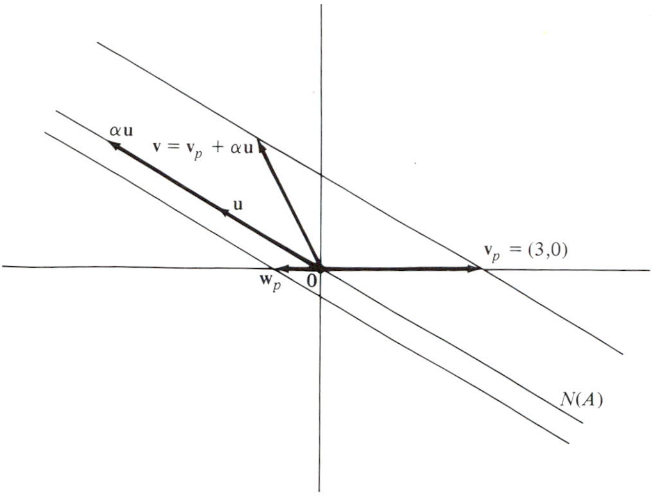


Figure 2.9

Since $\mathbf{v}_p = (2, 0, 0)$ and $\mathbf{w}_p = (3, 0, 0)$ are solutions of $A\mathbf{x} = \mathbf{b}$ and $A\mathbf{x} = \mathbf{c}$, respectively, the solution sets of $A\mathbf{x} = \mathbf{b}$ and $A\mathbf{x} = \mathbf{c}$ are the planes

$$\mathbf{x} = \mathbf{v}_p + \beta_1 \mathbf{u}_1 + \beta_2 \mathbf{u}_2$$
$$\mathbf{x} = \mathbf{w}_p + \gamma_1 \mathbf{u}_1 + \gamma_2 \mathbf{u}_2.$$

Both planes are parallel to $N(A)$. The first plane is obtained by translating the nullspace of $A$ by the solution $\mathbf{v}_p$ of the equation $A\mathbf{x} = \mathbf{b}$. Similarly, the second plane is obtained by translating $N(A)$ by the solution $\mathbf{w}_p$ of $A\mathbf{x} = \mathbf{c}$ (see Figure 2.10).

We proved that $R^m$ cannot be spanned by fewer than $m$ vectors and that it cannot contain more than $m$ independent vectors. The reason is that $R^m$ is $m$-dimensional. In the next theorem we prove that both of these facts remain true for subspaces of $R^m$.

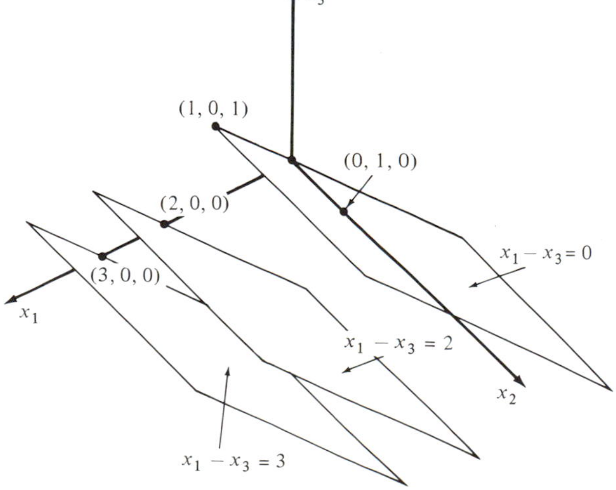


Figure 2.10

***Theorem 7*** *Let $V$ be an $n$-dimensional subspace of $R^m$.*

*(a) $V$ cannot be spanned by fewer than $n$ vectors.*
*(b) $V$ cannot contain more than $n$ independent vectors.*

**Proof** Suppose that $V$ has dimension $n$. Since $V$ has a basis consisting of $n$ vectors, it contains $n$ independent vectors. If $V$ were spanned by fewer than $n$ vectors, we would then have a contradiction to Theorem 2. This proves part (a). To prove part (b), we observe that $V$ is spanned by the $n$ vectors in any basis for $V$ and hence by Theorem 2 cannot contain more than $n$ independent vectors.

Thus the dimension of a subspace $V$ of $R^m$ is the minimum number of vectors needed to span $V$ and also the maximum number of independent vectors in $V$. Hence *a basis for a subspace $V$ is simultaneously a spanning set for $V$ of minimal size and an independent set of maximal size in $V$.*

To determine whether a given set of vectors forms a basis for a subspace $V$ we must determine whether the vectors span $V$ *and* are independent. However, if we somehow know that the dimension of $V$ is $n$ and our given set of vectors contains $n$ vectors, the following theorem states that they span $V$ if and only if they are independent. Thus it suffices in this case to determine whether the vectors span $V$ *or* are independent.

***Theorem 8*** *Let $V$ be an $n$-dimensional subspace of $R^m$.*

*(a) Any set of $n$ vectors that span $V$ form a basis for $V$.*
*(b) Any set of $n$ independent vectors in $V$ form a basis for $V$.*

**Proof** To prove part (a), suppose that $\mathbf{v}_1, \mathbf{v}_2, \ldots, \mathbf{v}_n$ span $V$. If these vectors were dependent, then by Theorem 7 of Section 2.4 one of them would be a linear combination of the remaining $n-1$ vectors and these remaining vectors would span $V$. This is impossible by Theorem 7(a). Thus $\mathbf{v}_1, \mathbf{v}_2, \ldots, \mathbf{v}_n$ are independent, which proves part (a).

Now for the proof of part (b). Suppose that the vectors $\mathbf{v}_1, \mathbf{v}_2, \ldots, \mathbf{v}_n$ in $V$ are independent but do not span $V$. Then there is some vector $\mathbf{v}$ in $V$ that is not a linear combination of $\mathbf{v}_1, \mathbf{v}_2, \ldots, \mathbf{v}_n$. It follows from Theorem 4 of Section 2.4 that the $n+1$ vectors $\mathbf{v}_1, \mathbf{v}_2, \ldots, \mathbf{v}_n, \mathbf{v}$ are independent. This is impossible by Theorem 7(b). Thus the $n$ vectors $\mathbf{v}_1, \mathbf{v}_2, \ldots, \mathbf{v}_n$ span $V$, which proves part (b).

## EXAMPLE 5

The three vectors $\mathbf{v}_1 = (3,0,0)$, $\mathbf{v}_2 = (1,2,0)$, and $\mathbf{v}_3 = (5,1,4)$ are independent because all columns of the matrix

$$A = \begin{bmatrix} 3 & 1 & 5 \\ 0 & 2 & 1 \\ 0 & 0 & 4 \end{bmatrix}$$

are pivot columns. Since dim $R^3 = 3$, it follows from Theorem 8 that $\mathbf{v}_1, \mathbf{v}_2, \mathbf{v}_3$ is a basis for $R^3$.

We remarked earlier in this section that the dimension of a subspace is a measure of its size. For example, all two-dimensional subspaces of $R^3$ are planes in $R^3$ through the origin. Similarly, any two-dimensional subspace of $R^7$ is a plane in $R^7$. If we disregard the fact that a plane in $R^7$ lives in $R^7$, then it is geometrically the same as a plane in $R^3$. In fact, any plane looks like $R^2$. This is no accident. Any basis for a plane consists of two vectors. Therefore, if $\mathbf{v}_1$ and $\mathbf{v}_2$ form a basis for $V$ and $\mathbf{v}$ is a vector in the plane, there is one and only one way to choose two real numbers $\alpha_1$ and $\alpha_2$ such that $\mathbf{v} = \alpha_1\mathbf{v}_1 + \alpha_2\mathbf{v}_2$. These numbers can be thought of as the name for $\mathbf{v}$ with respect to the basis $\mathbf{v}_1, \mathbf{v}_2$.

More generally, suppose that $V$ is an $n$-dimensional subspace of $R^m$, and let $\mathbf{v}_1, \mathbf{v}_2, \ldots, \mathbf{v}_n$ be a basis for $V$. Then by Theorem 2 of Section 2.4 any vector $\mathbf{u}$ in $V$ can be expressed as a linear combination

$$\mathbf{u} = \alpha_1\mathbf{v}_1 + \alpha_2\mathbf{v}_2 + \cdots + \alpha_n\mathbf{v}_n$$

of the basis vectors in a unique way. The vector $(\alpha_1, \alpha_2, \ldots, \alpha_n)$ uniquely specifies the vector $\mathbf{u}$ with respect to the reference frame determined by the basis $\mathbf{v}_1, \mathbf{v}_2, \ldots, \mathbf{v}_n$.

***Definition*** The vector $\mathbf{c}_\mathbf{v}(\mathbf{u}) = (\alpha_1, \alpha_2, \ldots, \alpha_n)$ is called the **coordinate vector** of $\mathbf{u}$ with respect to the basis $\mathbf{v}_1, \mathbf{v}_2, \ldots, \mathbf{v}_n$ (or simply the **v-coordinate vector** of $\mathbf{u}$). The scalars $\alpha_1, \alpha_2, \ldots, \alpha_n$ are called the **coordinates** of $\mathbf{u}$ with respect to the basis $\mathbf{v}_1, \mathbf{v}_2, \ldots, \mathbf{v}_n$ (or simply the **v-coordinates** of $\mathbf{u}$).

Thus the coordinates of a vector name a vector with respect to a basis. The concept also gives a precise meaning to our intuitive argument that any plane looks like $R^2$. We simply associate a vector in a plane with its coordinate vector with respect to a basis for the plane. In a similar way, any $n$-dimensional subspace $V$ of $R^m$ looks just like $R^n$. We simply choose a basis for $V$ and associate any vector in $V$ with its coordinate vector with respect to the basis.

Not only do coordinate vectors allow us to associate vectors in an $n$-dimensional subspace $V$ of $R^m$ with $R^n$ itself, but they allow us to do linear algebra in $V$ by doing linear algebra in $R^n$. What we mean here is that forming linear combinations of vectors in $V$ is the same as forming the same linear combination of the coordinate vectors. The proof is straightforward. Let $\mathbf{v}_1, \mathbf{v}_2, \ldots, \mathbf{v}_n$ be a basis for $V$, let $\mathbf{x}$ and $\mathbf{y}$ be in $V$, and let $\mathbf{c}_\mathbf{v}(\mathbf{x}) = (\alpha_1, \alpha_2, \ldots, \alpha_n)$ and $\mathbf{c}_\mathbf{v}(\mathbf{y}) = (\beta_1, \beta_2, \ldots, \beta_n)$ be their v-coordinate vectors. Then

$$\mathbf{x} = \alpha_1\mathbf{v}_1 + \alpha_2\mathbf{v}_2 + \cdots + \alpha_n\mathbf{v}_n,$$
$$\mathbf{y} = \beta_1\mathbf{v}_1 + \beta_2\mathbf{v}_2 + \cdots + \beta_n\mathbf{v}_n,$$

and hence

$$\mathbf{x} + \mathbf{y} = (\alpha_1 + \beta_1)\mathbf{v}_1 + (\alpha_2 + \beta_2)\mathbf{v}_2 + \cdots + (\alpha_n + \beta_n)\mathbf{v}_n.$$

This proves that

$$\begin{aligned}\mathbf{c_v}(\mathbf{x}) + \mathbf{c_v}(\mathbf{y}) &= (\alpha_1, \alpha_2, \ldots, \alpha_n) + (\beta_1, \beta_2, \ldots, \beta_n) \\ &= (\alpha_1 + \beta_1, \alpha_2 + \beta_2, \ldots, \alpha_n + \beta_n) = \mathbf{c_v}(\mathbf{x} + \mathbf{y})\end{aligned}$$

is the **v**-coordinate vector of $\mathbf{x} + \mathbf{y}$. Similarly, if $\gamma$ is a scalar, then since

$$\gamma\mathbf{x} = \gamma\alpha_1\mathbf{v}_1 + \gamma\alpha_2\mathbf{v}_2 + \cdots + \gamma\alpha_n\mathbf{v}_n,$$

the **v**-coordinate vector of $\gamma\mathbf{x}$ is

$$\mathbf{c_v}(\gamma\mathbf{x}) = (\gamma\alpha_1, \gamma\alpha_2, \ldots, \gamma\alpha_n) = \gamma(\alpha_1, \alpha_2, \ldots, \alpha_n) = \gamma\mathbf{c_v}(\mathbf{x}).$$

Since our assertion holds for sums and scalar multiples, it holds for arbitrary linear combinations.

We have proved that any $n$-dimensional subspace is just like $R^n$, both geometrically and algebraically. Moreover, it does not matter where the $n$-dimensional subspace lives. An $n$-dimensional subspace $V$ of $R^m$ and an $n$-dimensional subspace $U$ of $R^p$ both look like $R^n$ and hence must look like each other. In this respect the Euclidean spaces $R^n$ play a very special role in linear algebra. Every other subspace that we have studied is just a disguised form of $R^n$ for some $n$.

Because of the central role which the Euclidean spaces play in linear algebra, we need to pay special attention to coordinates of vectors in $R^m$. If $\mathbf{v}_1, \mathbf{v}_2, \ldots, \mathbf{v}_m$ is a basis for $R^m$, then the **v**-coordinate vector of $\mathbf{x} = (x_1, x_2, \ldots, x_m)$ is the unique vector $\mathbf{c_v}(\mathbf{x}) = (x'_1, x'_2, \ldots, x'_m)$ such that

$$\mathbf{x} = x'_1\mathbf{v}_1 + x'_2\mathbf{v}_2 + \cdots + x'_m\mathbf{v}_m.$$

This equation is begging to be written in terms of matrices. Letting $M$ be the matrix whose columns are $\mathbf{v}_1, \mathbf{v}_2, \ldots, \mathbf{v}_m$, it becomes

$$\mathbf{x} = M\mathbf{c_v}(\mathbf{x}).$$

As vectors, $\mathbf{x}$ and $\mathbf{c_v}(\mathbf{x})$ are different (unless of course $M$ is the identity matrix). *But they both name the same vector in* $R^m$: $(x_1, x_2, \ldots, x_m)$ is the name for $\mathbf{x}$ with respect to the standard basis, $(x'_1, x'_2, \ldots, x'_m)$ is the name for $\mathbf{x}$ with respect to the new basis $\mathbf{v}_1, \mathbf{v}_2, \ldots, \mathbf{v}_m$. The matrix $M$ determines how names are changed, and hence is called the **change of coordinates matrix**. $M$ changes the new name for a vector back to its standard name: $\mathbf{x} = M\mathbf{c_v}(\mathbf{x})$. Consequently, $M^{-1}$ changes the standard name of a vector to its new name: $\mathbf{c_v}(\mathbf{x}) = M^{-1}\mathbf{x}$.

**Result 1** Let $\mathbf{v}_1, \mathbf{v}_2, \ldots, \mathbf{v}_m$ be a basis for $R^m$. The matrix $M$ whose columns are $\mathbf{v}_1, \mathbf{v}_2, \ldots, \mathbf{v}_m$ is nonsingular and for any $\mathbf{x}$ in $R^m$,

$$\mathbf{x} = M\mathbf{c_v}(\mathbf{x}) \qquad \text{and} \qquad \mathbf{c_v}(\mathbf{x}) = M^{-1}\mathbf{x}.$$

That is, $M$ is the change of coordinates matrix that changes the **v**-coordinates of a vector to its standard coordinates.

## EXAMPLE 6

Consider the basis $\mathbf{v}_1 = (1, 2)$ and $\mathbf{v}_2 = (3, 5)$ for $R^2$. The change of coordinates matrix $M$ that changes $\mathbf{v}$-coordinates to standard coordinates is

$$M = \begin{bmatrix} 1 & 3 \\ 2 & 5 \end{bmatrix}.$$

The matrix

$$M^{-1} = \begin{bmatrix} -5 & 3 \\ 2 & -1 \end{bmatrix}$$

changes standard coordinates to $\mathbf{v}$-coordinates. For instance, the $\mathbf{v}$-coordinate vector of the vector $\mathbf{x} = (3, 4)$ is

$$\mathbf{c}_\mathbf{v}(\mathbf{x}) = M^{-1}\mathbf{x} = \begin{bmatrix} -5 & 3 \\ 2 & -1 \end{bmatrix}\begin{bmatrix} 3 \\ 4 \end{bmatrix} = \begin{bmatrix} -3 \\ 2 \end{bmatrix}.$$

The standard coordinate vector $\mathbf{x}$ of the vector whose $\mathbf{v}$-coordinate vector is $\mathbf{c}_\mathbf{v}(\mathbf{x}) = (-1, 1)$ is

$$\mathbf{x} = M\mathbf{c}_\mathbf{v}(\mathbf{x}) = \begin{bmatrix} 1 & 3 \\ 2 & 5 \end{bmatrix}\begin{bmatrix} -1 \\ 1 \end{bmatrix} = \begin{bmatrix} 2 \\ 3 \end{bmatrix}.$$

The coordinate vector of a vector $\mathbf{x} = (x_1, x_2, \ldots, x_m)$ in $R^m$ with respect to the standard basis is just $(x_1, x_2, \ldots, x_m)$ because

$$\mathbf{x} = x_1\mathbf{e}_1 + x_2\mathbf{e}_2 + \cdots + x_m\mathbf{e}_m.$$

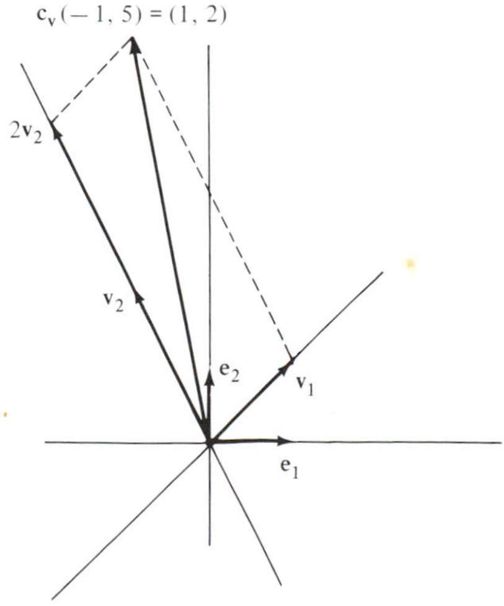


Figure 2.11

On the other hand, the coordinate vector of a basis vector $\mathbf{v}_i$ with respect to the basis $\mathbf{v}_1, \mathbf{v}_2, \ldots, \mathbf{v}_m$ is $\mathbf{e}_i$ because

$$\mathbf{v}_i = 0 \cdot \mathbf{v}_1 + \cdots + 0 \cdot \mathbf{v}_{i-1} + 1 \cdot \mathbf{v}_i + 0 \cdot \mathbf{v}_{i+1} + \cdots + 0 \cdot \mathbf{v}_m.$$

You are probably wondering why in the world we would want to work with any basis for $R^m$ other than the standard basis. It is a wonderful basis. Every vector is its own coordinate vector with respect to the standard basis. If we change bases, then vectors like $(1, 2)$ no longer name themselves. If the new basis for $R^2$ is $\mathbf{v}_1 = (1, 1)$, $\mathbf{v}_2 = (-1, 2)$, then $(1, 2)$ names the vector (see Figure 2.11)

$$1 \cdot \mathbf{v}_1 + 2 \cdot \mathbf{v}_2 = 1(1, 1) + 2(-1, 2) = (-1, 5).$$

Why introduce all this confusion?

Have faith! You will see in subsequent chapters that many results are greatly simplified by choosing a basis different from the standard basis. For now, we offer the following example.

### EXAMPLE 7

Let $\theta$ be a fixed angle. Let us rotate the standard basis vectors $\mathbf{e}_1$ and $\mathbf{e}_2$ of $R^2$ through the angle $\theta$ to obtain the vectors

$$\mathbf{v}_1 = (\cos\theta, \sin\theta) \qquad \text{and} \qquad \mathbf{v}_2 = (-\sin\theta, \cos\theta).$$

These two vectors are independent and hence form a basis for $R^2$. The matrix that changes standard coordinates to $\mathbf{v}$-coordinates is

$$M = \begin{bmatrix} \cos\theta & -\sin\theta \\ \sin\theta & \cos\theta \end{bmatrix}.$$

Thus when $\theta = \pi/4$,

$$M = \frac{1}{\sqrt{2}}\begin{bmatrix} 1 & -1 \\ 1 & 1 \end{bmatrix}. \tag{2}$$

Now suppose that we are asked to graph the equation

$$5x^2 - 6xy + 5y^2 = 8. \tag{3}$$

First we translate this equation into an equation in the coordinate system determined by $\mathbf{v}_1 = (\cos\theta, \sin\theta)$ and $\mathbf{v}_2 = (-\sin\theta, \cos\theta)$ when $\theta = \pi/4$. The change of coordinates matrix $M$ is given in (2) and the equation relating the standard coordinates $(x, y)$ to the $\mathbf{v}$-coordinates $(x', y')$ is

$$\begin{bmatrix} x \\ y \end{bmatrix} = \frac{1}{\sqrt{2}}\begin{bmatrix} 1 & -1 \\ 1 & 1 \end{bmatrix}\begin{bmatrix} x' \\ y' \end{bmatrix}$$

$$= \frac{1}{\sqrt{2}}\begin{bmatrix} x' - y' \\ x' + y' \end{bmatrix}.$$

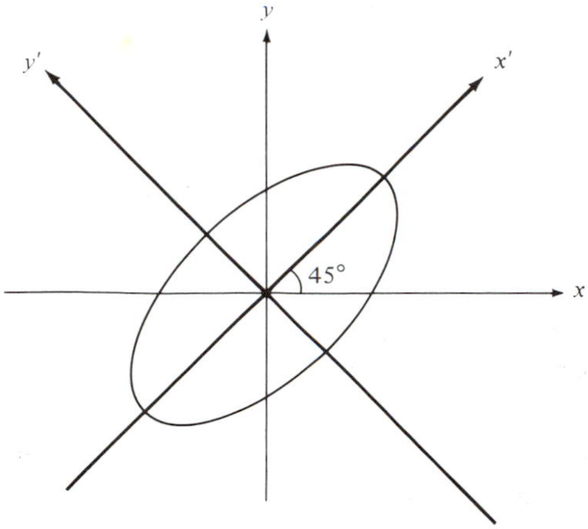


Figure 2.12

Substituting

$$x = \frac{1}{\sqrt{2}}(x' - y') \qquad \text{and} \qquad y = \frac{1}{\sqrt{2}}(x' + y')$$

into (3), we obtain (after a little algebra)

$$\frac{x'^2}{4} + y'^2 = 1.$$

This is, of course, the equation of an ellipse (see Figure 2.12).

How did we know that we should rotate the coordinate axes? Even if we knew the answer to that question, how did we know to rotate by $\pi/4$? Both of these questions are answered in Section 8.7.

## PROBLEMS 2.5

**1.** Define each of the following.
   (a) The dimension of a subspace.
   (b) A change of coordinates matrix.
   (c) The coordinates of a vector with respect to a basis.

**2.** In each part, let $V$ be the subspace of $R^m$ spanned by the given vectors. Find a basis for $V$ and the dimension of $V$.
   *(a) $(-2, 3, 1), (0, 1, 2), (-1, 0, 3)$
   (b) $(1, 1, 1, 1), (1, 2, 3, 4), (2, 3, 4, 5), (0, 1, 1, 0)$
   *(c) $(1, -2, 1, 1), (2, 0, 3, 1), (4, -4, 5, 3), (3, 1, 0, 1)$
   (d) $(-1, 0, 1, 1, 0), (2, 1, 3, 0, 1), (0, 1, 5, 2, 1), (2, 2, 8, 2, 2)$

**3.** In each part, let $V$ be the subspace spanned by the given vectors. Find the dimension of $V$ and a basis for $V$ by deleting some of the vectors if necessary.
   *(a) $(1, -1, 1), (3, 5, 1), (5, 3, 3)$
   (b) $(0, 3, 1), (1, 1, 1), (-3, 0, -2), (1, 0, 5)$

*(c) $(1,-1,2,0)$, $(2,-2,4,0)$, $(0,1,1,1)$, $(4,-3,9,1)$
(d) $(1,2,3,1,5)$, $(0,1,1,0,2)$, $(-1,0,1,2,-4)$, $(0,1,1,2,0)$, $(1,1,2,1,3)$ 2,

*4. Find the dimension of $C(A)$, $N(A)$, and $R(A)$ for each of the matrices in Problem 9 of Section 2.4.

*5. Find the dimension of $C(A)$, $N(A)$, and $R(A)$ for each of the matrices in Problem 5 of Section 2.3.

6. In each part, do the following.
(i) Find the dimension of $C(A)$, $N(A)$, and $R(A)$.
(ii) Find the constraints that a vector $\mathbf{b}$ must satisfy to be in $C(A)$.
(iii) Find a matrix $B$ such that $N(B) = C(A)$.
(iv) Find a matrix $B$ such that $C(B) = N(A)$.

*(a) $\begin{bmatrix} 1 & 2 & -1 & 1 & -1 & 1 \\ -1 & -2 & 1 & 1 & 2 & -3 \end{bmatrix}$

(b) $\begin{bmatrix} 1 & 2 & 1 & -1 \\ 3 & 6 & -3 & 3 \end{bmatrix}$

*(c) $\begin{bmatrix} 1 & 2 \\ 0 & 1 \\ 1 & 1 \end{bmatrix}$

(d) $\begin{bmatrix} 1 & 0 & 3 \\ 2 & 1 & -2 \\ -4 & -3 & 12 \end{bmatrix}$

*(e) $\begin{bmatrix} 1 & 2 & 3 \\ 2 & 3 & 4 \\ 5 & 6 & 7 \\ 8 & 9 & 10 \\ 10 & 11 & 12 \end{bmatrix}$

(f) $\begin{bmatrix} 2 & 3 & -1 & 2 \\ 4 & 6 & 0 & 3 \\ -2 & 5 & 3 & -1 \\ 0 & -1 & -2 & 1 \end{bmatrix}$

7. Show that the vectors $\mathbf{w}_1 = (-2,4,-2)$ and $\mathbf{w}_2 = (6,3,-4)$ form a basis for the nullspace of the matrix.

$$\begin{bmatrix} 1 & 2 & 3 \\ 2 & 4 & 6 \\ -3 & -6 & -9 \end{bmatrix}$$

8. In each part, do the following.
(i) Show that the given vectors form a basis for $R^m$ for an appropriate $m$.
(ii) Find the coordinates of the given vector $\mathbf{v}$ with respect to this basis.
(iii) Find the change of coordinates matrix that changes standard coordinates to $\mathbf{v}$-coordinates.

*(a) $(1,-3,2)$, $(1,2,4)$, $(3,-1,0)$, $\mathbf{v} = (11,-16,-8)$
(b) $(1,2,3)$, $(1,1,-1)$, $(1,4,1)$, $\mathbf{v} = (4,15,7)$
*(c) $(1,1,3)$, $(1,2,0)$, $(2,1,6)$, $\mathbf{v} = (2,4,3)$.
(d) $(1,0,1,0)$, $(0,1,1,1)$, $(1,-1,0,2)$, $(2,0,0,1)$, $\mathbf{v} = (1,2,3,4)$
(e) $(0,0,2,0)$, $(0,5,0,0)$, $(1,0,0,1)$, $(-2,0,0,3)$, $\mathbf{v} = (10,5,2,5)$

*9. Find a basis for and the dimension of the subspace $V$ spanned by the vectors $\mathbf{v}_1 = (1,-1,0,2)$, $\mathbf{v}_2 = (-2,2,0,-4)$, $\mathbf{v}_3 = (3,1,0,1)$, $\mathbf{v}_4 = (3,5,0,-4)$, and $\mathbf{v}_5 = (1,7,0,-8)$. Then find the coordinates of each of these vectors relative to the basis.

10. Find all $5 \times 3$ matrices whose nullspace is $R^3$.

11. Suppose that $\mathbf{v}_1, \mathbf{v}_2$ are independent and that $(\alpha, \gamma) \neq \mathbf{0}$. If $\alpha\beta\mathbf{v}_1 + \gamma\delta\mathbf{v}_2 = \alpha\mathbf{v}_1 + \gamma\mathbf{v}_2$, does it follow that $\beta = \delta = 1$?

12. Let $V$ be an $n$-dimensional subspace of $R^m$. Show that $V$ has subspaces of dimension $k$ for $k = 0, 1, \ldots, n$. (*Hint:* Let $\mathbf{v}_1, \mathbf{v}_2, \ldots, \mathbf{v}_n$ be a basis for $V$ and consider the subspaces spanned by $\mathbf{v}_1, \mathbf{v}_2, \ldots, \mathbf{v}_k$, $k = 1, \ldots, n$.)

13. Let $A$ be an $m \times n$ matrix and let $I$ be the $n \times n$ identity matrix. Show that the rank of $A$ is equal to $n$ if and only if there is an $n \times m$ matrix $B$ such that $BA = I$.

14. Let $A$ be an $m \times n$ matrix and $B$ be an $n \times r$ matrix. Show that $\text{rank}(AB) \leq$

rank $(B)$ and that rank $(AB) \leq$ rank $(A)$. (*Hint:* Use Problem 24 in Section 2.2.)

**15.** Suppose that $A$ is a nonsingular $m \times m$ matrix and $B$ is an $m \times n$ matrix. Show that rank $(AB) =$ rank $(B)$. (*Hint:* Use Problem 24 in Section 2.2.)

**16.** Suppose that $A$ is an $m \times n$ matrix and $B$ is a nonsingular $n \times n$ matrix. Show that rank $(AB) =$ rank $(A)$. (*Hint:* Use Problem 24 in Section 2.2.)

**17.** Let $A$ be an $m \times n$ matrix and let $B$ be an $n \times n$ matrix.
(a) Prove that if rank $(AB) =$ rank $(A)$, then rank $(A) \leq$ rank $(B)$. (*Hint:* Use Problem 14.)
(b) Prove that if rank $(AB) = n$, then rank $(A) =$ rank $(B) = n$.

***18.** Find the dimension of the subspace spanned by

$$(1, 1, \alpha), (1, \alpha, 1), (\alpha, 1, 1).$$

Your answer should depend on the value of $\alpha$.

**19.** Show that an $m \times n$ matrix $A$ has rank 1 if and only if there is a nonzero $m \times 1$ matrix $B$ and a nonzero $1 \times n$ matrix $C$ such that $A = BC$.

***20.** Let $V$ be a subspace of $R^m$. Prove that $\dim V \leq m$ and that $\dim V = m$ if and only if $V = R^m$.

**21.** Suppose that each of the vectors $\mathbf{w}_1, \mathbf{w}_2, \mathbf{w}_3, \mathbf{w}_4$ in $R^n$ $(n \geq 4)$ is a linear combination of the vectors $\mathbf{v}_1, \mathbf{v}_2, \mathbf{v}_3$ in $R^n$. Prove that the vectors $\mathbf{w}_1, \mathbf{w}_2, \mathbf{w}_3, \mathbf{w}_4$ are dependent.

**22.** Let $V$ be a subspace of $R^m$.
(a) Show that $V$ has dimension $n$ if and only if $n$ is the maximum number of independent vectors in $V$.
(b) Show that $V$ has dimension $n$ if and only if $n$ is the minimum number of vectors in any spanning set for $V$.

**23.** Let $V$ be a subspace of $R^m$ and let $W$ be a subspace of $V$. Show that $\dim W \leq \dim V$ and that $\dim W = \dim V$ if and only if $W = V$.

**24.** Suppose that $H$ and $K$ are subspaces of $R^7$, $\dim H = 4$ and $\dim K = 5$. What are the possible dimensions $H \cap K$ could have? Give an argument supporting your answer. [*Hint:* Here are two possible approaches.

1. How many independent homogeneous equations does it take to define $H$? How many to define $K$? Now how could you write down a set of equations to define $H \cap K$? What are the largest and smallest numbers of these equations that could be indpendent?
2. Choose a basis for $H \cap K$. Extend it first to a basis for $H$; then start over and extend it to a basis for $K$. How many independent vectors in $R^7$ does this give you altogether?]

**25.** (a) Show that every hyperplane (i.e., a subspace of dimension $n - 1$) in $R^n$ satisfies an equation of the form

$$a_1x_1 + a_2x_2 + \cdots + a_nx_n = 0.$$

(b) Prove that any $k$-dimensional subspace of $R^n$ is the nullspace of an $(n - k) \times n$ matrix.

(c) Show that any $k$-dimensional subspace of $R^n$ is the intersection of $n$–$k$ hyperplanes.

**26.** Let $U$ and $V$ be subspaces of $R^n$.
(a) Show that if the zero vector is the only vector common to $U$ and $V$, then

$$\dim (U + V) = \dim U + \dim V.$$

(b) Show in general that

$$\dim (U + V) = \dim U + \dim V - \dim (U \cap V).$$

(*Hint:* Begin with a basis for $U \cap V$, extend it to a basis for $U$, then extend it again to a basis for $V$. Now count.)

**27.** Find all possible price vectors for the closed economy described in Problem 5 of Section 1.2. Explain why all price vectors for this economy are essentially the same.

**28.** For each of the circuits in Problem 3 of Section 1.2, find its incidence matrix and determine how many currents must be measured before all currents can be calculated.

***29.** The price vectors $\mathbf{p} = (p_1, p_2, p_3, p_4)$ in a certain closed economy satisfy the equation $A\mathbf{p} = \mathbf{p}$, where

$$A = \begin{bmatrix} \frac{1}{2} & 0 & \frac{1}{4} & 0 \\ 0 & \frac{1}{3} & 0 & \frac{2}{3} \\ \frac{1}{2} & 0 & \frac{3}{4} & 0 \\ 0 & \frac{2}{3} & 0 & \frac{1}{3} \end{bmatrix}.$$

Find all possible price vectors for this economy. Are they all essentially the same? Can you interpret your answer in terms of noninteracting subeconomies?

## SUPPLEMENTARY PROBLEMS

**1.** Mark each of the following true or false and give a brief explanation.

*(a) If $V$ is a subspace of $R^n$, then $V$ contains a line through the origin.

(b) The dimension of a subspace is the maximum number of independent vectors that can be contained in the subspace.

*(c) The zero vector is a linear combination of any collection of vectors.

(d) Every set of independent vectors that span a subspace form a basis for the subspace.

*(e) If $\dim V = n$, then $n$ is the maximum number of independent vectors in $V$.

(f) If $\dim V = n$, then $V$ cannot be spanned by more than $n$ vectors.

*(g) If $\mathbf{u}_1, \mathbf{u}_2, \ldots, \mathbf{u}_n$ are independent, then $\mathbf{u}_2, \mathbf{u}_3, \ldots, \mathbf{u}_n$ are independent.

(h) Any independent set of vectors in $R^m$ is part of a basis for $R^m$.

*(i) It is possible to find five vectors that span $R^6$.

(j) If $\mathbf{v}$ is a vector in $R^m$, then there is a basis for $R^m$ including $\mathbf{v}$.

*(k) If $U$ and $V$ are both two-dimensional subspace of $R^3$, then there must be a nonzero vector $\mathbf{w}$ that is in both $U$ and $V$.

(l) If $U$ and $V$ are subspaces of $R^m$ and $\dim U = \dim V$, then $U = V$.

**2.** Let $A$ be an $m \times n$ matrix of rank $r$. Let $E$ be an echelon form of $A$. Mark each of the following true or false and give a brief explanation

*(a) If $m > n$, the rows of $A$ are dependent.

(b) If $m < n$, then $A\mathbf{x} = \mathbf{b}$ has at most one solution.

*(c) If $m = n$ and the rows of $A$ are independent, $A$ is nonsingular.

(d) If the rows of $A$ are independent, then $C(A) = R^m$.

*(e) $\dim N(A) \leq n - m$.

(f) If $n < m$, $N(A) \neq \{\mathbf{0}\}$.

*(g) If $m < n$, then $A\mathbf{x} = \mathbf{b}$ has at least one solution for every $\mathbf{b}$ in $R^m$.

(h) If $r < m = n$, then $A$ is singular.

*(i) If $r = n$, $C(A) = R^m$.

(j) $R(A) = R(E)$.

*(k) $C(A) = C(E)$.

(l) $N(A) = N(E)$.

*(m) $\dim C(A) = r$.

(n) If $E = I$, then $A$ is nonsingular.

*(o) $\dim C(E) = r$.

(p) If $A\mathbf{x} = \mathbf{b}$ has a unique solution for every $\mathbf{b}$ in $R^m$, then $r = m = n$.

*(q) If $A\mathbf{x} = \mathbf{b}$ has at least one solution for every $\mathbf{b}$ in $R^m$, then $r = m$.

(r) If $R(A) = R^n$, then $N(A) = \{\mathbf{0}\}$.

*(s) $N(A) = \{\mathbf{0}\}$ if and only if the columns of $A$ are independent.

(t) If $r = n = m$, then $A^2$ has independent columns.

*(u) Suppose that $r = m$. If $A\mathbf{x} = A\mathbf{y}$, then $\mathbf{x} = \mathbf{y}$.

(v) $A\mathbf{x} = \mathbf{b}$ is consistent if and only if rank $[A : \mathbf{b}] = r$.

*(w) If $m = n$ and each row and each column of $A$ has exactly one 1, and all other entries in $A$ are 0, then $A$ is nonsingular.

(x) If $m = n$, then $A$ has independent rows if and only if $A$ has independent columns.

*(y) If $m = n$ and $B$ is an $n \times r$ matrix of rank $r$, then $N(AB) = N(B)$.

(z) If $\alpha_1\mathbf{v}_1 + \alpha_2\mathbf{v}_2 + \alpha_3\mathbf{v}_3 = \mathbf{0}$, then $\mathbf{v}_1, \mathbf{v}_2, \mathbf{v}_3$ are dependent.

**3.** (a) Complete the following definition. The vectors $\mathbf{v}_1, \mathbf{v}_2, \ldots, \mathbf{v}_n$ are called linearly independent if....

(b) Determine whether the vectors $\mathbf{u} = (1,2,1)$, $\mathbf{v} = (0,1,2)$, $\mathbf{w} = (2,1,-4)$ are linearly independent.

*(c) Determine whether the vectors $(1,0,0,1)$, $(1,1,0,1)$, $(1,1,1,2)$, $(1,1,1,3)$ are linearly independent.

**4.** (a) Complete the following definition. The vectors $\mathbf{v}_1, \mathbf{v}_2, \ldots, \mathbf{v}_n$ in a subspace $V$ are called a basis for $V$ if....

(b) Determine whether the vectors $\mathbf{u} = (1,0,0,0)$, $\mathbf{v} = (1,1,0,0)$, $\mathbf{w} = (0,1,1,0)$ form a basis for $R^4$.

*(c) Determine whether the vectors $(1,1,1,1)$, $(9,12,12,15)$, $(1,7,9,15)$ form a basis for the subspace that they span.

**5.** (a) Complete the following definitions:

(i) A collection $V$ of $m$-vectors is a subspace of $R^m$ if....

(ii) The dimension of $V$ is....

(b) Show that the collection $V$ of all 3-vectors $\mathbf{x} = (x_1, x_2, x_3)$ whose components satisfy $x_1 + 2x_2 = 3x_3$ is a subspace. Then find a basis for $V$ and the dimension of $V$.

(c) Show that the collection of all 4-vectors $\mathbf{x}$ such that $x_3 = x_4$ is a subspace. Then find a basis for $V$ and the dimension of $V$.

*(d) Show that the collection of all 4-vectors $\mathbf{x}$ such that $\mathbf{x} = \mathbf{u} + \mathbf{v}$ where the components of $\mathbf{u}$ satisfy $u_1 + u_2 + u_3 = 0$ and those of $\mathbf{v}$ satisfy $2v_1 + v_3 = 0$ is a subspace. Then find a basis for $V$ and the dimension of $V$.

***6.** Do the vectors $(1,2,3)$, $(4,5,6)$, $(7,8,9)$ span $R^3$? If so, show that they do. If not, find an explicit vector in $R^3$ that is not in their span, and show that it is not.

**7.** Repeat Problem 6 for the vectors $(1,1,1)$, $(1,1,-1)$, and $(1,-1,-1)$.

**8.** Show that the vectors $(1,1,0)$, $(0,1,1)$, $(1,1,1)$, $(-1,0,2)$ span $R^3$. Do they form a basis?

**9.** Is $\mathbf{b} = (1,1,1)$ a linear combination of $(1,1,0)$, $(4,1,-3)$, $(2,2,0)$? Why or why not?

**10.** Suppose that the $m \times n$ matrix $A$ has been reduced to an echelon matrix $E$. Can one tell by looking at $E$ whether $C(A) = R^m$. How?

***11.** Let

$$A = \begin{bmatrix} 0 & 1 & 1 & 1 & 2 \\ 0 & 0 & 3 & 2 & 1 \\ 0 & 0 & -6 & -4 & -3 \\ 0 & 1 & -2 & -1 & 0 \end{bmatrix}.$$

(a) Find the dimension of $C(A)$, $N(A)$, and $R(A)$.

(b) Is there a scalar $\alpha$ such that $(1,1,1,\alpha)$ is in $C(A)$?

**12.** Let

$$A = \begin{bmatrix} 0 & 1 & 2 & 2 & 1 \\ 0 & 0 & 1 & 0 & 1 \\ 0 & 0 & 1 & 1 & 1 \\ 0 & 0 & 0 & 0 & 0 \\ 0 & 0 & 0 & 0 & 0 \end{bmatrix}.$$

(a) Find a basis for $C(A)$, $N(A)$, and $R(A)$.

(b) Are there numbers $\alpha$, $\beta$, and $\gamma$ such that $\mathbf{b} = (1,1,\alpha,\beta,\gamma)$ is in $C(A)$?

**13.** Let

$$A = \begin{bmatrix} 1 & 2 & 3 & 4 & 5 \\ 2 & 4 & 6 & 8 & 10 \\ 1 & 1 & 1 & 1 & 1 \\ 3 & 5 & 7 & 9 & 11 \\ 1 & 0 & 1 & 0 & 1 \end{bmatrix}.$$

(a) Find a basis for $C(A)$, $N(A)$, and $R(A)$.
(b) Find the general solution of $A\mathbf{x} = \mathbf{b}$.

***14.** (a) Let $A$ be an $n \times n$ matrix. Show that the set $V$ of all $n$-vectors $\mathbf{x}$ such that $A\mathbf{x} = 2\mathbf{x}$ is a subspace of $R^n$.
(b) If $A = \begin{bmatrix} 2 & 1 \\ 0 & 2 \end{bmatrix}$, find a basis for $V$.

**15.** (a) Let $\alpha$ be a scalar and $A$ be an $n \times n$ matrix. Show that the set $V$ of all $n$-vectors $\mathbf{v}$ such that $A\mathbf{v} = \alpha\mathbf{v}$ is a vector space.
(b) If

$$A = \begin{bmatrix} 3 & 2 & 3 \\ 2 & 6 & 6 \\ 0 & 0 & 2 \end{bmatrix}$$

and $\alpha = 2$, fina a basis for $V$ and the dimension of $V$.

**16.** Show that if $\mathbf{u}_1, \mathbf{u}_2, \mathbf{u}_3$ are in a subspace $V$ of $R^m$, $m \geq 3$, then the subspace spanned by $\mathbf{u}_1, \mathbf{u}_2, \mathbf{u}_3$ is contained in $V$.

**17.** Let $\mathbf{v}_1, \ldots, \mathbf{v}_n$ be dependent vectors in $R^m$, and let $V$ be the subspace spanned by these vectors. How would you find an independent set of vectors that span $V$?

**18.** Define a subspace and give an example of a three-dimensional subspace of $R^4$.

**19.** What is the geometric meaning of an inconsistent system?

**20.** Suppose that $\mathbf{v}_1, \mathbf{v}_2, \ldots, \mathbf{v}_n$ are vectors in $R^m$.
(a) Describe a practical method by which you can determine whether the $\mathbf{v}$'s are independent.
(b) Describe a practical method by which you can determine whether the $\mathbf{v}$'s span $R^m$.

**21.** Let $\mathbf{u}$ and $\mathbf{v}$ be vectors in $R^m$ that are not both zero. Show that $\mathbf{u}$ and $\mathbf{v}$ are dependent if and only if one of them is a scalar multiple of the other. Hence they are dependent if and only if they lie on a line in $R^m$ through the origin.

**22.** For each of the following subsets of $R^3$, determine whether the subset is a subspace and whether it is empty, a point, a line, a plane or all of $R^3$. Give a justification for each answer.
*(a) The subset spanned by $(0,0,0)$, $(1,1,1)$, $(2,2,5)$.
(b) The collection of all vectors $(x,y,z)$ such that $x - 3z = 0$.
*(c) The collection of all vectors $(x,y,z)$ such that

$$\begin{aligned} x + y + z &= 2 \\ y + 2z &= 3 \\ x + 3y + 5z &= 5. \end{aligned}$$

(d) The collection of all vectors $(x,y,z)$ such that

$$\begin{aligned} x - y - z &= 3 \\ x + y + z &= 4 \\ y + 3z &= 6. \end{aligned}$$

(e) The collection of all vectors $(x,y,z)$ such that $x + y + z = 0$ and $x + 2y + z = 0$.

***23.** In each part, decide whether the two collections of vectors span the same subspace. Interpret your answer geometrically.
(a) $(1,1), (-1,1), (1,2)$ and $(1,5), (0,1)$
(b) $(1,2,3)$, $(2,4,6)$, $(-2,-3,-4)$, $(3,4,5)$ and $(1,0,0), (0,1,0)$
(c) $(1,2,-2,3)$, $(2,4,-3,4)$, $(3,6,-4,5)$ and $(1,2,0,-1), (0,0,1,-2)$

**24.** In each of the following, determine whether the given plane contains the given line.
*(a) Plane $x_1 + x_2 - x_3 = 0$, line $\mathbf{x} = \alpha(5,-2,3)$.
(b) Plane $3x_1 + x_2 + 2x_3 = 1$, line $\mathbf{x} = (3,2,-5) + \alpha(-1,-3,3)$.

*(c) Plane $2x_1 + 3x_2 - 2x_3 + x_4 = 4$, line $\mathbf{x} = (7, 3, 9, 5) + \alpha(-3, 2, 1, 2)$.

(d) Plane $\mathbf{x} = (1, 1, 1, 1) + \alpha(-1, 1, 2, -2) + \beta(0, 2, 4, 1)$, line $\mathbf{x} = (0, 6, 11, 1) + \alpha(2, 4, 8, 7)$.

**25.** Determine conditions on the vectors $\mathbf{u}_1$, $\mathbf{u}_2$, $\mathbf{a}$, $\mathbf{v}_1$, and $\mathbf{v}_2$ in $R^4$ which guarantee that the plane $\mathbf{x} = \alpha_1\mathbf{u}_1 + \alpha_2\mathbf{u}_2$ is parallel to the plane $\mathbf{x} = \mathbf{a} + \beta_1\mathbf{v}_1 + \beta_2\mathbf{v}_2$.

**26.** How many hyperplanes in $R^5$ must one intersect to define a line in $R^5$?

**27.** In each of the following, find $3 \times 3$ matrices $A$ and $B$ such that the stated condition holds.

(a) Rank$(A) = 2$ and rank$(A^2) = 2$.

(b) Rank$(A) = 2$ and rank$(A^2) = 1$.

(c) Rank$(AB) = 2$ and rank$(BA) = 1$.

**28.** Suppose that $A\mathbf{x} = \mathbf{0}$ has a nontrivial solution and that $x_3$ is a free variable. Explain why the third column of $A$ is a linear combination of the other columns of $A$.

***29.** Assume that $A$, $B$, and $C$ are $n \times n$ matrices.

(a) $(A - I)(A + I) = A^2 - I$. (Sometimes/Always/Never)

(b) If $AB = C$ and $C$ is nonsingular, then $A$ must be nonsingular. (True/False)

(c) Given that $A$ and $B$ are singular, $A + B$ is singular. (Sometimes/Always/Never)

***30.** Prove that $A$ and $5A$ have the same column space, nullspace, and row space. Is there anything special about the number 5?

**31.** Let $\mathbf{v}_1, \mathbf{v}_2, \ldots, \mathbf{v}_n$ and $\mathbf{w}_1, \mathbf{w}_2, \ldots, \mathbf{w}_k$ be vectors in $R^m$, and let $A$ and $B$ be the matrices having these vectors as columns. Show that the $\mathbf{v}_i$'s span the same subspace as the $\mathbf{w}_j$'s if and only if $A^T$ and $B^T$ have the same reduced echelon form.

***32.** Find a single $4 \times 3$ matrix $A$ whose row space is spanned by the vectors $(1, 2, 3)$ and $(1, 1, -1)$ and whose column space is spanned by the vectors $(2, 2, 0, 1)$ and $(3, 5, 1, 3)$. Explain why your candidate for $A$ meets the requirements.

**33.** Let $A$ and $B$ be $3 \times 3$ matrices, and let $C = AB$. Make a table of all the possible combinations of rank$(A)$, rank$(B)$, and rank$(C)$, and give an example for each. For instance, an example of rank$(A) = 3$, rank$(B) = 2$, rank$(C) = 2$ is $A = I$,

$$B = \begin{bmatrix} 1 & 0 & 0 \\ 0 & 1 & 0 \\ 0 & 0 & 0 \end{bmatrix},$$

$C = B$. On the other hand, rank$(A) = 3$, rank$(B) = 2$, rank$(C) = 0$ cannot occur. So we get the following two entries:

| Rank ($A$) | Rank ($B$) | Rank ($C$) | Example |
|---|---|---|---|
| 3 | 2 | 2 | $A = I, \quad B = C = \begin{bmatrix} 1 & 0 & 0 \\ 0 & 1 & 0 \\ 0 & 0 & 0 \end{bmatrix}$ |
| 3 | 2 | 0 | Impossible |

# 3

# Inconsistent Systems, Inner Products, and Projections

In Chapter 1 we saw that Gaussian elimination not only solves any consistent system of equations, but that it also identifies any inconsistent system by producing a false equation. In Chapter 2 we saw how to characterize consistent and inconsistent systems of equations geometrically. If the vector of constant terms lies in the column space of the coefficient matrix, the system is consistent. Otherwise, it is inconsistent. The algebraic question of whether a system of equations is consistent is equivalent to the geometric question of whether a certain vector lies in a subspace. In this chapter we use this geometric characterization of consistent systems together with the concepts of length and angle to investigate the question of what can be done with inconsistent systems.

## 3.1 Inconsistent Systems of Equations

We begin our discussion of inconsistent systems with several examples of situations where these systems arise naturally.

**EXAMPLE 1**

In Section 1.2 we studied the problem of finding the equation of a line $y = a + bx$ passing through the $m$ points $(x_1, y_1), \ldots, (x_m, y_m)$, with $x_1 < x_2 < \cdots < x_m$. The coefficients $a$ and $b$ must satisfy the system of $m$ equations

$$\begin{array}{ccc} a + bx_1 & = & y_1 \\ \vdots & & \vdots \\ a + bx_m & = & y_m \end{array} \qquad \text{or} \qquad \begin{bmatrix} 1 & x_1 \\ \vdots & \vdots \\ 1 & x_m \end{bmatrix} \begin{bmatrix} a \\ b \end{bmatrix} = \begin{bmatrix} y_1 \\ \vdots \\ y_m \end{bmatrix}.$$

When $m = 2$ these equations always have a unique solution. Geometrically, the reason for this is that any two distinct points determine a straight line. However, when $m$ is greater than 2, the system may not have a solution. Three or more distinct points do not always lie on a line. In fact, the more points we have, the more likely it is that they do not lie on a straight line.

## EXAMPLE 2

Suppose that an object is traveling along a straight line and that we have reason to believe that it is moving with a constant velocity. To determine its velocity $v$, we measure the position of the object at times $t_1, \ldots, t_m$. If $y_i$ denotes the position at time $t_i$ and $y_0$ denotes the initial position of the object (which of course is unknown), then the following system of equations must be satisfied:

$$\begin{array}{ccc} y_0 + vt_1 & = & y_1 \\ \vdots & & \vdots \\ y_0 + vt_m & = & y_m \end{array} \qquad \text{or} \qquad \begin{bmatrix} 1 & t_1 \\ \vdots & \vdots \\ 1 & t_m \end{bmatrix} \begin{bmatrix} y_0 \\ v \end{bmatrix} = \begin{bmatrix} y_1 \\ \vdots \\ y_m \end{bmatrix}.$$

Since measurements of physical quantities are *never* exact, we *must expect* this system of equations to be inconsistent. We could, of course, use only two of the measurements and thereby be assured that the equations have a unique solution. But this is usually not a good idea. The line $L_1$ in Figure 3.1 is determined by two of the data points, but it certainly does not "fit the data" very well. On the other hand, even though the line $L_2$ does not pass through *any* of the data points, it does "fit the data" well. When we use only two of the data points, the errors in the individual measurements will usually affect our results significantly. By using more data points than are needed, we reduce the effects on our results of the errors in individual measurements. But when we do this we must be prepared to deal with an inconsistent system of equations.

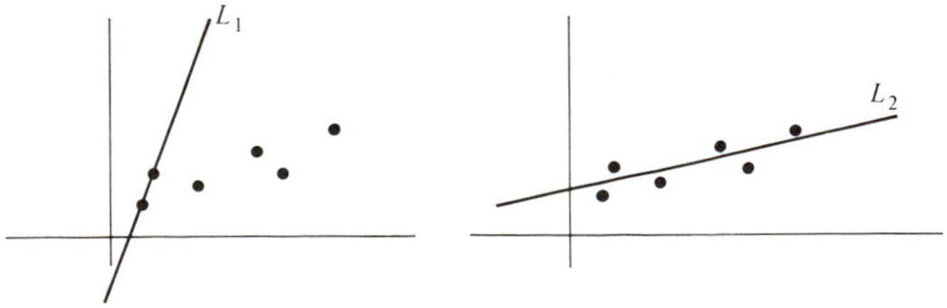


Figure 3.1

The problems that we considered in Examples 1 and 2 are mathematically identical. Both examples led to an $m \times 2$ system of equations with $m > 2$. In the general case we consider a system of equations $A\mathbf{x} = \mathbf{b}$, where $A$ is an $m \times n$ matrix of rank $r < m$ and $\mathbf{b}$ is an $m$-vector. We have seen in Chapter 2 that this system will have a solution if and only if the components of the vector $\mathbf{b}$ satisfy $m - r$ constraint equations (one for each zero row of the echelon matrix). Obviously, when $m - r$ is large, it will be difficult for the components of an $m$-vector $\mathbf{b}$ to satisfy all the constraints. In other words, the larger the value of $m - r$, the more likely it is that the system $A\mathbf{x} = \mathbf{b}$ is inconsistent. From an algebraic point of view, then, the question of the consistency of the system $A\mathbf{x} = \mathbf{b}$ is equivalent to the question of whether the vector $\mathbf{b}$ satisfies the constraint equations.

Now consider the question of the consistency of the system $A\mathbf{x} = \mathbf{b}$ ($A$ an $m \times n$ matrix of rank $r < m$) from a geometric point of view. In Chapter 2 we saw that the equation $A\mathbf{x} = \mathbf{b}$ is consistent if and only if the vector $\mathbf{b}$ lies in the column space of $A$, an $r$-dimensional subspace of $R^m$. If $m - r$ is large, this $r$-dimensional subspace is a rather "thin" subspace of $R^m$ and most $m$-vectors will not lie in it. Moreover, the larger $m - r$ is, the "thinner" the subspace is. (A line is a "thin" subspace of $R^2$ and an even "thinner" subspace of $R^3$.) Thus the larger $m - r$ is, the more likely it is that the vector $\mathbf{b}$ will not be in the column space of $A$ and hence the more likely it is that the system is inconsistent.

We now turn to the question of what can be done with inconsistent systems. As above, let $A$ be an $m \times n$ matrix of rank $r < m$ and let $\mathbf{b}$ be an $m$-vector. Suppose that the system $A\mathbf{x} = \mathbf{b}$ is inconsistent. In this case the $m$-vector $\mathbf{b}$ does not lie in the column space of $A$. We cannot solve $A\mathbf{x} = \mathbf{b}$ as it stands. Perhaps by changing the system slightly we can obtain a system that can be solved. One way to do this is to replace $\mathbf{b}$ by some vector $\mathbf{p}$ that *is* in the column space of $A$. The new system $A\mathbf{x} = \mathbf{p}$ will be consistent and any solution $\bar{\mathbf{x}}$ to $A\mathbf{x} = \mathbf{p}$ can be considered in some sense an "approximate solution" to the original inconsistent system $A\mathbf{x} = \mathbf{b}$. But for this process to be reasonable, we must choose the "best possible" replacement for $\mathbf{b}$. Since our objective is to obtain something that is close to being a solution of $A\mathbf{x} = \mathbf{b}$, it is reasonable to choose $\mathbf{p}$ to be as close to $\mathbf{b}$ as possible (while still being in the column space of $A$). This $\mathbf{p}$ is the "best possible" replacement for $\mathbf{b}$, the consistent system $A\mathbf{x} = \mathbf{p}$ is the "best possible" replacement for $A\mathbf{x} = \mathbf{b}$, and the "best" we can obtain as an approximate solution to $A\mathbf{x} = \mathbf{b}$ is a vector $\bar{\mathbf{x}}$ such that $A\bar{\mathbf{x}} = \mathbf{p}$.

It is clear from the discussion above that we must develop a procedure for finding the vector $\mathbf{p}$ in the column space of $A$ that is closest to $\mathbf{b}$. Let us first consider this problem in the familiar setting of three-dimensional space. Suppose that $A$ is a $3 \times 2$ matrix and that the rank of $A$ is either 1 or 2. Hence the column space of $A$ is a line or plane in $R^3$. From Figure 3.2(a) and (b) it is clear that the vector $\mathbf{p}$ that is closest to $\mathbf{b}$ is found by dropping a perpendicular from $\mathbf{b}$ to the line or plane.

The geometry of three-dimensional space has suggested a method that can be used to attack the problem of "solving" an inconsistent system of equations. This method depends on the concepts of length and perpendicularity. Hence to extend this method to the general case of an $m \times n$ system, we must first extend the notions of length and perpendicularity from $R^3$ to $R^m$. This is the topic of the next section.

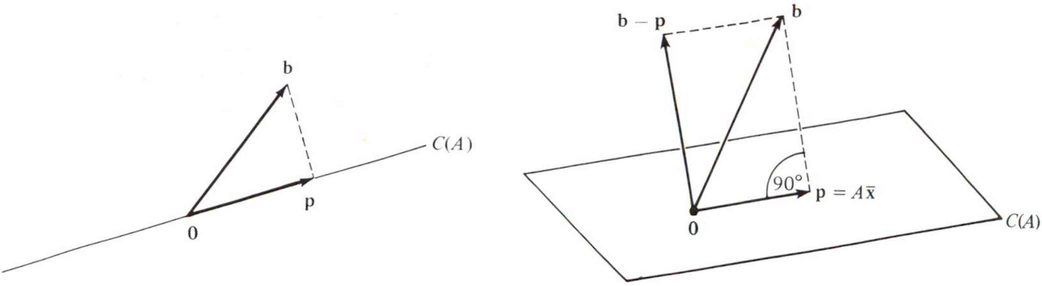


(a) (b)

**Figure 3.2**

# PROBLEMS 3.1

***1.** Suppose we know that an object is traveling with constant velocity. The position of the object is measured at various times and the following data are obtained.

| $t$ | 1 | 2 | 3 | 4 | 5 |
|---|---|---|---|---|---|
| $d$ | 3 | 5 | 9 | 11 | 12 |

What equations must the velocity and the initial position satisfy? Show that the equations are inconsistent.

***2.** Suppose that we know from certain theoretical considerations that the variables $t$ and $y$ satisfy a relationship of the form

$$y = a + bt + ct^2.$$

An experiment is conducted that gives the data in the following table. What equations must the vector $\mathbf{x} = (a, b, c)$ satisfy? Show that the equations are inconsistent.

| $t$ | −1 | 0 | 1 | 2 |
|---|---|---|---|---|
| $y$ | −2 | −1 | 0 | 3 |

**3.** Repeat Problem 2 when the relationship is of the form:
(a) $y = at + bt^2 + ct^3$
(b) $y = a + bt^2$

**4.** Let $(x_1, y_1), \ldots, (x_m, y_m)$ be $m$ points in $R^2$ with $x_1 < x_2 < \cdots < x_m$, and let $p(x) = a_0 + a_1x + \cdots + a_nx^n$ be a polynomial of degree $n$.
(a) Show that the matrix form of the system of equations which the coefficients $a_0, a_1, \ldots, a_n$ of the polynomial $p$ must satisfy in order that

$$p(x_i) = y_i, \qquad i = 1, \ldots, m,$$

is $A\mathbf{x} = \mathbf{b}$, where

$$A = \begin{bmatrix} 1 & x_1 & \cdots & x_1^n \\ \vdots & \vdots & & \vdots \\ 1 & x_m & \cdots & x_m^n \end{bmatrix},$$

$$\mathbf{x} = \begin{bmatrix} a_0 \\ \vdots \\ a_n \end{bmatrix}, \qquad \mathbf{b} = \begin{bmatrix} y_1 \\ \vdots \\ y_m \end{bmatrix}.$$

(b) Show that if the rank of the matrix $A$ in part (a) is $m$, then there is a solution to the equation $A\mathbf{x} = \mathbf{b}$. Hence there is a polynomial of degree $n$ which passes through the given points $(x_i, y_i)$, $i = 1, \ldots, m$.
(c) Show that there is one and only one polynomial of degree $m - 1$ which passes through the given set of $m$ points. (*Hint:* Show that if $n = m - 1$ then the rank of $A$ is $m$. Do this for $m = 2$, 3, and 4.)
(d) Show that if the rank of $A$ is $n + 1$,

then $m \geq n+1$, and the equation has a solution if and only if **b** lies in the column space of $A$. Therefore, there may not be a polynomial of degree $n$ which passes through the given points. As we have suggested, the larger $m$ gets, the less likely it is that the equations have a solution.

# 3.2 Length and Orthogonality

In this section we use elementary geometric arguments based on the Pythagorean theorem to derive algebraic characterizations of length and perpendicularity in $R^2$ and $R^3$. These algebraic formulas will provide us with the means to generalize these concepts to $R^m$ when $m > 3$ in such a way that the definitions are consistent with the usual definitions when $m = 2$ or $m = 3$.

The Pythagorean theorem states that the square of the length of the hypotenuse of a right triangle is equal to the sum of the squares of the lengths of the other two sides. If $\mathbf{x} = (x_1, x_2)$ is a vector in $R^2$, we find that [see Figure 3.3(a)]

$$\text{length of } \mathbf{x} = \sqrt{x_1^2 + x_2^2}.$$

Similarly, if $\mathbf{x} = (x_1, x_2, x_3)$ is a vector in $R^3$, then two applications of the Pythagorean theorem give [Figure 3.3(b)]

$$\begin{aligned}\text{length of } \mathbf{x} &= \sqrt{\left(\sqrt{x_1^2 + x_2^2}\right)^2 + x_3^2} \\ &= \sqrt{x_1^2 + x_2^2 + x_3^2}.\end{aligned}$$

These formulas suggest how we should define the length of a vector in $R^m$ when $m > 3$ so that the definition agrees with our findings when $m = 2$ or $m = 3$.

***Definition*** If $\mathbf{x}$ is a vector in $R^m$, then the **length** (or **norm**) of $\mathbf{x}$, denoted by $\|\mathbf{x}\|$, is defined by the equation

$$\|\mathbf{x}\| = \sqrt{x_1^2 + x_2^2 + \cdots + x_m^2}.$$

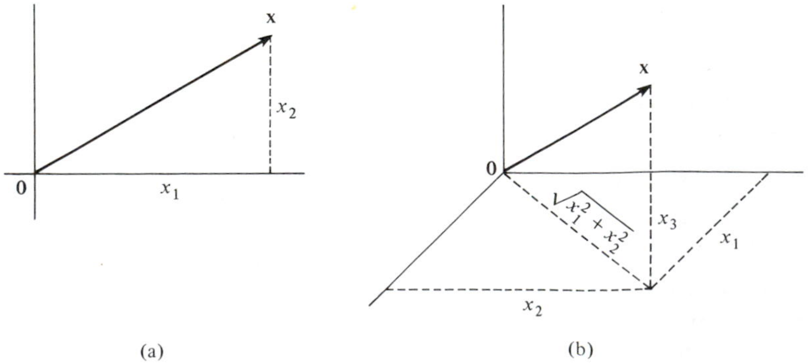


Figure 3.3

### EXAMPLE 1

$$\|(0,1,2)\| = \sqrt{0^2 + 1^2 + 2^2} = \sqrt{5}$$
$$\|(1,2,3,-2)\| = \sqrt{1^2 + 2^2 + 3^2 + (-2)^2} = \sqrt{18} = 3\sqrt{2}$$

If $\mathbf{e}_1, \ldots, \mathbf{e}_m$ are the standard basis vectors for $R^m$, then $\|\mathbf{e}_j\| = 1, j = 1, \ldots, m$. Vectors of length 1 are called **unit vectors**. Given any nonzero vector $\mathbf{x}$, the vector $\mathbf{y} = \mathbf{x}/\|\mathbf{x}\|$ is a unit vector in the direction of $\mathbf{x}$.

Since the length of a vector in $R^m$ is the square root of a sum of squares, it is nonnegative. In other words, $\|\mathbf{x}\| \geq 0$ for all $\mathbf{x}$. Since $\|0\|^2 = 0^2 + \cdots + 0^2$, the zero vector has zero length. It is the only vector with zero length. To prove this, suppose that $\mathbf{x} = (x_1, \ldots, x_m)$ is a nonzero vector. At least one of its components is not zero, say $x_i$. Then

$$0 < x_i^2 \leq x_1^2 + \cdots + x_i^2 + \cdots + x_m^2 = \|\mathbf{x}\|^2,$$

so that $\|\mathbf{x}\| > 0$. Thus we have shown that for all $\mathbf{x}$,

$$\|\mathbf{x}\| \geq 0, \quad \text{and} \quad \|\mathbf{x}\| = 0 \quad \text{if and only if} \quad \mathbf{x} = 0. \tag{1}$$

Again motivated by the geometry of $R^2$ and $R^3$, we define the distance between two vectors $\mathbf{x}$ and $\mathbf{y}$ in $R^m$ to be the length of the vector $\mathbf{x} - \mathbf{y}$ (see Figure 3.4).

***Definition*** If $\mathbf{x} = (x_1, \ldots, x_m)$ and $\mathbf{y} = (y_1, \ldots, y_m)$ are two vectors in $R^m$, then the **distance** between $\mathbf{x}$ and $\mathbf{y}$ is defined to be

$$\|\mathbf{x} - \mathbf{y}\| = \sqrt{(x_1 - y_1)^2 + (x_2 - y_2)^2 + \cdots + (x_m - y_m)^2}.$$

Note that the length of a vector is equal to its distance from the origin (the zero vector).

### EXAMPLE 2

The distance between $(0, 2, -2, 1)$ and $(-2, 0, -2, 2)$ is 3 since $[0 - (-2)]^2 + (2 - 0)^2 + [-2 - (-2)]^2 + (1 - 2)^2 = 9$.

Having generalized the notions of length and distance to $R^m$, we now turn to the notion of perpendicularity.

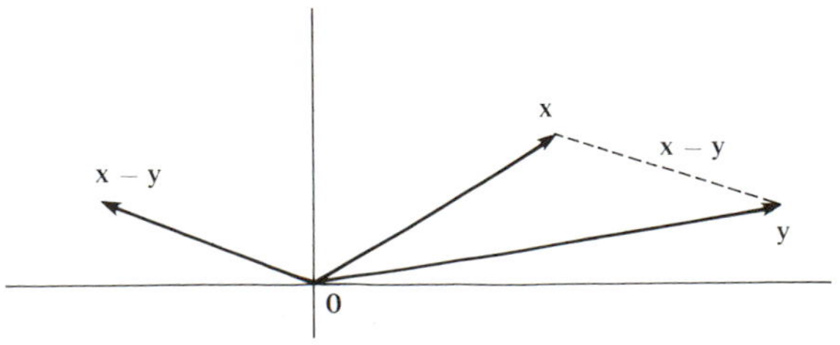


Figure 3.4

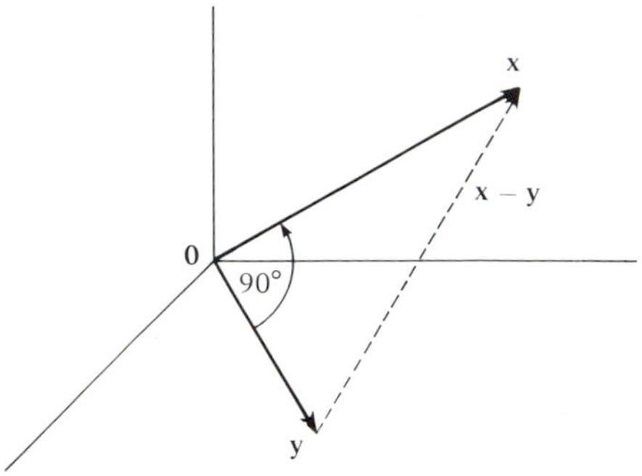


Figure 3.5

Suppose that $\mathbf{x} = (x_1, x_2, x_3)$ and $\mathbf{y} = (y_1, y_2, y_3)$ are two nonzero perpendicular vectors in $R^3$. Then the triangle in Figure 3.5 is a right triangle and we conclude from the Pythagorean theorem that

$$\|\mathbf{x} - \mathbf{y}\|^2 = \|\mathbf{x}\|^2 + \|\mathbf{y}\|^2.$$

Since

$$\begin{aligned}\|\mathbf{x} - \mathbf{y}\|^2 &= (x_1 - y_1)^2 + (x_2 - y_2)^2 + (x_3 - y_3)^2 \\ &= \|\mathbf{x}\|^2 + \|\mathbf{y}\|^2 - 2(x_1y_1 + x_2y_2 + x_3y_3),\end{aligned}$$

the equality $\|\mathbf{x} - \mathbf{y}\|^2 = \|\mathbf{x}\|^2 + \|\mathbf{y}\|^2$ holds if and only if

$$\langle \mathbf{x}, \mathbf{y} \rangle = x_1y_1 + x_2y_2 + x_3y_3 = 0. \tag{2}$$

Thus $\mathbf{x}$ and $\mathbf{y}$ are perpendicular if and only if equation (2) is true. The reader should use a similar argument to check that the vectors $\mathbf{x} = (x_1, x_2)$ and $\mathbf{y} = (y_1, y_2)$ are perpendicular if and only if

$$\langle \mathbf{x}, \mathbf{y} \rangle = x_1y_1 + x_2y_2 = 0.$$

Thus two vectors $\mathbf{x}$ and $\mathbf{y}$ in $R^2$ or $R^3$ are perpendicular if and only if $\langle \mathbf{x}, \mathbf{y} \rangle = 0$. The beauty of this algebraic characterization of the geometric notion of perpendicularity is that it allows us to extend the notion of perpendicularity to $R^m$ when $m > 3$.

***Definition*** Two $m$-vectors $\mathbf{x} = (x_1, \ldots, x_m)$ and $\mathbf{y} = (y_1, \ldots, y_m)$ are called **orthogonal** (or **perpendicular**) if

$$\langle \mathbf{x}, \mathbf{y} \rangle = x_1y_1 + \cdots + x_my_m = 0.$$

**EXAMPLE 3**

The 4-vectors $(1, 1, 0, 1)$ and $(1, -1, 3, 0)$ are orthogonal since $1 \cdot 1 + 1 \cdot (-1) + 0 \cdot 3 + 1 \cdot 0 = 0$. Similarly, $(1, 0, 0)$ and $(0, 5, 6)$ are orthogonal in $R^3$.

The inner product and the norm are related by the equation

$$\|\mathbf{x}\| = \sqrt{\langle \mathbf{x}, \mathbf{x} \rangle}.$$

In this notation (1) becomes

$$\langle \mathbf{x}, \mathbf{x} \rangle \geq 0, \quad \text{and} \quad \langle \mathbf{x}, \mathbf{x} \rangle = 0 \quad \text{if and only if} \quad \mathbf{x} = \mathbf{0}. \tag{3}$$

In terms of orthogonality the fact that

$$\langle \mathbf{x}, \mathbf{x} \rangle = 0 \quad \text{if and only if} \quad \mathbf{x} = \mathbf{0}$$

says that *the zero vector is orthogonal to itself and it is the only vector with this property*. Moreover, since

$$\langle \mathbf{0}, \mathbf{x} \rangle = 0 \quad \text{for all} \quad \mathbf{x} \text{ in } R^m,$$

the zero vector is orthogonal to every vector in $R^m$. It is also the only vector with this property. For if a vector $\mathbf{u}$ in $R^m$ is orthogonal to every vector in $R^m$, then in particular it is orthogonal to itself and hence must be the zero vector.

**Result 1** Let $\mathbf{u}$ be a vector in $R^m$. Then $\langle \mathbf{u}, \mathbf{x} \rangle = 0$ for all $\mathbf{x}$ in $R^m$ if and only if $\mathbf{u} = \mathbf{0}$.

In Theorem 1 we list some important properties of the inner product. The first three are simple consequences of the definition. The fourth is simply a restatement of (3).

***Theorem 1*** *Let* $\mathbf{x}$, $\mathbf{y}$, and $\mathbf{z}$ *be m-vectors and* $\alpha$ *a scalar.*

(*a*) $\langle \mathbf{x}, \mathbf{y} \rangle = \langle \mathbf{y}, \mathbf{x} \rangle$.
(*b*) $\langle \mathbf{x}, \mathbf{y} + \mathbf{z} \rangle = \langle \mathbf{x}, \mathbf{y} \rangle + \langle \mathbf{x}, \mathbf{z} \rangle$.
(*c*) $\langle \alpha\mathbf{x}, \mathbf{y} \rangle = \alpha\langle \mathbf{x}, \mathbf{y} \rangle$.
(*d*) $\langle \mathbf{x}, \mathbf{x} \rangle \geq 0$, *and* $\langle \mathbf{x}, \mathbf{x} \rangle = 0$ *if and only if* $\mathbf{x} = \mathbf{0}$.

## EXAMPLE 4

In this example we will show that the Pythagorean theorem holds in $R^m$. This should come as no surprise for two reasons. First, we used the Pythagorean theorem to motivate our definition of length in $R^m$, $m > 3$. Second, the Pythagorean theorem is concerned only with the relationship between the lengths of the sides of a right triangle, and hence it should be independent of the dimension of the space in which the triangle is situated.

Let us now prove the Pythagorean theorem. Suppose that $\mathbf{x}$ and $\mathbf{y}$ are $m$-vectors. Using (a) and (b) of Theorem 1, we obtain

$$\begin{aligned} \|\mathbf{x} + \mathbf{y}\|^2 &= \langle \mathbf{x} + \mathbf{y}, \mathbf{x} + \mathbf{y} \rangle \\ &= \langle \mathbf{x}, \mathbf{x} \rangle + \langle \mathbf{x}, \mathbf{y} \rangle + \langle \mathbf{y}, \mathbf{x} \rangle + \langle \mathbf{y}, \mathbf{y} \rangle \\ &= \|\mathbf{x}\|^2 + 2\langle \mathbf{x}, \mathbf{y} \rangle + \|\mathbf{y}\|^2. \end{aligned}$$

Thus $\|\mathbf{x} + \mathbf{y}\|^2 = \|\mathbf{x}\|^2 + \|\mathbf{y}\|^2$ if and only if $\langle \mathbf{x}, \mathbf{y} \rangle = 0$, that is, if and only if $\mathbf{x}$ and $\mathbf{y}$ are orthogonal. Thus the square of the length of the hypotenuse ($\|\mathbf{x} + \mathbf{y}\|^2$) is equal to the sum of the squares of the lengths of the other two sides ($\|\mathbf{x}\|^2 + \|\mathbf{y}\|^2$) if and only if the triangle in Figure 3.6 is a right triangle.

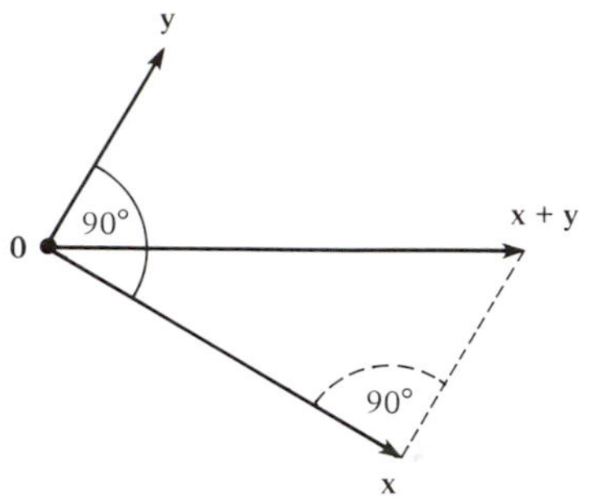


Figure 3.6

***Definition*** A set $\mathbf{v}_1, \ldots, \mathbf{v}_k$ of $m$-vectors is called **orthogonal** if $\langle \mathbf{v}_i, \mathbf{v}_j \rangle = 0$ whenever $i \neq j$.

That is, a set of vectors is orthogonal if every two distinct vectors in the set are orthogonal to each other. We also say that the vectors $\mathbf{v}_1, \ldots, \mathbf{v}_k$ are mutually orthogonal. If, in addition, all the vectors in the set have unit length, then the set of vectors is called **orthonormal**.

## EXAMPLE 5

If $\mathbf{v}_1, \mathbf{v}_2, \mathbf{v}_3$ are vectors in $R^m$, they are orthogonal if and only if

$$\langle \mathbf{v}_1, \mathbf{v}_2 \rangle = \langle \mathbf{v}_1, \mathbf{v}_3 \rangle = \langle \mathbf{v}_2, \mathbf{v}_3 \rangle = 0. \tag{4}$$

They are orthonormal if and only if (4) holds and

$$\|\mathbf{v}_1\| = \|\mathbf{v}_2\| = \|\mathbf{v}_3\| = 1.$$

If $\mathbf{v}_1, \ldots, \mathbf{v}_k$ is an orthogonal set of nonzero vectors, and if

$$\mathbf{w}_1 = \mathbf{v}_1/\|\mathbf{v}_1\|, \ldots, \mathbf{w}_k = \mathbf{v}_k/\|\mathbf{v}_k\|,$$

then $\mathbf{w}_1, \ldots, \mathbf{w}_k$ is an orthonormal set of vectors. Thus it is always possible to *normalize* an orthogonal set of nonzero vectors to obtain an orthonormal set by simply dividing each vector in the set by its length.

The standard basis vectors $\mathbf{e}_1, \ldots, \mathbf{e}_m$ in $R^m$ are the most familiar example of an orthonormal set of vectors. They provide us with a picture of $R^m$ which suggests that $R^m$ cannot contain more than $m$ mutually orthogonal nonzero vectors, one orthogonal vector for each dimension. The geometry of $R^m$ also suggests the more general result that nonzero orthogonal vectors must be independent.

***Theorem 2*** *If $\mathbf{v}_1, \ldots, \mathbf{v}_k$ are nonzero orthogonal vectors in $R^m$, then they are linearly independent.*

**Proof** Suppose that

$$\alpha_1 \mathbf{v}_1 + \cdots + \alpha_k \mathbf{v}_k = \mathbf{0}. \tag{5}$$

Taking the inner product of $\mathbf{v}_1$ with both sides of this equation, we obtain

$$\alpha_1\langle \mathbf{v}_1, \mathbf{v}_1\rangle + \cdots + \alpha_k\langle \mathbf{v}_1, \mathbf{v}_k\rangle = \langle \mathbf{v}_1, \mathbf{0}\rangle = 0.$$

Since the $\mathbf{v}$'s are orthogonal, $\langle \mathbf{v}_1, \mathbf{v}_j\rangle = 0$ for $j = 2, \ldots, k$ and the equation reduces to $\alpha_1\langle \mathbf{v}_1, \mathbf{v}_1\rangle = 0$. Finally, since $\mathbf{v}_1 \neq 0$, $\langle \mathbf{v}_1, \mathbf{v}_1\rangle = \|\mathbf{v}_1\|^2 \neq 0$ [Theorem 1(d)], and hence $\alpha_1 = 0$. Thus equation (5) reduces to $\alpha_2\mathbf{v}_2 + \cdots + \alpha_k\mathbf{v}_k = 0$. Now form the inner product of $\mathbf{v}_2$ with both sides of this equation and conclude in the same way that $\alpha_2 = 0$. Repeating this argument we find that each of the $\alpha_i$'s is equal to zero. Hence $\mathbf{v}_1, \ldots, \mathbf{v}_k$ are linearly independent.

We need to assume that the vectors are nonzero in this theorem because any set of vectors that contains the zero vector is dependent. Also, it follows from this theorem that $R^m$ cannot contain more than $m$ mutually orthogonal nonzero vectors because such a set of vectors is independent and $R^m$ cannot contain more than $m$ independent vectors.

### EXAMPLE 6

Each of the following collections of vectors is independent because they are nonzero orthogonal vectors.

(a) $(1, 1, 1)$, $(1, -1, 0)$, and $(1, 1, -2)$
(b) $(1, 1, 1, -1)$, $(3, -1, -1, 1)$, and $(0, 1, 1, 2)$

### EXAMPLE 7

We can find a vector $\mathbf{x}$ orthogonal to a given vector $\mathbf{u}$ by merely solving the equation

$$\langle \mathbf{u}, \mathbf{x}\rangle = u_1x_1 + u_2x_2 + \cdots + u_mx_m = 0.$$

For instance, if $\mathbf{u} = (1, 2, -1, 2)$, then a vector $\mathbf{x}$ is orthogonal to $\mathbf{u}$ if and only if

$$\langle \mathbf{u}, \mathbf{x}\rangle = x_1 + 2x_2 - x_3 + 2x_4 = 0.$$

The general solution of this equation is

$$\mathbf{x} = x_2(-2, 1, 0, 0) + x_3(1, 0, 1, 0) + x_4(-2, 0, 0, 1).$$

Thus $\mathbf{x}$ is orthogonal to $\mathbf{u}$ if and only if $\mathbf{x}$ is in the subspace spanned by

$$(-2, 1, 0, 0), \quad (1, 0, 1, 0), \quad (-2, 0, 0, 1).$$

## PROBLEMS 3.2

**1.** Find the length of each of the following vectors.

*(a) $(1, -1, 2)$
*(b) $(\frac{1}{3}, -\frac{1}{3}, \frac{1}{6}, -\frac{1}{12})$
(c) $(\sqrt{2}, 5, 0, -2)$
(d) $(\frac{1}{2}, \frac{1}{3}, \frac{1}{4}, \frac{2}{3})$
*(e) $(-1, 0, \pi, 0)$
*(f) $(1/a, b)$
(g) $(5, 0, 1, 0, 1, 3)$
(h) $(\frac{1}{2}, \frac{5}{8})$

2. Find the distance between each of the following pairs of vectors.
   *(a) $(1,0,1)$ and $(0,-1,0)$
   (b) $(1,1,\sqrt{3},1)$ and $(1,0,0,1)$
   *(c) $(1,\frac{1}{2},\frac{1}{3},\frac{2}{3})$ and $(2,\frac{1}{3},\frac{1}{6},-\frac{1}{3})$
   (d) $(-1,2,0,1,1)$ and $(-1,-2,0,\frac{1}{2},0)$

*3. Which of the following pairs of vectors are orthogonal?
   (a) $(1,2)$ and $(2,1)$
   (b) $(1,2,3)$ and $(3,0,-1)$
   (c) $(\sqrt{2},5,0,-1)$ and $(\sqrt{2},-1,7,-3)$
   (d) $(1,0,0,0,1)$ and $(0,3,7,-2,0)$

4. Expand each of the following using Theorem 1.
   (a) $\langle \mathbf{x} + \beta\mathbf{y}, \mathbf{z}\rangle$
   (b) $\langle \mathbf{x} + \beta\mathbf{y}, \mathbf{x} + \beta\mathbf{y}\rangle$
   (c) $\langle \alpha\mathbf{x} + \beta\mathbf{y}, \alpha\mathbf{x} + \beta\mathbf{y}\rangle$

5. Use Theorem 1 to prove that $\mathbf{x} \neq \mathbf{0}$ if and only if $\langle \mathbf{x},\mathbf{x}\rangle > 0$.

*6. Determine which of the following sets of vectors are independent.
   (a) $(1,1)$ and $(-1,1)$
   (b) $(-1,2,3)$ and $(6,0,2)$
   (c) $(0,0,1,0,0)$, $(0,0,0,1,0)$, $(0,0,0,0,1)$ and $(0,0,0,0,0)$

7. Find a nonzero vector that is orthogonal to:
   *(a) $(1,0,1)$ and $(1,1,-1)$
   (b) $(1,-1,2)$ and $(2,1,-1)$

8. Find two nonzero 3-vectors that are orthogonal to each other and to the vector:
   *(a) $(1,1,1)$ (b) $(1,1,0)$ (c) $(1,2,1)$

9. Find two nonzero 4-vectors that are orthogonal to each other and to the vector:
   *(a) $(1,2,1,-3)$
   (b) $(1,1,0,0)$
   (c) $(1,0,2,1)$

10. Let $\mathbf{a} = (1,3,5)$ and $\mathbf{b} = (1,1,1)$. Find vectors $\mathbf{x}$ and $\mathbf{y}$ such that $\mathbf{a} = \mathbf{x} + \mathbf{y}$, $\mathbf{x}$ is collinear with $\mathbf{b}$, and $\mathbf{y}$ is orthogonal to $\mathbf{b}$. Interpret this procedure geometrically.

11. Verify properties (a), (b), and (c) of Theorem 1.

12. Let $\mathbf{x}$ and $\mathbf{y}$ be vectors in $R^m$. Prove each of the following.
   *(a) $\|\mathbf{x}+\mathbf{y}\| = \|\mathbf{x}-\mathbf{y}\|$ if and only if $\mathbf{x}$ is orthogonal to $\mathbf{y}$.
   (b) $\mathbf{x}-\mathbf{y}$ is orthogonal to $\mathbf{x}+\mathbf{y}$ if and only if $\|\mathbf{x}\| = \|\mathbf{y}\|$. What does this say about the parallelogram determined by $\mathbf{x}$ and $\mathbf{y}$?
   *(c) If $\mathbf{x}$ and $\mathbf{y}$ are orthonormal vectors, then $\|\mathbf{x}-\mathbf{y}\| = \sqrt{2}$.
   (d) If $\mathbf{x}$ and $\mathbf{y}$ are orthogonal, then $\|\mathbf{x}-\mathbf{y}\|^2 = \|\mathbf{x}\|^2 + \|\mathbf{y}\|^2$.
   (e) If $\mathbf{v}$ is orthogonal to $\mathbf{x}$ and $\mathbf{y}$, then $\mathbf{v}$ is orthogonal to any linear combination of $\mathbf{x}$ and $\mathbf{y}$.
   (f) If $\mathbf{x}$ and $\mathbf{y}$ are independent and $\mathbf{v}$ is orthogonal to $\mathbf{x}$ and $\mathbf{y}$, then $\mathbf{x}$, $\mathbf{y}$, and $\mathbf{v}$ are independent.
   (g) Find and prove a generalization of part (f).
   *(h) Show that $\langle \mathbf{x},\mathbf{y}\rangle = \frac{1}{4}\{\|\mathbf{x}+\mathbf{y}\|^2 - \|\mathbf{x}-\mathbf{y}\|^2\}$.

*13. If $\mathbf{x}$ is a unit vector in $R^m$, show that the matrix $A = \mathbf{x}\mathbf{x}^T$ is symmetric and has the property that $A^2 = A$.

14. Let $\mathbf{a} = (a_1,a_2,a_3)$ and $\mathbf{b} = (b_1,b_2,b_3)$ be two noncollinear vectors.
   (a) Show that there is a nonzero vector $\mathbf{c} = (c_1,c_2,c_3)$ that is orthogonal to both $\mathbf{a}$ and $\mathbf{b}$. One way of showing this is to solve the system of equations

$$\langle \mathbf{a},\mathbf{c}\rangle = 0$$
$$\langle \mathbf{b},\mathbf{c}\rangle = 0$$

   or

$$\begin{bmatrix} \text{—}\mathbf{a}\text{—} \\ \text{—}\mathbf{b}\text{—} \end{bmatrix} \begin{bmatrix} | \\ \mathbf{c} \\ | \end{bmatrix} = 0.$$

   (b) Show that $(a_2b_3 - a_3b_2,\, a_3b_1 - a_1b_3,\, a_1b_2 - a_2b_1)$ is a solution to these equations. This particular solution is called the **cross product** of **a** and **b**, and is denoted by $\mathbf{a} \times \mathbf{b}$.

(c) Show that any vector orthogonal to **a** and **b** is a scalar multiple of $\mathbf{a} \times \mathbf{b}$.

***15.** Show that

$$||\mathbf{x} + \mathbf{y}||^2 + ||\mathbf{x} - \mathbf{y}||^2 = 2||\mathbf{x}||^2 + 2||\mathbf{y}||^2$$

for any two $m$-vectors **x** and **y**. Give a geometric interpretation of this result in terms of parallelograms when **x** and **y** are vectors in $R^3$. It is because of this interpretation that this equation is called the *parallelogram law*.

**16.** Let $A$ be an $m \times n$ matrix. Prove each of the following.

*(a) If **u** and **v** are vectors in $R^m$ and $\langle \mathbf{u}, \mathbf{x} \rangle = \langle \mathbf{v}, \mathbf{x} \rangle$ for all **x** in $R^m$, then $\mathbf{u} = \mathbf{v}$. (*Hint:* Use Result 1.)

(b) If $\langle A\mathbf{x}, \mathbf{y} \rangle = 0$ for all **x** in $R^n$ and **y** in $R^m$, then $A = 0$.

(c) If $B$ is also an $m \times n$ matrix and $\langle A\mathbf{x}, \mathbf{y} \rangle = \langle B\mathbf{x}, \mathbf{y} \rangle$ for all **x** in $R^n$ and **y** in $R^m$, then $A = B$.

**17.** Let $A$ be a $n \times n$ matrix. Then $A$ is called an **orthogonal matrix** if its columns form an orthonormal set of vectors in $R^n$.

*(a) Show that $A$ is orthogonal if and only if $A^TA = I$. Therefore, $A$ is orthogonal if and only if $A^{-1} = A^T$.

*(b) Show that $A$ is orthogonal if and only if its rows form an orthonormal set of vectors in $R^n$. This proves the remarkable fact that the rows of a matrix form an orthonormal set if and only if the columns form an orthonormal set.

(c) Show that $A$ is orthogonal if and only if $A^{-1}$ is orthogonal.

(d) Show that if $A$ and $B$ are orthogonal, then $AB$ and $BA$ are orthogonal.

*(e) Show that if $A$ is an orthogonal matrix, then

$$\begin{aligned} ||A\mathbf{x}|| &= ||\mathbf{x}|| \quad \text{for all } \mathbf{x} \\ \langle A\mathbf{x}, A\mathbf{y} \rangle &= \langle \mathbf{x}, \mathbf{y} \rangle \quad \text{for all } \mathbf{x} \text{ and } \mathbf{y}. \end{aligned}$$

Thus orthogonal matrices preserve length and orthogonality.

(f) Show that if $\langle A\mathbf{x}, A\mathbf{y} \rangle = \langle \mathbf{x}, \mathbf{y} \rangle$ for all **x** and **y**, then $A$ is orthogonal.

(g) Show that if $||A\mathbf{x}|| = ||\mathbf{x}||$ for all **x**, then $A$ is orthogonal. [*Hint:* Use Problem 12(h) and part (f).]

*(h) Show that if **x** is a unit vector, then $I - 2\mathbf{x}\mathbf{x}^T$ is an orthogonal matrix.

(i) Show that if $A$ is orthogonal, then $A$ is symmetric if and only if $A^2 = I$.

(j) Show that if $A$ is symmetric, then $A$ is orthogonal if and only if $A^2 = I$.

*(k) Find all orthogonal matrices that are triangular.

(l) Show that if $A$ is orthogonal and $B$ is any matrix with orthonormal columns, then $AB$ has orthonormal columns.

(m) Find all $3 \times 3$ orthogonal matrices whose first two columns are:

(1) $\begin{bmatrix} 1 & 0 \\ 0 & 1 \\ 0 & 0 \end{bmatrix}$

(2) $\begin{bmatrix} 1/\sqrt{2} & 1/\sqrt{3} \\ 0 & 1/\sqrt{3} \\ 1/\sqrt{2} & -1/\sqrt{3} \end{bmatrix}$

(3) $\begin{bmatrix} 1/\sqrt{2} & 0 \\ 1/\sqrt{3} & 1/\sqrt{3} \\ -1/\sqrt{6} & 2/\sqrt{6} \end{bmatrix}$

(n) Find an orthogonal matrix whose first column is:

(1) $(1, 0, 0)$
(2) $(0, 1, 0)$
(3) $(1/\sqrt{2}, 0, 1/\sqrt{2})$
(4) $(1/\sqrt{3}, -1/\sqrt{3}, 1/\sqrt{3})$

(o) How many $2 \times 2$ orthogonal matrices are there all of whose entries are zeros and ones? List them.

(p) How many $3 \times 3$ orthogonal matrices are there all of whose entries are zeros and ones? List them.

(q) Generalize parts (o) and (p).

# 3.3 The Cauchy–Schwarz Inequality and Angles

In this section we discuss some deeper properties of the inner product. We begin with the Cauchy–Schwarz inequality, which is undoubtedly one of the most important inequalities in higher mathematics. Its proof, however, requires only those properties of the inner product given in Theorem 1 of Section 3.2 and the quadratic formula.

***Theorem 1*** *(Cauchy–Schwarz Inequality) If* $\mathbf{x}$ *and* $\mathbf{y}$ *are vectors in* $R^m$, *then*

$$|\langle \mathbf{x}, \mathbf{y} \rangle| \leq \|\mathbf{x}\| \cdot \|\mathbf{y}\|.$$

*Moreover, equality holds if and only if one of the vectors is a scalar multiple of the other.*

**Proof** If either $\mathbf{x}$ or $\mathbf{y}$ are zero, the result is obvious. Hence let us suppose that both $\mathbf{x}$ and $\mathbf{y}$ are nonzero. If one of these vectors is a scalar multiple of the other, then again the result holds. For instance, if $\mathbf{x} = \alpha\mathbf{y}$, then

$$\begin{aligned} |\langle \mathbf{x}, \mathbf{y} \rangle| &= |\langle \alpha\mathbf{y}, \mathbf{y} \rangle| = |\alpha| \cdot \|\mathbf{y}\|^2 \\ &= (|\alpha| \cdot \|\mathbf{y}\|) \cdot \|\mathbf{y}\| = \|\mathbf{x}\| \cdot \|\mathbf{y}\|. \end{aligned}$$

Finally, suppose that neither vector is a scalar multiple of the other. Then $\mathbf{x} - \alpha\mathbf{y} \neq \mathbf{0}$ for all scalars $\alpha$ and hence

$$\begin{aligned} 0 < \|\mathbf{x} - \alpha\mathbf{y}\|^2 &= \langle \mathbf{x} - \alpha\mathbf{y}, \mathbf{x} - \alpha\mathbf{y} \rangle \\ &= \langle \mathbf{x}, \mathbf{x} \rangle - 2\alpha\langle \mathbf{x}, \mathbf{y} \rangle + \alpha^2\langle \mathbf{y}, \mathbf{y} \rangle \\ &= \|\mathbf{x}\|^2 - 2\alpha\langle \mathbf{x}, \mathbf{y} \rangle + \alpha^2\|\mathbf{y}\|^2. \end{aligned}$$

The right side is a quadratic in $\alpha$ which has no real solutions. Therefore, its discriminant must be negative:

$$b^2 - 4ac = 4\langle \mathbf{x}, \mathbf{y} \rangle^2 - 4\|\mathbf{x}\|^2\|\mathbf{y}\|^2 < 0.$$

Thus $|\langle \mathbf{x}, \mathbf{y} \rangle| < \|\mathbf{x}\| \cdot \|\mathbf{y}\|$ and the proof is complete.

Without vector notation, this inequality does not look quite so simple. It becomes

$$|x_1y_1 + \cdots + x_my_m| \leq \sqrt{(x_1^2 + \cdots + x_m^2)}\sqrt{(y_1^2 + \cdots + y_m^2)}.$$

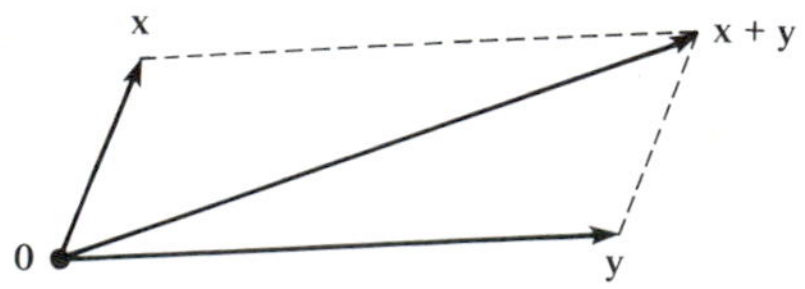


Figure 3.7

Using the Cauchy–Schwarz inequality we can prove another important inequality, the *triangle inequality*. It states that the length of any one side of a triangle cannot exceed the sum of the lengths of the other two (see Figure 3.7).

***Theorem 2*** *(Triangle Inequality) For any two vectors* $\mathbf{x}$ *and* $\mathbf{y}$ *in* $R^m$, $\|\mathbf{x} + \mathbf{y}\| \leq \|\mathbf{x}\| + \|\mathbf{y}\|$.

**Proof** To prove this inequality we expand the quantity $\|\mathbf{x} + \mathbf{y}\|^2$ and use the Cauchy–Schwarz inequality to obtain

$$\begin{aligned}\|\mathbf{x} + \mathbf{y}\|^2 &= \langle \mathbf{x} + \mathbf{y}, \mathbf{x} + \mathbf{y} \rangle \\ &= \|\mathbf{x}\|^2 + 2\langle \mathbf{x}, \mathbf{y} \rangle + \|\mathbf{y}\|^2 \\ &\leq \|\mathbf{x}\|^2 + 2\|\mathbf{x}\| \cdot \|\mathbf{y}\| + \|\mathbf{y}\|^2 \\ &= (\|\mathbf{x}\| + \|\mathbf{y}\|)^2.\end{aligned}$$

Taking the square root of both sides of this inequality completes the proof.

The Cauchy–Schwarz inequality also enables us to extend the concept of angle to $m$-dimensional space. Every calculus student has encountered the formula

$$\langle \mathbf{x}, \mathbf{y} \rangle = \|\mathbf{x}\| \cdot \|\mathbf{y}\| \cos\theta, \tag{1}$$

where $\mathbf{x}$ and $\mathbf{y}$ are vectors in $R^2$ and $\theta$ is the angle between these vectors. It is an immediate consequence of the law of cosines (see Problem 7). It also follows from the addition formula for the cosine. From Figure 3.8 we obtain

$$\begin{aligned}\cos\theta &= \cos(\alpha - \beta) \\ &= \cos\alpha\cos\beta + \sin\alpha\sin\beta \\ &= \frac{x_1 y_1 + x_2 y_2}{\|\mathbf{x}\| \cdot \|\mathbf{y}\|} \\ &= \frac{\langle \mathbf{x}, \mathbf{y} \rangle}{\|\mathbf{x}\| \cdot \|\mathbf{y}\|}.\end{aligned}$$

Equation (1) suggests that we define the cosine of the angle $\theta$ between two nonzero vectors $\mathbf{x}$ and $\mathbf{y}$ by the equation

$$\cos\theta = \frac{\langle \mathbf{x}, \mathbf{y} \rangle}{\|\mathbf{x}\| \cdot \|\mathbf{y}\|}.$$

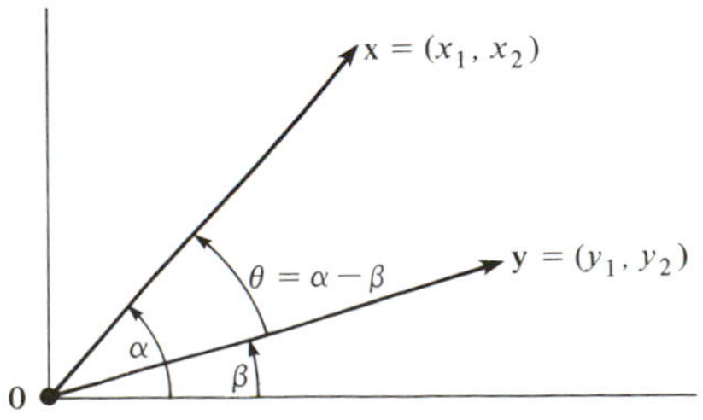


Figure 3.8

Of course, this definition makes sense if and only if

$$-1 \leq \frac{\langle \mathbf{x}, \mathbf{y} \rangle}{\|\mathbf{x}\| \cdot \|\mathbf{y}\|} \leq 1.$$

But this is exactly the statement of the Cauchy–Schwarz inequality. Therefore, given two vectors $\mathbf{x}$ and $\mathbf{y}$ in $R^m$, it follows from the Cauchy–Schwarz inequality that there is a unique number $\theta$, $0 \leq \theta \leq \pi$, such that

$$\langle \mathbf{x}, \mathbf{y} \rangle = \|\mathbf{x}\| \cdot \|\mathbf{y}\| \cos \theta.$$

$\theta$ is called the **angle** between $\mathbf{x}$ and $\mathbf{y}$.

This definition of angle in $R^m$ is consistent with our definition of orthogonality in $R^m$. For if $\mathbf{x}$ and $\mathbf{y}$ are nonzero vectors, then it follows directly from the equation $\langle \mathbf{x}, \mathbf{y} \rangle = \|\mathbf{x}\| \cdot \|\mathbf{y}\| \cos \theta$ that $\langle \mathbf{x}, \mathbf{y} \rangle = 0$ if and only if $\cos \theta = 0$, and this holds if and only if $\theta = \pi/2$.

## PROBLEMS 3.3

**1.** Find the angle between each of the following pairs of vectors.

*(a) $(2, 1)$ and $(1, 1)$

(b) $(1, 1, 0)$ and $(0, 1, 0)$

*(c) $(1, 1, 0, 1)$ and $(1, 1, 1, 1)$

(d) $(-1, 0, 2, -2, 0)$ and $(0, 1, 1, 0, 5)$

*(e) $(1, 1, 1)$ and $(1, 1, 0)$

(f) $(1, 0, 1, 1, 1)$ and $(2, 3, 1, 1, 1)$

**2.** Find all unit vectors that make an angle of $\pi/3$ with:

*(a) $(1, 0)$ (b) $(1, 1)$ (c) $(1, 2)$

**3.** Find all unit vectors that make an angle of $\pi/4$ with:

*(a) $(1, 0)$ (b) $(1, 1)$ (c) $(1, 2)$

**4.** Find all unit vectors that make an angle of $\pi/3$ with:

*(a) $(1, 0, 0)$ and $(0, 1, 0)$

(b) $(1, 0, 1)$ and $(0, 1, 0)$

**5.** Let $\mathbf{v} = (1, 4, 2)$. Let $\alpha$, $\beta$, and $\gamma$ be the angles that this vector makes with each of the coordinate vectors $\mathbf{e}_1$, $\mathbf{e}_2$, and $\mathbf{e}_3$. The numbers $\cos \alpha$, $\cos \beta$, and $\cos \gamma$ are called the *direction cosines* of $\mathbf{v}$. Find the direction cosines of the vector $\mathbf{v}$ and show that the vector

$$(\cos \alpha, \cos \beta, \cos \gamma)$$

is a unit vector in the direction of $\mathbf{v}$.

***6.** Show that if $\mathbf{x}$ and $\mathbf{y}$ are both unit vectors in $R^m$ with $\langle \mathbf{x}, \mathbf{y} \rangle = 1$, then $\mathbf{x} = \mathbf{y}$. What can you conclude when $\langle \mathbf{x}, \mathbf{y} \rangle = -1$?

**7.** Verify formula (1) in this section using the law of cosines. In case you have forgotten (as one of the authors did), the law of cosines states that $c^2 = a^2 + b^2 - 2ab \cos \theta$.

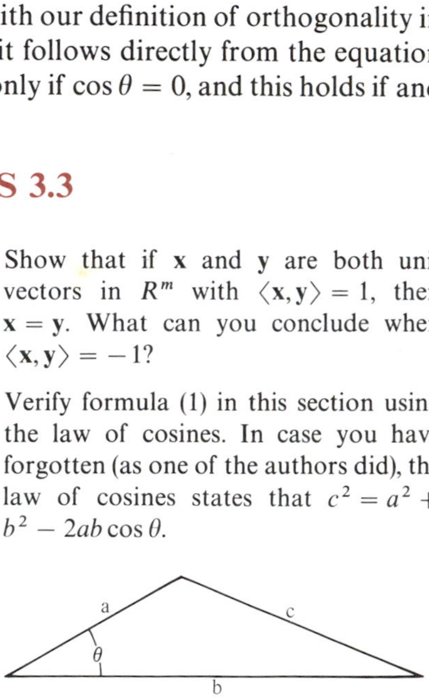


**8.** Show that $\big| \|\mathbf{x}\| - \|\mathbf{y}\| \big| \leq \|\mathbf{x} + \mathbf{y}\|$. (*Hint:* Apply the triangle inequality to $\mathbf{x} = \mathbf{x} + \mathbf{y} - \mathbf{y}$ and $\mathbf{y} = \mathbf{y} + \mathbf{x} - \mathbf{x}$.)

**9.** Let $\mathbf{v} = (1, 2, 3)$. Choose a vector $\mathbf{w}$ so that both of the following conditions are satisfied:

1. $\mathbf{w}$ is orthogonal to the vector $(1, 1, 1)$.
2. The angle between $\mathbf{v}$ and $\mathbf{w}$ is greater than 90 degrees.

***10.** Let $\alpha$ be the angle between the lines $y = 2x$ and $y = 3x$. Let $\beta$ be the angle between the lines $y = x$ and $y = \frac{4}{3}x$. Show that $\alpha = \beta$.

*11. Consider the triangle in $R^2$ with vertices $(2, 3)$, $(7, 7)$, and $(12, 1)$. Are any of its angles obtuse (i.e., greater than 90 degrees)? If so, find one. If not, show that they are all acute. Answer the same question for the triangle in $R^4$ with vertices $(0, 0, 0, 0)$, $(1, 1, 0, 0)$, and $(0, 1, 1, -1)$.

12. If **u**, **v**, and **w** are three unit vectors in $R^2$ such that $\mathbf{u} + \mathbf{v} + \mathbf{w} = \mathbf{0}$, then show that each vector makes an angle of $2\pi/3$ with the other two vectors.

13. Let **x** and **y** be vectors in $R^m$ and let $\theta$ be the angle between **x** and **y**. Prove that

$$\|\mathbf{x} + \mathbf{y}\|^2 = \|\mathbf{x}\|^2 + \|\mathbf{y}\|^2 + 2\|\mathbf{x}\| \cdot \|\mathbf{y}\| \cos \theta.$$

(*Hint:* Expand $\|\mathbf{x}+\mathbf{y}\|^2 = \langle \mathbf{x}+\mathbf{y}, \mathbf{x}+\mathbf{y} \rangle$.)

# 3.4 Orthogonal Complements

In our previous discussion of orthogonality we have restricted ourselves to talking about what it means for two vectors to be orthogonal. We discuss next what it means for a vector to be orthogonal to an entire subspace.

The following definition is clearly motivated by the geometry of $R^2$ and $R^3$ (see Figure 3.9).

***Definition*** A vector **u** in $R^m$ is **orthogonal to a subspace** $V$ of $R^m$, written $\mathbf{u} \perp V$, if **u** is orthogonal to every vector in $V$, that is,

$$\langle \mathbf{u}, \mathbf{v} \rangle = 0 \qquad \text{for all } \mathbf{v} \text{ in } V. \tag{1}$$

Given a vector **u** and a subspace $V$, it is difficult to verify condition (1) directly since it requires that we check an infinite number of equations. However, the situation is actually not as bad as it seems; we need only check that the vector **u** is orthogonal to every vector in a spanning set for $V$.

**Result 1** Let **u** be a vector in $R^m$ and let $V$ be a subspace of $R^m$. Then **u** is orthogonal to $V$ if and only if **u** is orthogonal to a spanning set for $V$. That is, if $\mathbf{v}_1, \ldots, \mathbf{v}_n$ span $V$, then

$$\mathbf{u} \perp V \quad \text{if and only if} \quad \mathbf{u} \perp \mathbf{v}_i, \qquad i = 1, \ldots, n.$$

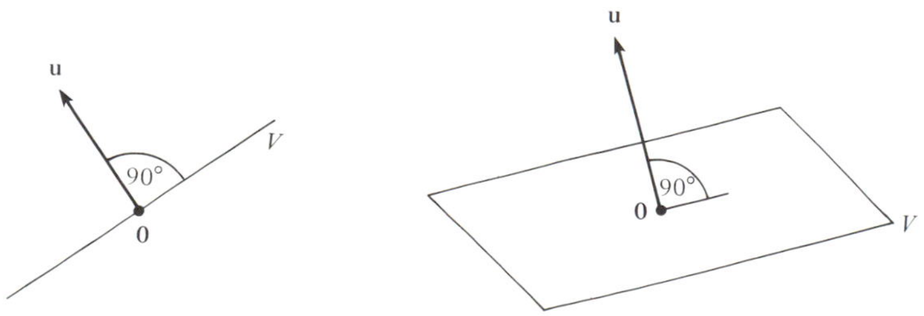


Figure 3.9

**Proof** First suppose that $\mathbf{u}$ is orthogonal to $V$. Then $\mathbf{u}$ is orthogonal to every vector in $V$ and hence to each of the vectors $\mathbf{v}_1, \ldots, \mathbf{v}_n$.

Conversely, suppose that $\mathbf{u}$ is orthogonal to each of the vectors $\mathbf{v}_1, \ldots, \mathbf{v}_n$. If $\mathbf{v}$ is any vector in $V$, then $\mathbf{v} = \alpha_1\mathbf{v}_1 + \cdots + \alpha_n\mathbf{v}_n$. Since $\langle \mathbf{u}, \mathbf{v}_1 \rangle = \cdots = \langle \mathbf{u}, \mathbf{v}_n \rangle = 0$,

$$\begin{aligned}\langle \mathbf{u}, \mathbf{v} \rangle &= \alpha_1 \langle \mathbf{u}, \mathbf{v}_1 \rangle + \cdots + \alpha_n \langle \mathbf{u}, \mathbf{v}_n \rangle \\ &= 0.\end{aligned}$$

Thus $\mathbf{u}$ is orthogonal to $\mathbf{v}$. Since $\mathbf{v}$ is an arbitrary vector in $V$, it follows that $\mathbf{u}$ is orthogonal to $V$.

Since a basis is a spanning set, it follows immediately that a vector is orthogonal to a subspace if and only if it is orthogonal to every vector in a basis for the subspace.

### EXAMPLE 1

The vector $\mathbf{x} = (2, 1, 1, -1)$ is orthogonal to the column space of the matrix

$$\begin{bmatrix} 1 & -1 & -1 & 1 \\ 0 & 0 & 2 & 1 \\ 0 & 2 & 0 & 1 \\ 2 & 0 & 0 & 4 \end{bmatrix}$$

since it is orthogonal to each column of $A$. Similarly, the vector $(-1, 1, 2)$ is orthogonal to the subspace $x - y - 2z = 0$ because the subspace is spanned by

$$(1, 1, 0) \quad \text{and} \quad (2, 0, 1)$$

and $(-1, 1, 2)$ is orthogonal to each of these vectors.

We can use this last result to find all vectors that are orthogonal to a given subspace. Let $V$ be an $r$-dimensional subspace of $R^m$ and suppose that $\mathbf{v}_1, \ldots, \mathbf{v}_n$ span $V$. By the previous result, an $m$-vector $\mathbf{x}$ is orthogonal to $V$ if and only if the system of equations

$$\langle \mathbf{v}_i, \mathbf{x} \rangle = 0, \qquad i = 1, \ldots, n,$$

is satisfied. If $A$ is the matrix whose columns are the vectors $\mathbf{v}_1, \ldots, \mathbf{v}_n$, then $A$ has rank $r$ and this system of equations becomes $A^T\mathbf{x} = \mathbf{0}$. Thus an $m$-vector $\mathbf{x}$ is orthogonal to $V$ if and only if $\mathbf{x}$ is in the nullspace of the matrix $A^T$. Since the nullspace of $A^T$ is a subspace of $R^m$ of dimension $m - r$, the set of all vectors orthogonal to $V$ is a subspace of $R^m$ of dimension $m - r$. We call this subspace the **orthogonal complement** of $V$ and denote it by $V^\perp$ (read $V$ perp). Since $V^\perp$ contains *all* vectors in $R^m$ that are orthogonal to $V$, any vector contained in both $V$ and $V^\perp$ would have to be orthogonal to itself. By Theorem 1(d) of Section 3.2, the only vector that is orthogonal to itself is the zero vector. Hence the only vector contained in both $V$ and $V^\perp$ is the zero vector. We summarize our findings in parts (a), (b), and (c) of the following theorem.

***Theorem 1*** *Let $V$ be a subspace of $R^m$.*

(*a*) *The orthogonal complement $V^\perp$ of $V$ is a subspace of $R^m$.*
(*b*) *If the dimension of $V$ is $r$, then the dimension of $V^\perp$ is $m - r$.*
(*c*) *The only vector in $R^m$ that lies in both $V$ and $V^\perp$ is the zero vector.*
(*d*) *If $\mathbf{v}_1, \ldots, \mathbf{v}_r$ form a basis for $V$ and $\mathbf{w}_1, \ldots, \mathbf{w}_{m-r}$ form a basis for $V^\perp$, then $\mathbf{v}_1, \ldots, \mathbf{v}_r, \mathbf{w}_1, \ldots, \mathbf{w}_{m-r}$ form a basis for $R^m$.*

**Proof of (d)** Since there are $m$ vectors in this collection, it suffices to show that these vectors are independent. Suppose that

$$\mathbf{0} = \alpha_1 \mathbf{v}_1 + \cdots + \alpha_r \mathbf{v}_r + \beta_1 \mathbf{w}_1 + \cdots + \beta_{m-r} \mathbf{w}_{m-r}.$$

Then

$$\alpha_1 \mathbf{v}_1 + \cdots + \alpha_r \mathbf{v}_r = -(\beta_1 \mathbf{w}_1 + \cdots + \beta_{m-r} \mathbf{w}_{m-r}).$$

The vector on the left-hand side of this equation is in $V$, and the vector on the right-hand side is in $V^\perp$. Thus both vectors (being equal) must be in both $V$ and $V^\perp$. Thus by part (c)

$$\alpha_1 \mathbf{v}_1 + \cdots + \alpha_r \mathbf{v}_r = \mathbf{0} \qquad \text{and} \qquad \beta_1 \mathbf{w}_1 + \cdots + \beta_{m-r} \mathbf{w}_{m-r} = \mathbf{0}.$$

Since $\mathbf{v}_1, \ldots, \mathbf{v}_r$ are independent, all of the $\alpha_i$'s are 0. Similarly, all of the $\beta_j$'s are 0. This completes the proof.

Note that the proof of parts (a) and (b) of Theorem 1 provides us with a computational procedure for actually determining the orthogonal complement of a subspace. First we express the subspace $V$ as the column space of a matrix $A$ (the columns of $A$ consist of vectors that span $V$). Then the orthogonal complement of $V$ is the nullspace of $A^T$. Since any subspace can be expressed as the column space of a matrix in this way, we can summarize our procedure in terms of matrices.

***Theorem 2*** *The orthogonal complement of the column space $C(A)$ of a matrix $A$ is the nullspace of its transpose $A^T$. In symbols,*

$$C(A)^\perp = N(A^T).$$

## EXAMPLE 2

To find the orthogonal complement of the subspace $V$ spanned by $(1, 2, 2, 1)$ and $(1, 0, 2, 0)$, we form the matrix

$$A = \begin{bmatrix} 1 & 1 \\ 2 & 0 \\ 2 & 2 \\ 1 & 0 \end{bmatrix}$$

and compute the nullspace of $A^T$. The reduced echelon form of $A^T$ is $E = \begin{bmatrix} 1 & 0 & 2 & 0 \\ 0 & 1 & 0 & \frac{1}{2} \end{bmatrix}$ and hence a basis for the nullspace is $(-2, 0, 1, 0)$ and $(0, -1, 0, 2)$.

Thus the orthogonal complement $V^\perp$ is the two-dimensional subspace of $R^4$ spanned by these two vectors.

Let us find the orthogonal complements of the various subspaces of $R^3$. The orthogonal complement of the trivial subspace $\{\mathbf{0}\}$ is all of $R^3$, since every vector in $R^3$ is orthogonal to the zero vector. The orthogonal complement of a line through the origin is the plane through the origin that is perpendicular to the given line [see Figure 3.10(a)]. Why? Because by Theorem 1 the dimension of the orthogonal complement is 2 and a two-dimensional subspace of $R^3$ is a plane. A similar dimension argument proves that the orthogonal complement of a plane through the origin is the line through the origin perpendicular to the plane [see Figure 3.10(b)]. Finally, the orthogonal complement of $R^3$ is the zero vector, since the only vector orthogonal to every vector is the zero vector (Result 1 of Section 3.2).

Let $V$ be a plane in $R^3$ through the origin. Then $V^\perp$ is a one-dimensional subspace of $R^3$. If $\mathbf{n}$ is a basis for $V^\perp$, then

$$\langle \mathbf{n}, \mathbf{v} \rangle = 0 \tag{2}$$

if and only if $\mathbf{v}$ is in $V$. Therefore, (2) is an equation of the plane $V$. If we let $\mathbf{n} = (a, b, c)$ and $\mathbf{v} = (x, y, z)$, then we can write (2) as

$$ax + by + cz = 0. \tag{3}$$

Therefore, any plane in $R^3$ has an equation of this form. Conversely, any equation of the form (2), or equivalently of the form (3), is the equation of a plane (if $\mathbf{n} \neq \mathbf{0}$), for $\mathbf{n}$ spans a one-dimensional subspace of $R^3$ and hence by Theorem 1 its orthogonal complement is a two-dimensional subspace (i.e., a plane). This constitutes a simple proof of Result 1 of Section 1.10. Note also that by Theorem 2 it is a simple matter to find a vector normal to a plane: if the plane is spanned by $\mathbf{v}_1$ and $\mathbf{v}_2$, then $\mathbf{n}$ can be any nonzero vector in $N(A^T)$, where $A$ is the matrix whose columns are $\mathbf{v}_1$ and $\mathbf{v}_2$. Similarly, given a nonzero vector $\mathbf{n}$, the plane whose equation is

$$\langle \mathbf{n}, \mathbf{v} \rangle = 0$$

is $N(A^T)$, where $A$ is the matrix whose single column is $\mathbf{n}$.

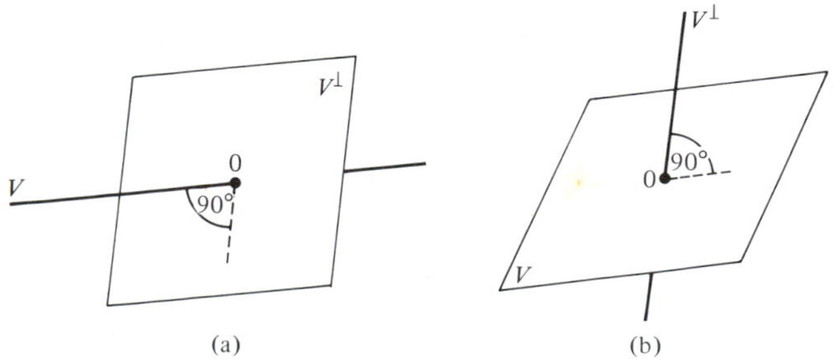


Figure 3.10

### EXAMPLE 3

Consider the plane

$$x + 2y + 8z = 0.$$

The vector $\mathbf{n} = (1, 2, 8)$ is orthogonal to the plane. If

$$A = \begin{bmatrix} 1 \\ 2 \\ 8 \end{bmatrix}, \qquad \text{then} \quad A^T = [1 \quad 2 \quad 8]$$

and the plane is $C(A)^\perp = N(A^T)$. A basis for the plane is $(-2, 1, 0)$ and $(-8, 0, 1)$.

### EXAMPLE 4

Consider the plane $V$ spanned by

$$(1, 3, 4) \quad \text{and} \quad (2, 1, 2).$$

To find a nonzero vector normal to the plane, form the matrix $A$ having these vectors as its columns. Then $V = C(A)$ and $V^\perp = C(A)^\perp = N(A^T)$. Since

$$A^T = \begin{bmatrix} 1 & 3 & 4 \\ 2 & 1 & 2 \end{bmatrix} \quad \text{reduces to} \quad \begin{bmatrix} 1 & 0 & \frac{2}{5} \\ 0 & 1 & \frac{6}{5} \end{bmatrix},$$

a basis for $N(A^T)$ is $(-2, -6, 5)$. Thus a Cartesian equation of the plane $V$ is

$$-2x - 6y + 5z = 0.$$

## PROBLEMS 3.4

**1.** Find a basis for the orthogonal complement of the subspace spanned by:
- *(a) $(1, 1)$
- (b) $(1, 0, 1)$
- *(c) $(1, 1, 3)$ and $(1, 1, 2)$
- (d) $(1, 1, 0, 0)$
- *(e) $(1, 1, 0, 0)$ and $(0, 2, 4, 5)$
- (f) $(2, 2, 1, 0)$, $(2, 4, 0, 1)$, and $(4, -2, 1, -1)$
- *(g) $(1, -1, 1, -1, 1)$
- (h) $(1, 1, 1, 1, 1)$ and $(1, 0, 1, 0, 1)$
- *(i) $(1, -1, 1, 1, 1)$, $(2, 0, 3, -2, 0)$, and $(3, -3, 4, 5, 2)$
- (j) $(1, 1, 5, 3)$ and $(2, 0, 1, -3)$
- *(k) $(0, 1, 2, 2)$ and $(0, 1, 1, 1)$
- (l) $(3, 0, -3, 0)$ and $(3, 2, -5, 2)$
- *(m) $(1, 2, 0, 1)$ and $(3, 1, 2, 2)$

**2.** For each subspace in Problem 1, verify parts (b) and (d) of Theorem 1.

**3.** Find the orthogonal complement of each of the following subspaces.
- *(a) The line spanned by $(1, -3, 4)$.
- (b) The line spanned by $(-1, 2, -5)$.
- *(c) The plane spanned by $(1, 2, 3)$ and $(2, 5, 3)$.
- (d) The plane spanned by $(1, 1, 1)$ and $(1, -1, 0)$.
- *(e) The plane $3x + 2y - 5z = 0$.
- (f) The plane $7x - 6y + z = 0$.

**4.** Verify that the vector $(4, 2, -1)$ is orthogonal to the column space of the matrix

$$A = \begin{bmatrix} 1 & 0 & -1 \\ -1 & 1 & 2 \\ 2 & 2 & 0 \end{bmatrix}.$$

What is the orthogonal complement of $C(A)$?

*5. Let $V$ be the subspace of $R^5$ consisting of all vectors of the form

$$(a, a+b, b+c, c, c).$$

(a) Find a basis for $V$.
(b) Find a basis for $V^\perp$.
(c) Write down equations for $V$; that is, write down a linear system whose solutions are exactly the vectors in $V$.
(d) Write down equations for $V^\perp$.

6. Repeat Problem 5 for the subspace of $R^8$ consisting of all vectors of the form

$$(a, a, a+b, b, b, b-c, b-c, c).$$

7. Let $\mathbf{v}_1 = (1, 0, 1)$ and $\mathbf{v}_2 = (1, 1, -1)$.
(a) Show that $\mathbf{v}_1$ is orthogonal to $\mathbf{v}_2$.
(b) Find a vector $\mathbf{v}_3$ orthogonal to $\mathbf{v}_1$ and $\mathbf{v}_2$.
(c) Do the vectors $\mathbf{v}_1, \mathbf{v}_2, \mathbf{v}_3$ form a basis for $R^3$?
(d) If possible, express $(1, 4, 1)$ as a linear combination of $\mathbf{v}_1, \mathbf{v}_2, \mathbf{v}_3$.

8. Determine the orthogonal complement of each of the subspaces of $R^2$. Also, when the subspace is a straight line, determine the equation of the orthogonal complement in terms of the equation of the line.

9. Given a straight line in $R^m$ through the origin, determine the equation of the orthogonal complement in terms of the equation of the line.

10. Let $S$ be a subset of $R^m$.
*(a) Show that the set $S^\perp$ of all vectors in $R^m$ which are orthogonal to $S$ is a subspace of $R^m$.
(b) Let $V$ be the subspace of $R^m$ that is spanned by the vectors in $S$. Show that $V^\perp = S^\perp$.
*(c) Let $S^{\perp\perp}$ be the set of all vectors that are orthogonal to $S^\perp$. Show that $S$ is a subset of $S^{\perp\perp}$.
(d) If $S$ is a subset of $T$, show that $T^\perp$ is a subspace of $S^\perp$.

11. Show that if $V$ is a subspace of $R^m$, then $V = V^{\perp\perp}$. (*Hint:* What is the dimension of $V^{\perp\perp}$?)

12. Let $A$ be an $m \times n$ matrix.
(a) Show that the nullspace of $A$ is the orthogonal complement of the row space of $A$.
(b) What is the orthogonal complement of the nullspace of $A^T$?
(c) What is the orthogonal complement of the nullspace of $A$?

13. Find the orthogonal complement of the column space and the row space of each of the following matrices.

*(a) $\begin{bmatrix} 1 & -3 \\ -2 & 6 \end{bmatrix}$ *(b) $\begin{bmatrix} 1 & -1 \\ 0 & 2 \\ 2 & 1 \end{bmatrix}$

(c) $\begin{bmatrix} 1 & 1 & 1 \\ 0 & 2 & 1 \\ 0 & 0 & 3 \end{bmatrix}$ (d) $\begin{bmatrix} 1 \\ 2 \\ 3 \\ 4 \end{bmatrix}$

(e) $\begin{bmatrix} 1 & 1 \\ -1 & -2 \\ 1 & -1 \\ -1 & 1 \end{bmatrix}$

14. Is there a matrix whose row space contains $(1, 2, -3)$ and whose nullspace contains $(1, 1, 1)$? If there is such a matrix, explain why and give an example. If not, explain why not.

15. Let $\mathbf{v}_1, \mathbf{v}_2, \ldots, \mathbf{v}_k$ and $\mathbf{w}_1, \mathbf{w}_2, \ldots, \mathbf{w}_k$ be two collections of vectors in $R^m$ with the property that

$$\langle \mathbf{v}_i, \mathbf{w}_j \rangle = 0 \quad \text{when} \quad i \neq j \quad \text{and} \quad \langle \mathbf{v}_i, \mathbf{w}_i \rangle = 1. \qquad (*)$$

(a) Prove that both sets of vectors $\mathbf{v}_1, \mathbf{v}_2, \ldots, \mathbf{v}_k$ and $\mathbf{w}_1, \mathbf{w}_2, \ldots, \mathbf{w}_k$ are independent.
(b) Prove that if $k = m$ and $\mathbf{v}_1, \mathbf{v}_2, \ldots, \mathbf{v}_m$ form an orthonormal basis for $R^m$, then $\mathbf{v}_i = \mathbf{w}_i$ for all $i$. (*Hint:* $\mathbf{w}_i$ is orthogonal to the subspace spanned by $\mathbf{v}_1, \ldots, \mathbf{v}_{i-1}, \mathbf{v}_{i+1}, \ldots, \mathbf{v}_m$.)
(c) Prove that if $k = m$ and $\mathbf{v}_1, \mathbf{v}_2, \ldots, \mathbf{v}_m$ form a basis for $R^m$, then there is a unique set of vectors $\mathbf{w}_1, \mathbf{w}_2, \ldots, \mathbf{w}_m$ such that property $(*)$ is satisfied. (*Hint:* By the hint given in part (b), $\mathbf{w}_i$

must belong to a one-dimensional subspace and there is a unique vector in this subspace whose inner product with $\mathbf{v}_i$ equals 1.)

(d) Find the vectors $\mathbf{w}_1, \mathbf{w}_2, \mathbf{w}_3, \mathbf{w}_4$ such that property (∗) is satisfied when $\mathbf{v}_1 = (1, 1, 1, 1)$, $\mathbf{v}_2 = (0, 1, 1, 1)$, $\mathbf{v}_3 = (0, 0, 1, 1)$, $\mathbf{v}_4 = (0, 0, 0, 1)$.

**16.** Two subspaces $V$ and $W$ of $R^m$ are called *orthogonal subspaces* if every vector in $V$ is orthogonal to every vector in $W$. Thus $V$ and $W$ are orthogonal subspaces if and only if $\langle \mathbf{v}, \mathbf{w} \rangle = 0$ for every $\mathbf{v}$ in $V$ and $\mathbf{w}$ in $W$.

(a) Show that the subspace $V$ spanned by $(-5, 0, 0, 0, 1)$ and $(0, -2, 0, 1, 0)$ is orthogonal to the subspace $W$ spanned by $(1, 2, 3, 4, 5)$ and $(1, -2, 3, -4, 5)$.

*(b) Is the subspace $W$ in part (a) the orthogonal complement of the subspace $V$?

(c) Show that a subspace $V$ and its orthogonal complement $V^\perp$ are orthogonal subspaces.

*(d) Show that if $V$ and $W$ are orthogonal subspaces of $R^m$ with the property that every vector $\mathbf{x}$ in $R^m$ can be written as a sum of a vector in $V$ and a vector in $W$, then $W = V^\perp$.

(e) Show that if $V$ and $W$ are orthogonal subspaces of $R^m$ with the property that $\dim V + \dim W = m$, then $W = V^\perp$.

**17.** Let $A$ and $B$ be $n \times n$ matrices. Show that if $A^T B = 0$, then $C(A)$ and $C(B)$ are orthogonal subspaces.

# 3.5 Projections

In Section 3.1 we discussed the question of what can be done with an inconsistent system of equations. We saw that one way to approach such a system was to replace it by a consistent system which could then be solved. The consistent system that is the "best possible" replacement for the inconsistent system $A\mathbf{x} = \mathbf{b}$ is the system $A\mathbf{x} = \mathbf{p}$, where $\mathbf{p}$ is the vector in the column space of $A$ which is closest to $\mathbf{b}$. In this section we develop a method for computing $\mathbf{p}$.

Let $V$ be a straight line or plane through the origin in $R^3$. If a vector $\mathbf{b}$ does not belong to the subspace $V$, then from geometry we know that the vector $\mathbf{p}$ in the subspace that is closest to $\mathbf{b}$ is found by dropping a perpendicular from $\mathbf{b}$ to the subspace (see Figure 3.11). This means that $\mathbf{p}$ is the vector in the subspace $V$ with the property that the vector $\mathbf{b} - \mathbf{p}$ is orthogonal to $V$. Therefore, $\mathbf{b} - \mathbf{p}$ is in $V^\perp$ because

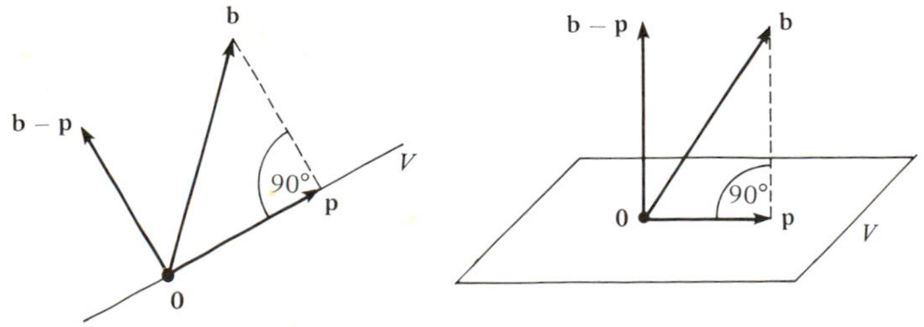


Figure 3.11

$V^{\perp}$ contains all vectors orthogonal to $V$. Since

$$\mathbf{b} = \mathbf{p} + (\mathbf{b} - \mathbf{p}),$$

we see that $\mathbf{b}$ can be expressed algebraically as a sum of a vector in $V$ and a vector in $V^{\perp}$. Thus the subspace $V$ and its orthogonal complement $V^{\perp}$ yield an interesting (and, as we shall see, very useful) algebraic description of $R^3$. Specifically, we are able to decompose any vector in $R^3$ into a sum of two vectors, one in $V$ and the other in $V^{\perp}$.

In our next theorem, we generalize this discussion to higher-dimensional spaces.

***Theorem 1*** *Let $V$ be a subspace of $R^m$ and let $V^{\perp}$ be its orthogonal complement. Given any vector $\mathbf{b}$ in $R^m$, there is a unique vector $\mathbf{p}$ in $V$ such that $\mathbf{b} - \mathbf{p}$ is in $V^{\perp}$. Since*

$$\mathbf{b} = \mathbf{p} + (\mathbf{b} - \mathbf{p}),$$

*the vectors $\mathbf{p}$ and $\mathbf{b} - \mathbf{p}$ give a unique decomposition of $\mathbf{b}$ into a sum of two vectors, one in $V$ and the other in $V^{\perp}$.*

**Proof** By Theorem 1(d) of Section 3.4, if $\mathbf{v}_1, \ldots, \mathbf{v}_n$ is a basis for $V$ and $\mathbf{w}_1, \ldots, \mathbf{w}_{m-n}$ is a basis for $V^{\perp}$, then $\mathbf{v}_1, \ldots, \mathbf{v}_n, \mathbf{w}_1, \ldots, \mathbf{w}_{m-n}$ is a basis for $R^m$. If $\mathbf{b}$ is an $m$-vector and

$$\mathbf{b} = \alpha_1 \mathbf{v}_1 + \cdots + \alpha_n \mathbf{v}_n + \beta_1 \mathbf{w}_1 + \cdots + \beta_{m-n} \mathbf{w}_{m-n}$$

is the expansion of $\mathbf{b}$ in terms of this basis, then

$$\begin{aligned} \mathbf{p} &= \alpha_1 \mathbf{v}_1 + \cdots + \alpha_n \mathbf{v}_n && \text{is in } V, \qquad \text{and} \\ \mathbf{q} &= \beta_1 \mathbf{w}_1 + \cdots + \beta_{m-n} \mathbf{w}_{m-n} && \text{is in } V^{\perp}. \end{aligned}$$

Thus there is a vector $\mathbf{p}$ in $V$ such that $\mathbf{b} - \mathbf{p} = \mathbf{q}$ is in $V^{\perp}$.

It remains to show that the vector $\mathbf{p}$ is unique. Suppose that there is another vector $\mathbf{p}_1$ in $V$ such that $\mathbf{b} - \mathbf{p}_1$ is also in $V^{\perp}$. Since $V^{\perp}$ is a subspace, the vector

$$(\mathbf{b} - \mathbf{p}_1) - (\mathbf{b} - \mathbf{p}) = \mathbf{p} - \mathbf{p}_1$$

is in $V^{\perp}$. But it is also in $V$ since $\mathbf{p}$ and $\mathbf{p}_1$ are both in $V$. By Theorem 1(c) of Section 3.4, $\mathbf{p} = \mathbf{p}_1$. Thus $\mathbf{p}$ is unique.

***Definition*** Let $V$ be a subspace of $R^m$ and $\mathbf{b}$ a vector in $R^m$. Then the **projection of b onto** $V$ is the unique vector $\mathbf{p}$ in $V$ such that $\mathbf{b} - \mathbf{p}$ is orthogonal to $V$. The vector $\mathbf{b} - \mathbf{p}$ is called the **projection of b onto** $V^{\perp}$. Thus a vector $\mathbf{p}$ is the projection of $\mathbf{b}$ onto $V$ if and only if $\mathbf{p}$ is in $V$ and $\mathbf{b} - \mathbf{p}$ is in $V^{\perp}$.

Recall that our procedure for dealing with an inconsistent system $A\mathbf{x} = \mathbf{b}$ requires that we find the vector in the column space of $A$ which is closest to $\mathbf{b}$. The following result shows that this vector is the projection of $\mathbf{b}$ onto the column space of $A$. This result also shows that in $R^3$ the projection of a vector onto a subspace corresponds to the geometrical construction of dropping a perpendicular from a vector to a subspace.

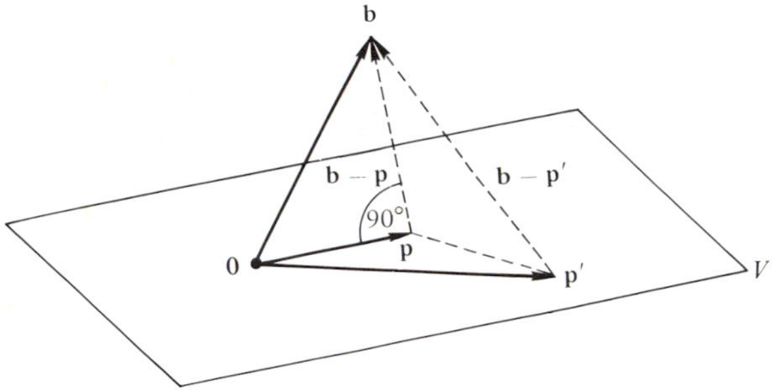


Figure 3.12

***Theorem 2*** *Let $V$ be a subspace of $R^m$ and* $\mathbf{b}$ *a vector in $R^m$. The projection of* $\mathbf{b}$ *onto $V$ is the vector in $V$ that is closest to* $\mathbf{b}$.

**Proof** Let $\mathbf{p}$ be the projection of $\mathbf{b}$ onto $V$. If $\mathbf{p}'$ is any other vector in $V$, $\mathbf{p}' - \mathbf{p}$ is in $V$ and hence is orthogonal to $\mathbf{b} - \mathbf{p}$ (see Figure 3.12). By the Pythagorean theorem

$$\|\mathbf{b} - \mathbf{p}'\|^2 = \|\mathbf{b} - \mathbf{p}\|^2 + \|\mathbf{p} - \mathbf{p}'\|^2.$$

Since $\|\mathbf{p} - \mathbf{p}'\| \geq 0$, it follows that $\|\mathbf{b} - \mathbf{p}\| \leq \|\mathbf{b} - \mathbf{p}'\|$. This completes the proof.

The proof of Theorem 1 provides us with a method for computing the projection of a vector onto a subspace. Given an $n$-dimensional subspace $V$ of $R^m$, first determine a basis for $V$, say $\mathbf{v}_1, \ldots, \mathbf{v}_n$. As we have seen in Section 3.4, we can use this basis for $V$ to find a basis for $V^\perp$. We form the matrix $A$ having the vectors $\mathbf{v}_1, \ldots, \mathbf{v}_n$ as its columns. Then $V^\perp$ is the nullspace of $A^T$, and hence a basis $\mathbf{w}_1, \ldots, \mathbf{w}_{m-n}$ for $V^\perp$ can be read off from the echelon form of $A^T$. The vectors $\mathbf{v}_1, \ldots, \mathbf{v}_n, \mathbf{w}_1, \ldots, \mathbf{w}_{m-n}$ form a basis for $R^m$. Now given any vector $\mathbf{b}$ in $R^m$, express $\mathbf{b}$ in terms of this basis,

$$\mathbf{b} = \alpha_1\mathbf{v}_1 + \cdots + \alpha_n\mathbf{v}_n + \beta_1\mathbf{w}_1 + \cdots + \beta_{m-n}\mathbf{w}_{m-n}.$$

The vector

$$\mathbf{p} = \alpha_1\mathbf{v}_1 + \cdots + \alpha_n\mathbf{v}_n$$

is the projection of $\mathbf{b}$ onto $V$ and the vector

$$\mathbf{b} - \mathbf{p} = \beta_1\mathbf{w}_1 + \cdots + \beta_{m-n}\mathbf{w}_{m-n}$$

is the projection of $\mathbf{b}$ onto $V^\perp$.

## EXAMPLE 1

Let $V$ be the subspace of $R^3$ spanned by the vectors $(2, 0, -1)$ and $(1, 1, 0)$. Since these vectors are independent, they form a basis for $V$. To find a basis for $V^\perp$, we let $A$ be the matrix

$$A = \begin{bmatrix} 2 & 1 \\ 0 & 1 \\ -1 & 0 \end{bmatrix}$$

and find the nullspace of $A^T$. The reduced echelon form of $A^T$ is

$$E = \begin{bmatrix} 1 & 0 & -\frac{1}{2} \\ 0 & 1 & \frac{1}{2} \end{bmatrix}.$$

Hence a basis for the nullspace of $A^T$ is $(1, -1, 2)$, so this vector is a basis for $V^\perp$. To express the 3-vector $\mathbf{b} = (7, -1, 5)$ in terms of the basis $(2, 0, -1)$, $(1, 1, 0)$, and $(1, -1, 2)$ of $R^3$, we use Gaussian elimination to reduce the augmented matrix

$$\begin{bmatrix} 2 & 1 & 1 & : & 7 \\ 0 & 1 & -1 & : & -1 \\ -1 & 0 & 2 & : & 5 \end{bmatrix}$$

to the echelon form

$$\begin{bmatrix} 1 & 0 & 0 & : & 1 \\ 0 & 1 & 0 & : & 2 \\ 0 & 0 & 1 & : & 3 \end{bmatrix}.$$

Thus the projection of $(7, -1, 5)$ onto $V$ is

$$\mathbf{p} = 1(2, 0, -1) + 2(1, 1, 0)$$

and the projection of $(7, -1, 5)$ onto $V^\perp$ is

$$(7, -1, 5) - \mathbf{p} = 3(1, -1, 2).$$

This procedure is very long. However, if we look closely at it we see that some improvements can be made. We began with a subspace $V$ of $R^m$ and a basis $\mathbf{v}_1, \ldots, \mathbf{v}_n$ for $V$. We then formed the matrix $A$ whose columns are $\mathbf{v}_1, \ldots, \mathbf{v}_n$. The subspaces $V$ and $V^\perp$ and the matrix $A$ are related as follows:

1. $V$ is the column space of $A$.
2. $V^\perp$ is the nullspace of $A^T$.

Now $\mathbf{p}$ is the projection of $\mathbf{b}$ onto $V$ if and only if $\mathbf{p}$ is in $V$ and $\mathbf{b} - \mathbf{p}$ is in $V^\perp$. Using this and 1 and 2 above we see that $\mathbf{p}$ is the projection of $\mathbf{b}$ onto $V$ if and only if the following two conditions hold.

1. $\mathbf{p} = A\bar{\mathbf{x}}$ for some $\bar{\mathbf{x}}$ in $R^n$.
2. $A^T(\mathbf{b} - \mathbf{p}) = A^T(\mathbf{b} - A\bar{\mathbf{x}}) = \mathbf{0}$.

Finally, observe that our argument depends only on the fact that the columns of $A$ span $V$ and does not require that the columns of $A$ be independent. Thus we have proved the following result.

***Theorem 3*** *Let $V$ be a subspace of $R^m$ and let $A$ be a matrix whose columns span $V$. A vector $\mathbf{p}$ is the projection of an m-vector $\mathbf{b}$ onto $V$ if and only if the following two conditions hold.*

(*a*) $\mathbf{p} = A\bar{\mathbf{x}}$ *for some* $\bar{\mathbf{x}}$ *in* $R^n$.
(*b*) $A^TA\bar{\mathbf{x}} = A^T\mathbf{b}$.

The equations $A^TA\bar{\mathbf{x}} = A^T\mathbf{b}$ in condition (b) of Theorem 3 are called the **normal equations**. A solution to this system gives us a vector $\bar{\mathbf{x}}$ such that $\mathbf{p} = A\bar{\mathbf{x}}$ is the projection of $\mathbf{b}$ onto $V$. In the next section we will have much more to say about the solution $\bar{\mathbf{x}}$ to this system.

The normal equations are obtained by simply multiplying both sides of the equation $A\bar{\mathbf{x}} = \mathbf{b}$ by $A^T$. In view of this remark it should be impossible to forget them.

Given a subspace $V$ of $R^m$ and a vector $\mathbf{b}$ in $R^m$, Theorem 3 gives us a method for finding the projection of $\mathbf{b}$ onto $V$.

1. Let $A$ be a matrix whose columns span $V$.
2. Use Gaussian elimination to find a solution $\bar{\mathbf{x}}$ of the normal equations
$$A^TA\bar{\mathbf{x}} = A^T\mathbf{b}.$$
3. The projection of $\mathbf{b}$ onto $V$ is
$$\mathbf{p} = A\bar{\mathbf{x}}.$$

## EXAMPLE 2

Consider the problem of finding the projection of $(4, -1, 1)$ onto the subspace $V$ of $R^3$ spanned by $(1, 0, 1)$ and $(1, 1, 0)$. Letting $A$ be the matrix whose columns are these vectors, we have

$$A = \begin{bmatrix} 1 & 1 \\ 0 & 1 \\ 1 & 0 \end{bmatrix}, \qquad A^TA = \begin{bmatrix} 2 & 1 \\ 1 & 2 \end{bmatrix}, \qquad A^T\mathbf{b} = \begin{bmatrix} 5 \\ 3 \end{bmatrix}.$$

Gaussian elimination applied to the augmented matrix $[A^TA : A^T\mathbf{b}]$ yields the echelon matrix

$$\begin{bmatrix} 1 & 0 & : & \frac{7}{3} \\ 0 & 1 & : & \frac{1}{3} \end{bmatrix}.$$

Hence $\bar{\mathbf{x}} = (\frac{7}{3}, \frac{1}{3})$ and the projection is given by

$$\mathbf{p} = A\bar{\mathbf{x}} = \begin{bmatrix} 1 & 1 \\ 0 & 1 \\ 1 & 0 \end{bmatrix} \begin{bmatrix} \frac{7}{3} \\ \frac{1}{3} \end{bmatrix} = \begin{bmatrix} \frac{8}{3} \\ \frac{1}{3} \\ \frac{7}{3} \end{bmatrix} = \frac{1}{3} \begin{bmatrix} 8 \\ 1 \\ 7 \end{bmatrix}.$$

The projection onto $V^\perp$ is

$$\mathbf{b} - \mathbf{p} = (4, -1, 1) - \tfrac{1}{3}(8, 1, 7) = (\tfrac{4}{3}, -\tfrac{4}{3}, -\tfrac{4}{3}).$$

## EXAMPLE 3

Suppose that we wish to find the projection of one $m$-vector $\mathbf{b}$ onto a nonzero $m$-vector $\mathbf{v}$. This problem is equivalent to finding the projection of $\mathbf{b}$ onto the one-dimensional subspace $V$ of $R^m$ that is spanned by $\mathbf{v}$. Let $A$ be the matrix whose single column is $\mathbf{v}$. Then

$$A^TA = \mathbf{v}^T\mathbf{v} = \langle \mathbf{v}, \mathbf{v} \rangle,$$
$$A^T\mathbf{b} = \mathbf{v}^T\mathbf{b} = \langle \mathbf{v}, \mathbf{b} \rangle,$$

and hence the normal equations $A^TA\bar{\mathbf{x}} = A^T\mathbf{b}$ become

$$\langle \mathbf{v}, \mathbf{v} \rangle \bar{\mathbf{x}} = \langle \mathbf{v}, \mathbf{b} \rangle.$$

Thus

$$\bar{\mathbf{x}} = \frac{\langle \mathbf{v}, \mathbf{b} \rangle}{\langle \mathbf{v}, \mathbf{v} \rangle}$$

and

$$\mathbf{p} = \frac{\langle \mathbf{v}, \mathbf{b} \rangle}{\langle \mathbf{v}, \mathbf{v} \rangle}\mathbf{v}.$$

## EXAMPLE 4

If $\mathbf{v} = (1, 1, 1, 1)$ and $\mathbf{b} = (2, 1, 6, 3)$, then $\langle \mathbf{v}, \mathbf{b} \rangle = 12$ and $\langle \mathbf{v}, \mathbf{v} \rangle = 4$. Thus the projection of $\mathbf{b}$ onto $\mathbf{v}$ is

$$\mathbf{p} = \frac{\langle \mathbf{v}, \mathbf{b} \rangle}{\langle \mathbf{v}, \mathbf{v} \rangle}\mathbf{v} = \frac{12}{4}(1, 1, 1, 1) = (3, 3, 3, 3).$$

## EXAMPLE 5

If $\mathbf{p}$ is the projection of $\mathbf{b}$ onto $\mathbf{v}$ and $\theta$ is the angle between $\mathbf{v}$ and $\mathbf{b}$, then

$$\|\mathbf{p}\| = \|\mathbf{b}\|\,|\cos\theta| \qquad \text{and} \qquad \mathbf{p} = \|\mathbf{b}\|\cos\theta\frac{\mathbf{v}}{\|\mathbf{v}\|}.$$

For obvious reasons the quantity $\|\mathbf{b}\|\cos\theta$ is called the component of $\mathbf{b}$ in the direction of $\mathbf{v}$ (see Figure 3.13).

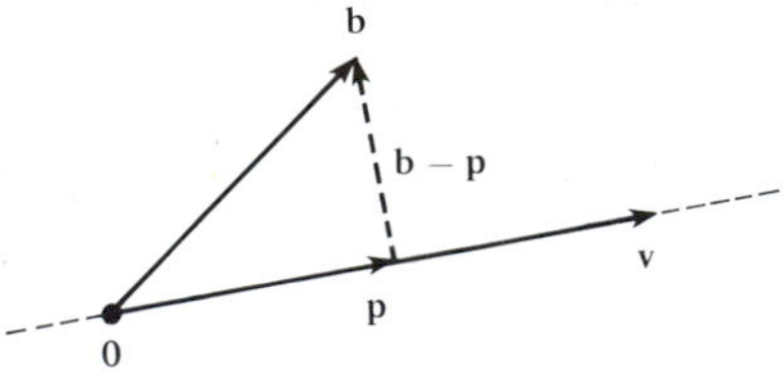


Figure 3.13

Up to this point in our discussion of projections we have been careful to point out that the columns of the matrix $A$ are only required to span the subspace $V$; they are not required to be independent. If they are independent, then the matrix $A^TA$ is nonsingular. This follows immediately from the following theorem.

***Theorem 4*** *For any matrix $A$, rank* $(A^TA) = \text{rank}(A)$.

**Proof** Let $A$ be an $m \times n$ matrix. Then $A^TA$ is an $n \times n$ matrix. Since

$$\text{nullity}(A) = n - \text{rank}(A)$$

and

$$\text{nullity}(A^TA) = n - \text{rank}(A^TA),$$

it follows that $\text{rank}(A^TA) = \text{rank}(A)$ if and only if $\text{nullity}(A^TA) = \text{nullity}(A)$. Thus the theorem follows from the fact that $N(A^TA) = N(A)$, which we now prove. If $\mathbf{x}$ is in $N(A)$, then

$$A^TA\mathbf{x} = A^T\mathbf{0} = \mathbf{0},$$

so that $\mathbf{x}$ is in $N(A^TA)$. Thus $N(A)$ is a subspace of $N(A^TA)$. Conversely, suppose that $\mathbf{x}$ is in $N(A^TA)$. Then $A^TA\mathbf{x} = \mathbf{0}$ and hence

$$0 = \langle \mathbf{x}, A^TA\mathbf{x} \rangle = \langle A\mathbf{x}, A\mathbf{x} \rangle = \|A\mathbf{x}\|^2,$$

from which we conclude that $A\mathbf{x} = \mathbf{0}$. Thus $\mathbf{x}$ is in $N(A)$, and hence $N(A^TA)$ is also a subspace of $N(A)$. Consequently, $N(A^TA) = N(A)$ and the theorem is proved.

It follows that when the columns of an $m \times n$ matrix $A$ are independent, then $A^TA$ is an $n \times n$ matrix with rank $n$. Thus $A^TA$ is nonsingular and the normal equations $A^TA\bar{\mathbf{x}} = A^T\mathbf{b}$ have the unique solution

$$\bar{\mathbf{x}} = (A^TA)^{-1}A^T\mathbf{b}.$$

The projection of $\mathbf{b}$ onto $V$ is given by

$$\mathbf{p} = A\bar{\mathbf{x}} = A(A^TA)^{-1}A^T\mathbf{b}.$$

Thus if $P$ is the matrix

$$P = A(A^TA)^{-1}A^T,$$

it follows that for every $\mathbf{b}$ in $R^m$,

$$\mathbf{p} = P\mathbf{b}.$$

Hence applying $P$ to any vector $\mathbf{b}$ gives the projection of $\mathbf{b}$ onto $V$. For this reason, $P$ is called a **projection matrix** for the projection of $R^m$ onto $V$. Since

$$(I - P)\mathbf{b} = \mathbf{b} - \mathbf{p},$$

$I - P$ is the projection matrix for the projection of $R^m$ onto $V^{\perp}$.

The important point here is that when the columns of $A$ are independent, the normal equations $A^TA\bar{\mathbf{x}} = A^T\mathbf{b}$ have a unique solution. However, the solution should not be computed by a straightforward application of the formula

$$\bar{\mathbf{x}} = (A^TA)^{-1}A^T\mathbf{b}.$$

That is, it should not be computed by finding $(A^TA)^{-1}$ and then forming the product $(A^TA)^{-1}A^T\mathbf{b}$. This procedure requires us to find the inverse of a matrix, a time-consuming and hence expensive task. It should be computed by solving the normal equations. But hang on, there is something to be learned here. Solving the normal equations is accomplished by reducing the augmented matrix

$$[A^TA \; : \; A^T\mathbf{b}] \tag{1}$$

to reduced echelon form. When the columns of $A$ are independent, $A^TA$ is an $n \times n$ nonsingular matrix and hence its reduced echelon form is $I$, where $I$ is the $n \times n$ identity matrix. Thus the reduced echelon form of (1) is

$$[I \; : \; \bar{\mathbf{x}}] = [I \; : \; (A^TA)^{-1}A^T\mathbf{b}].$$

Now let's apply this last observation to computing a projection matrix. A straightforward application of the formula

$$P = A(A^TA)^{-1}A^T$$

would require us to compute $(A^TA)^{-1}$, a task to be avoided if possible. However, $(A^TA)^{-1}$ is only required to find $(A^TA)^{-1}A^T$ and then $A[(A^TA)^{-1}A^T]$. Forming the product $(A^TA)^{-1}A^T$ is equivalent to forming the augmented matrix.

$$[A^TA \; : \; A^T] \tag{2}$$

and reducing it to reduced echelon form. Since $A^TA$ is nonsingular, its reduced echelon form is the $n \times n$ identity matrix $I$. Thus the reduced echelon form of (2) is

$$[I \; : \; (A^TA)^{-1}A^T].$$

Finally, we find the projection matrix with one final multiplication by $A$:

$$P = A[(A^TA)^{-1}A^T].$$

### EXAMPLE 6

Let us find the projection matrix for the projection of $R^3$ onto the subspace spanned by $(1, 0, -1)$ and $(2, 1, 0)$. Let $A$ be the matrix

$$A = \begin{bmatrix} 1 & 2 \\ 0 & 1 \\ -1 & 0 \end{bmatrix}.$$

Then

$$A^TA = \begin{bmatrix} 2 & 2 \\ 2 & 5 \end{bmatrix}$$

and the augmented matrix

$$[A^TA : A^T] = \begin{bmatrix} 2 & 2 & : & 1 & 0 & -1 \\ 2 & 5 & : & 2 & 1 & 0 \end{bmatrix}$$

reduces to

$$[I : (A^TA)^{-1}A^T] = \begin{bmatrix} 1 & 0 & : & \frac{1}{6} & -\frac{1}{3} & -\frac{5}{6} \\ 0 & 1 & : & \frac{1}{3} & \frac{1}{3} & \frac{1}{3} \end{bmatrix}.$$

Thus

$$P = A[(A^TA)^{-1}A^T] = \begin{bmatrix} 1 & 2 \\ 0 & 1 \\ -1 & 0 \end{bmatrix} \frac{1}{6}\begin{bmatrix} 1 & -2 & -5 \\ 2 & 2 & 2 \end{bmatrix} = \frac{1}{6}\begin{bmatrix} 5 & 2 & -1 \\ 2 & 2 & 2 \\ -1 & 2 & 5 \end{bmatrix}.$$

## EXAMPLE 7

Suppose that $V$ is a five-dimensional subspace of $R^7$. Then $V^\perp$ is a two-dimensional subspace of $R^7$. If $P$ is the projection matrix for $V$ and $Q$ is the projection matrix for $V^\perp$, then

$$Q = I - P \qquad \text{and hence} \qquad P = I - Q.$$

Thus $Q$ is determined by $P$, and vice versa. Which one is easier to compute? Let's see. To find $P$ we begin with a matrix $A$ whose columns form a basis for $V$. Thus $A$ is a $7 \times 5$ matrix and hence $A^TA$ is a $5 \times 5$ matrix. Reducing

$$[A^TA : A^T] \qquad \text{to} \qquad [I : (A^TA)^{-1}A^T]$$

requires us to reduce a $5 \times 5$ matrix to the identity $I$. $(A^TA)^{-1}A^T$ is a $5 \times 7$ matrix and finally

$$P = A(A^TA)^{-1}A^T$$

is a $7 \times 7$ matrix.

On the other hand, to find $Q$ we begin with a matrix $B$ whose columns form a basis for $V^\perp$. Thus $B$ is a $7 \times 2$ matrix and hence $B^TB$ is a $2 \times 2$ matrix. Reducing

$$[B^TB : B^T] \qquad \text{to} \qquad [I : (B^TB)^{-1}B^T]$$

only requires us to reduce a $2 \times 2$ matrix to $I$! $(B^TB)^{-1}B^T$ is a $2 \times 7$ matrix and

$$Q = B(B^TB)^{-1}B^T$$

is a $7 \times 7$ matrix. $P$ is determined from $Q$ by merely subtracting $Q$ from $I$:

$$P = I - Q.$$

I think (hope!) you will agree that it is much easier to compute $P$ from the equation $P = I - Q$ than to compute $P$ directly.

The moral of this example is that the larger the dimension of the subspace $V$, the more difficult it is to compute the projection matrix for $V$ via the formula

$P = A(A^TA)^{-1}A^T$. When $V$ is a subspace of $R^m$ and $\dim V > m/2$, it is easier to compute $P$ from the formula $P = I - Q$, where $Q$ is the projection matrix for $V^\perp$.

### EXAMPLE 8

Let $V$ be a subspace of $R^m$, let $\mathbf{v}_1, \ldots, \mathbf{v}_n$ be a basis for $V$, and let $A$ be the $m \times n$ matrix having $\mathbf{v}_1, \ldots, \mathbf{v}_n$ as its columns. Since the columns of $A$ are independent, $A^TA$ is nonsingular (Theorem 4) and hence the normal equations $A^TA\bar{\mathbf{x}} = A^T\mathbf{b}$ have a unique solution for all $m$-vectors $\mathbf{b}$. Now let $\mathbf{p}$ be the projection of $\mathbf{b}$ onto $V$. Since $\mathbf{p} = A\bar{\mathbf{x}}$, it follows that $\bar{\mathbf{x}}$ is the coordinate vector of $\mathbf{p}$ with respect to the basis we have chosen. If $\mathbf{b}$ happens to be in the subspace $V$, then clearly $\mathbf{p} = \mathbf{b}$ (i.e., the projection of $\mathbf{b}$ onto $V$ is $\mathbf{b}$ itself). Hence in this case $\bar{\mathbf{x}}$ is the coordinate vector of $\mathbf{b}$ itself with respect to the basis $\mathbf{v}_1, \ldots, \mathbf{v}_n$.

### EXAMPLE 9

Let $\mathbf{v}$ be a nonzero vector in $R^m$. To find the projection matrix $P$ for the projection of $R^m$ onto $\mathbf{v}$ (i.e., onto the subspace $V$ spanned by $\mathbf{v}$), we let $A$ be the matrix whose single column is $\mathbf{v}$ and compute:

$$\begin{aligned} P &= A(A^TA)^{-1}A^T \\ &= \mathbf{v}(\langle \mathbf{v}, \mathbf{v} \rangle)^{-1}\mathbf{v}^T \\ &= \frac{1}{\langle \mathbf{v}, \mathbf{v} \rangle}\mathbf{v}\mathbf{v}^T. \end{aligned}$$

Since $\langle \mathbf{v}, \mathbf{v} \rangle = \|\mathbf{v}\|^2$,

$$P = \mathbf{w}\mathbf{w}^T, \qquad \text{where} \qquad \mathbf{w} = \frac{\mathbf{v}}{\|\mathbf{v}\|}.$$

For instance, let $V$ be the subspace of $R^4$ spanned by $\mathbf{v} = (0, 1, 0, 1)$. If $A$ is the matrix whose single column is $\mathbf{v}$, then $A^TA = \langle \mathbf{v}, \mathbf{v} \rangle = 2$ and

$$P = A(A^TA)^{-1}A^T = \frac{1}{2}\begin{bmatrix} 0 \\ 1 \\ 0 \\ 1 \end{bmatrix}\begin{bmatrix} 0 & 1 & 0 & 1 \end{bmatrix} = \frac{1}{2}\begin{bmatrix} 0 & 0 & 0 & 0 \\ 0 & 1 & 0 & 1 \\ 0 & 0 & 0 & 0 \\ 0 & 1 & 0 & 1 \end{bmatrix}.$$

## PROBLEMS 3.5

**1.** Find the projection of $\mathbf{b}$ onto $\mathbf{v}$ for each of the following.

*(a) $\mathbf{v} = (0, 1)$, $\quad\mathbf{b} = (2, 1)$
(b) $\mathbf{v} = (1, 1, 1)$, $\quad\mathbf{b} = (5, 1, 1)$
(c) $\mathbf{v} = (1, 2, -1)$, $\quad\mathbf{b} = (0, 1, 2)$
*(d) $\mathbf{v} = (1, -1, 0, 1)$, $\quad\mathbf{b} = (3, -2, 8, 5)$
(e) $\mathbf{v} = (1, 0, 2, 0, 3)$, $\quad\mathbf{b} = (4, 5, 5, 5, 0)$

**2.** Find the projection of $\mathbf{b} = (1, 1, 1)$ onto the subspace $V$ of $R^3$ spanned by $(1, 0, 1)$ and $(0, 1, 1)$. Also find the projection of $\mathbf{b}$ onto $V^\perp$.

*3. (a) Find the projection of $\mathbf{u} = (1, 2, 3, 1, 1)$ onto the subspace $V$ spanned by $(1, 0, 1, 0, 1)$ and $(0, 1, 0, 1, 0)$. Also find the projection of $\mathbf{u}$ onto $V^{\perp}$.
(b) What is the distance from the vector $\mathbf{u}$ to the subspace $V$?
(c) What vector in the subspace $V$ is nearest to $\mathbf{u}$?

4. Find the projection of the vector $\mathbf{u} = (2, 1, 0, 1)$ onto the subspace $V$ of $R^4$ consisting of all vectors $(x_1, x_2, x_3, x_4)$ such that $x_1 + x_2 + x_3 + x_4 = 0$. Also answer parts (b) and (c) of Problem 3.

*5. Find the projection of $\mathbf{u} = (1, 1, 1)$ onto the subspace $V$ of $R^3$ spanned by $(0, 2, 3)$, $(1, 0, 1)$, $(-1, 2, 2)$. Also answer parts (b) and (c) of Problem 3.

6. Find the projection of the vector $(1, -1, 4)$ onto the plane $x - 2y + z = 0$. Find the perpendicular distance from the vector to the plane.

7. Let $\mathbf{u} \neq \mathbf{0}$ and $\mathbf{a}$ be $m$-vectors. The equation of the line in $R^m$ through $\mathbf{a}$ in the direction of $\mathbf{u}$ is $\mathbf{x} = \mathbf{a} + t\mathbf{u}$ where $t$ is any real number.
(a) If $m = 2$, show that the equation of the line through $\mathbf{a} = (x_0, y_0)$ in the direction of $\mathbf{u} = (-b, a)$ can be written in the form

$$a(x - x_0) + b(y - y_0) = 0.$$

(b) If $\mathbf{b}$ is an $m$-vector that is not on the line $\mathbf{x} = \mathbf{a} + t\mathbf{u}$, the distance from $\mathbf{b}$ to the line is defined to be the distance from $\mathbf{b} - \mathbf{a}$ to the subspace spanned by $\mathbf{u}$. Illustrate this definition with a picture when $m = 2$ and $m = 3$.
(c) If $\mathbf{p}$ is the projection of $\mathbf{b} - \mathbf{a}$ onto the subspace spanned by $\mathbf{u}$, show that the point on the line nearest the point $\mathbf{b}$ is $\mathbf{a} + \mathbf{p}$.

8. Use Problem 7 to find the distance from the point $\mathbf{b}$ to the line $\mathbf{x} = \mathbf{a} + t\mathbf{u}$ and the point on the line that is nearest to $\mathbf{b}$ if:
*(a) $\mathbf{b} = (1, 1)$, $\mathbf{a} = (-1, 1)$, and $\mathbf{u} = (-3, 2)$
(b) $\mathbf{b} = (0, 1)$, $\mathbf{a} = (0, -1)$, and $\mathbf{u} = (-1, 1)$
*(c) $\mathbf{b} = (2, 4)$, $\mathbf{a} = (4, 1)$, and $\mathbf{u} = (1, 1)$
(d) $\mathbf{b} = (3, 1, 1)$, $\mathbf{a} = (1, 1, 1)$, and $\mathbf{u} = (1, 0, 1)$
*(e) $\mathbf{b} = (1, 1, 1)$, $\mathbf{a} = (0, 1, 2)$, and $\mathbf{u} = (1, 2, 3)$
(f) $\mathbf{b} = (1, 2, 3)$, $\mathbf{a} = (0, 2, 1)$, and $\mathbf{u} = (2, 1, 2)$
*(g) $\mathbf{b} = (1, 1, 0, 1)$, $\mathbf{a} = (1, 0, 1, 0)$, and $\mathbf{u} = (0, 1, 1, 1)$
(h) $\mathbf{b} = (1, 2, -1, 0, 1)$, $\mathbf{a} = (0, 2, 1, 1, 1)$, and $\mathbf{u} = (2, 1, 1, -1, 1)$

9. Let $\mathbf{n} \neq \mathbf{0}$ and $\mathbf{a}$ be $m$-vectors. The equation of the plane in $R^m$ through $\mathbf{a}$ and orthogonal to $\mathbf{n}$ is

$$\langle \mathbf{x} - \mathbf{a}, \mathbf{n} \rangle = 0.$$

(a) If $m = 3$, show that if $\mathbf{a} = (x_0, y_0, z_0)$ and $\mathbf{n} = (a, b, c)$, then the equation of the plane can be written in the form

$$a(x - x_0) + b(y - y_0) + c(z - z_0) = 0.$$

(b) If $\mathbf{b}$ is an $m$-vector that is not on the plane, the distance from $\mathbf{b}$ to the plane is defined to be the distance from $\mathbf{b} - \mathbf{a}$ to the orthogonal complement of the subspace spanned by $\mathbf{n}$. Illustrate this definition with a picture when $m = 3$.
(c) Show that the distance from $\mathbf{b}$ to the plane $\langle \mathbf{x} - \mathbf{a}, \mathbf{n} \rangle = 0$ is

$$\frac{|\langle \mathbf{b} - \mathbf{a}, \mathbf{n} \rangle|}{\|\mathbf{n}\|}.$$

(d) If $\mathbf{p}$ is the projection of $\mathbf{b} - \mathbf{a}$ onto the orthogonal complement of the subspace spanned by $\mathbf{n}$, show that the point on the plane nearest to $\mathbf{b}$ is

$$\mathbf{b} - \frac{\langle \mathbf{b} - \mathbf{a}, \mathbf{n} \rangle}{\langle \mathbf{n}, \mathbf{n} \rangle}\mathbf{n}.$$

10. Use Problem 9 to find the distance from the point $\mathbf{b}$ to the plane $\langle \mathbf{x} - \mathbf{a}, \mathbf{u} \rangle = 0$ and the point on the plane that is nearest to $\mathbf{b}$ if:
*(a) $\mathbf{b} = (2, 1, -1)$, $\mathbf{a} = (0, 2, 2)$, and $\mathbf{n} = (5, -2, 3)$

(b) $\mathbf{b} = (1, 1, 1)$, $\mathbf{a} = (1, 0, 1)$, and $\mathbf{n} = (1, -1, 1)$

*(c) $\mathbf{b} = (1, -4, -3)$, $\mathbf{a} = (1, -1, -1)$, and $\mathbf{n} = (2, -3, 6)$

(d) $\mathbf{b} = (0, 1, 6)$, $\mathbf{a} = (0, 1, 0)$, and $\mathbf{n} = (1, 2, -1)$

*(e) $\mathbf{b} = (2, 3, -1)$, $\mathbf{a} = (1, 0, 0)$, and $\mathbf{n} = (1, 1, 1)$

(f) $\mathbf{b} = (1, 2, 3, 4)$, $\mathbf{a} = (1, 2, 3, 0)$, and $\mathbf{n} = (1, 0, 1, 0)$

*(g) $\mathbf{b} = (-1, 1, 1, -1)$, $\mathbf{a} = (2, 2, 1, 2)$, and $\mathbf{n} = (3, 0, 1, 3)$

(h) $\mathbf{b} = (-3, -1, 0, 1, 2, 1)$, $\mathbf{a} = (0, 1, 0, 1, 0, 1)$, and $\mathbf{n} = (1, 1, 1, 0, 1, 1)$

*(i) $\mathbf{b} = (4, 3, 2, 1, 0, -1)$, $\mathbf{a} = (3, 2, 1, 0, -1, -2)$, and $\mathbf{n} = (2, -1, 1, 0, 0, 1)$

***11.** Let $V$ be the subspace spanned by $(1, 0, 1, 0)$, $(1, 1, -1, 0)$, and $(1, -2, -1, 1)$.

(a) Find the projection matrix $P$ for the projection onto the subspace $V$.

(b) Find the projection of $(1, 0, 0, 0)$ and $(1, 1, 0, 0)$ onto the subspace $V$.

(c) What is the nullspace of $P$?

(d) What is the column space of $P$?

**12.** Find the projection matrix for each of the following subspaces.

*(a) The subspace in Problem 2.

(b) The subspace in Problem 3.

*(c) The subspace in Problem 4.

(d) The subspace in Problem 5.

(e) The subspace in Problem 6.

**13.** Let $A$ be an $m \times n$ matrix. If $A^TA = 0$, explain why $A = 0$. (*Hint:* Use Theorem 4).

**14.** Show that if $A$ is an $n \times n$ matrix of rank $n$, then

$$(A^TA)^{-1} = A^{-1}(A^T)^{-1}.$$

**15.** Let $V$ be a subspace of $R^m$ and $\mathbf{b}$ an $m$-vector.

(a) Show that $\mathbf{b}$ is in $V$ if and only if the projection of $\mathbf{b}$ onto $V$ is $\mathbf{b}$ itself.

(b) Show that $\mathbf{b}$ is orthogonal to $V$ if and only if the projection of $\mathbf{b}$ onto $V$ is the zero vector.

**16.** Let $\mathbf{v}$ be any nonzero $m$-vector.

(a) Show that if $\mathbf{b}$ is an $m$-vector and $\mathbf{p}$ is the projection of $\mathbf{b}$ onto $\mathbf{v}$, then $\mathbf{p} = \alpha\mathbf{v}$ for some scalar $\alpha$.

(b) Use the fact that $(\mathbf{b} - \mathbf{p}) \perp \mathbf{v}$ to show that $\alpha = \langle \mathbf{b}, \mathbf{v} \rangle / \langle \mathbf{v}, \mathbf{v} \rangle$. Hence

$$\mathbf{p} = \frac{\langle \mathbf{b}, \mathbf{v} \rangle}{\langle \mathbf{v}, \mathbf{v} \rangle}\mathbf{v}.$$

Compare this proof with the one given in the text.

**17.** Let $\mathbf{u}$ be a nonzero $m$-vector. Show that any $m$-vector $\mathbf{x}$ can be written in one and only one way as $\mathbf{x} = \mathbf{a} + \mathbf{b}$, where $\mathbf{a}$ and $\mathbf{u}$ are collinear and $\mathbf{b}$ is orthogonal to $\mathbf{u}$.

**18.** Let $V$ be a subspace of $R^m$ and let $\mathbf{b}_1, \mathbf{b}_2$ be $m$-vectors. If $\mathbf{p}_1$ and $\mathbf{p}_2$ are the projections of $\mathbf{b}_1$ and $\mathbf{b}_2$ onto $V$, what are the projections of $\mathbf{b}_1 + \mathbf{b}_2$ and $\alpha\mathbf{b}_1$ onto $V$?

**19.** Let $V$ be a subspace of $R^m$ and let $P$ be the projection matrix for the projection of $R^m$ onto $V$.

*(a) Show that $P^2 = P$.

(b) Show that $P = P^T$.

(c) Show that $I - P$ has properties (a) and (b).

(d) Show that $I - P$ is the projection matrix for the projection of $R^m$ onto $V^\perp$.

(e) Show that $C(P) = V$ and $C(I - P) = V^\perp$.

(f) Show that $N(P) = V^\perp$ and $N(I - P) = V$.

(g) Find a geometric interpretation for each of the matrices $I - 2P$ and $2P - I$.

**20.** Let $P$ be a symmetric matrix with the property that $P^2 = P$.

(a) Show that $P$ is the projection matrix for the projection onto the column space of $P$. For this reason a symmetric matrix $P$ with $P^2 = P$ is called a **projection matrix**. (*Hint:* Show that for any $\mathbf{x}$, $\mathbf{x} - P\mathbf{x}$ is orthogonal to the column space of $P$.)

(b) What is the nullspace of $P$?

**21.** Let $P_1$ and $P_2$ be projection matrices (i.e., $P_i^2 = P_i$ and $P_i^T = P_i$ for $i = 1, 2$).
(a) Show that $P_1 P_2$ is a projection matrix if and only if $P_1 P_2 = P_2 P_1$.
(b) Show that if $P_1 P_2 = 0$, then $P_1 + P_2$ is a projection matrix.

**22.** Let $\mathbf{v}_1, \mathbf{v}_2, \ldots, \mathbf{v}_n$ be orthonormal vectors in $R^m$ and let $P_i = \mathbf{v}_i \mathbf{v}_i^T$, $i = 1, 2, \ldots, n$. Verify each of the following.
(a) $P_i^2 = P_i$, $i = 1, 2, \ldots, n$.
(b) $P_i P_j = 0$ if $i \neq j$.
(c) If $A = \alpha_1 P_1 + \alpha_2 P_2 + \cdots + \alpha_n P_n$, then $A^k = \alpha_1^k P_1 + \alpha_2^k P_2 + \cdots + \alpha_n^k P_n$ for any positive integer $k$.
(d) If $A = \alpha_1 P_1 + \alpha_2 P_2 + \cdots + \alpha_n P_n$, then $A^T = A$.
(e) Use parts (c) and (d) to show that $P = P_1 + P_2 + \cdots + P_n$ is a projection matrix.

**23.** Find all $n \times n$ nonsingular projection matrices.

**24.** Let $A$ be a square matrix. If for all $\mathbf{x}$, $(\mathbf{x} - A\mathbf{x}) \perp C(A)$, show that $A$ is a projection matrix.

***25.** Let $A$ be an $n \times n$ matrix, and let $V$ be the set of all vectors $\mathbf{x}$ in $R^n$ such that $A\mathbf{x} = \mathbf{x}$. Show that if $V = N(A)^\perp$, then $A$ is a projection onto $V$.

**26.** Show that if $A$ is a square matrix such that $A^2 = A$ and $C(A)^\perp = N(A)$, then $A$ is symmetric and hence $A$ is a projection matrix.

**27.** Show that a vector $\mathbf{u}$ is orthogonal to a subspace $V$ if and only if the distance from $\mathbf{u}$ to $V$ is $\|\mathbf{u}\|$.

**28.** The Cauchy–Schwarz inequality states that for any two $m$-vectors $\mathbf{u}$ and $\mathbf{v}$, $|\langle \mathbf{u}, \mathbf{v} \rangle| \leq \|\mathbf{u}\| \cdot \|\mathbf{v}\|$. Prove this inequality by computing the length of $\mathbf{u} - \mathbf{p}$, where $\mathbf{p}$ is the projection of $\mathbf{u}$ onto $\mathbf{v}$. Can you now give a reason for the choice of $\alpha$ in the proof of the Cauchy–Schwarz inequality?

**29.** Use Theorem 3 to show that for every $m \times n$ matrix $A$ and every $m$-vector $\mathbf{b}$, the normal equations $A^T A \bar{\mathbf{x}} = A^T \mathbf{b}$ always have a solution.

**30.** Let $A$ be an $m \times n$ matrix.
(a) Show that $C(A^T) = C(A^T A)$.
(b) Use part (a) to show that the normal equations always have a solution.

**31.** In this problem we outline a different proof of Theorem 3. Let $A$ be an $m \times n$ matrix and $\mathbf{b}$ an $m$-vector.
(a) Show that $\mathbf{p}$ is the projection of $\mathbf{b}$ onto $C(A)$ if and only if $\mathbf{p} = A\bar{\mathbf{x}}$ and $(\mathbf{b} - A\bar{\mathbf{x}}) \perp C(A)$.
(b) Show that $(\mathbf{b} - A\bar{\mathbf{x}}) \perp C(A)$ if and only if $\langle \mathbf{x}, A^T\mathbf{b} - A^T A\bar{\mathbf{x}} \rangle = 0$ for all $\mathbf{x}$ in $R^n$.
(c) Now use Result 1 of Section 3.2 to show that $\langle \mathbf{x}, A^T\mathbf{b} - A^T A\bar{\mathbf{x}} \rangle = 0$ for all $\mathbf{x}$ in $R^n$ if and only if $A^T A\bar{\mathbf{x}} = A^T\mathbf{b}$.
(d) Use parts (a), (b), and (c) to prove Theorem 3.

**32.** Let $A$ be an $m \times n$ matrix. Show that if $\mathbf{b}$ is any vector in the column space of $A$, then there is a unique vector $\mathbf{x}$ in the row space of $A$ such that $A\mathbf{x} = \mathbf{b}$. [*Hint:* Use Problem 12(a) of Section 3.4 and Theorem 1].

# 3.6
# The Least-Squares Solution of Inconsistent Systems

In this section we return to and complete our discussion of the problem of finding the best possible solution of an inconsistent system of equations. Suppose that $A\mathbf{x} = \mathbf{b}$ is an inconsistent system. Then $\mathbf{b}$ does not lie in the column space of $A$ and (as we have seen in Section 3.5) the best we can do is to replace $\mathbf{b}$ by its projection $\mathbf{p}$

onto the column space of $A$. The system $A\bar{\mathbf{x}} = \mathbf{p}$ is consistent and a solution $\bar{\mathbf{x}}$ to this system is called (for reasons to be discussed below) a least-squares solution to the inconsistent system $A\mathbf{x} = \mathbf{b}$. We have seen in Section 3.5 the best way to find $\mathbf{p}$ is to solve the normal equations

$$A^T A\bar{\mathbf{x}} = A^T\mathbf{b}.$$

But since it is $\bar{\mathbf{x}}$ and not the projection $\mathbf{p} = A\bar{\mathbf{x}}$ that we are interested in now, we do not need to find $\mathbf{p}$!

***Definition*** A **least-squares solution** to an inconsistent system $A\mathbf{x} = \mathbf{b}$ is a solution to the normal equations $A^T A\bar{\mathbf{x}} = A^T\mathbf{b}$.

### EXAMPLE 1

The system of equations

$$\begin{aligned} x_1 - x_2 &= -3 \\ 2x_1 + x_2 &= 0 \\ 3x_1 + 2x_2 &= 6 \end{aligned} \qquad \text{or} \qquad \begin{bmatrix} 1 & -1 \\ 2 & 1 \\ 3 & 2 \end{bmatrix}\begin{bmatrix} x_1 \\ x_2 \end{bmatrix} = \begin{bmatrix} -3 \\ 0 \\ 6 \end{bmatrix}$$

is inconsistent. Since

$$A^T A = \begin{bmatrix} 1 & 2 & 3 \\ -1 & 1 & 2 \end{bmatrix}\begin{bmatrix} 1 & -1 \\ 2 & 1 \\ 3 & 2 \end{bmatrix} = \begin{bmatrix} 14 & 7 \\ 7 & 6 \end{bmatrix}$$

and

$$A^T\mathbf{b} = \begin{bmatrix} 1 & 2 & 3 \\ -1 & 1 & 2 \end{bmatrix}\begin{bmatrix} -3 \\ 0 \\ 6 \end{bmatrix} = \begin{bmatrix} 15 \\ 15 \end{bmatrix},$$

the normal equations are

$$\begin{bmatrix} 14 & 7 \\ 7 & 6 \end{bmatrix}\begin{bmatrix} x_1 \\ x_2 \end{bmatrix} = \begin{bmatrix} 15 \\ 15 \end{bmatrix}.$$

The solution of this system is $\bar{\mathbf{x}} = (-\frac{3}{7}, 3)$, which is the least-squares solution of the inconsistent system.

### EXAMPLE 2

Consider the inconsistent system

$$\begin{bmatrix} 1 & 0 & 2 \\ 0 & 1 & -2 \\ 1 & 1 & 0 \\ 1 & 2 & -2 \end{bmatrix}\begin{bmatrix} x_1 \\ x_2 \\ x_3 \end{bmatrix} = \begin{bmatrix} -1 \\ -1 \\ 1 \\ 1 \end{bmatrix}.$$

Notice that in this example the third column is a combination of the first two. Since

$$A^TA = \begin{bmatrix} 3 & 3 & 0 \\ 3 & 6 & -6 \\ 0 & -6 & 12 \end{bmatrix} \quad \text{and} \quad A^T\mathbf{b} = \begin{bmatrix} 1 \\ 2 \\ -2 \end{bmatrix}$$

the normal equations are

$$\begin{bmatrix} 3 & 3 & 0 \\ 3 & 6 & -6 \\ 0 & -6 & 12 \end{bmatrix} \begin{bmatrix} x_1 \\ x_2 \\ x_3 \end{bmatrix} = \begin{bmatrix} 1 \\ 2 \\ -2 \end{bmatrix}.$$

The reduced echelon form of this system is

$$\begin{bmatrix} 1 & 0 & 2 & : & 0 \\ 0 & 1 & -2 & : & \frac{1}{3} \\ 0 & 0 & 0 & : & 0 \end{bmatrix}.$$

The solution is $(\bar{x}_1, \bar{x}_2, \bar{x}_3) = (0, \frac{1}{3}, 0) + \bar{x}_3(-2, 2, 1)$. There are infinitely many solutions to the normal equations in this case. Any of these solutions is a least-squares solution to the inconsistent system.

Let $A$ be an $m \times n$ matrix ($m \geq n$). If $A$ has rank $n$, the normal equations have a unique solution. If $A$ has rank less than $n$, then there are infinitely many solutions (see Example 2). In most applications this situation will not arise. For a discussion of how to choose the optimal solution when it does arise, we refer the reader to Problem 15.

We now discuss the reason for calling a solution $\bar{\mathbf{x}}$ to the normal equations a least-squares solution. Since the distance between two vectors in $R^m$ is given by the length of their difference, the projection $\mathbf{p}$ has the property that

$$\|\mathbf{b} - \mathbf{p}\| \leq \|\mathbf{b} - \mathbf{y}\|$$

for all $\mathbf{y}$ in the column space of $A$. Equivalently, since $\mathbf{p} = A\bar{\mathbf{x}}$ for some $\bar{\mathbf{x}}$ in $R^n$, the vector $\bar{\mathbf{x}}$ has the property that

$$\|\mathbf{b} - A\bar{\mathbf{x}}\| \leq \|\mathbf{b} - A\mathbf{x}\| \qquad \text{for all } \mathbf{x} \text{ in } R^n.$$

If we let $\mathbf{r}(\mathbf{x}) = \mathbf{b} - A\mathbf{x}$, then we see that the minimum of the function $\|\mathbf{r}(\mathbf{x})\|$ occurs at $\bar{\mathbf{x}}$; that is, $\|\mathbf{r}(\bar{\mathbf{x}})\| \leq \|\mathbf{r}(\mathbf{x})\|$ for all $\mathbf{x}$ in $R^n$. Let $(A\mathbf{x})_i$ denote the $i$th component of the vector $A\mathbf{x}$ and $r_i(\mathbf{x})$ denote the $i$th component of the vector $\mathbf{r}(\mathbf{x})$. Then we have $r_i(\mathbf{x}) = b_i - (A\mathbf{x})_i$. Since

$$\|\mathbf{r}(\mathbf{x})\|^2 = r_1(\mathbf{x})^2 + \cdots + r_m(\mathbf{x})^2.$$

and since $r_i(\mathbf{x})$ measures the error in the $i$th component, *the vector $\bar{\mathbf{x}}$ has the property that the sum of the squares of the errors $r_i(\bar{\mathbf{x}})$ in each of the components is as small as possible.* For this reason $\bar{\mathbf{x}}$ is called a least-squares solution.

We have two equivalent ways of viewing a least-squares solution to an inconsistent system. From the geometric point of view, a least-squares solution to an inconsistent system $A\mathbf{x} = \mathbf{b}$ is a vector $\bar{\mathbf{x}}$ with the property that $\mathbf{p} = A\bar{\mathbf{x}}$ is the projection of $\mathbf{b}$ onto the column space of $A$. From the analytic point of view, *a least-squares solution to an inconsistent system $A\mathbf{x} = \mathbf{b}$ is a vector $\bar{\mathbf{x}}$ with the property that the sum of the squares of the errors in each of the components of $\mathbf{b} - A\bar{\mathbf{x}}$ is as small as possible.*

### EXAMPLE 3

In the examples in Section 3.1 we considered the problem of passing a straight line $y = a + bx$ through a collection of points $(x_1, y_1), \ldots, (x_m, y_m)$ with $x_1 < x_2 < \cdots < x_m$. In the unlikely event that all these points lie on a straight line, the answer to our problem is obvious; the straight line that gives the best possible fit is this straight line. When the data points do not lie on a straight line, then we are faced with the inconsistent system of equations

$$A\mathbf{x} = \begin{bmatrix} 1 & x_1 \\ 1 & x_2 \\ \vdots & \vdots \\ 1 & x_m \end{bmatrix} \begin{bmatrix} a \\ b \end{bmatrix} = \begin{bmatrix} y_1 \\ y_2 \\ \vdots \\ y_m \end{bmatrix}.$$

Let $\bar{\mathbf{x}} = (\bar{a}, \bar{b})$ be the least-squares solution to this problem. Then $A\bar{\mathbf{x}}$ is the vector in the column space of $A$ which is nearest to $\mathbf{b}$ and the line $y = \bar{a} + \bar{b}x$ gives the best possible fit to the data in the sense of least-squares. It is called the *regression line* in economics and statistics. From our discussion above,

$$\mathbf{r}(\mathbf{x}) = (y_1 - (a + bx_1), \ldots, y_m - (a + bx_m))$$

and the least-squares solution has the property that the sum of the squares of the errors

$$r_i(\bar{\mathbf{x}}) = y_i - (\bar{a} + \bar{b}x_i) = y_i - (A\bar{\mathbf{x}})_i$$

in each of the components is as small as possible. The errors $r_i(\bar{\mathbf{x}})$ are the differences between the observed values $y_i$ and the corresponding values of $y$ that lie on the line. In Figure 3.14 these errors are represented by the vertical-line segments joining the points to the line. The regression line has the property that the sum of the squares of the length of each of these segments is as small as possible.

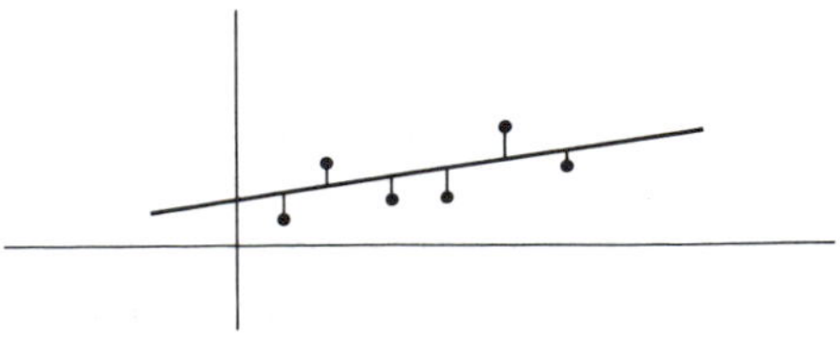

Figure 3.14

## EXAMPLE 4

In Example 2 of Section 3.1 we supposed that we had good reason to believe that the object was traveling with a constant velocity and hence that the variables satisfied a linear relationship $y = y_0 + vt$. Finding the coefficients of the best straight-line fit to the data will give us an estimate for the velocity of the object. It will also give us a way of predicting the position of the object at future times.

Suppose now that we do not know what relationship (if any) exists between the variables with which we are concerned. For example, we may be looking for a relationship between the price of a product and the amount sold, or high school grade-point average and college grade-point average, or the amount of fertilizer applied and the yield of a crop, or smoking and cancer. The usual procedure in this situation is to collect data and plot them. We hope that the graph of the data will suggest a relationship between the variables. For example, the graph in Figure 3.15(a) suggests that the variables are in a linear relationship $y = a + bx$ and the graph in Figure 3.15(b) suggests a quadratic relationship $y = a + bx + cx^2$. Figure 3.15(c), on the other hand, suggests that there is no relationship between the variables. In the first two cases, we are led to the inconsistent systems

$$\begin{bmatrix} 1 & x_1 \\ 1 & x_2 \\ \vdots & \vdots \\ 1 & x_m \end{bmatrix} \begin{bmatrix} a \\ b \end{bmatrix} = \begin{bmatrix} y_1 \\ y_2 \\ \vdots \\ y_m \end{bmatrix} \quad \text{and} \quad \begin{bmatrix} 1 & x_1 & x_1^2 \\ 1 & x_2 & x_2^2 \\ \vdots & \vdots & \vdots \\ 1 & x_m & x_m^2 \end{bmatrix} \begin{bmatrix} a \\ b \\ c \end{bmatrix} = \begin{bmatrix} y_1 \\ y_2 \\ \vdots \\ y_m \end{bmatrix},$$

respectively. The least-squares solutions to these inconsistent systems will give us the coefficients of the best possible linear or quadratic fit to the data. In the third case, we might decide that even though the variables appear to be unrelated, we want to find the *best* possible linear (or quadratic,...) fit to the data. Again, we will generate an inconsistent system of equations.

## EXAMPLE 5

*Multiple regression* involves finding the linear equation

$$y = a_0 + a_1x_1 + \cdots + a_nx_n$$

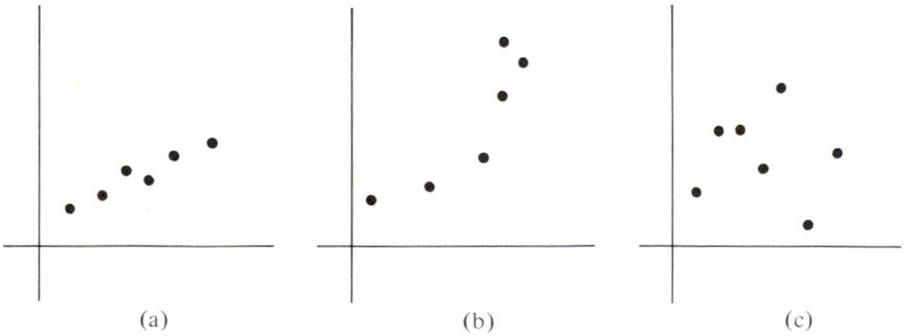

Figure 3.15

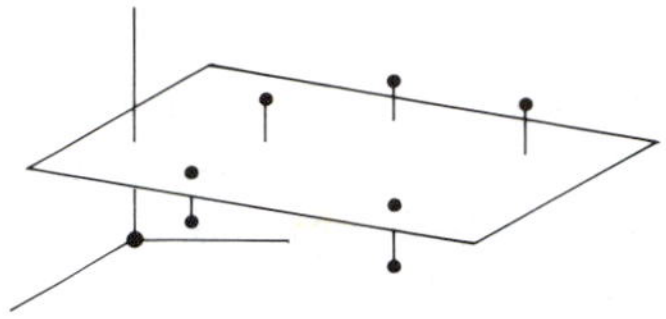

Figure 3.16

which gives the best fit (in the sense of least-squares) to the data

$$(x_{11}, x_{12}, \ldots, x_{1n}, y_1), \ldots, (x_{m1}, x_{m2}, \ldots, x_{mn}, y_m).$$

This problem is equivalent to finding the least-squares solution $\bar{\mathbf{x}} = (\bar{a}_0, \bar{a}_1, \ldots, \bar{a}_n)$ to the (usually) inconsistent system $A\mathbf{x} = \mathbf{b}$, where

$$A = \begin{bmatrix} 1 & x_{11} & \cdots & x_{1n} \\ 1 & x_{21} & \cdots & x_{2n} \\ \vdots & \vdots & & \vdots \\ 1 & x_{m1} & \cdots & x_{mn} \end{bmatrix}, \quad \mathbf{x} = \begin{bmatrix} a_0 \\ a_1 \\ \vdots \\ a_n \end{bmatrix}, \quad \text{and} \quad \mathbf{b} = \begin{bmatrix} y_1 \\ y_2 \\ \vdots \\ y_m \end{bmatrix}.$$

In the case where $y$ depends only on two parameters, $x_1$ and $x_2$, that is,

$$y = a_0 + a_1 x_1 + a_2 x_2,$$

the problem becomes one of finding the plane in $R^3$ that best fits the data $(x_{11}, x_{12}, y_1), \ldots, (x_{m1}, x_{m2}, y_m)$. In this case,

$$\mathbf{r}(\mathbf{x}) = (y_1 - (a_0 + a_1 x_{11} + a_2 x_{12}), \ldots, y_m - (a_0 + a_1 x_{m1} + a_2 x_{m2}))$$

and the least-squares solution $\bar{\mathbf{x}} = (\bar{a}_0, \bar{a}_1, \bar{a}_2)$ has the property that the sum of the squares of the errors

$$r_i(\bar{\mathbf{x}}) = y_i - (\bar{a}_0 + \bar{a}_1 x_{i1} + \bar{a}_2 x_{i2})$$

in each of the components is a minimum. In Figure 3.16 these errors are represented by the vertical line segments joining the points to the plane.

## PROBLEMS 3.6

**1.** Find a least-squares solution to each of the following inconsistent systems of equations.

*(a)
$$\begin{aligned} x_1 + x_2 &= 1 \\ x_1 - x_2 &= 0 \\ 2x_1 &= 2 \end{aligned}$$

*(b)
$$\begin{aligned} x_1 + 2x_2 \phantom{{}+2x_3} &= 0 \\ 2x_1 + 3x_2 + 2x_3 &= 1 \\ x_1 + 3x_2 - 2x_3 &= 1 \\ x_2 - 2x_3 &= 0 \end{aligned}$$

(c)
$$\begin{aligned} -x_1 + 2x_2 - x_3 &= 1 \\ 2x_1 - 4x_2 + 3x_3 &= 1 \\ x_2 + x_3 &= 2 \\ 2x_1 - 4x_2 + x_3 &= 1 \end{aligned}$$

(d)
$$\begin{aligned} x_1 \phantom{{}+2x_2} + 2x_3 &= -1 \\ x_2 - 2x_3 &= -2 \\ x_1 + x_2 + x_3 &= 1 \\ x_1 + 2x_2 - 2x_3 &= 1 \end{aligned}$$

***2.** Find the least-squares solutions to Problems 1, 2, and 3 in Section 3.1.

**3.** Find the best fit by a line of the form $y = a$ to the data

| $t$ | 1 | 2 | 3 | 4 | 5 |
|---|---|---|---|---|---|
| $y$ | 7 | 4 | 5 | 3 | 8 |

**4.** In each part, find the best fit to the data (in the sense of least-squares) by a line $y = a + bx$.
*(a) Data: $(-3, 2), (0, 2), (3, 4)$
(b) Data: $(-1, 0), (0, 1), (1, 2), (2, 4)$
(c) Data: $(1, 1), (1, 2), (1, 5)$

**5.** In each part, find the best fit (in the sense of least-squares) to the data by a quadratic function $y = a + bx + cx^2$.
*(a) Data: $(-1, 1), (0, 1), (1, 4), (2, 10)$
(b) Data: $(-1, -1), (0, 1), (1, 1), (2, -1)$
(c) Data: $(-2, 4), (-1, 2), (1, -6), (2, -2)$

***6.** Find the best fit to the data

| | | | | | |
|---|---|---|---|---|---|
| $x_1$ | 0 | 1 | 1 | 0 | $-1$ |
| $x_2$ | 0 | 1 | 0 | 1 | 1 |
| $y$ | 2 | 3 | $-4$ | 1 | 1 |

by a linear equation of the form:
(a) $y = a_0 + a_1x_1 + a_2x_2$
(b) $y = a_1x_1 + a_2x_2$

**7.** Find the best fit to the data

| | | | | |
|---|---|---|---|---|
| $x_1$ | 1 | $-1$ | $-1$ | 1 |
| $x_2$ | 1 | 1 | $-1$ | $-1$ |
| $y$ | 5 | 1 | 1 | 1 |

by a function of the form $y = a_0 + a_1x_1 + a_2x_2$.

**8.** Find the best fit to the data

| | | | |
|---|---|---|---|
| $x_1$ | 1 | 2 | 1 |
| $x_2$ | 1 | 3 | 5 |
| $y$ | $-2$ | 13 | 0 |

by a function of the form $y = a_1x_1 + a_2x_2$.

**9.** Repeat Problem 8 for the data

| | | | |
|---|---|---|---|
| $x_1$ | 3 | 1 | 1 |
| $x_2$ | 2 | 1 | 1 |
| $y$ | 12 | 4 | 6 |

**10.** Find the best fit to the data

| | | | |
|---|---|---|---|
| $x$ | 0 | 1 | 1 |
| $y$ | 1 | 2 | 0 |
| $z$ | 4 | 8 | 1 |

by a function of the form $z = ax^2 + by^2$.

**11.** By solving the normal equations, show that the coefficients $(\bar{a}, \bar{b})$ of the least-squares line $y = \bar{a} + \bar{b}x$ for the data $(x_1, y_1), \ldots, (x_m, y_m)$ are given by

$$\bar{b} = \frac{(x_1y_1 + \cdots + x_my_m) - m\bar{x}\bar{y}}{(x_1^2 + \cdots + x_m^2) - m\bar{x}^2}$$

$$a = \bar{y} - \bar{b}\bar{x},$$

where $\bar{x} = (1/m)(x_1 + \cdots + x_m)$ and $\bar{y} = (1/m)(y_1 + \cdots + y_m)$ are the averages of the $x_i$'s and $y_i$'s, respectively. Therefore, the least-squares line is

$$y = \bar{y} + \bar{b}(x - \bar{x}).$$

**12.** Using calculus, show that the minimum of the expression

$$\mathbf{r}(a, b) = (y_1 - a - bx_1)^2 + \cdots + (y_m - a - bx_m)^2$$

is $(\bar{a}, \bar{b})$, where $\bar{a}$ and $\bar{b}$ are given in Problem 11.

***13.** Suppose that a plot of the data $(x_1, y_1), \ldots, (x_m, y_m)$ indicates that the data can be best represented by an exponential curve of the form

$$y = \alpha\beta^x.$$

Taking logarithms (to the base $e$) of both sides, we obtain

$$\log y = \log \alpha + (\log \beta)x.$$

This is a linear equation of the form

$$z = a + bx,$$

where $z = \log y$, $a = \log \alpha$, and $b = \log \beta$. If we find the coefficients $\bar{a}$ and $\bar{b}$ of the regression line for the data $(x_1, z_1), \ldots, (x_m, z_m)$, where $z_i = \log y_i$, then we can determine $\alpha$ and $\beta$ from the equations

$$\alpha = e^{\bar{a}} \qquad \text{and} \qquad \beta = e^{\bar{b}}.$$

Carry out this procedure for the following data:

| $x$ | 0 | 1 | 2 |
|---|---|---|---|
| $y$ | $e^2$ | $e^5$ | $e^{13}$ |

**14.** The Malthusian law of population growth predicts that under ideal conditions the population $N$ at time $t$ is

$$N = N_0 e^{\lambda t}$$

where $N_0$ is the population at time $t = 0$ and $\lambda$ is a constant (called the population growth rate). Suppose that a population begins with 6 members and that observations are made to collect the following data:

| $t$ | 3 | 5 | 6 |
|---|---|---|---|
| $N$ | 15 | 25 | 40 |

(a) What is the population growth rate?

(b) What is the predicted population when $t = 8$?

**15.** If $A$ is an $m \times n$ matrix with rank less than $n$, then the normal equations $A^TA\bar{\mathbf{x}} = A^T\mathbf{b}$ do not have a unique solution. But all the solutions $\bar{\mathbf{x}}$ have the property that $\mathbf{p} = A\bar{\mathbf{x}}$ is the projection of $\mathbf{b}$ onto the column space of $A$.

(a) Use Theorem 1 of Section 3.5 and Problem 12 of Section 3.4 to show that any solution $\bar{\mathbf{x}}$ can be decomposed into

$$\bar{\mathbf{x}} = \bar{\mathbf{x}}_1 + \bar{\mathbf{x}}_2$$

where $\bar{\mathbf{x}}_1$ is in the nullspace of $A$ and $\bar{\mathbf{x}}_2$ is in the row space of $A$.

(b) Use Problem 32 of Section 3.5 to show that for any $m$-vector $\mathbf{b}$ there is a unique vector $\bar{\mathbf{x}}_2$ in the row space of $A$ such that $A\bar{\mathbf{x}}_2$ is the projection of $\mathbf{b}$ onto the column space of $A$.

(c) Show that if $\bar{\mathbf{y}}$ is any other solution to the normal equations $A^TA\bar{\mathbf{x}} = A^T\mathbf{b}$, then $\|\bar{\mathbf{x}}_2\| < \|\bar{\mathbf{y}}\|$. Thus among all solutions to the normal equations there is one with smallest length. This solution is called the *optimal solution.*

# 3.7
# Orthogonal Bases and Projections

In Section 3.5 we saw that if the columns of the matrix $A$ form a basis for a subspace $V$ (and not simply a spanning set for $V$), then the normal equations $A^TA\bar{\mathbf{x}} = A^T\mathbf{b}$ have a unique solution. In this section we investigate what follows from the assumption that the columns of $A$ not only form a basis for $V$ but form an orthonormal basis for $V$. This naturally leads us to investigate the possibility of constructing an orthonormal basis for a subspace.

Suppose that $V$ is a subspace of $R^m$ and that $\mathbf{w}_1, \ldots, \mathbf{w}_n$ is an orthonormal basis for $V$. As always, we let $A$ be the matrix whose columns are these basis vectors. Since the columns of $A$ are independent, given any $m$-vector $\mathbf{b}$ the normal equations

$$A^TA\bar{\mathbf{x}} = A^T\mathbf{b} \tag{1}$$

have a unique solution. This much we already know. But the columns of $A$ are not only independent vectors, they are also orthonormal vectors. Hence

$$A^TA = \begin{bmatrix} -\mathbf{w}_1- \\ \vdots \\ -\mathbf{w}_n- \end{bmatrix} \begin{bmatrix} | & & | \\ \mathbf{w}_1 & \cdots & \mathbf{w}_n \\ | & & | \end{bmatrix} = I,$$

and the normal equations reduce to

$$\bar{\mathbf{x}} = A^T\mathbf{b}.$$

Thus the normal equations not only have a unique solution, they have a solution that is trivial to find. It is

$$\bar{\mathbf{x}} = A^T\mathbf{b} = \begin{bmatrix} \langle \mathbf{w}_1, \mathbf{b} \rangle \\ \vdots \\ \langle \mathbf{w}_n, \mathbf{b} \rangle \end{bmatrix}.$$

Therefore, the projection of an $m$-vector $\mathbf{b}$ onto $V$ is given by

$$\mathbf{p} = A\bar{\mathbf{x}} = \langle \mathbf{w}_1, \mathbf{b} \rangle \mathbf{w}_1 + \cdots + \langle \mathbf{w}_n, \mathbf{b} \rangle \mathbf{w}_n. \tag{2}$$

and the projection matrix $P$ is given by

$$P = AA^T.$$

We recognize the terms $\langle \mathbf{w}_i, \mathbf{b} \rangle \mathbf{w}_i$ in the sum on the right-hand side of equation (2) as the projections of $\mathbf{b}$ onto the unit vectors $\mathbf{w}_i$ (see Example 3 of Section 3.5). Thus $\mathbf{p}$ is the sum of the projections of $\mathbf{b}$ onto the orthonormal basis $\mathbf{w}_1, \ldots, \mathbf{w}_n$. In particular, if $\mathbf{b}$ happens to be a vector in the subspace $V$, then $\mathbf{p} = \mathbf{b}$ and (2) gives the expansion of $\mathbf{b}$ in terms of this orthonormal basis for $V$. The projection matrix for the projection of $R^m$ onto $\mathbf{w}_i$ is $P_i = \mathbf{w}_i\mathbf{w}_i^T$ (see Example 9 of Section 3.5). Thus the projection matrix $P$ for the projection of $R^m$ onto $V$ decomposes into a sum of the $n$ projection matrices $P_1, P_2, \ldots, P_n$:

$$\begin{aligned} P &= P_1 + P_2 + \cdots + P_n \\ &= \mathbf{w}_1\mathbf{w}_1^T + \mathbf{w}_2\mathbf{w}_2^T + \cdots + \mathbf{w}_n\mathbf{w}_n^T. \end{aligned}$$

We summarize our findings in the following theorem.

***Theorem 1*** *Let $\mathbf{w}_1, \ldots, \mathbf{w}_n$ be an orthonormal basis for a subspace $V$ of $R^m$, let $A$ be the matrix whose columns are $\mathbf{w}_1, \ldots, \mathbf{w}_n$, and let $\mathbf{b}$ be a vector in $R^m$.*

*(a) The projection of $\mathbf{b}$ onto $V$ is*

$$\mathbf{p} = A\bar{\mathbf{x}} = \langle \mathbf{w}_1, \mathbf{b} \rangle \mathbf{w}_1 + \cdots + \langle \mathbf{w}_n, \mathbf{b} \rangle \mathbf{w}_n.$$

*Thus $\mathbf{p}$ is the sum of the projections of $\mathbf{b}$ onto the orthonormal vectors $\mathbf{w}_1, \ldots, \mathbf{w}_n$.*

(b) *If* **b** *is a vector in the subspace V, then the expansion of* **b** *in terms of the orthonormal basis* $\mathbf{w}_1, \ldots, \mathbf{w}_n$ *is given by*

$$\mathbf{b} = A\bar{\mathbf{x}} = \langle \mathbf{w}_1, \mathbf{b} \rangle \mathbf{w}_1 + \cdots + \langle \mathbf{w}_n, \mathbf{b} \rangle \mathbf{w}_n.$$

*That is,* **b** *is the sum of its projections onto each vector in the orthonormal basis. The coefficients* $\langle \mathbf{w}_1, \mathbf{b} \rangle, \ldots, \langle \mathbf{w}_n, \mathbf{b} \rangle$ *give the coordinates of* **b** *with respect to this orthonormal basis.*

(c) *The projection matrix* $P = AA^T$ *decomposes into the sum of the n projection matrices for the projection of* $R^m$ *onto each of the orthonormal basis vectors* $\mathbf{w}_1, \mathbf{w}_2, \ldots, \mathbf{w}_n$:

$$P = A^T A = \mathbf{w}_1 \mathbf{w}_1^T + \mathbf{w}_2 \mathbf{w}_2^T + \cdots + \mathbf{w}_n \mathbf{w}_n^T.$$

Note that the formulas in (a) and (b) are exactly the same. If **b** is not in the subspace $V$, then the formula gives the expansion of the projection of **b** onto $V$ in terms of the orthonormal basis. If **b** is in the subspace, then the formula gives the expansion of **b** itself in terms of the orthonormal basis. In particular, if $V = R^m$, then:

***Theorem 2*** *Let* $\mathbf{w}_1, \mathbf{w}_2, \ldots, \mathbf{w}_m$ *be an orthonormal basis for* $R^m$. *If* **v** *is any vector in* $R^m$, *then the expansion of* **v** *in terms of this basis is*

$$\mathbf{v} = \langle \mathbf{w}_1, \mathbf{v} \rangle \mathbf{w}_1 + \cdots + \langle \mathbf{w}_m, \mathbf{v} \rangle \mathbf{w}_m.$$

*The coordinate vector of* **v** *with respect to this basis is*

$$\mathbf{c}_{\mathbf{w}}(\mathbf{v}) = (\langle \mathbf{w}_1, \mathbf{v} \rangle, \ldots, \langle \mathbf{w}_m, \mathbf{v} \rangle).$$

## EXAMPLE 1

Suppose that $V$ is a two-dimensional subspace of $R^m$ and that $\mathbf{w}_1$ and $\mathbf{w}_2$ form an orthonormal basis for $V$. If **b** is any vector in $V$, it is geometrically obvious [see

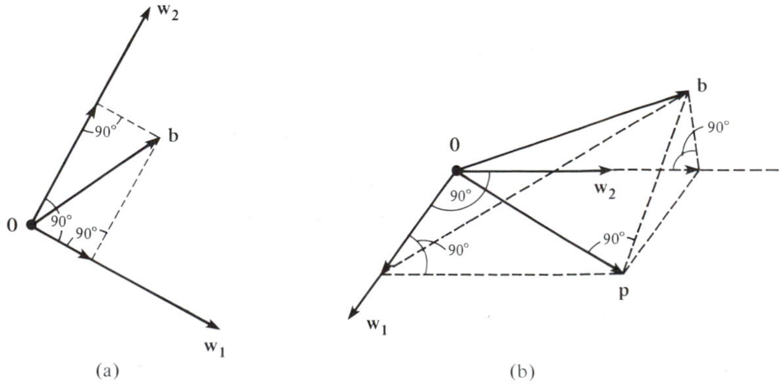


Figure 3.17

Figure 3.17(a)] that the expansion of $\mathbf{b}$ in terms of this basis is given by the formula

$$\begin{aligned}\mathbf{b} &= (\text{proj of } \mathbf{b} \text{ onto } \mathbf{w}_1) + (\text{proj of } \mathbf{b} \text{ onto } \mathbf{w}_2)\\ &= \langle \mathbf{w}_1, \mathbf{b}\rangle \mathbf{w}_1 + \langle \mathbf{w}_2, \mathbf{b}\rangle \mathbf{w}_2.\end{aligned}$$

Similarly, if $\mathbf{b}$ is a vector not in $V$, then the projection of $\mathbf{b}$ onto the subspace $V$ is obviously given by [Figure 3.17(b)]

$$\begin{aligned}\mathbf{p} &= \text{proj of } \mathbf{b} \text{ onto } V\\ &= (\text{proj of } \mathbf{b} \text{ onto } \mathbf{w}_1) + (\text{proj of } \mathbf{b} \text{ onto } \mathbf{w}_2)\\ &= \langle \mathbf{w}_1, \mathbf{b}\rangle \mathbf{w}_1 + \langle \mathbf{w}_2, \mathbf{b}\rangle \mathbf{w}_2.\end{aligned}$$

The only difference between an orthonormal basis and an orthogonal basis is that in the former all vectors have unit length. The vectors in an orthogonal basis can always be normalized to form an orthonormal basis by dividing each of the vectors by its length. However, it is often easier to perform hand calculations with an orthogonal basis than an orthonormal basis, especially if the vectors have integer components. In this case the components of orthonormal vectors usually involve square roots which result from the division of the vector by its length. For example, the vectors

$$(1, 1, 0), \quad (1, -1, 1) \quad \text{and} \quad (-1, 1, 2)$$

form an orthogonal basis for $R^3$. If we normalize these vectors to obtain an orthonormal basis, we obtain

$$\left(\frac{1}{\sqrt{2}}, \frac{1}{\sqrt{2}}, 0\right), \quad \left(\frac{1}{\sqrt{3}}, -\frac{1}{\sqrt{3}}, \frac{1}{\sqrt{3}}\right), \quad \text{and} \quad \left(-\frac{1}{\sqrt{6}}, \frac{1}{\sqrt{6}}, \frac{2}{\sqrt{6}}\right).$$

What happens to the results in Theorem 1 if we replace the orthonormal basis $\mathbf{w}_1, \ldots, \mathbf{w}_n$ by an orthogonal basis $\mathbf{v}_1, \ldots, \mathbf{v}_n$? Very little. The columns of $A$ become $\mathbf{v}_1, \ldots, \mathbf{v}_n$. Since they are orthogonal,

$$A^T A = \begin{bmatrix} -\mathbf{v}_1- \\ \vdots \\ -\mathbf{v}_n- \end{bmatrix} \begin{bmatrix} | & & | \\ \mathbf{v}_1 & \cdots & \mathbf{v}_n \\ | & & | \end{bmatrix} = \begin{bmatrix} \langle \mathbf{v}_1, \mathbf{v}_1\rangle & & \\ & \ddots & \\ & & \langle \mathbf{v}_n, \mathbf{v}_n\rangle \end{bmatrix}.$$

Thus, although $A^T A$ is no longer the identity matrix, it is a diagonal matrix. Since

$$A^T \mathbf{b} = \begin{bmatrix} -\mathbf{v}_1- \\ \vdots \\ -\mathbf{v}_n- \end{bmatrix} \mathbf{b} = \begin{bmatrix} \langle \mathbf{v}_1 \cdot \mathbf{b}\rangle \\ \vdots \\ \langle \mathbf{v}_n, \mathbf{b}\rangle \end{bmatrix},$$

the normal equations $A^T A \bar{\mathbf{x}} = A^T \mathbf{b}$ become

$$\begin{bmatrix} \langle \mathbf{v}_1, \mathbf{v}_1\rangle & & \\ & \ddots & \\ & & \langle \mathbf{v}_n, \mathbf{v}_n\rangle \end{bmatrix} \bar{\mathbf{x}} = \begin{bmatrix} \langle \mathbf{v}_1, \mathbf{b}\rangle \\ \vdots \\ \langle \mathbf{v}_n, \mathbf{b}\rangle \end{bmatrix}.$$

Hence

$$\bar{\mathbf{x}} = \begin{bmatrix} \langle \mathbf{v}_1, \mathbf{b}\rangle / \langle \mathbf{v}_1, \mathbf{v}_1\rangle \\ \vdots \\ \langle \mathbf{v}_n, \mathbf{b}\rangle / \langle \mathbf{v}_n, \mathbf{v}_n\rangle \end{bmatrix}.$$

Therefore,

$$\mathbf{p} = A\bar{\mathbf{x}} = \begin{bmatrix} | & & | \\ \mathbf{v}_1 & \cdots & \mathbf{v}_n \\ | & & | \end{bmatrix} \begin{bmatrix} \langle \mathbf{v}_1, \mathbf{b}\rangle / \langle \mathbf{v}_1, \mathbf{v}_1\rangle \\ \vdots \\ \langle \mathbf{v}_n, \mathbf{b}\rangle / \langle \mathbf{v}_n, \mathbf{v}_n\rangle \end{bmatrix}$$

$$= \frac{\langle \mathbf{v}_1, \mathbf{b}\rangle}{\langle \mathbf{v}_1, \mathbf{v}_1\rangle}\mathbf{v}_1 + \cdots + \frac{\langle \mathbf{v}_n, \mathbf{b}\rangle}{\langle \mathbf{v}_n, \mathbf{v}_n\rangle}\mathbf{v}_n.$$

Thus when we have an orthogonal basis for $V$, Theorem 1 becomes:

***Theorem 3*** *Let $\mathbf{v}_1, \mathbf{v}_2, \ldots, \mathbf{v}_n$ be an orthogonal basis for a subspace $V$ of $R^m$, let $A$ be the matrix whose columns are $\mathbf{v}_1, \mathbf{v}_2, \ldots, \mathbf{v}_n$, and let $\mathbf{b}$ be a vector in $R^m$.*

(*a*) *The projection of $\mathbf{b}$ onto $V$ is*

$$\mathbf{p} = A\bar{\mathbf{x}} = \frac{\langle \mathbf{v}_1, \mathbf{b}\rangle}{\langle \mathbf{v}_1, \mathbf{v}_1\rangle}\mathbf{v}_1 + \cdots + \frac{\langle \mathbf{v}_n, \mathbf{b}\rangle}{\langle \mathbf{v}_n, \mathbf{v}_n\rangle}\mathbf{v}_n.$$

*Thus $\mathbf{p}$ is the sum of the projections of $\mathbf{b}$ onto the orthogonal vectors $\mathbf{v}_1, \ldots, \mathbf{v}_n$.*

(*b*) *If $\mathbf{b}$ is a vector in the subspace $V$, then the expansion of $\mathbf{b}$ in terms of the orthogonal basis $\mathbf{v}_1, \ldots, \mathbf{v}_n$ is given by*

$$\mathbf{b} = A\bar{\mathbf{x}} = \frac{\langle \mathbf{v}_1, \mathbf{b}\rangle}{\langle \mathbf{v}_1, \mathbf{v}_1\rangle}\mathbf{v}_1 + \cdots + \frac{\langle \mathbf{v}_n, \mathbf{b}\rangle}{\langle \mathbf{v}_n, \mathbf{v}_n\rangle}\mathbf{v}_n.$$

*Thus $\mathbf{b}$ is the sum of its projections onto each vector in the orthogonal basis. The coefficients $\langle \mathbf{v}_1, \mathbf{b}\rangle / \langle \mathbf{v}_1, \mathbf{v}_1\rangle, \ldots, \langle \mathbf{v}_n, \mathbf{b}\rangle / \langle \mathbf{v}_n, \mathbf{v}_n\rangle$ give the coordinates of $\mathbf{b}$ with respect to this orthogonal basis.*

(*c*) *The projection matrix $P = A(A^TA)^{-1}A^T$ decomposes into the sum of the $n$ projection matrices for the projection of $R^m$ onto each of the orthogonal basis vectors $\mathbf{v}_1, \mathbf{v}_2, \ldots, \mathbf{v}_n$:*

$$P = \frac{1}{\langle \mathbf{v}_1, \mathbf{v}_1\rangle}\mathbf{v}_1\mathbf{v}_1^T + \cdots + \frac{1}{\langle \mathbf{v}_n, \mathbf{v}_n\rangle}\mathbf{v}_n\mathbf{v}_n^T.$$

If we now normalize the $\mathbf{v}$'s to obtain an orthonormal basis

$$\mathbf{w}_1 = \frac{\mathbf{v}_1}{\|\mathbf{v}_1\|}, \ldots, \mathbf{w}_n = \frac{\mathbf{v}_n}{\|\mathbf{v}_n\|},$$

then Theorem 3 reduces to Theorem 1. Thus given an orthogonal basis, it makes no difference whether we first normalize it and then compute the projection or whether we first compute the projection and then normalize. The result is exactly the same.

### EXAMPLE 2

The three vectors (1, 1, 0), (1, −1, 1), and (−1, 1, 2) are orthogonal vectors in $R^3$. Hence they form a basis for $R^3$, and any 3-vector $\mathbf{x} = (x_1, x_2, x_3)$ is given (uniquely) by the sum of its projections onto these vectors.

$$\mathbf{x} = (x_1, x_2, x_3) = \frac{x_1 + x_2}{2}(1, 1, 0) + \frac{x_1 - x_2 + x_3}{3}(1, -1, 1) + \frac{(-x_1 + x_2 + 2x_3)}{6}(-1, 1, 2).$$

It we had first normalized the vectors to obtain an orthonormal basis, our result would have been

$$\mathbf{x} = (x_1, x_2, x_3) = \frac{x_1 + x_2}{\sqrt{2}}\left(\frac{1}{\sqrt{2}}, \frac{1}{\sqrt{2}}, 0\right) + \frac{x_1 - x_2 + x_3}{\sqrt{3}}\left(\frac{1}{\sqrt{3}}, -\frac{1}{\sqrt{3}}, \frac{1}{\sqrt{3}}\right) + \frac{-x_1 + x_2 + 2x_3}{\sqrt{6}}\left(-\frac{1}{\sqrt{6}}, \frac{1}{\sqrt{6}}, \frac{2}{\sqrt{6}}\right).$$

Of course, these two expressions for $\mathbf{x}$ are the same.

### EXAMPLE 3

Let $V$ be the subspace of $R^3$ spanned by the two vectors (1, 1, 0) and (1, −1, 1). Since these vectors are orthogonal, the projection of (1, 1, 1) onto $V$ is

$$\mathbf{p} = \frac{2}{2}(1, 1, 0) + \frac{1}{3}(1, -1, 1) = \left(\tfrac{4}{3}, \tfrac{2}{3}, \tfrac{1}{3}\right).$$

### EXAMPLE 4

Consider the following inconsistent system of equations:

$$\begin{bmatrix} 1 & 4 \\ -1 & 2 \\ 1 & 0 \\ -1 & 2 \end{bmatrix}\begin{bmatrix} x_1 \\ x_2 \end{bmatrix} = \begin{bmatrix} -3 \\ -3 \\ 0 \\ -3 \end{bmatrix}.$$

Since the columns of $A$ are orthogonal, the projection of (−3, −3, 0, −3) onto the column space of $A$ is given by

$$\mathbf{p} = \frac{3}{4}(1, -1, 1, -1) + \frac{-24}{24}(4, 2, 0, 2).$$

Hence the least-squares solution to the inconsistent system is $\bar{\mathbf{x}} = (\tfrac{3}{4}, -1)$.

## PROBLEMS 3.7

**1.** Find the least-squares solution to the inconsistent system $A\mathbf{x} = \mathbf{b}$ if $A$ and $\mathbf{b}$ are given by:

*(a) $\begin{bmatrix} 1 & 1 \\ 1 & -1 \\ 1 & 0 \end{bmatrix}$ and $\begin{bmatrix} 1 \\ 0 \\ 3 \end{bmatrix}$

(b) $\begin{bmatrix} 1 & 1 \\ -1 & 1 \\ 0 & 2 \end{bmatrix}$ and $\begin{bmatrix} -3 \\ 2 \\ 5 \end{bmatrix}$

*(c) $\begin{bmatrix} 1 & 2 \\ -1 & 3 \\ 1 & 1 \end{bmatrix}$ and $\begin{bmatrix} -1 \\ 0 \\ 2 \end{bmatrix}$

(d) $\begin{bmatrix} 1 & -4 \\ -1 & 1 \\ 1 & 5 \end{bmatrix}$ and $\begin{bmatrix} 1 \\ 1 \\ 1 \end{bmatrix}$

*(e) $\begin{bmatrix} 4 & 1 \\ 2 & -2 \\ 2 & 2 \\ 1 & -4 \end{bmatrix}$ and $\begin{bmatrix} 0 \\ 1 \\ 0 \\ 1 \end{bmatrix}$

(f) $\begin{bmatrix} -2 & 1 \\ 4 & -1 \\ 1 & 2 \\ 2 & 2 \end{bmatrix}$ and $\begin{bmatrix} 1 \\ 2 \\ 1 \\ 1 \end{bmatrix}$

*(g) $\begin{bmatrix} -2 & -2 & 1 \\ -1 & 4 & -2 \\ 4 & 1 & 2 \\ 2 & -2 & -4 \end{bmatrix}$ and $\begin{bmatrix} 1 \\ 0 \\ 0 \\ 0 \end{bmatrix}$

(h) $\begin{bmatrix} 1 & 0 & 0 \\ 0 & 1 & 0 \\ 0 & 0 & 1 \\ 0 & 0 & 1 \end{bmatrix}$ and $\begin{bmatrix} 3 \\ 2 \\ 1 \\ 0 \end{bmatrix}$

*(i) $\begin{bmatrix} 1 & -1 & 1 \\ 1 & 1 & -1 \\ 0 & 2 & 1 \\ 0 & 0 & 3 \end{bmatrix}$ and $\begin{bmatrix} 1 \\ 1 \\ 1 \\ 4 \end{bmatrix}$

***2.** The vectors $(1,1,1,1)$, $(1,-1,1,-1)$, and $(1,2,-1,-2)$ are orthogonal. Express $(6,5,0,-7)$ as a linear combination of these vectors.

**3.** (a) Show that $(1,1,1,1)$, $(1,1,-2,0)$, $(-2,4,1,-3)$, $(-3,1,-1,3)$ form an orthogonal basis of $R^4$.
(b) Find the coordinate vector of $(1,2,3,4)$ with respect to this basis.
(c) Find the projection of $(1,2,3,4)$ onto the subspace $V$ spanned by $(1,1,1,1)$ and $(1,1,-2,0)$.

**4.** Show that $(\frac{1}{5},\frac{2}{5},\frac{2}{5},\frac{4}{5})$, $(\frac{4}{5},-\frac{2}{5},-\frac{2}{5},\frac{1}{5})$, $(\frac{2}{5},-\frac{1}{5},\frac{4}{5},-\frac{2}{5})$, $(\frac{2}{5},\frac{4}{5},-\frac{1}{5},-\frac{2}{5})$ form an orthonormal basis for $R^4$. Find the coordinate vector of $(1,2,2,1)$ with respect to this basis. Write $(1,2,2,1)$ as a linear combination of the basis vectors.

***5.** Show that the vector $(1,1,0,0)$ is not in the subspace $V$ spanned by $(1,1,1,1)$, $(1,-1,1,-1)$, and $(1,2,-1,-2)$, and then find its projection onto $V$.

**6.** Suppose that we wish to find the best straight-line fit to the data $(t_1, y_1), \ldots, (t_m, y_m)$. Show that if the $t$-coordinates of the data are symmetric about 0, then the columns of the matrix $A$ are orthogonal.

***7.** Find the best straight-line fit to the data

| $t$ | $-5$ | $-4$ | $-2$ | $2$ | $4$ | $5$ |
|---|---|---|---|---|---|---|
| $y$ | $7$ | $5$ | $1$ | $0$ | $-2$ | $-5$ |

Carefully plot the data and draw the straight line (see Problem 6).

**8.** Suppose that $A$ has orthonormal columns and that $A\mathbf{x} = \mathbf{b}$ is consistent. Show that $\mathbf{x} = A^T\mathbf{b}$ is the unique solution of $A\mathbf{x} = \mathbf{b}$.

**9.** Let $\mathbf{v}_1, \mathbf{v}_2, \ldots, \mathbf{v}_n$ be an orthonormal basis for a subspace $V$ of $R^m$. If $\mathbf{v}$ is in $V$ and

$$\mathbf{v} = \alpha_1\mathbf{v}_1 + \alpha_2\mathbf{v}_2 + \cdots + \alpha_n\mathbf{v}_n,$$

then $\alpha_i = \langle \mathbf{v}_i, \mathbf{v} \rangle$. Prove this without using Theorem 1.

**10.** Let $\mathbf{w}_1, \ldots, \mathbf{w}_m$ be an orthonormal basis for $R^m$. If $\mathbf{u}$ and $\mathbf{v}$ are in $R^m$, show that

*(a) $\langle \mathbf{u}, \mathbf{v} \rangle = \langle \mathbf{u}, \mathbf{w}_1 \rangle \langle \mathbf{v}, \mathbf{w}_1 \rangle + \cdots + \langle \mathbf{u}, \mathbf{w}_m \rangle \langle \mathbf{v}, \mathbf{w}_m \rangle$

(b) $\|\mathbf{u}\|^2 = \langle \mathbf{u}, \mathbf{w}_1 \rangle^2 + \cdots + \langle \mathbf{u}, \mathbf{w}_m \rangle^2$

**11.** Suppose that $\mathbf{w}_1, \ldots, \mathbf{w}_k$ are orthonormal vectors in $R^m$ and that for every $\mathbf{u}$ in $R^m$

$$\|\mathbf{u}\|^2 = \langle \mathbf{u}, \mathbf{w}_1 \rangle^2 + \cdots + \langle \mathbf{u}, \mathbf{w}_k \rangle^2.$$

Show that $\mathbf{w}_1, \ldots, \mathbf{w}_k$ form an orthonormal basis for $R^m$. (*Hint:* Let $V$ be the subspace of $R^m$ spanned by $\mathbf{w}_1, \ldots, \mathbf{w}_k$, and let $P$ be the projection matrix for $V$. Show that $P\mathbf{u} = \mathbf{u}$ for all $\mathbf{u}$ in $R^m$ by showing that $\|P\mathbf{u} - \mathbf{u}\|^2 = 0$ for all $\mathbf{u}$ in $R^m$.)

# 3.8 The Gram–Schmidt Process

In the preceding section we saw some of the advantages that result from using an orthogonal basis for a subspace. Such bases substantially simplify the algebraic computations involved in finding the representation of a vector in terms of the basis and the projection of a vector onto a subspace. Such a basis also corresponds to our mental image of the geometry of space. In view of these remarks it is certainly worthwhile to investigate the following two questions.

1. Does a subspace always have an orthogonal basis?
2. If so, is there a reasonable way to actually construct an orthogonal basis?

Of course, an affirmative answer to the second question implies an affirmative answer to the first. Hence we will show that the answer to both questions is affirmative if we exhibit a procedure for constructing an orthogonal basis for a subspace.

Suppose that $V$ is a subspace of $R^m$ and that $\mathbf{u}_1, \mathbf{u}_2, \ldots, \mathbf{u}_n$ form a basis for $V$. We are going to use this basis to construct an orthogonal basis $\mathbf{v}_1, \mathbf{v}_2, \ldots, \mathbf{v}_n$ for $V$. The first step of our construction is easy. Let $\mathbf{v}_1 = \mathbf{u}_1$. The vector $\mathbf{u}_2$ is not in the subspace spanned by $\mathbf{v}_1 = \mathbf{u}_1$ and in general it is not orthogonal to $\mathbf{v}_1$. Our second step, then, is to let $\mathbf{v}_2$ be the projection of $\mathbf{u}_2$ onto the orthogonal complement of the subspace spanned by $\mathbf{v}_1$.

$$\mathbf{v}_2 = \mathbf{u}_2 - \frac{\langle \mathbf{v}_1, \mathbf{u}_2 \rangle}{\langle \mathbf{v}_1, \mathbf{v}_1 \rangle} \mathbf{v}_1. \tag{1}$$

The vectors $\mathbf{v}_1, \mathbf{v}_2$ are nonzero (why?) orthogonal vectors. Moreover, they span the same subspace as $\mathbf{u}_1, \mathbf{u}_2$ because $\mathbf{v}_1 = \mathbf{u}_1$ and by (1), $\mathbf{v}_2$ is a linear combination of $\mathbf{u}_1$ and $\mathbf{u}_2$. The vector $\mathbf{u}_3$ is not in the subspace spanned by $\mathbf{v}_1, \mathbf{v}_2$ and in general it is not orthogonal to this subspace. The third step, then, is to let $\mathbf{v}_3$ be the projection of $\mathbf{u}_3$ onto the orthogonal complement of the subspace spanned by $\mathbf{v}_1, \mathbf{v}_2$:

$$\mathbf{v}_3 = \mathbf{u}_3 - \frac{\langle \mathbf{v}_1, \mathbf{u}_3 \rangle}{\langle \mathbf{v}_1, \mathbf{v}_1 \rangle} \mathbf{v}_1 - \frac{\langle \mathbf{v}_2, \mathbf{u}_3 \rangle}{\langle \mathbf{v}_2, \mathbf{v}_2 \rangle} \mathbf{v}_2.$$

Then $\mathbf{v}_1, \mathbf{v}_2, \mathbf{v}_3$ are nonzero (why?) orthogonal vectors that span the same subspace as $\mathbf{v}_1, \mathbf{v}_2, \mathbf{u}_3$ (which in turn span the same subspace as $\mathbf{u}_1, \mathbf{u}_2, \mathbf{u}_3$). Continue the construction. The final step is to let $\mathbf{v}_n$ be the projection of $\mathbf{u}_n$ onto the orthogonal complement of the subspace spanned by $\mathbf{v}_1, \mathbf{v}_2, \ldots, \mathbf{v}_{n-1}$:

$$\mathbf{v}_n = \mathbf{u}_n - \frac{\langle \mathbf{v}_1, \mathbf{u}_n \rangle}{\langle \mathbf{v}_1, \mathbf{v}_1 \rangle} \mathbf{v}_1 - \cdots - \frac{\langle \mathbf{v}_{n-1}, \mathbf{u}_n \rangle}{\langle \mathbf{v}_{n-1}, \mathbf{v}_{n-1} \rangle} \mathbf{v}_{n-1}.$$

The vectors $\mathbf{v}_1, \mathbf{v}_2, \ldots, \mathbf{v}_n$ form an orthogonal basis for $V$. Letting $\mathbf{w}_1 = \mathbf{v}_1/\|\mathbf{v}_1\|$, $\mathbf{w}_2 = \mathbf{v}_2/\|\mathbf{v}_2\|, \ldots, \mathbf{w}_n = \mathbf{v}_n/\|\mathbf{v}_n\|$, we obtain an orthonormal basis for $V$.

This procedure for converting a basis to an orthonormal basis is called the **Gram–Schmidt process**. Since every nontrivial subspace $V$ of $R^m$ has a basis, we have proved:

> ***Theorem 1*** *(Gram–Schmidt) Every nontrivial subspace $V$ of $R^m$ has an orthogonal basis, and hence an orthonormal basis.*

### EXAMPLE 1

The three vectors (1, 1, 0), (2, 1, 0), and (1, 1, 1) form a basis for $R^3$. If we apply the Gram–Schmidt process to these vectors, we obtain

$$\begin{aligned}
\mathbf{v}_1 &= (1, 1, 0), \\
\mathbf{v}_2 &= (2, 1, 0) - \frac{3}{2}(1, 1, 0) = (1/2, -1/2, 0), \\
\mathbf{v}_3 &= (1, 1, 1) - \frac{2}{2}(1, 1, 0) - \frac{0}{1/2}(1/2, -1/2, 0) \\
&= (0, 0, 1).
\end{aligned}$$

### EXAMPLE 2

Consider the subspace $V$ of $R^3$ spanned by $\mathbf{u}_1 = (1, -1, 3)$, $\mathbf{u}_2 = (0, 1, 2)$, and $\mathbf{u}_3 = (-1, 2, -1)$. If we apply the Gram–Schmidt process to these vectors, we obtain

$$\begin{aligned}
\mathbf{v}_1 &= (1, -1, 3), \\
\mathbf{v}_2 &= (0, 1, 2) - \frac{5}{11}(1, -1, 3) = \frac{1}{11}(-5, 16, 7), \\
\mathbf{v}_3 &= (-1, 2, -1) - \frac{-6}{11}(1, -1, 3) - \frac{1}{11}(-5, 16, 7) = (0, 0, 0).
\end{aligned}$$

What happened? That is, why is $\mathbf{v}_3 = \mathbf{0}$? Since $\mathbf{v}_3$ is the projection of $\mathbf{u}_3$ onto the orthogonal complement of the subspace spanned by $\mathbf{v}_1, \mathbf{v}_2$, it follows that $\mathbf{u}_3$ is in the subspace spanned by $\mathbf{v}_1, \mathbf{v}_2$. This subspace is the same as the subspace spanned by $\mathbf{u}_1, \mathbf{u}_2$, and hence $\mathbf{u}_1, \mathbf{u}_2, \mathbf{u}_3$ are dependent vectors. Thus a basis for the subspace $V$ is $\mathbf{u}_1$, $\mathbf{u}_2$, and $\mathbf{v}_1$, $\mathbf{v}_2$ is an orthogonal basis for $V$ (see Problem 13).

### EXAMPLE 3

Let us find an orthogonal basis for $R^3$ which contains the vector $(1, 1, 1)$. We begin with $\mathbf{v}_1 = (1, 1, 1)$. Clearly, the other two vectors must come from the orthgonal complement of the subspace spanned by $\mathbf{v}_1$. If $A$ is the matrix whose single column is $\mathbf{v}_1$, then the orthogonal complement is $C(A)^{\perp} = N(A^T)$. A basis for $N(A^T)$ is $(-1, 1, 0)$ and $(-1, 0, 1)$. Applying the Gram–Schmidt process to the vectors $(-1, 1, 0)$ and $(-1, 0, 1)$, we obtain $(-1, 1, 0)$ and $(-\frac{1}{2}, -\frac{1}{2}, 1)$. Thus an orthogonal basis for $R^3$ that contains the vector $(1, 1, 1)$ is $(1, 1, 1)$, $(-1, 1, 0)$, $(-\frac{1}{2}, -\frac{1}{2}, 1)$.

In general, we can extend an orthogonal basis for a subspace $V$ of $R^m$ to an orthogonal basis for $R^m$ by simply adding to the orthogonal basis for $V$ an orthogonal basis for $V^{\perp}$.

Finally, it should be clear from the proof of the Gram–Schmidt process that it can be used to replace any set of independent vectors with a set of orthogonal vectors that span the same subspace.

**Result 1** Let $\mathbf{u}_1, \ldots, \mathbf{u}_k$ be independent vectors in $R^n$. Then there is an orthogonal set of vectors $\mathbf{v}_1, \ldots, \mathbf{v}_k$ with the property that for each $i$, $1 \le i \le k$, $\mathbf{v}_1, \ldots, \mathbf{v}_i$ and $\mathbf{u}_1, \ldots, \mathbf{u}_i$ span the same subspace.

## PROBLEMS 3.8

**1.** For each of the following, find an orthogonal basis for $R^3$ containing the given vectors.
*(a) $(1, 1, -1), (-1, 2, 1)$
(b) $(1, 1, 1), (1, 0, -1)$
*(c) $(1, 2, 3)$
(d) $(1, -2, 2)$

**2.** Find an orthogonal basis of $R^4$ containing the vectors $(1, 1, 1, 0)$ and $(-1, 1, 0, 1)$.

**3.** Find an orthogonal basis for the nullspace of each of the following matrices.
*(a) $A = \begin{bmatrix} 1 & -1 & 1 & 2 \end{bmatrix}$
(b) $A = \begin{bmatrix} 1 & 0 & -1 & 0 \\ 2 & 1 & -3 & 1 \end{bmatrix}$

**4.** Find an orthogonal basis for the subspace $V$ spanned by the given vectors.
(a) $(2, 1, 3)$ and $(1, 2, 0)$
*(b) $(1, 1, 1)$, $(0, 1, 1)$, and $(0, 0, 1)$
(c) $(1, 1, 0, 0)$, $(1, -2, 1, 0)$, and $(0, 1, 1, 1)$
*(d) $(-1, 2, 0, 2)$, $(2, -4, 1, -4)$, and $(-1, 3, 1, 1)$
(e) $(1, 1, 2, 1)$, $(-1, 1, 0, 1)$, and $(2, 1, 1, 0)$

**5.** Apply Gram–Schmidt to change $(1, -1, 2)$, $(1, 1, 1)$, and $(2, -1, 1)$ to an orthogonal set of vectors spanning the same subspace of $R^3$.

***6.** (a) Find an orthonormal basis for the subspace of $R^4$ which is spanned by $(1, 1, 0, 0)$, $(1, 1, 1, 0)$, $(1, 1, 1, 1)$.
(b) Find the projection matrix for this subspace.
(c) Find the projection of $(1, 0, 0, 0)$ onto this subspace.

**7.** Find an orthonormal basis for the subspace $V$ of $R^4$ spanned by the vectors $(-1, 2, 0, 2)$, $(2, -4, 1, -4)$, $(-1, 3, 1, 1)$. Then find the projection matrix for $V$.

**8.** Let $V$ be the subspace of $R^3$ spanned by the vectors $(1, 2, 1)$ and $(4, 2, 4)$.

(a) Find an orthogonal basis for $V$.
(b) Find the projection matrix for $V$.
(c) Find the projection of the vector $(2, 6, 0)$ onto $V$.
(d) Find the distance of the vector $(2, 6, 0)$ from $V$.

**9.** Let $V$ be the subspace of $R^4$ spanned by $(1, 1, 1, 0)$, $(0, 1, 1, 0)$, $(0, 0, 1, 0)$.
(a) Find an orthogonal basis for $V$.
(b) Find an orthonormal basis for $V$.
(c) Find the projection of $(1, 0, 1, 1)$ onto $V$.

**10.** Let $V$ be the subspace of $R^4$ spanned by $(1, -1, 1, -1)$ and $(1, 1, 5, 1)$.
(a) Find an orthogonal basis for $V$.
(b) What point in $V$ is closest to $(1, 1, 1, -1)$?
(c) What is the distance from $(1, 1, 1, -1)$ to $V$?

**11.** Find the projection matrix $P$ for the subspace spanned by $(1, 0, 1, 0)$, $(1, 1, -1, 0)$, and $(1, -2, -1, 1)$. Then use $P$ to extend these vectors to an orthogonal basis of $R^4$.

**12.** Show that if the Gram–Schmidt process is applied to the vectors $\mathbf{u}_1$, $\mathbf{u}_2$, $\mathbf{u}_3$, and $\mathbf{u}_4$, where $\mathbf{u}_4$ is a linear combination of $\mathbf{u}_1$, $\mathbf{u}_2$, and $\mathbf{u}_3$, then $\mathbf{v}_4 = \mathbf{0}$.

**13.** Describe what happens when the Gram–Schmidt process is applied to a dependent set of vectors.

**14.** Let $A$ be an $m \times n$ matrix with independent columns $\mathbf{u}_1, \mathbf{u}_2, \ldots, \mathbf{u}_n$. Apply the Gram–Schmidt process to these vectors to obtain the orthonormal vectors $\mathbf{w}_1, \mathbf{w}_2, \ldots, \mathbf{w}_n$, where

$$\begin{aligned}\mathbf{v}_1 &= \mathbf{u}_1,\\ &\vdots\\ \mathbf{v}_n &= \mathbf{u}_n - \frac{\langle \mathbf{v}_1, \mathbf{u}_n\rangle}{\langle \mathbf{v}_1, \mathbf{v}_1\rangle}\mathbf{v}_1 - \cdots\\ &\quad - \frac{\langle \mathbf{v}_{n-1}, \mathbf{u}_n\rangle}{\langle \mathbf{v}_{n-1}, \mathbf{v}_{n-1}\rangle}\mathbf{v}_{n-1},\end{aligned}$$

and

$$\mathbf{w}_i = \frac{\mathbf{v}_i}{\|\mathbf{v}_i\|}, \qquad i = 1, \ldots, n.$$

Let $Q$ be the matrix whose columns are $\mathbf{w}_1, \ldots, \mathbf{w}_n$.
(a) Show that

$$\begin{aligned}\mathbf{u}_1 &= \|\mathbf{v}_1\|\mathbf{w}_1,\\ \mathbf{u}_2 &= \langle \mathbf{w}_1, \mathbf{u}_2\rangle\mathbf{w}_1 + \|\mathbf{v}_2\|\mathbf{w}_2,\\ \vdots &= \\ \mathbf{u}_n &= \langle \mathbf{w}_1, \mathbf{u}_n\rangle\mathbf{w}_1 + \cdots\\ &\quad + \langle \mathbf{w}_{n-1}, \mathbf{u}_n\rangle\mathbf{w}_{n-1} + \|\mathbf{v}_n\|\mathbf{w}_n\end{aligned}$$

(*Hint:* Show that $\|\mathbf{v}_i\| = \langle \mathbf{w}_i, \mathbf{u}_i\rangle$.)
(b) Find a triangular matrix $R$ such that $A = QR$. This decomposition of the matrix $A$ is called its *QR decomposition.*
(c) Show that $R$ is nonsingular.
(d) Show that if $A$ is square, then $Q$ is an orthogonal matrix.
(e) Show that if $A$ is square, then the normal equations become $R\bar{\mathbf{x}} = Q^T\mathbf{b}$.
(f) Suppose that $m > n$, so that $A$ is not square. Extend the orthonormal vectors $\mathbf{w}_1, \ldots, \mathbf{w}_n$ to an orthonormal basis of $R^m$, and let $Q_1$ be the matrix whose columns are this orthonormal basis. Thus $Q_1$ is an orthogonal matrix. Can the matrix $R$ be replaced by a matrix $R_1$ so that $A = Q_1R_1$?

***15.** For each part of Problem 4, let $A$ be the matrix whose columns are the given vectors. Find the $QR$ decomposition of each of the matrices.

**16.** Suppose that $\mathbf{u}_1, \mathbf{u}_2, \ldots, \mathbf{u}_m$ form a basis for $R^m$. Show that the following procedure produces an orthonormal basis for $R^m$.
1. Let $\mathbf{w}_1 = \mathbf{u}_1/\|\mathbf{u}_1\|$ and $\mathbf{v}_i^{(1)} = \mathbf{u}_i - \langle \mathbf{u}_i, \mathbf{w}_1\rangle\mathbf{w}_1$, $i = 2, 3, \ldots, m$.
2. Let $\mathbf{w}_2 = \mathbf{v}_2^{(1)}/\|\mathbf{v}_2^{(1)}\|$ and $\mathbf{v}_i^{(2)} = \mathbf{v}_i^{(1)} - \langle \mathbf{v}_i^{(1)}, \mathbf{w}_2\rangle\mathbf{w}_2$, $i = 3, 4, \ldots, m$.
3. Continue.

This procedure is called the **modified Gram–Schmidt process.** It is superior to Gram–Schmidt for computer implementation.

## SUPPLEMENTARY PROBLEMS

**1.** Let $V$ be a subspace of $R^m$ and let $P$ be a projection matrix for the projection of $V$ onto $R^m$. Mark each of the following true or false and given a brief explanation.

*(a) rank $P = \dim V$

(b) $N(P) = V^\perp$

*(c) $\dim V + \dim V^\perp < m$

(d) If $\mathbf{x}$ and $\mathbf{y}$ are orthogonal, then $P\mathbf{x}$ and $P\mathbf{y}$ are orthogonal.

*(e) $C(P) = V$

(f) If $\mathbf{b}$ is not in $V$, there is a unique vector in $V$ such that $\mathbf{b} - \mathbf{p}$ is orthogonal to $V$.

*(g) For any $\mathbf{b}$ in $R^m$, $P\mathbf{b}$ is orthogonal to $\mathbf{b}$.

(h) If $W$ is a subspace orthogonal to $V$, then $W = V^\perp$.

*(i) $\{\mathbf{0}\}^\perp = R^m$.

(j) $V$ has an orthonormal basis.

**2.** Let $A$ be an $m \times n$ matrix of rank $r$. Mark each of the following true or false and give a brief explanation.

*(a) The normal equations always have a solution.

(b) The normal equations always have a unique solution if and only if $r = n$.

*(c) If $r = n$ and $P = A(A^TA)^{-1}A^T$, then rank $P = r$.

(d) $N(A)^\perp = C(A^T)$

*(e) It is possible for $R(A)$ to contain $(1, 2, 1, 0)$ and $N(A)$ to contain $(0, 0, 1, 0)$.

(f) $\langle A\mathbf{x}, A\mathbf{y}\rangle = \langle \mathbf{x}, \mathbf{y}\rangle$ if $A^T = A^{-1}$.

*(g) If $r = m$ and $A^TA\bar{\mathbf{x}} = A^T\mathbf{b}$, then $A\bar{\mathbf{x}} = \mathbf{b}$.

(h) If the columns of $A$ are orthogonal, then $r = n$.

*(i) If $A^TA$ is nonsingular, then $A$ is nonsingular.

**3.** For each of the following subspaces $V$ and vectors $\mathbf{b}$:

(i) Find the projection of $\mathbf{b}$ onto $V$.

(ii) Find the vector in $V$ closest to $\mathbf{b}$.

(iii) Find the distance from $\mathbf{b}$ to $V$.

*(a) $V$ is the plane $x - 2y + z = 0$, $\mathbf{b} = (1, -1, 4)$.

(b) $V$ is the plane $4x - 5y - 3z = 0$, $\mathbf{b} = (1, 1, 1)$.

*(c) $V$ is the plane $x + y - z = 0$, $\mathbf{b} = (1, 1, 0)$.

(d) $V$ is the plane $x + y + z = 0$, $\mathbf{b} = (1, 1, 2)$.

*(e) $V$ is the plane $x + 2y + 2z = 0$, $\mathbf{b} = (1, 1, 1)$.

(f) $V$ is the plane $2x + y + z = 0$, $\mathbf{b} = (3, 1, 2)$.

*(g) $V$ is the plane $z = 0$, $\mathbf{b} = (1, 2, 3)$.

(h) $V$ is the plane $2x + y - z = 0$, $\mathbf{b} = (3, 1, 1)$.

*(i) $V$ is the subspace spanned by $(3, 1, 1)$ and $(2, 1, 1)$, $\mathbf{b} = (5, 3, -7)$.

(j) $V$ is the line spanned by $(1, 1, 2)$, $\mathbf{b} = (2, -1, 3)$.

*(k) $V$ is the subspace spanned by $(1, 0, 1, 0)$, $(1, 1, -1, 0)$, $(1, -2, -1, 1)$, $\mathbf{b} = (1, 1, 1, 1)$.

(l) $V$ is the subspace spanned by $(1, 0, 0, 0)$, $(0, 1, 0, 0)$, $(0, 1, 1, 1)$, $\mathbf{b} = (0, 0, 0, 1)$.

*(m) $V$ is the subspace spanned by $(1, 0, 0, 1)$, $(2, 1, 0, 1)$, $\mathbf{b} = (1, 1, 1, 1)$.

(n) $V$ is the orthogonal complement of the subspace spanned by $(1, 0, 1, 0)$, $\mathbf{b} = (1, 1, 1, 1)$.

**4.** (a) Find the projection of $(1, -2, 1)$ onto $(1, 0, 3)$.

(b) What is the angle between these two vectors?

***5.** Find the least-squares solution $\bar{x}$ to the system

$$\begin{aligned} 3x &= 10 \\ 4x &= 5 \\ 2x &= 1. \end{aligned}$$

**6.** Find the best fit to the data $(-\frac{3}{2}, -1)$, $(-\frac{1}{2}, 1)$, $(\frac{1}{2}, 1)$, $(\frac{3}{2}, -1)$ by a quadratic $y = a + bx + cx^2$.

**7.** Find an orthonormal basis for the subspace of $R^4$ spanned by $(-1, 2, 0, 2)$, $(2, -4, 1, -4)$, and $(-1, 3, 1, 1)$.

**8.** Let $A = [1 \quad -1 \quad -2 \quad 6]$.

(a) Find an orthogonal basis for the nullspace $N(A)$ of $A$.

(b) Find a vector $\mathbf{u}$ in $N(A)$ and a vector $\mathbf{w}$ orthogonal to $N(A)$ such that $\mathbf{u} + \mathbf{w} = (1, 1, 3, -1)$.

**9.** Let

$$A = \begin{bmatrix} 1 & 1 & 1 & 1 \\ 0 & 1 & 1 & 0 \\ 0 & 0 & 1 & -1 \\ 0 & 0 & 1 & -1 \end{bmatrix}.$$

(a) Find a basis for $R(A)$, $C(A)$, and $N(A)$.

(b) What constraints must the components of a vector $\mathbf{b}$ satisfy so that $A\mathbf{x} = \mathbf{b}$ is consistent?

(c) Find a matrix $B$ so that $N(B) = C(A)$.

(d) Find the general solution to $A\mathbf{x} = \mathbf{b}$ if $\mathbf{b} = (-1, -1, 1, 1)$.

(e) Does $A$ have an inverse? Why?

(f) Find a basis for $C(A)^{\perp}$.

(g) Find the projection of $(0, 0, 1, 2)$ onto $C(A)$ and onto $C(A)^{\perp}$.

(h) Find all least-squares solutions to $A\mathbf{x} = \mathbf{b}$, where $\mathbf{b} = (0, 0, 1, 2)$.

**10.** Let $V$ be the subspace of $R^4$ spanned by $(1, 0, 2, 1)$, $(1, 1, 1, 0)$, and $(0, 1, -1, -1)$.

(a) Find a basis for $V$.

(b) Find a basis for $V^{\perp}$.

(c) Write $\mathbf{w} = (1, 1, 1, 1)$ as a sum of two of two vectors, one from $V$ and the other from $V^{\perp}$.

(d) Find the projection of $\mathbf{w} = (1, 1, 1, 1)$ onto $V$ and onto $V^{\perp}$.

(e) Find the vector in $V$ that minimizes the expression $\|\mathbf{v} - \mathbf{w}\|$, where $\mathbf{w} = (1, 1, 1, 1)$.

**11.** Let $A$ be an $m \times n$ matrix. How would you find $R(A)^{\perp}$?

**12.** Explain why the vector $(a, b, c)$ is orthogonal to the plane $ax_1 + bx_2 + cx_3 = 0$.

**13.** Let $\mathbf{v}_1$ and $\mathbf{v}_2$ be two independent vectors in $R^5$, and let $V$ be the subspace spanned by these vectors.

(a) What is the dimension of $V^{\perp}$? Why?

(b) How would you find a basis for $V^{\perp}$?

(c) How would you find the projection matrix for $V^{\perp}$?

**14.** Let $\mathbf{v}_1, \mathbf{v}_2, \ldots, \mathbf{v}_k$ be vectors in $R^m$ and suppose that each of these vectors is orthogonal to a nonzero vector $\mathbf{x}$ in $R^m$.

(a) Show that any linear combination of the $\mathbf{v}_i$'s is orthogonal to $\mathbf{x}$.

(b) Show that if $\mathbf{v}_1, \mathbf{v}_2, \ldots, \mathbf{v}_k$ are independent, then so are $\mathbf{v}_1, \mathbf{v}_2, \ldots, \mathbf{v}_k, \mathbf{x}$.

**15.** Let $P$ be a projection matrix. Show that if $R = I - 2P$, then $\|R\mathbf{x}\| = \|\mathbf{x}\|$ for all $\mathbf{x}$.

***16.** Let $U$ be a two-dimensional subspace of $R^3$ (i.e., $U$ is a plane through the origin). Let $P$ be the projection matrix for $U$.

(a) Describe (in terms of $U$) $C(P)$.

(b) Describe (in terms of $U$) $N(P)$.

(c) What is the rank of $P$?

**17.** Let $\mathbf{a} = (-2, 1, 1)$ and $\mathbf{b} = (1, 0, -1)$.

(a) Find the projection of $\mathbf{a}$ onto $\mathbf{b}$.

(b) Find the angle between $\mathbf{a}$ and $\mathbf{b}$.

(c) Find the equation of the line through $\mathbf{a}$ and $\mathbf{b}$.

(d) Find the equation of the plane through $\mathbf{a}$, $\mathbf{b}$, and $\mathbf{c} = (0, 2, 3)$.

**18.** Find the angle between the planes $x - 2y + 4z = 7$ and $3x - y + 2z + 5 = 0$.

**19.** Find the equation of the plane passing through the point $(2, 0, 1)$ and perpendicular to the line $x = 2t$, $y = 5 - t$, $z = 3t - 4$.

**20.** Find the distance between the point $(1, -4, -3)$ and the plane $2x - 3y + 6z = -1$.

***21.** Find the distance between the two lines $\mathbf{x} = (1, 5, -1) + t(1, -1, 2)$ and $\mathbf{u} = (2, 4, 5) + s(1, -3, 1)$.

**22.** Find the distance from the point $(1, 0, 0)$ to the line $\mathbf{x} = (1, 1, 1) + t(-1, 1, 2)$.

***23.** Find the equation of the plane that is parallel to the plane $2x - y + 2z + 4 = 0$ and such that the point $(3, 2, -1)$ is equidistant from both planes.

**24.** Find the distance from $(3, 3, 1)$ to the line $\mathbf{x} = (1, 2, 3) + t(1, -2, 2)$.

***25.** Find the distance from the line $\mathbf{x} = (1, 0, 0) + \alpha(1, -2, 0)$ to the plane through $(0, 1, 1), (1, -1, 1), (2, 3, 4)$.

**26.** Find the point on the plane $x + y + z = 1$ closest to $(2, 3, -1)$. Find the distance from this point to the plane.

**27.** Show that the planes $2x + y + z + 5 = 0$ and $-3x - 4y + 10z + 4 = 0$ are perpendicular.

***28.** Find the distance between the two lines $\mathbf{x} = (1, 0, -4) + s(2, -1, 1)$ and $x = -1 + t, y = 3 - 2t, z = 2$.

**29.** Let $\mathbf{u}, \mathbf{v}, \mathbf{w}$ be nonzero orthogonal vectors. Prove that $\mathbf{w}$ is not a linear combination of $\mathbf{u}$ and $\mathbf{v}$.

**30.** Find the distance from the line $x = t$, $y = 1 + t$, $z = 1 + t$ to the $y$-axis.

**31.** Let $\mathbf{u}_1, \mathbf{u}_2, \ldots, \mathbf{u}_n$ and $\mathbf{v}_1, \mathbf{v}_2, \ldots, \mathbf{v}_n$ be two sets of orthonormal vectors in $R^m$. Suppose that for every $k = 1, 2, \ldots n$ the vectors

$$\mathbf{u}_1, \mathbf{u}_2, \ldots, \mathbf{u}_k \quad \text{and} \quad \mathbf{v}_1, \mathbf{v}_2, \ldots, \mathbf{v}_k$$

span the same subspace. Prove that $\mathbf{u}_i = \pm\mathbf{v}_i$ for every $i = 1, 2, \ldots, n$.

# 4

# Function Spaces

In Chapter 2 we called $R^m$ together with the operations of addition and scalar multiplication that satisfied eight identities a vector space. We then defined a subspace as a collection of vectors in $R^m$ that contains the zero vector and is closed under addition and scalar multiplication. In this chapter we consider the set $C(I)$ of all continuous real-valued functions on an interval $I$. There is a natural way to define operations of addition and scalar multiplication on $C(I)$ that satisfy the same eight identities that are satisfied by vector addition and scalar multiplication (see Section 2.1). Thus $C(I)$ has an algebraic structure analogous to that of $R^m$, and for that reason we call $C(I)$ a function space. As you would expect, we define a subspace of $C(I)$ to be a collection of functions in $C(I)$ that contains the zero function and is closed under the operations of addition of functions and multiplication of a function by a scalar.

## 4.1 Function Spaces

We begin by defining the relation of equality, the zero function, the sum of two functions, and multiplication of a function by a scalar. Two functions $f$ and $g$ defined on an interval $I$ are **equal** if they have the same value at every point $x$ in $I$. That is,

$$f = g \quad \text{if} \quad f(x) = g(x) \quad \text{for all } x \text{ in } I.$$

The zero function, denoted by the symbol 0, is the function whose value at every point $x$ in $I$ is the number 0. That is,

$$0(x) = 0 \qquad \text{for all } x \text{ in } I.$$

Let $f$ and $g$ be two functions defined on an interval $I$ and let $\alpha$ be a scalar. The **sum** $f + g$ of the functions $f$ and $g$ is defined to be the function whose value at every point $x$ in $I$ is given by $f(x) + g(x)$; that is,

$$(f + g)(x) = f(x) + g(x) \qquad \text{for all } x \text{ in } I.$$

Similarly, the **scalar multiple** $\alpha f$ of a function $f$ is the function whose value at every point $x$ in $I$ is given by $\alpha f(x)$; that is,

$$(\alpha f)(x) = \alpha f(x) \qquad \text{for all } x \text{ in } I.$$

Notice that if we think of the values of a function as the "components of an infinite vector," then these definitions of equality, addition, and scalar multiplication are the natural infinite-dimensional generalizations of these corresponding concepts in $R^m$.

Let $C(I)$ denote the collection of all real-valued functions that are continuous on the interval $I$. We know from calculus that if $f$ and $g$ are continuous functions on $I$ and $\alpha$ is a scalar, then $f + g$ and $\alpha f$ are continuous functions on $I$. Thus $C(I)$ is closed under addition and scalar multiplication. We leave it as an exercise for you to verify that these operations satisfy the following eight conditions.

I. Conditions for addition
  (1) $f + g = g + f$
  (2) $f + (g + h) = (f + g) + h$
  (3) $f + 0 = f$
  (4) $f + (-1)f = f - f = 0$

II. Conditions for scalar multiplication
  (5) $\alpha(\beta f) = (\alpha\beta)f$
  (6) $1f = f$
  (7) $\alpha(f + g) = \alpha f + \alpha g$
  (8) $(\alpha + \beta)f = \alpha f + \beta f$

Motivated then by our work in Chapter 2, we call $C(I)$ together with the operations of addition and scalar multiplication a **function space**. And we define a **subspace** of $C(I)$ to be a collection of functions in $C(I)$ that contains the zero function and is closed under addition and scalar multiplication.

Observe that when $I_1$ and $I_2$ are different intervals, the function spaces $C(I_1)$ and $C(I_2)$ are different. (Why?) Also, when $I = (-\infty, \infty)$, we denote $C(I)$ by $C(R)$.

## EXAMPLE 1

The collection $C^n(I)$, $n \geq 1$, of all functions on an interval $I$ which have $n$ continuous derivatives is a function space. For any $n$, it is a subspace of $C(I)$. Notice that these spaces are not equal to $C(I)$. For example, if we consider the function $f(x) = |x|$ on the interval $I = (-1, 1)$, then $f$ is in $C(I)$, but since $f$ is not differentiable at $x = 0$, $f$ is not in $C^1(I)$. Thus $C^1(I)$ is a **proper** subspace of $C(I)$, that is, a subspace of $C(I)$ that is not equal to $C(I)$. It is also true that $C^2(I)$ is a proper subspace of $C^1(I)$, and so on.

### EXAMPLE 2

The collection $C^\infty(R)$ of all functions that are infinitely differentiable on $R$ is a subspace of $C(R)$. Many of the functions that you studied in calculus belong to this subspace: for instance, $f(x) = e^x$, $f(x) = \sin x$, and $f(x) = \cos x$.

### EXAMPLE 3

Recall that a *polynomial of degree n* is a function of the form

$$p(x) = a_n x^n + a_{n-1}x^{n-1} + \cdots + a_1 x + a_0 \qquad (a_n \neq 0).$$

The numbers $a_0, a_1, \ldots, a_n$ are called the *coefficients* of $p(x)$. The zero polynomial is not assigned a degree by this definition. We declare it to have degree 0 together with all the other constant polynomials. The sum of two polynomials is a polynomial and the product of a scalar and a polynomial is a polynomial. From calculus we know that polynomials are continuous functions. Thus the collection $P(I)$ of all polynomials defined on an interval $I$ is a subspace of $C(I)$.

### EXAMPLE 4

A *homogeneous linear differential equation* is an equation of the form

$$y^{(n)} + p_{n-1}(x)y^{(n-1)} + \cdots + p_1(x)y = 0,$$

where $p_1, \ldots, p_{n-1}$ are continuous functions and $y^{(k)}$ denotes the $k$th derivative of $y$. [The symbols $y'$, $y''$, and $y'''$ are usually used to denote $y^{(1)}$, $y^{(2)}$, and $y^{(3)}$.] The number $n$ is called the *order* of the equation. $y' + 2xy = 0$ and $y'' + 2x^2y' + y = 0$ are examples of homogeneous linear differential equations of order 1 and 2, respectively. By a *solution* to a differential equation we mean a function $y = y(x)$ which when substituted into the equation makes it an identity. For example, $y = \sin x$ in a solution to the second-order equation $y'' + y = 0$. So is $y = \cos x$. So are $y = \alpha \sin x$ and $y = \sin x + \cos x$. (Check this!) In fact, the set of all solutions to this differential equation is a subspace of $C(I)$ called the *solution space*. For the zero function is clearly a solution. Suppose that $y_1$ and $y_2$ are both solutions and $\alpha$ is a scalar. Then

$$\begin{aligned}(y_1 + y_2)'' + (y_1 + y_2) &= (y_1'' + y_1) + (y_2'' + y_2) = 0\\ (\alpha y_1)'' + \alpha y_1 &= \alpha(y_1'' + y_1) = \alpha 0 = 0.\end{aligned}$$

Thus $y_1 + y_2$ and $\alpha y_1$ are solutions, as asserted.

## PROBLEMS 4.1

*1. Determine whether the collection of all functions $f$ in $C(R)$ that satisfy the stated condition is a function space. If the collection fails to be a function space, list those conditions that are not satisfied.

(a) $f(4) = 0$
(b) $f(4) = 1$
(c) $f(4) = f(1)$

(d) $2f(1) + f(-1) = 0$
(e) $f$ is an *odd* function; that is, $f(-x) = -f(x)$ for all $x$.
(f) $f$ is an *even* function; that is, $f(x) = f(-x)$ for all $x$.
(g) $f$ is a nonnegative function; that is, $f(x) \geq 0$ for all $x$.

*2. Determine whether the collection of all polynomials in $C(R)$ that satisfy the stated condition is a function space. If the collection fails to be a function space, list those conditions that are not satisfied.
(a) All polynomials with integer coefficients.
(b) All polynomials with zero constant term.
(c) All polynomials with nonzero constant term.
(d) All polynomials with leading coefficient 1.
(e) All polynomials containing only odd powers of $x$.
(f) All polynomials containing only even powers of $x$.

3. In each part, determine whether the set of all functions $f$ in $C^2(R)$ that satisfy the stated condition is a function space.
(a) $f'(x) - f(x) = 0$ for all $x$ in $R$
(b) $f''(1) = 1$
(c) $f(x) = f(x)^2$ for all $x$ in $R$
(d) $f(0) + f(1) = f(2)$
(e) $\lim_{x \to \infty} f(x) = 0$
(f) $f(x) \geq 0$ for all $x$ in $R$
(g) $f(0) + f'(0) = 0$
(h) $(xf(x))' = 0$ for all $x$ in $R$
(i) $f''(x) = 0$ for all $x$ in $R$
(j) $f(0) = 0$ or $f(1) = 0$
(k) $f(0) = f(1)$ and $f''(2) = 0$

*4. Is the collection of all polynomials defined on $R$ with degree exactly 5, together with the zero polynomial, a subspace of $C(R)$?

5. Show that the set of all solutions to a homogeneous linear differential equation of order 3 is a subspace of $C^3(R)$.

6. Show that for any interval $I$, $P(I)$ is a subspace of $C(I)$.

7. Let $P_n(I)$ denote the collection of all polynomials of degree $\leq n$, where $n$ is a fixed nonnegative integer. Show that $P_n(I)$ is a subspace of $P(I)$ for any $n$.

*8. Let $I = [0, 1]$. Determine which of the following collections of functions are function spaces. If the collection fails to be a function space, list the conditions that are not satisfied.
(a) The collection of all functions $f$ in $C(I)$ such that $f(0) = f(1)$
(b) The collection of all functions $f$ in $C^2(I)$ such that $f''(1/2) = 0$
(c) The collection of all functions $f$ in $C^2(I)$ such that $f''(1/2) = 1$

*9. For each of the following differential equations, determine whether the set of its solutions is a function space. If it fails to be a function space, list those conditions that are not satisfied.
(a) $y'' + 2y' + y = 0$
(b) $y'' + xy' + 7y = 0$
(c) $y''' + xy'' + x^2y' + x^3y = 0$, $y(1) = 0$
(d) $y''' + xy'' + x^2y' + x^3y = 0$, $y(1) = 2$

# 4.2 Concepts of Linear Algebra in Function Spaces

Thus far we have seen that the function space $C(I)$ is defined in the same way that the vector space $R^m$ is defined. They are both collections of objects together with operations of addition and scalar multiplication that satisfy identities (1) through

(8). The only difference is that in one case the objects are $m$-tuples of real numbers and in the other case the objects are functions. Hence it is not surprising that the concepts of linear combination, spanning, linear independence, dimension, and basis in a function space are defined exactly the same way in which they are defined in $R^m$.

For example, if $f_1, \ldots, f_n$ and $f$ are functions in $C(I)$, then $f$ is a **linear combination** of $f_1, \ldots, f_n$ if there exist scalars $\alpha_1, \ldots, \alpha_n$ such that

$$f = \alpha_1 f_1 + \cdots + \alpha_n f_n. \tag{1}$$

We must be careful to understand exactly what equation (1) means. Two functions that are defined on the same interval are *equal* if they produce the same value at every point in the interval. Thus equation (1) means that

$$f(x) = \alpha_1 f_1(x) + \cdots + \alpha_n f_n(x) \qquad \text{for all } x \text{ in } I.$$

### EXAMPLE 1

Suppose that $f(x) = x^2 - 6$, $f_1(x) = (x - 1)^2$, $f_2(x) = x - 1$, and $f_3(x) = 1$ for all $x$. Then $f = f_1 + 2f_2 - 5f_3$; that is, $f(x) = f_1(x) + 2f_2(x) - 5f_3(x)$ for all $x$.

### EXAMPLE 2

Let $f(x) = 1$, $g(x) = \sin^2 x$, and $h(x) = \cos^2 x$ for all $x$. Then $f = g + h$.

As another example, consider the definition of linear independence in the function space setting. Let $f_1, \ldots, f_n$ be functions in $C(I)$. They are called **linearly independent** if the only solution of the equation

$$\alpha_1 f_1 + \cdots + \alpha_n f_n = 0 \qquad \text{is} \qquad \alpha_1 = \cdots = \alpha_n = 0.$$

The equation $\alpha_1 f_1 + \cdots + \alpha_n f_n = 0$ is a special case of equation (1). Thus the equation $\alpha_1 f_1 + \cdots + \alpha_n f_n = 0$ (the zero function) means that $\alpha_1 f_1(x) + \cdots + \alpha_n f_n(x) = 0$ (the number zero) for all $x$ in $I$.

The importance of the phrase "for all $x$ in $I$" cannot be overemphasized. The function $f(x) = x^2 - 1$ and the zero function have the same value when $x$ equals 1 or $-1$, but they are certainly not the same function. Two functions may have the same value at one or many points in the interval $I$ without being equal. To be equal, they must have the same value at *every* point of $I$.

### EXAMPLE 3

Consider the polynomials 1, $x$, and $x^2$ as functions in $C(R)$. We assert that these polynomials are independent. For suppose that

$$0 = \alpha_0 + \alpha_1 x + \alpha_2 x^2 \qquad \text{for all } x. \tag{2}$$

If we choose three distinct numbers in $I$, say $-1, 0$, and 1, and set $x = -1, 0$, and 1, we obtain three equations in three unknowns.

$$\begin{aligned} \alpha_0 - \alpha_1 + \alpha_2 &= 0 \\ \alpha_0 \qquad\qquad &= 0 \\ \alpha_0 + \alpha_1 + \alpha_2 &= 0 \end{aligned} \qquad \text{or} \qquad \begin{bmatrix} 1 & -1 & 1 \\ 1 & 0 & 0 \\ 1 & 1 & 1 \end{bmatrix} \begin{bmatrix} \alpha_0 \\ \alpha_1 \\ \alpha_2 \end{bmatrix} = \begin{bmatrix} 0 \\ 0 \\ 0 \end{bmatrix}$$

The only solution of this system is $\alpha_0 = \alpha_1 = \alpha_2 = 0$. Thus the functions 1, $x$ and $x^2$ are independent.

There is another useful technique for showing that a collection of functions is independent. Setting $x = 0$ in (2), we obtain $\alpha_0 = 0$. Differentiating both sides of (2) we obtain $0 = \alpha_1 + 2\alpha_2 x$ for all $x$ in $I$, and setting $x = 0$ in this equation gives $\alpha_1 = 0$. Differentiating again, we obtain $2\alpha_2 = 0$ or $\alpha_2 = 0$. Thus $\alpha_0 = \alpha_1 = \alpha_2 = 0$.

This technique gives us another occasion to stress the importance of understanding equation (2). If two functions are equal, they must have the same derivative. The equation

$$0 = \alpha_0 + \alpha_1 x + \alpha_2 x^2 \qquad \text{for all } x \text{ in } I$$

states that the function $\alpha_0 + \alpha_1 x + \alpha_2 x^2$ and the zero function are equal *as functions*. Thus it is legitimate to differentiate both sides to obtain $0 = \alpha_1 + 2\alpha_2 x$ for all $x$ in $I$. Contrast this with the following situation. We cannot differentiate both sides of the equation $x^2 - 1 = 0$. For if we do, we obtain $2x = 0$ and hence $x = 0$, an absurdity. The reason we cannot differentiate both sides of the equation is that it is not an identity in $x$; it does not assert that the function $x^2 - 1$ and the zero function are *equal functions.*

It should be clear that the same reasoning could be used to show that for any $n$ the polynomials $1, x, \ldots, x^n$ are independent in $C(R)$ (see Problem 1).

## EXAMPLE 4

The functions $e^x$ and $e^{2x}$ are independent in $C(R)$. For suppose that

$$\alpha_1 e^x + \alpha_2 e^{2x} = 0 \qquad \text{for all } x. \tag{3}$$

Setting $x = 0$, we obtain $\alpha_1 + \alpha_2 = 0$. Differentiating both sides of (3), we obtain $\alpha_1 e^x + 2\alpha_2 e^{2x} = 0$ for all $x$ and setting $x = 0$ in this equation gives $\alpha_1 + 2\alpha_2 = 0$. Thus

$$\begin{aligned} \alpha_1 + \alpha_2 &= 0 \\ \alpha_1 + 2\alpha_2 &= 0 \end{aligned} \qquad \text{or} \qquad \begin{bmatrix} 1 & 1 \\ 1 & 2 \end{bmatrix} \begin{bmatrix} \alpha_1 \\ \alpha_2 \end{bmatrix} = \begin{bmatrix} 0 \\ 0 \end{bmatrix}.$$

The only solution of this system is $\alpha_1 = \alpha_2 = 0$.

Example 4 suggests that if $a_1, a_2, \ldots, a_n$ are distinct numbers, then $e^{a_1 x}$, $e^{a_2 x}, \ldots, e^{a_n x}$ are independent in $C(R)$. This is true (see Problem 9).

## EXAMPLE 5

The method for showing the independence of functions based on differentiation works even if 0 is not in the interval. To prove that the functions 1, $x$, and $x^2$ are independent on $I = [1, 4]$, we proceed as follows. Suppose that

$$\alpha_0 + \alpha_1 x + \alpha_2 x^2 = 0 \qquad \text{for all } x \text{ in } I. \tag{4}$$

Setting $x = 3$ we obtain the equation $\alpha_0 + 3\alpha_1 + 9\alpha_2 = 0$. Again, differentiating both sides of (4) gives $\alpha_1 + 2\alpha_2 x = 0$ for all $x$ in $I$ and setting $x = 3$ we obtain $\alpha_1 + 6\alpha_2 = 0$. A final differentiation gives $2\alpha_2 = 0$. Thus we have the system of equations

$$\begin{aligned} \alpha_0 + 3\alpha_1 + 9\alpha_2 &= 0 \\ \alpha_1 + 6\alpha_2 &= 0 \\ 2\alpha_2 &= 0 \end{aligned} \qquad \text{or} \qquad \begin{bmatrix} 1 & 3 & 9 \\ 0 & 1 & 6 \\ 0 & 0 & 2 \end{bmatrix} \begin{bmatrix} \alpha_0 \\ \alpha_1 \\ \alpha_2 \end{bmatrix} = \begin{bmatrix} 0 \\ 0 \\ 0 \end{bmatrix}.$$

Hence $\alpha_0 = \alpha_1 = \alpha_2 = 0$.

***Definition*** Let $V$ be a subspace of $C(I)$. A collection $f_1, f_2, \ldots, f_n$ of functions in $V$ **spans** $V$ if every function in $V$ is a linear combination of these functions. If, in addition, the functions $f_1, f_2, \ldots, f_n$ are independent, then we say that they form a **basis** for $V$. When $V$ has a basis consisting of $n$ functions we say that $V$ is $n$-**dimensional**.

## EXAMPLE 6

The polynomials 1, $x$, and $x^2$ clearly span the space $P_2(R)$ of polynomials of degree $\leq 2$. In Example 3 we showed that they are independent. Hence they form a basis and $\dim P_2(R) = 3$.

## EXAMPLE 7

Consider the first-order differential equation $y' - \lambda y = 0$. A solution to this equation is a function $y = y(x)$ whose derivative is $\lambda$ times $y$. The function $y = Ce^{\lambda x}$ has this property for any constant $C$. Indeed, if

$$y = Ce^{\lambda x},$$

then

$$y' = \lambda C e^{\lambda x} = \lambda y.$$

These exponential functions are the only nonzero functions with this property, and hence they are the only solutions to the differential equation. To prove this, suppose that $y$ is a solution of $y' - \lambda y = 0$. We assert that $ye^{-\lambda x}$ is a constant. Let

$$z = ye^{-\lambda x}.$$

Then

$$\begin{aligned} z' &= y'e^{-\lambda x} - \lambda y e^{-\lambda x} \\ &= \lambda y e^{-\lambda x} - \lambda y e^{-\lambda x} \\ &= 0. \end{aligned}$$

Since the derivative of $z$ is the zero function, $z$ is a constant function. To determine the constant, we observe that $z(0) = y(0)$. Denoting $y(0)$ by $C$, it follows that $z(x) = C$ for all $x$ and hence that $y = Ce^{\lambda x}$. Thus the solution space is spanned by the function $e^{\lambda x}$; it is one-dimensional.

As Example 6 suggests, for any $n$ the polynomials $1, x, \ldots, x^n$ for a basis for $P_n(R)$. Hence $\dim P_n(R) = n + 1$ (see Problem 21).

Now consider the space $P(R)$ of all polynomials. For every $n$, $P_n(R)$ is a subspace of $P(R)$ of dimension $n + 1$. This implies that the dimension of $P(R)$ (if it is defined) would have to be greater than or equal to $n + 1$ for every $n$. Clearly, no finite number can satisfy this condition. Hence we say that the dimension of $P(R)$ is infinite.

We can also prove that the dimension of $P(R)$ is not finite by showing that no finite set of polynomials can possibly span the collection of all polynomials. To prove this, we note that any finite set of polynomials must contain a polynomial of largest degree, say $k$. It follows that no polynomial of degree larger than $k$ can be a linear combination of these polynomials. Hence $P(R)$ cannot have a finite basis, that is, a basis with a finite number of elements.

***Definition*** If a subspace $V$ of $C(I)$ has a basis consisting of a finite number of functions, we say that $V$ is **finite-dimensional**. If a subspace $V$ does not have a basis consisting of a finite number of functions, we say that $V$ is **infinite-dimensional**.

## PROBLEMS 4.2

**1.** Show that $1$, $x$, $x^2$, and $x^3$ form a basis for $P_3(R)$. If you are confident that you could also show that $1, x, x^2, x^3$, and $x^4$ form a basis for $P_4(R)$, stop. Otherwise, do it.

**2.** Show that $e^x$, $e^{2x}$, and $e^{3x}$ are independent in $C(R)$.

***3.** Determine which of the following collections of polynomials are independent in $P(R)$.

(a) $1, 1 - t, 1 + t$

(b) $t, t^2 + 1, t^2 - 1$

(c) $1, 2t, 1 + t, 3t^2, 4t^3$

(d) $1 + t, 1 + t + t^2, 1 + t + t^2 + t^3$

***4.** Determine which of the following collections of functions are independent in $C(R)$.

(a) $\sin x, \cos x, \sin 2x$

(b) $\sin x, \cos x, x$

(c) $e^x, \sin x$

(d) $\sin x, \sin x + \cos x, \sin x - \cos x$

**5.** For each of the following triples of functions in $C(R)$, either show that they're independent, or show that they are dependent.

(a) $f(x) = 2x + 3$, $g(x) = 5x - 1$, $h(x) = 7x - 15$

(b) $f(x) = \sin x$, $g(x) = \sin 2x$, $h(x) = \sin 3x$
(c) $f(x) = x^2 + 1$, $g(x) = x^2 - 1$, $h(x) = x - 1$

*6. In each part, determine whether the given functions in $C(I)$ are independent.
(a) $x^2, e^x, \log x$; $I = (0, \infty)$
(b) $x^2, e^x, \sqrt{x}$; $I = (0, \infty)$
(c) $x, e^x, \cos x$; $I = R$
(d) $e^x, x \sin x, \cos x$; $I = R$
(e) $x^3, \sqrt{x}, e^x$; $I = (0, \infty)$
(f) $x, x^{3/2}, e^x$; $I = (0, \infty)$

7. Show that $1$, $1 + t$, $1 + t^2$, and $1 + t^3$ form a basis for $P_3(R)$.

8. Find a basis for $P_0(R)$.

9. Show that if $a_1$, $a_2$, $a_3$ are distinct numbers, then $e^{a_1 x}$, $e^{a_2 x}$, and $e^{a_3 x}$ are independent in $C(R)$.

10. For each of the following, consider the subspace $V$ of $C(R)$ spanned by the given collection of functions. Find a basis for $V$ and the dimension of $V$.
*(a) $1, 1 - t, 1 + t$
(b) $t, t^2 + 1, t^2 - 1$
*(c) $\sin x, \cos x, \sin 2x$
(d) $\sin x, \cos x, x$
*(e) $e^x, \sin x$
(f) $1, e^x, xe^x$

11. In each part, $V$ is the collection of polynomials in $P_3(R)$ that satisfy the given condition.
(i) Show that $V$ is a function space.
(ii) Find a basis for $V$ and the dimension of $V$.
*(a) $\int_{-1}^{1} p(x)\,dx = 0$
(b) $p''(x) = 0$ for all $x$
*(c) $xp''(x) = 0$ for all $x$
(d) $p'(x) + p''(x) = 0$
*(e) $p(x) = a + bx + cx^2 + dx^3$, where $a - b = 0$
(f) $p(x) = a + bx + cx^2 + dx^3$, where $a + c = 0$ and $a + d = 0$.

12. In each part, $V$ is the collection of polynomials in $P_4(R)$ that satisfy the stated condition.
(i) Show that $V$ is a function space.
(ii) Find a basis for $V$ and the dimension of $V$.
*(a) $p(2) = 0$ and $p(5) = 0$ (*Hint:* If $p(a) = 0$, then $x - a$ is a factor of $p(x)$.)
(b) $p(0) = p(1) = p(2) = p(3) = 0$
*(c) $p''(x) = 0$ for all $x$
(d) $p(x) = a + bx + cx^2 + dx^3 + ex^4$, where $a + b + c + d + e = 0$, $b + d + e = 0$, $d - e = 0$
*(e) $\int_0^1 xp(x)\,dx = 0$

13. (a) Show that $1 + x$, $1 + 2x$, $1 + x^2$ form a basis for $P_2(R)$.
(b) Express $1 + 3x^2$ as a linear combination of these functions.

14. Let $V$ be the space of polynomials of degree $\leq 2$.
(a) Show that $(x - 1)(x - 2)/2$, $x(x - 2)/(-1)$, $x(x - 1)/2$ form a basis for $V$.
(b) If $p$ is any polynomial in $V$, find the coordinates of $p$ with respect to this basis.

15. Suppose that $a$ and $b$ are two distinct real numbers (i.e., $a \neq b$). Show that $(x - a)^2$, $(x - a)(x - b)$, and $(x - b)^2$ are linearly independent.

16. Let $f(x) = ae^{bx} + ce^{dx}$, where $a, b, c$, and $d$ are real numbers.
(a) Show that $f$, $f'$, and $f''$ are linearly dependent.
(b) Express $f''$ as a linear combination of $f$ and $f'$.

*17. Given that the solution space to the differential equation $y'' - y = 0$ has dimension 2, show that every solution to $y'' - y = 0$ can be written uniquely in the form $c_1 e^x + c_2 e^{-x}$.

18. Repeat Problem 17 for each of the following. In each case assume that the solution space has dimension 2.
(a) $y'' - y' - 6y = 0$; $c_1 e^{3x} + c_2 e^{-2x}$
(b) $y'' - 4y = 0$; $c_1 e^{2x} + c_2 e^{-2x}$
(c) $y'' + 2y' + y = 0$; $c_1 e^{-x} + c_2 xe^{-x}$
(d) $y'' - 6y' + 9y = 0$; $c_1 e^{3x} + c_2 xe^{3x}$

**19.** Find values of $\alpha, \beta, \gamma$ so that the formula

$$\int_{a-h}^{a+h} f(x)\,dx = \alpha f(a-h) + \beta f(a) + \gamma f(a+h)$$

is exact for polynomials of degree $\leq 2$. This formula is Simpson's rule. (*Hint:* It is sufficient to require this formula to be exact on a basis for $P_2$.)

**20.** Let $f$ be a polynomial of degree 4.
(a) Show that $f, f', f'', f''', f^{(\text{iv})}$ form a basis for $P_4(R)$.
(b) State and prove a generalization.

**21.** Explain why for any $n$, $1, x, \ldots, x^n$ form a basis for $P_n(R)$.

**22.** Let $f_1, f_2, \ldots, f_n$ be in $C(I)$. Suppose that $x_1, x_2, \ldots, x_n$ are $n$ points in $I$ and that the matrix

$$A = \begin{bmatrix} f_1(x_1) & f_2(x_1) & \cdots & f_n(x_1) \\ f_1(x_2) & f_2(x_2) & \cdots & f_n(x_2) \\ \vdots & \vdots & & \vdots \\ f_1(x_n) & f_2(x_n) & \cdots & f_n(x_n) \end{bmatrix}$$

has rank $n$. Show that the functions $f_1, f_2, \ldots, f_n$ are independent in $C(I)$. (*Hint:* Show that if the functions are dependent, then rank $A < n$.)

**23.** Let $I = R$. Let $f_1(x) = \sin x$ and $f_2(x) = \cos x$. Then $f_1$ and $f_2$ are independent.
(a) What is the rank of

$$A = \begin{bmatrix} \sin 0 & \cos 0 \\ \sin \pi & \cos \pi \end{bmatrix}?$$

(b) What does this say about Problem 22?

**24.** Let $f_1, f_2, \ldots, f_n$ be in $C^{(n-1)}(I)$ and let $x$ be in $I$. The matrix

$$\begin{bmatrix} f_1(x) & f_2(x) & \cdots & f_n(x) \\ f'_1(x) & f'_2(x) & \cdots & f'_n(x) \\ \vdots & \vdots & & \vdots \\ f_1^{(n-1)}(x) & f_2^{(n-1)}(x) & \cdots & f_n^{(n-1)}(x) \end{bmatrix}$$

is called the **Wronskian matrix** and is denoted by $W(x)$.
(a) Show that if for some $x$ in $I$, $W(x)$ is nonsingular, then the functions $f_1, f_2, \ldots, f_n$ are independent.
(b) Show that the converse is not true. (*Hint:* Let $I = [-1, 1]$, $f_1(x) = x^3$, $f_2(x) = |x^3|$.)

**25.** Let $x_1$, $x_2$, $x_3$ be three distinct points and define

$$l_1(x) = \frac{(x - x_2)(x - x_3)}{(x_1 - x_2)(x_1 - x_3)}$$
$$l_2(x) = \frac{(x - x_1)(x - x_3)}{(x_2 - x_1)(x_2 - x_3)}$$
$$l_3(x) = \frac{(x - x_1)(x - x_2)}{(x_3 - x_1)(x_3 - x_2)}$$

Clearly, $l_1$, $l_2$, $l_3$ are polynomials of degree 2. They are called **Lagrange polynomials**.
(a) Show that $l_i(x_j) = 1$ if $i = j$ and $l_i(x_j) = 0$ if $i \neq j$.
(b) Show that $l_1$, $l_2$, $l_3$ are independent and hence form a basis for $P_2(R)$.
(c) Given $p$ in $P_2(R)$, show that the coordinate vector of $p$ with respect to this basis is $(p(x_1), p(x_2), p(x_3))$. Thus $p(x) = p(x_1)l_1(x) + p(x_2)l_2(x) + p(x_3)l_3(x)$ for all $x$. The formula is called **Lagrange's interpolation formula**.
(d) Generalize parts (a), (b), and (c).
(e) In Problem 4(c) of Section 3.1 you were asked to show that there is one and only one polynomial of degree $m - 1$ which passes through $m$ distinct points $(x_1, y_1)$, $(x_2, y_2), \ldots, (x_m, y_m)$. What are the coordinates of this polynomial with respect to the basis $l_1, l_2, \ldots, l_m$ for $P_{m-1}(R)$?

# 4.3 Inner Products in Function Spaces

In this section we extend the concept of an inner product to function spaces. The properties of the inner product in $R^m$ which were necessary to define length and angle and to prove the Cauchy–Schwarz inequality are the following:

(1) $\langle \mathbf{x}, \mathbf{y} \rangle = \langle \mathbf{y}, \mathbf{x} \rangle$.

(2) $\langle \mathbf{x}, \mathbf{y} + \mathbf{z} \rangle = \langle \mathbf{x}, \mathbf{y} \rangle + \langle \mathbf{x}, \mathbf{z} \rangle$.

(3) $\langle \alpha\mathbf{x}, \mathbf{y} \rangle = \alpha\langle \mathbf{x}, \mathbf{y} \rangle$.

(4) $\langle \mathbf{x}, \mathbf{x} \rangle \geq 0$, and $\langle \mathbf{x}, \mathbf{x} \rangle = 0$ if and only if $\mathbf{x} = \mathbf{0}$.

All of these properties are consequences of the definition of the inner product in $R^m$ (see Theorem 1 of Section 3.2). Hence when attempting to define an inner product in function spaces we should demand that these properties be satisfied. The reason is that if these properties are satisfied, all the results of Chapter 3 will carry over to this new setting with no change.

***Definition*** Let $I$ be a closed and bounded interval, say $I = [a, b]$, and let $f$ and $g$ be two functions in $C(I)$. The **inner product** of $f$ and $g$, denoted by $\langle f, g \rangle$, is defined by the equation

$$\langle f, g \rangle = \int_a^b f(x)g(x)\,dx. \tag{1}$$

There is a way to view this definition which shows that it is closely related to the inner product in $R^m$. In $R^m$ the inner product $\langle \mathbf{x}, \mathbf{y} \rangle$ is defined to be the sum of products of corresponding components. If we think of the values of a function as the "components of a vector" and of the integral as a "sum," then

$$\langle f, g \rangle = \int_a^b f(x)g(x)\,dx$$

can also be thought of as a sum of products of corresponding components.

Properties (1)–(4) above follow from elementary properties of the integral (the student should verify this). Also, notice that we have not defined an inner product on $C(I)$ when $I$ is an interval of infinite length or when $I$ is an open interval $(a, b)$. Why?

The remainder of this section is devoted to defining the norm of a function and the relation of orthogonality in function spaces. In the next section we discuss projections, least-squares approximations, and the Gram–Schmidt process in the function space setting.

## Norm and Orthogonality

Let $I = [a, b]$. The **norm** of a function $f$ in $C(I)$ is

$$\|f\| = \langle f, f \rangle^{1/2} = \left[\int_a^b f(x)^2\,dx\right]^{1/2}.$$

The **distance** between two functions $f$ and $g$ in $C(I)$ is

$$\|f - g\| = \left[\int_a^b (f(x) - g(x))^2\,dx\right]^{1/2}.$$

Two functions $f$ and $g$ in $C(I)$ are **orthogonal** if

$$\langle f, g\rangle = \int_a^b f(x)g(x)\,dx = 0.$$

The reader should observe that if we think of the values of functions as their components and think of integrals as sums, then the definitions of length, distance, and orthogonality are similar to those given in $R^m$. Also observe that the norm of a function is not its arc length, and that orthogonality of functions has nothing whatsoever to do with their graphs intersecting orthogonally.

### EXAMPLE 1

Let $I = [0, 2\pi]$. Then

$$\|\sin x\|^2 = \int_0^{2\pi} \sin^2 x\,dx = \frac{x}{2} - \frac{\sin 2x}{4}\bigg|_0^{2\pi} = \pi.$$

The distance between the two functions $\sin x$ and $\cos x$ can be computed from the relation

$$\begin{aligned}\|\sin x - \cos x\|^2 &= \int_0^{2\pi} (\sin x - \cos x)^2\,dx \\ &= \int_0^{2\pi} (\sin^2 x - 2\sin x\cos x + \cos^2 x)\,dx \\ &= \int_0^{2\pi} (1 - 2\sin x\cos x)\,dx \\ &= (x - \sin^2 x)\bigg|_0^{2\pi} \\ &= 2\pi.\end{aligned}$$

The functions $\sin x$ and $\cos x$ are orthogonal since

$$\langle \sin x, \cos x\rangle = \int_0^{2\pi} \sin x\cos x\,dx = \frac{\sin^2 x}{2}\bigg|_0^{2\pi} = 0.$$

### EXAMPLE 2

If $I = [0, 2\pi]$, then the $2n + 1$ functions

$$1, \sin x, \cos x, \ldots, \sin nx, \cos nx$$

are orthogonal in $C(I)$. In fact,

$$\int_0^{2\pi} \sin mx \sin nx \, dx = \begin{cases} 0 & \text{if } m \neq n \\ \pi & \text{if } m = n \end{cases}$$

$$\int_0^{2\pi} \cos mx \cos nx \, dx = \begin{cases} 0 & \text{if } m \neq n \\ \pi & \text{if } m = n \end{cases}$$

$$\int_0^{2\pi} \sin mx \cos nx \, dx = 0 \qquad \text{for all positive integers } m \text{ and } n.$$

The norm of the constant function 1 is $\sqrt{2\pi}$ and the norm of each of the other functions is $\sqrt{\pi}$. We ask the reader to verify these facts in Problem 11.

## PROBLEMS 4.3

**1.** Let $I = [0, 1]$. Compute the norm of each of the following functions in $C(I)$.
*(a) 1 (b) $x$
(c) $x^2 + 2$ *(d) $\dfrac{1}{\sqrt{x^2 + 1}}$

**2.** Let $I = [-1, 1]$. Find the norm of each of the following functions.
*(a) $e^x$ (b) $\sin \pi x$ (c) $xe^{-x}$

**3.** Let $I = [0, 1]$. Find each of the following.
*(a) $\langle 1, x^2 \rangle$ (b) $\langle x, \sqrt{x} \rangle$
*(c) $\left\langle x, \dfrac{1}{1 + x^2} \right\rangle$

**4.** Let $I = [0, 1]$. Find two nonzero polynomials of degree $\leq 2$ that are orthogonal to:
*(a) $x$ (b) $x + 1$
*(c) $x^2$ (d) $x^3$

**5.** Compute the norm of $\sin x$ and the distance between $\sin x$ and $\cos x$ if $I = [0, \pi/2]$. Also, test the orthogonality of $\sin x$ and $\cos x$.

**6.** Interpret the norm of a function and the distance between two functions in terms of area. Also, show that if $I = [-1, 1]$, the functions $x$ and $x^2$ are orthogonal in $C(I)$. Then draw the graphs of these functions and observe that the graphs do not intersect orthogonally.

***7.** Show that the polynomials 1 and $x - \frac{1}{2}$ are orthogonal in $C(I)$, $I = [0, 1]$. Then find a nonzero polynomial of degree $\leq 2$ that is orthogonal to both of these polynomials.

**8.** Let $I = [0, 1]$. Show that the only polynomial of degree $\leq 2$ that is orthogonal to every polynomial of degree $\leq 2$ is the zero polynomial. Can you generalize this result?

**9.** Let $I = [-1, 1]$. Find a basis for the orthogonal complement of the subspace of $P_3(I)$ that is spanned by:
*(a) $x$ (b) $x^2$
*(c) 1 and $x$ (d) $x$ and $x^2$
*(e) $1 + x$ and $x^2$ (f) $1, x^2$, and $x^3$

***10.** For each of the subspaces and their orthogonal complements given in Problem 9, express the function $x^3$ as a sum of a vector in the subspace and a vector in the orthogonal complement. Do the same for $1 + x^2$ and $1 + x^3$.

**11.** Use the identities

$$\sin a \sin b = \frac{1}{2}[\cos(a-b) - \cos(a+b)]$$

$$\cos a \cos b = \frac{1}{2}[\cos(a-b) + \cos(a+b)]$$

$$\sin a \cos b = \frac{1}{2}[\sin(a-b) + \sin(a+b)]$$

$$\sin^2 mx = \frac{1}{2}(1 - \cos 2mx)$$

$$\cos^2 mx = \frac{1}{2}(1 + \cos 2mx)$$

to prove the assertions made in Example 2.

**12.** Show that the results of Example 2 remain valid if we replace the interval $I = [0, 2\pi]$ by the interval $I = [-\pi, \pi]$. Will they remain valid for any interval of length $2\pi$?

**13.** A function $f$ is orthogonal to a subspace $V$ of $C(I)$ if $f$ is orthogonal to every function in $V$.
(a) Let $I = [-1, 1]$. Show that the function $x$ is orthogonal to the subspace of $P(I)$ spanned by 1, $x^2$, $x^4$.
(b) Show that a function $f$ is orthogonal to a finite-dimensional subspace $V$ of $C(I)$ if and only if $f$ is orthogonal to every function in a basis for $V$.

**14.** (a) Show that if $f_1$, $f_2$, and $f_3$ are nonzero orthogonal functions in $C(I)$, $I = [a, b]$, then they are linearly independent.
(b) If $f_1, \ldots, f_n$ are nonzero orthogonal functions, are they independent? Why?

**15.** If $V$ is a subspace of $C(I)$, $V^\perp$ is defined to be the collection of all functions orthogonal to $V$.
(a) Show that $V^\perp$ is a subspace.
(b) Show that $V \subseteq V^{\perp\perp}$.

**16.** Let $I = [-1, 1]$ and let $V$ be the subspace of $P_n(I)$ consisting of all even polynomials. Show that $V^\perp$ consists of all odd polynomials.

**17.** It is important to observe that the definition of the inner product (1) depends on the interval $I$. Let $I_1 = [0, 1]$ and $I_2 = [0, 2]$. Then both $x$ and $x^2$ belong to $C(I_1)$ and $C(I_2)$. However, show that their inner product when they are viewed as elements of the function space $C(I_1)$ is different from their inner product when they are viewed as elements of the function space $C(I_2)$. Hence we should expect that those properties which are defined in terms of the inner product will depend on the interval $I$.

# 4.4
# Projections in Function Spaces

Having extended the notion of an inner product to function spaces, we now extend the other concepts and results of Chapter 3 to function spaces.

## Projections

Let $V$ be a subspace of $C(I)$ and let $f$ be a function in $C(I)$. Our work in Chapter 3 suggests that we define the projection of $f$ onto $V$ to be a function $h$ such that

1. $h$ is in $V$,
2. $f - h$ is orthogonal to $V$, that is, $f - h$ is in $V^\perp$.

It turns out that if $V$ is not finite-dimensional, such a function may not exist. We ask you to take our word for that. However, if $V$ is finite-dimensional, then such a function $h$ does exist. In fact, if we choose an orthogonal basis $g_1, g_2, \ldots, g_k$ for $V$ (see

Gram–Schmidt below), then the function

$$h = \frac{\langle f, g_1 \rangle}{\langle g_1, g_1 \rangle} g_1 + \frac{\langle f, g_2 \rangle}{\langle g_2, g_2 \rangle} g_2 + \cdots + \frac{\langle f, g_k \rangle}{\langle g_k, g_k \rangle} g_k$$

satisfies conditions 1 and 2. Certainly $h$ is in $V$. Since $g_1, g_2, \ldots, g_k$ are orthogonal,

$$\langle h, g_i \rangle = \langle f, g_i \rangle, \qquad i = 1, 2, \ldots, k,$$

and hence

$$\langle f - h, g_i \rangle = \langle f, g_i \rangle - \langle h, g_i \rangle = 0, \qquad i = 1, 2, \ldots, k.$$

Since $f - h$ is orthogonal to each element of a basis for $V$, it is orthogonal to $V$.

***Definition*** Let $V$ be a finite-dimensional subspace of $C(I)$ and let $g_1, g_2, \ldots, g_k$ be an orthogonal basis for $V$. If $f$ is in $C(I)$, the **projection of $f$ onto $V$** is

$$h = \frac{\langle f, g_1 \rangle}{\langle g_1, g_1 \rangle} g_1 + \frac{\langle f, g_2 \rangle}{\langle g_2, g_2 \rangle} g_2 + \cdots + \frac{\langle f, g_k \rangle}{\langle g_k, g_k \rangle} g_k. \tag{1}$$

The **projection of $f$ orthogonal to $V$** is $f - h$.

## EXAMPLE 1

The projection of $x$ onto $\cos x$ in $C(I)$, $I = [0, \pi]$, is the function

$$h(x) = \frac{\langle x, \cos x \rangle}{\langle \cos x, \cos x \rangle} \cos x.$$

Since

$$\langle x, \cos x \rangle = \int_0^\pi x \cos x \, dx = x \sin x + \cos x \Big|_0^\pi = -2,$$

$$\langle \cos x, \cos x \rangle = \int_0^\pi \cos^2 x \, dx = \frac{x}{2} + \frac{\sin 2x}{4} \Big|_0^\pi = \frac{\pi}{2},$$

we obtain

$$h(x) = \frac{-4}{\pi} \cos x.$$

## EXAMPLE 2

Let $I = [0, 2\pi]$. By equation (1), the projection of $x$ onto the subspace $V$ of $C(I)$ spanned by the three orthogonal functions 1, $\sin x$, $\cos x$ is

$$h(x) = \frac{\langle x, 1 \rangle}{\langle 1, 1 \rangle} 1 + \frac{\langle x, \sin x \rangle}{\langle \sin x, \sin x \rangle} \sin x + \frac{\langle x, \cos x \rangle}{\langle \cos x, \cos x \rangle} \cos x.$$

Since

$$\langle 1, 1\rangle = 2\pi,$$

$$\langle x, 1\rangle = \int_0^{2\pi} x\,dx = 2\pi^2,$$

$$\langle x, \sin x\rangle = \int_0^{2\pi} x \sin x\,dx = -x\cos x + \sin x\Big|_0^{2\pi} = -2\pi,$$

$$\langle x, \cos x\rangle = \int_0^{2\pi} x\cos x\,dx = x\sin x + \cos x\Big|_0^{2\pi} = 0,$$

$$\langle \sin x, \sin x\rangle = \langle \cos x, \cos x\rangle = \pi,$$

we have

$$h(x) = \pi - 2\sin x.$$

Now let us find the projection of $x$ onto the subspace $V_1$ of $C(I)$, $I = [0, 2\pi]$, spanned by the four orthogonal functions 1, $\sin x$, $\cos x$, and $\sin 2x$. The projection is given by

$$h_1(x) = \frac{\langle x, 1\rangle}{\langle 1, 1\rangle}1 + \frac{\langle x, \sin x\rangle}{\langle \sin x, \sin x\rangle}\sin x + \frac{\langle x, \cos x\rangle}{\langle \cos x, \cos x\rangle}\cos x + \frac{\langle x, \sin 2x\rangle}{\langle \sin 2x, \sin 2x\rangle}\sin 2x,$$

and since the first three terms have already been computed, we need only compute the last term.

$$\langle x, \sin 2x\rangle = \int_0^{2\pi} x\sin 2x\,dx = -\frac{x\cos 2x}{2} + \frac{\sin 2x}{4}\Big|_0^{2\pi} = -\pi$$

$$\langle \sin 2x, \sin 2x\rangle = \int_0^{2\pi} \sin^2 2x\,dx = \pi$$

Thus

$$h_1(x) = \pi - 2\sin x - \sin 2x.$$

## The Gram–Schmidt Process

If $V$ is a finite-dimensional subspace of $C(I)$ and $f_1, \ldots, f_k$ is a basis for $V$, we can construct an orthogonal basis $g_1, \ldots, g_k$ for $V$ using the **Gram–Schmidt** process. The procedure is exactly the same as it was in $R^m$. We begin by letting $g_1 = f_1$. Then let $g_2$

be the projection of $f_2$ onto the orthogonal complement of the subspace spanned by $g_1$.

$$g_2 = f_2 - \frac{\langle f_2, g_1 \rangle}{\langle g_1, g_1 \rangle} g_1$$

In general, having constructed $g_1, \ldots, g_i$ $(1 \le i < k)$, let $g_{i+1}$ be the projection of $f_{i+1}$ onto the orthogonal complement of the subspace spanned by the orthogonal vectors $g_1, \ldots, g_i$. Thus

$$g_{i+1} = f_{i+1} - \frac{\langle f_{i+1}, g_1 \rangle}{\langle g_1, g_1 \rangle} g_1 - \cdots - \frac{\langle f_{i+1}, g_i \rangle}{\langle g_i, g_i \rangle} g_i.$$

### EXAMPLE 3

Let $I = [-1, 1]$. The function space $P_3(I)$ has $1, x, x^2$, and $x^3$ as a basis. Let us apply the Gram–Schmidt orthogonalization process to find an orthogonal basis. Let $g_1(x) = 1$. Then since $\langle x, 1 \rangle = 0$ and $\langle 1, 1 \rangle = 2$, we have $g_2(x) = x - (0/2) \cdot 1 = x$. Similarly, since $\langle x^2, 1 \rangle = \frac{2}{3}$, $\langle x^2, x \rangle = 0$, and $\langle x, x \rangle = \frac{2}{3}$, we obtain

$$\begin{aligned} g_3(x) &= x^2 - \frac{\langle x^2, 1 \rangle}{\langle 1, 1 \rangle} 1 - \frac{\langle x^2, x \rangle}{\langle x, x \rangle} x \\ &= x^2 - \frac{\frac{2}{3}}{2} - \frac{0}{\frac{2}{3}} x \\ &= x^2 - \tfrac{1}{3}. \end{aligned}$$

Since $\langle x^3, 1 \rangle = 0$, $\langle x^3, x \rangle = \frac{2}{5}$, $\langle x^3, x^2 - \frac{1}{3} \rangle = 0$, and $\langle x^2 - \frac{1}{3}, x^2 - \frac{1}{3} \rangle = \frac{8}{45}$, we obtain

$$\begin{aligned} g_4(x) &= x^3 - \frac{\langle x^3, 1 \rangle}{\langle 1, 1 \rangle} 1 - \frac{\langle x^3, x \rangle}{\langle x, x \rangle} x - \frac{\langle x^3, x^2 - \frac{1}{3} \rangle}{\langle x^2 - \frac{1}{3}, x^2 - \frac{1}{3} \rangle} (x^2 - \tfrac{1}{3}) \\ &= x^3 - \tfrac{3}{5} x. \end{aligned}$$

Thus the polynomials $1$, $x$, $x^2 - \frac{1}{3}$, and $x^3 - \frac{3}{5}x$ form an orthogonal basis for the space of polynomials of degree $\le 3$ on $[-1, 1]$. If these polynomials are adjusted so that $g_1(1) = g_2(1) = g_3(1) = g_4(1) = 1$, then they are called the *Legendre polynomials*. The first four Legendre polynomials are $1$, $x$, $\frac{3}{2}(x^2 - \frac{1}{3})$, $\frac{5}{2}(x^3 - \frac{3}{5}x)$.

## Least-Squares Approximations

In Chapter 3 we saw that the projection of a vector **b** onto a subspace gives the vector in the subspace which is closest to **b**. Thus we can think of the projection **p** of **b** onto a subspace as the best possible approximation to **b** by vectors in the subspace. It is this last interpretation of the projection that is most useful in the function space setting. Given a function $f$ in $C(I)$ that is not in a finite-dimensional subspace $V$ of $C(I)$, the projection $h$ of $f$ onto $V$ gives the best possible approximation to $f$ by functions in $V$ in the sense that $\|f - h\|$ is as small as possible. It is called the **least-squares approximation** to $f$ by a function in $V$.

In many of the applications of this result, the subspace $V$ is one of the following subspaces of $C(I)$:

1. The polynomials of degree $\leq n$ for some $n$.
2. The *trigonometric polynomials* of degree $\leq n$; that is, $V$ is the subspace of $C(I)$ spanned by $1, \sin x, \cos x, \sin 2x, \cos 2x, \ldots, \sin nx, \cos nx$ for some $n$.

The main reason for the interest in approximating a function by a polynomial is that the function may be complicated and difficult to compute. Polynomials, on the other hand, are simple, both conceptually and computationally. The reason for the interest in approximating a function by a trigonometric polynomial is that these approximations are important in any branch of physics or engineering that deals with oscillatory phenomena, heat conduction, or potential theory.

## EXAMPLE 4

Consider the problem of finding the polynomial of degree $\leq 1$ which is nearest $\sin \pi x$ on the interval $[-1, 1]$. The solution is the projection of $\sin \pi x$ onto the subspace $P_1(I)$ of $C(I)$, where $I = [-1, 1]$. Since 1 and $x$ form an orthogonal basis for $P_1$, the projection is

$$h(x) = \frac{\langle \sin \pi x, 1 \rangle}{\langle 1, 1 \rangle} 1 + \frac{\langle \sin \pi x, x \rangle}{\langle x, x \rangle} x.$$

Since

$$\begin{aligned}
\langle \sin \pi x, 1 \rangle &= \int_{-1}^{1} \sin \pi x \, dx = 0, \\
\langle 1, 1 \rangle &= 2, \\
\langle \sin \pi x, x \rangle &= \int_{-1}^{1} x \sin \pi x \, dx \\
&= -\frac{1}{\pi} x \cos \pi x + \frac{1}{\pi^2} \sin \pi x \Big|_{-1}^{1} \\
&= \frac{2}{\pi}, \\
\langle x, x \rangle &= \int_{-1}^{1} x^2 \, dx = \frac{2}{3},
\end{aligned}$$

we obtain

$$h(x) = \frac{2/\pi}{2/3} x = \frac{3}{\pi} x.$$

### EXAMPLE 5

Consider now the problem of finding the least-squares approximation to $\sin \pi x$ on $I=[-1, 1]$ by a polynomial of degree $\leq 2$. In Example 3 we found that the functions 1, $x$, and $x^2 - \frac{1}{3}$ form an orthogonal basis for $P_2(I)$. Hence the least-squares approximation is

$$h_1(x) = \frac{\langle \sin \pi x, 1\rangle}{\langle 1, 1\rangle} 1 + \frac{\langle \sin \pi x, x\rangle}{\langle x, x\rangle} x + \frac{\langle \sin \pi x, x^2 - \frac{1}{3}\rangle}{\langle x^2 - \frac{1}{3}, x^2 - \frac{1}{3}\rangle}(x^2 - \tfrac{1}{3}).$$

Notice that the first two terms have already been computed in Example 4. Since $\langle \sin \pi x, x^2 - \frac{1}{3}\rangle = 0$,

$$h_1(x) = \frac{3}{\pi} x = h(x).$$

### EXAMPLE 6

Let us find the least-squares approximation to the function $x^2$ on $[-\pi, \pi]$ by a trigonometric polynomial of degree $\leq 1$. We have seen that the trigonometric functions

$$1, \quad \sin x, \quad \cos x$$

are orthogonal. Hence the projection is

$$\begin{aligned} h &= \frac{\langle x^2, 1\rangle}{\langle 1, 1\rangle} 1 + \frac{\langle x^2, \sin x\rangle}{\langle \sin x, \sin x\rangle} \sin x + \frac{\langle x^2, \cos x\rangle}{\langle \cos x, \cos x\rangle} \cos x \\ &= \frac{\langle x^2, 1\rangle}{2\pi} 1 + \frac{\langle x^2, \sin x\rangle}{\pi} \sin x + \frac{\langle x^2, \cos x\rangle}{\pi} \cos x. \end{aligned}$$

Since

$$\langle x^2, 1\rangle = \int_{-\pi}^{\pi} x^2\, dx = \frac{2\pi^3}{3},$$

$$\langle x^2, \sin x\rangle = \int_{-\pi}^{\pi} x^2 \sin x\, dx = 0,$$

$$\langle x^2, \cos x\rangle = \int_{-\pi}^{\pi} x^2 \cos x\, dx = -4\pi,$$

we obtain

$$h(x) = \frac{\pi^2}{3} - 4\cos x.$$

In general, if $f$ is a continuous function on $[-\pi, \pi]$, then the least-squares approximation to $f$ by a trigonometric polynomial of degree $\leq n$ is

$$\begin{aligned} h_n(x) = a_0 + a_1 \cos x + b_1 \sin x + a_2 \cos 2x + b_2 \sin 2x \\ + \cdots + a_n \cos nx + b_n \sin nx, \end{aligned}$$

where

$$a_0 = \frac{\langle f(x), 1\rangle}{\langle 1, 1\rangle} = \frac{1}{2\pi}\int_{-\pi}^{\pi} f(x)\,dx,$$

and for $k = 1, 2, \ldots, n$,

$$a_k = \frac{\langle f(x), \cos kx\rangle}{\langle \cos kx, \cos kx\rangle} = \frac{1}{\pi}\int_{-\pi}^{\pi} f(x)\cos kx\,dx,$$

$$b_k = \frac{\langle f(x), \sin kx\rangle}{\langle \sin kx, \sin kx\rangle} = \frac{1}{\pi}\int_{-\pi}^{\pi} f(x)\sin kx\,dx.$$

As $n$ increases we expect the approximation to get better. Why? Because the subspaces get bigger and hence there is more room to get close to the function $f$. In fact, it can be shown that $\lim_{n\to\infty} \|f - h_n\| = 0$, which says that

$$f(x) = a_0 + (a_1 \cos x + b_1 \sin x) + \cdots + (a_n \cos nx + b_n \sin nx) + \cdots.$$

This is called the **Fourier series** expansion of $f$.

## PROBLEMS 4.4

**1.** Let $I = [0, 1]$. Find the projection of:
(a) $1 + x$ onto $x^2$
*(b) $x + x^2$ onto $x$
(c) $\sin \pi x$ onto 1

***2.** Find the projection of $x$ onto $\sin x$ in $C(I)$, where $I = [0, \pi]$.

***3.** Let $I = [0, 1]$. Use the result of Problem 7 of Section 4.3 to find the projection of $x^2$ onto $P_1(I)$.

***4.** Let $I = [0, 2\pi]$. Find the projection of $x^2$ onto the subspace of $C(I)$ spanned by 1, $\sin x$, and $\cos x$. Do the same for $\sin^2 x$ and $\sin x \cos x$.

***5.** Let $I = [0, 2\pi]$. Find the projections of $x$, $x^2$, and $\sin x$ onto the subspace of $C(I)$ spanned by 1 and $\sin 2x$.

**6.** Find an orthogonal basis for the subspace of $C(I)$ spanned by
*(a) 1, $x$, and $x^2$ when $I = [0, 1]$
(b) $e^x$ and $e^{-x}$ when $I = [0, 1]$
*(c) $x$ and $e^x$ when $I = [0, 1]$
(d) $x$ and $\sin x$ when $I = [0, 2\pi]$
*(e) $x$, $x^2$, and $x^3$ when $I = [-1, 1]$
(f) $x$ and $\sqrt{x}$ when $I = [0, 1]$

**7.** Let $I = [-1, 1]$. Find all polynomials in $P_2(I)$ that are orthogonal to:
(a) $x + 1$ (b) $x$ and $1 + x^2$

**8.** Let $I = [0, 1]$ and let $V$ be the subspace of $C(I)$ spanned by $f_1(x) = x$, $f_2(x) = x^2$, $f_3(x) = \sqrt{x}$. Find an orthogonal basis for $V$.

**9.** Let $V$ be the subspace of $C(I)$, $I = [0, 2\pi]$, which is spanned by $1 + \sin x$, $\cos x$.
(a) Find the dimension of $V$ and a basis for $V$.
(b) Find the projection of $\sin x \cos x$ onto $V$.
(c) Is $\sin x$ in $V$? Why?

***10.** Find the polynomial of degree $\leq 3$ which is nearest $\sin \pi x$ on the interval $[-1, 1]$. Having done this, how would you find the least-squares approximation to $\sin \pi x$ on $[-1, 1]$ by a polynomial of degree $\leq 5$?

*11. Let $I = [0, 2\pi]$. Find the function in the subspace of $C(I)$ spanned by 1 and $\cos 2x$ that is nearest $x$. Do the same for $\sin x$.

12. Find the polynomial of degree $\le 2$ that is nearest:
*(a) $\sin x$ on the interval $[-\pi, \pi]$
(b) $\sqrt[3]{x}$ on the interval $[-1, 1]$
*(c) $x^3$ on the interval $[-1, 1]$
(d) $e^x$ on the interval $[0, 1]$
*(e) $\cos x$ on the interval $[-\pi, \pi]$
(f) $\sqrt{x}$ on the interval $[0, 1]$

13. Find the least-squares approximation to $x^3$ by a polynomial of degree $\le 1$ on $I = [0, 1]$. Then find the least-squares approximation to $x^3$ by a polynomial of degree $\le 2$ on $I = [0, 1]$.

*14. Let $I = [-1, 1]$. Find the polynomial of degree $\le 1$ that is nearest to $e^x$.

15. Find the least-squares approximation to $e^{2x}$ on $I = [0, 1]$ by a polynomial of degree $\le 1$.

*16. Let $V$ be the subspace of $C(I)$, $I = [0, 1]$, spanned by 1 and $e^x$. Find the least-squares approximation to $f(x) = x$ by functions in this subspace.

17. What polynomial of degree 2 is closest to $y = x^4$ in the sense of least-squares over the interval $-1 \le x \le 1$?

*18. Find the value of $a$ for which the expression $\int_0^1 (x - a \sin \pi x)^2\, dx$ is a minimum.

19. Find the polynomial of degree $\le 3$ that minimizes the expression

$$\int_{-1}^{1} [p(x) - e^x]^2\, dx$$

over all polynomials $p(x)$ of degree $\le 3$.

20. Let $f$ be continuous on $[0, 2\pi]$. Show that the trigonometric polynomial of degree $n$

$$g(x) = \tfrac{1}{2}a_0 + a_1 \cos x + b_1 \sin x + \cdots + a_n \cos nx + b_n \sin nx$$

which minimizes the least-squares error

$$\int_0^{2\pi} [f(x) - g(x)]^2\, dx$$

has coefficients

$$a_j = \frac{1}{\pi} \int_0^{2\pi} f(x) \cos jx\, dx, \quad j = 0, 1, 2, \ldots, n$$

$$b_j = \frac{1}{\pi} \int_0^{2\pi} f(x) \sin jx\, dx, \quad j = 1, \ldots, n.$$

21. Let $g_1, \ldots, g_k$ be a basis for a subspace $V$ of $C(I)$, $I = [a, b]$. The least-squares approximation to a function $f$ in $C(I)$ by a function in $V$ is that function $h$ in $V$ such that

$$\|h - f\|^2 = \int_a^b [h(x) - f(x)]^2\, dx$$

is as small as possible. To find $h$ we must find scalars $\alpha_1, \ldots, \alpha_k$ such that

$$h = \alpha_1 g_1 + \cdots + \alpha_k g_k$$

and

$$e = \int_a^b [\alpha_1 g_1(x) + \cdots + \alpha_k g_k(x) - f(x)]^2\, dx$$

is as small as possible. If $e$ is viewed as a function of $\alpha_1, \ldots, \alpha_k$, then $e$ will be minimized only if the equations

$$\frac{\partial e}{\partial \alpha_j} = 0, \qquad j = 1, \ldots, k, \tag{1}$$

are satisfied.
(a) Show that for $j = 1, \ldots, k$,

$$\frac{\partial e}{\partial \alpha_j} = 2 \int_a^b g_j(x)[\alpha_1 g_1(x) + \cdots + \alpha_k g_k(x) - f(x)]\, dx.$$

(b) Using the inner product notation, show that equations (1) can be written as

$$\alpha_1 \langle g_j, g_1 \rangle + \cdots + \alpha_k \langle g_j, g_k \rangle = \langle g_j, f \rangle, \qquad j = 1, \ldots, k.$$

(c) Show that these last equations can be written in matrix notation as

$$\begin{bmatrix} \langle g_1, g_1 \rangle \cdots \langle g_1, g_k \rangle \\ \vdots \quad\quad \vdots \\ \langle g_k, g_1 \rangle \cdots \langle g_k, g_k \rangle \end{bmatrix} \begin{bmatrix} \alpha_1 \\ \vdots \\ \alpha_k \end{bmatrix} = \begin{bmatrix} \langle g_1, f \rangle \\ \vdots \\ \langle g_k, f \rangle \end{bmatrix}.$$

(d) Show that if $g_1, \ldots, g_k$ are thought of as the "columns" of a matrix $A$ and $f$ is thought of as a "column vector" **b**, then this last equation can be written as

$$\begin{bmatrix} -g_1- \\ \vdots \\ -g_k- \end{bmatrix} \begin{bmatrix} | & & | \\ g_1 & \cdots & g_k \\ | & & | \end{bmatrix} \begin{bmatrix} \alpha_1 \\ \vdots \\ \alpha_k \end{bmatrix} = \begin{bmatrix} -g_1- \\ \vdots \\ -g_k- \end{bmatrix} \begin{bmatrix} | \\ f \\ | \end{bmatrix}$$

or

$$A^T A\bar{\mathbf{x}} = A^T \mathbf{b}.$$

***22.** Clearly, the function $\sin x$ is not in $P_2(I)$, $I = [-\pi, \pi]$. Thus the equation

$$\sin x = a + bx + cx^2 \qquad \text{for all } x \text{ in } I$$

is certainly inconsistent. As above, if we view the functions 1, $x$, and $x^2$ as the columns of a matrix $A$ and $\sin x$ as a column vector, this equation becomes

$$\begin{bmatrix} | & | & | \\ 1 & x & x^2 \\ | & | & | \end{bmatrix} \begin{bmatrix} a \\ b \\ c \end{bmatrix} = \begin{bmatrix} | \\ \sin x \\ | \end{bmatrix}.$$

Find a least-squares solution to this problem.

**23.** Interpret Problems 11 and 12 as inconsistent systems.

**24.** Investigate the difference between a least-squares approximation of $\sin x$ by a polynomial of degree $\leq 2$ and the second-degree Taylor polynomial approximation of $\sin x$ about $x = 0$.

# 5

# Linear Transformations

In Chapters 1 through 3 we regarded the equation $A\mathbf{x} = \mathbf{b}$ as something to be solved for $\mathbf{x}$. In this chapter we view $A$ as transforming a vector $\mathbf{x}$ into a new vector $A\mathbf{x}$. From this point of view, an $m \times n$ matrix $A$ defines a function from $R^n$ into $R^m$. Functions $T$ from $R^n$ into $R^m$ that can be defined by a matrix via the equation

$$T(\mathbf{x}) = A\mathbf{x}$$

satisfy two important properties:

$$T(\mathbf{x} + \mathbf{y}) = T(\mathbf{x}) + T(\mathbf{y}) \qquad \text{and} \qquad T(\alpha\mathbf{x}) = \alpha T(\mathbf{x}).$$

Our goal is to study functions from $R^n$ into $R^m$ that satisfy these properties.

## 5.1 Linear Transformations

The first problem posed in Chapter 1 was this: Given an $m \times n$ matrix $A$ and a vector $\mathbf{b}$ in $R^m$, solve $A\mathbf{x} = \mathbf{b}$. This problem soon gave way to the problem of existence and uniqueness.

*Existence:* For what vectors $\mathbf{b}$ does the system $A\mathbf{x} = \mathbf{b}$ have a solution?
*Uniqueness:* Given a consistent system $A\mathbf{x} = \mathbf{b}$, when is the solution unique?

Posing these two questions represented a subtle but important change in perspective which set the stage for introducing the column space and the nullspace in Chapter 2. The column space of $A$ is the collection of all vectors $\mathbf{b}$ in $R^m$ such that $A\mathbf{x} = \mathbf{b}$ has a solution . It is also the collection of all possible linear combinations of the columns of $A$. That is, $C(A)$ is just the collection of all possible vectors $A\mathbf{x}$ for $\mathbf{x}$ in $R^n$. This change in perspective suggests that we view $A$ as defining a function $T$ from $R^n$ to $R^m$.

This function $T$ transforms a vector $\mathbf{x}$ in $R^n$ to the vector $T(\mathbf{x}) = A\mathbf{x}$ in $R^m$. The vector $T(\mathbf{x})$ is called the **image** of $\mathbf{x}$ under $T$. The collection of all possible images of vectors in $R^n$ is called the **image of $R^n$ under** $T$ and is denoted by $T(R^n)$. Thus the image of $R^n$ under $T$ is the collection of all vectors $\mathbf{y}$ in $R^m$ such that $\mathbf{y} = T(\mathbf{x})$ for some vector $\mathbf{x}$ in $R^n$. Since $T(\mathbf{x}) = A\mathbf{x}$ for all $\mathbf{x}$ in $R^n$, the image of $R^n$ under $T$ is the collection of all vectors $\mathbf{y}$ in $R^m$ such that $\mathbf{y} = A\mathbf{x}$ for some $\mathbf{x}$ in $R^n$. From our current perspective then, *the image of $R^n$ under the function $T$ defined by $T(\mathbf{x}) = A\mathbf{x}$ is the column space of $A$.*

## EXAMPLE 1

Consider the function from $R^3$ into $R^2$ defined by the equation $T(\mathbf{x}) = A\mathbf{x}$, where $A$ is the matrix

$$A = \begin{bmatrix} 1 & 1 & 2 \\ 0 & 2 & 3 \end{bmatrix}.$$

The image of $\mathbf{x} = (1, 1, 1)$ under $T$ is the vector

$$T(\mathbf{x}) = A\mathbf{x} = \begin{bmatrix} 1 & 1 & 2 \\ 0 & 2 & 3 \end{bmatrix}\begin{bmatrix} 1 \\ 1 \\ 1 \end{bmatrix} = \begin{bmatrix} 4 \\ 5 \end{bmatrix}.$$

In general, the image of an arbitrary vector $\mathbf{x} = (x_1, x_2, x_3)$ under $T$ is

$$T(\mathbf{x}) = A\mathbf{x} = \begin{bmatrix} 1 & 1 & 2 \\ 0 & 2 & 3 \end{bmatrix}\begin{bmatrix} x_1 \\ x_2 \\ x_3 \end{bmatrix} = \begin{bmatrix} x_1 + x_2 + 2x_3 \\ 2x_2 + 3x_3 \end{bmatrix}.$$

Thus the image of $R^3$ under $T$ is the collection of all vectors of the form

$$T(\mathbf{x}) = A\mathbf{x} = \begin{bmatrix} x_1 + x_2 + 2x_3 \\ 2x_2 + 3x_3 \end{bmatrix} = x_1\begin{bmatrix} 1 \\ 0 \end{bmatrix} + x_2\begin{bmatrix} 1 \\ 2 \end{bmatrix} + x_3\begin{bmatrix} 2 \\ 3 \end{bmatrix},$$

which of course is the column space of $A$.

Given an $m \times n$ matrix $A$, the function $T$ from $R^n$ into $R^m$ defined by $A$ via the equation $T(\mathbf{x}) = A\mathbf{x}$ has two very important properties. For any two vectors $\mathbf{x}$ and $\mathbf{y}$ in $R^n$ and any scalar $\alpha$,

$$\begin{aligned} T(\mathbf{x} + \mathbf{y}) &= T(\mathbf{x}) + T(\mathbf{y}) \qquad &&[\text{since } A(\mathbf{x} + \mathbf{y}) = A\mathbf{x} + A\mathbf{y}], \\ T(\alpha\mathbf{x}) &= \alpha T(\mathbf{x}) &&[\text{since } A(\alpha\mathbf{x}) = \alpha A\mathbf{x}]. \end{aligned}$$

Functions from $R^n$ into $R^m$ that have these two properties are the most important functions in linear algebra because they preserve the most important operations in linear algebra: vector addition and scalar multiplication.

To indicate that a function $T$ maps vectors in $R^n$ into vectors in $R^m$, we write $T\colon R^n \to R^m$ and say that $T$ maps $R^n$ into $R^m$.

***Definition*** A function $T: R^n \to R^m$ is called a **linear transformation** if for all vectors $\mathbf{x}$ and $\mathbf{y}$ in $R^n$ and for all scalars $\alpha$,

(a) $T(\mathbf{x} + \mathbf{y}) = T(\mathbf{x}) + T(\mathbf{y})$,
(b) $T(\alpha\mathbf{x}) = \alpha T(\mathbf{x})$.

A linear transformation $T: R^n \to R^m$ is also called a **linear map**, a **linear mapping**, and a **linear function**.

Thus any function $T: R^n \to R^m$ defined by an $m \times n$ matrix $A$ via the equation $T(\mathbf{x}) = A\mathbf{x}$ is a linear transformation. Here are some other examples.

### EXAMPLE 2

Consider the function $T: R^2 \to R^3$ defined by $T(x_1, x_2) = (x_1 + x_2, 3x_2, 2x_1 - x_2)$. We will prove that $T$ is linear. Let $\mathbf{x} = (x_1, x_2)$ and $\mathbf{y} = (y_1, y_2)$ be any two vectors in $R^2$ and let $\alpha$ be any scalar. Then

$$\begin{aligned} T(\mathbf{x} + \mathbf{y}) &= T(x_1 + y_1, x_2 + y_2) \\ &= (x_1 + y_1 + x_2 + y_2, 3(x_2 + y_2), 2(x_1 + y_1) - (x_2 + y_2)) \\ &= (x_1 + x_2, 3x_2, 2x_1 - x_2) + (y_1 + y_2, 3y_2, 2y_1 - y_2) \\ &= T(\mathbf{x}) + T(\mathbf{y}), \\ T(\alpha\mathbf{x}) &= T(\alpha x_1, \alpha x_2) \\ &= (\alpha x_1 + \alpha x_2, 3\alpha x_2, 2\alpha x_1 - \alpha x_2) \\ &= \alpha(x_1 + x_2, 3x_2, 2x_1 - x_2) \\ &= \alpha T(\mathbf{x}). \end{aligned}$$

Thus $T$ is linear.

A linear transformation $T: R^n \to R^m$ preserves vector addition and scalar multiplication. If $\mathbf{x}$ and $\mathbf{y}$ are two vectors in $R^n$, then first adding $\mathbf{x}$ and $\mathbf{y}$ and then applying $T$ to their sum gives the same result as first applying $T$ to $\mathbf{x}$ and $\mathbf{y}$ separately and then adding (see Figure 5.1). Similarly, first multiplying $\mathbf{x}$ by a scalar $\alpha$ and then applying $T$ results in the same vector as multiplying the image of $\mathbf{x}$ under $T$ by $\alpha$.

### EXAMPLE 3

The transformation $T: R^2 \to R^2$ defined by $T(x_1, x_2) = (x_1x_2, x_2)$ is not linear because $T$ does not preserve vector addition. For instance, if $\mathbf{x} = (2, 3)$ and $\mathbf{y} = (3, 2)$, then $T(\mathbf{x} + \mathbf{y}) = T(5, 5) = (25, 5)$ but $T(\mathbf{x}) + T(\mathbf{y}) = (6, 3) + (6, 2) = (12, 5) \neq (25, 5)$.

### EXAMPLE 4

The function $I: R^m \to R^m$ defined by $I(\mathbf{x}) = \mathbf{x}$ maps every vector in $R^m$ onto itself. It is clearly a linear transformation and is called the **identity transformation**. The image of $R^m$ under $I$ is $R^m$.

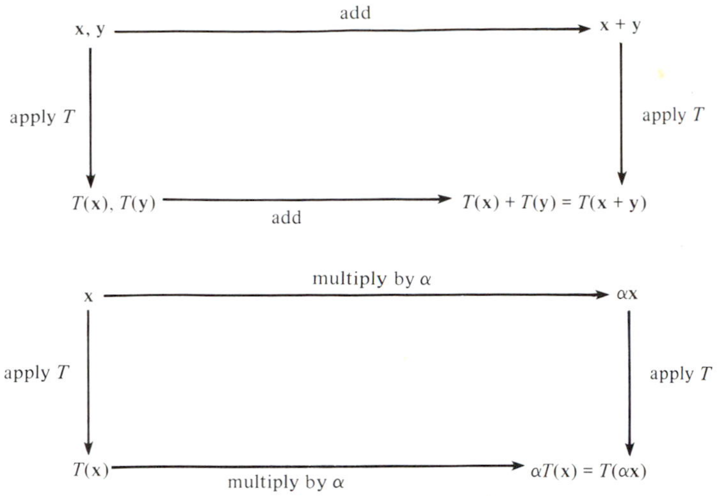


**Figure 5.1**

### EXAMPLE 5

The function 0: $R^n \to R^m$ defined by $0(\mathbf{x}) = \mathbf{0}$ maps every vector in $R^n$ onto the zero vector in $R^m$. It, too, is clearly a linear transformation and is called the **zero transformation**. The image of $R^n$ under the zero transformation is the trivial subspace $\{\mathbf{0}\}$.

***Theorem 1*** *Let* $T$: $R^n \to R^m$ *be a linear transformation. Then*:

(*a*) $T(\mathbf{0}) = \mathbf{0}$.
(*b*) $T(-\mathbf{x}) = -T(\mathbf{x})$ *for all* $\mathbf{x}$ in $R^n$.
(*c*) $T(\mathbf{x} - \mathbf{y}) = T(\mathbf{x}) - T(\mathbf{y})$ *for all* $\mathbf{x}$ and $\mathbf{y}$ in $R^n$.
(*d*) $T(\alpha\mathbf{x} + \beta\mathbf{y}) = \alpha T(\mathbf{x}) + \beta T(\mathbf{y})$ *for all* $\mathbf{x}$ and $\mathbf{y}$ *in* $R^n$.
(*e*) *The image of* $R^n$ *under* $T$ *is a subspace of* $R^m$.

**Proof** Part (a) is easy once you see the trick: $T(\mathbf{0}) = T(\mathbf{0}+\mathbf{0}) = T(\mathbf{0}) + T(\mathbf{0})$ and subtracting $T(\mathbf{0})$ from both sides yields $\mathbf{0} = T(\mathbf{0})$. Parts (b) and (c) are special cases of part (d) obtained, respectively, from $\alpha = -1$, $\beta = 0$ and $\alpha = 1$, $\beta = -1$. For part (d) we have

$$\begin{aligned} T(\alpha\mathbf{x} + \beta\mathbf{y}) &= T(\alpha\mathbf{x}) + T(\beta\mathbf{y}) && \text{(since } T \text{ preserves addition)} \\ &= \alpha T(\mathbf{x}) + \beta T(\mathbf{y}) && \text{(since } T \text{ preserves scalar multiplication).} \end{aligned}$$

Finally, since the image of $R^n$ under $T$ consists of all vectors $T(\mathbf{x})$ for $\mathbf{x}$ in $R^n$, the proof of (e) follows directly from part (a) and the definition of a linear transformation. For by part (a), $\mathbf{0}$ is in $T(R^n)$. If $T(\mathbf{x})$ and $T(\mathbf{y})$ are in $T(R^n)$ and

$\alpha$ is a scalar, then

$$T(\mathbf{x}) + T(\mathbf{y}) = T(\mathbf{x} + \mathbf{y}) \qquad \text{and} \qquad \alpha T(\mathbf{x}) = T(\alpha \mathbf{x}).$$

Therefore, $T(\mathbf{x}) + T(\mathbf{y})$ and $\alpha T(\mathbf{x})$ are also in $T(R^n)$.

Clearly, part (d) extends to any linear combination of vectors in $R^n$. If $\mathbf{v}_1, \mathbf{v}_2, \ldots, \mathbf{v}_k$ are vectors in $R^n$ and $\alpha_1, \alpha_2, \ldots, \alpha_k$ are scalars, then

$$T(\alpha_1 \mathbf{v}_1 + \alpha_2 \mathbf{v}_2 + \cdots + \alpha_k \mathbf{v}_k) = \alpha_1 T(\mathbf{v}_1) + \alpha_2 T(\mathbf{v}_2) + \cdots + \alpha_k T(\mathbf{v}_k).$$

Thus linear transformations preserve linear combinations, which is why they are the natural functions to study in linear algebra.

In Example 2, we showed that the function $T: R^2 \to R^3$ defined by

$$T(x_1, x_2) = (x_1 + x_2, 3x_2, 2x_1 - x_2)$$

is a linear transformation. The vector on the right-hand side specifies the column vector

$$\begin{bmatrix} x_1 + x_2 \\ 3x_2 \\ 2x_1 - x_2 \end{bmatrix},$$

which in turn is just

$$A\mathbf{x} = \begin{bmatrix} 1 & 1 \\ 0 & 3 \\ 2 & -1 \end{bmatrix} \begin{bmatrix} x_1 \\ x_2 \end{bmatrix}.$$

Thus $T(\mathbf{x}) = A\mathbf{x}$ and we see that the linear transformation $T$ can be defined by the matrix $A$. Notice that the first column of $A$ is just $T(\mathbf{e}_1)$ and the second column of $A$ is $T(\mathbf{e}_2)$.

In general, given any linear transformation $T: R^n \to R^m$, there is an $m \times n$ matrix $A$ such that $T(\mathbf{x}) = A\mathbf{x}$. To see this, let $\mathbf{e}_1, \mathbf{e}_2, \ldots, \mathbf{e}_n$ be the standard basis vectors for $R^n$ and let $A$ be the matrix whose columns are $T(\mathbf{e}_1), T(\mathbf{e}_2), \ldots, T(\mathbf{e}_n)$. If $\mathbf{x}$ is in $R^n$, then $\mathbf{x} = x_1\mathbf{e}_1 + x_2\mathbf{e}_2 + \cdots + x_n\mathbf{e}_n$ and since $T$ is linear, we have

$$\begin{aligned} T(\mathbf{x}) &= T(x_1\mathbf{e}_1 + x_2\mathbf{e}_2 + \cdots + x_n\mathbf{e}_n) \\ &= x_1 T(\mathbf{e}_1) + x_2 T(\mathbf{e}_2) + \cdots + x_n T(\mathbf{e}_n) \\ &= A\mathbf{x}. \end{aligned}$$

***Definition*** Let $\mathbf{e}_1, \mathbf{e}_2, \ldots, \mathbf{e}_n$ be the standard basis for $R^n$. If $T: R^n \to R^m$ is a linear transformation, then the $m \times n$ matrix $A$ having the vectors $T(\mathbf{e}_1), T(\mathbf{e}_2), \ldots, T(\mathbf{e}_n)$ as its columns is called the **standard matrix** for $T$.

In this termonology we have proved

***Theorem 2*** *If $T: R^n \to R^m$ is a linear transformation and $A$ is the standard matrix for $T$, then $T(\mathbf{x}) = A\mathbf{x}$ for all $\mathbf{x}$ in $R^n$.*

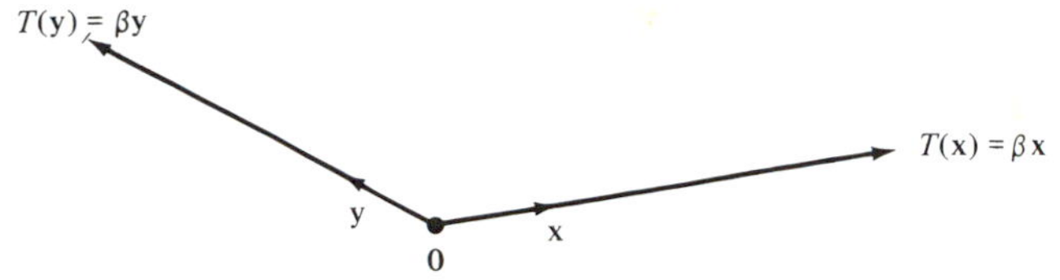


Figure 5.2

When $A$ is an $m \times n$ matrix and $T: R^n \to R^m$ is defined by $T(\mathbf{x}) = A\mathbf{x}$, what is the standard matrix for $T$? It is the matrix $A$ itself because

$$T(\mathbf{e}_i) = A\mathbf{e}_i = \text{the } i\text{th column of } A.$$

Every matrix defines a linear transformation, and conversely every linear transformation defines a matrix. *Thus linear transformations correspond to matrices.* The power of this result is that it enables us to translate questions about linear transformations to questions about matrices.

### EXAMPLE 6

Let $\beta$ be a fixed scalar. The function $T: R^m \to R^m$ defined by $T(\mathbf{x}) = \beta\mathbf{x}$ is a linear transformation because

$$\begin{aligned} T(\mathbf{x} + \mathbf{y}) &= \beta(\mathbf{x} + \mathbf{y}) = \beta\mathbf{x} + \beta\mathbf{y} = T(\mathbf{x}) + T(\mathbf{y}) \\ T(\alpha\mathbf{x}) &= \beta(\alpha\mathbf{x}) = \alpha(\beta\mathbf{x}) = \alpha T(\mathbf{x}). \end{aligned}$$

The standard matrix for $T$ is $\beta I$, where $I$ is the $m \times m$ identity matrix. If $\beta > 1$, $T$ stretches vectors by a factor $\beta$ and hence is called a **dilation**. If $0 < \beta < 1$, $T$ contracts vectors by a factor $\beta$ and is called a **contraction** (see Figure 5.2). If $\beta = 0$, $T$ is the zero transformation and if $\beta = 1$, $T$ is the identity transformation. As long as $\beta \neq 0$, the image of $T$ is $R^m$. For if $\mathbf{z}$ is any vector in $R^m$, then $(1/\beta)\mathbf{z}$ is a vector in $R^m$ such that

$$T\left(\frac{1}{\beta}\mathbf{z}\right) = \frac{1}{\beta}T(\mathbf{z}) = \frac{1}{\beta}\beta\mathbf{z} = \mathbf{z}.$$

### EXAMPLE 7

Let $\theta$ be a fixed angle and let $T: R^2 \to R^2$ be the function that rotates every vector in $R^2$ counterclockwise through the angle $\theta$ (see Figure 5.3). It is geometrically obvious that if $\mathbf{x}$ and $\mathbf{y}$ are vectors in $R^2$ and $\alpha$ is a scalar, then $T(\mathbf{x} + \mathbf{y}) = T(\mathbf{x}) + T(\mathbf{y})$ and $T(\alpha\mathbf{x}) = \alpha T(\mathbf{x})$ (see Figure 5.4). From geometry then $T$ is a linear transformation, called a **rotation**. To verify algebraically that $T$ is linear, we need a formula for $T$. That's easy using polar coordinates. Let

$$\mathbf{x} = \begin{bmatrix} x_1 \\ x_2 \end{bmatrix} = \begin{bmatrix} r\cos\phi \\ r\sin\phi \end{bmatrix}.$$

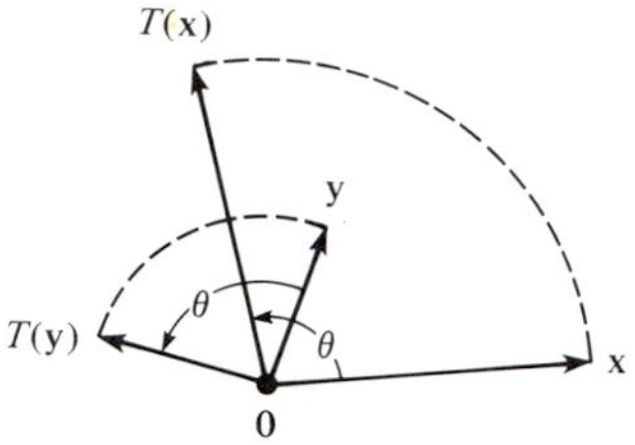


Figure 5.3

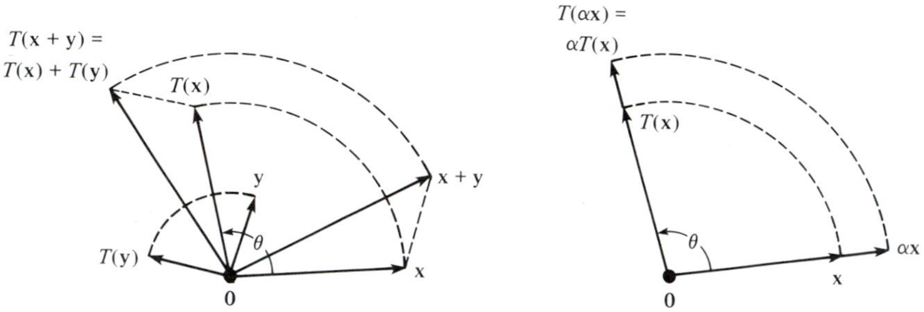


Figure 5.4

Then (see Figure 5.5)

$$T(\mathbf{x}) = \begin{bmatrix} r\cos(\theta+\phi) \\ r\sin(\theta+\phi) \end{bmatrix} = \begin{bmatrix} r\cos\theta\cos\phi - r\sin\theta\sin\phi \\ r\sin\theta\cos\phi + r\cos\theta\sin\phi \end{bmatrix}$$
$$= \begin{bmatrix} x_1\cos\theta - x_2\sin\theta \\ x_1\sin\theta + x_2\cos\theta \end{bmatrix} = \begin{bmatrix} \cos\theta & -\sin\theta \\ \sin\theta & \cos\theta \end{bmatrix}\begin{bmatrix} x_1 \\ x_2 \end{bmatrix}.$$

Thus $T$ is given by a matrix and hence is linear. The image of $T$ is the column space

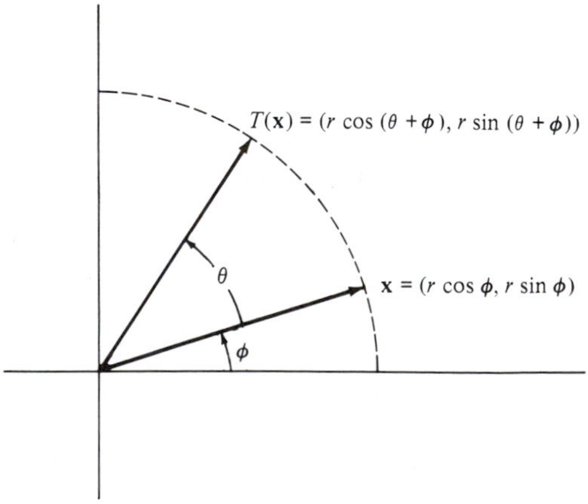


Figure 5.5

of its standard matrix

$$A = \begin{bmatrix} \cos\theta & -\sin\theta \\ \sin\theta & \cos\theta \end{bmatrix}.$$

Since the columns of $A$ are nonzero orthonormal vectors, $A$ has rank 2. Therefore, $T(R^2) = C(A) = R^2$.

In the following example we visit an old friend from Chapter 3 which we trust will be warmly remembered.

## EXAMPLE 8

Let $V$ be a subspace of $R^m$ and let $T\colon R^m \to R^m$ be the function that projects all vectors in $R^m$ onto $V$ (see Figure 5.6). $T$ is called the **projection** of $R^m$ onto $V$. Let $A$ be a matrix whose columns form a basis for $V$. Then $P = A(A^TA)^{-1}A^T$ is the projection matrix for the subspace $V$ (see Section 3.5). If $\mathbf{x}$ is any vector in $R^m$, then $P\mathbf{x}$ is the projection of $\mathbf{x}$ onto $V$ and hence $T(\mathbf{x}) = P\mathbf{x}$. Hence $T$ is linear and $P$ is the standard matrix for $T$. The image of every vector in $R^m$ under $T$ is in $V$. Since every vector $\mathbf{v}$ in $V$ is also in the image of $T$ (because $T(\mathbf{v}) = \mathbf{v}$ when $\mathbf{v}$ is in $V$), the image of $R^m$ under $T$ is $V$.

In Example 8 we showed that if $V$ is a subspace of $R^m$ and $T\colon R^m \to R^m$ is the projection onto $V$, then the standard matrix for $T$ is the projection matrix for $V$. Let $A$ be a matrix whose columns form a basis for $V$. The computation of a projection matrix from a straightforward application of the formula

$$P = A(A^TA)^{-1}A^T$$

is not simple; it involves finding the inverse of $A^TA$. Since the standard matrix for $T$ is the matrix whose columns are $T(\mathbf{e}_1)$, $T(\mathbf{e}_2), \ldots, T(\mathbf{e}_m)$, these vectors are the columns of $P$. Moreover, since $T(\mathbf{e}_i)$ is the projection of $\mathbf{e}_i$ onto $V$, we can find $T(\mathbf{e}_i)$ by solving the normal equations

$$A^TA\bar{\mathbf{x}}_i = A^T\mathbf{e}_i, \qquad i = 1, 2, \ldots, m.$$

The $\bar{\mathbf{x}}_i$ are found by reducing the augmented matrix

$$[A^TA : A^T\mathbf{e}_1 \quad A^T\mathbf{e}_2 \quad \cdots \quad \mathbf{A}^T\mathbf{e}_m] = [A^TA : A^T] \tag{1}$$

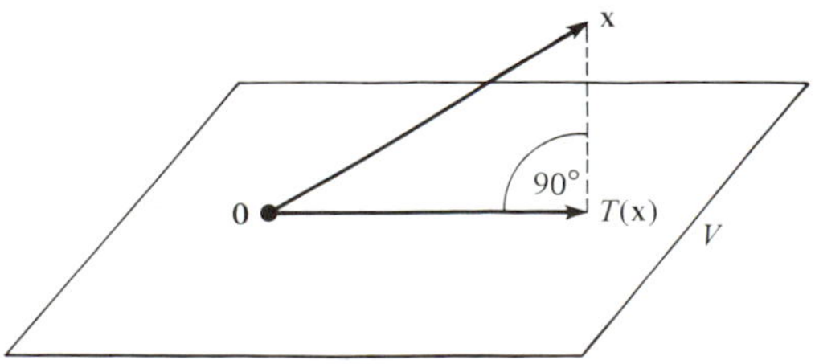


Figure 5.6

to reduced echelon form

$$[I : \bar{\mathbf{x}}_1 \quad \bar{\mathbf{x}}_2 \quad \cdots \quad \bar{\mathbf{x}}_m] = [I : (A^TA)^{-1}A^T].$$

Thus the standard matrix for $T$ is

$$P = A[\bar{\mathbf{x}}_1 \quad \bar{\mathbf{x}}_2 \quad \cdots \quad \mathbf{x}_m] = A[(A^TA)^{-1}A^T].$$

## EXAMPLE 9

Let $T$: $R^4 \to R^4$ be the projection of $R^4$ onto the subspace $x_1 + x_2 + x_3 + x_4 = 0$. A basis for this subspace is

$$(-1, 1, 0, 0), \quad (-1, 0, 1, 0), \quad (-1, 0, 0, 1).$$

Let $A$ be the matrix whose columns are these vectors. Then

$$A^TA = \begin{bmatrix} -1 & 1 & 0 & 0 \\ -1 & 0 & 1 & 0 \\ -1 & 0 & 0 & 1 \end{bmatrix} \begin{bmatrix} -1 & -1 & -1 \\ 1 & 0 & 0 \\ 0 & 1 & 0 \\ 0 & 0 & 1 \end{bmatrix} = \begin{bmatrix} 2 & 1 & 1 \\ 1 & 2 & 1 \\ 1 & 1 & 2 \end{bmatrix}.$$

Form the matrix

$$[A^TA : A^T] = \begin{bmatrix} 2 & 1 & 1 & : & -1 & 1 & 0 & 0 \\ 1 & 2 & 1 & : & -1 & 0 & 1 & 0 \\ 1 & 1 & 2 & : & -1 & 0 & 0 & 1 \end{bmatrix}$$

and use Gaussian elimination to reduce it to

$$[I : (A^TA)^{-1}A^T] = \begin{bmatrix} 1 & 0 & 0 & : & -\frac{1}{4} & \frac{3}{4} & -\frac{1}{4} & -\frac{1}{4} \\ 0 & 1 & 0 & : & -\frac{1}{4} & -\frac{1}{4} & \frac{3}{4} & -\frac{1}{4} \\ 0 & 0 & 1 & : & -\frac{1}{4} & -\frac{1}{4} & -\frac{1}{4} & \frac{3}{4} \end{bmatrix}.$$

Thus the standard matrix for $T$ is

$$P = A[(A^TA)^{-1}A^T] = \frac{1}{4}\begin{bmatrix} 3 & -1 & -1 & -1 \\ -1 & 3 & -1 & -1 \\ -1 & -1 & 3 & -1 \\ -1 & -1 & -1 & 3 \end{bmatrix}.$$

## EXAMPLE 10

A close relative of a projection is a reflection. Let $V$ be a line in $R^2$ through the origin. Let $T$: $R^2 \to R^2$ be the function that reflects every vector in $R^2$ across the line $V$. It is clear geometrically that $T$ is a linear transformation (see Figure 5.7). To verify this fact algebraically, we develop a formula for $T$. Let $P$ be the projection matrix for $V$. If $\mathbf{p}$ is the projection of $\mathbf{x}$ onto the line $V$, then (see Figure 5.8)

$$\begin{aligned} T(\mathbf{x}) &= \mathbf{p} + (\mathbf{p} - \mathbf{x}) \\ &= 2\mathbf{p} - \mathbf{x} \\ &= (2P - I)\mathbf{x}. \end{aligned}$$

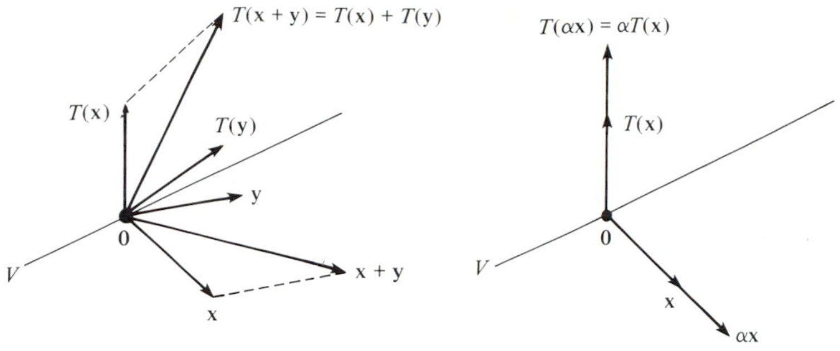


Figure 5.7

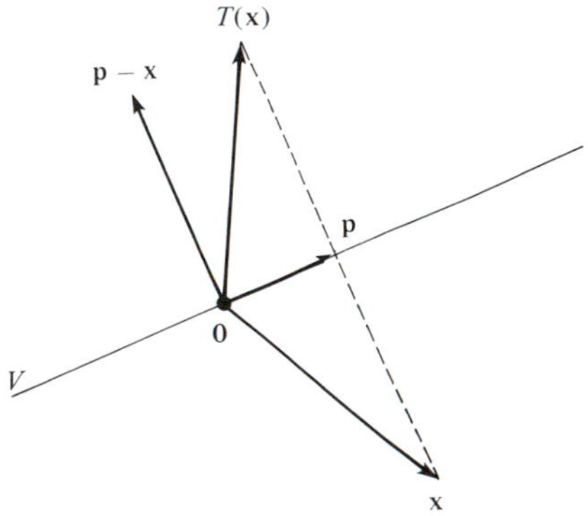


Figure 5.8

So we have a formula for $T$; $T$ is multiplication by the matrix $2P - I$. Thus $T$ is linear and its standard matrix is $2P - I$. Since every vector in $R^2$ is the reflection across $V$ of some vector, the image of $T$ is $R^2$.

In general, if $V$ is a subspace of $R^m$ and $P$ is the projection matrix for $V$, the linear transformation $T: R^m \to R^m$ defined by

$$T(\mathbf{x}) = (2P - I)\mathbf{x}$$

is called the **reflection** of $R^m$ across $V$ (see Figure 5.9).

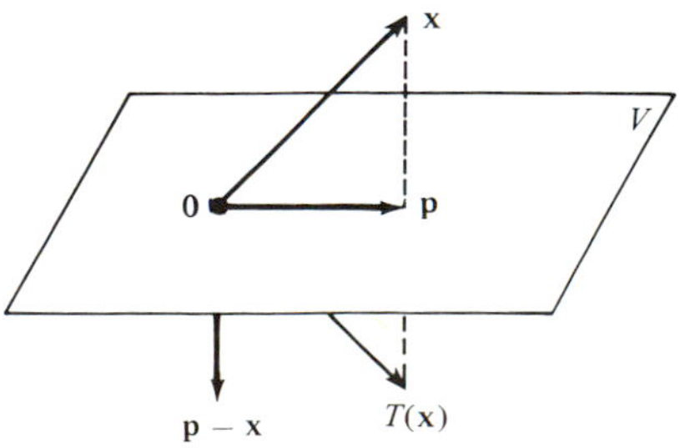


Figure 5.9

As the previous examples indicate, the relationship between linear transformations and matrices is a fruitful one. Representing a linear transformation by a matrix enables us to translate questions about a linear transformation to corresponding questions about a matrix. We are now going to investigate further the relationship between linear transformations and matrices. On the collection of matrices we defined the operations of addition of matrices, multiplication of a matrix by a scalar, and multiplication of two matrices. Are there operations on the collection of linear transformations that correspond to these operations on matrices? Yes! We begin with addition and scalar multiplication.

***Definition*** Let $S: R^n \to R^m$ and $T: R^n \to R^m$ be two linear transformations and let $\alpha$ be a scalar.

(a) The **sum** of $S$ and $T$ is the function $S + T: R^n \to R^m$ defined by

$$(S + T)(\mathbf{x}) = S(\mathbf{x}) + T(\mathbf{x})$$

for all $\mathbf{x}$ in $R^n$.

(b) The **scalar product** of $\alpha$ and $S$ is the function $\alpha S: R^n \to R^m$ defined by

$$(\alpha S)(\mathbf{x}) = \alpha S(\mathbf{x})$$

for all $\mathbf{x}$ in $R^n$.

Note that this definition of the sum and scalar product of two linear transformations is just the definition of the sum and scalar product of two functions (see Section 4.1). Thus we add two linear transformations in the same way that we add any two functions and similarly for multiplication by a scalar.

***Theorem 3*** *Let $S: R^n \to R^m$ and $T: R^n \to R^m$ be two linear transformations, and let $\alpha$ be a scalar.*

*(a) $S + T$ and $\alpha S$ are linear transformations.*

*(b) If $A$ and $B$ are the standard matrices for $S$ and $T$, respectively, then $A + B$ and $\alpha A$ are the standard matrices for $S + T$ and $\alpha S$.*

**Proof of (a)** Let $\mathbf{x}$ and $\mathbf{y}$ be in $R^n$ and let $\beta$ be a scalar. Then applying the definitions and the assumption that $S$ and $T$ are linear, we obtain

$$\begin{aligned}
(S + T)(\mathbf{x} + \mathbf{y}) &= S(\mathbf{x} + \mathbf{y}) + T(\mathbf{x} + \mathbf{y}) && \text{(definition)}\\
&= S(\mathbf{x}) + S(\mathbf{y}) + T(\mathbf{x}) + T(\mathbf{y}) && \text{(linearity)}\\
&= S(\mathbf{x}) + T(\mathbf{x}) + S(\mathbf{y}) + T(\mathbf{y}) && (R^m \text{ is a vector space})\\
&= (S + T)(\mathbf{x}) + (S + T)(\mathbf{y}) && \text{(definition)}\\
(S + T)(\beta\mathbf{x}) &= S(\beta\mathbf{x}) + T(\beta\mathbf{x}) && \text{(definition)}\\
&= \beta S(\mathbf{x}) + \beta T(\mathbf{x}) && \text{(linearity)}\\
&= \beta(S(\mathbf{x}) + T(\mathbf{x})) && (R^m \text{ is a vector space})\\
&= \beta(S + T)(\mathbf{x}). && \text{(definition).}
\end{aligned}$$

Thus $S + T: R^n \to R^m$ is a linear transformation. We now verify that $\alpha S$ is a linear transformation.

$$\begin{aligned}
(\alpha S)(\mathbf{x}+\mathbf{y}) &= \alpha S(\mathbf{x}+\mathbf{y}) && \text{(definition)}\\
&= \alpha(S(\mathbf{x})+S(\mathbf{y})) && \text{(linearity)}\\
&= \alpha S(\mathbf{x})+\alpha S(\mathbf{y}) && (R^m \text{ is a vector space})\\
&= (\alpha S)(\mathbf{x})+(\alpha S)(\mathbf{y}) && \text{(definition)}\\
(\alpha S)(\beta\mathbf{x}) &= \alpha S(\beta\mathbf{x}) && \text{(definition)}\\
&= \alpha\beta S(\mathbf{x}) && \text{(linearity)}\\
&= \beta\alpha S(\mathbf{x}) && (R^m \text{ is a vector space})\\
&= \beta(\alpha S)(\mathbf{x}) && \text{(definition).}
\end{aligned}$$

Thus $\alpha S\colon R^n \to R^m$ is also a linear transformation. This completes the proof of (a).

**Proof of (b)** The standard matrix of $S + T$ is the matrix whose columns are

$$(S+T)(\mathbf{e}_1),\quad (S+T)(\mathbf{e}_2),\quad \ldots,\quad (S+T)(\mathbf{e}_n).$$

Since $(S+T)(\mathbf{e}_i) = S(\mathbf{e}_i) + T(\mathbf{e}_i)$ for $i = 1, 2, \ldots, n$, the matrix of $S + T$ is the matrix whose columns are

$$S(\mathbf{e}_1)+T(\mathbf{e}_1),\quad S(\mathbf{e}_2)+T(\mathbf{e}_2),\quad \ldots,\quad S(e_n)+T(\mathbf{e}_n),$$

and these vectors are just the columns of $A + B$. Thus $A + B$ is the standard matrix for $S + T$. Similarly, since for $i = 1, 2, \ldots, n$ the $i$th column of the standard matrix for $\alpha S$ is $(\alpha S)(\mathbf{e}_i) = \alpha S(\mathbf{e}_i)$, the standard matrix for $\alpha S$ is $\alpha A$.

Well, we couldn't ask for much more. The sum of two transformations corresponds to the sum of the matrices that represent the transformations; a scalar multiple of a transformation is represented by the same scalar multiple of the matrix that represents the transformation. Moreover, the operations of addition and scalar multiplication on the collection of linear transformations from $R^n$ into $R^m$ satisfy the same identities as the corresponding operations on the $m \times n$ matrices.

## EXAMPLE 11

Let $V$ be a subspace of $R^m$ and let $T\colon R^m \to R^m$ be the projection of $R^m$ onto $V$. Let $P$ be the standard matrix for $T$. We defined the reflection of $R^m$ across $V$ to the linear transformation $S\colon R^m \to R^m$ defined by the equation

$$S(\mathbf{x}) = (2P - I)\mathbf{x}.$$

Thus $S = 2T - I$. Using this equation we can also define a projection transformation in terms of a reflection transformation: $T = \frac{1}{2}(S + I)$ (see Figure 5.10).

The operation on linear transformations that corresponds to matrix multiplication is composition.

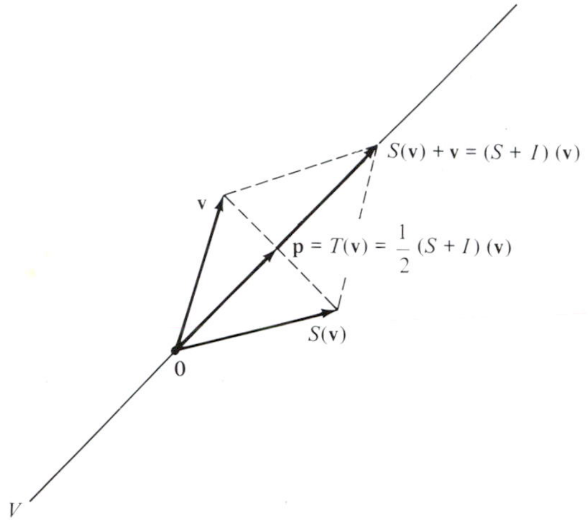


Figure 5.10

***Definition*** Let $T: R^n \to R^m$ and $S: R^m \to R^k$ be two linear transformations. The composition of $S$ and $T$ is the function $S \circ T: R^n \to R^k$ defined by

$$S \circ T(\mathbf{x}) = S(T(\mathbf{x}))$$

for all $\mathbf{x}$ in $R^n$.

Notice that as with addition and scalar multiplication of linear transformations, the composition of two linear transformations is just their composition as functions.

***Theorem 4*** *Let $T: R^n \to R^m$ and $S: R^m \to R^k$ be two linear transformations.*

*(a) $S \circ T$ is a linear transformation.*
*(b) If $A$ and $B$ are the standard matrices for $S$ and $T$, respectively, then $AB$ is the standard matrix for $S \circ T$.*

**Proof of (a)** Let $\mathbf{x}$ and $\mathbf{y}$ be vectors in $R^n$ and let $\alpha$ be a scalar. Then

$$\begin{aligned} S \circ T(\mathbf{x} + \mathbf{y}) &= S(T(\mathbf{x} + \mathbf{y})) && \text{(definition)} \\ &= S(T(\mathbf{x}) + T(\mathbf{y})) && \text{(linearity of } T) \\ &= S(T(\mathbf{x})) + S(T(\mathbf{y})) && \text{(linearity of } S) \\ &= S \circ T(\mathbf{x}) + S \circ T(\mathbf{y}) && \text{(definition)} \end{aligned}$$

and

$$\begin{aligned} S \circ T(\alpha\mathbf{x}) &= S(T(\alpha\mathbf{x})) && \text{(definition)} \\ &= S(\alpha T(\mathbf{x})) && \text{(linearity of } T) \\ &= \alpha S(T(\mathbf{x})) && \text{(linearity of } S) \\ &= \alpha(S \circ T)(\mathbf{x}) && \text{(definition)} \end{aligned}$$

Thus $S \circ T: R^n \to R^k$ is indeed a linear transformation.

**Proof of (b)** The columns of the standard matrix for $S \circ T$ are the vectors

$$S \circ T(\mathbf{e}_1), S \circ T(\mathbf{e}_2), \ldots, S \circ T(\mathbf{e}_n).$$

Since $A$ and $B$ are the standard matrices for $S$ and $T$, respectively,

$$\begin{aligned}(S \circ T)(\mathbf{e}_i) &= S(T(\mathbf{e}_i)) = A(B\mathbf{e}_i) = (AB)\mathbf{e}_i \\ &= i\text{th column of } AB\end{aligned}$$

for $i = 1, 2, \ldots, n$. Thus $AB$ is indeed the standard matrix for $S \circ T$.

## EXAMPLE 12

Let $T$: $R^2 \rightarrow R^2$ be the linear transformation defined as follows. If $\mathbf{x}$ is a vector in $R^2$, then $T(\mathbf{x})$ is the vector obtained by first projecting $\mathbf{x}$ onto the line spanned by $(1, -1)$ and then reflecting this projection across the line spanned by $(0, 1)$. The definition of $T$ should make it clear that $T = R \circ S$, where $S$: $R^2 \rightarrow R^2$ is the projection onto the subspace $V$ spanned by $(1, -1)$ and $R$: $R^2 \rightarrow R^2$ is the reflection of $R^2$ across the subspace spanned by $(0, 1)$. Let us first compute the standard matrix for $T$ and then verify that it is the product of the standard matrix for $R$ with the standard matrix for $S$. From elementary geometry (see Figure 5.11),

$$\begin{aligned}T(\mathbf{e}_1) &= (-1/2, -1/2) \\ T(\mathbf{e}_2) &= (1/2, 1/2)\end{aligned}$$

Thus the standard matrix for $T$ is

$$C = \begin{bmatrix} -1/2 & 1/2 \\ -1/2 & 1/2 \end{bmatrix}.$$

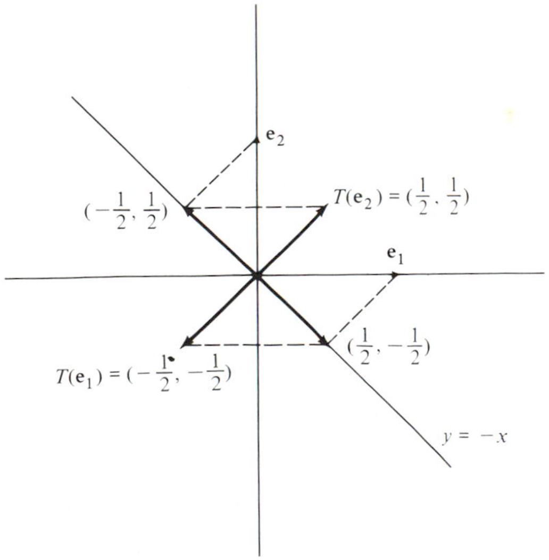


Figure 5.11

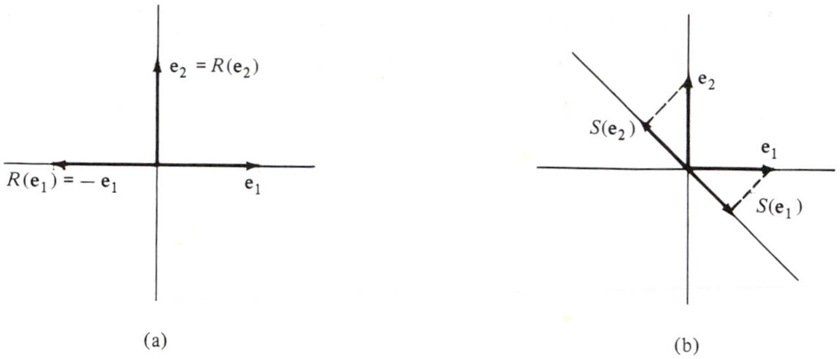


**Figure 5.12**

Again from elementary geometry we see that [see Figure 5.12 (a) and (b)]

$$R(\mathbf{e}_1) = (-1, 0), \qquad R(\mathbf{e}_2) = (0, 1)$$

and

$$S(\mathbf{e}_1) = (1/2, -1/2), \qquad S(\mathbf{e}_2) = (-1/2, 1/2).$$

Thus the standard matrices for $R$ and $S$ are, respectively,

$$A = \begin{bmatrix} -1 & 0 \\ 0 & 1 \end{bmatrix} \quad \text{and} \quad B = \begin{bmatrix} 1/2 & -1/2 \\ -1/2 & 1/2 \end{bmatrix}.$$

Finally,

$$AB = \begin{bmatrix} -1 & 0 \\ 0 & 1 \end{bmatrix}\begin{bmatrix} 1/2 & -1/2 \\ -1/2 & 1/2 \end{bmatrix} = \begin{bmatrix} -1/2 & 1/2 \\ -1/2 & 1/2 \end{bmatrix} = C.$$

When $T$ is a linear transformation from $R^n$ into itself, we define $T^2 = T \circ T$, $T^3 = T \circ T \circ T$, and in general we define $T^n$ to be $T$ composed with itself $n$ times. For instance, if $T$: $R^3 \to R^3$ is the linear transformation defined by

$$T(x_1, x_2, x_3) = (x_1 + x_2, x_3, x_2 + x_3),$$

then

$$T(1, 2, 3) = (3, 3, 5),$$
$$T^2(1, 2, 3) = T(3, 3, 5) = (6, 5, 8)$$

and

$$T^3(1, 2, 3) = T(6, 5, 8) = (11, 8, 13).$$

**EXAMPLE 13**

If $T$: $R^2 \to R^2$ is a rotation through an angle $\theta$, then $T^2$ is a rotation through an angle $2\theta$. For any nonnegative integer $n$, $T^n$ is a rotation through the angle $n\theta$.

## EXAMPLE 14

Let $V$ be a subspace of $R^m$. If $T: R^m \to R^m$ is the projection of $R^m$ onto $V$, then $T^2 = T$. If $S: R^m \to R^m$ is the reflection of $R^m$ across $V$, then $S^2 = I$.

# PROBLEMS 5.1

**1.** Complete each of the following definitions.

(a) The image of $R^n$ under $T$ is....

(b) A function $T: R^n \to R^m$ is called a linear transformation if....

(c) The standard matrix of a linear transformation $T: R^n \to R^m$ is....

(d) A rotation is....

(e) The sum of two linear transformations is....

(f) The composition of two linear transformations is....

**2.** Which of the following functions are linear? If the function is linear, find its standard matrix.

*(a) $T: R^3 \to R^2$ defined by $T(x_1, x_2, x_3) = (x_2, x_1)$

(b) $T: R^2 \to R^2$ defined by $T(x_1, x_2) = (x_2, x_1 x_2)$

*(c) $T: R^1 \to R^3$ defined by $T(x_1) = (x_1, 0, 1)$

(d) $T: R^2 \to R^1$ defined by $T(x_1, x_2) = 2x_1 + 3x_2$

*(e) $T: R^n \to R^1$ defined by $T(\mathbf{x}) = \|\mathbf{x}\|$

(f) $T: R^3 \to R^3$ defined by $T(x_1, x_2, x_3) = (x_1, x_2, x_3) + (1, 0, 1)$

*(g) $T: R^2 \to R^2$ defined by $T(x_1, x_2) = (1 + x_1^2, x_1 + x_2)$

**3.** In each part, find the standard matrix for the given linear transformation.

*(a) $T: R^3 \to R^2$ defined by $T(x_1, x_2, x_3) = (2x_1 + x_2, x_2 - x_3)$

(b) $T: R^4 \to R^1$ defined by $T(x_1, x_2, x_3, x_4) = x_1 + 2x_2 - x_3 + 5x_4$

*(c) $T: R^2 \to R^4$ defined by $T(x_1, x_2) = (0, x_2, x_1, x_1 + x_2)$

(d) $T: R^3 \to R^3$ defined by $T(x_1, x_2, x_3) = (x_1 + x_2 + x_3, x_1 + x_2, x_1)$

*(e) $T: R^3 \to R^3$ defined by $T(x_1, x_2, x_3) = (x_1 + x_2, x_2 + x_3, x_3 + x_1)$

(f) $T: R^3 \to R^2$ defined by $T(x_1, x_2, x_3) = (-x_2, x_1)$

**4.** Let $T: R^2 \to R^2$ be a rotation through an angle $\theta$. Find $T(1, 2)$ if $\theta$ is

*(a) $\dfrac{\pi}{6}$ (b) $\dfrac{5\pi}{6}$ *(c) $\dfrac{\pi}{4}$ (d) $-\dfrac{\pi}{4}$

**5.** Let $T: R^2 \to R^2$ be the projection of $R^2$ onto the subspace spanned by $(1, 3)$.

(a) Find the standard matrix of $T$.

(b) Find $T(5, 2)$.

(c) What is the image of $R^2$ under $T$?

(d) What is the standard matrix for $T^2$? $T^3$? $T^n$?

***6.** Let $T: R^3 \to R^3$ be the projection of $R^3$ onto the plane spanned by the vectors $(1, 0, 0)$ and $(0, 1, 1)$.

(a) Find the standard matrix for $T$.

(b) Find $T(1, 2, 1)$.

(c) What is the image of $R^3$ under $T$?

(d) What is the standard matrix for $T^2$? $T^3$? $T^n$?

**7.** Consider the reflection $T$ of $R^3$ across the plane spanned by $(1, 0, 0)$ and $(0, 1, 1)$.

(a) Find the standard matrix for $T$.

(b) Compute $T(1, 2, 3)$ and $T(2, -1, 2)$.

(c) What is the image of $R^3$ under $T$?

(d) What is the standard matrix for $T^2$? $T^3$? $T^n$?

**8.** In Example 9 we found the standard matrix for the projection of $R^4$ onto the subspace $x_1 + x_2 + x_3 + x_4 = 0$. Compute this matrix again as follows.

(a) Find the standard matrix for the projection onto the orthogonal complement of the given subspace.

(b) Use part (a) to find the standard matrix for the projection onto $V$.
(c) When will this method be less work than the method used in Example 9?

*9. Let $T$ be the projection of $R^3$ onto the subspace $x_1 + 2x_2 - 2x_3 = 0$. Compute the standard matrix for $T$ in two ways.

10. Repeat Problem 9 for the projection of $R^4$ onto the subspace spanned by $(1, 0, 1, 0)$ and $(1, 1, 0, 1)$.

11. Repeat Problem 9 for the projection of $R^4$ onto the subspace $x_1 + 2x_2 + 3x_3 + 4x_4 = 0$.

12. Let $V$ be a subspace of $R^m$.
(a) If $V = R^m$, what is the standard matrix for the projection of $R^m$ onto $V$?
(b) If $V = R^m$, what is the standard matrix for the reflection of $R^m$ across $V$?
(c) If $V = \{\mathbf{0}\}$, what is the standard matrix for the projection of $R^m$ onto $V$?
(d) If $V = \{\mathbf{0}\}$, what is the standard matrix for the reflection of $R^m$ across $V$?

*13. Find a formula for the sum of the linear transformations in Problem 3(a) and (f). Use this formula to find the standard matrix for the sum. Verify that this matrix is the sum of the standard matrices found in Problem 3(a) and (f).

14. Repeat Problem 13 for the linear transformations in Problem 3(d) and (e).

*15. Find a formula for the composition of the linear transformations in Problem 3(a) and (c). Use this formula to find the standard matrix for the composition. Verify that this matrix is the product of the matrices in Problem 3(a) and (c).

16. Repeat Problem 15 for each of the following linear transformations.
*(a) The linear transformations in Problem 3(d) and (a).
(b) The linear transformations in Problem 3(e) and (a).
*(c) The linear transformations in Problem 3(d) and (e).
(d) The linear transformations in Problem 3(e) and (d).
*(e) The linear transformations in Problem 3(d) and (f).
(f) The linear transformations in Problem 3(e) and (f).
*(g) The linear transformations in Problem 3(f) and (c).
(h) The linear transformations in Problem 3(c) and (b).
*(i) The linear transformations in Problem 3(d) and (d).
(j) The linear transformations in Problem 3(e) and (e).

In Problems 17 through 24, answer each of the following questions about the linear transformation $T: R^2 \to R^2$.
(a) Find the standard matrix for $T$.
(b) Express $T$ as a composition of two linear transformations, find the standard matrix of each of these transformations, and finally compute the matrix for $T$ as a product of these two matrices. Compare this matrix with the matrix you found in part (a).
(c) Find $T(5, 8)$.
(d) Describe $T(R^2)$.
(e) Find all vectors $\mathbf{x}$ in $R^2$ for which $T(\mathbf{x}) = \mathbf{x}$.
(f) Find all vectors $\mathbf{x}$ in $R^2$ for which $T(\mathbf{x}) = \mathbf{0}$.

*17. $T(\mathbf{x})$ is the vector obtained by first reflecting $\mathbf{x}$ across the $x$-axis and then projecting this reflection of $\mathbf{x}$ onto the line spanned by the vector $(1, 1)$.

18. $T(\mathbf{x})$ is the vector obtained by first projecting $\mathbf{x}$ onto the line spanned by the vector $(1, 1)$ and then reflecting this projection across the $x$-axis.

*19. $T(\mathbf{x})$ is the vector obtained by first projecting $\mathbf{x}$ onto the line $y = -x$ and then rotating this projection through an angle $3\pi/2$ (counterclockwise).

***20.** $T(\mathbf{x})$ is the vector obtained by first rotating $\mathbf{x}$ through an angle of $3\pi/2$ and then projecting onto the line $y = -x$.

**21.** $T(\mathbf{x})$ is the vector obtained by first projecting $\mathbf{x}$ onto the line $y = 2x$ and then rotating this projection through an angle of $2\pi/3$.

***22.** $T(\mathbf{x})$ is the vector obtained by first rotating $\mathbf{x}$ by an angle of $2\pi/3$ and then projecting this rotation onto the line $y = 2x$.

**23.** $T(\mathbf{x})$ is the vector obtained by first projecting $\mathbf{x}$ onto the line spanned by $(1, 1)$ and then reflecting this projection across the line spanned by $(1, -1)$.

**24.** $T(\mathbf{x})$ is the vector obtained by first reflecting $\mathbf{x}$ across the line spanned by $(1, -1)$ and then projecting this reflection onto the line spanned by $(1, 1)$.

***25.** Let $S$ be the projection of $R^3$ onto the plane $x_1 + x_2 + x_3 = 0$ and let $T$ be the projection of $R^3$ onto the plane $x_1 - x_2 + x_3 = 0$. Find the standard matrices of $S \circ T$ and $T \circ S$.

**26.** Let $S$ be the projection of $R^4$ onto the $x_1x_2$-plane and $T$ be the projection of $R^4$ onto the plane $x_3 = x_4$. Find the standard matrices for $S \circ T$ and $T \circ S$.

**27.** Let $S$ be a counterclockwise rotation of $R^2$ through an angle $\theta$ and $T$ a counterclockwise rotation through an angle $\phi$.
(a) Find the standard matrices for $S$ and $T$.
(b) Find a formula for $S \circ T$ and then find its standard matrix.
(c) Verify that the matrix in part (b) is a product of the matrices in part (a).
(d) Find the standard matrices for a rotation through the angle $-\theta$ and the angle $-\phi$. What is the relationship between these matrices and those found in part (a)?

**28.** Let $L$ be the straight line in $R^2$ defined by the equation $ax + by = 0$.
(a) Show that the projection matrix for $L$ is

$$\frac{1}{a^2 + b^2}\begin{bmatrix} b^2 & -ab \\ -ab & a^2 \end{bmatrix}.$$

(b) Show that the reflection matrix for $L$ is

$$\frac{1}{a^2 + b^2}\begin{bmatrix} b^2 - a^2 & -2ab \\ -2ab & a^2 - b^2 \end{bmatrix}.$$

(c) Let $\phi$ be the angle which $L$ makes with the positive $x_1$-axis. Show that the matrix in part (a) can be written as

$$\begin{bmatrix} \cos^2\phi & \sin\phi\cos\phi \\ \sin\phi\cos\phi & \sin^2\phi \end{bmatrix}.$$

(d) Show that the matrix in part (c) can be written as a product of three matrices, the right most a rotation through an angle $-\phi$, the middle a projection, the left most a rotation through an angle $\phi$. Interpret this geometrically.
(e) Show that the matrix in part (b) can be written as

$$\begin{bmatrix} \cos\theta & \sin\theta \\ \sin\theta & -\cos\theta \end{bmatrix},$$

where $\theta = 2\phi$. Thus a matrix of this form represents a reflection in the line making an angle $\theta/2$ with the $x_1$-axis.

**29.** Let $T$ be the rotation of $R^2$ through the angle $\theta$. Show that the standard matrix for $T$ is orthogonal.

**30.** Let $A$ be a $2 \times 2$ orthogonal matrix. Show that $A$ is either a rotation or a rotation followed by a reflection.

**31.***(a) Show that the matrix

$$A = \begin{bmatrix} \cos\theta & -\sin\theta & 0 \\ \sin\theta & \cos\theta & 0 \\ 0 & 0 & 1 \end{bmatrix}$$

represents a rotation of $R^3$ about the $x_3$-axis through an angle $\theta$.

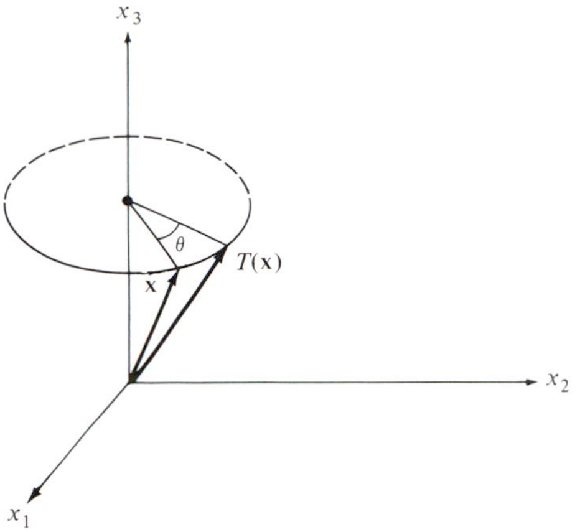


(b) Show that the matrix

$$\begin{bmatrix} \cos\theta & 0 & -\sin\theta \\ 0 & 1 & 0 \\ \sin\theta & 0 & \cos\theta \end{bmatrix}$$

represents a rotation of $R^3$ about the $x_2$-axis through an angle $\theta$.

(c) Show that the matrix

$$\begin{bmatrix} 1 & 0 & 0 \\ 0 & \cos\theta & -\sin\theta \\ 0 & \sin\theta & \cos\theta \end{bmatrix}$$

represents a rotation of $R^3$ about the $x_1$-axis through an angle $\theta$.

**32.** Let $S$ be the projection of $R^3$ onto the plane $x_1 = x_2$. Find the standard matrices for $S \circ T$ and $T \circ S$ when $T$ is rotation of $R^3$ through an angle $\theta$ about:
*(a) The $x_1$-axis.
(b) The $x_2$-axis.
(c) The $x_3$-axis.

**33.** Repeat Problem 32 when $S$ is the projection of $R^3$ onto the line spanned by $(1, 1, 1)$.

**34.** Let $V$ be a subspace of $R^m$ and let $T$ be the projection of $R^m$ onto $V$. Find the standard matrices for $S \circ T$ and $T \circ S$ for each of the following linear transformations $S$. Explain each answer geometrically.
(a) $S$ is the reflection of $R^m$ across $V$.
(b) $S$ is the reflection of $R^m$ across $V^\perp$.
(c) $S$ is the projection of $R^m$ onto $V^\perp$.

**35.** Show that conditions (a) and (b) in the definition of a linear transformation can be replaced by the single condition that

$$T(\alpha\mathbf{x} + \beta\mathbf{y}) = \alpha T(\mathbf{x}) + \beta T(\mathbf{y})$$

holds for all vectors $\mathbf{x}$ and $\mathbf{y}$ and scalars $\alpha$ and $\beta$.

**36.** Let $\mathbf{y}$ be an $m$-vector. Show that the transformation $T: R^m \to R^1$ defined by $T(\mathbf{x}) = \langle \mathbf{x}, \mathbf{y} \rangle$ is a linear transformation. What is the standard matrix for $T$?

**37.** Let $\mathbf{v}_1, \mathbf{v}_2, \ldots, \mathbf{v}_n$ be vectors in $R^m$. Show that the transformation $T: R^n \to R^m$ defined by $T(x_1, x_2, \ldots, x_n) = x_1\mathbf{v}_1 + x_2\mathbf{v}_2 + \cdots + x_n\mathbf{v}_n$ is linear.

**38.** A linear transformation $T: R^n \to R^n$ is called an **isometry** if $\langle T(\mathbf{x}), T(\mathbf{y}) \rangle = \langle \mathbf{x}, \mathbf{y} \rangle$ for all vectors $\mathbf{x}$ and $\mathbf{y}$ in $R^n$. Let $T$ be an isometry.
(a) Show that $\|T(\mathbf{x})\| = \|\mathbf{x}\|$ for every vector $\mathbf{x}$ in $R^n$.
*(b) Show that the standard matrix for $T$ is orthogonal and that $T(\mathbf{e}_1)$, $T(\mathbf{e}_2), \ldots, T(\mathbf{e}_n)$ is an orthonormal basis for $R^n$.
(c) Show that the standard matrix for $T$ is nonsingular.

**39.** Show that each of the following linear transformations is an isometry.
(a) A rotation of $R^2$.
(b) A rotation of $R^3$ about a coordinate axis.

**40.** Show that if the standard matrix $A$ of a linear transformation $T: R^n \to R^n$ is orthogonal, then $T$ is an isometry.

***41.** Let $V$ be a subspace of $R^n$ and let $T$ be the reflection of $R^n$ across $V$. Let $A$ be the standard matrix for $T$.
(a) Show that $A$ is a symmetric orthogonal matrix.
(b) Show that $T$ is an isometry.
(c) Show that $T^2 = I$.

**42.** Let $T: R^n \to R^n$ be a linear transformation. Show that if

$$\langle T(\mathbf{x}), \mathbf{y}\rangle = \langle \mathbf{x}, T(\mathbf{y})\rangle$$

for all $\mathbf{x}$, $\mathbf{y}$ in $R^n$, then the standard matrix for $T$ is symmetric. Is the converse true?

**43.** Let $T: R^n \to R^n$ be the function defined by

$$T(x_1, x_2, \ldots, x_n) = (0, x_1, x_2, \ldots, x_{n-1}).$$

(a) Show that $T$ is linear and find its standard matrix.
(b) Find $T^2$.
(c) Find $T^3$.
(d) Deduce a consequence from parts (b) and (c). For obvious reasons, $T$ is called a **shift operator**.

**44.** A linear transformation $T: R^n \to R^n$ is called **invertible** if there is a function $S: R^n \to R^n$ such that $S \circ T(\mathbf{x}) = \mathbf{x}$ and $T \circ S(\mathbf{x}) = \mathbf{x}$ for all $\mathbf{x}$ in $R^n$.
(a) Show that $S$ is unique. It is called the **inverse of** $T$ and is denoted by $T^{-1}$
(b) Show that $T^{-1}$ is linear.
*(c) Show that if $A$ is the standard matrix for $T$, then $A^{-1}$ is the standard matrix for $T^{-1}$.
(d) Show that if the standard matrix of a linear transformation $T: R^n \to R^n$ is nonsingular, then $T$ is invertible.
(e) Conclude from parts (c) and (d) that a linear transformation $T: R^n \to R^n$ is invertible if and only if its standard matrix is nonsingular.

**45.** Show that each of the following linear transformations is invertible.
(a) A rotation of $R^2$.
(b) A reflection of $R^m$ across a subspace $V$.
(c) A rotation of $R^3$ about a coordinate axis.

**46.** A function $S: R^n \to R^m$ is called an **affine** transformation if there is a linear transformation $T: R^n \to R^m$ and a vector $\mathbf{b}$ in $R^m$ such that

$$S(\mathbf{x}) = T(\mathbf{x}) + \mathbf{b}.$$

*(a) Show that a function $S: R^n \to R^m$ is affine if and only if there is a vector $\mathbf{b}$ in $R^m$ such that $S(\mathbf{x}) - \mathbf{b}$ is a linear transformation.
(b) Show that a function $S: R^n \to R^m$ is affine if and only if there is an $m \times n$ matrix $A$ and a vector $\mathbf{b}$ in $R^m$ such that

$$S(\mathbf{x}) = A\mathbf{x} + \mathbf{b}.$$

(c) Let $T: R^n \to R^m$ be a linear transformation and let $\mathbf{a}$ be a fixed vector in $R^n$. Show that the transformation $S: R^n \to R^m$ defined by

$$S(\mathbf{x}) = T(\mathbf{x} + \mathbf{a})$$

is affine.
(d) If $T: R^2 \to R^2$ is a rotation through an angle $\theta$ and $\mathbf{a}$ is a fixed vector in $R^2$, describe $S(\mathbf{x}) = T(\mathbf{x} + \mathbf{a})$.
(e) If $Q: R^n \to R^m$ and $R: R^m \to R^p$ are affine, show that their composition $S = R \circ Q$ is affine. Find a matrix $C$ and a vector $\mathbf{c}$ such that $S(\mathbf{x}) = C\mathbf{x} + \mathbf{c}$. Determine $C$ and $\mathbf{c}$ in terms of the corresponding quantities for $Q$ and $R$.

# 5.2
# Properties of Linear Transformations

In the preceding section we saw that the concept of existence of solutions of a linear system corresponds to the concept of the image of a linear transformation. We proved that the image of $R^n$ under a linear transformation $T: R^n \to R^m$ defined by an $m \times n$ matrix $A$ is the column space of the matrix $A$. We then proved that any linear transformation $T: R^n \to R^m$ can be defined by a matrix, namely its standard matrix. Therefore,

***Theorem 1*** *Let $T: R^n \to R^m$ be a linear transformation and let $A$ be its standard matrix.*

(a) *$T(R^n) = C(A)$. In particular, $T(R^n)$ is a subspace.*
(b) *$T(R^n)$ is spanned by $T(\mathbf{e}_1), T(\mathbf{e}_2), \ldots, T(\mathbf{e}_n)$.*
(c) *$\dim T(R^n) = \operatorname{rank} A \leq n$.*
(d) *$T(R^n) = R^m$ if and only if $\operatorname{rank} A = m$.*

When $T(R^n) = R^m$, we say that $T$ maps $R^n$ **onto** $R^m$. Since $\dim T(R^n) \leq n$, $T$ cannot map $R^n$ onto $R^m$ when $n < m$.

We now turn to the question of uniqueness. A linear system $A\mathbf{x} = \mathbf{b}$ has a unique solution for every vector $\mathbf{b}$ in $C(A)$ if and only if $A$ has a trivial nullspace: $N(A) = \{\mathbf{0}\}$. Since a linear transformation $T: R^n \to R^m$ corresponds to an $m \times n$ matrix $A$ (its standard matrix) via the equation $T(\mathbf{x}) = A\mathbf{x}$, the set of vectors $\mathbf{x}$ in $R^n$ such that $T(\mathbf{x}) = \mathbf{0}$ corresponds to the nullspace of $A$. We call this subset of $R^n$ the **kernel** of $T$ and denote it by $\ker(T)$. Since $\ker(T) = N(A)$, the kernel of $T$ is a subspace of $R^n$. This fact also follows directly from the properties of a linear transformation. The zero vector is in $\ker(T)$ since $T(\mathbf{0}) = \mathbf{0}$ (Theorem 1 in Section 5.1). If $\mathbf{u}$ and $\mathbf{v}$ are in $\ker(T)$ and $\alpha$ is a scalar, then

$$T(\mathbf{u} + \mathbf{v}) = T(\mathbf{u}) + T(\mathbf{v}) = \mathbf{0} + \mathbf{0} = \mathbf{0},$$
$$T(\alpha\mathbf{u}) = \alpha T(\mathbf{u}) = \alpha\mathbf{0} = \mathbf{0}.$$

Hence $\mathbf{u} + \mathbf{v}$ and $\alpha\mathbf{u}$ are in $\ker(T)$.

***Theorem 2*** *Let $T: R^n \to R^m$ be a linear transformation.*

(a) *The kernel of $T$ is a subspace of $R^n$.*
(b) *If $A$ is the standard matrix of $T$, then $\ker(T) = N(A)$.*

## EXAMPLE 1

If $T$ is the rotation transformation of $R^2$ through an angle $\theta$, then $\ker(T) = \{\mathbf{0}\}$. $T$ cannot rotate a nonzero vector onto the zero vector.

## EXAMPLE 2

Let $V$ be a $k$-dimensional subspace of $R^n$, $k < n$, and let $T$ be the projection of $R^n$ onto $V$. Then $T(\mathbf{v}) = \mathbf{0}$ if and only if $\mathbf{v}$ is orthogonal to $V$. Hence the kernel of $T$ is the orthogonal complement of the subspace $V$.

## EXAMPLE 3

The standard matrix for the linear transformation $T: R^3 \to R^2$ defined by

$$T(x_1, x_2, x_3) = (x_1 - x_3, x_2 + x_3) \text{ is } A = \begin{bmatrix} 1 & 0 & -1 \\ 0 & 1 & 1 \end{bmatrix}.$$

The vector $(1, -1, 1)$ forms a basis for the nullspace of $A$. Hence $\ker(T)$ is the straight line in $R^3$ spanned by this vector.

In terms of a linear transformation $T: R^n \to R^m$ the uniqueness property becomes: For every vector $\mathbf{y}$ in the image of $T$, the equation $T(\mathbf{x}) = \mathbf{y}$ has a unique solution. Thus if $T$ has the uniqueness property and if $\mathbf{x}_1$ and $\mathbf{x}_2$ are in $R^n$ with $\mathbf{x}_1 \neq \mathbf{x}_2$, then $T(\mathbf{x}_1) \neq T(\mathbf{x}_2)$. For if $T(\mathbf{x}_1) = T(\mathbf{x}_2)$ and we denote their common value by $\mathbf{y}$, then $\mathbf{x}_1$ and $\mathbf{x}_2$ are both solutions of $T(\mathbf{x}) = \mathbf{y}$, contradicting uniqueness.

***Definition*** A linear transformation $T: R^n \to R^m$ is called **one-to-one** if whenever $\mathbf{x}_1$ and $\mathbf{x}_2$ are in $R^n$ with $\mathbf{x}_1 \neq \mathbf{x}_2$, then $T(\mathbf{x}_1) \neq T(\mathbf{x}_2)$. That is, $T$ is one-to-one if $T$ maps any two distinct vectors in $R^n$ onto distinct vectors in $R^m$.

The transformation in Example 3 is not one-to-one because $T(1, 1, 0) = (1, 1)$ and $T(2, 0, 1) = (1, 1)$. Neither is the projection in Example 2. There are certainly many different vectors in $R^n$ which have the same projection onto the subspace $V$. In fact, if $\mathbf{v}$ is any vector in $V$, then every vector of the form $\mathbf{v} + \mathbf{w}$, where $\mathbf{w}$ is in $V^\perp$, is mapped onto $\mathbf{v}$ (see Figure 5.13). On the positive side, we note that every rotation of $R^2$ is one-to-one. Rotating two different vectors through the same angle surely cannot result in the same vector.

In view of the relationship between transformations and matrices, you should be expecting the following theorem.

***Theorem 3*** *A linear transformation $T: R^n \to R^m$ is one-to-one if and only if its kernel is trivial: $\ker(T) = \{\mathbf{0}\}$.*

**Proof** First, assume that $T$ is one-to-one. If $\mathbf{x}$ is in $R^n$ and $\mathbf{x} \neq \mathbf{0}$, then $\mathbf{x}$ is not in $\ker(T)$ because $T(\mathbf{x}) \neq T(\mathbf{0}) = \mathbf{0}$. Therefore, $\ker(T)$ contains no nonzero vectors. Conversely, suppose that $\ker(T) = \{\mathbf{0}\}$, and let $\mathbf{x}_1$ and $\mathbf{x}_2$ be in $R^n$. If $\mathbf{x}_1 \neq \mathbf{x}_2$, then $\mathbf{x}_1 - \mathbf{x}_2 \neq \mathbf{0}$ and hence $\mathbf{x}_1 - \mathbf{x}_2$ is not in $\ker(T)$. Therefore $T(\mathbf{x}_1) - T(\mathbf{x}_2) = T(\mathbf{x}_1 - \mathbf{x}_2) \neq \mathbf{0}$. Consequently, $T(\mathbf{x}_1) \neq T(\mathbf{x}_2)$ and $T$ is one-to-one.

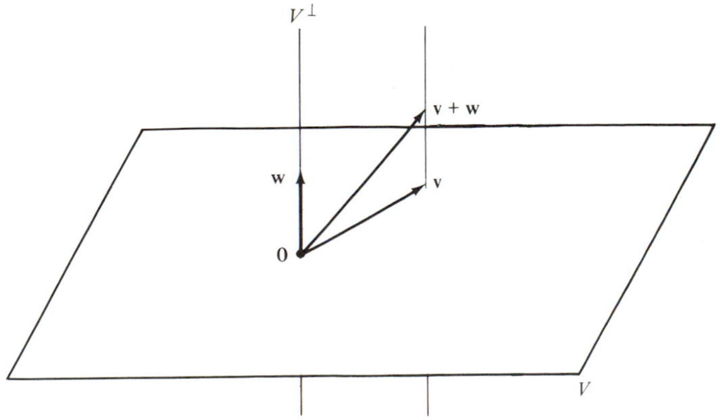


Figure 5.13

Note that Theorem 3 says that a linear transformation maps any two distinct vectors in $R^n$ onto distinct vectors in $R^m$ if and only if it maps every nonzero vector onto a nonzero vector. Note also that if we let $A$ be the standard matrix for $T$, then the theorem follows immediately from Theorem 1 of Section 2.3.

### EXAMPLE 4

Let us prove that the reflection $T$ of $R^m$ across a subspace $V$ of $R^m$ is one-to-one. Let $P$ be the projection matrix for $V$. Then $T(\mathbf{x}) = (2P - I)\mathbf{x}$ for all $\mathbf{x}$ in $R^m$ (see Example 10 of Section 5.1). Suppose that $\mathbf{u}$ is a vector in the nullspace of $T$. Then $\mathbf{0} = T(\mathbf{u}) = 2P\mathbf{u} - \mathbf{u}$, or $\mathbf{u} = 2P\mathbf{u}$. Since $P\mathbf{u}$, the projection of $\mathbf{u}$ onto $V$, is in $V$, it follows that $\mathbf{u}$ is in $V$. But this means that $P\mathbf{u} = \mathbf{u}$. Therefore, $\mathbf{u} = 2\mathbf{u}$ and hence $\mathbf{u} = \mathbf{0}$. We conclude that $\ker(T) = \{\mathbf{0}\}$.

We now state and prove a very important theorem: the **rank plus nullity** theorem for linear transformations. If $T: R^n \to R^m$ is a linear transformation, then the **nullity** of $T$ is the dimension of its kernel, and the **rank** of $T$ is the dimension of its image. Thus if $A$ is the standard matrix for $T$, then:

$$\text{nullity of } T = \dim \ker(T) = \dim N(A) = \text{nullity of } A.$$
$$\text{rank of } T = \dim T(R^n) = \dim C(A) = \text{rank of } A.$$

Thus from the rank plus nullity theorem for matrices, we have proved the rank plus nullity theorem for linear transformations.

***Theorem 4*** *(Rank Plus Nullity Theorem) Let $T: R^n \to R^m$ be a linear transformation. Then the rank of $T$ plus the nullity of $T$ is $n$:*

$$\dim T(R^n) + \dim \ker(T) = n.$$

***Theorem 5*** *Let $T: R^n \to R^m$ be a linear transformation and let $\mathbf{v}_1, \mathbf{v}_2, \ldots, \mathbf{v}_n$ be a basis for $R^n$. The following conditions are equivalent.*

*(a) $T$ is one-to-one.*
*(b) The vectors $T(\mathbf{v}_1), T(\mathbf{v}_2), \ldots, T(\mathbf{v}_n)$ form a basis for $T(R^n)$.*
*(c) $\dim T(R^n) = n$.*

**Proof** The proof follows immediately from the rank plus nullity theorem. Condition (a) is equivalent to the nullity of $T$ being 0. Conditions (b) and (c) are equivalent to the rank of $T$ being $n$.

It follows immediately from Theorem 5 that a linear transformation $T: R^n \to R^m$ cannot be one-to-one when $m < n$.

The nature of a linear transformation $T: R^n \to R^m$ cannot be fully understood without studying how $T$ transforms subsets of $R^n$ and, in particular, subspaces of $R^n$. For example, a rotation of $R^2$ through an angle $\theta$ carries each straight line onto a straight line and carries the whole plane onto itself. The dimension of a subspace of $R^2$ does not change under rotation. In contrast to this, projecting $R^2$ onto, say, the

$x$-axis collapses the whole plane onto a one-dimensional subspace. All one-dimensional subspaces of $R^2$ with the exception of the $y$-axis are carried onto the $x$-axis. The $y$-axis is projected onto the origin.

Let $T: R^n \to R^m$ be a linear transformation and let $U$ be a set of vectors in $R^n$. Then $T$ transforms $U$ into a set of vectors in $R^m$, namely the set $Z$ of all images (under $T$) of the vectors in $U$. This subset $Z$ of $R^m$ is called the **image of *U* under *T*** and we say that $T$ maps $U$ **onto** $Z$. When convenient we will denote the image of $U$ under $T$ by $T(U)$.

## EXAMPLE 5

Let $T: R^n \to R^m$ be a linear transformation. Let $\mathbf{u}$ be a nonzero vector in $R^n$ and let $V$ be the line through the origin determined by $\mathbf{u}$. An equation for this line is

$$\mathbf{x} = \alpha\mathbf{u}, \qquad \alpha \text{ a scalar.}$$

Since $T$ is a linear transformation, the image of any vector $\mathbf{x} = \alpha\mathbf{u}$ in $V$ is

$$T(\mathbf{x}) = \alpha T(\mathbf{u}).$$

Thus the image of the line $V$ under $T$ is the subspace spanned by the vector $T(\mathbf{u})$. If $T(\mathbf{u}) = \mathbf{0}$, then the image of $V$ under $T$ consists of the origin. If $T(\mathbf{u}) \neq \mathbf{0}$, then the image of $V$ under $T$ is the line through the origin determined by the vector $T(\mathbf{u})$. Thus the image of a line through the origin under a linear transformation $T$ is either the origin or a line through the origin (see Figure 5.14).

## EXAMPLE 6

Let $T: R^n \to R^m$ be a linear transformation. Suppose that $n \geq 2$ and let $\mathbf{v}$ and $\mathbf{w}$ be two noncollinear vectors in $R^n$. If $V$ is the plane through the origin determined by $\mathbf{u}$

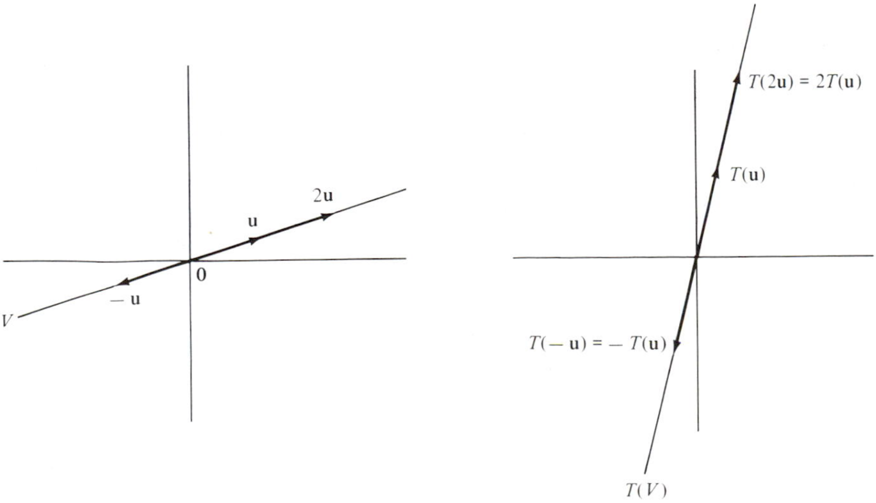


Figure 5.14

and $\mathbf{v}$, then an equation of $V$ is

$$\mathbf{x} = \alpha\mathbf{u} + \beta\mathbf{v}, \qquad \text{where } \alpha \text{ and } \beta \text{ are scalars.}$$

Since $T$ is a linear transformation, the image of any vector $\mathbf{x} = \alpha\mathbf{u} + \beta\mathbf{v}$ under $T$ is

$$T(\mathbf{x}) = \alpha T(\mathbf{u}) + \beta T(\mathbf{v}).$$

Thus the image of the plane $V$ under $T$ is the subspace of $R^m$ spanned by $T(\mathbf{u})$ and $T(\mathbf{v})$. If both $T(\mathbf{u}) = \mathbf{0}$ and $T(\mathbf{v}) = \mathbf{0}$, then the image is the trivial subspace $\{\mathbf{0}\}$. If $T(\mathbf{u})$ and $T(\mathbf{v})$ are dependent (i.e., collinear) and at least one of them is not zero, say $T(\mathbf{u})$, then the image of $V$ under $T$ is the line determined by $T(\mathbf{u})$. Finally, if $T(\mathbf{u})$ and $T(\mathbf{v})$ are independent (i.e., noncollinear), then the image of the plane $V$ is the plane spanned by $T(\mathbf{u})$ and $T(\mathbf{v})$. Thus we have shown that the image of a plane through the origin is a point (namely $\mathbf{0}$), a line through the origin, or a plane through the origin (see Figure 5.15).

We turn now to the generalization of (a), (b), and (c) of Theorem 1 to the image of a subspace under a linear transformation.

***Theorem 6*** *Let $T: R^n \to R^m$ be a linear transformation. Suppose that $V$ is a subspace of $R^n$ and that $\mathbf{v}_1, \mathbf{v}_2, \ldots, \mathbf{v}_k$ span $V$. Then*

*(a) The image $T(V)$ of $V$ is a subspace of $R^m$.*
*(b) The vectors $T(\mathbf{v}_1), T(\mathbf{v}_2), \ldots, T(\mathbf{v}_k)$ span $T(V)$.*
*(c) $\dim T(V) \le \dim V$.*

**Proof** Since $\mathbf{v}_1, \mathbf{v}_2, \ldots, \mathbf{v}_k$ are vectors that span $V$, every vector $\mathbf{v}$ in $V$ is a linear combination

$$\mathbf{v} = \alpha_1\mathbf{v}_1 + \alpha_2\mathbf{v}_2 + \cdots + \alpha_k\mathbf{v}_k$$

of the $\mathbf{v}_i$'s. Since

$$\begin{aligned} T(\mathbf{v}) &= T(\alpha_1\mathbf{v}_1 + \alpha_2\mathbf{v}_2 + \cdots + \alpha_k\mathbf{v}_k) \\ &= \alpha_1 T(\mathbf{v}_1) + \alpha_2 T(\mathbf{v}_2) + \cdots + \alpha_k T(\mathbf{v}_k), \end{aligned}$$

we see that the image of $V$ under $T$ is the set of all linear combinations of the vectors $T(\mathbf{v}_1), T(\mathbf{v}_2), \ldots, T(\mathbf{v}_k)$. This proves (a) and (b). If the $\mathbf{v}_i$'s form a basis for $V$ (i.e., $\dim V = k$), then $\dim T(V) \le k$ because $T(V)$ is spanned by $k$ vectors. This proves (c).

Part (b) of the theorem provides a method of identifying the image of a subspace as the column space of a matrix.

**Result 1** Let $T: R^n \to R^m$ be a linear transformation and let $A$ be its standard matrix. If $V$ is a subspace of $R^n$ spanned by the vectors $\mathbf{v}_1, \mathbf{v}_2, \ldots, \mathbf{v}_k$ and if $B$ is the matrix having these vectors as its columns, then the image of $V$ under $T$ is the column space of the matrix $AB$.

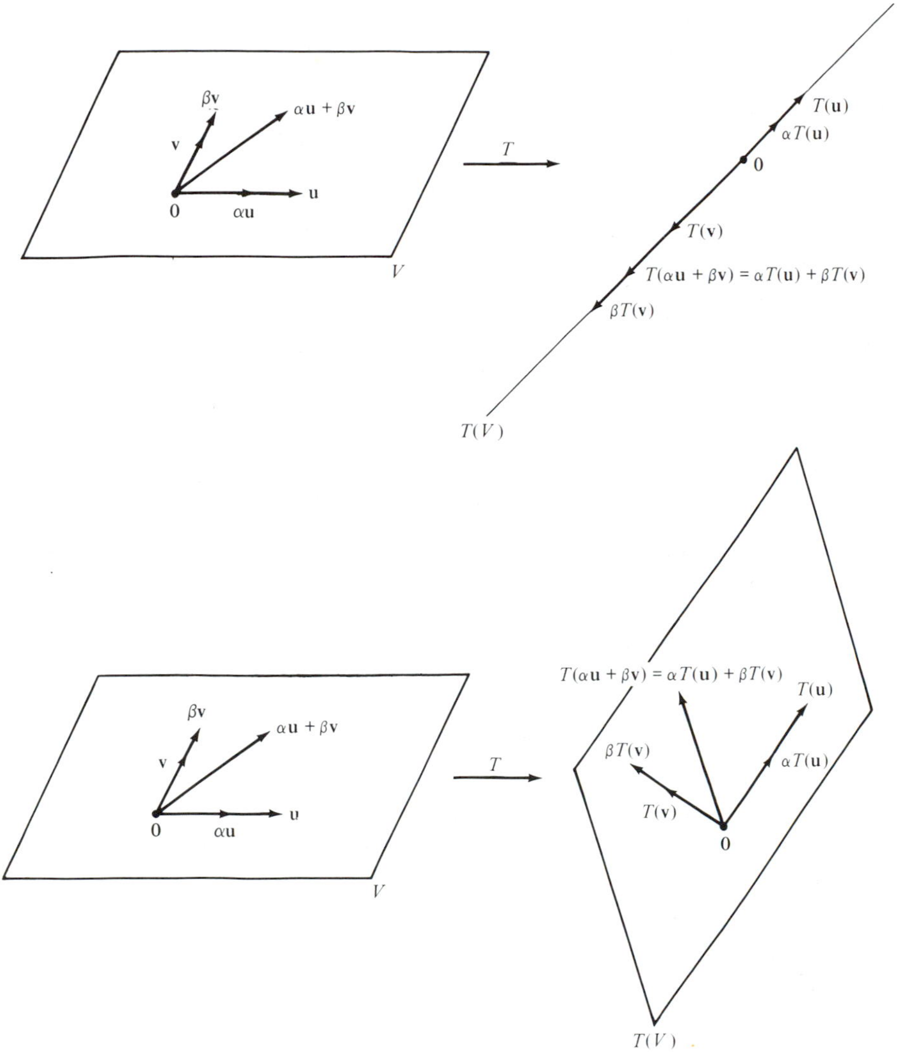


Figure 5.15

**Proof** The columns of the matrix $AB$ are the vectors $A\mathbf{v}_1, A\mathbf{v}_2, \ldots, A\mathbf{v}_k$, which are the vectors $T(\mathbf{v}_1), T(\mathbf{v}_2), \ldots, T(\mathbf{v}_k)$. Since $T(V)$ is spanned by these latter vectors, it is the column space of $AB$.

## EXAMPLE 7

Let $T\colon R^3 \to R^3$ be the transformation defined by $T(\mathbf{x}) = A\mathbf{x}$, where

$$A = \begin{bmatrix} 1 & 1 & 3 \\ 0 & 1 & 1 \\ 2 & 0 & 4 \end{bmatrix}.$$

Reducing $A$ to an echelon matrix shows that the first two columns of $A$ form a basis for the column space of $A$. Thus $T$ maps $R^3$ onto the plane spanned by the vectors $(1,0,2)$ and $(1,1,0)$. $T$ maps the plane $V$ spanned by $(2,1,1)$ and $(1,1,1)$ onto the column space of the matrix

$$\begin{bmatrix} 1 & 1 & 3 \\ 0 & 1 & 1 \\ 2 & 0 & 4 \end{bmatrix}\begin{bmatrix} 2 & 1 \\ 1 & 1 \\ 1 & 1 \end{bmatrix} = \begin{bmatrix} 6 & 5 \\ 2 & 2 \\ 8 & 6 \end{bmatrix}.$$

The two columns of this matrix are independent. Hence $T$ maps $V$ onto a plane. This plane is a subspace of the plane $T(R^3)$ and hence $T(V) = T(R^3)$. On the other hand, $T$ maps the plane $W$ spanned by $(1,1,0)$ and $(0,1,1)$ onto the column space of the matrix

$$\begin{bmatrix} 1 & 1 & 3 \\ 0 & 1 & 1 \\ 2 & 0 & 4 \end{bmatrix}\begin{bmatrix} 1 & 0 \\ 1 & 1 \\ 0 & 1 \end{bmatrix} = \begin{bmatrix} 2 & 4 \\ 1 & 2 \\ 2 & 4 \end{bmatrix}.$$

The columns of this matrix are dependent. Hence $T$ maps $W$ onto the line spanned by $(2,1,2)$.

It follows from Theorem 6 that a linear transformation $T\colon R^n \to R^m$ cannot map a subspace $V$ onto a subspace of dimension greater than the dimension of $V$. But the dimension of the image of $V$ under $T$ may be less that the dimension of $V$. The following theorem indicates when these dimensions are equal.

***Theorem 7*** *Let $T\colon R^n \to R^m$ be a linear transformation and let $V$ be a subspace of $R^n$. Let $\mathbf{v}_1, \mathbf{v}_2, \ldots, \mathbf{v}_k$ be a basis for $V$. Then the following conditions are equivalent.*

*(a) $T(\mathbf{v}) \neq \mathbf{0}$ for all nonzero vectors $\mathbf{v}$ in $V$.*
*(b) The vectors $T(\mathbf{v}_1), T(\mathbf{v}_2), \ldots, T(\mathbf{v}_k)$ form a basis for $T(V)$.*
*(c) $\dim T(V) = \dim V$.*

**Proof** First we show that conditions (a) and (b) are equivalent. Suppose that condition (a) holds. If there are scalars $\alpha_1, \alpha_2, \ldots, \alpha_k$ such that

$$\alpha_1 T(\mathbf{v}_1) + \alpha_2 T(\mathbf{v}_2) + \cdots + \alpha_k T(\mathbf{v}_k) = \mathbf{0}$$

then

$$T(\alpha_1 \mathbf{v}_1 + \alpha_2 \mathbf{v}_2 + \cdots + \alpha_k \mathbf{v}_k) = \mathbf{0}.$$

Since $T(\mathbf{v}) \neq \mathbf{0}$ for all nonzero vectors in $V$, it follows that

$$\alpha_1 \mathbf{v}_1 + \alpha_2 \mathbf{v}_2 + \cdots + \alpha_k \mathbf{v}_k = \mathbf{0}.$$

Thus $\alpha_1 = \alpha_2 = \cdots = \alpha_k = 0$ because the $\mathbf{v}_i$'s are independent. This proves that $T(\mathbf{v}_1), T(\mathbf{v}_2), \ldots, T(\mathbf{v}_k)$ are independent. Since these vectors span $T(V)$

(Theorem 6), they form a basis for $T(V)$. Conversely, suppose that

$$T(\mathbf{v}_1), T(\mathbf{v}_2), \ldots, T(\mathbf{v}_k)$$

form a basis for $T(V)$. If $\mathbf{v}$ is a nonzero vector in $V$ and

$$\mathbf{v} = \alpha_1 \mathbf{v}_1 + \alpha_2 \mathbf{v}_2 + \cdots + \alpha_k \mathbf{v}_k,$$

then at least one of the scalars, say $\alpha_i$, is not zero. Now

$$T(\mathbf{v}) = \alpha_1 T(\mathbf{v}_1) + \alpha_2 T(\mathbf{v}_2) + \cdots + \alpha_k T(\mathbf{v}_k).$$

Since $\alpha_i \neq 0$ and since the vectors $T(\mathbf{v}_1)$, $T(\mathbf{v}_2), \ldots, T(\mathbf{v}_k)$ are independent we conclude that $T(\mathbf{v}) \neq \mathbf{0}$. This shows that conditions (a) and (b) are equivalent.

Next we prove that conditions (b) and (c) are equivalent. If condition (b) holds, then clearly $\dim T(V) = k$. Conversely, suppose $\dim T(V) = k$. By Theorem 6

$$T(\mathbf{v}_1), T(\mathbf{v}_2), \ldots, T(\mathbf{v}_k)$$

span $T(V)$. Since the dimension of $T(V) = k$, we conclude from Theorem 8 of Section 2.5 that they form a basis for $T(V)$. Thus conditions (b) and (c) are equivalent.

Let $T\colon R^n \to R^m$ be a linear transformation. Theorem 7 says that $T$ maps a subspace $V$ onto a subspace of the same dimension as $V$ if and only if the zero vector is the only vector that is both in $V$ and in the kernel of $T$. Hence if $\ker T = \{\mathbf{0}\}$ (i.e., if $T$ is one-to-one), then

1. $T$ maps any subspace $V$ onto a subspace of the same dimension as $V$.
2. $T$ maps a basis for $V$ to a basis for $T(V)$.

If $\ker T \neq \{\mathbf{0}\}$ and $\mathbf{x}$ is a nonzero vector in the kernel of $T$, then $T$ maps every subspace containing $\mathbf{x}$ onto a subspace of smaller dimension.

## EXAMPLE 8

The linear transformation $T\colon R^2 \to R^3$ defined by $T(\mathbf{x}) = A\mathbf{x}$, where

$$A = \begin{bmatrix} 1 & 1 \\ 0 & 1 \\ 1 & 1 \end{bmatrix}.$$

is one-to-one because $\ker(T) = N(A) = \{\mathbf{0}\}$. Hence $T$ maps every one-dimensional subspace of $R^2$ onto a one-dimensional subspace of $R^3$. More precisely, if the vector $\mathbf{u}$ spans the straight line $U$ in $R^2$, then $T(\mathbf{u})$ spans the straight line $T(U)$ in $R^3$. $T$ maps $R^2$ onto the plane in $R^3$ spanned by $T(1,0) = (1,0,1)$ and $T(0,1) = (1,1,1)$.

## EXAMPLE 9

Let $T$ be the projection transformation of $R^3$ onto a plane $W$ (see Figure 5.16). Let $V$ be a plane in $R^3$. If $\mathbf{v}$ is a vector in $V$, then $T(\mathbf{v}) = \mathbf{0}$ if and only if $\mathbf{v}$ lies in

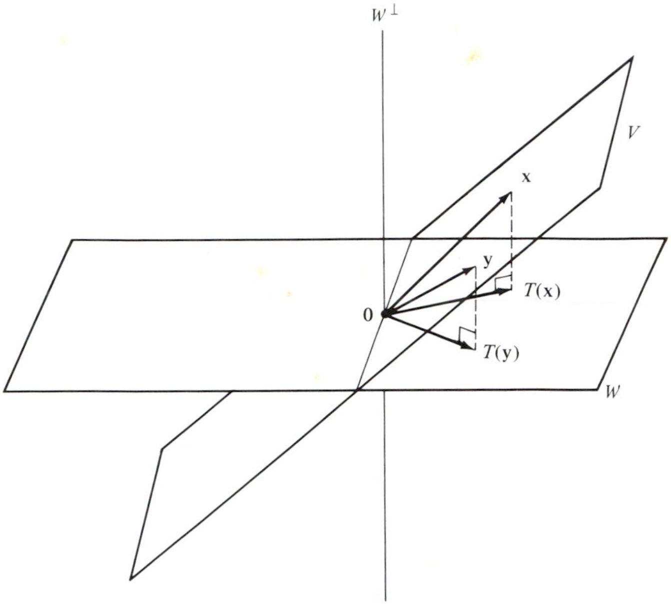


Figure 5.16

the orthogonal complement of $W$ (since $\ker T = W^{\perp}$). Thus by Theorem 7 $\dim T(V) = 2$ if and only if the plane $V$ and the straight line $W^{\perp}$ have no vectors in common besides the zero vector. If this is the case, then $T(V) = W$ because $T(V)$ is a two-dimensional subspace of the plane $W$. On the other hand, if the plane $V$ contains the line $W^{\perp}$, then we can choose a basis for $V$ consisting of a vector $\mathbf{v}_1$ in $W^{\perp}$ and a vector $\mathbf{v}_2$ not in $W^{\perp}$. By Theorem 6, $T(\mathbf{v}_1)$ and $T(\mathbf{v}_2)$ span $T(V)$. But $T(\mathbf{v}_1) = \mathbf{0}$ and $T(\mathbf{v}_2) \neq \mathbf{0}$. Consequently $T(\mathbf{v}_2)$ is a basis for $T(V)$ and $T(V)$ is one-dimensional.

## PROBLEMS 5.2

**1.** Complete each of the following definitions.

(a) $T: R^n \to R^m$ is onto if....

(b) If $V$ is a subspace, then $T(V)$ is....

(c) The kernel of $T$ is....

(d) $T$ is one-to-one if....

(e) The nullity of $T$ is....

**2.** Let $T: R^3 \to R^3$ be the linear transformation defined by $T(x_1, x_2, x_3) = (x_1 + x_2, 2x_1 + 3x_2 + x_3, x_2 + x_3)$. For each of ˙ɘ following subspaces, find a basis for ıʟs image under $T$.

*(a) $R^3$

(b) The plane spanned by $(1, 1, 0)$ and $(0, 0, 1)$

*(c) The plane spanned by $(0, 1, 1)$ and $(1, 1, 1)$

(d) The line spanned by $(5, 1, 3)$

***3.** Repeat Problem 2 with the linear transformation $T: R^3 \to R^2$ defined by $T(x_1, x_2, x_3) = (x_1 + x_3, x_2 + x_3)$.

***4.** Determine which of the following linear transformations are one-to-one.

(a) $T: R^2 \to R^2$ defined by
$T(x_1, x_2) = (x_1 + x_2, 2x_1 + 2x_2)$
(b) $T: R^2 \to R^3$ defined by
$T(x_1, x_2) = (x_1, x_1 + x_2, x_1 - x_2)$
(c) $T: R^3 \to R^2$ defined by
$T(x_1, x_2, x_3) = (x_1 - x_2, x_3)$
(d) $T: R^3 \to R^3$ defined by
$T(x_1, x_2, x_3) = (x_3, x_1, x_2)$
(e) $T: R^2 \to R^3$ defined by
$T(x_1, x_2) = (x_1, x_1, x_1)$

**5.** Let $T: R^3 \to R^2$ be the linear transformation defined by

$$T(x_1, x_2, x_3) = (x_1 + x_2 + x_3, x_2 - 5x_3).$$

(a) Show that $T(R^3) = R^2$.
(b) Find dim $T(V)$ where $V$ is the subspace of $R^3$ spanned by $\mathbf{v}_1 = (2, 1, 2)$, $\mathbf{v}_2 = (-4, 6, 3)$.

***6.** Let $T: R^4 \to R^3$ be the linear transformation defined by

$$T(x_1, x_2, x_3, x_4) = (x_1 + x_2, x_1 + x_3, x_1 + x_4).$$

(a) What is the rank of $T$? What is the nullity of $T$?
(b) If $U$ is the subspace defined by $x_2 = x_4$, find a basis for $T(U)$.

**7.** Let $T: R^4 \to R^4$ be the linear transformation defined by

$$T(x_1, x_2, x_3, x_4) = (x_1 + x_2 + x_3 + x_4, x_2 + x_3 + x_4, x_3 - x_4, x_3 - x_4).$$

Find a basis for each of the following.
(a) Ker$(T)$ (b) $T(R^4)$
(c) $T(U)$, where $U$ is the subspace spanned by $(0, 1, 0, 1)$ and $(0, 1, 1, 0)$
(d) $T(V)$, where $V$ is the subspace spanned by $(1, -2, 2, 1)$ and $(1, 0, 1, 0)$
(e) Why is the dimension of $T(U)$ different from that of $T(V)$?

***8.** Let $T: R^4 \to R^2$ be the linear transformation defined by

$$T(x, y, z, w) = (3x + 4y + 2z - w, 2x + y - 3z - w).$$

(a) Find a basis for ker$(T)$.
(b) Find a basis for $T(R^4)$.
(c) Find the general solution to $T(x, y, z, w) = (1, -1)$.

**9.** Which of the following linear transformations are one-to-one? Onto?
(a) $T: R^3 \to R^2$ defined by
$T(x, y, z) = (3x + y, x - 2y + z)$
(b) $T: R^2 \to R^3$ defined by
$T(x, y) = (x + y, x - y, 3x)$

**10.** Consider the transformation $T: R^3 \to R^2$ defined by

$$T(x_1, x_2, x_3) = (x_1 + x_2 + x_3, 2x_2 - 3x_3).$$

(a) Find a basis for the kernel of $T$. Is $T$ one-to-one?
(b) What is the image of $T$? Is $T$ onto?

**11.** For each of the following linear transformations, do the following.
(i) Find ker$(T)$ and $T(R^2)$.
(ii) If $L$ is a line in $R^2$ through the origin, find $T(L)$.
*(a) $T$ is the linear transformation in Example 12 of Section 5.1.
(b) $T$ is the linear transformation in Problem 17 of Section 5.1.
*(c) $T$ is the linear transformation in Problem 20 of Section 5.1.
(d) $T$ is the linear transformation in Problem 21 of Section 5.1.

**12.** Let $T: R^2 \to R^2$ be the linear transformation defined as follows. If $\mathbf{x}$ is a vector in $R^2$, then $T(\mathbf{x})$ is the vector obtained by projecting $\mathbf{x}$ onto the line spanned by $(1, 1)$ and then reflecting this projection across the $y$-axis.
(a) Find the standard matrix for $T$ and compute $T(3, 9)$.
(b) Describe the kernel of $T$ and describe $T(R^2)$.
(c) If $L$ is a line in $R^2$ through the origin, describe $T(L)$.

**13.** For each of the following matrices $A$, let $T$ be the linear transformation whose standard matrix is $A$. Do the following.
(i) Find a basis for ker$(T)$.
(ii) Find a basis for the image of $T$.
(iii) Give conditions that the components of a vector $\mathbf{x}$ must satisfy to

belong to the image of $T$.

*(a) $A = \begin{bmatrix} 1 & 1 & 1 & 1 \\ 0 & 1 & 1 & 0 \\ 0 & 0 & 1 & -1 \\ 0 & 0 & 1 & -1 \end{bmatrix}$

(b) $A = \begin{bmatrix} 1 & -2 & 3 & 4 & 5 \\ 2 & -4 & 5 & 9 & 11 \\ -1 & 1 & 0 & 0 & -3 \\ 0 & 1 & 1 & -8 & -6 \end{bmatrix}$

*(c) $A = \begin{bmatrix} 1 & 2 & 3 & 4 & 5 & 6 & 7 \\ 2 & 4 & 3 & 1 & 5 & 4 & 1 \\ 1 & 2 & 1 & 2 & 1 & 3 & 2 \\ 1 & 2 & 6 & 3 & 12 & 7 & 9 \end{bmatrix}$

(d) $A = \begin{bmatrix} 1 & 1 & 2 & 3 & 4 & 5 \\ 2 & 2 & 3 & 4 & 5 & 6 \\ 3 & 3 & 1 & -1 & -3 & -5 \\ 4 & 4 & 3 & 2 & 1 & 0 \end{bmatrix}$

**14.** Let $T: R^n \to R^n$ be a linear transformation and suppose that $T$ is one-to-one.
(a) Show that the image of a line through the origin is a line through the origin.
(b) Show that the image of a subspace $V$ of dimension $k$ is a subspace of dimension $k$.
(c) Show that the image of any line is a line and of any plane is a plane.
*(d) Show that the images of parallel lines are parallel lines.
(e) Show that the images of parallel planes are parallel planes.

**15.** For each of the following, find a linear transformation that satisfies the stated conditions.
*(a) $\ker(T)$ is spanned by $(2, 3, 1, 0)$ and $(4, 5, 0, 1)$ and $T(R^4)$ is the subspace $3x - y - 2z = 0$ of $R^3$.
(b) $\ker(T)$ is spanned by $(1, 1, 1)$ and $T(R^3)$ is the subspace $x = y$ of $R^3$.

***16.** Let $T: R^2 \to R^2$ be a linear transformation and let $\mathbf{v}_1 = (1, 2)$, $\mathbf{v}_2 = (3, 5)$. If $T(\mathbf{v}_1) = 6\mathbf{v}_1$ and $T(\mathbf{v}_2) = -\mathbf{v}_2$, find the standard matrix for $T$. (*Hint:* Write $\mathbf{e}_1$ and $\mathbf{e}_2$ as a linear combination of $\mathbf{v}_1$ and $\mathbf{v}_2$.)

**17.** Let $T: R^2 \to R^2$ be a linear transformation such that $T(3, 1) = (-2, 6)$ and $T(-1, 4) = (0, 5)$. Find the standard matrix for $T$.

***18.** Let $T$ be a linear transformation of $R^2$ into $R^3$ such that $T(1, 1) = (1, 2, 1)$ and $T(1, -1) = (1, 0, 1)$. What is the standard matrix of $T$?

**19.** Let $T: R^3 \to R^2$ be a linear transformation such that $T(1, 1, 1) = (3, 4)$, $T(0, 1, 1) = (1, 1)$, $T(2, 0, 1) = (0, 3)$. Find the standard matrix for $T$.

***20.** Let $A$ be a nonsingular matrix and let $T: R^n \to R^n$ be defined by $T(\mathbf{x}) = A\mathbf{x}$. Show that $T$ is one-to-one and maps $R^n$ onto $R^n$.

**21.** Let $T: R^2 \to R^2$ be a linear transformation and let $P$ be the parallelogram determined by two nonzero vectors $\mathbf{u}$ and $\mathbf{v}$ in $R^2$. Show that $T$ maps $P$ onto a parallelogram or onto a line segment (a degenerate parallelogram). (*Hint:* Every vector in $P$ is of the form $\alpha\mathbf{u} + \beta\mathbf{v}$, where $0 \le \alpha \le 1$ and $0 \le \beta \le 1$.)

***22.** Let $S: R^n \to R^m$ be a linear transformation. Show that the image of an affine subspace under $S$ is an affine subspace.

**23.** Let $T: R^n \to R^m$ be a linear transformation. Show that if for every $\mathbf{y}$ in $T(R^n)$ there is a unique $\mathbf{x}$ in $R^n$ such that $T(\mathbf{x}) = \mathbf{y}$, then $T$ is one-to-one.

**24.** Let $T: R^n \to R^m$ be a linear transformation and let $\mathbf{v}_1, \mathbf{v}_2, \ldots, \mathbf{v}_k$ be vectors in $R^n$. Let

$$\mathbf{w}_1 = T(\mathbf{v}_1), \mathbf{w}_2 = T(\mathbf{v}_2), \ldots, \mathbf{w}_k = T(\mathbf{v}_k).$$

Show that if $V$ is the subspace spanned by $\mathbf{v}_1, \mathbf{v}_2, \ldots, \mathbf{v}_k$ and $W$ is the subspace spanned by $\mathbf{w}_1, \mathbf{w}_2, \ldots, \mathbf{w}_k$, then $T(V) = W$. Is $T$ necessarily one-to-one?

***25.** Let $S$ and $T$ be two linear transformations from $R^n$ into $R^m$.

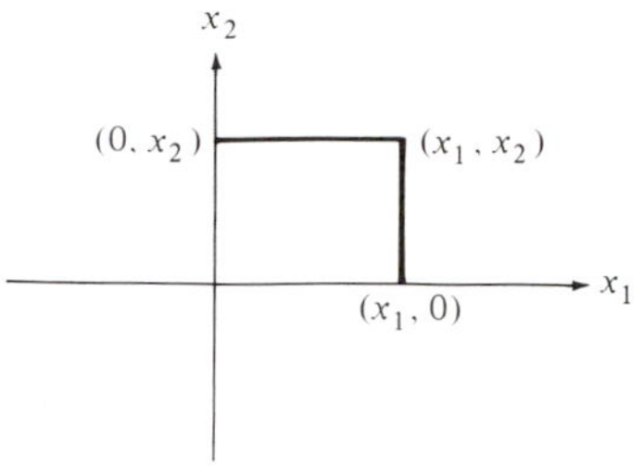


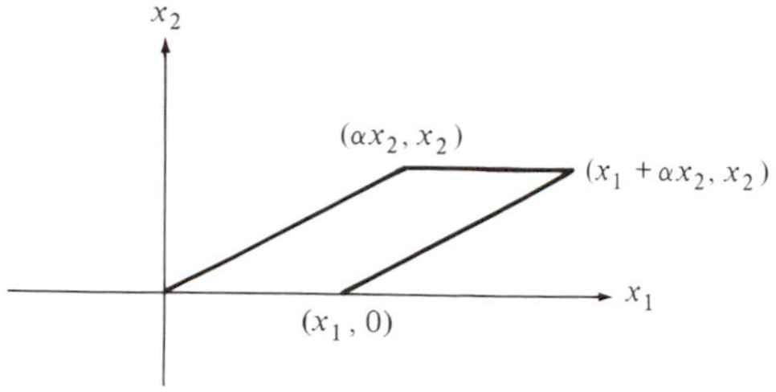


(a) Let $\mathbf{v}_1, \mathbf{v}_2, \ldots, \mathbf{v}_n$ be a basis for $R^n$. Show that if $S(\mathbf{v}_i) = T(\mathbf{v}_i)$ for $i = 1, 2, \ldots, n$, then $S = T$.

(b) Can $S(R^n) = T(R^n)$ and $S$ be different from $T$?

**26.** Let $T: R^n \to R^m$ be a linear transformation. Show that if $\dim T(R^n) < n$, then $T$ is not one-to-one.

**27.** Let $\mathbf{y}$ be a fixed vector in $R^m$, and let $T: R^m \to R$ be the linear transformation $T(\mathbf{x}) = \langle \mathbf{x}, \mathbf{y} \rangle$.

(a) Find the kernel of $T$.

(b) Find the image of $T$.

**28.** Let $T: R^n \to R^m$ be a linear transformation. Let $\mathbf{a}$ be a fixed vector in $R^n$ and let $\mathbf{b} = T(\mathbf{a})$.

(a) Show that $\mathbf{x}$ is a solution of the equation $T(\mathbf{x}) = \mathbf{b}$ if and only if

$$\mathbf{x} = \mathbf{a} + \mathbf{y},$$

where $\mathbf{y}$ is in $\ker(T)$.

(b) Use (a) to prove that $T$ is one-to-one if and only if $\ker(T)$ is the trivial subspace.

(c) Translate these statements into statements about linear systems.

**29.** Let $T: R^n \to R^m$ be a linear transformation. Show that $T(R^n)$ contains a line through the origin or else $T$ is the zero transformation.

**30.** Let $T: R^n \to R^m$ be a linear transformation and let $A$ be its standard matrix. If $V = R(A)$, show that

(a) $T(V) = C(A)$. (*Hint:* Use Problem 32 in Section 3.5.)

(b) $T(\mathbf{v}) \neq \mathbf{0}$ for all nonzero vectors $\mathbf{v}$ in $V$.

***31.** The linear transformation $T: R^2 \to R^2$ defined by

$$T(x_1, x_2) = (x_1 + \alpha x_2, x_2).$$

is called a **horizontal shear**. (See figure)

(a) Find the standard matrix of $T$.

(b) Show that each horizontal line is carried into itself under $T$.

(c) Show that the distance between $(x_1, x_2)$ and $T(x_1, x_2)$ is proportional to $x_2$.

(d) What is the image of a line $x_2 = mx_1 + b$ under $T$?

(e) What is the image of the unit square under $T$?

(f) What vectors are mapped into multiples of themselves by $T$?

**32.** Define a **vertical shear** and answer parts (a)–(f) of Problem 31.

**33.** Let $L$ be the line $y = 3x$ in $R^2$.

(a) Suppose that $T: R^2 \to R^2$ is a linear map with $T(R^2) = L$ *and* $\ker(T) = L$. Define $S: R^2 \to R^2$ by $S(\mathbf{v}) = T(T(\mathbf{v}))$. Find $S(R^2)$ and $\ker(S)$.

(b) Find one matrix $A$ such that the map $T: R^2 \to R^2$ defined by $T(\mathbf{x}) = A\mathbf{x}$ has *both* $T(R^2) = L$ and $\ker(T) = L$. Compute $A^2$. Explain how this agrees with your answer to part (a).

**34.** Let $V$ be a subspace of $R^m$ of dimension $n$, and let $T: R^m \to R^m$ be the projection of $R^m$ onto $V$.

(a) What is the image of $T$ and what is its dimension?

(b) What is the kernel of $T$ and what is its dimension?

(c) Can $T$ be one-to-one?

**35.** Let $T: R^n \to R^m$ be a linear transformation.
(a) Show that the image of $T$ is $R^m$ if and only if the rank of $T$ is $m$.
(b) Show that if $n < m$, then $T$ is not onto $R^m$. Can $T$ be one-to-one?
(c) Show that if $n > m$, then $T$ is not one-to-one. Can $T$ be onto?

**36.** Let $T: R^n \to R^n$ be a linear transformation.
(a) Show that $T$ is one-to-one if and only if its standard matrix is nonsingular.
(b) Show that $T$ is onto if and only if its standard matrix is nonsingular.
(c) Conclude that $T$ is one-to-one if and only if $T$ is onto.

**37.** Let $T: R^n \to R^m$ be a linear transformation. Show that $T$ is one-to-one if and only if $T$ sends independent vectors to independent vectors. That is, $T$ is one-to-one if and only if whenever $\mathbf{v}_1, \mathbf{v}_2, \ldots, \mathbf{v}_k$ are independent vectors in $R^n$, $T(\mathbf{v}_1), T(\mathbf{v}_2), \ldots, T(\mathbf{v}_k)$ are independent vectors in $R^m$.

**38.** Let $\mathbf{v}_1, \mathbf{v}_2, \ldots, \mathbf{v}_n$ be a basis for $R^n$ and let $\mathbf{w}_1, \mathbf{w}_2, \ldots, \mathbf{w}_n$ be any vectors in $R^m$.
(a) Show that there is a linear transformation $T: R^n \to R^m$ such that $T(\mathbf{v}_1) = \mathbf{w}_1$, $T(\mathbf{v}_2) = \mathbf{w}_2, \ldots, T(\mathbf{v}_n) = \mathbf{w}_n$. [*Hint:* If there is such a linear transformation, and $\mathbf{v}$ is in $R^n$, what must $T(\mathbf{v})$ be?]
(b) Show that there is only one such $T$.
(c) Let $V$ be a subspace of dimension $n-1$ of $R^n$ and let $\mathbf{v}$ be an $n$-vector not in $V$. Show that there is one and only one linear transformation $T: R^n \to R$ such that $\ker T = V$ and $T(\mathbf{v}) = 1$.

**39.** Let $V$ be a hyperplane in $R^n$ (i.e., an $n-1$ dimensional subspace of $R^n$). Let $\mathbf{v}_1, \mathbf{v}_2, \ldots, \mathbf{v}_{n-1}$ be a basis for $V$. Extend this basis for $V$ to a basis $\mathbf{v}_1, \mathbf{v}_2, \ldots, \mathbf{v}_{n-1}, \mathbf{v}_n$ for $R^n$. Define

$$T: R^n \to R$$

to be the linear transformation determined by $T(\mathbf{v}_i) = 0$, $i = 1, \ldots, n-1$, $T(\mathbf{v}_n) = 1$.
(a) Prove that $\ker(T) = V$.
(b) Show that if $S: R^n \to R^1$ is a linear transformation and $\ker(S) = V$, then $S = \alpha T$ for some scalar $\alpha$.
(c) Show that every hyperplane has an equation of the form

$$a_1x_1 + a_2x_2 + \cdots + a_nx_n = 0.$$

**40.** Let $V$ be an $(n-2)$-dimensional subspace of $R^n$. Let $\mathbf{v}_1, \ldots, \mathbf{v}_{n-2}$ be a basis for $V$ and expand it to a basis $\mathbf{v}_1, \ldots, \mathbf{v}_{n-2}, \mathbf{v}_{n-1}, \mathbf{v}_n$ for $R^n$. Define

$$T_1: R^n \to R \quad \text{and} \quad T_2: R^n \to R$$

to be the linear transformations defined by

$$\begin{aligned} T_1(\mathbf{v}_1) &= \cdots = T_1(\mathbf{v}_{n-2}) = 0, \\ &T_1(\mathbf{v}_{n-1}) = 1, \quad T_1(\mathbf{v}_n) = 0; \\ T_2(\mathbf{v}_1) &= \cdots = T_2(\mathbf{v}_{n-2}) = 0, \\ &T_2(\mathbf{v}_{n-1}) = 0, \quad T_2(\mathbf{v}_n) = 1 \end{aligned}$$

respectively.
(a) Show that $\ker(T_1) \cap \ker(T_2) = V$.
(b) Show that $V$ is the intersection of two hyperplanes. That is, show that $V$ is the nullspace of a $2 \times n$ matrix.
(c) Show that if $S: R^n \to R^1$ is a linear transformation with $\ker(S) = V$, then $S = \alpha_1 T_1 + \alpha_2 T_2$ for some scalars $\alpha_1$ and $\alpha_2$.

**41.** Generalize Problems 39 and 40.

**42.** Recall that a linear transformation $T: R^m \to R^m$ is called **invertible** if there is a linear transformation $S: R^m \to R^m$ such that $S \circ T = I = T \circ S$.
(a) Show that $T$ is invertible if and only if $T$ is one-to-one.
(b) Show that $T$ is invertible if and only if $T$ is onto.

**43.** Let $T: R^n \to R^n$ be an isometry (see Problem 38 in Section 5.1)
(a) Show that $T$ is one-to-one and onto.
(b) Show that $T$ is invertible.

**44.** Prove Theorem 4 without using matrices. (*Hint:* Begin with a basis for $\ker T$ and extend it to a basis for $R^n$.)

**45.** Prove Theorem 5 without using the rank plus nullity theorem. (*Hint:* Modify the proof of Theorem 7 given in the text.)

**46.** Prove Theorem 7 using matrix techniques. (*Hint:* Use Result 1.)

**47.** Let $T: R^n \to R^m$ be a linear transformation and let $V$ be a subspace of $R^n$. Prove that

$$\dim T(V) + \dim(\ker(T) \cap V) = \dim V.$$

(*Hint:* Begin with a basis for $\ker(T) \cap V$ and extend it to a basis for $V$.)

**48.** Prove the result in Problem 47 as follows. Let $A$ be the standard matrix for $T$ and let $B$ be a matrix whose columns form a basis for $V$. Then show that

$$\ker(T) \cap V = N(A) \cap C(B) = N(AB).$$

Now apply the rank plus nullity theorem for matrices to the matrix $AB$.

# 5.3 Matrix Representations of a Linear Transformation.

Let $T: R^n \to R^m$ be a linear transformation. The standard matrix for $T$ is the $m \times n$ matrix whose columns are $T(\mathbf{e}_1), T(\mathbf{e}_2), \ldots, T(\mathbf{e}_n)$. The standard matrix for $T$ represents $T$ by mapping the standard coordinate vector of $\mathbf{x}$ to the standard coordinate vector of $T(\mathbf{x})$.

**Question:** Given any basis $\mathbf{v}_1, \mathbf{v}_2, \ldots, \mathbf{v}_n$ for $R^n$ and basis $\mathbf{w}_1, \mathbf{w}_2, \ldots, \mathbf{w}_m$ for $R^m$, is there an $m \times n$ matrix $A$ that represents $T$ by mapping the $\mathbf{v}$-coordinate vector $\mathbf{c}_\mathbf{v}(\mathbf{x})$ of $\mathbf{x}$ to the $\mathbf{w}$-coordinate vector $\mathbf{c}_\mathbf{w}(T(\mathbf{x}))$ of $T(\mathbf{x})$? That is, is there an $m \times n$ matrix $A$ such that

$$A\mathbf{c}_\mathbf{v}(\mathbf{x}) = \mathbf{c}_\mathbf{w}(T(\mathbf{x}))$$

for all $\mathbf{x}$ in $R^n$?

If there is such a matrix, then certainly this last equation must hold for each of the basis vectors $\mathbf{v}_1, \mathbf{v}_2, \ldots, \mathbf{v}_n$. Thus we must have

$$A\mathbf{c}_\mathbf{v}(\mathbf{v}_i) = \mathbf{c}_\mathbf{w}(T(\mathbf{v}_i)), \qquad i = 1, 2, \ldots, n.$$

Since $\mathbf{c}_\mathbf{v}(\mathbf{v}_i) = \mathbf{e}_i$, the $i$th column of $A$ must be $\mathbf{c}_\mathbf{w}(T(\mathbf{v}_i))$:

$$i\text{th column of } A = A\mathbf{e}_i = A\mathbf{c}_\mathbf{v}(\mathbf{v}_i) = \mathbf{c}_\mathbf{w}(T(\mathbf{v}_i)).$$

Thus if there is such a matrix, then we have just found it:

$$A = [\mathbf{c}_\mathbf{w}(T(\mathbf{v}_1)) \quad \mathbf{c}_\mathbf{w}(T(\mathbf{v}_2)) \quad \cdots \quad \mathbf{c}_\mathbf{w}(T(\mathbf{v}_n))].$$

It remains to show that this matrix maps the $\mathbf{v}$-coordinate vector of any vector $\mathbf{x}$ to the $\mathbf{w}$-coordinate vector of $T(\mathbf{x})$. That is, we must show that

$$A\mathbf{c}_\mathbf{v}(\mathbf{x}) = \mathbf{c}_\mathbf{w}(T(\mathbf{x}))$$

for all $\mathbf{x}$ in $R^n$. Let $\mathbf{x}$ be in $R^n$, and let

$$\mathbf{c}_\mathbf{v}(\mathbf{x}) = (x'_1, x'_2, \ldots, x'_n)$$

be the **v**-coordinate vector of **x**. Then

$$\mathbf{x} = x'_1\mathbf{v}_1 + x'_2\mathbf{v}_2 + \cdots + x'_n\mathbf{v}_n$$

and hence

$$T(\mathbf{x}) = x'_1 T(\mathbf{v}_1) + x'_2 T(\mathbf{v}_2) + \cdots + x'_n T(\mathbf{v}_n).$$

In Section 2.5 we proved that forming a linear combination of vectors corresponds to forming the same linear combination of their coordinate vectors. Thus

$$\begin{aligned}\mathbf{c}_\mathbf{w}(T(\mathbf{x})) &= x'_1\mathbf{c}_\mathbf{w}(T(\mathbf{v}_1)) + x'_2\mathbf{c}_\mathbf{w}(T(\mathbf{v}_2)) + \cdots + x'_n\mathbf{c}_\mathbf{w}(T(\mathbf{v}_n))\\ &= [\mathbf{c}_\mathbf{w}(T(\mathbf{v}_1)) \quad \mathbf{c}_\mathbf{w}(T(\mathbf{v}_2)) \quad \cdots \quad \mathbf{c}_\mathbf{w}(T(\mathbf{v}_n))]\begin{bmatrix} x'_1 \\ x'_2 \\ \vdots \\ x'_n \end{bmatrix}\\ &= A\mathbf{c}_\mathbf{v}(\mathbf{x}).\end{aligned}$$

***Definition*** The matrix of a linear transformation $T\colon R^n \to R^m$ with respect to a basis $\mathbf{v}_1, \mathbf{v}_2, \ldots, \mathbf{v}_n$ of $R^n$ and a basis $\mathbf{w}_1, \mathbf{w}_2, \ldots, \mathbf{w}_m$ of $R^m$ is the $m \times n$ matrix $A$ whose $i$th column is the **w**-coordinate vector $\mathbf{c}_\mathbf{w}(T(\mathbf{v}_i))$ of $T(\mathbf{v}_i)$:

$$A = [\mathbf{c}_\mathbf{w}(T(\mathbf{v}_1)) \quad \mathbf{c}_\mathbf{w}(T(\mathbf{v}_2)) \quad \cdots \quad \mathbf{c}_\mathbf{w}(T(\mathbf{v}_n))].$$

### EXAMPLE 1

By our previous remarks, the matrix of a linear transformation $T\colon R^n \to R^m$ with respect to the standard bases for $R^n$ and $R^m$ is the standard matrix.

### EXAMPLE 2

Consider the linear transformation $T\colon R^3 \to R^2$ defined by

$$T(x_1, x_2, x_3) = (x_1 - x_2, x_2 + x_3).$$

Let's find the matrix of $T$ with respect to the basis

$$\mathbf{v}_1 = (2, 0, 1), \qquad \mathbf{v}_2 = (0, 2, 2), \qquad \mathbf{v}_3 = (0, 2, 3)$$

of $R^3$ and the basis

$$\mathbf{w}_1 = (1, 2), \qquad \mathbf{w}_2 = (0, 1)$$

of $R^2$. Our first problem, then, is to compute the **w**-coordinates of $T(\mathbf{v}_1)$, $T(\mathbf{v}_2)$, and $T(\mathbf{v}_3)$. These coordinates are found by forming the matrix $M$ whose columns are $\mathbf{w}_1$ and $\mathbf{w}_2$ and solving the three systems

$$M\mathbf{c}_\mathbf{w}(T(\mathbf{v}_1)) = T(\mathbf{v}_1) = \begin{bmatrix} 2 \\ 1 \end{bmatrix}, \qquad M\mathbf{c}_\mathbf{w}(T(\mathbf{v}_2)) = T(\mathbf{v}_2) = \begin{bmatrix} -2 \\ 4 \end{bmatrix},$$

$$M\mathbf{c}_\mathbf{w}(T(\mathbf{v}_3)) = T(\mathbf{v}_3) = \begin{bmatrix} -2 \\ 5 \end{bmatrix}.$$

Reduce the augmented matrix

$$\left[\begin{array}{cc:ccc} 1 & 0 & 2 & -2 & -2 \\ 2 & 1 & 1 & 4 & 5 \end{array}\right] \quad \text{to} \quad \left[\begin{array}{cc:ccc} 1 & 0 & 2 & -2 & -2 \\ 0 & 1 & -3 & 8 & 9 \end{array}\right].$$

Hence the coordinate vectors of $T(\mathbf{v}_1)$, $T(\mathbf{v}_2)$, and $T(\mathbf{v}_3)$ with respect to the basis $\mathbf{w}_1$ and $\mathbf{w}_2$ are

$$\mathbf{c}_{\mathbf{w}}(T(\mathbf{v}_1)) = \begin{bmatrix} 2 \\ -3 \end{bmatrix}, \qquad \mathbf{c}_{\mathbf{w}}(T(\mathbf{v}_2)) = \begin{bmatrix} -2 \\ 8 \end{bmatrix}, \qquad \mathbf{c}_{\mathbf{w}}(T(\mathbf{v}_3)) = \begin{bmatrix} -2 \\ 9 \end{bmatrix},$$

and the matrix of $T$ with respect to the bases $\mathbf{v}_1, \mathbf{v}_2, \mathbf{v}_3$ and $\mathbf{w}_1, \mathbf{w}_2$ is

$$A = \begin{bmatrix} 2 & -2 & -2 \\ -3 & 8 & 9 \end{bmatrix}.$$

The matrix $A$ represents $T$ by sending the $\mathbf{v}$-coordinates of a vector $\mathbf{x}$ to the $\mathbf{w}$-coordinates of $T(\mathbf{x})$. For instance, the $\mathbf{v}$-coordinate vector of $\mathbf{x} = (-6, 6, 7)$ is found by reducing

$$\left[\begin{array}{ccc:c} 2 & 0 & 0 & -6 \\ 0 & 2 & 2 & 6 \\ 1 & 2 & 3 & 7 \end{array}\right] \quad \text{to} \quad \left[\begin{array}{ccc:c} 1 & 0 & 0 & -3 \\ 0 & 1 & 0 & -1 \\ 0 & 0 & 1 & 4 \end{array}\right]$$

to be $\mathbf{c}_{\mathbf{v}}(\mathbf{x}) = (-3, -1, 4)$. Thus

$$A\mathbf{c}_{\mathbf{v}}(\mathbf{x}) = \begin{bmatrix} 2 & -2 & -2 \\ -3 & 8 & 9 \end{bmatrix} \begin{bmatrix} -3 \\ -1 \\ 4 \end{bmatrix} = \begin{bmatrix} -12 \\ 37 \end{bmatrix}$$

is supposed to be the $\mathbf{w}$-coordinate vector of

$$T(\mathbf{x}) = \begin{bmatrix} -12 \\ 13 \end{bmatrix}.$$

Since

$$(-12, 13) = -12(1, 2) + 37(0, 1),$$

$(-12, 37)$ is indeed the $\mathbf{w}$-coordinate vector of $T(\mathbf{x})$.

## EXAMPLE 3

Consider the linear transformation $T\colon R^2 \to R^2$ defined by

$$T(x_1, x_2) = (\tfrac{7}{2}x_1 - \tfrac{1}{2}x_2, \tfrac{3}{2}x_1 + \tfrac{3}{2}x_2)$$

To find the matrix of $T$ with respect to the same basis

$$\mathbf{v}_1 = (1, 1) \qquad \text{and} \qquad \mathbf{v}_2 = (1, 3)$$

for both copies of $R^2$, we compute

$$T(\mathbf{v}_1) = (3, 3) = 3(1, 1) \qquad \text{and} \qquad T(\mathbf{v}_2) = (2, 6) = 2(1, 3).$$

Thus

$$\mathbf{c}_{\mathbf{v}}(T(\mathbf{v}_1)) = (3, 0) \qquad \text{and} \qquad \mathbf{c}_{\mathbf{v}}(T(\mathbf{v}_2)) = (0, 2)$$

and the matrix of $T$ with respect to this basis is

$$\Lambda = \begin{bmatrix} 3 & 0 \\ 0 & 2 \end{bmatrix}.$$

Thus with respect to the basis we chose for $R^2$ the matrix for $T$ is a particularly nice matrix, namely a diagonal matrix. By contrast, the standard matrix for $T$ is not nearly as nice; it is

$$A = \frac{1}{2}\begin{bmatrix} 7 & -1 \\ 3 & 3 \end{bmatrix}.$$

As you might expect, some important information about the linear transformation $T$ is revealed by the diagonal matrix $\Lambda$ which is not revealed by the standard matrix $A$. For instance, $T$ has two favored directions, namely the directions specified by the basis vectors $\mathbf{v}_1$ and $\mathbf{v}_2$. Since

$$T(\mathbf{v}_1) = 3\mathbf{v}_1 \qquad \text{and} \qquad T(\mathbf{v}_2) = 2\mathbf{v}_2, \tag{1}$$

$T$ stretches vectors collinear with $\mathbf{v}_1$ by a factor of 3 and vectors collinear with $\mathbf{v}_2$ by a factor of 2. Now let $\mathbf{x}$ be any vector in $R^2$ and write $\mathbf{x}$ as a linear combination of $\mathbf{v}_1$ and $\mathbf{v}_2$:

$$\mathbf{x} = \alpha_1\mathbf{v}_1 + \alpha_2\mathbf{v}_2. \tag{2}$$

Then the $\mathbf{v}$-coordinate vector of $\mathbf{x}$ is $\mathbf{c}_{\mathbf{v}}(\mathbf{x}) = (\alpha_1, \alpha_2)$ and hence the $\mathbf{v}$-coordinate vector of $T(\mathbf{x})$ is

$$\Lambda\mathbf{c}_{\mathbf{v}}(\mathbf{x}) = \begin{bmatrix} 3 & 0 \\ 0 & 2 \end{bmatrix}\begin{bmatrix} \alpha_1 \\ \alpha_2 \end{bmatrix} = \begin{bmatrix} 3\alpha_1 \\ 2\alpha_2 \end{bmatrix}.$$

This means that

$$T(\mathbf{x}) = 3\alpha_1\mathbf{v}_1 + 2\alpha_2\mathbf{v}_2.$$

[Notice that we could have arrived at this equation directly by applying $T$ to both sides of (2) and then using (1).] $T$ weights the first $\mathbf{v}$-coordinate of $\mathbf{x}$ by a factor of 3 and the second by a factor of 2. For instance, if $\mathbf{x} = 2(1, 1) + (-1)(1, 3) = (1, -1)$, then

$$\mathbf{c}_{\mathbf{v}}(\mathbf{x}) = (2, -1)$$

and hence

$$\mathbf{c}_{\mathbf{v}}(T(\mathbf{x})) = \Lambda\mathbf{c}_{\mathbf{v}}(\mathbf{x}) = \begin{bmatrix} 3 & 0 \\ 0 & 2 \end{bmatrix}\begin{bmatrix} 2 \\ -1 \end{bmatrix} = \begin{bmatrix} 6 \\ -2 \end{bmatrix}.$$

Thus $T(\mathbf{x}) = 6\mathbf{v}_1 - 2\mathbf{v}_2$ (see Figure 5.17).

I think you will agree that none of these results can be seen directly from the definition of $T$ or from the standard matrix of $T$. Of course, it must seem like

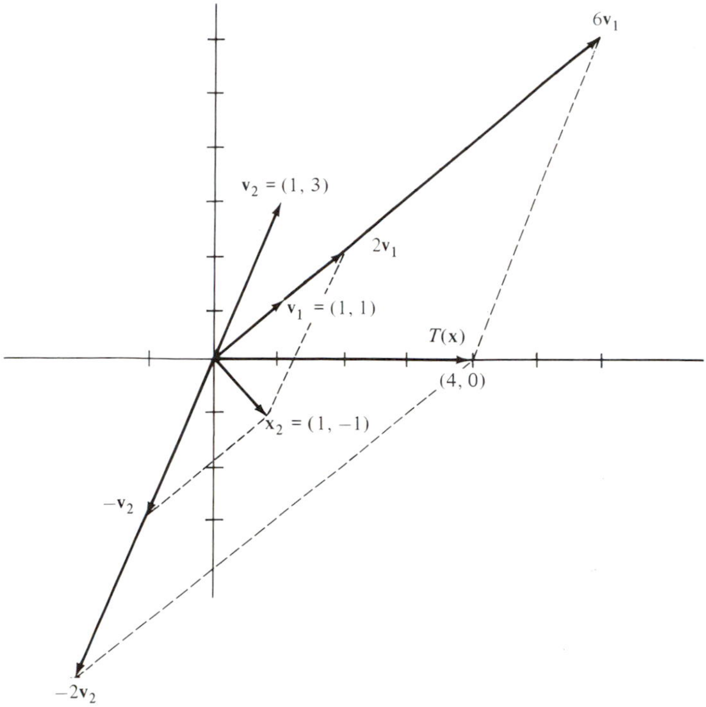


Figure 5.17

magic that we were able to find a basis that gives rise to such a simple matrix representation of $T$. We will address that question in Section 5.5 and Chapter 8. For now I hope that you are convinced that it is worthwhile to study the matrix of a transformation with respect to a basis other than the standard basis.

In constructing the matrix of a linear transformation $T$: $R^n \to R^m$ with respect to a basis $\mathbf{v}_1, \mathbf{v}_2, \ldots, \mathbf{v}_n$ of $R^n$ and a basis $\mathbf{w}_1, \mathbf{w}_2, \ldots, \mathbf{w}_m$ of $R^m$, all we have to know is what $T$ does to the basis $\mathbf{v}_1, \mathbf{v}_2, \ldots, \mathbf{v}_n$. That is, we need to know $T(\mathbf{v}_1)$, $T(\mathbf{v}_2), \ldots, T(\mathbf{v}_n)$. We then determine the $\mathbf{w}$-coordinate vectors $\mathbf{c}_\mathbf{w}(T(\mathbf{v}_1)), \mathbf{c}_\mathbf{w}(T(\mathbf{v}_2)), \ldots, \mathbf{c}_\mathbf{w}(T(\mathbf{v}_n))$ of these vectors and form the matrix $A$ with these vectors as its columns. Then for any vector $\mathbf{x}$ in $R^n$ we can determine $T(\mathbf{x})$ from its $\mathbf{w}$-coordinate vector

$$\mathbf{c}_\mathbf{w}(T(\mathbf{x})) = A\mathbf{c}_\mathbf{v}(\mathbf{x}).$$

### EXAMPLE 4

Suppose that we are given the basis

$$\mathbf{v}_1 = (1, 0), \qquad \mathbf{v}_2 = (1, 1)$$

of $R^2$ and the basis

$$\mathbf{w}_1 = (1, 0, 0), \qquad \mathbf{w}_2 = (1, 1, 0), \qquad \mathbf{w}_3 = (1, 1, 1)$$

of $R^3$. In addition, suppose we are given that $T\colon R^2 \to R^3$ is a linear transformation such that

$$T(\mathbf{v}_1) = (1, 0, 1), \qquad T(\mathbf{v}_2) = (1, 1, 2).$$

Finally, the questions we are asked are to find the matrix of $T$ with respect to the given bases and to find a formula for $T$. In order to find the matrix $A$ of $T$ with respect to the given bases of $R^2$ and $R^3$, we must first determine the $\mathbf{w}$-coordinate vectors of $T(\mathbf{v}_1)$ and $T(\mathbf{v}_2)$; they are found by reducing

$$\begin{bmatrix} 1 & 1 & 1 & : & 1 & 1 \\ 0 & 1 & 1 & : & 0 & 1 \\ 0 & 0 & 1 & : & 1 & 2 \end{bmatrix} \quad \text{to} \quad \begin{bmatrix} 1 & 0 & 0 & : & 1 & 0 \\ 0 & 1 & 0 & : & -1 & -1 \\ 0 & 0 & 1 & : & 1 & 2 \end{bmatrix}.$$

Thus

$$A = \begin{bmatrix} 1 & 0 \\ -1 & -1 \\ 1 & 2 \end{bmatrix}$$

and by our previous remarks,

$$\mathbf{c}_{\mathbf{w}}(T(\mathbf{x})) = A\mathbf{c}_{\mathbf{v}}(\mathbf{x}).$$

To complete our answer, then, we must compute $\mathbf{c}_{\mathbf{v}}(\mathbf{x})$ and $A\mathbf{c}_{\mathbf{v}}(\mathbf{x})$. Since

$$\begin{bmatrix} 1 & 1 & : & x_1 \\ 0 & 1 & : & x_2 \end{bmatrix} \quad \text{reduces to} \quad \begin{bmatrix} 1 & 0 & : & x_1 - x_2 \\ 0 & 1 & : & x_2 \end{bmatrix},$$

$$\mathbf{c}_{\mathbf{v}}(\mathbf{x}) = (x_1 - x_2, x_2).$$

Since

$$\mathbf{c}_{\mathbf{w}}(T(\mathbf{x})) = A\mathbf{c}_{\mathbf{v}}(\mathbf{x}) = \begin{bmatrix} 1 & 0 \\ -1 & -1 \\ 1 & 2 \end{bmatrix} \begin{bmatrix} x_1 - x_2 \\ x_2 \end{bmatrix} = \begin{bmatrix} x_1 - x_2 \\ -x_1 \\ x_1 + x_2 \end{bmatrix},$$

we have our formula for $T$:

$$\begin{aligned} T(\mathbf{x}) &= (x_1 - x_2)\mathbf{w}_1 - x_1\mathbf{w}_2 + (x_1 + x_2)\mathbf{w}_3 \\ &= (x_1 - x_2)(1, 0, 0) - x_1(1, 1, 0) + (x_1 + x_2)(1, 1, 1) \\ &= (x_1, x_2, x_1 + x_2). \end{aligned}$$

***Theorem 1*** *Let $\mathbf{v}_1, \mathbf{v}_2, \ldots, \mathbf{v}_n$ be a basis for $R^n$. If $T\colon R^n \to R^m$ is a linear transformation, then $T$ is completely determined by $T(\mathbf{v}_1), T(\mathbf{v}_2), \ldots, T(\mathbf{v}_n)$. That is, $T$ is completely determined by its values on a basis.*

**Proof** If $\mathbf{x}$ is any vector in $R^n$, then

$$\mathbf{x} = x'_1\mathbf{v}_1 + x'_2\mathbf{v}_2 + \cdots + x'_n\mathbf{v}_n. \tag{3}$$

Thus

$$T(\mathbf{x}) = x'_1 T(\mathbf{v}_1) + x'_2 T(\mathbf{v}_2) + \cdots + x'_n T(\mathbf{v}_n). \tag{4}$$

In view of our previous example, the proof of Theorem 1 probably felt too simple. What made that example complicated was determining $T$ via its coordinates. The vectors $T(\mathbf{v}_1), T(\mathbf{v}_2), \ldots, T(\mathbf{v}_n)$ are in $R^m$ and as soon as a basis for $R^m$ is selected, we can determine the coordinate vectors for these vectors and hence we can determine $A$. Then in terms of matrices, equation (4) becomes

$$\begin{aligned}\mathbf{c}_{\mathbf{w}}(T(\mathbf{x})) &= x_1'\mathbf{c}_{\mathbf{w}}(T(\mathbf{v}_1)) + x_2'\mathbf{c}_{\mathbf{w}}(T(\mathbf{v}_2)) + \cdots + x_n'\mathbf{c}_{\mathbf{w}}(T(\mathbf{v}_n)) \\ &= A\mathbf{c}_{\mathbf{v}}(\mathbf{x})\end{aligned}$$

where $\mathbf{c}_{\mathbf{v}}(\mathbf{x})$ is determined in (3).

## EXAMPLE 5

Let us find a formula and the standard matrix for the linear transformation $T\colon R^3 \to R^2$ defined by

$$T(1,0,3) = (1,0), \qquad T(1,2,5) = (1,2), \qquad T(1,3,2) = (1,-1).$$

The matrix of $T$ with respect to the basis $(1,0,3), (1,2,5), (1,3,2)$ for $R^3$ and the standard basis for $R^2$ is

$$A = \begin{bmatrix} 1 & 1 & 1 \\ 0 & 2 & -1 \end{bmatrix}.$$

Since the standard coordinate vector of $T(\mathbf{x})$ is

$$T(\mathbf{x}) = \mathbf{c}_{\mathbf{e}}(T(\mathbf{x})) = A\mathbf{c}_{\mathbf{v}}(\mathbf{x}),$$

we must compute $\mathbf{c}_{\mathbf{v}}(\mathbf{x})$ and multiply by $A$. If $M$ is the change of coordinates matrix

$$M = \begin{bmatrix} 1 & 1 & 1 \\ 0 & 2 & 3 \\ 3 & 5 & 2 \end{bmatrix},$$

then

$$M^{-1} = \frac{1}{8}\begin{bmatrix} 11 & -3 & -1 \\ -9 & 1 & 3 \\ 6 & 2 & -2 \end{bmatrix}.$$

Thus

$$\begin{aligned}A\mathbf{c}_{\mathbf{v}}(\mathbf{x}) = AM^{-1}\mathbf{x} &= \begin{bmatrix} 1 & 1 & 1 \\ 0 & 2 & -1 \end{bmatrix}\frac{1}{8}\begin{bmatrix} 11 & -3 & -1 \\ -9 & 1 & 3 \\ 6 & 2 & -2 \end{bmatrix}\begin{bmatrix} x_1 \\ x_2 \\ x_3 \end{bmatrix} \\ &= \begin{bmatrix} x_1 \\ -3x_1 + x_3 \end{bmatrix},\end{aligned}$$

and our formula for $T$ is

$$T(x_1, x_2, x_3) = (x_1, -3x_1 + x_3).$$

We conclude this section with an interpretation of a change of coordinates matrix. Let $\mathbf{v}_1, \mathbf{v}_2, \ldots, \mathbf{v}_m$ be a basis for $R^m$. The change of coordinates matrix that changes **v**-coordinates to standard coordinates is the matrix $M$ having these vectors as its columns. $M$ changes the name of a vector in the **v**-basis to its name in the standard basis: $\mathbf{x} = M\mathbf{c}_\mathbf{v}(\mathbf{x})$. Thus the vector stays fixed but its name changes, and this suggests that $M$ is the matrix of the identity transformation $I\colon R^m \to R^m$ with respect to some bases for $R^m$. The columns of $M$ are the vectors $\mathbf{v}_1, \mathbf{v}_2, \ldots, \mathbf{v}_m$, and these vectors are their own coordinate vectors with respect to the standard basis. Therefore, if the first copy of $R^m$ has the basis $\mathbf{v}_1, \mathbf{v}_2, \ldots \mathbf{v}_m$ and the second copy has the standard basis $\mathbf{e}_1, \mathbf{e}_2, \ldots, \mathbf{e}_m$, then the matrix of $I$ with respect to these bases is $M$.

$$\begin{array}{ccc} I\colon R^m & \longrightarrow & R^m \\ \mathbf{v}\text{-basis} & & \mathbf{e}\text{-basis} \end{array}$$

$$I(\mathbf{x}) = \mathbf{x} \text{ corresponds to } M\mathbf{c}_\mathbf{v}(\mathbf{x}) = \mathbf{x}$$

Now suppose that we have two arbitrary bases for $R^m$, say

$$\mathbf{v}_1, \mathbf{v}_2, \ldots, \mathbf{v}_m \qquad \text{and} \qquad \mathbf{w}_1, \mathbf{w}_2, \ldots, \mathbf{w}_m.$$

How do we change coordinates with respect to the **v**-basis to coordinates with respect to the **w**-basis? If **x** is a vector in $R^m$, then

$$\mathbf{c}_\mathbf{v}(\mathbf{x}) \qquad \text{and} \qquad \mathbf{c}_\mathbf{w}(\mathbf{x})$$

are simply different names for **x**. Thus changing from **v**-coordinates to **w**-coordinates simply changes the name for **x** from $\mathbf{c}_\mathbf{v}(\mathbf{x})$ to $\mathbf{c}_\mathbf{w}(\mathbf{x})$. The vector **x** doesn't change; its name changes. The linear transformation that leaves vectors fixed is the identity transformation. The matrix that represents the identity transformation $I$ from $R^m$ into $R^m$ with respect to the **v**-basis and the **w**-basis, respectively, maps the **v**-coordinates of a vector **x** to the **w**-coordinates of $I(\mathbf{x}) = \mathbf{x}$. Thus the matrix that changes **v**-coordinates to **w**-coordinates is the matrix of the identity transformation with respect to the **v**-basis and the **w**-basis respectively. It is the matrix

$$M = [\mathbf{c}_\mathbf{w}(\mathbf{v}_1) \quad \mathbf{c}_\mathbf{w}(\mathbf{v}_2) \quad \cdots \quad \mathbf{c}_\mathbf{w}(\mathbf{v}_m)].$$

$$\begin{array}{ccc} I\colon R^m & \longrightarrow & R^m \\ \mathbf{v}\text{-basis} & & \mathbf{w}\text{-basis} \end{array}$$

$$I(\mathbf{x}) = \mathbf{x} \text{ corresponds to } M\mathbf{c}_\mathbf{v}(\mathbf{x}) = \mathbf{c}_\mathbf{w}(\mathbf{x})$$

## EXAMPLE 6

Let us find the matrix which changes coordinates with respect to the basis $\mathbf{v}_1 = (1, 1)$, $\mathbf{v}_2 = (-1, 1)$ to coordinates with respect to the basis $\mathbf{w}_1 = (1, 2)$, $\mathbf{w}_2 = (2, 1)$. Since $M$ is the matrix of $I\colon R^2 \to R^2$ with respect to these bases, we must compute the coordinates of $I(\mathbf{v}_1) = \mathbf{v}_1$ and $I(\mathbf{v}_2) = \mathbf{v}_2$ with respect to the basis $\mathbf{w}_1, \mathbf{w}_2$. Reducing

$$\begin{bmatrix} 1 & 2 & : & 1 & -1 \\ 2 & 1 & : & 1 & 1 \end{bmatrix} \quad \text{to} \quad \begin{bmatrix} 1 & 0 & : & \frac{1}{3} & 1 \\ 0 & 1 & : & \frac{1}{3} & -1 \end{bmatrix}$$

we conclude that

$$M = \begin{bmatrix} \frac{1}{3} & 1 \\ \frac{1}{3} & -1 \end{bmatrix}.$$

## Alias vs Aliby

Given a nonsingular $m \times m$ matrix $A$, our standard interpretation of the transformation $T: R^m \to R^m$ defined by $T(\mathbf{x}) = A\mathbf{x}$ is a dynamic one: $A$ moves a vector $\mathbf{x}$ to the vector $A\mathbf{x}$ (Aliby). The static interpretation is that $A$ renames vectors. The columns $\mathbf{v}_1, \mathbf{v}_2, \ldots, \mathbf{v}_m$ of $A$ form a basis for $R^m$. Given a vector $\mathbf{x}$ in $R^m$, $A$ changes the name for $\mathbf{x}$ with respect to this basis to the name for $\mathbf{x}$ with respect to the standard basis: $\mathbf{x} = A\mathbf{c}_\mathbf{v}(\mathbf{x})$ (Alias). In this interpretation $A$ is the matrix of the identity transformation $I: R^m \to R^m$ with respect to the $\mathbf{v}$-basis and the standard basis, respectively. Hence any nonsingular matrix can always be viewed as a change of coordinates matrix; it changes the coordinates of a vector with respect to the basis of $R^m$ formed from its columns to the standard coordinates.

**Result 1** Let $A$ be an $m \times m$ nonsingular matrix, and let $\mathbf{v}_1, \mathbf{v}_2, \ldots, \mathbf{v}_m$ denote the columns of $A$. Then $A$ can be viewed as a change of coordinates matrix that changes coordinates with respect to the basis $\mathbf{v}_1, \mathbf{v}_2, \ldots, \mathbf{v}_m$ to standard coordinates. That is,

$$A\mathbf{c}_\mathbf{v}(\mathbf{x}) = \mathbf{x}.$$

### EXAMPLE 7

In Example 7 in Section 5.1 we discussed the linear transformation $T: R^2 \to R^2$ which rotates every vector through an angle $\theta$. This is a dynamic interpretation of $T$; vectors are being moved. There is also a static interpretation of $T$ which is based on a change of coordinates. The standard matrix of $T$ is

$$A = \begin{bmatrix} \cos\theta & -\sin\theta \\ \sin\theta & \cos\theta \end{bmatrix}.$$

The matrix $A$ is nonsingular (in fact, it is orthogonal). Hence the columns of $A$ form a basis for $R^2$. By our previous discussion, then, we can view $A$ as a change of coordinates matrix. The matrix $A$ changes the coordinates of a vector with respect to the basis $(\cos\theta, \sin\theta)$ and $(-\sin\theta, \cos\theta)$ to the coordinates of the vector with respect to the standard basis $\mathbf{e}_1$ and $\mathbf{e}_2$ of $R^2$. Under this interpretation vectors do not move but rather, their names are changed. Said differently, vectors stay fixed but the frame of reference changes. The standard basis vectors $\mathbf{e}_1$ and $\mathbf{e}_2$ are rotated through an angle $\theta$ to the new basis vectors $(\cos\theta, \sin\theta)$ and $(-\sin\theta, \cos\theta)$. Since coordinate

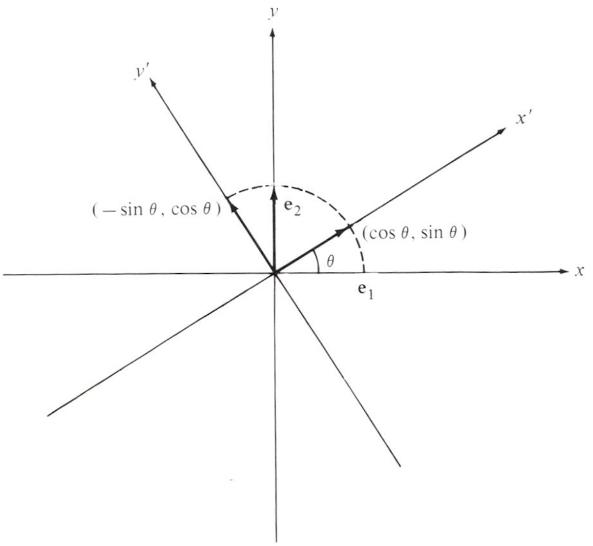


Figure 5.18

vectors determine coordinate axes, we can view $A$ as rotating the coordinate axes through an angle $\theta$ while leaving vectors fixed (see Figure 5.18).

## PROBLEMS 5.3

**1.** Complete each of the following definitions.
 (a) A change of coordinate matrix is....
 (b) The matrix of a linear transformation $T\colon R^n \to R^m$ with respect to the bases $\mathbf{v}_1, \mathbf{v}_2, \ldots, \mathbf{v}_n$ and $\mathbf{w}_1, \mathbf{w}_2, \ldots, \mathbf{w}_m$ of $R^n$ and $R^m$, respectively, is....

***2.** Find the coordinate vector of $(2, 1, 3)$ with respect to:
 (a) The basis $(1, 0, 1)$, $(1, 1, 0)$, $(1, 1, 1)$ of $R^3$
 (b) The basis $(0, 0, 1)$, $(0, 1, 1)$, $(1, 1, 1)$ of $R^3$

***3.** Let $T\colon R^3 \to R^2$ be defined by

$$T(x_1, x_2, x_3) = (3x_1 + 2x_2, x_2 - 2x_3).$$

Find the matrix for $T$ with respect to the bases $(1, 0, 1)$, $(1, 2, 1)$, $(0, 1, 2)$ for $R^3$ and $(1, 2)$, $(4, 0)$ for $R^2$.

***4.** Let $T$ be the linear transformation in Problem 3. Find the matrix for $T$ with respect to the bases $(0, 0, 1)$, $(0, 1, 1)$, $(1, 1, 1)$ for $R^3$ and $(1, 0)$, $(1, 1)$ for $R^2$.

**5.** Let $T\colon R^2 \to R^3$ be defined by $T(x, y) = (x - y, 2x + y, -x + 3y)$. Find the matrix of $T$ with respect to the bases $(3, 2)$, $(4, 1)$ for $R^2$ and $(1, 1, 1)$, $(0, 1, 1)$, $(0, 0, 1)$ for $R^3$.

**6.** Let $T\colon R^4 \to R^3$ be defined by

$$T(x_1, x_2, x_3, x_4) = (x_1 - x_2, x_2 + x_3 - x_4, x_3 - x_4).$$

 (a) Find the matrix of $T$ with respect to the basis $(1, 1, 1, 1)$, $(1, 1, 1, -1)$, $(1, 1, -1, -1)$, and $(1, -1, -1, -1)$ of $R^4$ and the basis $(2, 0, 0)$, $(0, 1, 0)$, $(0, 0, -2)$ of $R^3$.
 (b) Let $V$ be the subspace $x_4 = x_1 + x_3$ of $R^4$. What is $\dim T(V)$?

**7.** Let $T\colon R^2 \to R^3$ be the linear transformation defined by

$$T(x, y) = (x + y, 2x - y, 3x - 2y).$$

Find the matrix of $T$ with respect to the bases $(1,1)$, $(2,3)$ for $R^2$ and $(\frac{1}{3},\frac{2}{3},\frac{2}{3})$, $(-\frac{2}{3},\frac{2}{3},-\frac{1}{3})$, $(\frac{2}{3},\frac{1}{3},-\frac{2}{3})$ for $R^3$. (*Hint*: Use the fact that the given basis for $R^3$ is orthonormal.)

***8.** Let $T: R^3 \to R^2$ be defined by

$$T(x_1,x_2,x_3) = (3x_1 + 2x_2, x_2 - 2x_3).$$

(a) What is the kernel of $T$?
(b) What is the matrix of $T$ with respect to the bases $(1,0,1)$, $(1,2,1)$, $(0,1,2)$ for $R^3$ and $(1,2)$, $(1,1)$ of $R^2$?
(c) What is the nullspace of the matrix in part (b)?
(d) How are parts (a) and (c) related?

***9.** Suppose that the matrix of $T: R^3 \to R^2$ with respect to the standard basis for $R^3$ and the bases $(2,1)$, $(1,-1)$ of $R^2$ is

$$\begin{bmatrix} 1 & 1 & 0 \\ 1 & 0 & 1 \end{bmatrix}.$$

Find a formula for $T$ and find the standard matrix for $T$.

**10.** Suppose that the matrix of $T: R^3 \to R^4$ with respect to the bases $(1,0,1)$, $(1,1,0)$, $(0,0,1)$ for $R^3$ and $(1,0,0,0)$, $(1,1,0,0)$, $(1,1,1,0)$, $(1,1,1,1)$ for $R^4$ is

$$\begin{bmatrix} -2 & 1 & -1 \\ 1 & 2 & 0 \\ 0 & 3 & 1 \\ 1 & 4 & 0 \end{bmatrix}.$$

Find a formula for $T$ and find the standard matrix for $T$.

**11.** Find a formula and the standard matrix for the linear mapping $T: R^3 \to R^3$ defined in each part.

*(a) $T(1,0,0) = (1,1,0)$,
$T(0,1,0) = (2,0,1)$
$T(0,0,1) = (1,1,1)$

(b) $T(1,2,1) = (-1,2,-3)$,
$T(2,1,1) = (2,-1,1)$,
$T(0,0,1) = (0,0,3)$

(c) $T(1,2,3) = (1,1,-5)$,
$T(-1,0,-1) = (5,-1,1)$,
$T(1,3,2) = (0,1,0)$.

**12.** Let $\mathbf{v}_1,\mathbf{v}_2,\ldots,\mathbf{v}_n$ and $\mathbf{w}_1,\mathbf{w}_2,\ldots,\mathbf{w}_m$ be bases for $R^n$ and $R^m$, respectively, and let $A$ be an $m \times n$ matrix. Let $T: R^n \to R^m$ be the function defined by

$$\mathbf{c}_\mathbf{w}(T(\mathbf{x})) = A\mathbf{c}_\mathbf{v}(\mathbf{x}).$$

(a) Show that $T$ is a linear transformation.
(b) Find the matrix of $T$ with respect to the standard bases.

**13.** Let $\mathbf{v}_1,\mathbf{v}_2,\ldots \mathbf{v}_n$ be a basis for $R^n$.

(a) Show that $\mathbf{c}_\mathbf{v}(\mathbf{x}+\mathbf{y}) = \mathbf{c}_\mathbf{v}(\mathbf{x}) + \mathbf{c}_\mathbf{v}(\mathbf{y})$ for all $\mathbf{x}$ and $\mathbf{y}$ in $R^n$.
(b) Show that $\mathbf{c}_\mathbf{v}(\alpha\mathbf{x}) = \alpha\mathbf{c}_\mathbf{v}(\mathbf{x})$ for all $\mathbf{x}$ in $R^n$ and for all scalars $\alpha$.
(c) Show that $\mathbf{c}_\mathbf{v}(\mathbf{x}) = \mathbf{0}$ if and only if $\mathbf{x} = \mathbf{0}$.
(d) Show that for every $\mathbf{y}$ in $R^n$ there is a unique $\mathbf{x}$ in $R^n$ such that $\mathbf{c}_\mathbf{v}(\mathbf{x}) = \mathbf{y}$.

**14.** Let $T: R^n \to R^m$ be a linear transformation and let $A$ be the matrix of $T$ with respect to a basis $\mathbf{v}_1,\mathbf{v}_2,\ldots,\mathbf{v}_n$ of $R^n$ and a basis $\mathbf{w}_1,\mathbf{w}_2,\ldots,\mathbf{w}_m$ of $R^m$. Prove the following.

*(a) $T$ is one-to-one if and only if rank $A = n$.
(b) $T$ is onto if and only if rank $A = m$.
(c) $T(R^n)$ is spanned by the vectors whose $\mathbf{w}$-coordinates are the columns of $A$.
(d) Ker$(T)$ is spanned by the vectors whose $\mathbf{v}$-coordinates are vectors that span $N(A)$.

## 5.4 Similarity

When $T$ is a linear transformation from $R^m$ into itself, we usually use the same basis on both copies of $R^m$ to construct the matrix for $T$. That is, if $T: R^m \to R^m$ is a linear transformation and $\mathbf{v}_1,\mathbf{v}_2,\ldots,\mathbf{v}_m$ is a basis for $R^m$, then the matrix for $T$ with respect to this basis is the $m \times m$ matrix $A$ whose $j$th column is $\mathbf{c}_{\mathbf{v}}(T(\mathbf{v}_j))$:

$$A = [\mathbf{c}_{\mathbf{v}}(T(\mathbf{v}_1)) \quad \mathbf{c}_{\mathbf{v}}(T(\mathbf{v}_2)) \quad \cdots \quad \mathbf{c}_{\mathbf{v}}(T(\mathbf{v}_m))].$$

If $\mathbf{x}$ is a vector in $R^m$, then $\mathbf{c}_{\mathbf{v}}(\mathbf{x})$ is the coordinate vector of $\mathbf{x}$ with respect to the $\mathbf{v}$-basis and $A\mathbf{c}_{\mathbf{v}}(\mathbf{x})$ is the coordinate vector of $T(\mathbf{x})$ with respect to the $\mathbf{v}$-basis.

Since it is possible to represent a given linear transformation $T$: $R^m \to R^m$ with respect to different bases, it is natural to ask whether there is some basis for $R^m$ that gives rise to a particularly simple matrix representation of $T$. A complete answer to this question will have to wait until Section 5.5 and Chapter 8. For now we invite you to review Example 3 in Section 5.3 and to consider the following examples.

### EXAMPLE 1

Let $V$ be a line in $R^2$ through the origin and let $T$: $R^2 \to R^2$ be the projection of $R^2$ onto $V$. Let $\mathbf{v}_1$ be any nonzero vector on the line $V$ and let $\mathbf{v}_2$ be any nonzero vector orthogonal to $\mathbf{v}_1$. Then $\mathbf{v}_1$ and $\mathbf{v}_2$ form a basis for $R^2$. Since

$$\begin{aligned} T(\mathbf{v}_1) &= \mathbf{v}_1 = 1\cdot\mathbf{v}_1 + 0\cdot\mathbf{v}_2 \\ T(\mathbf{v}_2) &= \mathbf{0} = 0\cdot\mathbf{v}_1 + 0\cdot\mathbf{v}_2, \end{aligned}$$

if follows that

$$\mathbf{c}_{\mathbf{v}}(T(\mathbf{v}_1)) = (1,0) \qquad \text{and} \qquad \mathbf{c}_{\mathbf{v}}(T(\mathbf{v}_2)) = (0,0).$$

Hence the matrix of $T$ relative to the basis $\mathbf{v}_1$ and $\mathbf{v}_2$ is the diagonal matrix

$$\begin{bmatrix} 1 & 0 \\ 0 & 0 \end{bmatrix}.$$

### EXAMPLE 2

Let $V$, $\mathbf{v}_1$, and $\mathbf{v}_2$ be as in Example 1, but now let $T$ be the reflection of $R^2$ across $V$. Then $T(\mathbf{v}_1) = \mathbf{v}_1$ and $T(\mathbf{v}_2) = -\mathbf{v}_2$, so that the matrix of $T$ with respect to the basis $\mathbf{v}_1$ and $\mathbf{v}_2$ is the diagonal matrix

$$\begin{bmatrix} 1 & 0 \\ 0 & -1 \end{bmatrix}.$$

The general situation for projections (and reflections) is the same. If $V$ is an $n$-dimensional subspace of $R^m$ and $T$: $R^m \to R^m$ is the projection of $R^m$ onto $V$, then $T$ can be represented by a diagonal matrix whose first $n$ diagonal entries are 1 and whose remaining diagonal entries are 0. The basis for $R^m$ which gives this matrix is constructed as follows. Choose any basis $\mathbf{v}_1, \mathbf{v}_2, \ldots, \mathbf{v}_n$ for $V$ and any basis $\mathbf{v}_{n+1}, \ldots, \mathbf{v}_m$ for $V^{\perp}$. By Theorem 1 in Section 3.4 the vectors $\mathbf{v}_1, \mathbf{v}_2, \ldots, \mathbf{v}_m$ form a basis for $R^m$. Since $T(\mathbf{v}_i) = \mathbf{v}_i$ for $i = 1, \ldots, n$ and $T(\mathbf{v}_i) = \mathbf{0}$ for $i = n+1, \ldots, m$, the matrix of $T$ with respect to the basis is indeed a diagonal matrix whose first $n$ diagonal entries are 1 and whose remaining diagonal entries are 0. (See Problem 15 for the corresponding result about reflections.)

Here is an important point. In finding the diagonal matrix representation of a projection of $R^m$ onto a subspace $V$ we are free to choose *any* basis for $V$ and *any* basis for $V^\perp$. No matter what bases we choose for $V$ and $V^\perp$, we get a diagonal matrix. On the other hand, if we choose a basis for $R^m$ that does not consist of a basis for $V$ followed by a basis for $V^\perp$, the resulting matrix is not diagonal. For instance, suppose that the subspace $V$ in Example 1 is spanned by $(1, 1)$. Then the standard matrix for the projection $T$ of $R^2$ onto $V$ is the projection matrix. Since $T(1,0) = \frac{1}{2}(1,1)$ and $T(0,1) = \frac{1}{2}(1,1)$,

$$P = \frac{1}{2}\begin{bmatrix} 1 & 1 \\ 1 & 1 \end{bmatrix}, \tag{1}$$

which is certainly not a diagonal matrix. Some important mathematics is going on here! Let's investigate.

The key question would seem to be: *How are the various matrices that represent a given linear transformation related to one another*? Since matrices use coordinates of vectors with respect to a basis to represent the transformation with respect to the basis, the heart of the matter seems to be with the coordinates. Let's see.

Let $T: R^m \to R^m$ be a linear transformation and let $\mathbf{v}_1, \mathbf{v}_2, \ldots, \mathbf{v}_m$ and $\mathbf{w}_1, \mathbf{w}_2, \ldots, \mathbf{w}_m$ be two bases for $R^m$. Let $A$ and $B$ be the matrices that represent $T$ with respect to the $\mathbf{v}$-basis and the $\mathbf{w}$-basis, respectively. If $\mathbf{x}$ is in $R^m$, then

$$A\mathbf{c}_\mathbf{v}(\mathbf{x}) = \mathbf{c}_\mathbf{v}(T(\mathbf{x})) \qquad \text{and} \qquad B\mathbf{c}_\mathbf{w}(\mathbf{x}) = \mathbf{c}_\mathbf{w}(T(\mathbf{x})).$$

So $A\mathbf{c}_\mathbf{v}(\mathbf{x})$ and $B\mathbf{c}_\mathbf{w}(\mathbf{x})$ are simply different names for the same vector. Therefore if $M$ is the change of coordinates matrix that changes $\mathbf{v}$-coordinates to $\mathbf{w}$-coordinates, then

$$M\mathbf{c}_\mathbf{v}(\mathbf{x}) = \mathbf{c}_\mathbf{w}(\mathbf{x}). \tag{2}$$

Since $A\mathbf{c}_\mathbf{v}(\mathbf{x})$ and $B\mathbf{c}_\mathbf{w}(\mathbf{x})$ are coordinate vectors of the same vector $T(\mathbf{x})$ in the $\mathbf{v}$-basis and $\mathbf{w}$-basis, respectively,

$$\mathbf{c}_\mathbf{w}(A\mathbf{c}_\mathbf{v}(\mathbf{x})) = B\mathbf{c}_\mathbf{w}(\mathbf{x})$$

and hence

$$MA\mathbf{c}_\mathbf{v}(\mathbf{x}) = B\mathbf{c}_\mathbf{w}(\mathbf{x}).$$

Substituting equation (2) into the right hand side of this equation, we obtain

$$MA\mathbf{c}_\mathbf{v}(\mathbf{x}) = BM\mathbf{c}_\mathbf{v}(\mathbf{x}),$$

and since this holds for all $\mathbf{c}_\mathbf{v}(\mathbf{x})$, it follows from Result 1 in Section 1.5 that

$$MA = BM$$

or

$$A = M^{-1}BM.$$

It was a long road and we hope you made it. The result is significant!

***Theorem 1*** *Let $T: R^m \to R^m$ be a linear transformation, let $\mathbf{v}_1, \mathbf{v}_2, \ldots, \mathbf{v}_m$ and $\mathbf{w}_1, \mathbf{w}_2, \ldots, \mathbf{w}_m$ be two bases for $R^m$. Let $A$ be the matrix of $T$ with respect to the basis $\mathbf{v}_1, \mathbf{v}_2, \ldots, \mathbf{v}_m$ and let $B$ be the matrix of $T$ with respect to the basis $\mathbf{w}_1, \mathbf{w}_2, \ldots, \mathbf{w}_m$. Then*

$$A = M^{-1}BM,$$

*where $M$ is the change of coordinates matrix which changes* **v***-coordinates to* **w***-coordinates.*

Everything is as it ought to be. $A$ operates on the **v**-coordinates of $\mathbf{x}$ and produces the **v**-coordinates of the image $T(\mathbf{x})$. $M$ translates the **v**-coordinates of $\mathbf{x}$ to the **w**-coordinates of $\mathbf{x}$, $B$ operates on these **w**-coordinates of $\mathbf{x}$ and produces the **w**-coordinates of $T(\mathbf{x})$, and $M^{-1}$ translates these **w**-coordinates of $T(\mathbf{x})$ back to the **v**-coordinates of $T(\mathbf{x})$.

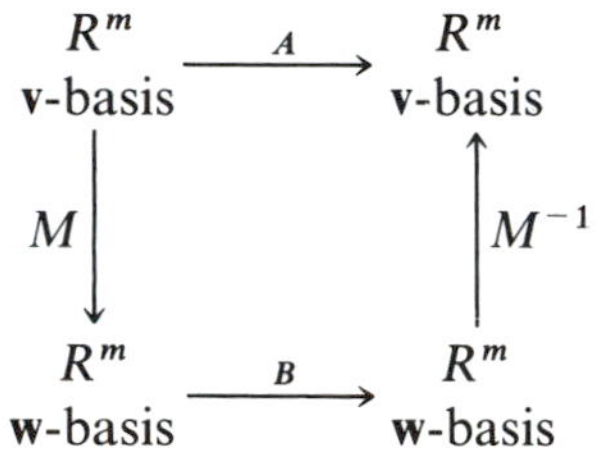


***Definition*** Let $A$ and $B$ be two $m \times m$ matrices. We say that $A$ and $B$ are **similar** if there is a nonsingular matrix $M$ such that

$$B = M^{-1}AM.$$

By Theorem 1, two $m \times m$ matrices $A$ and $B$ that represent the same linear transformation $T: R^m \to R^m$ are similar.

## EXAMPLE 3

We saw that if $V$ is the line spanned by $(1, 1)$, then the standard matrix of the projection $T$ of $R^2$ onto $V$ is [see equation (1)]

$$P = \frac{1}{2}\begin{bmatrix} 1 & 1 \\ 1 & 1 \end{bmatrix}.$$

The basis of $R^2$ consisting of $\mathbf{v}_1 = (1, 1)$ and $\mathbf{v}_2 = (-1, 1)$ gives rise to the diagonal matrix

$$\Lambda = \begin{bmatrix} 1 & 0 \\ 0 & 0 \end{bmatrix},$$

which represents $T$ with respect to the basis $\mathbf{v}_1, \mathbf{v}_2$. The matrix that changes the $\mathbf{v}$-coordinates of a vector to its standard coordinates is

$$M = \begin{bmatrix} 1 & -1 \\ 1 & 1 \end{bmatrix};$$

its inverse is

$$M^{-1} = \frac{1}{2}\begin{bmatrix} 1 & 1 \\ -1 & 1 \end{bmatrix}.$$

Thus

$$M^{-1}PM = \frac{1}{4}\begin{bmatrix} 1 & 1 \\ -1 & 1 \end{bmatrix}\begin{bmatrix} 1 & 1 \\ 1 & 1 \end{bmatrix}\begin{bmatrix} 1 & -1 \\ 1 & 1 \end{bmatrix} = \begin{bmatrix} 1 & 0 \\ 0 & 0 \end{bmatrix} = \Lambda,$$

just as Theorem 1 says.

## EXAMPLE 4

In Example 3 in Section 5.3 we studied the linear transformation defined by the equation

$$T(x_1, x_2) = (\tfrac{7}{2}x_1 - \tfrac{1}{2}x_2, \tfrac{3}{2}x_1 + \tfrac{3}{2}x_2).$$

We found that the matrix of $T$ with respect to the basis

$$\mathbf{v}_1 = (1, 1) \qquad \text{and} \qquad \mathbf{v}_2 = (1, 3)$$

is the diagonal matrix

$$\Lambda = \begin{bmatrix} 3 & 0 \\ 0 & 2 \end{bmatrix}.$$

The standard matrix of $T$ is

$$A = \frac{1}{2}\begin{bmatrix} 7 & -1 \\ 3 & 3 \end{bmatrix}.$$

It follows from Theorem 1 that if $M$ is the change of coordinates matrix which changes coordinates with respect to the basis $\mathbf{v}_1, \mathbf{v}_2$ to standard coordinates, then

$$\Lambda = M^{-1}AM.$$

Let's verify this equation. The matrix $M$ is

$$M = \begin{bmatrix} 1 & 1 \\ 1 & 3 \end{bmatrix}$$

and

$$M^{-1} = \frac{1}{2}\begin{bmatrix} 3 & -1 \\ -1 & 1 \end{bmatrix}.$$

Thus

$$M^{-1}AM = \frac{1}{4}\begin{bmatrix} 3 & -1 \\ -1 & 1 \end{bmatrix}\begin{bmatrix} 7 & -1 \\ 3 & 3 \end{bmatrix}\begin{bmatrix} 1 & 1 \\ 1 & 3 \end{bmatrix} = \begin{bmatrix} 3 & 0 \\ 0 & 2 \end{bmatrix}.$$

We are now in a position to interpret Theorem 1 of Section 3.7 terms of linear transformations. Let $\mathbf{w}_1, \mathbf{w}_2, \ldots, \mathbf{w}_n$ be an *orthonormal* basis for a subspace $V$ of $R^m$ and let $\mathbf{w}_{n+1}, \ldots, \mathbf{w}_m$ be an *orthonormal* basis for $V^\perp$. Then $\mathbf{w}_1, \mathbf{w}_2, \ldots, \mathbf{w}_m$ form an orthonormal basis for $R^m$. If $\mathbf{x}$ is in $R^m$, then by Theorem 2 of Section 3.7,

$$\mathbf{x} = \langle \mathbf{w}_1, \mathbf{x} \rangle \mathbf{w}_1 + \cdots + \langle \mathbf{w}_m, \mathbf{x} \rangle \mathbf{w}_m$$

and hence

$$\mathbf{c}_{\mathbf{w}}(\mathbf{x}) = (\langle \mathbf{w}_1, \mathbf{x} \rangle, \ldots, \langle \mathbf{w}_m, \mathbf{x} \rangle)$$

is the coordinate vector of $\mathbf{x}$ with respect to this orthonormal basis. Now let $T\colon R^m \to R^m$ be the projection of $R^m$ onto $V$. The matrix of $T$ with respect to the $\mathbf{w}$-basis is the diagonal matrix $\Lambda$ whose first $n$ diagonal entries are 1 and whose remaining diagonal entries are 0. $\Lambda$ represents $T$ by transforming the $\mathbf{w}$-coordinates of a vector $\mathbf{x}$ into the $\mathbf{w}$-coordinates of $\mathbf{p} = T(\mathbf{x})$. Thus the $\mathbf{w}$-coordinate vector of the projection $\mathbf{p} = T(\mathbf{x})$ of $\mathbf{x}$ onto $V$ is

$$\Lambda \mathbf{c}_{\mathbf{w}}(\mathbf{x}) = \begin{bmatrix} 1 & & & & & \\ & \ddots & & & & \\ & & 1 & & & \\ & & & 0 & & \\ & & & & \ddots & \\ & & & & & 0 \end{bmatrix}\begin{bmatrix} \langle \mathbf{w}_1, \mathbf{x} \rangle \\ \vdots \\ \langle \mathbf{w}_n, \mathbf{x} \rangle \\ \langle \mathbf{w}_{n+1}, \mathbf{x} \rangle \\ \vdots \\ \langle \mathbf{w}_m, \mathbf{x} \rangle \end{bmatrix} = \begin{bmatrix} \langle \mathbf{w}_1, \mathbf{x} \rangle \\ \vdots \\ \langle \mathbf{w}_n, \mathbf{x} \rangle \\ 0 \\ \vdots \\ 0 \end{bmatrix}.$$

Therefore, the expansion of $\mathbf{p}$ in terms of the basis $\mathbf{w}_1, \ldots, \mathbf{w}_m$ of $V$ is

$$\mathbf{p} = \langle \mathbf{w}_1, \mathbf{x} \rangle \mathbf{w}_1 + \cdots + \langle \mathbf{w}_n, \mathbf{x} \rangle \mathbf{w}_n.$$

This is exactly Theorem 1(a) of Section 3.7! But hang on, there is more to be learned here. If $A$ is the matrix whose columns are $\mathbf{w}_1, \ldots, \mathbf{w}_n$, then $A^TA = I$ and the projection matrix for $V$ is

$$P = A(A^TA)^{-1}A^T = AA^T.$$

$P$ is the standard matrix for $T$. By Theorem 1, then,

$$\Lambda = M^{-1}PM,$$

where $M$ is the change of coordinates matrix which changes $\mathbf{w}$-coordinates to standard coordinates. Thus the columns of $M$ are the orthonormal vectors $\mathbf{w}_1, \mathbf{w}_2, \ldots, \mathbf{w}_m$ and $M$ is an orthogonal matrix. Consequently, $M^{-1} = M^T$ and

$$\Lambda = M^TPM.$$

## EXAMPLE 5

From our work with projection matrices it follows that any projection matrix is similar to a diagonal matrix. And if we choose bases correctly (an orthonormal basis for the subspace followed by an orthonormal basis for its orthogonal complement), the matrix $M$ is an orthogonal matrix.

## EXAMPLE 6

In Example 4 we showed that the matrix

$$A = \frac{1}{2}\begin{bmatrix} 7 & -1 \\ 3 & 3 \end{bmatrix}$$

is similar to the diagonal matrix

$$\Lambda = \begin{bmatrix} 3 & \\ & 2 \end{bmatrix}.$$

The change of coordinates matrix $M$ which we computed in that example is not orthogonal. Is it possible to choose an orthogonal matrix $M$ for this matrix $A$ such that

$$\Lambda = M^T A M?$$

No! For if such a matrix existed, then taking transposes we obtain (since $\Lambda = \Lambda^T$)

$$\Lambda = M^T A^T M.$$

Thus we would have

$$M^T A^T M = M^T A M,$$

from which we would conclude that (verify!)

$$A^T = A,$$

which is not true.

In general, given a matrix $A$, if there is an orthogonal matrix $M$ such that

$$\Lambda = M^T A M$$

is a diagonal matrix, then $A$ must be symmetric (see Problem 16).

By Theorem 1, two matrices that represent the same linear transformation $T: R^m \to R^m$ are similar. The converse is also true. That is, similar matrices represent the same linear transformation with respect to appropriately chosen bases. Suppose that $A$ and $B$ are two $m \times m$ similar matrices. The matrix $A$ defines a linear transformation $T: R^m \to R^m$ via the equation $T(\mathbf{x}) = A\mathbf{x}$ and the standard matrix for $T$ is $A$ itself. If $\mathbf{v}_1, \mathbf{v}_2, \ldots, \mathbf{v}_m$ are the columns of $M$, the matrix $M$ changes coordinates of a vector with respect to the basis $\mathbf{v}_1, \mathbf{v}_2, \ldots, \mathbf{v}_m$ to its standard coordinates. The equation

$$B = M^{-1}AM$$

says that $B$ represents $T$ with respect to the basis $\mathbf{v}_1, \mathbf{v}_2, \ldots, \mathbf{v}_m$. For if $\mathbf{c}_\mathbf{v}(\mathbf{x})$ is the $\mathbf{v}$-coordinate vector of $\mathbf{x}$, then $\mathbf{x} = M\mathbf{c}_\mathbf{v}(\mathbf{x})$ and hence $B\mathbf{c}_\mathbf{v}(\mathbf{x}) = M^{-1}AM\mathbf{c}_\mathbf{v}(\mathbf{x}) = M^{-1}A\mathbf{x}$. But $A\mathbf{x}$ is the standard coordinate vector of $T(\mathbf{x})$, so $M^{-1}A\mathbf{x}$ is the $\mathbf{v}$-coordinate vector of $T(\mathbf{x})$. And that is exactly what it means to say that $B$ is the matrix of $T$ with respect to the basis $\mathbf{v}_1, \mathbf{v}_2, \ldots, \mathbf{v}_m$.

***Thoerem 2*** *Let $T: R^m \to R^m$ be a linear transformation and let $A$ and $B$ be two $m \times m$ matrices.*

(a) *If $A$ and $B$ both represent $T$, then $A$ and $B$ are similar.*
(b) *If $A$ and $B$ are similar, then $A$ and $B$ can be interpreted as representing the same linear transformation.*

In particular, we have proved the following theorem.

***Theorem 3*** *Let $T: R^m \to R^m$ be a linear transformation and let $A$ be its standard matrix.*

(a) *If an $m \times m$ matrix $B$ is the matrix of $T$ with respect to some basis $\mathbf{v}_1, \mathbf{v}_2, \ldots, \mathbf{v}_m$ of $R^m$, then*

$$B = M^{-1}AM,$$

*where $M$ is the matrix whose columns are $\mathbf{v}_1, \mathbf{v}_2, \ldots, \mathbf{v}_m$.*
(b) *If $M$ is a nonsingular $m \times m$ matrix, then the matrix*

$$B = M^{-1}AM$$

*is the matrix of $T$ with respect to the basis of $R^m$ consisting of the columns of $M$.*

Since similar matrices can be interpreted as representing the same linear transformation, they are bound to share many properties. In Problem 19 we ask you to verify some of these properties.

## PROBLEMS 5.4

**1.** Let $T: R^2 \to R^2$ be defined by

$$T(x_1, x_2) = (x_1 + x_2, 2x_2).$$

(a) Find the standard matrix $A$ for $T$.
(b) Find the matrix $B$ for $T$ with respect to the basis $(1, 0), (1, 1)$ for $R^2$.
(c) Find a nonsingular matrix $M$ such that

$$B = M^{-1}AM.$$

***2.** Find the matrix of a rotation of $R^2$ by $\pi/4$ with respect to each of the following bases.

(a) $(1, 0), (0, 1)$  (b) $(1, 1), (-1, 1)$

*3. Let $T: R^3 \to R^3$ be defined by

$$T(x_1, x_2, x_3) = (x_1 + x_2 + x_3, x_1 - x_2 + 3x_3, x_2).$$

(a) Find the standard matrix $A$ for $T$.
(b) Find the matrix $B$ for $T$ with respect to the basis $\mathbf{v}_1 = (3, 2, 1)$, $\mathbf{v}_2 = (1, 1, 1)$, $\mathbf{v}_3 = (0, 1, 0)$.
(c) Find the change of coordinates matrix $M$ that changes standard coordinates to $\mathbf{v}$-coordinates.
(d) Verify that $A = M^{-1}BM$.

*4. Let $T: R^3 \to R^3$ be defined by

$$T(x, y, z) = (x + y + z, x + y, z).$$

(a) Find the matrix of $T$ with respect to the standard basis for $R^3$.
(b) Find the matrix of $T$ with respect to the basis $(1, 1, 0)$, $(1, 0, 1)$, $(0, 1, 1)$ for $R^3$.
(c) Show that the matrix in part (a) is similar to the matrix in part (b).
(d) How do the matrices found in parts (a) and (b) represent the linear transformation $T$? Illustrate with an example for each.

5. Suppose that you are given a linear transformation $T: R^3 \to R^3$ such that

$$T(1, 1, 2) = (3, 4, 5)$$
$$T(1, 0, 1) = (4, -3, 2)$$
$$T(-3, 4, 5) = (1, 1, 2).$$

(a) What is $T(4, -3, 2)$? $T(1, 1, 1)$? $T(0, 0, 1)$?
(b) What is the matrix of $T$ with respect to the standard basis?
(c) What is the matrix of $T$ with respect to the basis $(4, -3, 2)$, $(1, 1, 1)$, $(0, 0, 1)$?
(d) Verify that the matrices in parts (b) and (c) are similar.
(e) What is the rank of $T$?
(f) Describe the kernel of $T$.
(g) Is $T$ invertible? Explain.

6. Let $T: R^3 \to R^3$ be the linear transformation defined by

$$T(x, y, z) = (x - y, y - z, z - x).$$

(a) Find the image of $R^3$ under $T$ and find a basis for it.
(b) Find the kernel of $T$ and find a basis for it.
(c) Is $T$ one-to-one?
(d) Is $T$ onto?
(e) Find the matrix $B$ of $T$ relative to the bases $(0, 1, 1)$, $(1, 0, 1)$, $(1, 1, 0)$ and $(1, 0, 0)$, $(0, 1, 0)$, $(0, 1, 1)$.
(f) What is the relation between $C(B)$, $N(B)$, and your answers to parts (a) and (b)?

*7. Let $V$ be the line in $R^2$ spanned by the vector $(1, 2)$. Find a basis of $R^2$ such that the matrix of the reflection transformation of $R^2$ across $V$ with respect to this basis is $\begin{bmatrix} 1 & 0 \\ 0 & -1 \end{bmatrix}$.

8. Let $T: R^3 \to R^3$ be the projection of $R^3$ onto the $xy$-plane. Verify that the standard matrix of $T$ and the matrix of $T$ with respect to the basis $(1, 1, 0)$, $(-1, 1, 0)$, $(0, 0, 1)$ are both equal to

$$\Lambda = \begin{bmatrix} 1 & & \\ & 1 & \\ & & 0 \end{bmatrix}.$$

Then find the change of coordinates matrix $M$ and verify that $\Lambda = M^{-1}\Lambda M$.

9. Repeat Problem 8 when $T: R^3 \to R^3$ is the reflection of $R^3$ across the $xy$-plane and

$$\Lambda = \begin{bmatrix} 1 & & \\ & 1 & \\ & & -1 \end{bmatrix}.$$

10. For each of the following matrices, find a basis for $R^3$ so that the matrix of $T(\mathbf{x}) = A\mathbf{x}$ has the form

$$B = \begin{bmatrix} 1 & 0 & 0 \\ 0 & a & b \\ 0 & c & d \end{bmatrix}.$$

Then find a change of coordinates

matrix $M$ such that $B = M^{-1}AM$.

*(a) $A = \begin{bmatrix} a & 0 & b \\ 0 & 1 & 0 \\ c & 0 & d \end{bmatrix}$

(b) $A = \begin{bmatrix} a & b & 0 \\ c & d & 0 \\ 0 & 0 & 1 \end{bmatrix}$

***11.** Let $T: R^3 \to R^2$ be given by

$$T(x, y, z) = (7x - 8y - 4z,\ 10x - 11y - 5z).$$

(a) Find the standard matrix $A$ of $T$.
(b) Show that $(1, 0, 1)$, $(1, 1, 0)$, $(0, -1, 2)$ and $(1, 2)$, $(2, 3)$ are bases for $R^3$ and $R^2$, respectively. Find the matrix $B$ of $T$ with respect to these bases.
(c) Find matrices $M_1$ and $M_2$ such that

$$A = M_1^{-1}BM_2.$$

**12.** Write out a proof of Theorem 3.

***13.** Let $T: R^n \to R^m$ be a linear transformation and let $\mathbf{v}_1, \mathbf{v}_2, \ldots, \mathbf{v}_n$ be a basis for $R^n$ and $\mathbf{w}_1, \mathbf{w}_2, \ldots, \mathbf{w}_m$ be an orthonormal basis for $R^m$. Find the matrix for $T$ with respect to these bases.

***14.** Let $T: R^n \to R^n$ be a linear transformation and let $\mathbf{v}_1, \mathbf{v}_2, \ldots, \mathbf{v}_n$ be a basis for $R^n$. Suppose that

$$T(\mathbf{v}_1) = \mathbf{v}_2,\ T(\mathbf{v}_2) = \mathbf{v}_3, \ldots, T(\mathbf{v}_{n-1}) = \mathbf{v}_n$$
$$T(\mathbf{v}_n) = a_1\mathbf{v}_1 + a_2\mathbf{v}_2 + \cdots + a_n\mathbf{v}_n.$$

Find the matrix of $T$ with respect to the basis $\mathbf{v}_1, \mathbf{v}_2, \ldots, \mathbf{v}_n$.

**15.** Show that any reflection matrix is similar to a diagonal matrix whose diagonal entries are 1's followed by $-1$'s.

**16.** Prove that if $A$ is similar to a diagonal matrix $\Lambda$ and there is an orthogonal matrix $M$ such that $\Lambda = M^TAM$, then $A$ is symmetric.

**17.** Let $T: R^n \to R^n$ be an isometry (see Problem 38 in Section 5.1).

(a) Show that if $\mathbf{v}_1, \mathbf{v}_2, \ldots, \mathbf{v}_n$ is an orthogonal (orthonormal) basis for $R^n$, then $T(\mathbf{v}_1), T(\mathbf{v}_2), \ldots, T(\mathbf{v}_n)$ also is an orthogonal (orthonormal) basis for $R^n$.
(b) Show that the matrix of $T$ with respect to an orthonormal basis for $R^n$ is orthogonal.

**18.** Let $T: R^n \to R^n$ be a linear transformation.

*(a) Show that $\operatorname{rank}(T) =$ rank of any matrix that represents $T$.
(b) Show that $\operatorname{nullity}(T) =$ nullity of any matrix that represents $T$.
(c) Show that $T$ is one-to-one if and only if the rank of any matrix that represents $T$ is $n$.
(d) Show that $T$ is onto if and only if the rank of any matrix that represents $T$ is $n$.
(e) Show that $T$ is invertible if and only if any matrix that represents $T$ is nonsingular.

**19.** Prove each of the following statements about similar matrices.

(a) Any matrix is similar to itself.
(b) If $A$ is similar to $B$, then $B$ is similar to $A$.
(c) If $A$ is similar to $B$ and $B$ is similar to $C$, then $A$ is similar to $C$.
(d) If $A$ is similar to $B$, then $\operatorname{rank}(A) = \operatorname{rank}(B)$.
(e) If $A$ is similar to $B$, then $A$ is nonsingular if and only if $B$ is nonsingular.
(f) If $A$ is similar to $B$, then $A$ is nilpotent (i.e., $A^k = 0$ for some positive integer $k$) if and only if $B$ is nilpotent.
*(g) If $A$ is similar to $B$, then $A$ is idempotent (i.e., $A^2 = A$) if and only if $B$ is idempotent.
(h) If $A$ is similar to $B$, then $A$ is similar to a diagonal matrix if and only if $B$ is similar to a diagonal matrix.
(i) If $A$ is similar to $B$, then $\operatorname{tr}(A) = \operatorname{tr}(B)$. [*Hint:* Use Problem 16(c) of Section 1.6.]

*20. Let $T: R^m \to R^m$ be a linear transformation. A subspace $V$ of $R^m$ is called **invariant** under $T$ if

$$T(V) \text{ is contained in } V.$$

Let $\mathbf{v}_1, \mathbf{v}_2, \ldots, \mathbf{v}_n$ be a basis for $V$. Extend this basis to a basis $\mathbf{v}_1, \ldots, \mathbf{v}_n, \mathbf{v}_{n+1}, \ldots, \mathbf{v}_m$ of $R^m$. What can you say about the matrix of $T$ with respect to this basis?

21. A linear transformation $T: R^n \to R^n$ is called **nilpotent** if $T^s$ is the zero transformation for some positive integer $s$. The **index of nilpotency** is the least positive integer $k$ such that:

1. $T^k(\mathbf{v}) = \mathbf{0}$ for all $\mathbf{v}$ in $R^n$.
2. $T^{k-1}(\mathbf{w}) \neq \mathbf{0}$ for some $\mathbf{w}$ in $R^n$.

*(a) Show that if $T$ is nilpotent and $T(\mathbf{v}) = \alpha\mathbf{v}$, then $\alpha = 0$ or $\mathbf{v} = \mathbf{0}$.

(b) Show that the right shift operator $T: R^n \to R^n$ defined by

$$T(x_1, x_2, \ldots, x_n) = (0, x_1, x_2, \ldots, x_{n-1})$$

is nilpotent. What is its index of nilpotency?

(c) Let $T: R^n \to R^n$ be nilpotent. Show that the matrix of $T$ with respect to any basis for $R^n$ is a nilpotent matrix. Also, show that if $T$ has index of nilpotency $k$, and $A$ is any matrix that represents $T$, then $A^k = 0$.

(d) Let $T: R^n \to R^n$ be nilpotent with index of nilpotency $k$.

(i) Show that if $T^{k-1}(\mathbf{w}) \neq \mathbf{0}$, then

$$\mathbf{w}, T(\mathbf{w}), \ldots, T^{k-1}(\mathbf{w})$$

are independent. [*Hint:* Suppose that

$$\alpha\mathbf{w} + \alpha_1 T(\mathbf{w}) + \cdots + \alpha_{k-1} T^{k-1}(\mathbf{w}) = \mathbf{0}$$

and apply $T^{k-1}$ to both sides.]

(ii) Show that $k \leq n$ and hence $T^n = 0$.

(iii) Show that the subspace spanned by the vectors in part (i) is invariant under $T$.

(iv) Show that if $k = n$, then the matrix of $T$ with respect to the basis in part (i) is the $n \times n$ matrix

$$\begin{bmatrix} 0 & 0 & 0 & \cdots & 0 & 0 \\ 1 & 0 & 0 & \cdots & 0 & 0 \\ 0 & 1 & 0 & \cdots & 0 & 0 \\ \vdots & \vdots & \vdots & & \vdots & \vdots \\ 0 & 0 & 0 & \cdots & 1 & 0 \end{bmatrix}.$$

(e) Let $A$ be an $n \times n$ nilpotent matrix. Show that $A^n = 0$.

(f) Show that a nilpotent linear transformation is not invertible.

22. Let $T: R^n \to R^m$ be a linear transformation, let $\mathbf{v}_1, \mathbf{v}_2, \ldots, \mathbf{v}_n$ and $\mathbf{v}'_1, \mathbf{v}'_2, \ldots, \mathbf{v}'_n$ be bases for $R^n$ and $\mathbf{w}_1, \mathbf{w}_2, \ldots, \mathbf{w}_m$ and $\mathbf{w}'_1, \mathbf{w}'_2, \ldots, \mathbf{w}'_m$ be bases for $R^m$. Let $A$ be the matrix for $T$ with respect to the $\mathbf{v}$-basis for $R^n$ and the $\mathbf{w}$-basis for $R^m$, and let $B$ be the matrix for $T$ with respect to the $\mathbf{v}'$-basis for $R^n$ and the $\mathbf{w}'$-basis for $R^m$. Show that there is an $n \times n$ nonsingular matrix $M_1$ and an $m \times m$ nonsingular matrix $M_2$ such that

$$A = M_2^{-1} B M_1.$$

23. Let $T: R^n \to R^n$ be a linear transformation of rank $k$. Show that there are bases $\mathbf{v}_1, \mathbf{v}_2, \ldots, \mathbf{v}_n$ and $\mathbf{w}_1, \mathbf{w}_2, \ldots, \mathbf{w}_n$ for $R^n$ such that the matrix of $T$ with respect to these bases is a diagonal matrix with the first $n - k$ diagonal entries 0 and the remaining $k$ diagonal entries 1. [*Hint:* Begin with a basis for $\ker(T)$ and extend to a basis for $R^n$.]

# 5.5
# Diagonal Representation of a Linear Transformation

Let $V$ be an $n$-dimensional subspace of $R^m$ and let $T\colon R^m \to R^m$ be the projection of $R^m$ onto $V$. As we saw in Section 5.4, if $\mathbf{v}_1, \mathbf{v}_2, \ldots, \mathbf{v}_n$ is a basis for $V$ and $\mathbf{v}_{n+1}, \ldots, \mathbf{v}_m$ is a basis for $V^\perp$, then $\mathbf{v}_1, \mathbf{v}_2, \ldots, \mathbf{v}_m$ is a basis for $R^m$ and the matrix of $T$ with respect to this basis is a diagonal matrix $\Lambda$. The first $n$ diagonal entries of $\Lambda$ are 1 [because $T(\mathbf{v}_i) = \mathbf{v}_i$ for $i = 1, \ldots, n$ and the $\mathbf{v}$-coordinate vector of $\mathbf{v}_i$ is $\mathbf{e}_i$]; the remaining diagonal entries of $\Lambda$ are 0 [because $T(\mathbf{v}_i) = \mathbf{0}$ for $i = n + 1, \ldots, m$ and the $\mathbf{v}$-coordinate vector of $\mathbf{0}$ is $\mathbf{0}$].

Similarly, if $S$ is the reflection of $R^m$ across $V$, then the matrix of $S$ with respect to this basis is also a diagonal matrix $\Lambda'$. The first $n$ diagonal entries of $\Lambda'$ are again 1 [because $S(\mathbf{v}_i) = \mathbf{v}_i$ for $i = 1, \ldots, n$ and the $\mathbf{v}$-coordinate vector of $\mathbf{v}_i$ is $\mathbf{e}_i$]; the remaining $m - n$ diagonal entries of $\Lambda'$ are $-1$ [because $S(\mathbf{v}_i) = -\mathbf{v}_i$ and the $\mathbf{v}$-coordinate vector of $-\mathbf{v}_i$ is $-\mathbf{e}_i$].

Both $T$ and $S$ behave in a particularly nice way on our basis. $T$ maps each basis vector onto a multiple of itself:

$$T(\mathbf{v}_i) = \lambda_i \mathbf{v}_i,$$

where $\lambda_i = 1$ for $i = 1, \ldots, n$ and $\lambda_i = 0$ for $i = n + 1, \ldots, m$. $S$ also maps each basis vector onto a multiple of itself:

$$S(\mathbf{v}_i) = \lambda_i \mathbf{v}_i,$$

where $\lambda_i = 1$ for $i = 1, \ldots, n$ and $\lambda_i = -1$ for $i = n + 1, \ldots, m$.

Now let's generalize. Let $T\colon R^m \to R^m$ be a linear transformation and let $\mathbf{v}_1, \mathbf{v}_2, \ldots, \mathbf{v}_m$ be a basis for $R^m$. Suppose that $T$ maps one of the basis vectors onto a multiple of itself. That is, suppose that there is a scalar $\lambda_i$ such that

$$T(\mathbf{v}_i) = \lambda_i \mathbf{v}_i.$$

Then the $i$th column of the matrix representing $T$ with respect to this basis is $\lambda_i \mathbf{e}_i$ (because $\lambda_i \mathbf{e}_i$ is the $\mathbf{v}$-coordinate vector of $\lambda_i \mathbf{v}_i$). Consequently, if $T$ maps every basis vector onto a multiple of itself, then there are $m$ scalars $\lambda_1, \lambda_2, \ldots, \lambda_m$ (not necessarily distinct) such that

$$T(\mathbf{v}_i) = \lambda_i \mathbf{v}_i, \qquad i = 1, \ldots, m.$$

Therefore, the matrix of $T$ with respect to this basis is the $m \times m$ diagonal matrix $\Lambda$ whose diagonal entries are $\lambda_1, \lambda_2, \ldots, \lambda_m$:

$$\Lambda = \begin{bmatrix} \lambda_1 & & & \\ & \lambda_2 & & \\ & & \ddots & \\ & & & \lambda_m \end{bmatrix}.$$

***Definition*** Let $T: R^m \to R^m$ be a linear transformation. A nonzero vector $\mathbf{v}$ in $R^m$ is called an **eigenvector** of $T$ if there is a scalar $\lambda$ such that

$$T(\mathbf{v}) = \lambda \mathbf{v}.$$

The scalar $\lambda$ is called an **eigenvalue** of $T$ associated with the eigenvector $\mathbf{v}$.

### EXAMPLE 1

Let $V$ be an $n$-dimensional subspace of $R^m$ and let $T: R^m \to R^m$ be the projection of $R^m$ onto $V$. Then any nonzero vector in $V$ is an eigenvector of $T$ associated with the eigenvalue 1 because

$$T(\mathbf{v}) = \mathbf{v} = 1 \cdot \mathbf{v} \qquad \text{for all } \mathbf{v} \text{ in } V.$$

Similarly, each nonzero vector in $V^\perp$ is an eigenvector of $T$ associated with the eigenvalue 0 because

$$T(\mathbf{v}) = \mathbf{0} = 0 \cdot \mathbf{v} \qquad \text{for all } \mathbf{v} \text{ in } V^\perp.$$

There are no other eigenvectors since no other nonzero vector is mapped onto a multiple of itself.

Using the terminology of eigenvectors and eigenvalues, we have proved:

***Theorem 1*** *Let $T: R^m \to R^m$ be a linear transformation. Suppose $\mathbf{v}_1, \mathbf{v}_2, \ldots, \mathbf{v}_m$ is a basis for $R^m$ consisting of eigenvectors of $T$. Then the matrix of $T$ with respect to this basis is the diagonal matrix $\Lambda$ whose diagonal entries $\lambda_1, \lambda_2, \ldots, \lambda_m$ are eigenvalues of $T$ associated with the eigenvectors $\mathbf{v}_1, \mathbf{v}_2, \ldots, \mathbf{v}_m$:*

$$\Lambda = \begin{bmatrix} \lambda_1 & & & \\ & \lambda_2 & & \\ & & \ddots & \\ & & & \lambda_m \end{bmatrix}.$$

### EXAMPLE 2

Let $T: R^2 \to R^2$ be the linear transformation considered in Example 3 in Section 5.3. $T$ is defined by the equation

$$T(x_1, x_2) = (\tfrac{7}{2}x_1 - \tfrac{1}{2}x_2, \tfrac{3}{2}x_1 + \tfrac{3}{2}x_2).$$

The vectors

$$\mathbf{v}_1 = (1, 1) \qquad \text{and} \qquad \mathbf{v}_2 = (1, 3)$$

form a basis for $R^2$ consisting of eigenvectors of $T$. The eigenvalue associated with the eigenvector $\mathbf{v}_1$ is $\lambda_1 = 3$ and the eigenvalue associated with eigenvector $\mathbf{v}_2$ is

$\lambda_2 = 2$. Thus by Theorem 1 the linear transformation $T$ is represented by the diagonal matrix

$$\Lambda = \begin{bmatrix} 3 & 0 \\ 0 & 2 \end{bmatrix}$$

with respect to this basis. This of course is exactly what we proved in Example 3 of Section 5.3.

The converse of Theorem 1 is also true and just as easy to prove.

***Theorem 2*** *Let $T: R^m \to R^m$ be a linear transformation. Suppose there is a basis $\mathbf{v}_1, \mathbf{v}_2, \ldots, \mathbf{v}_m$ of $R^m$ such that the matrix of $T$ with respect to this basis is a diagonal matrix $\Lambda$. If the diagonal entries of $\Lambda$ are $\lambda_1, \lambda_2, \ldots, \lambda_m$, then*

$$T(\mathbf{v}_i) = \lambda_i \mathbf{v}_i, \qquad i = 1, \ldots, m.$$

*Therefore, $\mathbf{v}_1, \mathbf{v}_2, \ldots, \mathbf{v}_m$ is a basis for $R^m$ consisting of eigenvectors of $T$.*

**Proof** The coordinate vector of $\mathbf{v}_i$ with respect to the basis $\mathbf{v}_1, \mathbf{v}_2, \ldots, \mathbf{v}_m$ is $\mathbf{e}_i$ and hence the coordinate vector of $T(\mathbf{v}_i)$ with respect to this basis is $\Lambda\mathbf{e}_i = \lambda_i \mathbf{e}_i$. Therefore, $T(\mathbf{v}_i) = \lambda_i \mathbf{v}_i$.

Putting these two theorems together we have proved:

***Theorem 3*** *A linear transformation $T: R^m \to R^m$ can be represented by a diagonal matrix if and only if there is a basis for $R^m$ consisting of eigenvectors of $T$.*

Thus we can obtain a beautifully simple matrix representation of a linear transformation *if we can find a basis consisting of eigenvectors of the transformation.* The "if" in the preceding sentence is a big if! Unfortunately, it is impossible to find such a basis for some linear transformations. That is, there are linear transformations for which there does not exist a basis consisting of eigenvectors of the transformation. A simple example is provided by the linear transformation that rotates the plane through an angle $\theta$, $0 < \theta < 2\pi$ and $\theta \neq \pi$. No nonzero vector is carried into a multiple of itself under such a rotation. Those linear transformations for which there is a basis consisting of eigenvectors present us with the problem of finding their eigenvectors. As with many concrete problems, the eigenvector problem is most easily solved using matrices. Of course, now we are required to translate the definition of an eigenvector and eigenvalue of a linear transformation into matrix terms. That's easy! If $T: R^m \to R^m$ is a linear transformation and $A$ is its standard matrix, then $T(\mathbf{x}) = A\mathbf{x}$. Thus if $\mathbf{v}$ is an eigenvector of $T$ and $\lambda$ is an associated eigenvalue, then

$$A\mathbf{v} = T(\mathbf{v}) = \lambda\mathbf{v}.$$

***Definition*** Let $A$ be an $m \times m$ matrix. A nonzero vector $\mathbf{v}$ in $R^m$ is called an **eigenvector** of $A$ if there is a scalar $\lambda$ such that

$$A\mathbf{v} = \lambda\mathbf{v}.$$

The scalar $\lambda$ is called an **eigenvalue** of $A$ corresponding to the eigenvector $\mathbf{v}$.

Thus the eigenvalues and eigenvectors of a linear transformation $T: R^m \to R^m$ are just the eigenvalues and eigenvectors of its standard matrix $A$. Conversely, the eigenvalues and eigenvectors of an $m \times m$ matrix $A$ are just the eigenvalues and eigenvectors of the transformation $T: R^m \to R^m$ defined by $T(\mathbf{x}) = A\mathbf{x}$. In fact, more is true. In Chapter 8 we will show that similar matrices have the same eigenvalues. Thus the eigenvalues of $T: R^m \to R^m$ are the same as the eigenvalues of any $m \times m$ matrix that represents $T$.

Every theorem about linear transformations corresponds to a theorem about matrices. For instance, the matrix version of Theorem 1 is:

***Theorem 4*** *Let $A$ be an $m \times m$ matrix. Suppose that $\mathbf{v}_1, \mathbf{v}_2, \ldots, \mathbf{v}_n$ is a basis for $R^m$ consisting of eigenvectors of $A$. Then $A$ is similar to the diagonal matrix $\Lambda$ whose diagonal entries $\lambda_1, \lambda_2, \ldots, \lambda_m$ are eigenvalues of $A$ associated with the eigenvectors $\mathbf{v}_1, \mathbf{v}_2, \ldots, \mathbf{v}_m$. Moreover, if $M$ is the matrix whose columns are $\mathbf{v}_1, \mathbf{v}_2, \ldots, \mathbf{v}_m$, then*

$$\Lambda = M^{-1}AM.$$

**Proof** Let $T: R^m \to R^m$ be the linear transformation defined by $T(\mathbf{x}) = A\mathbf{x}$. Then $A$ is the standard matrix for $T$ and hence the vectors $\mathbf{v}_1, \mathbf{v}_2, \ldots, \mathbf{v}_m$ form a basis for $R^m$ consisting of eigenvectors of $T$. By Theorem 1 the matrix of $T$ with respect to this basis is the diagonal matrix $\Lambda$ whose diagonal entries $\lambda_1, \lambda_2, \ldots, \lambda_m$ are eigenvalues of $T$ associated with the eigenvectors $\mathbf{v}_1, \mathbf{v}_2, \ldots, \mathbf{v}_m$. Thus $A$ is similar to the diagonal matrix $\Lambda$ because $A$ and $\Lambda$ represent the same linear transformation. Moreover, by Theorem 3(a) in Section 5.4 if $M$ is the matrix whose columns are $\mathbf{v}_1, \mathbf{v}_2, \ldots, \mathbf{v}_m$, then

$$\Lambda = M^{-1}AM.$$

## EXAMPLE 3

Consider the $3 \times 3$ matrix

$$A = \begin{bmatrix} 3 & 3 & -4 \\ 2 & 2 & -2 \\ 4 & 4 & -5 \end{bmatrix}.$$

It is easy to check that the vectors

$$\mathbf{v}_1 = (1, 0, 1), \qquad \mathbf{v}_2 = (-1, 1, 0), \qquad \mathbf{v}_3 = (1, 2, 2)$$

are independent, that they are eigenvectors of $A$, and that $\lambda_1 = 1, \lambda_2 = 0, \lambda_3 = 1$ are corresponding eigenvalues. Thus if $M$ is the matrix

$$M = \begin{bmatrix} 1 & -1 & 1 \\ 0 & 1 & 2 \\ 1 & 0 & 2 \end{bmatrix},$$

then by Theorem 4

$$\Lambda = \begin{bmatrix} -1 & & \\ & 0 & \\ & & 1 \end{bmatrix} = M^{-1}AM.$$

As an exercise, find $M^{-1}$ and verify this equation.

The matrix version of Theorem 2 is:

***Theorem 5*** *Let $A$ be an $m \times m$ matrix and suppose that $A$ is similar to an $m \times m$ diagonal matrix $\Lambda$. Then there is a basis for $R^m$ consisting of eigenvectors of $A$.*

**Proof** If $A$ is similar to a diagonal matrix $\Lambda$, then there is a nonsingular matrix $M$ such that

$$\Lambda = M^{-1}AM.$$

If $T: R^m \to R^m$ is defined by $T(\mathbf{x}) = A\mathbf{x}$, then $A$ is the standard matrix for $T$. By Theorem 3(b) in Section 5.4, $\Lambda$ is the matrix of $T$ with respect to the basis for $R^m$ consisting of the columns of $M$. By Theorem 2, the columns of $M$ are eigenvectors of $T$. Since $A$ is the standard matrix of $T$, the columns of $M$ are eigenvectors of $A$.

The matrix version of Theorem 3 is:

***Theorem 6*** *Let $A$ be an $m \times m$ matrix. $A$ is similar to a diagonal matrix if and only if there is a basis for $R^m$ consisting of eigenvectors of $A$.*

If it is known that $\lambda$ is an eigenvalue of a linear transformation $T: R^m \to R^m$, it is an easy matter to find the eigenvectors associated with $\lambda$. If

$$T(\mathbf{v}) = \lambda\mathbf{v},$$

then since $\lambda\mathbf{v} = \lambda I\mathbf{v}$,

$$(T - \lambda I)(\mathbf{v}) = \mathbf{0}.$$

Thus the eigenvectors of $T$ associated with the eigenvalue $\lambda$ are the nonzero vectors in the kernel of $T - \lambda I$. If $A$ is the standard matrix of $T$, then $\ker(T - \lambda I) = N(A - \lambda I)$ and the eigenvectors are simply the nonzero vectors in $N(A - \lambda I)$.

**Result 1** Let $T: R^m \to R^m$ be a linear transformation. If $\lambda$ is an eigenvalue of $T$, then the eigenvectors of $T$ associated with $\lambda$ are the nonzero vectors in the kernel of $T - \lambda I$.

## EXAMPLE 4

Let $T: R^3 \to R^3$ be the linear transformation defined by $T(\mathbf{x}) = A\mathbf{x}$, where

$$A = \begin{bmatrix} 5 & -3 & 2 \\ 6 & -4 & 4 \\ 4 & -4 & 5 \end{bmatrix}.$$

The eigenvalues of $T$ are $\lambda = 1$, $\lambda = 2$, and $\lambda = 3$. Since the standard matrix of $T - \lambda I$ is $A - \lambda I$, the eigenvectors associated with $\lambda = 1$, $\lambda = 2$, and $\lambda = 3$ are the nonzero vectors in

$$N(A - I), \quad N(A - 2I), \quad N(A - 3I),$$

respectively. By a routine computation we find that

$$\begin{aligned} &N(A - I) \text{ is spanned by } (1, 2, 1), \\ &N(A - 2I) \text{ is spanned by } (1, 1, 0), \\ &N(A - 3I) \text{ is spanned by } (1, 2, 2). \end{aligned}$$

Another routine computation will show you that these vectors are independent. Thus $T$ can be represented by the diagonal matrix

$$\Lambda = \begin{bmatrix} 1 & & \\ & 2 & \\ & & 3 \end{bmatrix}$$

with respect to the basis $(1, 2, 1)$, $(1, 1, 0)$, $(1, 2, 2)$.

We stop now. In Chapter 6 we study eigenvalues and eigenvectors in a more general setting. In Chapter 7 we develop some machinery that (at least in theory) will enable us to compute the eigenvalues (and hence the eigenvectors) of a matrix. Then in Chapter 8 we take up the topic of eigenvalues and eigenvectors afresh from the point of view of matrices. We also give a proof of Theorem 6 entirely in terms of matrices.

# PROBLEMS 5.5

**1.** In each part, do the following.

(i) Find the standard matrix $A$ of $T$.

(ii) Verify that the given numbers are eigenvalues.

(iii) If possible, find a basis for $R^m$ such that the matrix of $T$ with respect to this basis is a diagonal matrix $\Lambda$. You may assume that you have been given all eigenvalues of $T$.

(iv) Find the change of coordinates matrix $M$ and verify that $\Lambda = M^{-1}AM$.

*(a) $T(x, y) = (7x + 6y, -9x - 8y)$, $\lambda = 1, -2$
(b) $T(x, y) = (x + y, y)$, $\lambda = 1$
*(c) $T(x, y) = (3x + y, 4y)$, $\lambda = 3, 4$
(d) $T(x, y, z) = (x, -x + 2y, 4x + 2y + 3z)$, $\lambda = 1, 2, 3$
*(e) $T(x, y, z) = (3x - y, -x + 2y - z, -y + 3z)$, $\lambda = 1, 3, 4$
(f) $T(x, y, z) = (x - 2z, 0, -2x + 4z)$, $\lambda = 0, 5$
*(g) $T(x, y, z) = (x - 2y + 2z, -y + 2z, z)$ $\lambda = -1$
(h) $T(x, y, z) = (x - 2z, y + 2z, -z)$, $\lambda = -1, 1$

**2.** In each of the following, find the eigenvalues and eigenvectors of the linear transformation $T$. Then do parts (i)–(iv) of Problem 1.

*(a) $T$ is the linear transformation defined by

$$T(1, 1, 1) = (2, 2, 2),$$
$$T(2, 0, 1) = (4, 0, 2),$$
$$T(1, 0, 0) = (0, 0, 0).$$

(b) $T$ is the linear transformation defined by

$$T(-1, 2, 1) = (-3, 6, 3),$$
$$T(1, 1, 2) = (-1, -1, -2),$$
$$T(1, 2, 3) = (1, 2, 3).$$

(c) $T$ is the linear transformation defined by

$$T(2, 2, 1) = (2, 2, 1),$$
$$T(2, 0, 1) = (2, 0, 1),$$
$$T(0, 0, 1) = (0, 0, 9).$$

**3.** Find a linear transformation $T$ whose eigenvalues and eigenvectors are given. Then answer parts (i), (iii), and (iv) of Problem 1.

*(a) Eigenvalues $1, 1, 2$; eigenvectors $(1, 0, 0), (1, 1, 0), (1, 1, 1)$
(b) Eigenvalues $-1, 0, 1$; eigenvectors $(1, 1, 1), (-1, 1, 0), (1, 1, -2)$
*(c) Eigenvalues $-5, -1, 4$; eigenvectors $(1, 2, 3), (1, 3, 4), (1, 3, 5)$
(d) Eigenvalues $-1, 0, 1, 2$; eigenvectors $(1, 1, 1, 1), (1, -1, 1, -1), (1, 0, -1, 0), (0, 1, 0, -1)$

**4.** Let $T: R^2 \to R^2$ be a horizontal shear defined by

$$T(x_1, x_2) = (x_1 + \alpha x_2, x_2).$$

What are the eigenvalues and eigenvectors of $T$?

**5.** Let $T: R^2 \to R^2$ be defined by

$$T(x, y) = (5x - y, 3x + y).$$

Find a basis of $R^2$ so that the matrix of $T$ with respect to this basis is diagonal.

**6.** Let $\mathbf{v}_1$ be a fixed unit vector in $R^3$. Let $T: R^3 \to R^3$ be a rotation of $R^3$ about the line determined by $\mathbf{v}_1$ through an angle $\theta$. This line is called the **axis** of the rotation.

*(a) Show that $\lambda = 1$ is an eigenvalue of $T$.
(b) Extend $\mathbf{v}_1$ to an orthonormal basis $\mathbf{v}_1, \mathbf{v}_2, \mathbf{v}_3$ of $R^3$. Explain why $T(\mathbf{v}_1) = \mathbf{v}_1$ and $T(\mathbf{v}_1), T(\mathbf{v}_2), T(\mathbf{v}_3)$ are orthonormal vectors.
(c) Show that the matrix of $T$ with respect to the basis $\mathbf{v}_1, \mathbf{v}_2, \mathbf{v}_3$ is

$$\begin{bmatrix} 1 & 0 & 0 \\ 0 & \cos\theta & -\sin\theta \\ 0 & \sin\theta & \cos\theta \end{bmatrix}.$$

**7.** Find a matrix of the form in Problem 6(c) for each of the following rotations of $R^3$.

*(a) $T$ is a rotation about $\mathbf{v}_1 = (0, 0, 1)$ through an angle $\pi/2$.
(b) $T$ is a rotation about $\mathbf{v}_1 = (0, 1, 0)$ through an angle $\pi/6$.
*(c) $T$ is a rotation about $\mathbf{v}_1 = (1, 0, 0)$ through an angle $\pi$.
(d) $T$ is a rotation about $\mathbf{v}_1 = (1/\sqrt{2}, 1/\sqrt{2}, 0)$ through an angle $\pi/3$.
*(e) $T$ is a rotation about $\mathbf{v}_1 = (\frac{2}{7}, \frac{3}{7}, -\frac{6}{7})$ through an angle $\pi/4$.

(f) $T$ is a rotation about $\mathbf{v}_1 = (\frac{1}{3}, \frac{2}{3}, \frac{2}{3})$ through an angle $2\pi/3$.

(g) $T$ is a rotation about $\mathbf{v}_1 = (-1/\sqrt{3}, 1/\sqrt{3}, 1/\sqrt{3})$ through an angle $4\pi/3$.

**8.** Show that each of the following matrices represent rotations as follows.

(i) Show that $\lambda = 1$ is an eigenvalue of the matrix.

(ii) Find a unit eigenvector $\mathbf{v}_1$ for the eigenvalue $\lambda = 1$.

(iii) Extend $\mathbf{v}_1$ to an orthonormal basis for $R^3$ and find the matrix of $T(\mathbf{x}) = A\mathbf{x}$ with respect to this basis.

(iv) Find the axis and angle of the rotation.

*(a) $A = \frac{1}{3}\begin{bmatrix} 1 & -2 & 2 \\ 2 & -1 & -2 \\ 2 & 2 & 1 \end{bmatrix}$

(b) $A = \frac{1}{7}\begin{bmatrix} 3 & 2 & -6 \\ 2 & 6 & 3 \\ -6 & 3 & -2 \end{bmatrix}$

(c) $A = \begin{bmatrix} 1/2 & 1/2 & -\sqrt{2}/2 \\ 1/2 & 1/2 & \sqrt{2}/2 \\ \sqrt{2}/2 & -\sqrt{2}/2 & 0 \end{bmatrix}$

**9.** Let $T: R^3 \to R^3$ be the linear transformation defined by

$$T(\mathbf{e}_1) = -\mathbf{e}_3, \qquad T(\mathbf{e}_2) = \mathbf{e}_1, \qquad T(\mathbf{e}_3) = -\mathbf{e}_2$$

(a) Show that $\lambda = 1$ is an eigenvalue of $T$. Find a unit eigenvector $\mathbf{v}_1$ for the eigenvalue $\lambda = 1$.

(b) Extend $\mathbf{v}_1$ to an orthonormal basis for $R^3$ and find the matrix of $T$ with respect to this basis.

(c) Show that $T$ is a rotation. What is the axis and angle of the rotation?

**10.** Repeat Problem 9 for the linear transformation $T$ defined by

(a) $T(\mathbf{e}_1) = \mathbf{e}_2$, $T(\mathbf{e}_2) = \mathbf{e}_3$, $T(\mathbf{e}_3) = \mathbf{e}_1$

(b) $T(\mathbf{e}_1) = \mathbf{e}_3$, $T(\mathbf{e}_2) = \mathbf{e}_1$, $T(\mathbf{e}_3) = \mathbf{e}_2$

(c) $T(\mathbf{e}_1) = \mathbf{e}_2$, $T(\mathbf{e}_2) = -\mathbf{e}_3$, $T(\mathbf{e}_3) = -\mathbf{e}_1$

***11.** Show that a rotation of $R^3$ about any axis is an isometry.

**12.** Let $L$ be a line in $R^3$ through the origin and let $T$ be a rotation about $L$ through an angle $\theta$. Show that $\cos\theta = \frac{1}{2}(\mathrm{tr}(A) - 1)$, where $A$ is any matrix that represents $T$. [*Hint:* Use part (i) of Problem 19 in Section 5.4.]

**13.** Let $V$ be an $n$-dimensional subspace of $R^m$.

(a) Let $T: R^m \to R^m$ be the projection of $R^m$ onto $V$. Show that if $0 < n < m$, then both 0 and 1 are eigenvalues of $T$. What happens when $n = 0$?, $n = m$?

(b) Let $T: R^m \to R^m$ be the reflection of $R^m$ across $V$. Show that if $0 < n < m$, then both $-1$ and 1 are eigenvalues of $T$. What happens when $n = 0$?, $n = m$?

**14.** Let $T: R^n \to R^n$ be a linear transformation and let $\lambda_1, \lambda_2$ be distinct eigenvalues of $T$.

*(a) Show that if $\mathbf{v}_1$ and $\mathbf{v}_2$ are eigenvectors associated with $\lambda_1$ and $\lambda_2$, then $\mathbf{v}_1$ and $\mathbf{v}_2$ are independent.

(b) If $\lambda_3$ is another eigenvalue of $T$ and $\lambda_3$ is distinct from $\lambda_1$ and $\lambda_2$, show that if $\mathbf{v}_3$ is an eigenvector associated with $\lambda_3$, then $\mathbf{v}_1, \mathbf{v}_2, \mathbf{v}_3$ are independent.

(c) Find and prove a generalization.

**15.** Let $T: R^3 \to R^3$ have three distinct eigenvalues. Prove that $T$ can be represented by a diagonal matrix.

**16.** Find and prove a generalization of Problem 15.

**17.** Let $T: R^3 \to R^3$ have three distinct positive eigenvalues. Show that there is a linear transformation $S$ such the $S^2 = T$.

**18.** Find and prove a generalization of Problem 17.

***19.** Show that $T: R^n \to R^n$ is invertible if and only if 0 is not an eigenvalue of $T$.

**20.** Let $T: R^n \to R^n$ be invertible. What is the relation between the eigenvalues and eigenvectors of $T$ and $T^{-1}$?

**21.** Let $T: R^n \to R^n$ be a linear transformation and let $\mathbf{v}_1, \mathbf{v}_2, \ldots, \mathbf{v}_n$ be a basis for $R^n$. Show that if the matrix $A = [a_{ij}]$ of $T$ with respect to this basis is an upper triangular matrix, then the diagonal entries of the matrix are eigenvalues. (*Hint:* Show that $A - a_{ii}I$ is singular.)

**22.** Let $\lambda$ be an eigenvalue of a linear transformation $T: R^m \to R^m$.

*(a) Show that the collection of all eigenvectors of $\lambda$ together with the zero vector form a subspace of $R^m$. This subspace is called the **eigenspace** of $\lambda$ and is denoted by $E(\lambda)$.

(b) Show that if $\lambda = 0$, then the image of $E(\lambda)$ under $T$ is the trivial subspace $\{\mathbf{0}\}$.

(c) Show that if $\lambda \neq 0$, then the image of $E(\lambda)$ under $T$ is $E(\lambda)$.

(d) Show that if $T$ is a rotation of $R^3$ through an angle $\theta$, then $E(1)$ is the axis of the rotation.

**23.** Let $T: R^n \to R^n$ be a linear transformation. Suppose that $V$ is a one-dimensional subspace of $T$. Show that if $T(V)$ is contained in $V$, then every nonzero vector in $V$ is an eigenvector of $T$.

***24.** Let $T: R^n \to R^n$ satisfy $T^2 = T$. Show that 0 and 1 are the only possible eigenvalues of $T$.

**25.** Let $T: R^n \to R^n$ be nilpotent. Show that 0 is an eigenvalue of $T$ and that it is the only eigenvalue of $T$.

**26.** Prove that if $T: R^n \to R^n$ is nilpotent and $T$ can be represented by a triangular matrix $A$, then all diagonal entries of $A$ are 0.

**27.** Find all nilpotent transformations that can be represented by a diagonal matrix.

**28.** Let $T: R^n \to R^n$ be an isometry. Show that if $V$ is invariant under $T$, then so is $V^{\perp}$. (*Hint*: Show that the image of $V$ under $T$ is not only contained in $V$, but is $V$ itself.)

# 6

# Abstract Vector Spaces

The purpose of this brief chapter is to give the abstract definitions of a vector space and an inner product on a vector space. These abstract definitions specify those properties of $n$-vectors and inner products that are required to derive the results in Chapters 2 and 3. The advantage of this abstract approach is that if we know that a collection of objects has these properties, then all the results about vector spaces apply to that collection.

## 6.1 Abstract Vector Spaces

In Chapter 1 we pointed out that the operations of vector addition and scalar multiplication on $R^m$ satisfied eight identities. In Chapter 4 we pointed out that these same eight identities are satisfied in $C(I)$ by the operations of addition of functions and multiplication of a function by a scalar. $R^m$ with its two operations was called a vector space and $C(I)$ with its two operations was called a function space. The objects and operations in these spaces are different but the algebraic structure on them determined by their two operations is similar. We capture this algebraic structure with the following definition.

***Definition*** A **vector space** is a collection $V$ of objects, called vectors, together with two operations, called addition and scalar multiplication, which satisfy the following conditions. (In these conditions, $x$, $y$, and $z$ are vectors in $V$ and $\alpha$ and $\beta$ are scalars.)

I. Conditions for addition

(1) $x + y = y + x$

(2) $x + (y + z) = (x + y) + z$

(3) There is a unique vector 0 such that $x + 0 = x$ for all $x$ in $V$.
(4) For every vector $x$ in $V$ there is a unique vector $-x$ in $V$ such that $x + (-x) = 0$.

II. Conditions for scalar multiplication
(5) $\alpha(\beta x) = (\alpha\beta)x$
(6) $1x = x$
(7) $\alpha(x + y) = \alpha x + \alpha y$
(8) $(\alpha + \beta)x = \alpha x + \beta x$

This definition does not specify the particular nature of the objects that are called vectors. All that has been specified is that there must be two operations defined on the collection of these objects and that these operations must satisfy certain rules. In the case of $n$-tuples these rules follow easily from the definitions of addition of $n$-tuples, multiplication of an $n$-tuple by a scalar, and the properties of real numbers. In other situations the rules for vector operations may not be obvious and hence must be verified. Finally, since vectors in an abstract vector space can be objects other than $n$-tuples of real numbers, we do not use boldface letters to denote vectors in an abstract vector space.

### EXAMPLE 1

In Chapter 2 we showed that $R^n$ is a vector space with respect to the operations of adding $n$-tuples and multiplying an $n$-tuple by a scalar. The vectors in this vector space are $n$-tuples.

### EXAMPLE 2

In Chapter 4 we showed that the set $C(I)$ of all functions that are continuous on an interval $I$ is a vector space with respect to the usual operations of addition and scalar multiplication. The vectors in this vector space are continuous functions.

### EXAMPLE 3

Comparing these rules with the properties of $m \times n$ matrices given in Chapter 1 will show that the collection $M_{m,n}$ of all $m \times n$ matrices is a vector space with respect to the operations of addition of matrices and multiplying a matrix by a scalar. The vectors in this vector space are $m \times n$ matrices.

### EXAMPLE 4

In Chapter 5 we defined addition and scalar multiplication of linear transformations. It is easy to show that the collection $V$ of all linear transformations from $R^n$ to $R^m$ is a vector space with respect to these operations (see Problem 5). The vectors in this vector space are linear transformations.

## EXAMPLE 5

The collection of all infinite sequences $(a_1, a_2, \ldots)$ of real numbers can be thought of as an infinite-dimensional generalization of $R^n$. Thus it is natural to denote this collection by $R^\infty$. The natural generalizations of addition and scalar multiplication are

$$(a_1, a_2, \ldots) + (b_1, b_2, \ldots) = (a_1 + b_1, a_2 + b_2, \ldots)$$
$$\alpha(a_1, a_2, \ldots) = (\alpha a_1, \alpha a_2, \ldots).$$

It is easy to show that $R^\infty$ is a vector space with respect to these two operations (see Problem 3).

We define a **subspace** of a vector space $V$ to be a collection $W$ of vectors in $V$ that contains the zero vector and is closed with respect to the operations of addition and scalar multiplication defined on $V$. Thus a subspace of a vector space is a collection of vectors together with two operations. These operations satisfy all the defining conditions of a vector space because the vectors in $W$ are also in $V$ and hence (since $V$ is a vector space) satisfy the defining conditions of a vector space. This proves the following theorem.

***Theorem 1*** *A subspace $W$ of a vector space $V$ is a vector space with respect to the operations of addition and scalar multiplication in $V$.*

It follows from this theorem that all the subspaces that we studied in Chapters 2, 3, and 4 are themselves vector spaces.

The definitions of spanning, independence, basis, and dimension for an abstract vector space are the same as those given in Chapters 2 and 4. In particular, a vector space $V$ is called **finite-dimensional** if $V$ has a basis consisting of a finite number of vectors. Otherwise, $V$ is called **infinite-dimensional**. All the theorems in Chapter 2 dealing with the concepts of spanning, independence, basis, and dimension are true in any finite-dimensional vector space. However, many of the proofs that we presented in Chapter 2 depend on the choice of a basis for the vector space. This dependence was not noticed (except by your professor) because there is such a natural basis for $R^n$, namely the standard basis. It is, in fact, so natural that any $n$-tuple $\mathbf{x} = (x_1, x_2, \ldots, x_n)$ is its own coordinate vector with respect to the standard basis. In Section 6.3 we prove that if $V$ is a finite-dimensional vector space and if a basis for $V$ is specified, and if all $n$-tuples in Chapter 2 are assumed to be the coordinate vectors of the vectors in $V$ with respect to this basis, then all the proofs are valid.

However, it is not always convenient to be required to first choose a basis for $V$ in order to prove some result about a set of vectors in $V$ or to perform some calculations with the vectors in $V$. In addition, reasoning with the vectors themselves (instead of their coordinates with respect to some basis) often illuminates the structure of the proof more clearly. For these reasons it is desirable to give proofs of theorems about vector spaces that do not depend on the choice of a basis. Such proofs are called coordinate-free proofs. For instance, in Section 2.2 we proved that given a set of vectors $\mathbf{v}_1, \ldots, \mathbf{v}_n$ in $R^m$, the collection of all linear combinations of these vectors is a subspace of $R^m$. In our proof, we formed the matrix having these vectors as its columns. By doing this we have identified the vectors with their coordinate vectors with respect to the standard basis. For this reason, the proof is a coordinate proof. A coordinate-free proof might go like this.

**Proof of Theorem 2 of Section 2.2** The zero vector is obviously a linear combination of the $\mathbf{v}_i$'s. If $\mathbf{u}$ and $\mathbf{v}$ are linear combinations of the $\mathbf{v}_i$'s, say

$$\mathbf{u} = \beta_1\mathbf{v}_1 + \beta_2\mathbf{v}_2 + \cdots + \beta_n\mathbf{v}_n$$
$$\mathbf{v} = \gamma_1\mathbf{v}_1 + \gamma_2\mathbf{v}_2 + \cdots + \gamma_n\mathbf{v}_n$$

and $\alpha$ is any scalar, then $\mathbf{u} + \mathbf{v}$ and $\alpha\mathbf{u}$ are also linear combinations of the $\mathbf{v}_i$'s because

$$\mathbf{u} + \mathbf{v} = (\beta_1 + \gamma_1)\mathbf{v}_1 + (\beta_2 + \gamma_2)\mathbf{v}_2 + \cdots + (\beta_n + \gamma_n)\mathbf{v}_n$$

and

$$\alpha\mathbf{u} = (\alpha\beta_1)\mathbf{v}_1 + (\alpha\beta_2)\mathbf{v}_2 + \cdots + (\alpha\beta_n)\mathbf{v}_n.$$

This proof involves manipulating the vectors themselves without ever mentioning their coordinates. It is a coordinate-free proof. It applies equally well to a collection of functions in $C(I)$, or a collection of linear transformations from $R^n$ into $R^m$, etc.

## PROBLEMS 6.1

*1. Consider the set of all infinite sequences that have only finitely many nonzero terms. A sequence $(a_1, a_2, \ldots)$ is in this collection if and only if there is an $N$ such that $a_n = 0$ for all $n \geq N$. Is this set a vector space?

*2. Consider the set of all infinite sequences that have infinitely many terms equal to zero. A sequence $(a_1, a_2, \ldots)$ is in this set if and only if $a_n = 0$ for infinitely many $n$. [For example, $(1, 0, 1, 0, \ldots)$ is in this set.] Is this set a vector space?

3. Show that $R^\infty$ is a vector space, and show that it is infinite-dimensional.

4. Show that the set of all infinite sequences of the form $(a_1, a_2, \ldots)$ with $a_2 = 0$ is a vector space.

5. Show that the collection of all linear transformations from $R^n$ into $R^m$ is a vector space.

6. Let $\mathbf{v}_0$ be a fixed vector in $R^n$. Show that the set of all linear transformations from $R^n$ into $R^n$ with the property that $T(\mathbf{v}_0) = \mathbf{0}$ is a vector space.

7. Show that the collection of all trigonometric polynomials is a vector space, and show that it is infinite-dimensional.

8. Show that the collection of all trigonometric polynomials of degree $\leq n$ is a finite-dimensional vector space. What is its dimension?

9. (a) Show that the matrices $\begin{bmatrix} 1 & 0 \\ 0 & 0 \end{bmatrix}$, $\begin{bmatrix} 0 & 1 \\ 0 & 0 \end{bmatrix}$, $\begin{bmatrix} 0 & 0 \\ 1 & 0 \end{bmatrix}$, and $\begin{bmatrix} 0 & 0 \\ 0 & 1 \end{bmatrix}$ form a basis for the vector space of all $2 \times 2$ matrices.

   *(b) Let $V$ be the vector space of all $m \times n$ matrices. Find a basis for $V$. What is the dimension of $V$?

*(c) Is the set of all $m \times n$ echelon matrices a subspace of $V$?

***10.** Let $V$ be the vector space of all $n \times n$ matrices.

(a) Show that the set $W$ of all upper triangular matrices is a subspace of $V$. Find a basis for $W$ and the dimension of $W$.

(b) Show that the set $U$ of all diagonal $n \times n$ matrices is a subspace of $V$. Find a basis for $U$ and the dimension of $U$.

(c) Show that the set of all $n \times n$ symmetric matrices is a subspace of $V$. What is its dimension?

(d) Show that the set of all $n \times n$ skew-symmetric matrices is a subspace of $V$. What is its dimension?

***11.** Let $V$ be the vector space of all $n \times n$ matrices. Which of the following sets of matrices are subspaces of $V$?

(a) The matrices with trace equal to zero.

(b) The nonsingular matrices.

(c) The singular matrices.

(d) The matrices with all diagonal entries equal to zero.

(e) The orthogonal matrices.

**12.** Let $A$ be a $3 \times 3$ matrix. Consider the 10 matrices $I, A, A^2, \ldots, A^9$.

(a) Show that these 10 matrices are linearly dependent.

(b) Use part (a) to show that there exists an integer $k \leq 9$ and scalars $a_0, a_1, \ldots, a_{k-1}$ such that

$$A^k + a_{k-1}A^{k-1} + \cdots + a_1 A + a_0 I = 0.$$

(c) Generalize your argument to apply to the case where $A$ is an $n \times n$ matrix.

**13.** Identify those proofs in Chapter 2 that are coordinate proofs.

**14.** Show that the set of all convergent infinite sequences is a vector space.

**15.** Show that the set of all infinite sequences which converge to zero is a vector space.

***16.** Is the set of all infinite sequences that converge to 1 a vector space?

**17.** Let $V$ be the set of $2 \times 2$ matrices $A$ that satisfy *both* of the following properties simultaneously.

1. $A\begin{bmatrix} 1 \\ 2 \end{bmatrix}$ is a multiple of $\begin{bmatrix} 1 \\ 2 \end{bmatrix}$.

2. $A\begin{bmatrix} 2 \\ -3 \end{bmatrix}$ is a multiple of $\begin{bmatrix} 2 \\ -3 \end{bmatrix}$.

(a) Show that $V$ is a vector space.

(b) Find dim $V$, and give a basis for $V$.

**18.** Let $V$ be the collection of all polynomials $p$ of degree $\leq 4$ such that

$$p(0) = p(1) = 0.$$

(a) Show that $V$ is a vector space.

(b) Find a basis for $V$ and the dimension of $V$.

# 6.2
## Inner Product Spaces

In Chapter 4 we pointed out that all the theorems and results about inner products that were derived in Chapter 3 depended only on those properties of an inner product given in Theorem 1 of Section 3.2. When we defined an inner product in function spaces, all these properties were satisfied. Because of this, all the results from Chapter 3 carried over to function spaces unchanged.

Our motivation for the abstract definition of an inner product is exactly the same. We define an inner product on an abstract vector space to be a function that satisfies those properties given in Theorem 1 of Section 3.2. Then all the results and theorems about inner products that we derived in Chapter 3 will hold for a finite-dimensional vector space with an (abstract) inner product.

***Definition*** An **inner product** on a vector space $V$ is a function that assigns a real number $\langle u, v\rangle$ to every pair of vectors $u$ and $v$ in $V$ in such a way that for all vectors $u$ and $v$ and scalars $\alpha$,

(a) $\langle u, v\rangle = \langle v, u\rangle$.
(b) $\langle u, v + w\rangle = \langle u, v\rangle + \langle u, w\rangle$.
(c) $\langle \alpha u, v\rangle = \alpha\langle u, v\rangle$.
(d) $\langle u, u\rangle \geq 0$, and $\langle u, u\rangle = 0$ if and only if $u = 0$.

An **inner product space** is a vector space with an inner product.

### EXAMPLE 1

Let $\alpha_1, \alpha_2, \ldots, \alpha_m$ be fixed positive numbers. We claim that the formula

$$\langle \mathbf{u}, \mathbf{v}\rangle = \alpha_1 u_1 v_1 + \cdots + \alpha_m u_m v_m$$

defines an inner product on $R^m$. Condition (a) clearly holds since $\alpha_i u_i v_i = \alpha_i v_i u_i$ for $i = 1, \ldots, m$. Similarly, since $\alpha_i u_i(v_i + w_i) = \alpha_i u_i v_i + \alpha_i u_i w_i$ for $i = 1, \ldots, m$, condition (b) follows immediately. Condition (c) is just the distributive law. So far we have not used the assumption that the scalars $\alpha_i$ are positive. We do need this assumption to prove condition (d). Since

$$\langle \mathbf{u}, \mathbf{u}\rangle = \alpha_1 u_1^2 + \cdots + \alpha_m u_m^2$$

is a sum of nonnegative numbers, it is a nonnegative number and it is 0 if and only if $\mathbf{u} = \mathbf{0}$. This proves condition (d).

### EXAMPLE 2

The inner product on $C(I)$, $I = [a, b]$, corresponding to the inner product on $R^m$ in Example 1 is defined by the equation

$$\langle f, g\rangle = \int_a^b f(x)g(x)h(x)\,dx,$$

where $h$ is a fixed positive function in $C(I)$ [i.e., $h(x) > 0$ for all $x$ in $I$]. As in Example 1, conditions (a), (b), and (c) follow immediately. That $\langle f, f\rangle \geq 0$ and that if $f = 0$, then $\langle f, f\rangle = 0$ also follow immediately. The fact that $\langle f, f\rangle = 0$ implies that $f = 0$ is a more sophisticated fact from calculus. Suppose that $f \neq 0$, that is, suppose that $f(x_0) \neq 0$ for some $x_0$ in $[a, b]$. Since $h(x_0) > 0$, $f^2(x_0)h(x_0) > 0$ and continuity guarantees that there is a constant $\alpha > 0$ and an interval $[c, d]$ such that $a \leq c < x_0 < d \leq b$ and $f^2(x)h(x) \geq \alpha$ for all $x$ in $[c, d]$. Therefore,

$$\langle f, f\rangle = \int_a^b f^2(x)h(x)\,dx \geq \int_c^d f^2(x)h(x)\,dx \geq \alpha(d - c) > 0.$$

### EXAMPLE 3

Let $V$ be the vector space of all $2 \times 2$ matrices. If $A$ and $B$ are in $V$, define

$$\langle A, B \rangle = a_{11}b_{11} + a_{12}b_{12} + a_{21}b_{21} + a_{22}b_{22}.$$

If we think of a $2 \times 2$ matrix as a vector in $R^4$ [namely, the vector $(a_{11}, a_{12}, a_{21}, a_{22})$], then this is just the standard inner product on $R^4$. Thus this formula defines an inner product on $V$.

### EXAMPLE 4

An $m \times m$ matrix $A$ is called **positive definite** if $\mathbf{x}^T A \mathbf{x} > 0$ for all nonzero vectors $\mathbf{x}$ in $R^m$. Let $A$ be a positive definite symmetric matrix. The formula

$$\langle \mathbf{u}, \mathbf{v} \rangle = \mathbf{u}^T A \mathbf{v}$$

defines an inner product on $R^m$ (see Problem 6).

## PROBLEMS 6.2

**1.** The norm of a vector in $V$, denoted by $\|v\|$, is defined by the equation

$$\|v\| = \sqrt{\langle v, v \rangle}.$$

(a) Show that for any vector $v$ in $V$ and scalar $\alpha$,

$$\|\alpha v\| = |\alpha| \cdot \|v\|.$$

(b) Show that for any vector $v$ in $V$, $\|v\| \geq 0$, and $\|v\| = 0$ if and only if $v = 0$.

(c) State and prove the Cauchy–Schwarz inequality for vectors in $V$. (*Hint:* Use the proof given in Chapter 3.)

(d) State and prove the triangle inequality for vectors in $V$. (*Hint:* Use the proof given in Chapter 3.)

(e) How would you define the distance between two vectors in $V$?

***2.** Show that the formula

$$\langle \mathbf{u}, \mathbf{v} \rangle = u_1v_1 + 2u_1v_2 + 2u_2v_1 + u_2v_2$$

does not define an inner product on $R^2$. (*Hint:* Find a nonzero vector $\mathbf{u}$ such that $\langle \mathbf{u}, \mathbf{u} \rangle = 0$.)

***3.** Show that the formula

$$\langle \mathbf{u}, \mathbf{v} \rangle = u_1v_1 + u_2v_1 + u_1v_2 + 2u_2v_2$$

defines an inner product on $R^2$.

**4.** (a) Show that the formula

$$\langle \mathbf{u}, \mathbf{v} \rangle = u_1v_1 + u_1v_2 + u_2v_1 + u_2v_2$$

does not define an inner product on $R^2$. What fails?

(b) Show that the formula

$$\langle \mathbf{u}, \mathbf{v} \rangle = 2u_1v_1 + u_1v_2 + u_2v_1 + 2u_2v_2$$

does define an inner product on $R^2$.

**5.** Show that the formula

$$\langle (x_1, x_2, x_3), (y_1, y_2, y_3) \rangle = x_1y_1 + x_2y_2 + 2x_2y_3 + 2x_3y_2 + 3x_3y_3$$

does not define an inner product on $R^3$.

**6.** Verify that the formula given in Example 4 defines an inner product on $R^m$.

***7.** You are given that

$$\langle (a, b), (c, d) \rangle = 2ac - ad - bc + 4bd$$

is an inner product on $R^2$.

(a) What is the angle between $(1, 2)$ and $(3, 0)$?
(b) Is the angle between $(0, 1)$ and $(-1, 1)$ acute, right, or obtuse? Why?

**8.** Define two vectors $u$ and $v$ in $V$ to be orthogonal if $\langle u, v\rangle = 0$.
(a) Show that the zero vector in $V$ is orthogonal to every vector in $V$.
(b) Show that two vectors $u$ and $v$ in $V$ are orthogonal if and only if

$$\|u + v\| = \|u - v\|.$$

(c) Show that if $u$ and $v$ are orthogonal vectors in $V$, then

$$\|u + v\|^2 = \|u\|^2 + \|v\|^2.$$

**9.** Given the inner product

$$\langle \mathbf{u}, \mathbf{v}\rangle = 9u_1v_1 + u_2v_2$$

on $R^2$, find:
*(a) All vectors $\mathbf{v}$ orthogonal to $(1, 1)$.
(b) All vectors $\mathbf{w}$ orthogonal to $(1, -1)$.
(c) All values of $k$ such that the lines $y = kx$ and $y = -kx$ are perpendicular.
Draw pictures of the above!

**10.** Let $I = [1, e]$. For $f, g$ in $C(I)$, define

$$\langle f, g\rangle = \int_1^e (\log x) f(x) g(x)\, dx.$$

(a) Prove that this formula defines an inner product on $C(I)$.
*(b) Using the inner product in part (a), find a polynomial $p(x) = a + bx$ that is orthogonal to $f(x) = 1$.

**11.** Define a collection $v_1, \ldots, v_k$ of vectors in $V$ to be orthogonal if $\langle v_i, v_j\rangle = 0$ whenever $i \neq j$. Show that if $v_1, \ldots, v_k$ are nonzero orthogonal vectors in $V$, then they are linearly independent. (*Hint:* Use the proof given in Chapter 3.)

**12.** Let $V$ be a finite-dimensional inner product space and let $v_1, v_2, \ldots, v_n$ be an orthonormal basis for $V$.
(a) If $x = \alpha_1 v_1 + \alpha_2 v_2 + \cdots + \alpha_n v_n$ and

$$y = \beta_1 v_1 + \beta_2 v_2 + \cdots + \beta_n v_n,$$

show that

$$\langle x, y\rangle = \alpha_1\beta_1 + \alpha_2\beta_2 + \cdots + \alpha_n\beta_n.$$

(b) Show that if $x$ is in $V$, then

$$x = \langle x, v_1\rangle v_1 + \langle x, v_2\rangle v_2 + \cdots + \langle x, v_n\rangle v_n.$$

**13.** Let $S$ be a subset of $V$. Define a vector $v$ in $V$ to be orthogonal to the set $S$, written $v \perp S$, if $v$ is orthogonal to every vector in $S$.
(a) Show that the only vector in $V$ that is orthogonal to $V$ is the zero vector.
(b) Show that if $W$ is a subspace of $V$, then $v \perp W$ if and only if $v$ is orthogonal to a spanning set for $W$.

**14.** Let $S$ be a subset of $V$. Define the orthogonal complement of $S$, denoted by $S^{\perp}$, to be the set of all vectors in $V$ that are orthogonal to $S$.
(a) Show that $S^{\perp}$ is a subspace of $V$.
(b) Let $W$ be a subspace of $V$. Show that the only vector contained in both $W$ and $W^{\perp}$ is the zero vector.
(c) If $V$ is $n$-dimensional and $W$ is an $r$-dimensional subspace of $V$, show that the dimension of $W^{\perp}$ is $n - r$.
(d) Show that if $w_1, \ldots, w_r$ is a basis for $W$ and $w_{r+1}, \ldots, w_n$ is a basis for $W^{\perp}$, then $w_1, \ldots, w_r, w_{r+1}, \ldots, w_n$ is a basis for $V$. [*Hint:* See the proof of Theorem 1(d) in Section 3.4.]

**15.** Let $V$ be the vector space consisting of all $n \times n$ matrices.
(a) Show that

$$\langle A, B\rangle = \operatorname{tr}(AB^T)$$

defines an inner product on $V$. (*Hint:* Use Problem 16 in Section 1.6.)
(b) Show that $\|A\|^2$ is the sum of the squares of the entries in $A$.
(c) Show that

$$\begin{bmatrix} 1 & 0 \\ 0 & 0 \end{bmatrix}, \begin{bmatrix} 0 & 1 \\ 0 & 0 \end{bmatrix}, \begin{bmatrix} 0 & 0 \\ 1 & 0 \end{bmatrix}, \begin{bmatrix} 0 & 0 \\ 0 & 1 \end{bmatrix}$$

form an orthonormal basis for $V$ when $n = 2$.

(d) Find and prove a generalization of part (c).

(e) Show that if $\mathbf{v}_1, \mathbf{v}_2, \ldots, \mathbf{v}_n$ form an orthonormal basis for $R^n$, then the matrices $\mathbf{v}_i\mathbf{v}_j^T$, $i = 1, \ldots, n$, $j = 1, \ldots, n$, form an orthonormal basis for $V$.

(f) Let $W$ be the subspace consisting of all matrices in $V$ whose diagonal entries are 0. Show that $W^\perp$ consists of all diagonal matrices.

(g) Show that if $W$ is the subspace that consists of all symmetric matrices, then $W^\perp$ consists of all skew-symmetric matrices.

**16.** Prove Theorem 1 of Section 3.5 when $R^m$ is replaced by an abstract finite-dimensional inner product space $V$.

**17.** (a) Define the projection of a vector $b$ onto a finite-dimensional subspace $W$ of an abstract inner product space $V$.

(b) Show that the projection of $b$ onto $W$ is the vector in $W$ that is closest to $b$.

**18.** Show that the Gram–Schmidt process is valid in any finite-dimensional inner product space $V$.

**19.** If $p$ and $q$ are polynomials, define

$$\langle p, q\rangle = \int_0^\infty e^{-x}p(x)q(x)\,dx.$$

(a) Show that this formula defines an inner product on $P(R)$.

(b) Show that $\langle x^n, x^m\rangle = (n + m)!$

**20.** Let $V$ be the subset of $R^\infty$ consisting of all vectors $x = (x_1, x_2, \ldots)$ such that $x_1^2 + x_2^2 + \cdots$ converges.

(a) Show that $V$ is a subspace of $R^\infty$.

(b) If $x$ and $y$ are in $V$, define

$$\langle x, y\rangle = x_1y_1 + x_2y_2 + \cdots.$$

Prove that this series converges absolutely. (*Hint:* This is difficult. You will need to use the Cauchy–Schwarz inequality on the partial sums.)

(c) Prove that this formula defines an inner product on $V$.

**21.** Show that

$$\begin{bmatrix} a & b \\ b & d \end{bmatrix}$$

is positive definite if and only if $a > 0$ and $ad - b^2 > 0$. (*Hint:* Use the fact that if a quadratic $Ax^2 + Bx + C$ has no real roots, then $B^2 - 4AC < 0$.)

# 6.3 Linear Transformations

A **linear transformation** $T$ between two vector spaces $V$ and $W$ is a function $T\colon V \to W$ such that

(1) $T(x + y) = T(x) + T(y)$ for all $x, y$ in $V$.
(2) $T(\alpha x) = \alpha T(x)$ for all $x$ in $V$ and all scalars $\alpha$.

The **image** of $T$ is the set of all vectors $y$ in $W$ such that $y = T(x)$ for some $x$ in $V$ and is denoted by $T(V)$. When $T(V) = W$, we say that $T$ is **onto**. Theorem 1 of Section 5.1 and its proof are valid when $R^n$ is replaced by an $n$-dimensional vector space $V$, and $R^m$ is replaced by an $m$-dimensional vector space $W$. The **kernel** of $T$ is the set of all vectors $x$ in $V$ such that $T(x) = 0$ (the zero vector in $W$). It is a subspace of $V$ (verify!). $T$ is called **one-to-one** if $T(x) \neq T(y)$ whenever $x \neq y$. It is easy to prove that $T$ is one-to-one if and only if $\ker(T)=\{0\}$. For if $T$ is one-to-one and $T(x) = 0$, then since $T(0) = 0$, it follows that $T(x) = T(0)$. Thus $x = 0$ and $\ker(T) = \{0\}$. Conversely, suppose $\ker(T) = \{0\}$. If $T(x) = T(y)$, then $T(x - y) = 0$ and hence $x - y = 0$. Theorems 4 through 7 of Section 5.2 are true when $V$ and $W$ are finite-dimensional vector spaces (verify!). In Problem 48 we ask you to investigate the validity of the theorems in Chapter 5 when $V$ and $W$ are infinite-dimensional.

Let us return now to our discussion of coordinate versus coordinate-free proofs in Section 6.1. There we stated that if $V$ is an $n$-dimensional vector space and a basis for $V$ is specified, and if all $n$-tuples are interpreted as coordinate vectors of vectors in $V$, then all the coordinate proofs we gave for vectors in $R^n$ are valid for vectors in $V$. The reason this is true is that the linear transformation which sends vectors in $V$ to their coordinate vectors in $R^n$ is one-to-one and onto.

***Theorem 1*** *Let $V$ be an $n$-dimensional vector space and let $v_1, v_2, \ldots, v_n$ be a basis for $V$. The coordinate map*

$$\mathbf{c}_v: V \to R^n$$

*which maps every vector $x$ in $V$ to its $v$-coordinate vector $\mathbf{c}_v(x)$ is a linear transformation that is one-to-one and onto.*

**Proof** Let $x$ and $y$ be vectors in $V$, and let $\mathbf{c}_v(x) = (x_1, x_2, \ldots, x_n)$ and $\mathbf{c}_v(y) = (y_1, y_2, \ldots, y_n)$ be their coordinate vectors with respect to the basis $v_1, v_2, \ldots, v_n$. Then

$$\begin{aligned} x &= x_1 v_1 + x_2 v_2 + \cdots + x_n v_n \\ y &= y_1 v_1 + y_2 v_2 + \cdots + y_n v_n. \end{aligned}$$

Since

$$x + y = (x_1 + y_1)v_1 + (x_2 + y_2)v_2 + \cdots + (x_n + y_n)v_n$$

is the unique expansion of $x + y$ in terms of the basis,

$$\mathbf{c}_v(x + y) = (x_1 + y_1, x_2 + y_2, \ldots, x_n + y_n).$$

Thus $\mathbf{c}_v(x + y) = \mathbf{c}_v(x) + \mathbf{c}_v(y)$. Similarly, $\mathbf{c}_v(\alpha x) = \alpha \mathbf{c}_v(x)$ for any scalar $\alpha$. Thus the coordinate map is indeed linear. It is one-to-one since $\mathbf{c}_v(x) = 0$ clearly implies that $x = 0$. It is onto since any $n$-tuple $(x_1, x_2, \ldots, x_n)$ determines a unique vector

$$x = x_1 v_1 + x_2 v_2 + \cdots + x_n v_n$$

such that $\mathbf{c}_v(x) = (x_1, x_2, \ldots, x_n)$. This completes the proof.

A linear transformation $T: V \to W$ preserves linear combinations: if $x$ is a linear combination of $v_1, v_2, \ldots, v_n$, then $T(x)$ is the same linear combination of $T(v_1), T(v_2), \ldots, T(v_n)$. However, a linear transformation can involve some collapsing. This happens when $T$ sends a basis for $V$ into a dependent set of vectors. The condition on $T$ which prevents this collapsing is that $T$ is one-to-one. Then $T$ maps any independent set of vectors in $V$ to an independent set of vectors in $W$ and hence it maps any basis for $V$ to a basis for $T(V)$. Finally, the general situation is that the image $T(V)$ of $V$ is a subspace of $W$. When $T$ is onto, $T(V) = W$. Thus a linear transformation $T: V \to W$ which is both one-to-one and onto sets up a one-to-one correspondence between $V$ and $W$. Moreover, it not only preserves linear combinations, it preserves nonzero linear combinations. Therefore, it faithfully translates any linear combination in $V$ to an equivalent linear combination in $W$. Thus if $V$ and $W$ are vector spaces such that there is a linear transformation $T: V \to W$ that is one-to-one and onto, then $V$ and $W$ have the same algebraic structure and are called **isomorphic**. A one-to-one and onto linear transformation is called an **isomorphism**. In this terminology, the previous theorem becomes:

***Theorem 2*** *Let $V$ be an n-dimensional vector space, and let $v_1, v_2, \ldots, v_n$ be a basis for $V$. Then the coordinate map $\mathbf{c}_v$: $V \to R^n$ is an isomorphism, and $V$ is isomorphic to $R^n$.*

Therefore, any $n$-dimensional vector space is isomorphic to $R^n$, and the coordinate map establishes the isomorphism. Thus if $V$ is a finite-dimensional vector space and a basis for $V$ is specified, and if all vectors in Chapter 2 are assumed to be the coordinate vectors of vectors in $V$ with respect to this basis, then all theorems and their proofs in Chapter 2 are valid for $V$.

Now let's consider some examples of linear transformations.

## EXAMPLE 1

Let $I = [a, b]$ be a closed interval. If $f$ is a function in $C^1(I)$, then its derivative $f'$ is continuous on $I$ and hence belongs to $C(I)$. Let

$$D: C^1(I) \to C(I)$$

be the function that maps $f$ into its derivative, that is, $D(f) = f'$. For instance, if $f(x) = x^3$, then $D(f)(x) = 3x^2$. Similarly, if $g(x) = \sin x$, then $D(g)(x) = \cos x$. $D$ is a linear transformation because

$$D(f + g) = (f + g)' = f' + g' = D(f) + D(g)$$
$$D(\alpha f) = (\alpha f)' = \alpha f' = \alpha D(f).$$

The linear transformation $D$ is often called the **differential operator**. The kernel of $D$ is the set of all functions $f$ in $C^1(I)$ such that $D(f) = f' = 0$, and these functions are the constant functions. The image of $T$ is all of $C(I)$. To prove this fact we need to show that any continuous function on $I$ is equal to the derivative of another function. This is exactly what the *fundamental theorem of calculus* says. If $f$ is continuous on $I = [a, b]$, then the function

$$F(x) = \int_a^x f(t)\,dt$$

is differentiable on $I$ and $F'(x) = f(x)$. Thus $F$ is in $C^1(I)$ and $D(F) = f$.

We generalize this example as follows. Let

$$D^n\colon C^n(I) \to C(I)$$

be the function that maps $f$ onto its $n$th derivative; that is, $D^n(f) = f^{(n)}$. $D^n$ is a linear transformation since

$$\begin{aligned} D^n(f+g) &= (f+g)^{(n)} = f^{(n)} + g^{(n)} = D^n(f) + D^n(g), \\ D^n(\alpha f) &= (\alpha f)^{(n)} = \alpha f^{(n)} = \alpha D^n(f). \end{aligned}$$

We call $D^n$ a **differential operator of order $n$**. It is easy to show that the sum $D^n + D^m$ of two differential operators $D^n$ and $D^m$, and a scalar multiple of a differential operator $D^n$, define linear transformations from $C^n(I) \to C(I)$ if $m \le n$. The function $L\colon C^n(I) \to C(I)$ defined by

$$L = \alpha_n D^n + \alpha_{n-1} D^{n-1} + \cdots + \alpha_1 D + \alpha_0 I$$

is called a **linear differential operator of order $n$ with constant coefficients**. Its kernel is the solution space of the homogeneous linear differential equation

$$\alpha_n y^{(n)} + \alpha_{n-1} y^{(n-1)} + \cdots + \alpha_1 y + \alpha_0 y = 0.$$

See Example 4 of Section 4.1.

### EXAMPLE 2

Let $I = [a, b]$ be a closed interval, and let $T\colon C(I) \to R^1$ be defined by

$$T(f) = \int_a^b f(x)\,dx.$$

For instance, if $I = [0, 1]$, $f(x) = 2x$, and $g(x) = 3x^2 - 1$, then

$$T(f) = \int_0^1 2x\,dx = x^2\Big|_0^1 = 1,$$

$$T(g) = \int_0^1 (3x^2 - 1)\,dx = (x^3 - x)\Big|_0^1 = 0.$$

If $f$ and $g$ are any functions in $C(I)$ and $\alpha$ is any scalar, then

$$\begin{aligned} T(f+g) &= \int_a^b (f+g)(x)\,dx = \int_a^b (f(x) + g(x))\,dx \\ &= \int_a^b f(x)\,dx + \int_a^b g(x)\,dx = T(f) + T(g), \\ T(\alpha f) &= \int_a^b (\alpha f)(x)\,dx = \int_a^b \alpha f(x)\,dx \\ &= \alpha \int_a^b f(x)\,dx = \alpha T(f). \end{aligned}$$

Thus $T$ is a linear transformation.

## EXAMPLE 3

Let $T: P_3(R) \to P_3(R)$ be defined by $T(p)(x) = xp'(x)$. Since

$$T(p + q)(x) = x(p + q)'(x) = xp'(x) + xq'(x) = T(p) + T(q),$$
$$T(\alpha p)(x) = x(\alpha p)'(x) = \alpha xp'(x) = \alpha T(p)(x),$$

it follows that $T$ is a linear transformation. The kernel of $T$ consists of all $p$ in $P_3(R)$ such that $T(p) = 0$. If $p(x) = a + bx + cx^2 + dx^3$, then this equation becomes

$$T(p) = bx + 2cx^2 + 3dx^3 = 0.$$

Thus $b = c = d = 0$ and the kernel of $T$ consists of all constant polynomials. Thus $T$ is not one-to-one. What about the image of $T$? It consists of all polynomials $q$ in $P_3(R)$ such that $T(p) = q$ for some $p$ in $P_3(R)$. If $p(x) = a + bx + cx^2 + dx^3$ and $q(x) = a_1 + b_1x + c_1x^2 + d_1x^3$, then $T(p) = q$ if and only if

$$bx + 2cx^2 + 3dx^3 = a_1 + b_1x + c_1x^2 + d_1x^3.$$

Thus there is a polynomial $p$ such that $T(p) = q$ if and only if $a_1 = 0$. Consequently, the image of $T$ consists of all polynomials whose constant term is zero. $T$ is not onto.

Since any $n$-dimensional vector space is isomorphic to $R^n$ and the coordinate map establishes the isomorphism, it is reasonable to expect that any linear transformation between two finite-dimensional vector spaces is represented by a matrix that sends the coordinates of a vector to the coordinates of its image under the transformation. Indeed, suppose that $V$ and $W$ are finite-dimensional vector spaces and $T: V \to W$ is a linear transformation. Let $v_1, v_2, \ldots, v_n$ be a basis for $V$ and $w_1, w_2, \ldots, w_m$ be a basis for $W$. If $x$ is in $V$ anc $\mathbf{c}_v(x) = (x_1, x_2, \ldots, x_n)$, then

$$x = x_1v_1 + x_2v_2 + \cdots + x_nv_n$$

and hence

$$T(x) = x_1T(v_1) + x_2T(v_2) + \cdots + x_nT(v_n).$$

In terms of $w$-coordinates this equation becomes

$$\mathbf{c}_w(T(x)) = x_1\mathbf{c}_w(T(v_1)) + x_2\mathbf{c}_w(T(v_2)) + \cdots + x_n\mathbf{c}_w(T(v_n)).$$

Thus if $A$ is the matrix whose columns are

$$\mathbf{c}_w(T(v_1)), \quad \mathbf{c}_w(T(v_2)), \quad \ldots, \quad \mathbf{c}_w(T(v_n)),$$

then

$$\mathbf{c}_w(T(x)) = A\mathbf{c}_v(x). \tag{1}$$

Thus $T$ can be represented by a matrix. The matrix maps the $v$-coordinate vector of $x$ to the $w$-coordinate vector of $T(x)$.

Of course, the converse is also true. That is, given a basis $v_1, v_2, \ldots, v_n$ for $V$ and a basis $w_1, w_2, \ldots, w_m$ for $W$, any $m \times n$ matrix $A$ defines a linear transformation $T: V \to W$ by means of equation (1).

### EXAMPLE 4

Consider the differential operator $D$: $P_3(R) \to P_2(R)$. Choose the basis $v_1 = 1$, $v_2 = x$, $v_3 = x^2$, $v_4 = x^3$ for $P_3(R)$ and $w_1 = 1$, $w_2 = x$, $w_3 = x^2$ for $P_2(R)$. Since

$$\begin{aligned} D(1) &= 0 = 0\cdot 1 + 0\cdot x + 0\cdot x^2 \\ D(x) &= 1 = 1\cdot 1 + 0\cdot x + 0\cdot x^2 \\ D(x^2) &= 2x = 0\cdot 1 + 2\cdot x + 0\cdot x^2 \\ D(x^3) &= 3x^2 = 0\cdot 1 + 0\cdot x + 3\cdot x^2, \end{aligned}$$

we have

$$\mathbf{c}_w(D(1)) = (0,0,0), \qquad \mathbf{c}_w(D(x)) = (1,0,0)$$
$$\mathbf{c}_w(D(x^2)) = (0,2,0), \qquad \mathbf{c}_w(D(x^3)) = (0,0,3).$$

Thus the matrix of $D$ with respect to these bases is

$$A = \begin{bmatrix} 0 & 1 & 0 & 0 \\ 0 & 0 & 2 & 0 \\ 0 & 0 & 0 & 3 \end{bmatrix}.$$

The $v$-coordinate vector of $p(x) = a + bx + cx^2 + dx^3$ is $\mathbf{c}_v(p) = (a, b, c, d)$. Since

$$A\begin{bmatrix} a \\ b \\ c \\ d \end{bmatrix} = \begin{bmatrix} b \\ 2c \\ 3d \end{bmatrix},$$

$(b, 2c, 3d)$ is the $w$-coordinate vector of

$$D(a + bx + cx^2 + dx^3) = b + 2cx + 3dx^2.$$

As with transformations of $R^m$ into itself, when $T$ is a linear transformation of a vector space $V$ into itself, we usually use just one basis. Thus if $T$: $V \to V$ is a linear transformation and $v_1, v_2, \ldots, v_m$ is a basis for $V$, then the $m \times m$ matrix $A$ that represents $T$ with respect to this basis has the vectors $\mathbf{c}_v(T(v_1)), \ldots, \mathbf{c}_v(T(v_m))$ as its columns. $A$ represents $T$ by sending the $v$-coordinate of $x$ into the $v$-coordinates of $T(x)$:

$$\mathbf{c}_v(T(x)) = A\mathbf{c}_v(x).$$

### EXAMPLE 5

Consider the linear transformation $T$: $P_3(R) \to P_3(R)$ defined by $T(p)(x) = xp'(x)$. If we choose the basis $1, x, x^2, x^3$ for $P_3(R)$, then

$$\begin{aligned} T(1) &= 0 = 0\cdot 1 + 0\cdot x + 0\cdot x^2 + 0\cdot x^3 \\ T(x) &= x = 0\cdot 1 + 1\cdot x + 0\cdot x^2 + 0\cdot x^3 \\ T(x^2) &= 2x^2 = 0\cdot 1 + 0\cdot x + 2\cdot x^2 + 0\cdot x^3 \\ T(x^3) &= 3x^3 = 0\cdot 1 + 0\cdot x + 0\cdot x^2 + 3\cdot x^3. \end{aligned}$$

Thus the matrix for $T$ with respect to this basis is

$$\begin{bmatrix} 0 & 0 & 0 & 0 \\ 0 & 1 & 0 & 0 \\ 0 & 0 & 2 & 0 \\ 0 & 0 & 0 & 3 \end{bmatrix}.$$

If $p(x) = a + bx + cx^2 + dx^3$, then the coordinate vector of $T(p)$ with respect to this basis is

$$\begin{bmatrix} 0 & 0 & 0 & 0 \\ 0 & 1 & 0 & 0 \\ 0 & 0 & 2 & 0 \\ 0 & 0 & 0 & 3 \end{bmatrix}\begin{bmatrix} a \\ b \\ c \\ d \end{bmatrix} = \begin{bmatrix} 0 \\ b \\ 2c \\ 3d \end{bmatrix}.$$

Changing the basis of $V$ changes the matrix that represents a linear transformation $T: V \to V$. The exact relationship between the matrices is given in the following theorem.

***Theorem 3*** *Suppose that $T: V \to V$ is a linear transformation of an $m$-dimensional vector space $V$ into itself. Let $A$ be the matrix of $T$ with respect to a basis $v_1, v_2, \ldots, v_m$* for *$V$ and $B$ be the matrix of $T$ with respect to a basis $w_1, w_2, \ldots, w_m$ for $V$. Then*

$$A = M^{-1}BM,$$

*where $M$ is the change of coordinates matrix which changes $v$-coordinates to $w$-coordinates:*

$$M\mathbf{c}_v(x) = \mathbf{c}_w(x).$$

*The columns of $M$ are the vectors*

$$\mathbf{c}_w(v_1), \ldots, \mathbf{c}_w(v_m).$$

**Proof** The proof of this theorem is exactly the same as the proof of Theorem 1 in Section 5.4.

Thus if two matrices represent the same linear transformation, then they are similar. Conversely, similar matrices represent the same linear transformation (see Problem 37).

## EXAMPLE 6

Consider again the linear transformation $T: P_3(R) \to P_3(R)$ defined by $T(p)(x) = xp'(x)$ (see Example 5). This time we choose the basis $1$, $1 + x$, $1 + x + x^2$,

$1 + x + x^2 + x^3$ for $P_3(R)$. Since

$$T(1) = 0, \qquad T(1 + x) = x, \qquad T(1 + x + x^2) = x + 2x^2,$$
$$T(1 + x + x^2 + x^3) = x + 2x^2 + 3x^3,$$

the matrix of $T$ with respect to this basis is (verify!)

$$B = \begin{bmatrix} 0 & -1 & -1 & -1 \\ 0 & 1 & -1 & -1 \\ 0 & 0 & 2 & -1 \\ 0 & 0 & 0 & 3 \end{bmatrix}.$$

The change of coordinates matrix which changes coordinates with respect to the basis $1, x, x^2, x^3$ to coordinates with respect to the basis $1, 1 + x, 1 + x + x^2, 1 + x + x^2 + x^3$ is (verify!)

$$M = \begin{bmatrix} 1 & -1 & 0 & 0 \\ 0 & 1 & -1 & 0 \\ 0 & 0 & 1 & -1 \\ 0 & 0 & 0 & 1 \end{bmatrix}.$$

Since $M^{-1}$ is (verify!)

$$M^{-1} = \begin{bmatrix} 1 & 1 & 1 & 1 \\ 0 & 1 & 1 & 1 \\ 0 & 0 & 1 & 1 \\ 0 & 0 & 0 & 1 \end{bmatrix},$$

we should have $A = M^{-1}BM$. Carry out the computation to verify this fact.

There is no change in the definition of an eigenvalue and eigenvector of a linear transformation $T: V \to V$ when $V$ is an abstract vector space. A nonzero vector $v$ in $V$ is called an **eigenvector** of $T$ if there is a scalar $\lambda$ such that $T(v) = \lambda v$. The scalar $\lambda$ is called an **eigenvalue** of $T$ corresponding to the eigenvector $v$. This definition does not require $V$ to be finite-dimensional. When $V$ is a function space, the eigenvectors are usually called **eigenfunctions**.

## EXAMPLE 7

Recall that (see Example 2 in Section 4.1) $C^\infty(R)$ is the vector space of all functions that are infinitely differentiable on $R$. Since the derivative of an infinitely differentiable function is infinitely differentiable, the differential operator $D$ defines a linear transformation of $C^\infty(R)$ into itself. If $D(f) = \lambda f$, then $f'(x) = \lambda f(x)$. In Example 7 of Section 4.2 we showed that the solution space of this differential equation is the set of all multiples of the exponential function $f(x) = e^{\lambda x}$. Thus every real number $\lambda$ is an eigenvalue of $D$ and the corresponding eigenfunctions are the nonzero multiples of $e^{\lambda x}$.

When $V$ is a finite-dimensional vector space and $T: V \to V$ is a linear transformation, are Theorems 1, 2, and 3 of Section 5.5 still true? Not only are they still true, but their proofs are exactly the same! We summarize these theorems in this more general setting with the following theorem.

***Theorem 4*** *Let $V$ be a finite-dimensional vector space and let $T: V \to V$ be a linear transformation. Then $T$ can be represented by a diagonal matrix $\Lambda$ if and only if there is a basis for $V$ consisting of eigenvectors of $T$. In this case, the diagonal entries of $\Lambda$ are eigenvalues of $T$, the ith diagonal entry being an eigenvalue for the ith basis vector.*

## EXAMPLE 8

In Example 5 we found that the linear transformation $T: P_3(R) \to P_3(R)$ defined by $T(p)(x) = xp'(x)$ is represented by the diagonal matrix

$$\Lambda = \begin{bmatrix} 0 & & & \\ & 1 & & \\ & & 2 & \\ & & & 3 \end{bmatrix}$$

with respect to the basis $1, x, x^2, x^3$ of $P_3(R)$. Therefore, $\lambda = 0, 1, 2,$ and $3$ are eigenvalues of $T$ and $1, x, x^2$, and $x^3$ are corresponding eigenfunctions.

## EXAMPLE 9

Consider again the linear transformation $T: P_3(R) \to P_3(R)$ defined by $T(p)(x) = xp'(x)$. Let's pretend that we did not know that $P_3(R)$ has a basis of eigenfunctions of $T$ and that we are asked to find the eigenvalues and eigenfunctions of $T$. If $p(x) = a + bx + cx^2 + dx^3$, then the equation $T(p) = \lambda p$ becomes

$$xp'(x) = \lambda p(x)$$

or

$$bx + 2cx^2 + 3dx^3 = \lambda(a + bx + cx^2 + dx^3).$$

Equating coefficients, we obtain

$$\begin{aligned} \lambda a &= 0 \\ \lambda b &= b \\ \lambda c &= 2c \\ \lambda d &= 3d \end{aligned} \qquad \text{or} \qquad \begin{bmatrix} \lambda & & & \\ & \lambda - 1 & & \\ & & \lambda - 2 & \\ & & & \lambda - 3 \end{bmatrix} \begin{bmatrix} a \\ b \\ c \\ d \end{bmatrix} = \begin{bmatrix} 0 \\ 0 \\ 0 \\ 0 \end{bmatrix}.$$

If $\lambda = 3$ the matrix has rank 3 and a basis for its nullspace is $(0, 0, 0, 1)$. Thus $\lambda = 3$ is an eigenvalue of $T$ and the corresponding eigenfunctions are $p(x) = dx^3$, $d \neq 0$. Similarly, $\lambda = 0$, $\lambda = 1$, and $\lambda = 2$ are eigenvalues and the corresponding eigenfunctions are

$$\begin{aligned}\lambda = 0&: \quad p(x) = a,\ a \neq 0\\ \lambda = 1&: \quad p(x) = bx,\ b \neq 0\\ \lambda = 2&: \quad p(x) = cx^2,\ c \neq 0.\end{aligned}$$

## PROBLEMS 6.3

*1. Which of the following functions are linear?
   (a) $T: C^2(R) \to C(R)$ defined by $T(f) = f'' + 2f'$.
   (b) $T: C(R) \to R^1$ defined by $T(f) = f(0)$.
   (c) $T: P(R) \to R^1$ defined by $T(f) =$ degree of $f$.
   (d) $T: C^2(R) \to C(R)$ defined by $T(f)(x) = f''(x) + xf'(x) + (\sin x)f(x)$.

2. Let $V$ be the vector space of all $C^\infty$ functions and let $T: V \to V$ be the transformation defined by

$$T(f) = f' - f.$$

   (a) Show that $T$ is linear.
   *(b) Find all functions in the kernel of $T$.
   *(c) Is $T$ one-to-one?

3. Let $V$ be the vector space of all $C^\infty$ functions.
   (a) Determine which of the functions $f$, $g$, $h$ defined by $f(x) = 3x^2$, $g(x) = x^{1/3}$, $h(x) = \sin x$, are in $V$.
   (b) Determine whether $T: V \to V$ defined by $T(f)(x) = |f(x)|^2$ is linear.
   (c) Prove that $T: V \to V$ defined by $T(f) = f''$ is linear. Describe the kernel and the image of $T$. Is $T$ one-to-one? Is $T$ onto?
   (d) Consider the linear transformation $S: V \to R$ defined by $S(f) = \int_{-1}^{1} f(x)\,dx$. Find three different functions in the kernel of $S$. Also, prove that $S$ is onto.

*4. Let $T: P_3(R) \to P_3(R)$ be defined by

$$T(p)(x) = (x+1)^2 p''(x) + (x+1)p'(x) + p(x).$$

   (a) Find the matrix $A$ of $T$ with respect to the basis $1, x, x^2, x^3$.
   (b) Find the matrix $B$ of $T$ with respect to the basis $1$, $1 + x$, $(1 + x)^2$, $(1 + x)^3$.
   (c) Find the change of coordinates matrix $M$ and verify that $A = M^{-1}BM$.

5. In each of the following problems the functions are independent and span a finite-dimensional subspace $V$ of $C(R)$. Verify that the differential operator $D$ maps $V$ into itself and find its matrix with respect to the given basis for $V$.
   *(a) $\sin x$, $\cos x$
   (b) $1$, $e^x$
   *(c) $\sin x$, $\cos x$, $x \sin x$, $x \cos x$
   (d) $x$, $xe^x$
   (e) $e^x \sin x$, $e^x \cos x$

6. Let $T: P_2(R) \to R$ be defined by $T(p) = \int_0^1 p(x)\,dx$.
   (a) Show that $T$ is a linear transformation.
   (b) Find a basis for $\ker(T)$.

7. Let $V$ be the vector space of all polynomials and $T: V \to V$ be the linear transformation defined by

$$T(a_0 + a_1 x + a_2 x^2 + \cdots + a_n x^n) = a_0 x + \frac{a_1}{2}x^2 + \cdots + \frac{a_n}{n+1}x^{n+1}.$$

Describe the kernel and the image of $T$. Is $T$ one-to-one? Is $T$ onto?

8. Let $I = [a, b]$ be a closed interval, and let $c$ be a fixed number in $I$. Define a function $A: C(I) \to C(I)$ as follows: For $f$ in $C(I)$, $A(f)$ is the unique function $F$ in

$C(I)$ such that

$$F' = f \text{ and } F(c) = 0.$$

(a) Show that $A$ is a linear transformation.

(b) Show that it is one-to-one.

(c) Show that its image consists of all functions $g$ in $C^1(I)$ such that $g(c) = 0$.

*(d) What does "$A$" stand for?

***9.** Define $T\colon P_3 \to R^2$ by the equation $T(p) = (p(2), p(-5))$. Find a basis for $\ker(T)$. [*Hint:* If $p(x)$ is a polynomial, then $p(a) = 0$ if and only if $(x - a)$ is a factor of $p(x)$.]

**10.** Find the dimension of the image and the dimension of the kernel of each of the following linear transformations. Say whether each is onto, one-to-one, both or neither.

*(a) $T\colon R^3 \to R^2$ is defined by $T(x, y, z) = (x + 2y + 5z, 2x + 4y + 10z)$.

(b) $T\colon P_3 \to P_3$ is defined by $T(p)(x) = (x + 1)p''(x) + 3p'(x)$.

*(c) $T\colon P_1 \to V$, where $V$ is the plane $z = x + 2y$, is defined by $T(p) = p(2)(-3, 2, 1) + p'(5)(3, -1, 1)$.

(d) $T\colon M_{3,2} \to M_{3,2}$ is defined by $T(A) = A\begin{bmatrix} 1 & 1 \\ 2 & 2 \end{bmatrix}$.

**11.** Let $T\colon P_2(R) \to P_1(R)$ be defined by $T(p)(t) = p'(t)$. Let $t^2, t, 1$ and $t, 1$ be bases for $P_2(R)$ and $P_1(R)$, respectively.

(a) Find the matrix of $T$ with respect to these bases.

(b) How does this matrix represent $T$? Illustrate with an example.

**12.** Let $T\colon P_3(R) \to P_3(R)$ be the linear transformation defined by $T(p)(t) = p''(t) - 4p'(t) + p(t)$.

(a) Find the matrix of $T$ relative to the basis $1, t, t^2, t^3$.

(b) Explain how this matrix represents $T$. Illustrate with an example.

(c) Find the kernel of $T$.

(d) Is $T$ one-to-one? Onto?

***13.** For each of the linear transformations $T\colon V \to W$ below, find the matrix of $T$ with respect to the given bases.

(a) $T\colon M_{2,2} \to M_{2,2}$ is defined by

$$T(M) = M\begin{bmatrix} 1 & 2 \\ 3 & 4 \end{bmatrix} + 2M - M^T.$$

Basis: $\begin{bmatrix} 1 & 0 \\ 0 & 0 \end{bmatrix}, \begin{bmatrix} 0 & 1 \\ 0 & 0 \end{bmatrix}, \begin{bmatrix} 0 & 0 \\ 1 & 0 \end{bmatrix}, \begin{bmatrix} 0 & 0 \\ 0 & 1 \end{bmatrix}.$

(b) $T\colon P_3(R) \to P_2(R)$ is defined by

$$T(f) = f'.$$

Basis for $P_3(R)$: $1, x, x^2, x^3$; basis for $P_2(R)$: $1 + x, x + x^2, 1 + x^2$.

**14.** (a) Define $T\colon M_{3,2} \to M_{3,2}$ by

$$T(A) = A\begin{bmatrix} 1 & 1 \\ 2 & 3 \end{bmatrix}.$$

Find the rank of $T$. Find $\ker(T)$.

(b) Define $S\colon M_{2,2} \to M_{2,2}$ by

$$S(A) = \begin{bmatrix} 1 & 2 \\ 3 & 4 \end{bmatrix} A \begin{bmatrix} 1 & 2 \\ 3 & 4 \end{bmatrix}.$$

What is the matrix of $S$ with respect to the basis

$$\begin{bmatrix} 1 & 0 \\ 0 & 0 \end{bmatrix}, \begin{bmatrix} 0 & 1 \\ 0 & 0 \end{bmatrix}, \begin{bmatrix} 0 & 0 \\ 1 & 0 \end{bmatrix}, \begin{bmatrix} 0 & 0 \\ 0 & 1 \end{bmatrix}?$$

What is $\ker(S)$?

***15.** Let $T\colon P(R) \to P(R)$ be defined by $T(p)(x) = xp'(x)$. Show that for every nonnegative integer $n$, $\lambda = n$ is an eigenvalue of $T$ and that $p(x) = cx^n$ is an eigenfunction for any real number $c \neq 0$.

**16.** Find the eigenvalues and corresponding eigenfunctions of the linear transformation $D\colon P_3(R) \to P_3(R)$. What about when $D\colon P_4(R) \to P_4(R)$. Generalize.

**17.** Show that $\cos nx$ and $\sin nx$ are eigenfunctions of

$$D^2\colon C^\infty(R) \to C^\infty(R).$$

What are the corresponding eigenvalues?

*18. Let $V$ be the plane $2x - 7y + 3z = 0$, and define $T: R^3 \to V$ by $T(\mathbf{x})$ is the projection of $\mathbf{x}$ onto $V$. Find a basis of $R^3$ and a basis of $V$ so that the matrix of $T$ with respect to these bases is $\begin{bmatrix} 1 & 0 & 0 \\ 0 & 1 & 0 \end{bmatrix}$.

19. Let $V$ be the vector space of all $n \times n$ matrices and let $T: V \to V$ be defined by $T(A) = A^T$.
   (a) Show that $T$ is a linear transformation.
   (b) Show that $\lambda = \pm 1$ are the only eigenvalues of $T$.
   (c) Find the eigenspaces $E(1)$ and $E(-1)$.

20. Let $L$: $C^n(I) \to C(I)$ be a linear differential operator with constant coefficients. If $g$ is in $C(I)$, show that every solution of the equation $L(f) = g$ is of the form $f = f_1 + h$, where $h$ is in the $\ker(L)$ and $L(f_1) = g$.

21. Let $V$ be the subspace of $C(I)$, $I = [0, 2\pi]$, spanned by $1$, $x$, $x^2 \cos x$, $\sin x$. Let $T: V \to V$ be the linear map defined by $T(f) = f''$.
   (a) Show that the given functions form a basis for $V$.
   *(b) Find the matrix of $T$ with respect to this basis.
   (c) Find a basis for $\ker(T)$.
   (d) Find a basis for $T(V)$.
   *(e) Find the general solution of the equation $T(f)(x) = \sin x$.

22. Consider the differential operator $L$: $C^\infty(R) \to C^\infty(R)$ defined by $L(f) = f'' - f$.
   (a) Show that $e^x$ and $e^{-x}$ are in $\ker(L)$.
   *(b) Given that $\ker(L)$ has dimension 2, find the general solution of $L(f) = 0$.
   (c) Find the general solution of $L(f) = -x$. [*Hint:* What is $L(x)$?]

23. Consider the differential operator $L$: $C^\infty(R) \to C^\infty(R)$ defined by $L(f) = f'' + f$.
   (a) Show that $\sin x$ and $\cos x$ are solutions of $L(f) = 0$.
   (b) Given that $\ker(L)$ has dimension 2, find the general solution of $L(f) = 0$.
   (c) Find the general solution of $L(f) = e^x$.

24. Find a basis for the kernel of the differential operator $L$: $C^\infty(R) \to C^\infty(R)$.
   (a) $L(f) = f'$
   (b) $L(f) = f''$
   (c) $L(f) = f'''$
   (d) Generalize parts (a), (b), and (c).

25. Find the general solution of $L(f) = g$ if:
   (a) $L(f) = f'$ and $g(x) = \sin x$
   (b) $L(f) = f''$ and $g(x) = e^{-x}$
   (c) $L(f) = f'''$ and $g(x) = \cos x$

26. Let $g$ be a fixed function in $C(I)$ and define $T: C(I) \to C(I)$ by $T(f) = f \cdot g$, that is,

$$T(f)(x) = f(x) \cdot g(x) \qquad \text{for all } x \text{ in } I.$$

   (a) Show that $T$ is a linear transformation.
   (b) Find $\ker(T)$.

*27. Let $I = [-1, 1]$ and let $T: P_n(I) \to P_n(I)$ be defined by $T(f)(x) = f(-x)$. Show that $T$ is an isometry with respect to the standard inner product on $P_n(I)$.

28. Let $V$ be the plane $x + y - 2z = 0$ in $R^3$, and let $T: V \to R^4$ be defined by

$$T(x, y, z) = (x + z, 3x + y, y - 3z, 4x + y + z).$$

   (a) Find a basis for $T(V)$.
   (b) Find a basis for $\ker(T)$.

29. Suppose that $T: P_2(R) \to P_1(R)$ is the linear transformation whose matrix with respect to the basis $x^2 + 1$, $x + 1$, $x^2 - 5x - 2$ of $P_2(R)$ and $x + 3$, $2x - 1$ of $P_1(R)$ is

$$\begin{bmatrix} 1 & 2 & 0 \\ 1 & -1 & 3 \end{bmatrix}.$$

(a) Find $T(x^2 + x - 2)$.
(b) Find $\ker(T)$.

**30.** Define roll$(A)$ to be the matrix obtained from a matrix $A$ by "rotating" it 90 degrees clockwise; for example, if

$$A = \begin{bmatrix} a & b & c \\ d & e & f \\ g & h & k \end{bmatrix},$$

then

$$\text{roll}(A) = \begin{bmatrix} g & d & a \\ h & e & b \\ k & f & c \end{bmatrix}.$$

Define a square matrix $A$ to be **rollable** if $A = \text{roll}(A)$.

(a) Show that roll: $M_{n,n} \to M_{n,n}$ is a linear transformation.
(b) Show that $T$: $M_{n,n} \to M_{n,n}$ defined by $T(A) = \text{roll}(A) - A$ is a linear transformation.
(c) Show that $\ker(T)$ is the collection of the rollable matrices.
*(d) Let $V$ be the collection of all $3 \times 3$ rollable matrices. Find $\dim V$ and give a basis for $V$.
(e) Do the same for the collection of $4 \times 4$ rollable matrices.
(f) Do the same for $4 \times 4$ matrices that are both rollable and symmetric.
(g) Do the same for $4 \times 4$ rollable, skew-symmetric matrices.
(h) Do the same for $5 \times 5$ rollable, skew-symmetric matrices.

**31.** Show that if $T$: $V \to W$ is a one-to-one linear transformation, and $v_1, v_2, \ldots, v_k$ are linearly independent in $V$, then $T(v_1), T(v_2), \ldots, T(v_k)$ are linearly independent in $W$.

**32.** Suppose that $V$ and $W$ are finite-dimensional vector spaces. If $T$ is an isomorphism and $v_1, v_2, \ldots, v_k$ is a basis for $V$, show that $T(v_1), T(v_2), \ldots, T(v_k)$ is a basis for $W$.

**33.** Let $T$: $P_2(R) \to P_2(R)$ be defined by

$$T(p)(x) = p(x + 1).$$

(a) Find the matrix of $T$ with respect to the basis $1, x, x^2$ for $P_2(R)$.
(b) Find the matrix for the differential operator

$$1 + D + \frac{1}{2}D^2$$

with respect to the basis in part (a). The matrices in parts (a) and (b) should be equal.
(c) Generalize this problem to $P_3(R)$.
(d) Now generalize it to $P_n(R)$.

**34.** Let $V$ be a vector space of dimension $n$. Show that a linear transformation $T$: $V \to V$ is one-to-one if and only if $T$ maps $V$ onto $V$.

**35.** Let $v_1, v_2, \ldots, v_n$ be a basis for the vector space $V$ and let $A$ be the matrix for a linear transformation $T$: $V \to V$ with respect to this basis. Show that $A$ has rank $n$ if and only if $T$ is one-to-one.

**36.** Let $V$ be a vector space of dimension $n$ and let $T$: $V \to W$ be a linear transformation. Show that if $\dim W < n$, then $T$ is not one-to-one.

**37.** Suppose that $A$ and $B$ are $m \times m$ similar matrices. If $V$ is an $m$-dimensional vector space, show that it is possible to choose two different bases for $V$ so that $A$ and $B$ represent the same linear transformation $T$: $V \to V$ with respect to these bases.

**38.** Let $T$: $V \to W$ be a linear transformation and suppose that $T(x_0) = y_0$. Prove that the set of solutions of $T(x) = y_0$ is precisely the set of vectors of the form $x_0 + z$, where $z$ is in the kernel of $T$.

**39.** Prove the following:

(a) If $T$: $V \to W$ is linear, and $T(v_1), T(v_2), \ldots, T(v_k)$ are independent vectors in $W$, then $v_1, v_2, \ldots, v_k$ are independent vectors in $V$. (*Hint*: Start by assuming that $\alpha_1 v_1 + \alpha_2 v_2 + \cdots + \alpha_k v_k = 0$.)
(b) Use part (a) to prove that if $T$: $V \to W$ is a linear transforma-

tion of $V$ onto $W$ and $W$ is finite-dimensional, then $\dim V \geq \dim W$. (*Hint:* Start by choosing a basis for $W$. Since $T$ is onto, each of the basis vectors can be written as $T$ of something in $V$. Continue from there.)

**40.** Let $T: V \to W$ be a linear transformation. If $v_1, v_2, \ldots, v_k$ are dependent vectors in $V$, must $T(v_1), T(v_2), \ldots, T(v_k)$ be dependent?

**41.** (a) Let $S: V \to W$ and $T: V \to W$ be two linear transformations and let $\alpha$ be a scalar. We define $S + T: V \to W$ and $\alpha S: V \to W$ by the equations

$$(S + T)(x) = S(x) + T(x),$$
$$(\alpha S)(x) = \alpha S(x).$$

Show that $S + T$ and $\alpha S$ are linear.

(b) Let $v_1, v_2, \ldots, v_n$ and $w_1, w_2, \ldots, w_m$ be bases for $V$ and $W$, respectively. If $A$ and $B$ are the matrices of $S$ and $T$, respectively, with respect to the $v$-basis and the $w$-basis, show that $A + B$ and $\alpha A$ are the matrices of $S + T$ and $\alpha S$, respectively, with respect to the $v$-basis and $w$-basis.

**42.** Let $S: U \to V$ and $T: V \to W$ be two linear transformations.

(a) Show that the transformation $T \circ S: U \to W$ defined by

$$(T \circ S)(x) = T(S(x))$$

is linear.

(b) Let $u_1, u_2, \ldots, u_n$ be a basis for $U$, $v_1, v_2, \ldots, v_m$ be a basis for $V$, and $w_1, w_2, \ldots, w_k$ be a basis for $W$. Let $A$ be the matrix of $S$ with respect to the $u$-basis and the $v$-basis, and let $B$ be the matrix of $T$ with respect to the $v$-basis and the $w$-basis. Show that $BA$ is the matrix of $T \circ S$ with respect to the $u$-basis and the $w$-basis.

**43.** Let $V$ and $W$ be finite-dimensional vector spaces.

(a) Show that if $\dim V = \dim W$, then $V$ and $W$ are isomorphic.

(b) Show that if there is an isomorphism $T: V \to W$, then $\dim V = \dim W$.

**44.** Show that $T: R^{n+1} \to P_n(R)$ defined by

$$\begin{aligned} T(a_0, a_1, \ldots, a_n) &= a_1 + a_2 x + \cdots + a_{n-1} x^{n-2} \\ &\quad + a_n x^{n-1} + a_0 x^n \end{aligned}$$

is an isomorphism.

**45.** Let $V$ be an $m$-dimensional vector space and let $v_1, v_2, \ldots, v_m$ be a basis for $V$. Let $w_1, w_2, \ldots, w_m$ be arbitrary vectors in another vector space $W$. Show that there is one and only one linear transformation $T: V \to W$ such that $T(v_i) = w_i$ for $i = 1, 2, \ldots, m$.

**46.** Let $T: V \to W$ be a linear transformation. If $V$ is finite-dimensional, prove the rank plus nullity theorem:

$$\dim V = \dim T(V) + \dim \ker(T).$$

**47.** Let $V$ be an $n$-dimensional inner product space and let $T: V \to R$ be a linear transformation.

(a) Show that $T(V)$ is $\{0\}$ or $R$.

(b) Show that if $T \neq 0$, then the nullity of $T$ is $n - 1$.

(c) Show that there is a unique vector $y$ in $V$ such that $T(x) = \langle x, y \rangle$ for all $x$ in $V$. [*Hint:* Choose an orthonormal basis for $\ker(T)$ and extend it to an orthonormal basis for $V$. Write $x$ as a linear combination of this basis and then compute $T(x)$.]

**48.** Which theorems in Chapter 5 remain true when $V$ and $W$ are infinite-dimensional vector spaces?

# 7

# Determinants

In this chapter we define the determinant of a square matrix. The determinant function associates with every square matrix $A$ a unique *number* (denoted det $A$) called the **determinant** of $A$. The determinant of a $2 \times 2$ matrix $\begin{bmatrix} a & b \\ c & d \end{bmatrix}$ is defined by the formula $ad - bc$. It is possible similarly to define the determinant of an $n \times n$ matrix by an explicit formula. However, for $n > 4$ this formula is quite complicated and useless for computation. Therefore, we choose to define determinants by means of their basic properties. These basic properties can be derived in a formal manner, but their derivation is quite messy and not very instructive. Our approach has the advantage that it places emphasis on those properties of determinants that are most useful for computational purposes. This will enable us to derive a reasonably simple procedure for computing determinants that is based on the Gaussian elimination process.

## 7.1 Properties of the Determinant

As indicated above, we treat the determinant of a square matrix as a function. Although we shall not prove this, there are four simple properties that characterize the determinant function. That is, there is one and only one function defined on square matrices that satisfies Properties 1 through 4 listed below, and this function is called the determinant.

***Property 1*** If a scalar multiple of one row of a square matrix $A$ is added to another row of $A$, then the determinant of the resulting matrix is equal to the determinant of $A$.

For example,

$$\det\begin{bmatrix} 1 & 2 & -3 \\ -4 & 3 & 5 \\ 4 & 8 & -5 \end{bmatrix} = \det\begin{bmatrix} 1 & 2 & -3 \\ -4 & 3 & 5 \\ 2 & 4 & 1 \end{bmatrix}.$$

Here the first matrix is the result of adding twice the first row of the second matrix to its third row.

***Property 2*** If any two rows of a square matrix $A$ are interchanged, then the determinant of the resulting matrix is the negative of the determinant of $A$.

For example,

$$\det\begin{bmatrix} 1 & 2 & -3 \\ 2 & 4 & 1 \\ -4 & 3 & 5 \end{bmatrix} = (-1)\det\begin{bmatrix} 1 & 2 & -3 \\ -4 & 3 & 5 \\ 2 & 4 & 1 \end{bmatrix}.$$

***Property 3*** If any row of a square matrix $A$ is multiplied by a scalar $\alpha$, then the determinant of the resulting matrix is $\alpha$ times the determinant of $A$. Equivalently, if a matrix $A$ is the matrix obtained from a matrix $B$ by dividing a row of $B$ by the constant $\alpha$, then $\det B = \alpha \det A$.

For example,

$$\det\begin{bmatrix} 1 & 2 & -3 \\ -4 & 3 & 5 \\ 4 & 8 & 2 \end{bmatrix} = 2\det\begin{bmatrix} 1 & 2 & -3 \\ -4 & 3 & 5 \\ 2 & 4 & 1 \end{bmatrix}.$$

***Property 4*** The determinant of the $n \times n$ identity matrix $I$ is equal to 1.

For example, when $n = 3$ we have

$$\det\begin{bmatrix} 1 & 0 & 0 \\ 0 & 1 & 0 \\ 0 & 0 & 1 \end{bmatrix} = 1.$$

At this point we recommend that you verify each of these properties for the determinant of a $2 \times 2$ matrix using the formula

$$\det\begin{bmatrix} a & b \\ c & d \end{bmatrix} = ad - bc.$$

***Definition*** The **determinant function** is the unique real-valued function defined on $n \times n$ matrices that satisfies Properties 1 through 4.

***Theorem 1*** *If a square matrix has a zero row, then its determinant is equal to zero.*

**Proof** Suppose that a square matrix $A$ has a zero row. Then multiplying this zero row by 2 does not change $A$; hence (by Property 3) $\det A = 2\det A$. The only number that is equal to twice itself is the number 0.

We leave it to you to prove:

***Theorem 2*** *The determinant of a square matrix with two identical rows is equal to zero.*

Next suppose that we are dealing with an upper triangular matrix that has no zeros on the main diagonal, say

$$A = \begin{bmatrix} 2 & 5 & 8 & 4 \\ 0 & -1 & 4 & 7 \\ 0 & 0 & 3 & -2 \\ 0 & 0 & 0 & 2 \end{bmatrix}.$$

We can make all entries above the main diagonal zero without changing either the determinant or the diagonal entries. We do this by back substitution. We first subtract appropriate multiples of the last row from the preceding rows. Then we continue this process row by row until we obtain a diagonal matrix. Thus (by Property 1)

$$\det A = \det \begin{bmatrix} 2 & 0 & 0 & 0 \\ 0 & -1 & 0 & 0 \\ 0 & 0 & 3 & 0 \\ 0 & 0 & 0 & 2 \end{bmatrix}.$$

Now we use Property 3 successively to "factor out" the diagonal entries. Since $\det I = 1$ (Property 4),

$$\det A = 2 \cdot (-1) \cdot 3 \cdot 2 \cdot \det \begin{bmatrix} 1 & 0 & 0 & 0 \\ 0 & 1 & 0 & 0 \\ 0 & 0 & 1 & 0 \\ 0 & 0 & 0 & 1 \end{bmatrix} = -12.$$

This method works for any upper triangular matrix with no zero entries on the main diagonal. It follows that the determinant of an upper triangular matrix is equal to the product of the diagonal entries provided that these entries are nonzero. When one of the diagonal entries is zero, the back substitution procedure does not eliminate all entries above the main diagonal. But the last row with a zero diagonal entry will become a zero row and hence the determinant is zero. So in this case we also know that the determinant is equal to the product of the diagonal entries. For example,

$$\det\begin{bmatrix} 5 & 1 & 2 & 1 \\ 0 & 0 & 3 & -3 \\ 0 & 0 & 4 & 7 \\ 0 & 0 & 0 & 2 \end{bmatrix} = \det\begin{bmatrix} 5 & 1 & 0 & 0 \\ 0 & 0 & 0 & 0 \\ 0 & 0 & 4 & 0 \\ 0 & 0 & 0 & 2 \end{bmatrix} = 0 = 5\cdot 0\cdot 4\cdot 2.$$

***Theorem 3*** *The determinant of any upper triangular square matrix $A$ is equal to the product of the diagonal entries of $A$.*

We are now ready to develop a procedure to compute the determinant of an arbitrary square matrix.

## EXAMPLE 1

Let

$$A = \begin{bmatrix} 1 & 2 & -3 \\ 2 & 4 & 1 \\ -4 & 3 & 5 \end{bmatrix}.$$

Subtracting multiples of the first row from the other rows gives (by Property 1)

$$\det A = \det\begin{bmatrix} 1 & 2 & -3 \\ 0 & 0 & 7 \\ 0 & 11 & -7 \end{bmatrix}.$$

Now an interchange of rows gives us a triangular matrix. Hence by Property 2 and Theorem 3, we obtain

$$\det A = -\det\begin{bmatrix} 1 & 2 & -3 \\ 0 & 11 & -7 \\ 0 & 0 & 7 \end{bmatrix} = -1\cdot 11\cdot 7 = -77.$$

Example 1 illustrates how to proceed in general to compute the determinant of a square matrix. We reduce the matrix to upper triangular form by forward elimination. If we add multiples of a row to another row, the determinant does not change. Of course, we have to keep track of how often we interchange rows because each such interchange switches the sign of the determinant. Finally, we read off the determinant from the triangular form.

In this reduction it is sometimes convenient (although not necessary) to multiply a row by a nonzero scalar. Property 3 tells us how to compensate for this. For example, if

$$A = \begin{bmatrix} 19 & -38 & 95 \\ 7 & 91 & 14 \\ 5 & -8 & 9 \end{bmatrix},$$

it is convenient to divide the first row by 19 (i.e., multiply by $\frac{1}{19}$) and to divide the second row by 7. By Property 3 we have

$$\det A = 19\cdot 7\cdot \det\begin{bmatrix} 1 & -2 & 5 \\ 1 & 13 & 2 \\ 5 & -8 & 9 \end{bmatrix}.$$

In other words, when we have calculated the determinant of the new matrix, we have to multiply our result by $19\cdot 7 = 133$ to obtain the determinant of $A$.

### EXAMPLE 2

Suppose that at some stage in the reduction of a square matrix $A$ to a triangular form we obtain the matrix

$$\begin{bmatrix} 1 & 3 & 3 & 6 \\ 0 & 0 & -4 & 9 \\ 0 & 0 & 5 & 8 \\ 0 & 0 & -3 & -2 \end{bmatrix}.$$

Further reduction will not change the zero entry in the second position on the diagonal. Thus the final triangular matrix will have a zero on the main diagonal. Hence we can conclude without finishing the reduction that $\det A = 0$.

### EXAMPLE 3

Note that the fourth row of the matrix

$$A = \begin{bmatrix} 1 & 2 & -3 & 2 \\ 2 & 0 & -1 & 5 \\ 8 & -3 & 2 & 4 \\ 2 & 4 & -6 & 4 \end{bmatrix}$$

is a multiple of the first row. Thus we can obtain a zero row and it follows that $\det A = 0$. In general, if one row of a square matrix is a multiple of another row, then the determinant must be 0.

***Theorem 4*** *If $A$ is a square matrix, then $A$ and its transpose have the same determinant*; $\det A = \det A^T$.

We omit the proof of this theorem. In Problem 12 we ask you to give a proof when $A$ is a $2 \times 2$ matrix.

It follows from this theorem that Properties 1 through 4 and Theorems 1 and 2 remain valid if the word "row" is replaced by the word "column." Therefore, in reducing to triangular form we may add multiples of one column to another column, interchange columns, and multiply a column by a nonzero scalar. If a square matrix has a zero column or two identical columns, then its determinant is zero.

### EXAMPLE 4

The third column of the matrix

$$A = \begin{bmatrix} -3 & 1 & -2 \\ 2 & 5 & 7 \\ 4 & 1 & 5 \end{bmatrix}$$

is the sum of its first and second columns. Then subtracting these two columns from the third gives a zero column and hence $\det A = 0$.

Considering how involved matrix multiplication and the evaluation of determinants are, the following theorem is quite remarkable. We omit the proof but ask you to verify the theorem in a couple of special cases (see Problem 13).

***Theorem 5*** *If A and B are square matrices, then*

$$\det(AB) = (\det A)\cdot(\det B).$$

An important consequence of this theorem is the following theorem.

***Theorem 6*** *If A is nonsingular, then* $\det A \neq 0$ *and*

$$\det A^{-1} = \frac{1}{\det A}. \tag{1}$$

**Proof** Since $AA^{-1} = I$,

$$(\det A)(\det A^{-1}) = \det(AA^{-1}) = \det I = 1.$$

Therefore, $\det A \neq 0$ and dividing the equation by $\det A$ gives us (1).

Thus when $A$ is nonsingular, $\det A \neq 0$. The converse of this result is also true. Suppose that we have reduced $A$ to its reduced echelon form $R$. Then $\det R = \alpha \cdot \det A$, where $\alpha$ is a nonzero scalar. Thus if $\det A \neq 0$, then $\det R \neq 0$ and hence $R$ has no zero rows and $A$ is nonsingular. This proves the following theorem.

***Theorem 7*** *A square matrix is nonsingular if and only if its determinant is nonzero. Equivalently, a square matrix is singular if and only if its determinant is zero.*

We now verify that the formula for the determinant of a $2 \times 2$ matrix,

$$\det\begin{bmatrix} a & b \\ c & d \end{bmatrix} = ad - bc,$$

is a consequence of the properties of determinants given above. To see this, let us assume first that $a \neq 0$. Then

$$\det\begin{bmatrix} a & b \\ c & d \end{bmatrix} = a\det\begin{bmatrix} 1 & b/a \\ c & d \end{bmatrix}$$

$$= a\det\begin{bmatrix} 1 & b/a \\ 0 & d - bc/a \end{bmatrix}$$

$$= a \cdot 1 \cdot \left(d - \frac{bc}{a}\right)$$

$$= ad - bc.$$

If $a = 0$, then

$$\det\begin{bmatrix} 0 & b \\ c & d \end{bmatrix} = -\det\begin{bmatrix} c & d \\ 0 & b \end{bmatrix} = -bc = ad - bc.$$

We leave it to you to verify a following formula for the determinant of a $3 \times 3$ matrix (see Problem 15):

$$\det\begin{bmatrix} a_{11} & a_{12} & a_{13} \\ a_{21} & a_{22} & a_{23} \\ a_{31} & a_{32} & a_{33} \end{bmatrix} = a_{11}a_{22}a_{33} + a_{12}a_{23}a_{31} + a_{13}a_{21}a_{32}$$

$$- a_{13}a_{22}a_{31} - a_{11}a_{23}a_{32} - a_{12}a_{21}a_{33}. \qquad (2)$$

It is easy to remember this formula with the help of the following schematic device. Write the first two columns of the $3 \times 3$ matrix to the right of the matrix and compute the products as indicated by the arrows.

$$
\begin{array}{ccccc}
a_{11} & a_{12} & a_{13} & a_{11} & a_{12} \\
a_{21} & a_{22} & a_{23} & a_{21} & a_{22} \\
a_{31} & a_{32} & a_{33} & a_{31} & a_{32}
\end{array}
$$

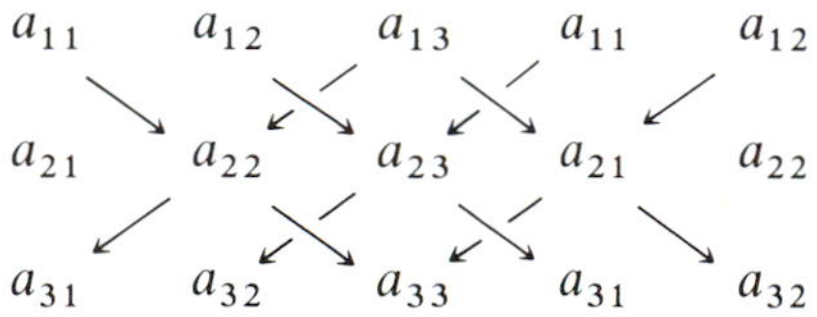

These six products are the products in (2). The products corresponding to the arrows pointing to the right have plus signs attached to them and products corresponding to the arrows pointing to the left have negative signs attached to them. This schematic device works only for $3 \times 3$ determinants and cannot be used for higher-order determinants.

## EXAMPLE 5

The determinant of the matrix

$$\begin{bmatrix} 2 & -1 & 4 \\ 3 & 1 & 2 \\ 5 & 2 & 3 \end{bmatrix} \quad \text{is} \quad
\begin{array}{ccccc}
2 & -1 & 4 & 2 & -1 \\
3 & 1 & 2 & 3 & 1 \\
5 & 2 & 3 & 5 & 2
\end{array}$$

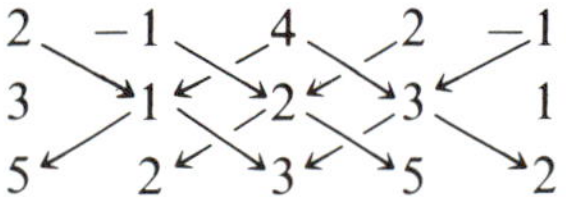

$$= 6 + (-10) + 24 - 20 - 8 - (-9) = 1.$$

# PROBLEMS 7.1

***1.** Compute the determinants of the following matrices. Then use Theorem 7 to determine which of these matrices are singular.

(a) $\begin{bmatrix} -3 & 1 \\ 2 & 5 \end{bmatrix}$ (b) $\begin{bmatrix} 1 & 0 & 0 \\ -2 & 3 & 0 \\ 5 & 9 & 4 \end{bmatrix}$

(c) $\begin{bmatrix} 2 & 0 & 1 \\ -1 & 1 & 2 \\ 2 & -2 & -2 \end{bmatrix}$

(d) $\begin{bmatrix} 5 & 1 & 2 \\ 8 & -7 & 9 \\ 5 & 1 & 2 \end{bmatrix}$

(e) $\begin{bmatrix} 3 & 5 & -6 \\ 1 & 9 & -2 \\ -2 & 8 & 4 \end{bmatrix}$

(f) $\begin{bmatrix} \frac{1}{2} & \frac{1}{3} & \frac{1}{4} \\ \frac{1}{4} & -\frac{1}{6} & \frac{1}{4} \\ 0 & 2 & \frac{1}{5} \end{bmatrix}$

(g) $\begin{bmatrix} 0 & 1 & 2 & 3 \\ 1 & 0 & -1 & 2 \\ 3 & 4 & 0 & -1 \\ -1 & 2 & 2 & 0 \end{bmatrix}$

(h) $\begin{bmatrix} 3 & 1 & 0 & 0 \\ 4 & 2 & 0 & 0 \\ 0 & 0 & 1 & 5 \\ 0 & 0 & 2 & 1 \end{bmatrix}$

(i) $\begin{bmatrix} 1 & -3 & 0 & 0 & 0 \\ 2 & 1 & 5 & 0 & 0 \\ 0 & 2 & -2 & 1 & 0 \\ 0 & 0 & 1 & 3 & 2 \\ 0 & 0 & 0 & 4 & 1 \end{bmatrix}$

**2.** Let

$$M = \begin{bmatrix} 7 & 8 & 9 & 9 \\ 11 & 12 & 13 & 13 \\ 0 & 5 & 1 & 1 \\ -2 & 4 & 3 & 3 \end{bmatrix}.$$

Is $M$ nonsingular? Give a reason for your answer. (*Hint:* No computation is necessary.)

**3.** Let

$$A = \begin{bmatrix} 1 & 2 & 3 \\ 1 & 1 & 2 \\ 0 & 1 & 2 \end{bmatrix} \quad \text{and} \quad B = \begin{bmatrix} 1 & 1 & 1 \\ 2 & 1 & 2 \\ 3 & 1 & 2 \end{bmatrix}.$$

(a) Find the determinant of $A$ and $B$.
(b) Find $A^{-1}$ and $B^{-1}$.
(c) Find $(AB)^{-1}$ and $(BA)^{-1}$.

***4.** Evaluate the determinants of the following matrices.

(a) $\begin{bmatrix} 0 & a \\ b & c \end{bmatrix}$ (b) $\begin{bmatrix} 0 & 0 & a \\ 0 & b & c \\ d & e & f \end{bmatrix}$

(c) $\begin{bmatrix} 0 & 0 & 0 & a \\ 0 & 0 & b & c \\ 0 & d & e & f \\ g & h & i & j \end{bmatrix}$

***5.** Find all values of $x$ for which the matrix

$$\begin{bmatrix} 1 & x & x^2 \\ x^2 & 1 & x \\ x & x^2 & 1 \end{bmatrix}$$

is singular.

**6.** Let

$$a = \begin{bmatrix} \lambda & 1 & 1 \\ 1 & \lambda & 1 \\ 1 & 1 & \lambda \end{bmatrix}.$$

For what values of $\lambda$ does rank $A = 1$? Rank $A = 2$? Rank $A = 3$?

**7.** Verify that each of the Properties 1 through 4 holds for the determinant of a

2 × 2 matrix by using the formula

$$\det\begin{bmatrix} a & b \\ c & d \end{bmatrix} = ad - bc.$$

***8.** Suppose that $A$, $B$, and $C$ are $3 \times 3$ matrices such that $\det A = 2$, $\det B = -\frac{5}{3}$, and $\det C = 0$. Calculate the following determinants.

(a) $\det A^{-1}$ (b) $\det(3B)$
(c) $\det(ABC)$ (d) $\det(A^T B^{-1})$

**9.** Suppose that $A$ is an $n \times n$ matrix and $\alpha$ is a scalar. Show that $\det(\alpha A) = \alpha^n \det A$. When is $\det(-A) = \det A$?

***10.** Let $A$ and $B$ be $n \times n$ matrices. We know that in general $AB \neq BA$. However, show that $AB$ and $BA$ have the same determinant.

**11.** Prove that the determinant of a lower triangular square matrix (i.e., a matrix with all zeros above the main diagonal) is equal to the product of the diagonal entries.

**12.** Prove Theorem 4 for 2 × 2 matrices.

**13.** Verify Theorem 5 when:

(a) $A$ and $B$ are 2 × 2 matrices.
(b) $A$ is an $n \times n$ diagonal matrix and $B$ is an arbitrary $n \times n$ matrix.

**14.** Verify the formula for the determinant of a 3 × 3 matrix given in equation (2) of the text.

***15.** Suppose that $A$ is a skew-symmetric $n \times n$ matrix (skew-symmetric means that $A^T = -A$). Show that if $n$ is odd, then $\det A = 0$. Given an example of a skew-symmetric 2 × 2 matrix with nonzero determinant.

**16.** Prove that if $A$ is an orthogonal matrix, then $\det A$ is equal to 1 or $-1$. (*Hint:* If $A$ is orthogonal, then $A^T A = I$.)

**17.** Let $T: R^2 \to R^2$ be a linear transformation and let $A$ be its standard matrix.

(a) Show that if $A$ is orthogonal and $\det A = 1$, then $A$ is a rotation.
(b) Show that if $A$ is orthogonal and $\det A = -1$, then $A$ is a rotation followed by a reflection.

(*Hint:* See Problem 30 in Section 5.1.)

***18.** What can you say about the determinant of the following matrix?

$$\begin{bmatrix} 0 & 0 & \cdots & 0 & a_{1n} \\ 0 & 0 & \cdots & a_{2,n-1} & a_{2n} \\ \vdots & & & & \vdots \\ 0 & a_{n-1,2} & \cdots & a_{n-1,n-1} & a_{n-1,n} \\ a_{n1} & a_{n2} & \cdots & a_{n,n-1} & a_{nn} \end{bmatrix}$$

**19.** The Vandermonde matrix is

$$V(x_1, x_2, \ldots, x_n) = \begin{bmatrix} 1 & x_1 & \cdots & x_1^{n-1} \\ 1 & x_2 & \cdots & x_2^{n-1} \\ \vdots & \vdots & & \vdots \\ 1 & x_n & \cdots & x_n^{n-1} \end{bmatrix}.$$

Show that:

(a) $\det V(x_1, x_2, x_3) = (x_3 - x_1)(x_3 - x_2)(x_2 - x_1)$
(b) $\det V(x_1, x_2, x_3, x_4) = (x_4 - x_3)(x_4 - x_2)(x_4 - x_1)(x_3 - x_2)(x_3 - x_1)(x_2 - x_1)$
(c) Generalize parts (a) and (b).

# 7.2 Expansion by Cofactors

Although the Gaussian elimination process provides us with a very efficient method for computing determinants, it is sometimes useful to use an alternative method. In this section we develop such a method, the expansion of a determinant by cofactors.

We mentioned in Section 7.1 a formula for the determinant of a $3 \times 3$ matrix $A$:

$$\det\begin{bmatrix} a_{11} & a_{12} & a_{13} \\ a_{21} & a_{22} & a_{23} \\ a_{31} & a_{32} & a_{33} \end{bmatrix} = a_{11}a_{22}a_{33} + a_{12}a_{23}a_{31} + a_{13}a_{21}a_{32}$$
$$- a_{13}a_{22}a_{31} - a_{11}a_{23}a_{32} - a_{12}a_{21}a_{33}.$$

Rearranging the terms in this formula and factoring out the quantities $a_{11}$, $a_{12}$, and $a_{13}$ leads to

$$\begin{aligned} \det A = a_{11}&(a_{22}a_{33} - a_{23}a_{32}) \\ - a_{12}&(a_{21}a_{33} - a_{23}a_{31}) \\ + a_{13}&(a_{21}a_{32} - a_{22}a_{31}). \end{aligned}$$

Hence

$$\det A = a_{11}A_{11} + a_{12}A_{12} + a_{13}A_{13}, \tag{1}$$

where

$$A_{11} = \det\begin{bmatrix} a_{22} & a_{23} \\ a_{32} & a_{33} \end{bmatrix}, \quad A_{12} = -\det\begin{bmatrix} a_{21} & a_{23} \\ a_{31} & a_{33} \end{bmatrix}, \quad A_{13} = \det\begin{bmatrix} a_{21} & a_{22} \\ a_{31} & a_{32} \end{bmatrix}.$$

The quantity $A_{11}$, which is called the cofactor of $a_{11}$, is the determinant of the matrix obtained from $A$ by omitting its first row and first column. Similarly, $A_{12}$, the cofactor of $a_{12}$, is the negative of the determinant of the matrix obtained from $A$ by omitting its first row and second column. Finally, $A_{13}$ is the determinant of the matrix obtained from $A$ by omitting its first row and third column.

In general, the evaluation of the determinant of an $n \times n$ matrix $A$ can be reduced to the evaluation of $n$ determinants of $(n-1) \times (n-1)$ matrices. Let

$$A = \begin{bmatrix} a_{11} & a_{12} & \cdots & a_{1n} \\ a_{21} & a_{22} & \cdots & a_{2n} \\ \vdots & \vdots & & \vdots \\ a_{n1} & a_{n2} & \cdots & a_{nn} \end{bmatrix}.$$

The **cofactor** of an entry $a_{ij}$ is defined to be the number

$$A_{ij} = (-1)^{i+j}D_{ij},$$

where $D_{ij}$ is the determinant of the $(n-1) \times (n-1)$ matrix obtained from $A$ by omitting the $i$th row and $j$th column of $A$. The sign $(-1)^{i+j}$ is $+1$ or $-1$, depending on whether $i + j$ is even or odd. If we replace the entries $a_{ij}$ of $A$ by the corresponding sign $(-1)^{i+j}$, we obtain a checkerboard pattern. The patterns for $n = 3$ and $n = 4$ are as follows.

$$\begin{bmatrix} + & - & + \\ - & + & - \\ + & - & + \end{bmatrix} \qquad \begin{bmatrix} + & - & + & - \\ - & + & - & + \\ + & - & + & - \\ - & + & - & + \end{bmatrix}$$

Note that the pattern always begins with a plus sign in the upper left corner and that the signs alternate.

**EXAMPLE 1**

Consider the matrix

$$A = \begin{bmatrix} 1 & -2 & 4 \\ 2 & -7 & 9 \\ -1 & 8 & -1 \end{bmatrix}.$$

The cofactor of the entry $a_{23} = 9$ is

$$A_{23} = (-1)^{2+3} \det \begin{bmatrix} 1 & -2 \\ -1 & 8 \end{bmatrix} = -(8 - 2) = -6.$$

The cofactor of the entry $a_{31} = -1$ is

$$A_{31} = (-1)^{3+1} \det \begin{bmatrix} -2 & 4 \\ -7 & 9 \end{bmatrix} = +(-18 + 28) = 10.$$

You should verify that the other cofactors are $A_{11} = -65$, $A_{12} = -7$, $A_{13} = 9$, $A_{21} = 30$, $A_{22} = 3$, $A_{32} = -1$, and $A_{33} = -3$.

The following result, whose proof we omit, gives us a method for computing determinants by means of cofactors.

**Result 1** Let $A$ be an $n \times n$ matrix. Then the sum of the products of the entries in any row (or column) of $A$ and their corresponding cofactors is equal to the determinant of $A$. In symbols, for any $i$ and any $j$,

$$\det A = a_{i1}A_{i1} + a_{i2}A_{i2} + \cdots + a_{in}A_{in}, \tag{2}$$
$$\det A = a_{1j}A_{1j} + a_{2j}A_{2j} + \cdots + a_{nj}A_{nj}. \tag{3}$$

Formula (2) is called the **cofactor expansion** of the determinant of $A$ along the $i$th row. Formula (3) is the cofactor expansion down the $j$th column. Thus equation (1) is the cofactor expansion of the determinant of the $3 \times 3$ matrix $A$ along the first row. If $A$ is a $4 \times 4$ matrix, then $\det A = a_{21}A_{21} + a_{22}A_{22} + a_{23}A_{23} + a_{24}A_{24}$ is the cofactor expansion of $\det A$ along the second row and $\det A = a_{13}A_{13} + a_{23}A_{23} + a_{33}A_{33} + a_{43}A_{43}$ is the cofactor expansion down the third column.

**EXAMPLE 2**

The cofactors of the matrix

$$A = \begin{bmatrix} 1 & -2 & 4 \\ 2 & -7 & 9 \\ -1 & 8 & -1 \end{bmatrix}$$

were calculated in Example 1. Expanding the determinant of $A$ along the third row gives

$$\begin{aligned}\det A &= (-1)\cdot A_{31} + 8\cdot A_{32} + (-1)\cdot A_{33}\\ &= (-1)\cdot 10 + 8\cdot(-1) + (-1)\cdot(-3) = -15.\end{aligned}$$

Expanding down the first column gives the same value,

$$\begin{aligned}\det A &= 1\cdot A_{11} + 2\cdot A_{21} + (-1)\cdot A_{31}\\ &= 1\cdot(-65) + 2\cdot 30 + (-1)\cdot 10 = -15.\end{aligned}$$

### EXAMPLE 3

Any $1 \times 1$ matrix $[a]$ is upper triangular and $a$ is its only diagonal entry. Hence $\det[a] = a$. In particular, if

$$A = \begin{bmatrix} a_{11} & a_{12} \\ a_{21} & a_{22} \end{bmatrix},$$

then the cofactor expansion along the first row yields the familiar formula $\det A = a_{11}A_{11} + a_{12}A_{12} = a_{11}a_{22} - a_{12}a_{21}$.

When we compute a determinant using cofactors, then for obvious reasons it is usually more efficient to choose the row or column of the matrix that has the most zero entries.

### EXAMPLE 4

Let

$$A = \begin{bmatrix} 2 & 1 & 0 & 5 \\ -2 & 0 & -1 & 1 \\ 3 & 4 & 3 & 0 \\ 1 & -2 & 0 & 6 \end{bmatrix}.$$

Expanding the determinant of $A$ down the third column gives

$$\det A = 0\cdot A_{13} + (-1)\cdot A_{23} + 3\cdot A_{33} + 0\cdot A_{43} = -A_{23} + 3\cdot A_{33},$$

where

$$A_{23} = -\det\begin{bmatrix} 2 & 1 & 5 \\ 3 & 4 & 0 \\ 1 & -2 & 6 \end{bmatrix} \quad \text{and} \quad A_{33} = +\det\begin{bmatrix} 2 & 1 & 5 \\ -2 & 0 & 1 \\ 1 & -2 & 6 \end{bmatrix}.$$

The expansion of $A_{23}$ down the third column gives

$$A_{23} = -5\cdot\det\begin{bmatrix} 3 & 4 \\ 1 & -2 \end{bmatrix} - 6\cdot\det\begin{bmatrix} 2 & 1 \\ 3 & 4 \end{bmatrix} = -5\cdot(-10) - 6\cdot 5 = 20.$$

The expansion of $A_{33}$ along the second row gives

$$A_{33} = 2\cdot\det\begin{bmatrix} 1 & 5 \\ -2 & 6 \end{bmatrix} - 1\cdot\det\begin{bmatrix} 2 & 1 \\ 1 & -2 \end{bmatrix} = 2\cdot 16 - (-5) = 37.$$

Thus $\det A = -20 + 3\cdot 37 = 91$.

Sometimes we can combine elementary row operations with expansion by cofactors and significantly reduce the computation required. We illustrate the procedure for the matrix in Example 4.

## EXAMPLE 5

Let $A$ be the matrix of Example 4. The determinant of $A$ is equal to the determinant of the matrix

$$B = \begin{bmatrix} 2 & 1 & 0 & 5 \\ -2 & 0 & -1 & 1 \\ -3 & 4 & 0 & 3 \\ 1 & -2 & 0 & 6 \end{bmatrix}$$

obtained from $A$ by adding three times the second row to the third row. Expanding $\det B$ down the third column gives

$$\det A = \det B = -(-1)\det\begin{bmatrix} 2 & 1 & 5 \\ -3 & 4 & 3 \\ 1 & -2 & 6 \end{bmatrix}.$$

Using row operations on the $3 \times 3$ matrix (adding suitable multiples of the third row to each of the others) gives

$$\det A = \det\begin{bmatrix} 0 & 5 & -7 \\ 0 & -2 & 21 \\ 1 & -2 & 6 \end{bmatrix}.$$

Now expanding down the first column yields

$$\det A = \det\begin{bmatrix} 5 & -7 \\ -2 & 21 \end{bmatrix} = 91.$$

We now have two methods for computing the determinant of an $n \times n$ matrix: reduction to triangular form and cofactor expansion. The reduction to triangular form is usually much faster than the cofactor expansion unless $n = 2$ or $n = 3$. For example, the computation of the determinant of a $5 \times 5$ matrix by cofactors involves in general more than 200 multiplications, whereas the reduction to triangular form can be done with fewer than 50 multiplications. The difference between the two methods increases dramatically as the size of the matrix increases. For a $10 \times 10$ matrix the reduction to triangular form involves fewer than 350 multiplications, but the cofactor method involves more than 6 million multiplications! No computer in the world could ever compute the determinant of a $100 \times 100$ matrix by cofactor expansion because this would involve about $10^{158}$ multiplications. However, a modern computer could calculate such a determinant by reduction to triangular form in a few seconds because fewer than 350,000 multiplications are needed.

We conclude this section with an example illustrating the typical situation that arises in Chapter 8.

### EXAMPLE 6

Consider the matrix

$$A = \begin{bmatrix} \lambda & -1 & 0 \\ 0 & \lambda - 3 & -2 \\ 2 & 1 & \lambda + 2 \end{bmatrix},$$

where $\lambda$ is an unknown quantity. Let us calculate the determinant of $A$. In this case it will not be a number but an expression in terms of $\lambda$. Expanding down the first column, we obtain

$$\begin{aligned} \det A &= \lambda \cdot \det \begin{bmatrix} \lambda - 3 & -2 \\ 1 & \lambda + 2 \end{bmatrix} + 2 \cdot \det \begin{bmatrix} -1 & 0 \\ \lambda - 3 & -2 \end{bmatrix} \\ &= \lambda[(\lambda - 3)(\lambda + 2) + 2] + 2 \cdot 2 \\ &= \lambda(\lambda^2 - \lambda - 4) + 4 \\ &= \lambda^3 - \lambda^2 - 4\lambda + 4 \\ &= (\lambda - 1)(\lambda - 2)(\lambda + 2). \end{aligned}$$

It follows that the matrix $A$ is singular ($\det A = 0$) if and only if $\lambda = 1, 2$, or $-2$.

## PROBLEMS 7.2

*1. Calculate the determinants of the following matrices.

(a) $\begin{bmatrix} -2 & 2 & 9 \\ 3 & -4 & 1 \\ 5 & 0 & 8 \end{bmatrix}$

(b) $\begin{bmatrix} 0 & 2 & -3 \\ -2 & 0 & 3 \\ 2 & -3 & 0 \end{bmatrix}$

(c) $\begin{bmatrix} \frac{1}{3} & 0 & \frac{1}{3} \\ \frac{1}{2} & 1 & \frac{1}{2} \\ -1 & \frac{1}{3} & \frac{2}{3} \end{bmatrix}$

(d) $\begin{bmatrix} 1 & -1 & 1 & 0 \\ -1 & 1 & 0 & 1 \\ 1 & 0 & 1 & -1 \\ 0 & 1 & -1 & 1 \end{bmatrix}$

(e) $\begin{bmatrix} 1 & 2 & 3 \\ 1 & 1 & 2 \\ 2 & 3 & 1 \end{bmatrix}$

(f) $\begin{bmatrix} 7 & 0 & 2 & 1 \\ 2 & -2 & 1 & 0 \\ 0 & 1 & 0 & -2 \\ 5 & 1 & -1 & 3 \end{bmatrix}$

**2.** Verify that the determinant of the matrix $\begin{bmatrix} a-\lambda & b \\ c & d-\lambda \end{bmatrix}$ is $\lambda^2 - (a+d)\lambda + ad - bc$.

***3.** Determine all scalars $\lambda$ for which the matrix $\begin{bmatrix} 3-\lambda & 1 \\ -6 & -4-\lambda \end{bmatrix}$ is singular.

***4.** Repeat Problem 3 with the matrices

(a) $\begin{bmatrix} 1-\lambda & -2 \\ 1 & -1-\lambda \end{bmatrix}$

(b) $\begin{bmatrix} -\lambda & 0 & -1 \\ -2 & 2-\lambda & -3 \\ -1 & 2 & -2-\lambda \end{bmatrix}$

**5.** Verify that if $a$, $b$, $c$, and $d$ are integers, then the determinant of the matrix $\begin{bmatrix} a & b \\ c & d \end{bmatrix}$ is an integer.

**6.** Suppose that $A$ is an $n \times n$ matrix and all entries of $A$ are integers. Explain why det $A$ also is an integer.

**7.** Verify that

$$\det \begin{bmatrix} 1 & x & x^2 \\ 1 & y & y^2 \\ 1 & z & z^2 \end{bmatrix} = (x-y)(y-z)(z-x).$$

**8.** Let $d_n$ be the determinant of the $n \times n$ matrix

$$D_n = \begin{bmatrix} x & y & 0 & 0 & 0 & 0 & \cdots & 0 \\ y & x & y & 0 & 0 & 0 & \cdots & 0 \\ 0 & y & x & y & 0 & 0 & \cdots & 0 \\ \vdots & & & & & & & \vdots \\ 0 & \cdots & & 0 & 0 & 0 & y \; x & y \\ 0 & \cdots & & 0 & 0 & 0 & 0 \; y & x \end{bmatrix}.$$

Show that $d_n = xd_{n-1} - y^2 d_{n-2}$.

**9.** (a) Show that the determinants of the matrices

$$\begin{bmatrix} 1 & x \\ x & 1 \end{bmatrix}, \quad \begin{bmatrix} 1 & x & x^2 \\ x^2 & 1 & x \\ x & x^2 & 1 \end{bmatrix},$$

and

$$\begin{bmatrix} 1 & x & x^2 & x^3 \\ x^3 & 1 & x & x^2 \\ x^2 & x^3 & 1 & x \\ x & x^2 & x^3 & 1 \end{bmatrix}$$

are $1 - x^2$, $(1 - x^3)^2$, and $(1 - x^4)^3$, respectively.

(b) On the basis of part (a), give a formula for the determinant of the matrix

$$\begin{bmatrix} 1 & x & x^2 & \cdots & x^n \\ x^n & 1 & x & \cdots & x^{n-1} \\ \vdots & & & & \vdots \\ x & x^2 & x^3 & \cdots & 1 \end{bmatrix}.$$

# 7.3 The Cross Product

In this section we define the cross product of two vectors in 3-space and then use this product to give a geometric interpretation of $2 \times 2$ and $3 \times 3$ determinants.

***Definition*** For any two vectors $\mathbf{a} = (a_1, a_2, a_3)$ and $\mathbf{b} = (b_1, b_2, b_3)$ in $R^3$, their **cross product a × b** is defined to be the vector

$$\mathbf{a} \times \mathbf{b} = (a_2 b_3 - a_3 b_2, a_3 b_1 - a_1 b_3, a_1 b_2 - a_2 b_1).$$

Note that the components of $\mathbf{a} \times \mathbf{b}$ are $2 \times 2$ determinants,

$$\mathbf{a} \times \mathbf{b} = \left( \det \begin{bmatrix} a_2 & a_3 \\ b_2 & b_3 \end{bmatrix}, -\det \begin{bmatrix} a_1 & a_3 \\ b_1 & b_3 \end{bmatrix}, \det \begin{bmatrix} a_1 & a_2 \\ b_1 & b_2 \end{bmatrix} \right). \tag{1}$$

This formula makes it somewhat easier to remember the definition of the cross product.

### EXAMPLE 1

If $\mathbf{a} = (2, 3, -1)$ and $\mathbf{b} = (-2, 1, 5)$, then

$$\begin{aligned} \mathbf{a} \times \mathbf{b} &= \left( \det \begin{bmatrix} 3 & -1 \\ 1 & 5 \end{bmatrix}, -\det \begin{bmatrix} 2 & -1 \\ -2 & 5 \end{bmatrix}, \det \begin{bmatrix} 2 & 3 \\ -2 & 1 \end{bmatrix} \right) \\ &= (16, -8, 8). \end{aligned}$$

If $\mathbf{a} = (a_1, a_2, a_3)$ and $\mathbf{b} = (b_1, b_2, b_3)$ are two collinear vectors in $R^3$, say $\mathbf{b} = \alpha \mathbf{a}$, then

$$\det \begin{bmatrix} a_i & a_j \\ b_i & b_j \end{bmatrix} = \det \begin{bmatrix} a_i & a_j \\ \alpha a_i & \alpha a_j \end{bmatrix} = \alpha \det \begin{bmatrix} a_i & a_j \\ a_i & a_j \end{bmatrix} = 0$$

and hence $\mathbf{a} \times \mathbf{b} = \mathbf{0}$. In particular, $\mathbf{a} \times \mathbf{a} = \mathbf{0}$.

The cross product, the inner product, and $3 \times 3$ determinants are related by the formula

$$\det \begin{bmatrix} a_1 & a_2 & a_3 \\ b_1 & b_2 & b_3 \\ c_1 & c_2 & c_3 \end{bmatrix} = \langle \mathbf{a} \times \mathbf{b}, \mathbf{c} \rangle, \tag{2}$$

where $\mathbf{a} = (a_1, a_2, a_3)$, $\mathbf{b} = (b_1, b_2, b_3)$, and $\mathbf{c} = (c_1, c_2, c_3)$. Indeed, expanding the determinant by cofactors along the third row gives

$$c_1 \det \begin{bmatrix} a_2 & a_3 \\ b_2 & b_3 \end{bmatrix} - c_2 \det \begin{bmatrix} a_1 & a_3 \\ b_1 & b_3 \end{bmatrix} + c_3 \det \begin{bmatrix} a_1 & a_2 \\ b_1 & b_2 \end{bmatrix},$$

which by (1) is equal to $\langle \mathbf{c}, \mathbf{a} \times \mathbf{b} \rangle = \langle \mathbf{a} \times \mathbf{b}, \mathbf{c} \rangle$.

Since the determinant of a square matrix with two identical rows is equal to zero, it follows from (2) that $\langle \mathbf{a} \times \mathbf{b}, \mathbf{a} \rangle = 0$ and $\langle \mathbf{a} \times \mathbf{b}, \mathbf{b} \rangle = 0$. Consequently, the vector $\mathbf{a} \times \mathbf{b}$ is orthogonal to both $\mathbf{a}$ and $\mathbf{b}$. Suppose that $\mathbf{a}$ and $\mathbf{b}$ are not collinear. Then we can find a vector $\mathbf{c} = (c_1, c_2, c_3)$ in $R^3$ such that $\mathbf{a}, \mathbf{b}, \mathbf{c}$ form a basis for $R^3$. Hence the matrix

$$\begin{bmatrix} a_1 & a_2 & a_3 \\ b_1 & b_2 & b_3 \\ c_1 & c_2 & c_3 \end{bmatrix}$$

is nonsingular and therefore by (2) we have $\langle \mathbf{a} \times \mathbf{b}, \mathbf{c} \rangle \neq 0$. This implies that $\mathbf{a} \times \mathbf{b} \neq \mathbf{0}$. The vectors $\mathbf{a}$ and $\mathbf{b}$ span a plane in $R^3$ and the orthogonal complement

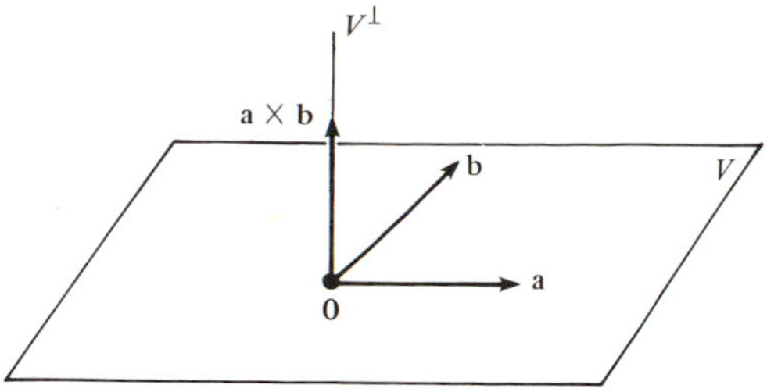


Figure 7.1

of this plane is a line through the origin. The vector $\mathbf{a} \times \mathbf{b} \neq \mathbf{0}$, being orthogonal to $\mathbf{a}$ and $\mathbf{b}$, is on this line and hence spans this line (see Figure 7.1).

We summarize this discussion in the following result.

**Result 1** Let $\mathbf{a}$ and $\mathbf{b}$ be two vectors in $R^3$. Then their cross product $\mathbf{a} \times \mathbf{b}$ is orthogonal to both $\mathbf{a}$ and $\mathbf{b}$. If $\mathbf{a}$ and $\mathbf{b}$ are not collinear, then $\mathbf{a} \times \mathbf{b} \neq \mathbf{0}$ and a vector $\mathbf{c}$ in $R^3$ is orthogonal to both $\mathbf{a}$ and $\mathbf{b}$ if and only if $\mathbf{c}$ is a scalar multiple of $\mathbf{a} \times \mathbf{b}$.

## EXAMPLE 2

The vectors $\mathbf{a} = (1, 2, 0)$ and $\mathbf{b} = (-1, 3, 1)$ are not collinear and hence span a plane $V$ in $R^3$. The orthogonal complement of $V$ is the straight line spanned by the vector $\mathbf{a} \times \mathbf{b} = (2, -1, 5)$.

Now we wish to compute the length of the cross product of two vectors $\mathbf{a} = (a_1, a_2, a_3)$ and $\mathbf{b} = (b_1, b_2, b_3)$. We have

$$\|\mathbf{a} \times \mathbf{b}\|^2 = (a_2 b_3 - a_3 b_2)^2 + (a_3 b_1 - a_1 b_3)^2 + (a_1 b_2 - a_2 b_1)^2.$$

A straightforward algebraic manipulation (which the reader should perform) shows that this quantity is equal to

$$(a_1^2 + a_2^2 + a_3^2)(b_1^2 + b_2^2 + b_3^2) - (a_1 b_1 + a_2 b_2 + a_3 b_3)^2,$$

which is simply

$$\|\mathbf{a}\|^2 \|\mathbf{b}\|^2 - \langle \mathbf{a}, \mathbf{b} \rangle^2.$$

Since $\langle \mathbf{a}, \mathbf{b} \rangle = \|\mathbf{a}\| \|\mathbf{b}\| \cos\theta$, where $\theta$ is the angle between $\mathbf{a}$ and $\mathbf{b}$, it follows that

$$\begin{aligned} \|\mathbf{a} \times \mathbf{b}\|^2 &= \|\mathbf{a}\|^2 \|\mathbf{b}\|^2 - \|\mathbf{a}\|^2 \|\mathbf{b}\|^2 \cos^2\theta \\ &= \|\mathbf{a}\|^2 \|\mathbf{b}\|^2 (1 - \cos^2\theta) \\ &= \|\mathbf{a}\|^2 \|\mathbf{b}\|^2 \sin^2\theta. \end{aligned}$$

Taking square roots yields

$$\|\mathbf{a} \times \mathbf{b}\| = \|\mathbf{a}\| \cdot \|\mathbf{b}\| \sin\theta.$$

This formula has a nice geometric interpretation. Consider the parallelogram determined by $\mathbf{a}$ and $\mathbf{b}$, that is, the parallelogram with vertices $\mathbf{0}$, $\mathbf{a}$, $\mathbf{b}$, and $\mathbf{a} + \mathbf{b}$ (see

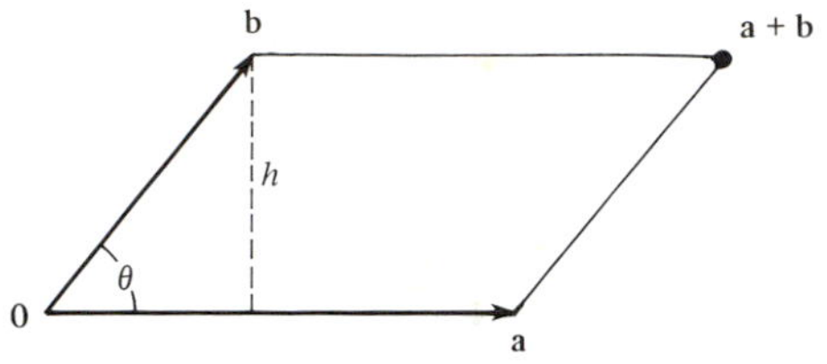


Figure 7.2

Figure 7.2). The area of the parallelogram is equal to the base $\|\mathbf{a}\|$ times the height $h = \|\mathbf{b}\| \sin\theta$. Hence the area is equal to $\|\mathbf{a}\| \cdot \|\mathbf{b}\| \sin\theta = \|\mathbf{a} \times \mathbf{b}\|$. This proves:

**Result 2** If **a** and **b** are two vectors in $R^3$ and $\theta$ is the angle between **a** and **b**, then $\|\mathbf{a} \times \mathbf{b}\| = \|\mathbf{a}\| \cdot \|\mathbf{b}\| \sin\theta$ is equal to the area of the parallelogram determined by **a** and **b**.

### EXAMPLE 3

The area of the parallelogram determined by the vectors $\mathbf{a} = (1, 1, 1)$ and $\mathbf{b} = (2, 0, 3)$ is $\|\mathbf{a} \times \mathbf{b}\| = \|(3, -1, -2)\| = \sqrt{14}$.

We conclude this section with two examples that give geometric interpretations for $2 \times 2$ and $3 \times 3$ determinants.

### EXAMPLE 4

Let $A = \begin{bmatrix} a_1 & a_2 \\ b_1 & b_2 \end{bmatrix}$. We will show that the absolute value of det $A$ is equal to the area of the parallelogram determined by $\mathbf{a} = (a_1, a_2)$ and $\mathbf{b} = (b_1, b_2)$ (see Figure 7.3).

First note that this area is equal to the area of the parallelogram determined by the vectors $\bar{\mathbf{a}} = (a_1, a_2, 0)$ and $\bar{\mathbf{b}} = (b_1, b_2, 0)$ in $R^3$. Hence, by Result 2, the area is

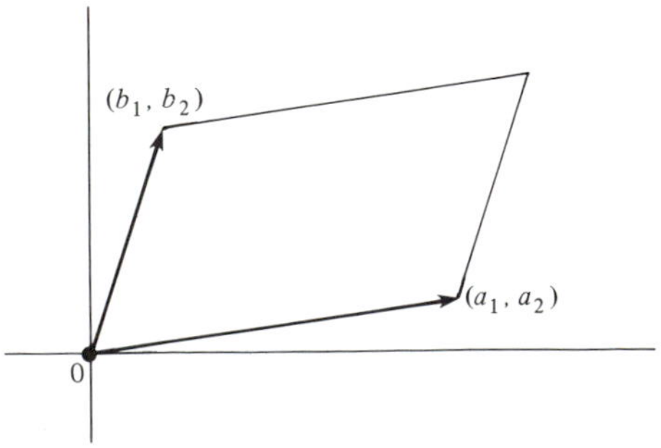


Figure 7.3

equal to

$$\|\bar{\mathbf{a}} \times \bar{\mathbf{b}}\| = \|(0, 0, a_1 b_2 - a_2 b_1)\| = \left|\det \begin{bmatrix} a_1 & a_2 \\ b_1 & b_2 \end{bmatrix}\right|.$$

### EXAMPLE 5

Let

$$A = \begin{bmatrix} a_1 & a_2 & a_3 \\ b_1 & b_2 & b_3 \\ c_1 & c_2 & c_3 \end{bmatrix}.$$

We will show that the absolute value of $\det A$ is equal to the volume of the parallelopiped determined by $\mathbf{a} = (a_1, a_2, a_3)$, $\mathbf{b} = (b_1, b_2, b_3)$, and $\mathbf{c} = (c_1, c_2, c_3)$ (see Figure 7.4). The volume is equal to the area of the base times the height $h$. By Result 2 the area of the base is $\|\mathbf{a} \times \mathbf{b}\|$. Since $\mathbf{a} \times \mathbf{b}$ is perpendicular to $\mathbf{a}$ and $\mathbf{b}$, the height $h$ is equal to the length of the projection of $\mathbf{c}$ onto $\mathbf{a} \times \mathbf{b}$. Hence

$$\begin{aligned} h &= \left\| \frac{\langle \mathbf{a} \times \mathbf{b}, \mathbf{c} \rangle}{\langle \mathbf{a} \times \mathbf{b}, \mathbf{a} \times \mathbf{b} \rangle} \mathbf{a} \times \mathbf{b} \right\| \\ &= \frac{|\langle \mathbf{a} \times \mathbf{b}, \mathbf{c} \rangle|}{\|\mathbf{a} \times \mathbf{b}\|^2} \|\mathbf{a} \times \mathbf{b}\| = \frac{|\langle \mathbf{a} \times \mathbf{b}, \mathbf{c} \rangle|}{\|\mathbf{a} \times \mathbf{b}\|}. \end{aligned}$$

Thus the volume is

$$\|\mathbf{a} \times \mathbf{b}\| \cdot \frac{|\langle \mathbf{a} \times \mathbf{b}, \mathbf{c} \rangle|}{\|\mathbf{a} \times \mathbf{b}\|} = |\langle \mathbf{a} \times \mathbf{b}, \mathbf{c} \rangle|$$

and $|\langle \mathbf{a} \times \mathbf{b}, \mathbf{c} \rangle|$ is equal to $|\det A|$ by equation (2).

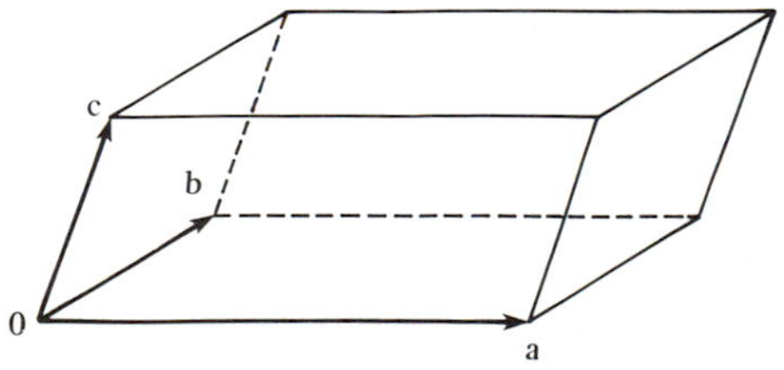


Figure 7.4

## PROBLEMS 7.3

*1. For the given vectors **a** and **b**, find a vector orthogonal to **a** and **b** and find the area of the parallelogram determined by **a** and **b**.

(a) $\mathbf{a} = (1, -1, 0)$, $\mathbf{b} = (3, 2, 3)$
(b) $\mathbf{a} = (-1, 2, 3)$, $\mathbf{b} = (5, 1, 2)$
(c) $\mathbf{a} = (2, 6, -1)$, $\mathbf{b} = (1, 1, 2)$

*2. Find the area of the parallelogram determined by the vectors:
(a) $(1, 2), (-2, 5)$
(b) $(2, 3), (3, 1)$

*3. Find the volume of the parallelopiped determined by the vectors:
(a) $(1, 1, 1), (1, 0, 2), (3, -1, 2)$
(b) $(2, -1, 2), (1, 1, -1), (1, 5, 5)$

*4. Find the area of the triangle with vertices:
(a) $(0, 0), (4, 3), (-2, 1)$
(b) $(0, 0, 0), (1, 2, 3), (2, -2, 4)$

5. Let $\mathbf{e}_1, \mathbf{e}_2, \mathbf{e}_3$ be the standard basis of $R^3$. Calculate $\mathbf{e}_1 \times \mathbf{e}_2$, $\mathbf{e}_2 \times \mathbf{e}_3$, and $\mathbf{e}_3 \times \mathbf{e}_1$.

6. Verify that for all vectors $\mathbf{a}$, $\mathbf{b}$, and $\mathbf{c}$ in $R^3$ and for all scalars $\alpha$:
(a) $\mathbf{a} \times \mathbf{0} = \mathbf{0} \times \mathbf{b} = \mathbf{0}$
(b) $(\alpha\mathbf{a}) \times \mathbf{b} = \mathbf{a} \times (\alpha\mathbf{b}) = \alpha(\mathbf{a} \times \mathbf{b})$
(c) $(\mathbf{a} + \mathbf{b}) \times \mathbf{c} = (\mathbf{a} \times \mathbf{c}) + (\mathbf{b} \times \mathbf{c})$
(d) $\mathbf{a} \times (\mathbf{b} + \mathbf{c}) = (\mathbf{a} \times \mathbf{b}) + (\mathbf{a} \times \mathbf{c})$
(e) $\mathbf{a} \times \mathbf{b} = -(\mathbf{b} \times \mathbf{a})$

7. Verify that for all vectors $\mathbf{a}$, $\mathbf{b}$, and $\mathbf{c}$ in $R^3$:
(a) $(\mathbf{a} \times \mathbf{b}) \times \mathbf{c} + (\mathbf{b} \times \mathbf{c}) \times \mathbf{a} + (\mathbf{c} \times \mathbf{a}) \times \mathbf{b} = \mathbf{0}$
(b) $(\mathbf{a} \times \mathbf{b}) \times \mathbf{c} = \langle \mathbf{a}, \mathbf{c} \rangle \mathbf{b} - \langle \mathbf{a}, \mathbf{b} \rangle \mathbf{c}$
Is it always true that $(\mathbf{a} \times \mathbf{b}) \times \mathbf{c} = \mathbf{a} \times (\mathbf{b} \times \mathbf{c})$? Justify your answer.

8. (Chapter 5 required) Let $T: R^2 \to R^2$ be a linear transformation and let $A$ be the standard matrix for $T$. Let $Q$ be the rectangle determined by two nonzero orthogonal vectors $\mathbf{a}$ and $\mathbf{b}$. Then $T$ maps $Q$ onto a parallelogram (see Problem 21 of Section 5.2). Show that if $A$ is nonsingular, then the area of this parallelogram is equal to $|\det A| \cdot$(area of $Q$). Does this formula hold if $A$ is singular?

# 7.4 The Adjoint Matrix

In Section 1.11 we discussed a method (based on the Gaussian elimination process) for computing the inverse of a nonsingular square matrix. We now present another method for calculating inverses.

Consider an $n \times n$ matrix

$$A = \begin{bmatrix} a_{11} & a_{12} & \cdots & a_{1n} \\ a_{21} & a_{22} & \cdots & a_{2n} \\ \vdots & \vdots & & \vdots \\ a_{n1} & a_{n2} & \cdots & a_{nn} \end{bmatrix}.$$

For each entry $a_{ij}$ let $A_{ij}$ be the cofactor of $a_{ij}$. The **adjoint matrix** of $A$ is defined to be the $n \times n$ matrix whose entry in the $i$th row and $j$th column is $A_{ji}$. Thus the adjoint matrix is the transpose of the matrix of cofactors. We will denote this matrix by adj $A$.

$$\operatorname{adj} A = \begin{bmatrix} A_{11} & A_{21} & \cdots & A_{n1} \\ A_{12} & A_{22} & \cdots & A_{n2} \\ \vdots & \vdots & & \vdots \\ A_{1n} & A_{2n} & \cdots & A_{nn} \end{bmatrix}$$

Note that, for instance, the entry in the first row and second column of adj $A$ is $A_{21}$, *not* $A_{12}$.

## EXAMPLE 1

The adjoint matrix of

$$A = \begin{bmatrix} 1 & -2 & 4 \\ 2 & -7 & 9 \\ -1 & 8 & -1 \end{bmatrix} \quad \text{is} \quad \operatorname{adj} A = \begin{bmatrix} -65 & 30 & 10 \\ -7 & 3 & -1 \\ 9 & -6 & -3 \end{bmatrix}.$$

(The cofactors of $A$ were calculated in Example 1 of Section 7.2.)

Now consider the product of $A$ with its adjoint,

$$P = \begin{bmatrix} a_{11} & a_{12} & \cdots & a_{1n} \\ a_{21} & a_{22} & \cdots & a_{2n} \\ \vdots & \vdots & & \vdots \\ a_{n1} & a_{n2} & \cdots & a_{nn} \end{bmatrix} \begin{bmatrix} A_{11} & A_{21} & \cdots & A_{n1} \\ A_{12} & A_{22} & \cdots & A_{n2} \\ \vdots & \vdots & & \vdots \\ A_{1n} & A_{2n} & \cdots & A_{nn} \end{bmatrix}.$$

The entry in the $i$th row and $j$th column of $P$ is

$$p_{ij} = a_{i1}A_{j1} + a_{i2}A_{j2} + \cdots + a_{in}A_{jn}.$$

If $i = j$, then $p_{ii} = a_{i1}A_{i1} + a_{i2}A_{i2} + \cdots + a_{in}A_{in}$ is the expansion of the determinant of $A$ along the $i$th row of $A$ (i.e., $p_{ii} = \det A$). Thus all entries on the main diagonal of $P$ are equal to $\det A$. Next suppose that $i \neq j$. We claim that in this case $p_{ij} = 0$. To see this, replace the $j$th row of $A$ by the $i$th row and call the resulting matrix $B$. Now the $i$th row and the $j$th row of $B$ are identical, and consequently $\det B = 0$. Note that the changing of the $j$th row of $A$ has no effect on the cofactors belonging to this row because these cofactors do not involve the entries of the $j$th row. It follows that $A_{j1} = B_{j1}$, $A_{j2} = B_{j2}, \ldots, A_{jn} = B_{jn}$. Since the entries in the $j$th row of $B$ are $a_{i1}, a_{i2}, \ldots, a_{in}$ we conclude that $p_{ij} = a_{i1}A_{j1} + a_{i2}A_{j2} + \cdots + a_{in}A_{jn}$ is the cofactor expansion of $\det B$ along the $j$th row (i.e., $p_{ij} = \det B = 0$). Hence all entries of $P$ off the main diagonal are zero. It follows that

$$P = \begin{bmatrix} \det A & 0 & \cdots & 0 \\ 0 & \det A & \cdots & 0 \\ \vdots & \vdots & & \vdots \\ 0 & 0 & \cdots & \det A \end{bmatrix} = \det A \begin{bmatrix} 1 & 0 & \cdots & 0 \\ 0 & 1 & \cdots & 0 \\ \vdots & \vdots & & \vdots \\ 0 & 0 & \cdots & 1 \end{bmatrix}.$$

This gives us the equation

$$A \operatorname{adj} A = (\det A)I, \tag{1}$$

where $I$ is the $n \times n$ identity matrix. Now suppose that $A$ is nonsingular. Then $\det A \neq 0$ and dividing the equation (1) by $\det A$ yields

$$I = \frac{1}{\det A} \cdot (A \operatorname{adj} A) = A\left(\frac{1}{\det A} \cdot \operatorname{adj} A\right).$$

It follows that the matrix $\dfrac{1}{\det A} \cdot \operatorname{adj} A$ must be the inverse of $A$.

**Result 1** If $A$ is a nonsingular square matrix, then

$$A^{-1} = \frac{1}{\det A} \cdot \operatorname{adj} A.$$

**EXAMPLE 2**

The determinant of $A = \begin{bmatrix} a & b \\ c & d \end{bmatrix}$ is $ad - bc$ and the adjoint of $A$ is $\begin{bmatrix} d & -b \\ -c & a \end{bmatrix}$. A direct calculation shows that

$$\begin{bmatrix} a & b \\ c & d \end{bmatrix}\begin{bmatrix} d & -b \\ -c & a \end{bmatrix} = \begin{bmatrix} ad - bc & 0 \\ 0 & ad - bc \end{bmatrix} = (ad - bc)\begin{bmatrix} 1 & 0 \\ 0 & 1 \end{bmatrix}.$$

If $A$ is nonsingular, then $ad - bc \neq 0$ and we obtain the familiar formula

$$A^{-1} = \frac{1}{ad - bc}\begin{bmatrix} d & -b \\ -c & a \end{bmatrix}.$$

**EXAMPLE 3**

The determinant of the matrix

$$A = \begin{bmatrix} 1 & -2 & 4 \\ 2 & -7 & 9 \\ -1 & 8 & -1 \end{bmatrix}$$

is $\det A = -15 \neq 0$. Thus $A$ is nonsingular and (see Example 1)

$$A^{-1} = \frac{1}{\det A} \cdot \operatorname{adj} A = \frac{1}{-15}\begin{bmatrix} -65 & 30 & 10 \\ -7 & 3 & -1 \\ 9 & -6 & -3 \end{bmatrix}.$$

Result 1 gives us a precise formula for the inverse of a nonsingular square matrix $A$ in terms of its determinant and cofactors. However, from a computational point of view this result is practically useless. To compute the inverse of an $n \times n$ matrix $A$ we would have to calculate $n^2 + 1$ determinants, the determinant of $A$, and $n^2$ cofactors [each cofactor being the determinant of an $(n - 1) \times (n - 1)$ matrix]. For example, if $n = 4$, this would involve the evaluation of the determinant of a $4 \times 4$ matrix and of the determinants of sixteen $3 \times 3$ matrices, a formidable task. The method for computing inverses by Gaussian elimination is much faster. Only in the case $n = 2$ do we obtain the useful and familiar formula for the inverse of a nonsingular $2 \times 2$ matrix (see Example 2).

Using Result 1, we can derive a rule for representing the solution of the linear system $A\mathbf{x} = \mathbf{b}$ provided that $A$ is a nonsingular square matrix.

**Result 2** (Cramer's Rule) Let $A$ be a nonsingular $n \times n$ matrix and let $\mathbf{b}$ be a vector in $R^n$. Then the components of the unique solution $\mathbf{x} = (x_1, x_2, \ldots, x_n)$ of $A\mathbf{x} = \mathbf{b}$ are

$$x_i = \frac{\det A_i}{\det A}, \qquad i = 1, 2, \ldots, n,$$

where $A_i$ is the matrix obtained by replacing the $i$th column of $A$ by $\mathbf{b}$.

**Proof** Since $\mathbf{x} = A^{-1}\mathbf{b} = \dfrac{1}{\det A}(\text{adj } A)\mathbf{b}$, it follows that

$$x_i = \frac{1}{\det A}(b_1 A_{1i} + b_2 A_{2i} + \cdots + b_n A_{ni}),$$

where $\mathbf{b} = (b_1, b_2, \ldots, b_n)$. But $b_1 A_{1i} + b_2 A_{2i} + \cdots + b_n A_{ni}$ is the expansion of the determinant of $A_i$ down the $i$th column (which is $\mathbf{b}$).

## EXAMPLE 4

Consider the linear system

$$\begin{bmatrix} 1 & 1 & 2 \\ 0 & 2 & -1 \\ 3 & 1 & 2 \end{bmatrix} \begin{bmatrix} x_1 \\ x_2 \\ x_3 \end{bmatrix} = \begin{bmatrix} 1 \\ 1 \\ 2 \end{bmatrix}.$$

The determinant of the coefficient matrix $A$ is equal to $-10$ and hence $A$ is nonsingular. Now (using the notation of Result 2) we have

$$\det A_1 = \det \begin{bmatrix} 1 & 1 & 2 \\ 1 & 2 & -1 \\ 2 & 1 & 2 \end{bmatrix} = -5,$$

$$\det A_2 = \det \begin{bmatrix} 1 & 1 & 2 \\ 0 & 1 & -1 \\ 3 & 2 & 2 \end{bmatrix} = -5,$$

and

$$\det A_3 = \det \begin{bmatrix} 1 & 1 & 1 \\ 0 & 2 & 1 \\ 3 & 1 & 2 \end{bmatrix} = 0.$$

Hence by Cramer's rule the unique solution is $x_1 = \dfrac{-5}{-10} = \dfrac{1}{2}$, $x_2 = \dfrac{-5}{-10} = \dfrac{1}{2}$, $x_3 = \dfrac{0}{-10} = 0$.

Cramer's rule can be applied only to linear systems with nonsingular coefficient matrices. If the system has $n$ equations, we must evaluate the determinants of $n + 1$ matrices of size $n \times n$. One attempt to apply Cramer's rule to a $6 \times 6$ system will convince the reader that the Gaussian elimination process is a much more efficient method for solving the system.

## PROBLEM 7.4

***1.** Use adjoint matrices to find the inverses of the following nonsingular matrices.

(a) $\begin{bmatrix} -1 & 2 & 2 \\ 3 & 0 & 1 \\ 5 & 3 & -2 \end{bmatrix}$

(b) $\begin{bmatrix} 2 & -1 & 0 \\ 0 & 2 & -1 \\ -5 & -1 & 2 \end{bmatrix}$

(c) $\begin{bmatrix} \frac{1}{3} & 0 & \frac{1}{3} \\ \frac{1}{2} & 1 & \frac{1}{2} \\ -1 & \frac{1}{3} & \frac{2}{3} \end{bmatrix}$

***2.** Consider the upper triangular matrix

$$A = \begin{bmatrix} a & b & c \\ 0 & d & e \\ 0 & 0 & f \end{bmatrix},$$

where $adf \neq 0$. Find the inverse of $A$. Can you explain why in general the inverse of a nonsingular upper triangular matrix must also be upper triangular?

**3.** Let $A$ be a square matrix whose entries are integers. Suppose that $\det A = \pm 1$. Explain why all entries of $A^{-1}$ are also integers (see Problem 6 in Section 7.2).

***4.** Use Cramer's rule to solve the following systems.

(a) $\begin{bmatrix} 2 & -5 \\ 7 & 1 \end{bmatrix} \begin{bmatrix} x_1 \\ x_2 \end{bmatrix} = \begin{bmatrix} 3 \\ 4 \end{bmatrix}$

(b) $\begin{bmatrix} 1 & 1 & -3 \\ 2 & 1 & 2 \\ -1 & -2 & 1 \end{bmatrix} \begin{bmatrix} x_1 \\ x_2 \\ x_3 \end{bmatrix} = \begin{bmatrix} 1 \\ 0 \\ 2 \end{bmatrix}$

# 8

# Eigenvalues and Eigenvectors

This chapter is all about the equation $A\mathbf{v} = \lambda\mathbf{v}$, where $A$ is a square matrix. In Chapter 1 we saw that determining prices in a stable economy is equivalent to solving the equation $A\mathbf{p} = \mathbf{p}$. In Section 8.4 we develop the Leslie population distribution model. We will find that the equilibrium population distribution $\mathbf{p}$ is a solution of $A\mathbf{p} = \lambda\mathbf{p}$, where $\lambda$ is the "natural" growth rate. Those of you who have studied Section 5.5 know that the equation $A\mathbf{v} = \lambda\mathbf{v}$ is all tied up with the diagonal representation of a linear transformation. Whether or not you studied Section 5.5 you will learn in Section 8.3 that this equation is intimately related to converting a matrix into a diagonal matrix.

## 8.1 Eigenvalues and Eigenvectors

***Definition*** Let $A$ be an $n \times n$ matrix. A nonzero vector $\mathbf{v}$ is called an **eigenvector** of $A$ if there is a number $\lambda$ such that

$$A\mathbf{v} = \lambda\mathbf{v}.$$

The number $\lambda$ is called an **eigenvalue** associated with the eigenvector $\mathbf{v}$.

Since eigenvalues and eigenvectors are defined only for square matrices, *we will assume in this chapter that all matrices are square matrices.*

### EXAMPLE 1

Let $A = \begin{bmatrix} 1 & 2 \\ 0 & 3 \end{bmatrix}$. Since

$$\begin{bmatrix} 1 & 2 \\ 0 & 3 \end{bmatrix}\begin{bmatrix} 1 \\ 1 \end{bmatrix} = \begin{bmatrix} 3 \\ 3 \end{bmatrix} = 3\begin{bmatrix} 1 \\ 1 \end{bmatrix},$$

the vector $(1, 1)$ is an eigenvector of $A$ associated with the eigenvalue 3. On the other hand, since

$$\begin{bmatrix} 1 & 2 \\ 0 & 3 \end{bmatrix}\begin{bmatrix} 1 \\ 2 \end{bmatrix} = \begin{bmatrix} 5 \\ 6 \end{bmatrix} \neq \lambda\begin{bmatrix} 1 \\ 2 \end{bmatrix}$$

for any scalar $\lambda$, the vector $(1, 2)$ is not an eigenvector.

If $\lambda$ is an eigenvalue of a matrix $A$ and $\mathbf{v}$ is an associated eigenvector, then $A\mathbf{v}$ is a multiple of $\mathbf{v}$, that is, $A\mathbf{v} = \lambda\mathbf{v}$. $A\mathbf{v}$ and $\mathbf{v}$ are collinear vectors and the length of $A\mathbf{v}$ is $|\lambda|$ times the length of $\mathbf{v}$. Thus the eigenvectors of $A$ associated with the eigenvalue $\lambda$ are those vectors in $R^n$ that when multiplied by $A$ remain collinear to themselves but whose length is changed by the eigenvalue $\lambda$. Geometrically, then, the eigenvectors determine those directions in space that are unchanged ($\lambda > 0$), reversed ($\lambda < 0$), or annihilated ($\lambda = 0$) by $A$.

The definition of an eigenvector specifically excludes the zero vector from being an eigenvector. Without this restriction any scalar $\lambda$ would be an eigenvalue of $A$ having the zero vector as an eigenvector and there would be nothing special about eigenvalues. In addition, by imposing this restriction it follows that an eigenvector of a square matrix $A$ is associated with exactly one eigenvalue. In other words, distinct eigenvalues have different eigenvectors. For suppose that $\mathbf{v}$ is a nonzero vector and that $A\mathbf{v} = \lambda_1\mathbf{v}$ and $A\mathbf{v} = \lambda_2\mathbf{v}$. Then $\lambda_1\mathbf{v} = \lambda_2\mathbf{v}$, and we have

$$\mathbf{0} = \lambda_1\mathbf{v} - \lambda_2\mathbf{v} = (\lambda_1 - \lambda_2)\mathbf{v}.$$

Since $\mathbf{v}$ is an eigenvector, $\mathbf{v} \neq \mathbf{0}$ and it follows that $\lambda_1 - \lambda_2 = 0$. Hence $\lambda_1 = \lambda_2$.

### EXAMPLE 2

Although the zero vector is excluded from being an eigenvector, the number zero is not excluded from being an eigenvalue. For example, the vector $\mathbf{v} = (-2, 1, 0)$ is an eigenvector of the matrix

$$A = \begin{bmatrix} 1 & 2 & 3 \\ 0 & 0 & 1 \\ 0 & 0 & 2 \end{bmatrix}$$

with eigenvalue 0 since $A\mathbf{v} = \mathbf{0} = 0\mathbf{v}$.

Let $\lambda$ be an eigenvalue of an $n \times n$ matrix $A$. Then the collection of all eigenvectors associated with $\lambda$ *almost* form a subspace of $R^n$. For suppose that $\mathbf{v}$ and

$\mathbf{w}$ are eigenvectors and that $\alpha$ is a scalar. Then

$$A(\mathbf{v} + \mathbf{w}) = A\mathbf{v} + A\mathbf{w} = \lambda\mathbf{v} + \lambda\mathbf{w} = \lambda(\mathbf{v} + \mathbf{w})$$
$$A(\alpha\mathbf{v}) = \alpha A\mathbf{v} = \alpha\lambda\mathbf{v} = \lambda(\alpha\mathbf{v}).$$

Thus the sum of two eigenvectors is an eigenvector and a nonzero scalar multiple of an eigenvector is an eigenvector. But the zero vector is specifically excluded from being an eigenvector, so the collection of all eigenvectors associated with $\lambda$ is not a subspace. We define the **eigenspace** of $\lambda$ to be the collection of all eigenvectors associated with $\lambda$ together with the zero vector. We denote the eigenspace of $\lambda$ by $E(\lambda)$.

Now let us examine some of the consequences that follow from the equation

$$A\mathbf{v} = \lambda\mathbf{v}.$$

Let $A$ be an $n \times n$ matrix and $I$ be the $n \times n$ identity matrix. Since $\lambda I\mathbf{v} = \lambda\mathbf{v}$, the equation $A\mathbf{v} = \lambda\mathbf{v}$ is equivalent to the equation

$$(A - \lambda I)\mathbf{v} = \mathbf{0}.$$

Therefore, $\lambda$ is an eigenvalue of $A$ with an associated eigenvector $\mathbf{v}$ if and only if

$$(A - \lambda I)\mathbf{v} = \mathbf{0} \qquad \text{and} \qquad \mathbf{v} \neq \mathbf{0}.$$

This proves the following theorem.

***Theorem 1*** *Let $A$ be a square matrix.*

(*a*) *$\lambda$ is an eigenvalue of $A$ if and only if the nullspace of $A - \lambda I$ is nontrivial.*
(*b*) *If $\lambda$ is an eigenvalue, then $\mathbf{v}$ is an eigenvector associated with $\lambda$ if and only if $\mathbf{v}$ is a nonzero vector in the nullspace of $A - \lambda I$.*

It follows from (b) that $E(\lambda) = N(A - \lambda I)$. Since the nullspace of a matrix is a subspace, we have another proof of the fact that the eigenspace $E(\lambda)$ is a subspace.

Perhaps the most important consequence of (b) is that it provides us with a method for finding the eigenvectors associated with an eigenvalue $\lambda$. We simply compute a basis for $N(A - \lambda I)$ using the method presented in Chapter 2. The eigenvectors associated with $\lambda$ are then the nontrivial linear combinations of these basis vectors.

## EXAMPLE 3

$\lambda = 2$ is an eigenvalue of the matrix

$$A = \begin{bmatrix} 2 & -5 & 5 \\ 0 & 3 & -1 \\ 0 & -1 & 3 \end{bmatrix}.$$

To find the associated eigenvectors, we compute the nullspace of

$$A - 2I = \begin{bmatrix} 0 & -5 & 5 \\ 0 & 1 & -1 \\ 0 & -1 & 1 \end{bmatrix}.$$

The reduced echelon form of this matrix is

$$\begin{bmatrix} 0 & 1 & -1 \\ 0 & 0 & 0 \\ 0 & 0 & 0 \end{bmatrix}.$$

Thus a basis for the eigenspace $E(2)$ is $(1, 0, 0)$ and $(0, 1, 1)$. Any eigenvector associated with the eigenvalue 2 can be expressed as a nontrivial linear combination of these two eigenvectors. Thus the eigenvectors are

$$\alpha(1,0,0) + \beta(0,1,1) = (\alpha, \beta, \beta),$$

where at least one of $\alpha$ and $\beta$ is not zero. $\lambda = 4$ is also an eigenvalue of the matrix $A$. Since $N(A - 4I)$ is spanned by $(5, -1, 1)$ (verify!) the eigenvectors associated with the eigenvalue 4 are precisely the nonzero multiples of $(5, -1, 1)$.

## PROBLEMS 8.1

**1.** Let $A = \begin{bmatrix} 3 & 1 \\ -3 & 7 \end{bmatrix}$. Show that $\lambda = 4$ and $\lambda = 6$ are eigenvalues of $A$ with eigenvectors $(1, 1)$ and $(1, 3)$.

**2.** Let $A = \begin{bmatrix} 2 & 0 \\ 0 & 3 \end{bmatrix}$. Show that $\lambda = 2$ and $\lambda = 3$ are eigenvalues of $A$ with eigenvectors $\mathbf{v}_1 = (1, 0)$ and $\mathbf{v}_2 = (0, 1)$.

***3.** Show that $(1, -1)$ and $(-1, 6)$ are eigenvectors of $A = \begin{bmatrix} 3 & 1 \\ -6 & -4 \end{bmatrix}$. What are the associated eigenvalues?

***4.** Show that $(1, 0, 1)$, $(-1, 3, 0)$, and $(1, 0, 2)$ are eigenvectors of each of the following matrices.

(a) $\begin{bmatrix} -12 & -5 & 9 \\ 0 & 3 & 0 \\ -18 & -6 & 15 \end{bmatrix}$

(b) $\begin{bmatrix} -4 & -1 & 3 \\ 0 & -1 & 0 \\ -6 & -2 & 5 \end{bmatrix}$

What are the associated eigenvalues?

**5.** Show that $(1, 0, 0)$, $(2, 1, 0)$, and $(3, 4, 1)$ are eigenvectors of each of the following matrices.

(a) $\begin{bmatrix} 2 & -2 & 2 \\ 0 & 1 & -4 \\ 0 & 0 & 0 \end{bmatrix}$

(b) $\begin{bmatrix} -1 & 0 & 6 \\ 0 & -1 & 8 \\ 0 & 0 & 1 \end{bmatrix}$

What are the associated eigenvalues?

***6.** Show that $(1, 0, 0, 0)$, $(1, 1, 0, 0)$, $(1, 1, 1, 0)$, and $(1, 1, 1, 1)$ are eigenvectors of each of the following matrices.

(a) $\begin{bmatrix} -1 & -1 & -1 & -1 \\ 0 & -2 & -1 & -1 \\ 0 & 0 & -3 & -1 \\ 0 & 0 & 0 & -4 \end{bmatrix}$

(b) $\begin{bmatrix} 1 & 0 & 1 & 0 \\ 0 & 1 & 1 & 0 \\ 0 & 0 & 2 & 0 \\ 0 & 0 & 0 & 2 \end{bmatrix}$

What are the associated eigenvalues?

**7.** In each part, find a basis for the eigenspace associated with the given eigenvalues.

*(a) $A = \begin{bmatrix} 2 & 3 \\ 4 & 6 \end{bmatrix}, \quad \lambda = 0, \quad \lambda = 8$

(b) $A = \begin{bmatrix} 2 & 0 & 1 \\ 0 & 1 & 0 \\ 1 & 0 & 2 \end{bmatrix}, \quad \lambda = 1, \quad \lambda = 3$

*(c) $A = \begin{bmatrix} 1 & 1 & 1 \\ 0 & 1 & 0 \\ 0 & 0 & 2 \end{bmatrix}, \quad \lambda = 1, \quad \lambda = 2$

(d) $A = \begin{bmatrix} 1 & 1 & 1 \\ 0 & 1 & 1 \\ 0 & 1 & 1 \end{bmatrix}, \quad \lambda=0, \ \lambda=1, \ \lambda=2$

(e) $A = \begin{bmatrix} 0 & 2 & 1 \\ -2 & 3 & 0 \\ 0 & -3 & -1 \end{bmatrix}, \quad \lambda = 2$

**8.** Show that if $\lambda$ is an eigenvalue of $A$ with eigenvector $\mathbf{v}$, then

$$\lambda = \frac{\langle \mathbf{v}, A\mathbf{v} \rangle}{\langle \mathbf{v}, \mathbf{v} \rangle}.$$

***9.** Let $\lambda$ be an eigenvalue of a matrix $A$. Show that if $\mathbf{v}$ is any eigenvector in $E(\lambda)$, then $A\mathbf{v}$ is also in $E(\lambda)$.

**10.** Let $\mathbf{v}$ be an eigenvector of a matrix $A$. Show that $\mathbf{v}$ is an eigenvector of the matrix $B = A - \alpha I$ for any number $\alpha$. What is the corresponding eigenvalue?

***11.** Show that the standard basis vectors $\mathbf{e}_1, \ldots, \mathbf{e}_n$ for $R^n$ are eigenvectors for any $n \times n$ diagonal matrix. What are the associated eigenvalues?

**12.** Show that 0 is an eigenvalue of a matrix $A$ if and only if $A$ is singular.

***13.** Let $A$ be an invertible matrix. Show that $\lambda$ is an eigenvalue of $A$ if and only if $\lambda^{-1}$ is an eigenvalue of $A^{-1}$. How are the eigenvectors related?

**14.** Let $\lambda$ be an eigenvalue of $A$.

(a) Show that $\lambda^2$ is an eigenvalue of $A^2$, $\lambda^3$ is an eigenvalue of $A^3$, and so on.

(b) If $\mathbf{v}$ is an eigenvector of $A$ associated with $\lambda$, show that $\mathbf{v}$ is an eigenvector of $A^2, A^3, A^4, \ldots$ associated with $\lambda^2, \lambda^3, \lambda^4, \ldots$.

**15.** Let $A$ be an $n \times n$ matrix whose rows all add up to 2. Show that $\lambda = 2$ is an eigenvalue of $A$. What is an associated eigenvector? Is there anything special about the number 2?

**16.** Show that if $\lambda_1$ and $\lambda_2$ are eigenvalues of a matrix $A$, and if $E(\lambda_1) \cap E(\lambda_2)$ is not the trivial subspace, then $\lambda_1 = \lambda_2$.

**17.** Suppose that $A$ is idempotent, that is, $A^2 = A$.

*(a) Prove that if $\lambda$ is an eigenvalue of $A$, then $\lambda = 0$ or $\lambda = 1$.

*(b) Prove that if $A \neq 0$, then $\lambda = 1$ is an eigenvalue of $A$. (*Hint:* If $A \neq 0$, then there is a vector $\mathbf{v}$ in $R^n$ such that $A\mathbf{v} \neq \mathbf{0}$.)

(c) Prove that $E(1) = C(A)$.

(d) Prove that if $A \neq I$, then $\lambda = 0$ is an eigenvalue of $A$. (*Hint:* If $A \neq I$, then there is a vector $\mathbf{v}$ in $R^n$ such that $A\mathbf{v} \neq \mathbf{v}$. Thus $A\mathbf{v} - \mathbf{v} \neq \mathbf{0}$.)

(e) Prove that $E(0) = N(A)$.

**18.** A matrix $A$ is called *nilpotent* if there is a positive integer $n$ such that $A^n = 0$. Show that 0 is the only eigenvalue of a nilpotent matrix.

**19.** Let $A$ be an orthogonal matrix (i.e., $A^{-1} = A^T$). Show that if $\lambda$ is an eigenvalue of $A$, then $|\lambda| = 1$. (*Hint:* Use Problem 17(e) of Section 3.2.)

# 8.2 The Characteristic Equation

Theorem 1(b) of Section 8.1 provides us with a method for finding the eigenvectors associated with an eigenvalue $\lambda$. Once it is known that $\lambda$ is an eigenvalue, the eigenvectors associated with $\lambda$ are simply the nonzero vectors in the nullspace of $A - \lambda I$. We now address the question of how to find the eigenvalues. Theorem 1(a) of Section 8.1 provides the answer. $\lambda$ is an eigenvalue of $A$ if and only if the nullspace of $A - \lambda I$ is nontrivial, and this holds if and only if $A - \lambda I$ is singular. Since singular matrices are exactly those matrices whose determinant is zero (Theorem 7 of Section 7.1), we have proved the following theorem.

***Theorem 1*** *$\lambda$ is an eigenvalue of $A$ if and only if*

$$\det(A - \lambda I) = 0.$$

Before we can use this result to find the eigenvalues of a matrix, we must study the equation

$$\det(A - \lambda I) = 0.$$

In view of Theorem 1, the eigenvalues are precisely the zeros of the function $p(\lambda) = \det(A - \lambda I)$. Let us compute $p(\lambda)$ when $A$ is a $2 \times 2$ matrix and when $A$ is a $3 \times 3$ matrix.

First suppose that $A$ is a $2 \times 2$ matrix.

$$A = \begin{bmatrix} a_{11} & a_{12} \\ a_{21} & a_{22} \end{bmatrix}$$

Then

$$A - \lambda I = \begin{bmatrix} a_{11} & a_{12} \\ a_{21} & a_{22} \end{bmatrix} - \lambda \begin{bmatrix} 1 & 0 \\ 0 & 1 \end{bmatrix} = \begin{bmatrix} a_{11} - \lambda & a_{12} \\ a_{21} & a_{22} - \lambda \end{bmatrix}$$

and

$$\begin{aligned} \det(A - \lambda I) &= (a_{11} - \lambda)(a_{22} - \lambda) - a_{12}a_{21} \\ &= \lambda^2 - (a_{11} + a_{22})\lambda + (a_{11}a_{22} - a_{12}a_{21}). \end{aligned}$$

Thus when $A$ is a $2 \times 2$ matrix, $p(\lambda) = \det(A - \lambda I)$ is a polynomial of degree 2 in $\lambda$.

Now suppose that $A$ is a $3 \times 3$ matrix.

$$A = \begin{bmatrix} a_{11} & a_{12} & a_{13} \\ a_{21} & a_{22} & a_{23} \\ a_{31} & a_{32} & a_{33} \end{bmatrix}$$

Then

$$A - \lambda I = \begin{bmatrix} a_{11} & a_{12} & a_{13} \\ a_{21} & a_{22} & a_{23} \\ a_{31} & a_{32} & a_{33} \end{bmatrix} - \lambda \begin{bmatrix} 1 & 0 & 0 \\ 0 & 1 & 0 \\ 0 & 0 & 1 \end{bmatrix}$$

$$= \begin{bmatrix} a_{11} - \lambda & a_{12} & a_{13} \\ a_{21} & a_{22} - \lambda & a_{23} \\ a_{31} & a_{32} & a_{33} - \lambda \end{bmatrix}$$

and

$$\det(A - \lambda I) = (a_{11} - \lambda)\det\begin{bmatrix} a_{22} - \lambda & a_{23} \\ a_{32} & a_{33} - \lambda \end{bmatrix}$$

$$- a_{12}\det\begin{bmatrix} a_{21} & a_{23} \\ a_{31} & a_{33} - \lambda \end{bmatrix} + a_{13}\det\begin{bmatrix} a_{21} & a_{22} - \lambda \\ a_{31} & a_{32} \end{bmatrix}.$$

The first term is a polynomial of degree 3 in $\lambda$ and the second and third terms are polynomials of degree 1 in $\lambda$. Thus when $A$ is a $3 \times 3$ matrix, $p(\lambda) = \det(A - \lambda I)$ is a polynomial of degree 3 in $\lambda$.

These examples prove the following result when $n = 2$ and $n = 3$. A similar argument works for an $n \times n$ matrix.

**Result 1** Let $A$ be an $n \times n$ matrix. The function $p(\lambda) = \det(A - \lambda I)$ is a polynomial of degree $n$ in $\lambda$.

The polynomial $p(\lambda) = \det(A - \lambda I)$ is called the **characteristic polynomial** of $A$ and the equation $\det(A - \lambda I) = 0$ is called the **characteristic equation** of $A$. In this terminology Theorem 1 becomes:

***Theorem 2*** *$\lambda$ is an eigenvalue of a matrix $A$ if and only if $\lambda$ is a zero of the characteristic equation* $\det(A - \lambda I) = 0$.

Theorem 1(b) of Section 8.1 showed that finding the eigenspace $E(\lambda)$ of an eigenvalue $\lambda$ is equivalent to finding the nullspace of $A - \lambda I$. Theorem 2 shows that finding the eigenvalues of a matrix is equivalent to finding the zeros of the characteristic polynomial.

## EXAMPLE 1

Let us find the eigenvalues and the associated eigenvectors of the matrix

$$A = \begin{bmatrix} 1 & 0 & 1 \\ 0 & 1 & 0 \\ 1 & 2 & 1 \end{bmatrix}.$$

The characteristic polynomial of $A$ is

$$\det(A-\lambda I)=\det\begin{bmatrix}1-\lambda & 0 & 1\\ 0 & 1-\lambda & 0\\ 1 & 2 & 1-\lambda\end{bmatrix}=\lambda(1-\lambda)(\lambda-2).$$

The zeros of the characteristic equation are 0, 1, and 2 and by Theorem 2 these are the eigenvalues of $A$. To find the eigenvectors associated with these eigenvalues, we compute the nullspace of $A-0I$, $A-1I$, and $A-2I$. Using Gaussian elimination we find that the reduced echelon forms of $A$, $A-I$, and $A-2I$ are

$$\begin{bmatrix}1 & 0 & 1\\ 0 & 1 & 0\\ 0 & 0 & 0\end{bmatrix},\quad \begin{bmatrix}1 & 2 & 0\\ 0 & 0 & 1\\ 0 & 0 & 0\end{bmatrix},\quad \text{and}\quad \begin{bmatrix}1 & 0 & -1\\ 0 & 1 & 0\\ 0 & 0 & 0\end{bmatrix}.$$

Thus the eigenspace $E(0)$ is spanned by $(-1,0,1)$, the eigenspace $E(1)$ is spanned by $(-2,1,0)$, and the eigenspace $E(2)$ is spanned by $(1,0,1)$.

| **Eigenvalue** | **Eigenvectors** |
|---|---|
| 0 | $\alpha(-1,0,1)$, $\alpha \neq 0$ |
| 1 | $\alpha(-2,1,0)$, $\alpha \neq 0$ |
| 2 | $\alpha(1,0,1)$, $\alpha \neq 0$ |

### EXAMPLE 2

Let us verify that $\lambda = 2$ and $\lambda = 4$ are the eigenvalues of the matrix

$$A=\begin{bmatrix}2 & -5 & 5\\ 0 & 3 & -1\\ 0 & -1 & 3\end{bmatrix}$$

considered in Example 3 of Section 8.1. The characteristic polynomial of $A$ is

$$\begin{aligned}\det(A-\lambda I)&=\det\begin{bmatrix}2-\lambda & -5 & 5\\ 0 & 3-\lambda & -1\\ 0 & -1 & 3-\lambda\end{bmatrix}\\ &=(2-\lambda)\det\begin{bmatrix}3-\lambda & -1\\ -1 & 3-\lambda\end{bmatrix}\\ &=(2-\lambda)(\lambda-4)(\lambda-2).\end{aligned}$$

Thus $\lambda = 2$ and $\lambda = 4$ are the zeros of the characteristic polynomial. (Since $\lambda = 2$ is a double root of the characteristic equation, we say that it has **multiplicity** 2.) In Example 3 of Section 8.1 we also saw that a basis for the eigenspace $E(2)$ is $(0,1,1)$ and $(1,0,0)$, and a basis for the eigenspace $E(4)$ is $(5,-1,1)$.

| Eigenvalue | Eigenvectors |
|---|---|
| 2 | $\alpha(0,1,1)+\beta(1,0,0)$, $\alpha$ or $\beta \neq 0$ |
| 4 | $\alpha(5,-1,1)$, $\alpha \neq 0$ |

### EXAMPLE 3

Let

$$A = \begin{bmatrix} -1 & 1 & 0 \\ 0 & 5 & 0 \\ 4 & -2 & 5 \end{bmatrix}.$$

The characteristic polynomial of $A$ is

$$\begin{aligned} \det(A-\lambda I) &= \det \begin{bmatrix} -1-\lambda & 1 & 0 \\ 0 & 5-\lambda & 0 \\ 4 & -2 & 5-\lambda \end{bmatrix} \\ &= (-1-\lambda)(5-\lambda)^2, \end{aligned}$$

so the eigenvalues of $A$ are $\lambda = -1$ and $\lambda = 5$. The reduced echelon form of

$$A + I = \begin{bmatrix} 0 & 1 & 0 \\ 0 & 6 & 0 \\ 4 & -2 & 6 \end{bmatrix} \quad \text{is} \quad \begin{bmatrix} 1 & 0 & \frac{3}{2} \\ 0 & 1 & 0 \\ 0 & 0 & 0 \end{bmatrix},$$

and of

$$A - 5I = \begin{bmatrix} -6 & 1 & 0 \\ 0 & 0 & 0 \\ 4 & -2 & 0 \end{bmatrix} \quad \text{is} \quad \begin{bmatrix} 1 & 0 & 0 \\ 0 & 1 & 0 \\ 0 & 0 & 0 \end{bmatrix}.$$

Thus the eigenspace $E(-1)$ is spanned by $(-\frac{3}{2}, 0, 1)$ and the eigenspace $E(5)$ is spanned by $(0, 0, 1)$.

| Eigenvalue | Eigenvectors |
|---|---|
| $-1$ | $\alpha(-\frac{3}{2},0,1)$, $\alpha \neq 0$ |
| 5 | $\alpha(0,0,1)$, $\alpha \neq 0$ |

All of the examples above were carefully arranged so that it was easy to factor the characteristic polynomial. However, you are no doubt well aware that it is difficult, if not impossible, to find the zeros of most polynomials. Polynomials of degree 2 are simple; we have the quadratic formula. It is also possible to find the zeros of any polynomial of degree 3 or 4 in the sense that there are formulas, similar to the quadratic formula although much more complicated, which will give us the zeros. But this is as far as it goes. Evariste Galois (1811–1832) showed that there is no analogue of the quadratic formula for polynomials of degree 5 or higher. For such polynomials we must resort to ad hoc methods to find their zeros. In Chapter 9 we discuss some numerical methods for finding the eigenvalues of a matrix that are not based on finding the zeros of a characteristic polynomial.

The discussion thus far has left open the extremely important question: Do polynomials always have zeros? If we insist that all zeros of a polynomial be real numbers, then the answer is definitely no! The polynomial $\lambda^2 + 1$ $\left(\text{which is the characteristic polynomial of the matrix } \begin{bmatrix} 0 & -1 \\ 0 & 0 \end{bmatrix}\right)$ has no real zeros. It does, however, have the two complex zeros $\pm i$, where $i = \sqrt{-1}$. In fact, if we allow complex zeros, then the following theorem answers our question. *Every polynomial of degree $n \geq 1$ has $n$ zeros.* [Note that these zeros need not be distinct. If the polynomial has a repeated zero (i.e., a zero with multiplicity greater than 1), then this zero must be counted the number of times it is repeated. For example, the three zeros of $(\lambda - 1)^2(\lambda + 1)$ are $\lambda = 1$, $\lambda = 1$, and $\lambda = -1$. The multiplicity of a zero is the number of times the zero is repeated.] This result, called the *fundamental theorem of algebra*, was first proved by Gauss in 1799. Its proof is very difficult and will not be given here.

Thus if we allow complex numbers to be eigenvalues, any $n \times n$ matrix $A$ has exactly $n$ eigenvalues. They are the $n$ zeros of the characteristic equation $\det(A - \lambda I) = 0$. As we have just pointed out, this equation is in general very difficult to solve, and it may have complex roots. When an eigenvalue $\lambda$ is a complex number, the eigenvectors associated with $\lambda$ will have complex entries. Since the eigenspace $E(\lambda)$ is the nullspace of $A - \lambda I$, the procedure for finding the eigenspace associated with a complex eigenvalue $\lambda$ is simply that of applying Gaussian elimination to a matrix with complex entries. We refer the reader to Chapter 10 where we develop the necessary arithmetic of complex numbers and give several examples. *In the remainder of this chapter we deal only with matrices all of whose eigenvalues are real.*

One situation where it is exceptionally easy to find the eigenvalues is described in the following result.

***Theorem 3*** *If $A$ is a triangular matrix, then the eigenvalues of $A$ are its diagonal entries.*

This result applies to either upper or lower triangular matrices. In particular, it applies to diagonal matrices.

**Proof** If $A$ is upper triangular, say

$$A = \begin{bmatrix} a_{11} & a_{12} & \cdots & a_{1n} \\ 0 & a_{21} & \cdots & a_{2n} \\ \vdots & \vdots & & \vdots \\ 0 & 0 & \cdots & a_{nn} \end{bmatrix},$$

then

$$A - \lambda I = \begin{bmatrix} a_{11} - \lambda & a_{12} & \cdots & a_{1n} \\ 0 & a_{22} - \lambda & \cdots & a_{2n} \\ \vdots & \vdots & & \vdots \\ 0 & 0 & \cdots & a_{nn} - \lambda \end{bmatrix}$$

is also upper triangular. Consequently,

$$\det(A - \lambda I) = (a_{11} - \lambda)(a_{22} - \lambda)\cdots(a_{nn} - \lambda)$$

and the zeros of this polynomial are $a_{11}, a_{22}, \ldots, a_{nn}$. The essence of this proof is the following. If $A$ is triangular, then $A - \lambda I$ is triangular and the determinant of a triangular matrix is the product of its diagonal entries.

We conclude this section with two useful facts that relate the eigenvalues of a matrix to its diagonal entries and its determinant.

**Result 2** Let $A$ be an $n \times n$ matrix with the eigenvalues $\lambda_1, \ldots, \lambda_n$. Then the sum of the eigenvalues is the sum of the diagonal entries of $A$,

$$\lambda_1 + \lambda_2 + \cdots + \lambda_n = a_{11} + a_{22} + \cdots + a_{nn},$$

and the product of the eigenvalues is the determinant of $A$,

$$\lambda_1 \lambda_2 \cdots \lambda_n = \det A.$$

The proof of this result will be omitted. It is useful as a numerical check on the eigenvalues. If one of these equations fails, there must be a mistake. However, these equations are not sufficient to guarantee that the eigenvalues are correct. That is, these equations may be satisfied by incorrect eigenvalues. We illustrate this with an example.

## EXAMPLE 4

In Example 2 we showed that the eigenvalues of the matrix

$$A = \begin{bmatrix} 2 & -5 & 5 \\ 0 & 3 & -1 \\ 0 & -1 & 3 \end{bmatrix}$$

are $\lambda_1 = 2$, $\lambda_2 = 2$, and $\lambda_3 = 4$. Thus $\lambda_1 + \lambda_2 + \lambda_3 = 8$ and the sum of the diagonal entries of $A$ is $2 + 3 + 3 = 8$. Similarly, $\lambda_1 \lambda_2 \lambda_3 = 16 = \det A$. However, suppose that we had erroneously obtained $1 + \sqrt{5}$, $1 + \sqrt{5}$, and $6 - 2\sqrt{5}$ for the eigenvalues. Since $(1 + \sqrt{5}) + (1 + \sqrt{5}) + (6 - 2\sqrt{5}) = 8$ and $(1 + \sqrt{5})(1 + \sqrt{5})(6 - 2\sqrt{5}) = 16$, these equations would not have detected the error.

# PROBLEMS 8.2

1. Find the eigenvalues and a basis for each of the corresponding eigenspaces.

*(a) $\begin{bmatrix} 2 & 0 \\ -1 & 3 \end{bmatrix}$  (b) $\begin{bmatrix} 1 & -2 \\ -1 & 2 \end{bmatrix}$

*(c) $\begin{bmatrix} 0 & -1 \\ -3 & 2 \end{bmatrix}$

(d) $\begin{bmatrix} 1 & 0 & 1 \\ 1 & 1 & 1 \\ 0 & 0 & -3 \end{bmatrix}$

*(e) $\begin{bmatrix} 1 & 1 & 0 \\ 1 & 5 & -2 \\ 1 & 3 & -1 \end{bmatrix}$

(f) $\begin{bmatrix} 1 & 1 & 1 \\ 1 & -1 & 3 \\ 0 & 1 & 0 \end{bmatrix}$

*(g) $\begin{bmatrix} 0 & 0 & 0 \\ 0 & 1 & 0 \\ 1 & 0 & 1 \end{bmatrix}$

(h) $\begin{bmatrix} 2 & 0 & 1 \\ 0 & 1 & 2 \\ 0 & 0 & 1 \end{bmatrix}$

*(i) $\begin{bmatrix} 0 & 1 & 1 \\ 0 & 0 & 2 \\ 0 & 0 & 1 \end{bmatrix}$

(j) $\begin{bmatrix} 0 & 0 & 2 \\ 0 & 0 & 0 \\ 2 & 0 & 0 \end{bmatrix}$

*(k) $\begin{bmatrix} 1 & -1 & 0 \\ -1 & 2 & -1 \\ 0 & -1 & 1 \end{bmatrix}$

(l) $\begin{bmatrix} -2 & -2 & -4 \\ 2 & 3 & 2 \\ 3 & 2 & 5 \end{bmatrix}$

*(m) $\begin{bmatrix} 3 & 1 & 0 \\ 0 & 2 & 1 \\ 0 & 1 & 2 \end{bmatrix}$

(n) $\begin{bmatrix} 2 & 0 & 0 \\ 0 & 2 & 7 \\ 0 & 0 & -5 \end{bmatrix}$

*(o) $\begin{bmatrix} 1 & -1 & 4 \\ 3 & 2 & -1 \\ 2 & 1 & -1 \end{bmatrix}$

(p) $\begin{bmatrix} 1 & 1 & 0 & 0 \\ 0 & 1 & 0 & 0 \\ 0 & 0 & 1 & 1 \\ 0 & 0 & 0 & 1 \end{bmatrix}$

*(q) $\begin{bmatrix} 1 & 1 & 0 & 0 \\ 0 & 1 & 0 & 0 \\ 0 & 0 & -2 & 0 \\ 0 & 0 & 0 & -2 \end{bmatrix}$

(r) $\begin{bmatrix} 2 & 1 & 0 & 0 \\ 0 & 2 & 1 & 0 \\ 0 & 0 & 2 & 0 \\ 0 & 0 & 0 & -1 \end{bmatrix}$

2. Verify that Result 2 holds for each of the matrices in Problem 1.

3. Prove Result 2 for $2 \times 2$ matrices.

4. Verify Result 1 for $4 \times 4$ matrices.

*5. Suppose that $A$ is a $2 \times 2$ matrix with characteristic polynomial $\lambda^2 + \lambda - 6$. Find the characteristic polynomial of the matrix $A^2$.

6. Show that $A$ and $A^T$ have the same eigenvalues. (*Hint:* Show that $A$ and $A^T$ have the same characteristic equation.) Do they have the same eigenvectors? If so, give a proof. If not, give a counterexample.

*7. Two matrices $A$ and $B$ are **similar** if there is a nonsingular matrix $M$ such that $A = M^{-1}BM$. Show that similar matrices have the same characteristic equation and hence the same eigenvalues. How are the eigenvectors related?

# 8.3 Diagonalization

We have seen a couple of examples where a proper choice of a basis for $R^m$ simplifies computations and leads to a deeper understanding of the mathematics involved. For example, by choosing an orthonormal basis for a subspace $V$ of $R^m$, the normal equations $A^TA\bar{\mathbf{x}} = A^T\mathbf{b}$ simplify to $\bar{\mathbf{x}} = A^T\mathbf{b}$ (since $A^TA = I$) and the projection matrix for $V$ simplifies from $P = A(A^TA)^{-1}A^T$ to $P = AA^T$. Moreover, with respect to an orthonormal basis, the projection matrix decomposes into a sum of $n$ projection matrices ($n = \dim V$), where each summand is the projection of $R^m$ onto one of the basis vectors. Specifically, if $\mathbf{w}_1, \mathbf{w}_2, \ldots, \mathbf{w}_n$ is an orthonormal basis for $V$, then

$$P = \mathbf{w}_1\mathbf{w}_1^T + \mathbf{w}_2\mathbf{w}_2^T + \cdots + \mathbf{w}_n\mathbf{w}_n^T,$$

where $\mathbf{w}_i\mathbf{w}_i^T$ is the projection of $R^m$ onto $\mathbf{w}_i$. This decomposition is a special case of a very important theorem called the spectral theorem for symmetric matrices (see Section 8.6).

Those of you who studied Section 5.5 saw that if $R^m$ has a basis consisting of eigenvectors of a linear transformation $T$: $R^m \to R^m$, then $T$ can be represented by a diagonal matrix with respect to this basis.

We are now going to investigate the consequences of the assumption that $R^n$ has a basis consisting of eigenvectors of an $n \times n$ matrix $A$.

Let $A$ be an $n \times n$ matrix and suppose that $\mathbf{v}_1, \mathbf{v}_2, \ldots, \mathbf{v}_n$ form a basis of $R^n$ consisting of eigenvectors of $A$. Let $\lambda_1, \lambda_2, \ldots, \lambda_n$ be the corresponding eigenvalues. If $\mathbf{x}$ is in $R^n$ and

$$\mathbf{x} = \alpha_1\mathbf{v}_1 + \alpha_2\mathbf{v}_2 + \cdots + \alpha_n\mathbf{v}_n$$

is its expansion in terms of the basis, then

$$\begin{aligned} A\mathbf{x} &= \alpha_1 A\mathbf{v}_1 + \alpha_2 A\mathbf{v}_2 + \cdots + \alpha_n A\mathbf{v}_n \\ &= \lambda_1\alpha_1\mathbf{v}_1 + \lambda_2\alpha_2\mathbf{v}_2 + \cdots + \lambda_n\alpha_n\mathbf{v}_n. \end{aligned} \tag{1}$$

This equation makes it clear that the eigenvectors determine the important directions of the matrix $A$; $A$ sends each $\mathbf{v}_i$ onto a multiple of itself: $A\mathbf{v}_i = \lambda_i\mathbf{v}_i$. Equation (1) also makes it clear that the eigenvalue $\lambda_i$ determines the relative importance of the corresponding eigenvector $\mathbf{v}_i$ because the coefficient $\alpha_i$ of $\mathbf{v}_i$ is multiplied by the corresponding eigenvalue $\lambda_i$. For instance, suppose that $A$ is a $2 \times 2$ matrix and that $\mathbf{v}_1, \mathbf{v}_2$ are independent eigenvectors of $A$. If the corresponding eigenvalues are $\lambda_1 = 10$ and $\lambda_2 = 2$ and if

$$\mathbf{x} = \alpha_1\mathbf{v}_1 + \alpha_2\mathbf{v}_2,$$

then

$$A\mathbf{x} = 10\alpha_1\mathbf{v}_1 + 2\alpha_2\mathbf{v}_2.$$

Thus the first coordinate of $A\mathbf{x}$ with respect to the basis $\mathbf{v}_1, \mathbf{v}_2$ is 10 times the first coordinate of $\mathbf{x}$ with respect to this basis. On the other hand, the second coordinate of $A\mathbf{x}$ is just 2 times the second coordinate of $\mathbf{x}$. The eigenvector $\mathbf{v}_1$ has much more influence on $A\mathbf{x}$ than the eigenvector $\mathbf{v}_2$ (see Figure 8.1).

Now for the interpretation of (1) in terms of matrices. Equation (1),

$$A\mathbf{x} = \lambda_1\alpha_1\mathbf{v}_1 + \lambda_2\alpha_2\mathbf{v}_2 + \cdots + \lambda_n\alpha_n\mathbf{v}_n,$$

is clearly equivalent to the matrix equation

$$A\mathbf{x} = \begin{bmatrix} | & | & & | \\ \mathbf{v}_1 & \mathbf{v}_2 & \cdots & \mathbf{v}_n \\ | & | & & | \end{bmatrix} \begin{bmatrix} \lambda_1 & & & \\ & \lambda_2 & & \\ & & \ddots & \\ & & & \lambda_n \end{bmatrix} \begin{bmatrix} \alpha_1 \\ \alpha_2 \\ \vdots \\ \alpha_n \end{bmatrix}. \tag{2}$$

Let

$$M = \begin{bmatrix} | & | & & | \\ \mathbf{v}_1 & \mathbf{v}_2 & \cdots & \mathbf{v}_n \\ | & | & & | \end{bmatrix} \quad \text{and} \quad \Lambda = \begin{bmatrix} \lambda_1 & & & \\ & \lambda_2 & & \\ & & \ddots & \\ & & & \lambda_n \end{bmatrix}.$$

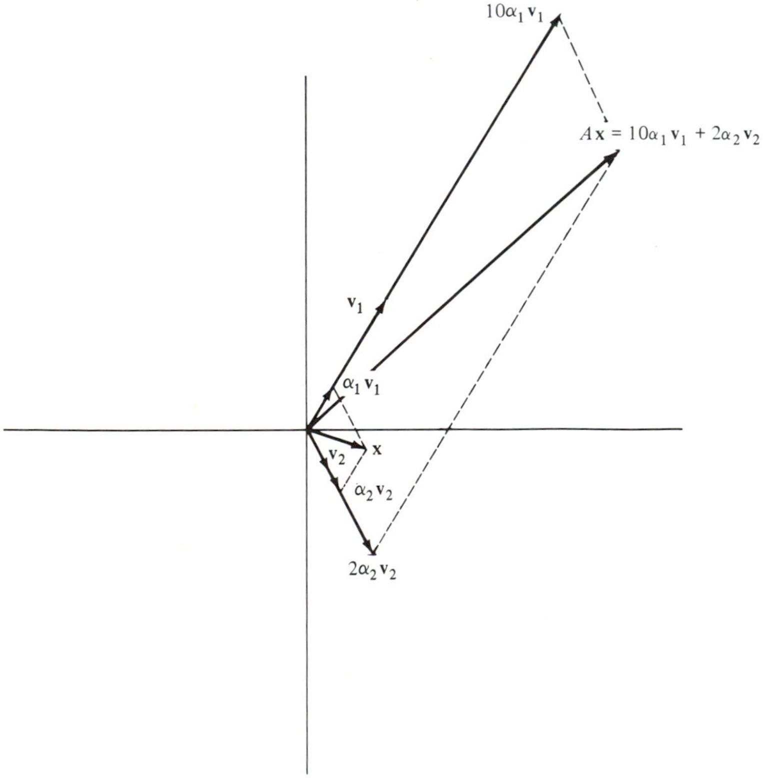


Figure 8.1

Since $\mathbf{x} = \alpha_1\mathbf{v}_1 + \alpha_2\mathbf{v}_2 + \cdots + \alpha_n\mathbf{v}_n$,

$$\mathbf{x} = M\begin{bmatrix}\alpha_1\\ \alpha_2\\ \vdots\\ \alpha_n\end{bmatrix} \qquad \text{so that} \qquad \begin{bmatrix}\alpha_1\\ \alpha_2\\ \vdots\\ \alpha_n\end{bmatrix} = M^{-1}\mathbf{x}.$$

Consequently, equation (2) becomes

$$A\mathbf{x} = M\Lambda M^{-1}\mathbf{x}$$

and since this equation holds for all $\mathbf{x}$ in $R^n$, it follows from Result 1 in Section 1.5 that

$$A = M\Lambda M^{-1} \qquad \text{or} \qquad \Lambda = M^{-1}AM.$$

***Definition*** A square matrix $A$ is said to be **diagonalizable** if there is a nonsingular matrix $M$ such that

$$\Lambda = M^{-1}AM$$

is a diagonal matrix. The matrix $M$ is called a **diagonalizing matrix** for $A$.

We have proved that if $A$ is an $n \times n$ matrix that has $n$ independent eigenvectors, then $A$ is diagonalizable. Actually, we have proved much more. If $\mathbf{v}_1, \mathbf{v}_2, \ldots, \mathbf{v}_n$ are independent eigenvectors of $A$ and $\lambda_1, \lambda_2, \ldots, \lambda_n$ are associated eigenvalues, then:

1. The matrix $M$ whose columns are these independent eigenvectors is a diagonalizing matrix for $A$.
2. The diagonal matrix $\Lambda = M^{-1}AM$ has the eigenvalues of $A$ as its diagonal entries, the $i$th diagonal entry being an eigenvalue for the eigenvector in the $i$th column of $M$.

Is the converse of this result true? That is, if $A$ is an $n \times n$ diagonalizable matrix, must $A$ have $n$ independent eigenvectors? Yes! For suppose that $A$ is an $n \times n$ diagonalizable matrix. Then there is a nonsingular matrix $M$ such that $\Lambda = M^{-1}AM$ is a diagonal matrix. Thus

$$AM = M\Lambda,$$

where $M$ is a nonsingular matrix and $\Lambda$ is a diagonal matrix. Let

$$M = \begin{bmatrix} | & & | \\ \mathbf{w}_1 & \cdots & \mathbf{w}_n \\ | & & | \end{bmatrix} \qquad \text{and} \qquad \Lambda = \begin{bmatrix} \mu_1 & & \\ & \ddots & \\ & & \mu_n \end{bmatrix}.$$

Then

$$AM = \begin{bmatrix} | & & | \\ A\mathbf{w}_1 & \cdots & A\mathbf{w}_n \\ | & & | \end{bmatrix}$$

and

$$M\Lambda = \begin{bmatrix} | & & | \\ \mathbf{w}_1 & \cdots & \mathbf{w}_n \\ | & & | \end{bmatrix} \begin{bmatrix} \mu_1 & & \\ & \ddots & \\ & & \mu_n \end{bmatrix} = \begin{bmatrix} | & & | \\ \mu_1\mathbf{w}_1 & \cdots & \mu_n\mathbf{w}_n \\ | & & | \end{bmatrix}.$$

Since $AM = M\Lambda$, we obtain

$$A\mathbf{w}_1 = \mu_1\mathbf{w}_1, \ldots, A\mathbf{w}_n = \mu_n\mathbf{w}_n.$$

Because $M$ is nonsingular, its columns $\mathbf{w}_1, \ldots, \mathbf{w}_n$ are independent and nonzero. Thus the vectors $\mathbf{w}_1, \ldots, \mathbf{w}_n$ are independent eigenvectors of $A$ and $\mu_1, \ldots, \mu_n$ are the associated eigenvalues.

***Theorem 1*** *An $n \times n$ matrix $A$ is diagonalizable if and only if it has $n$ independent eigenvectors. Any diagonalizing matrix $M$ has independent eigenvectors of $A$ as its columns and the diagonal matrix $\Lambda$ has the eigenvalues of $A$ as its diagonal entries, the* ith *diagonal entry of $\Lambda$ being the eigenvalue for the eigenvector in the* ith *column of $M$.*

## EXAMPLE 1

Let $A$ be the matrix

$$A = \begin{bmatrix} 1 & 0 & 1 \\ 0 & 1 & 0 \\ 1 & 2 & 1 \end{bmatrix}.$$

In Example 1 of Section 8.2 we found that 0, 1, and 2 are the eigenvalues of $A$ and that the eigenspaces $E(0)$, $E(1)$, and $E(2)$ are spanned by the vectors $(-1, 0, 1)$, $(-2, 1, 0)$, and $(1, 0, 1)$, respectively. These eigenvectors are independent (verify!). Thus $A$ is a diagonalizable matrix and if

$$M = \begin{bmatrix} -1 & -2 & 1 \\ 0 & 1 & 0 \\ 1 & 0 & 1 \end{bmatrix}, \quad \text{then} \quad \Lambda = \begin{bmatrix} 0 & & \\ & 1 & \\ & & 2 \end{bmatrix}.$$

Let us verify for this example that $A = M\Lambda M^{-1}$. Since

$$M^{-1} = \frac{1}{2}\begin{bmatrix} -1 & -2 & 1 \\ 0 & 2 & 0 \\ 1 & 2 & 1 \end{bmatrix},$$

$$\begin{aligned} M\Lambda M^{-1} &= \frac{1}{2}\begin{bmatrix} -1 & -2 & 1 \\ 0 & 1 & 0 \\ 1 & 0 & 1 \end{bmatrix}\begin{bmatrix} 0 & & \\ & 1 & \\ & & 2 \end{bmatrix}\begin{bmatrix} -1 & -2 & 1 \\ 0 & 2 & 0 \\ 1 & 2 & 1 \end{bmatrix} \\ &= \frac{1}{2}\begin{bmatrix} 0 & -2 & 2 \\ 0 & 1 & 0 \\ 0 & 0 & 2 \end{bmatrix}\begin{bmatrix} -1 & -2 & 1 \\ 0 & 2 & 0 \\ 1 & 2 & 1 \end{bmatrix} = \begin{bmatrix} 1 & 0 & 1 \\ 0 & 1 & 0 \\ 1 & 2 & 1 \end{bmatrix} \\ &= A. \end{aligned}$$

## EXAMPLE 2

In Example 2 of Section 8.2 we considered the matrix

$$A = \begin{bmatrix} 2 & -5 & 5 \\ 0 & 3 & -1 \\ 0 & -1 & 3 \end{bmatrix}.$$

We found that $A$ has two eigenvalues $\lambda = 2$ and $\lambda = 4$. The eigenvectors $(0, 1, 1)$ and$(1, 0, 0)$ form a basis for the eigenspace $E(2)$ and the eigenvector $(5, -1, 1)$ forms a basis for the eigenspace $E(4)$. Since the vectors $(0, 1, 1)$, $(1, 0, 0)$, and $(5, -1, 1)$ are independent (verify!), the matrix $A$ is diagonalizable. If $M$ is the matrix

$$M = \begin{bmatrix} 0 & 1 & 5 \\ 1 & 0 & -1 \\ 1 & 0 & 1 \end{bmatrix}, \qquad \text{then} \qquad \Lambda = \begin{bmatrix} 2 & & \\ & 2 & \\ & & 4 \end{bmatrix}.$$

## EXAMPLE 3

Consider now the matrix

$$A = \begin{bmatrix} -1 & 1 & 0 \\ 0 & 5 & 0 \\ 4 & -2 & 5 \end{bmatrix}$$

of Example 3 in 8.2. The eigenvalues of $A$ are $\lambda = -1$ and $\lambda = 5$ and both of the eigenspaces $E(-1)$ and $E(5)$ are one-dimensional. Therefore, it is impossible for $A$ to have three independent eigenvectors, which means that $A$ is not diagonalizable.

A natural question to ask at this point is whether there are any "nice" conditions which guarantee that an $n \times n$ matrix $A$ has $n$ independent eigenvectors. We begin to answer this question with the following theorem.

***Theorem 2*** *Let $A$ be an $n \times n$ matrix. If $A$ has $n$ distinct eigenvalues, then $A$ has $n$ independent eigenvectors. Hence $A$ is diagonalizable.*

This theorem follows immediately from the following more general theorem.

***Theorem 3*** *Let $A$ be an $n \times n$ matrix and let $\lambda_1, \ldots, \lambda_r$ be distinct eigenvalues of $A$ with associated eigenvectors $\mathbf{v}_1, \ldots, \mathbf{v}_r$. Then $\mathbf{v}_1, \ldots, \mathbf{v}_r$ are independent eigenvectors.*

In other words, eigenvectors associated with distinct eigenvalues are independent. (If the zero vector were allowed to be an eigenvector, this would not be true. Why?) However, the converse is not true! *An $n \times n$ matrix may have $n$ independent eigenvectors without having $n$ distinct eigenvalues* (see Example 2). In the extreme, the identity matrix has $n$ independent eigenvectors $\mathbf{e}_1, \mathbf{e}_2, \ldots, \mathbf{e}_n$ but has only one eigenvalue $\lambda = 1$.

**Proof** First we prove that $\mathbf{v}_1$ and $\mathbf{v}_2$ are independent. Suppose that there are scalars $\alpha_1$ and $\alpha_2$ such that

$$\alpha_1\mathbf{v}_1 + \alpha_2\mathbf{v}_2 = \mathbf{0}. \tag{3}$$

Multiplying by $A$ we obtain (since $A\mathbf{v}_i = \lambda_i\mathbf{v}_i$, $i = 1, 2$)

$$\alpha_1\lambda_1\mathbf{v}_1 + \alpha_2\lambda_2\mathbf{v}_2 = \mathbf{0}.$$

Now subtract $\lambda_2$ times the previous equation from this equation to obtain

$$\alpha_1(\lambda_1 - \lambda_2)\mathbf{v}_1 = \mathbf{0}.$$

Since $\mathbf{v}_1$ is an eigenvector, $\mathbf{v}_1 \neq \mathbf{0}$. Hence $\alpha_1(\lambda_1 - \lambda_2) = 0$ and because $\lambda_1 \neq \lambda_2$ we must have $\alpha_1 = 0$. From (3) we now find that $\alpha_2 = 0$ since $\mathbf{v}_2 \neq \mathbf{0}$. Therefore, $\alpha_1 = \alpha_2 = 0$ and $\mathbf{v}_1$ and $\mathbf{v}_2$ are independent.

Next we show that $\mathbf{v}_1$, $\mathbf{v}_2$, and $\mathbf{v}_3$ are independent. Suppose that there are scalars $\alpha_1$, $\alpha_2$, and $\alpha_3$ such that

$$\alpha_1\mathbf{v}_1 + \alpha_2\mathbf{v}_2 + \alpha_3\mathbf{v}_3 = \mathbf{0}. \tag{4}$$

As above, we multiply this equation by $A$ to obtain

$$\alpha_1\lambda_1\mathbf{v}_1 + \alpha_2\lambda_2\mathbf{v}_2 + \alpha_3\lambda_3\mathbf{v}_3 = \mathbf{0}.$$

Now subtract $\lambda_3$ times equation (4) from this equation and obtain

$$\alpha_1(\lambda_1 - \lambda_3)\mathbf{v}_1 + \alpha_2(\lambda_2 - \lambda_3)\mathbf{v}_2 = \mathbf{0}.$$

Because $\mathbf{v}_1$ and $\mathbf{v}_2$ are independent it follows that

$$\alpha_1(\lambda_1 - \lambda_3) = \alpha_2(\lambda_2 - \lambda_3) = 0.$$

Since the $\lambda$'s are distinct, $\alpha_1 = \alpha_2 = 0$. From (4) we now find that $\alpha_3 = 0$. Thus $\mathbf{v}_1$, $\mathbf{v}_2$, and $\mathbf{v}_3$ are independent. Repeating this argument proves the result.

## EXAMPLE 4

Since the matrix $A$ in Example 1 is a $3 \times 3$ matrix with three distinct eigenvalues, Theorem 2 guarantees that it is diagonalizable. However, we cannot conclude anything from this theorem about the diagonalizability of the matrices in Example 2 or 3.

## EXAMPLE 5

When $A$ is diagonalizable, finding powers of $A$ is simple. Since $\Lambda = M^{-1}AM$,

$$\begin{aligned} A &= M\Lambda M^{-1}, \\ A^2 &= (M\Lambda M^{-1})(M\Lambda M^{-1}) = M\Lambda^2 M^{-1}, \\ A^3 &= AA^2 = (M\Lambda M^{-1})(M\Lambda^2 M^{-1}) = M\Lambda^3 M^{-1}. \end{aligned}$$

It is clear that by repeating this process we obtain

$$A^k = M\Lambda^k M^{-1}.$$

For instance, in Example 1 we found that the matrix

$$A = \begin{bmatrix} 1 & 0 & 1 \\ 0 & 1 & 0 \\ 1 & 2 & 1 \end{bmatrix}$$

can be written as

$$A = \begin{bmatrix} -1 & -2 & 1 \\ 0 & 1 & 0 \\ 1 & 0 & 1 \end{bmatrix} \begin{bmatrix} 0 & & \\ & 1 & \\ & & 2 \end{bmatrix} \begin{bmatrix} -\frac{1}{2} & -1 & \frac{1}{2} \\ 0 & 1 & 0 \\ \frac{1}{2} & 1 & \frac{1}{2} \end{bmatrix}$$
$$= M\Lambda M^{-1}.$$

Thus for $m \geq 1$

$$A^m = \begin{bmatrix} -1 & -2 & 1 \\ 0 & 1 & 0 \\ 1 & 0 & 1 \end{bmatrix} \begin{bmatrix} 0 & & \\ & 1 & \\ & & 2^m \end{bmatrix} \begin{bmatrix} -\frac{1}{2} & -1 & \frac{1}{2} \\ 0 & 1 & 0 \\ \frac{1}{2} & 1 & \frac{1}{2} \end{bmatrix}$$
$$= \begin{bmatrix} 2^{m-1} & 2^m - 2 & 2^{m-1} \\ 0 & 1 & 0 \\ 2^{m-1} & 2^m & 2^{m-1} \end{bmatrix}.$$

# PROBLEMS 8.3

*1. Determine which of the matrices in Problem 1 of Section 8.2 are diagonalizable. For each matrix $A$ that is diagonalizable, find a nonsingular matrix $M$ and a diagonal matrix $\Lambda$ such that $\Lambda = M^{-1}AM$.

*2. Find a $2 \times 2$ matrix $A$ that has an eigenvalue $\lambda = 1$ with associated eigenvector $(1, 3)$ and an eigenvalue $\lambda = -1$ with associated eigenvector $(1, 4)$.

3. Find a $3 \times 3$ matrix $A$ such that $1, -1, 0$ are its eigenvalues and $(1, -1, 1), (1, 1, 0), (1, -1, 0)$ are its corresponding eigenvectors.

4. Find a matrix with eigenvalues 0, 1, and 3 and corresponding eigenvectors $(1, 0, 0)$, $(1, 1, 0)$, and $(1, 1, 1)$.

*5. Is the converse of Theorem 3 true?

6. Prove Theorem 2 using Theorem 3.

*7. If $A$ is diagonalizable, is the diagonalizing matrix unique? If so, why? If not, how are the diagonalizing matrices related (if at all)?

*8. If $A$ is diagonalizable, is the diagonal matrix unique? If so, why? If not, how are the diagonal matrices related (if at all)?

*9. Suppose that $A$ is an $n \times n$ matrix with exactly one eigenvalue $\lambda$ with multiplicity $n$. Show that if $A$ is diagonalizable, then $A$ is a diagonal matrix. Find $A$.

10. Let $A$ be diagonalizable. Explain why the rank of $A$ is equal to the number of nonzero eigenvalues of $A$.

11. Show that if $A$ and $B$ are diagonalizable and both have the same eigenvectors, then $AB = BA$. That is, $A$ and $B$ must commute.

**12.** Show that if $A$ has distinct eigenvalues, and $B$ is a matrix that commutes with $A$, then $A$ and $B$ have the same eigenvectors. (*Hint:* Let $\mathbf{v}$ be an eigenvector of $A$ and $\lambda$ be an associated eigenvalue. Then $AB\mathbf{v} = BA\mathbf{v} = B\lambda\mathbf{v} = \lambda B\mathbf{v}$. Now use the fact that the eigenspaces are one-dimensional.)

**13.** Show that if $A$ has distinct eigenvalues and $B$ is a matrix that commutes with $A$, then $B$ is diagonalizable.

**14.** Let $A$ be diagonalizable and $B$ be a nonsingular matrix that commutes with $A$. Show that if $M$ is a diagonalizing matrix for $A$, then $BM$ is also.

***15.** Find all $2 \times 2$ matrices $A$ satisfying *all* of the following three conditions:

1. $A\begin{bmatrix}1\\2\end{bmatrix}$ is a multiple of $\begin{bmatrix}1\\2\end{bmatrix}$.
2. $A\begin{bmatrix}3\\5\end{bmatrix}$ is a multiple of $\begin{bmatrix}3\\5\end{bmatrix}$.
3. $A\begin{bmatrix}7\\2\end{bmatrix}$ is a multiple of $\begin{bmatrix}7\\2\end{bmatrix}$.

Support your answer. (*Hint:* Show that $A$ is diagonalizable)

***16.** Let $\mathbf{v}_1 = (1, 2)$ and $\mathbf{v}_2 = (3, 5)$. Find all $2 \times 2$ matrices $A$ such that $\mathbf{v}_1$ and $\mathbf{v}_2$ are the eigenvectors of $A$.

***17.** In Problem 7 in Section 8.2 you were asked to prove that similar matrices have the same eigenvalues. Is the converse true? That is, if two matrices have the same eigenvalues each with the same multiplicity, are they necessarily similar?

***18.** If $A$ is $n \times n$ and if $A$ is similar to a diagonal matrix, must it follow that $A$ has $n$ independent eigenvectors?

***19.** Are the matrices

$$\begin{bmatrix}1 & & \\ & 5 & \\ & & -2\end{bmatrix}$$

and

$$\begin{bmatrix}5 & & \\ & 1 & \\ & & -2\end{bmatrix}$$

similar?

**20.** Show that any two diagonal matrices which have the same diagonal entries (counting multiplicity) are similar.

**21.** If $A$ and $B$ are diagonalizable and have the same eigenvalues, show that $A$ and $B$ are similar matrices.

***22.** What matrices are similar to the identity matrix?

***23.** How are the eigenvectors of similar matrices related?

**24.** Show that if $A$ and $B$ are similar, then $\operatorname{tr} A = \operatorname{tr} B$. (*Hint:* Use Problem 16 of Section 1.6.)

**25.** Let $A$ and $B$ be $n \times n$ matrices. Suppose that one of them, say $A$, is nonsingular. Show that $AB$ and $BA$ are similar and hence have the same eigenvalues.

**26.** Show that if $A$ is similar to $B$, then $A^T$ is similar to $B^T$.

**27.** Show that if $A$ is similar to $B$ and $B$ is diagonalizable, then $A$ is diagonalizable.

**28.** Find all $3 \times 3$ matrices that are diagonalizable and whose only eigenvalue is 5.

**29.** Can an idempotent matrix be diagonalized? (*Hint:* Use Problem 17 of Section 8.1.)

**30.** Show that if $A$ is diagonalizable, then the eigenvalues of $A^2$ are exactly the squares of the eigenvalues of $A$ and that $A$ and $A^2$ have the same eigenvectors.

**31.** If $A \neq 0$ and $A$ is nilpotent, is $A$ similar to a diagonal matrix? In other words, can $A$ be diagonalized? (*Hint:* Use Problem 18 of Section 8.1.)

# 8.4 Applications

## Predicting Population (One Species)

When one studies a particular animal species (e.g., bacteria, fish, deer, human beings) one frequently wishes to make reasonable predictions about the future population of that species. All changes in the population of an isolated species are caused by births and deaths. It is not too difficult to determine the present birth rate and the present death rate of the species. But it is clearly impossible to determine the rates of births and deaths in the future. Therefore, if we wish to make predictions about the future population we are forced to make some reasonable assumptions about these rates in the future. We will assume that the future rates of births and deaths will be the same as the present rates. (This may or may not be true. If it is true, our predictions will be correct. If not, our predictions may be in error.) Let us express these ideas in appropriate symbols.

Choose some basic time interval as the unit of time. This may be 1 year, 1 minute, 3 days, and so on, depending on the species being studied. Let $p^{(0)}$ be the initial population and $p^{(k)}$ be the population after $k$ units of time have passed. Let $i$ be the rate of change (increase if positive, decrease if negative) of the population over one unit of time. (We can interpret $i$ as the birth rate minus the death rate. It is normally positive unless the species is dying out.) Then we have the following **difference equation**:

$$p^{(k+1)} - p^{(k)} = ip^{(k)} \tag{1}$$

or, equivalently,

$$p^{(k+1)} = (1 + i)p^{(k)}. \tag{2}$$

The two sides of equation (1) are the two different ways to compute the change in population during the $(k + 1)$st time interval. Equations (1) and (2) hold when $k$ is a nonnegative integer.

It is possible to obtain a simple formula for $p^{(k)}$. We know that $p^{(k)} = (1 + i)p^{(k-1)}$. But we also know that $p^{(k-1)} = (1 + i)p^{(k-2)}$ because (2) holds for every positive integer $k$. These two equalities can be combined to give $p^{(k)} = (1 + i)^2p^{(k-2)}$. Continuing this process, we eventually conclude that

$$p^{(k)} = (1 + i)^k p^{(0)}. \tag{3}$$

We have now accomplished our goal. Given the present population $p^{(0)}$ of a species and the (assumed constant) rate of change $i$ of the species, we can predict the population $p^{(k)}$ after $k$ time intervals have elapsed.

## Predicting Population (More Than One Species)

The ideas discussed above can also be used to predict the future populations of two or more species. If the species are independent (i.e., if they do not interact in any way), then there is nothing new to be done. But the possibility of interaction (e.g., competition, predation, symbiosis, scavenging) creates the need for a more complex model of population growth. Again, let us express our ideas in appropriate symbols for the case of two different species.

Let $p_1^{(k)}$ and $p_2^{(k)}$ be the population of species 1 and 2 after $k$ time units have passed. Under the same assumptions made above, we would expect $p_1^{(k+1)} - p_1^{(k)} = ip_1^{(k)}$ for some rate of change $i$. But we are considering the case of interacting species, and the presence of the second species must also influence the population of the first. We will assume that this added influence is proportional to the population of the second species. (Of course, this assumption may not always be valid. When it is not, our predictions may be in error.) We are thus assuming that future changes in the population of the first species satisfy the equation

$$p_1^{(k+1)} - p_1^{(k)} = ip_1^{(k)} + bp_2^{(k)}$$

for some numbers $i$ and $b$. A similar equation holds for the population of the second species. Changing notation so that $a_{ij}$ represents the rate of change of species $i$ due to the presence of species $j$, we have the following equations:

$$\begin{aligned} p_1^{(k+1)} - p_1^{(k)} &= a_{11}p_1^{(k)} + a_{12}p_2^{(k)} \\ p_2^{(k+1)} - p_2^{(k)} &= a_{21}p_1^{(k)} + a_{22}p_2^{(k)}. \end{aligned} \tag{4}$$

If $\mathbf{p}^{(k)}$ is the vector $(p_1^{(k)}, p_2^{(k)})$ and $A$ is the matrix $[a_{ij}]$, then (4) can be written as the **difference equation**

$$\mathbf{p}^{(k+1)} - \mathbf{p}^{(k)} = A\mathbf{p}^{(k)}$$

or, equivalently,

$$\mathbf{p}^{(k+1)} = (I + A)\mathbf{p}^{(k)}, \tag{5}$$

where $I$ is the identity matrix.

This matrix equation is the matrix analogue of equation (2) and can be solved in a similar fashion, giving

$$\mathbf{p}^{(k)} = (I + A)^k\mathbf{p}^{(0)}. \tag{6}$$

We will refer to $\mathbf{p}^{(k)}$ as the population vector and $A$ as the growth matrix.

### EXAMPLE 1

Suppose that we have two species, the first of which is increasing and is not influenced by the second, and the second of which scavenges on the dead of the first. If the scavenging animal cannot live except on the dead of the other species, then the entries in the growth matrix $A$ would be as follows:

$a_{11} > 0$: The first species is increasing.
$a_{12} = 0$: The second species does not influence the first.

$a_{21} > 0$: The second scavenges on the first.
$a_{22} < 0$: The second species will die out if it cannot scavenge upon the first.

For instance, suppose that $A$ is the matrix

$$A = \begin{bmatrix} 0.2 & 0 \\ 0.4 & -1.2 \end{bmatrix}.$$

To find the population after $k$ time intervals requires us to compute

$$(I + A)^k.$$

The eigenvalues of the matrix

$$I + A = \begin{bmatrix} 1.2 & 0 \\ 0.4 & -0.2 \end{bmatrix}$$

are $\lambda = 1.2$ and $\lambda = -0.2$. The corresponding eigenvectors are $\mathbf{v}_1 = (7, 2)$ and $\mathbf{v}_2 = (0, 1)$. Thus $I + A$ is diagonalizable and

$$\begin{aligned} I + A &= M\Lambda M^{-1} \\ &= \begin{bmatrix} 7 & 0 \\ 2 & 1 \end{bmatrix}\begin{bmatrix} 1.2 & \\ & -0.2 \end{bmatrix}\frac{1}{7}\begin{bmatrix} 1 & 0 \\ -2 & 7 \end{bmatrix} \end{aligned}$$

Thus

$$\mathbf{p}^{(k)} = (I + A)^k \mathbf{p}^{(0)} = M\Lambda^k M^{-1}\mathbf{p}^{(0)}.$$

Setting

$$\begin{bmatrix} \alpha \\ \beta \end{bmatrix} = M^{-1}\mathbf{p}^{(0)} = M^{-1}\begin{bmatrix} p_1^{(0)} \\ p_2^{(0)} \end{bmatrix},$$

we obtain

$$\mathbf{p}^{(k)} = M\Lambda^k \begin{bmatrix} \alpha \\ \beta \end{bmatrix} = \alpha\lambda_1^k \mathbf{v}_1 + \beta\lambda_2^k \mathbf{v}_2.$$

For large $k$, $\lambda_2^k = (-0.2)^k$ is very close to zero. Hence the vector $\mathbf{p}^{(k)}$ is very close to the vector $\alpha\lambda_1^k\mathbf{v}_1$. Thus for large $k$ the population behaves like $\alpha\lambda_1^k\mathbf{v}_1$. Both species grow without bound and the ratio $p_1^{(k)}/p_2^{(k)}$ tends to $\frac{7}{2}$. (Notice that the limiting ratio is independent of the initial population distribution.)

## Predicting the Age Distribution of a Population

There are situations where we are interested in the numbers of animals alive in certain age groups. For example, to determine appropriate Social Security tax levels, there must be some predictions of the age distribution of the U.S. population. Let us develop a model (the **Leslie model**) to make such predictions.

Let the basic unit of time be 5 years. We wish to estimate the size of each age group of the population a certain number of time units in the future. Define variables as follows:

$p_1^{(k)}$: Number of people 0–4 years old after $k$ intervals.
$p_2^{(k)}$: Number of people 5–9 years old after $k$ intervals.
$p_3^{(k)}$: Number of people 10–14 years old after $k$ intervals.
$\vdots$
$p_{20}^{(k)}$: Number of people 95–99 years old after $k$ intervals.

(We assume that the number of persons 100 and over is negligible.) Let us relate the population at a certain time to the population 5 years later. A certain fraction of people in the $i$th age group will survive to become members of the $(i + 1)$st age group. In addition, the people in the $i$th age group have been responsible for births in proportion to their numbers. Let $s_i$ be the fraction of the $i$th age group that survives after 5 years and $b_i$ be the birth rate for the $i$th age group. If we assume that future births and deaths will continue in the same proportion, then we have the following relations:

$$\begin{aligned} p_{i+1}^{(k+1)} &= s_i p_i^{(k)}, \qquad i = 1,\ldots,19 \\ p_1^{(k+1)} &= b_1 p_1^{(k)} + b_2 p_2^{(k)} + \cdots + b_{20} p_{20}^{(k)}. \end{aligned} \tag{7}$$

The values for $b_1,\ldots,b_{20}$ and $s_1,\ldots,s_{19}$ will of course be obtained from studies of the present behavior of the population.

The system of equations (7) can be rewritten as a matrix equation. If $A$ is the matrix

$$A = \begin{bmatrix} b_1 & b_2 & b_3 & \cdots & b_{19} & b_{20} \\ s_1 & 0 & 0 & \cdots & 0 & 0 \\ 0 & s_2 & 0 & \cdots & 0 & 0 \\ 0 & 0 & s_3 & \cdots & 0 & 0 \\ & \vdots & & & \vdots & \\ 0 & 0 & 0 & \cdots & s_{19} & 0 \end{bmatrix}$$

and $\mathbf{p}^{(k)}$ the vector $(p_1^{(k)}, p_2^{(k)},\ldots,p_{20}^{(k)})$, then (7) becomes

$$\mathbf{p}^{(k+1)} = A\mathbf{p}^{(k)}.$$

This equation has a solution

$$\mathbf{p}^{(k)} = A^k\mathbf{p}^{(0)}.$$

Again we see that a complete solution of our problem is obtained by computing powers of a matrix.

In the Leslie model for predicting population distributions, the basic time unit and the time span of each population group must be equal. By choosing age groups that span a time interval equal to our basic unit of time, we have succeeded in obtaining a special kind of matrix $A$. Matrices of this form are called **Leslie matrices** after P. H. Leslie, who was one of the first to investigate this method for predicting population distributions.

This model was developed to study the age distribution of a population. But there are other questions that can be investigated once we know the Leslie matrix, $A$, for the population. Is there a distribution of the population that is stable under the birth and death processes described by the matrix $A$? (In symbols, does there exist a vector $\mathbf{p}$ such that $A\mathbf{p} = \lambda\mathbf{p}$ for some constant $\lambda$?) What is the natural rate of growth of this population? How can we determine this growth rate from the matrix $A$? Is this population likely to grow in a stable manner, approaching some fixed age distribution, or will there be wild fluctuations in the age distribution?

## EXAMPLE 2

In this example we apply the process of diagonalization to the problem of predicting the age distribution of a certain fish population. This population has been divided into three age groups, the group of fish in their first year of life, their second year of life, and their third year of life, respectively. Each year $\frac{1}{32}$ of the fish in the first group survive to become members of the second group. The fish in the second group have an average of 26 offspring per individual and $\frac{1}{5}$ of the fish in the second group survive to become members of the third group. The fish in the third group have an average of 30 offspring per individual and then die. Let us denote the number of fish in the first, second, and third groups after $k$ years by $a_k$, $b_k$, and $c_k$, respectively. Let $N_k = a_k + b_k + c_k$ be the total number of fish after $k$ years. Then $a_{k+1} = 26b_k + 30c_k$, $b_{k+1} = a_k/32$, and $c_{k+1} = b_k/5$. Setting $\mathbf{x}_k = (a_k, b_k, c_k)$ and

$$A = \begin{bmatrix} 0 & 26 & 30 \\ \frac{1}{32} & 0 & 0 \\ 0 & \frac{1}{5} & 0 \end{bmatrix},$$

we obtain the difference equation

$$\mathbf{x}_{k+1} = A\mathbf{x}_k, \qquad k = 0, 1, 2, \ldots, \tag{8}$$

where $\mathbf{x}_0 = (a_0, b_0, c_0)$ is the initial age distribution of the population. The matrix $A$ is the Leslie matrix for the population. The solution to (8) is

$$\mathbf{x}_k = A^k\mathbf{x}_0, \qquad k = 0, 1, 2, \ldots.$$

We are interested in the long-range behavior of the population (i.e., in $\mathbf{x}_k$ as $k$ becomes very large). The characteristic equation of $A$ is

$$0 = \det(A - \lambda I) = -\lambda^3 + \tfrac{13}{16}\lambda + \tfrac{3}{16}.$$

This equation has three distinct solutions: $\lambda_1 = 1$, $\lambda_2 = -\frac{1}{4}$, and $\lambda_3 = -\frac{3}{4}$. Hence $A$ is diagonalizable. The vectors $\mathbf{v}_1 = (160, 5, 1)$, $\mathbf{v}_2 = (40, -5, 4)$, and $\mathbf{v}_3 = (360, -15, 4)$ are eigenvectors associated with the eigenvalues 1, $-\frac{1}{4}$, and $-\frac{3}{4}$, respectively. Thus $A = M\Lambda M^{-1}$, where

$$M = \begin{bmatrix} 160 & 40 & 360 \\ 5 & -5 & -15 \\ 1 & 4 & 4 \end{bmatrix} \quad \text{and} \quad \Lambda = \begin{bmatrix} 1 & 0 & 0 \\ 0 & -\frac{1}{4} & 0 \\ 0 & 0 & -\frac{3}{4} \end{bmatrix}.$$

This yields $\mathbf{x}_k = A^k\mathbf{x}_0 = M\Lambda^k M^{-1}\mathbf{x}_0$. Setting

$$\begin{bmatrix}\alpha\\ \beta\\ \gamma\end{bmatrix} = M^{-1}\mathbf{x}_0 = M^{-1}\begin{bmatrix}a_0\\ b_0\\ c_0\end{bmatrix},$$

we obtain

$$\mathbf{x}_k = M\Lambda^k\begin{bmatrix}\alpha\\ \beta\\ \gamma\end{bmatrix} = M\begin{bmatrix}\alpha\\ (-\frac{1}{4})^k\beta\\ (-\frac{3}{4})^k\gamma\end{bmatrix} = \alpha\mathbf{v}_1 + (-\tfrac{1}{4})^k\beta\mathbf{v}_2 + (-\tfrac{3}{4})^k\gamma\mathbf{v}_3.$$

For large $k$, $(-\frac{1}{4})^k$ and $(-\frac{3}{4})^k$ are very close to zero. Therefore, $\mathbf{x}_k$ approaches

$$\mathbf{x}_\infty = \alpha\mathbf{v}_1 = \alpha(160, 5, 1)$$

and the number of fish $N_k$ approaches

$$N_\infty = \alpha(160 + 5 + 1) = 166\alpha.$$

If we solve

$$M\begin{bmatrix}\alpha\\ \beta\\ \gamma\end{bmatrix} = \begin{bmatrix}a_0\\ b_0\\ c_0\end{bmatrix}$$

we obtain $\alpha = (a_0 + 32b_0 + 30c_0)/350$. We conclude that in the long run the population approaches a stable state. In this state the total number of fish is $166\alpha$ and of these fish about 96.4% belong to the first group, 3.0% belong to the second group, and 0.6% belong to the third group. Notice that in contrast to the number $N_\infty = 166\alpha$, these percentages do not depend upon the initial values $a_0$, $b_0$, and $c_0$.

## Markov Processes

Suppose that we are studying the geographical distribution of the U.S. population. What we are interested in are the population shifts between different regions. Let us divide the United States into two regions: the Sun Belt and everywhere else. Suppose that the population movement between these two regions has been studied over several years and the following figures determined: percentage of Sun Belt population moving out, 15% per year; percentage of other population moving into Sun Belt, 20% per year. (These figures were invented by the authors and have no connection with reality.) Let the Sun Belt be region 1 and everywhere else be region 2. In addition, let $f_1^{(k)}$ ($f_2^{(k)}$) be the fraction of the population living in region 1 (region 2) after $k$ years. [Note that $f_1^{(k)} + f_2^{(k)} = 1$ for all $k \geq 0$.] Then the following equation holds for all $k \geq 0$:

$$\begin{bmatrix}f_1^{(k+1)}\\ f_2^{(k+1)}\end{bmatrix} = \begin{bmatrix}0.85 & 0.20\\ 0.15 & 0.80\end{bmatrix}\begin{bmatrix}f_1^{(k)}\\ f_2^{(k)}\end{bmatrix}$$

or, letting $A$ be the coefficient matrix and $\mathbf{f}^{(k)}$ the distribution vector,

$$\mathbf{f}^{(k+1)} = A\mathbf{f}^{(k)}.$$

The process described by this difference equation is called a **Markov process** (after the Russian mathematician A. A. Markov, 1856–1922, who first studied processes of this type). In a Markov process there are a certain number of states (two in our example). The entries in the coefficient (or **transition**) matrix indicate how the distribution among the states changes over one time interval; it gives information about the transitions between states. In particular, the entry in the $i$th row and the $j$th column of the transition matrix is the fraction of objects in state $j$ which move to state $i$ (0.85 of the Sun Belt people stay while 0.15 move out). In general, a Markov process is described by a difference equation

$$\mathbf{f}^{(k+1)} = A\mathbf{f}^{(k)}, \tag{9}$$

where the transition matrix $A$ has columns that add up to 1 and the **state vectors** $\mathbf{f}^{(k)}$ all have coordinates that sum to 1.

By this time it should be obvious that equation (9) has a solution given by

$$\mathbf{f}^{(k)} = A^k\mathbf{f}^{(0)}.$$

So in our example we can predict the population distribution $k$ years from now by computing

$$\begin{bmatrix} 0.85 & 0.20 \\ 0.15 & 0.80 \end{bmatrix}^k \begin{bmatrix} 0.15 \\ 0.85 \end{bmatrix}.$$

Again we have the problem of raising a matrix to a power. The eigenvalues and corresponding eigenvectors of the matrix are

$$\begin{aligned} \lambda &= 1, & \mathbf{v}_1 &= (\tfrac{4}{7}, \tfrac{3}{7}) \\ \lambda &= 0.65, & \mathbf{v}_2 &= (-1, 1). \end{aligned}$$

Therefore,

$$A = M\Lambda M^{-1} = \begin{bmatrix} \frac{4}{7} & -1 \\ \frac{3}{7} & 1 \end{bmatrix} \begin{bmatrix} 1 & \\ & 0.65 \end{bmatrix} \begin{bmatrix} 1 & 1 \\ -\frac{3}{7} & \frac{4}{7} \end{bmatrix}.$$

If

$$\begin{bmatrix} \alpha \\ \beta \end{bmatrix} = M^{-1}\mathbf{p}^{(0)},$$

then

$$A^k\mathbf{p}^{(0)} = M\Lambda^k M^{-1}\mathbf{p}^{(0)} = \alpha\mathbf{v}_1 + \beta(0.65)^k\mathbf{v}_2.$$

When $k$ is large, $(0.65)^k$ is close to zero and hence $A^k\mathbf{p}^{(0)}$ is close to $\alpha\mathbf{v}_1$.

# PROBLEMS 8.4

*1. Suppose that an amount $x^{(0)}$ is invested at $r$ percent per year and is compounded monthly. Determine a formula for the amount of money after $k$ months.

2. What are the signs of the entries in the growth matrix $A$ when we have two competing species (see Example 1)?

3. Study the long-range behavior of the populations whose growth matrices are:

*(a) $\begin{bmatrix} -0.8 & 0.4 \\ -0.8 & 0.4 \end{bmatrix}$ (b) $\begin{bmatrix} -0.4 & 0.5 \\ -0.1 & 0.2 \end{bmatrix}$

4. Consider a population divided into three age groups with the following Leslie matrix.

$$\begin{bmatrix} 0 & 2 & 0 \\ 0.72 & 0 & 0 \\ 0 & 0.5 & 0 \end{bmatrix}$$

(a) Show that if the initial population distribution is (200, 120, 50), the population in each age group increases by $\frac{1}{5}$ after one time interval has passed.

(b) In part (a) it was shown that the relative proportion of animals in each age group remained the same after one time interval had passed. Will this continue indefinitely?

5. Suppose that the Leslie matrix for a population satisfies $A\mathbf{v} = \lambda\mathbf{v}$ for some vector $v$.

(a) Show that $A^2\mathbf{v} = \lambda^2\mathbf{v}$, $A^3\mathbf{v} = \lambda^3\mathbf{v}$, and $A^4\mathbf{v} = \lambda^4\mathbf{v}$.

(b) If the initial population distribution for this population is given by $\mathbf{v}$, describe the future population distribution.

*(c) What is the natural growth rate of this population?

*(d) Are there any stipulations that you should make as to which values of $\lambda$ are acceptable or meaningful?

*6. A certain fish population has been studied. Of the eggs that hatch each year, $\frac{1}{100}$ survive. Of the 1-year-old fish, $\frac{1}{5}$ survive. Of the 2-year-old fish, $\frac{1}{2}$ survive. All of the 3-year-old fish spawn and then die. The number of eggs that hatch averages 1100 per 3-year-old fish. What is the Leslie matrix $A$ for this population if the basic unit of time is 1 year? Compute $A^4$. What does your answer tell you about the population?

7. A certain fish population has the following Leslie matrix. (The basic time interval is 1 year.)

$$A = \begin{bmatrix} 0 & 0 & 110 \\ 0.05 & 0 & 0 \\ 0 & 0.2 & 0 \end{bmatrix}$$

If the initial population is 1000 in each age group, what will be the population of each age group after 1 year? 2 years? 3 years? 4 years?

Compute $A^3$. How does this help you predict the future distribution of the fish population? Describe the way this fish population changes from year to year.

*8. Referring to Example 2, express $N_\infty$ in terms of $N_0$ when:

(a) $a_0 = b_0 = c_0 = N_0/3$

(b) $a_0 = N_0, b_0 = c_0 = 0$

(c) $a_0 = c_0 = 0, b_0 = N_0$

*9. Referring to Example 2, show that when $\mathbf{x}_0$ is a scalar multiple of (160, 5, 1), then $\mathbf{x}_k = \mathbf{x}_0$ for all $k$ (i.e., the population is completely stable).

10. In each part the Leslie matrix for a certain population is given. Do the following.

(i) Find the eigenvalues.

(ii) Describe how the population grows.

*(a) $\begin{bmatrix} 0 & \frac{13}{8} & 3 \\ \frac{1}{2} & 0 & 0 \\ 0 & \frac{1}{8} & 0 \end{bmatrix}$ (b) $\begin{bmatrix} 0 & \frac{3}{2} & 6 \\ \frac{1}{2} & 0 & 0 \\ 0 & \frac{1}{12} & 0 \end{bmatrix}$

**11.** Using the notation of Example 2, suppose that the Leslie matrix for a certain fish population is

$$A = \begin{bmatrix} 0 & 19 & 15 \\ \frac{1}{16} & 0 & 0 \\ 0 & \frac{1}{2} & 0 \end{bmatrix}.$$

(a) Show that the eigenvalues of $A$ are $\frac{5}{4}$, $-\frac{1}{2}$, and $-\frac{3}{4}$.

(b) Show that for large $k$, $\mathbf{x}_k$ is approximately $\alpha(\frac{5}{4})^k(100, 5, 2)$, where $\alpha$ is a constant.

(c) Show that in the long run about 93% of the fish belong to the first group, about 5% belong to the second group, and about 2% belong to the third group.

(d) Describe how this population grows.

**12.** The sequence of numbers

$$0, 1, 1, 2, 3, 5, 8, 13, 21, \ldots$$

is a famous sequence called the **Fibonacci sequence**. It is so famous that there is even a journal called **Fibonacci Quarterly**. The first two terms of the sequence are 0 and 1 and each successive term is the sum of the previous two terms:

$$s_0 = 0, \qquad s_1 = 1, \qquad s_{k+2} = s_{k+1} + s_k.$$

(a) Show that if $\mathbf{x}_k = (s_{k+1}, s_k)$ and

$$A = \begin{bmatrix} 1 & 1 \\ 1 & 0 \end{bmatrix},$$

then $\mathbf{x}_{k+1} = A\mathbf{x}_k$.

(b) Diagonalize $A$.

(c) Use part (b) to show that

$$s_k = \frac{1}{\sqrt{5}}\left[\left(\frac{1+\sqrt{5}}{2}\right)^k - \left(\frac{1-\sqrt{5}}{2}\right)^k\right].$$

In view of the fact that $s_k$ is an integer, this is a rather surprising formula.

(d) What happens to $s_k$ when $k$ gets large?

**13.** Consider the sequence defined as follows:

$$a_0 = 4, \quad a_1 = 7, \quad a_{k+2} = 5a_{k+1} - 6a_k.$$

Find a "closed-form" formula for $a_k$ using the method in Problem 12.

**14.** Repeat Problem 13 for the sequence

$$b_0 = 1, \qquad b_1 = 2, \qquad b_2 = 3,$$
$$b_{k+3} = b_{k+2} + 2b_{k+1} - 2b_k.$$

**15.** For each of the following matrices $A$ and vectors $\mathbf{x}_0$, do the following.

(i) Solve the difference equation

$$\mathbf{x}_{k+1} = A\mathbf{x}_k.$$

(ii) Determine what $\mathbf{x}_k$ does as $k$ gets large. Does your answer depend on the initial vector $\mathbf{x}_0$?

*(a) $A = \begin{bmatrix} \frac{1}{4} & \frac{5}{8} \\ \frac{3}{4} & \frac{3}{8} \end{bmatrix}$, $\mathbf{x}_0 = \begin{bmatrix} 3 \\ 2 \end{bmatrix}$

(b) $A = \begin{bmatrix} \frac{1}{3} & \frac{5}{6} \\ \frac{2}{3} & \frac{1}{6} \end{bmatrix}$, $\mathbf{x}_0 = \begin{bmatrix} 1 \\ 2 \end{bmatrix}$

*(c) $A = \begin{bmatrix} \frac{1}{3} & \frac{1}{6} \\ \frac{2}{3} & \frac{5}{6} \end{bmatrix}$, $\mathbf{x}_0 = \begin{bmatrix} a_0 \\ b_0 \end{bmatrix}$

(d) $A = \begin{bmatrix} 0.8 & 0.2 & 0.1 \\ 0.1 & 0.7 & 0.3 \\ 0.1 & 0.1 & 0.6 \end{bmatrix}$, $\mathbf{x}_0 = \begin{bmatrix} a_0 \\ b_0 \\ c_0 \end{bmatrix}$

*(e) $A = \begin{bmatrix} 1 & 0 & 0 \\ 1 & 0 & -2 \\ -1 & 1 & -3 \end{bmatrix}$, $\mathbf{x}_0 = \begin{bmatrix} a_0 \\ b_0 \\ c_0 \end{bmatrix}$

***16.** Immigration statistics for the United Function States (UFS) and the Union of Linear Republics (ULR) show that each year $\frac{1}{3}$ of the people in the UFS emmigrate to the ULR while $\frac{1}{6}$ of the people in the ULR emmigrate to the UFS; the rest of the people stay in their respective countries. Let $a_k$ and $b_k$ be the number of people in the UFS and ULR, respectively, after $k$ years.

(a) Find expressions for $a_k$ and $b_k$ in terms of the initial values $a_0$ and $b_0$ and $k$.

(b) What is your long-range prediction for the populations of the two countries? In what sense does this depend on $a_0$ and $b_0$?

**17.** There are two cities $a$ and $b$. City $a$ keeps $\frac{3}{4}$ of its currency in each period and sends

$\frac{1}{4}$ to city $b$. City $b$ keeps $\frac{1}{2}$ and sends $\frac{1}{2}$ to city $a$. Find the transition matrix and the steady-state $\mathbf{u}_\infty$. Diagonalize the transition matrix and find an expression for $\mathbf{u}_k$ in terms of the eigenvalues.

*18. Suppose that a city has two banks and that everyone in the city does business at one of these banks. (Assume also that the city's population remains constant.) Suppose that each year $\frac{1}{10}$ of bank $B$'s customers switch to bank $A$, and $\frac{2}{10}$ of bank $A$'s customers switch to bank $B$. Currently, $\frac{9}{10}$ of the population is with bank $B$. In the long run, how does the population distribute itself between the two banks? Would this be different if $B$ began with $\frac{6}{10}$ of the population?

19. Apply the Markov process idea to the distribution of sales among three supermarkets. What is the interpretation of the coefficients in the transition matrix? What questions are of interest, and what is their mathematical formulation?

20. Suppose that in a certain plant population, each plant is fertilized with a plant with its own genotype. The fraction of offspring in each genotype is given in the table. Let

$a_n$ = fraction of plants of genotype AA in the $n$th generation.
$b_n$ = fraction of plants of genotype Aa in the $n$th generation
$c_n$ = fraction of plants of genotype aa in the $n$th generation.

Let $a_0, b_0, c_0$ denote the initial distribution of genotypes.

*(a) Find a formula to determine the genotype distribution of each generation from the genotype distribution of the preceding generation.

(b) What happens in the long run?

| | | Genotype of parents | | |
|---|---|---|---|---|
| | | AA–AA | Aa–Aa | aa–aa |
| Genotype of offspring | AA | 1 | $\frac{1}{4}$ | 0 |
| | Aa | 0 | $\frac{1}{2}$ | 0 |
| | aa | 0 | $\frac{1}{4}$ | 1 |

21. Let $A$ be an $n \times n$ diagonalizable matrix. If $\lambda_1, \lambda_2, \ldots, \lambda_n$ are the eigenvalues of $A$ and $\mathbf{v}_1, \mathbf{v}_2, \ldots, \mathbf{v}_n$ are the corresponding eigenvectors, show that

$$A^k\mathbf{v} = \alpha_1\lambda_1^k\mathbf{v}_1 + \alpha_2\lambda_2^k\mathbf{v}_2 + \cdots + \alpha_n\lambda_n^k\mathbf{v}_n,$$

where $\mathbf{v} = \alpha_1\mathbf{v}_1 + \alpha_2\mathbf{v}_2 + \cdots + \alpha_n\mathbf{v}_n$.

22. Let $A$ be a transition matrix and let $\lambda$ be an eigenvalue of $A$ with $\lambda \neq 1$. Prove that if $\mathbf{v}$ is a corresponding eigenvector, the components of $\mathbf{v}$ add to 0. [*Hint:* $\mathbf{v}^TA = \mathbf{v}^T$, where $\mathbf{v} = (1, 1, \ldots, 1)$.]

*23. Show that any transition matrix has $\lambda = 1$ as an eigenvalue.

24. Let $P$ be a transition matrix all of whose columns are the same. If $\mathbf{x}$ is any state vector, show that

$$P\mathbf{x} = \mathbf{x}.$$

25. Let $A$ and $B$ be transition matrices.
(a) Show that $AB$ is a transition matrix.
(b) Show that $A^n$ is a transition matrix for any positive integer $n$.
(c) If $\mathbf{x}$ is a vector whose components are nonnegative and sum to 1, show that the same is true of $A\mathbf{x}$.

26. Let $A$ be the matrix

$$A = \begin{bmatrix} p & q \\ 1-p & 1-q \end{bmatrix},$$

where $0 \leq p \leq 1$ and $0 \leq q \leq 1$.
(a) Show that the eigenvalues of $A$ are 1 and $p - q$.
(b) Let $\mathbf{v}_1$ and $\mathbf{v}_2$ be eigenvectors associated with 1 and $p - q$, respectively. Show that if $|p - q| < 1$ and if $\mathbf{v} = \alpha_1\mathbf{v}_1 + \alpha_2\mathbf{v}_2$, then

$$\mathbf{v}, A\mathbf{v}, A^2\mathbf{v}, \ldots$$

tends toward $\alpha_1\mathbf{v}_1$.
(c) Show that the sum of the components of $\alpha_1\mathbf{v}_1$ is the same as the sum of the components of $\mathbf{v}$.

# 8.5 Matrices with Repeated Eigenvalues

In Section 8.3 we proved that if the eigenvalues of an $n \times n$ matrix $A$ are all distinct, then $A$ is diagonalizable. This is the best result we can obtain from just looking at the eigenvalues of $A$. Matrices with repeated eigenvalues may or may not be diagonalizable. In Example 2 of Section 8.3 we saw an example of a $3 \times 3$ matrix with a repeated eigenvalue that is diagonalizable, and in Example 3 of Section 8.3 we saw an example of a $3 \times 3$ matrix with a repeated eigenvalue that is not diagonalizable. *Is there a necessary and sufficient condition that a matrix with multiple eigenvalues be diagonalizable?*

The heart of the matter lies in the relationship between the multiplicity of $\lambda$ as a zero of the characteristic polynomial and the number of independent eigenvectors associated with $\lambda$, that is, the dimension of the eigenspace $E(\lambda)$. The multiplicity of $\lambda$ as a zero of the characteristic polynomial is called the **algebraic multiplicity** of $\lambda$. The dimension of the eigenspace $E(\lambda)$ is called the **geometric multiplicity** of $\lambda$.

***Theorem 1*** *The geometric multiplicity of $\lambda$ is always less than or equal to the algebraic multiplicity of $\lambda$.*

**Proof** Suppose that $A$ is an $n \times n$ matrix and that $\lambda$ is an eigenvalue of $A$ with geometric multiplicity $k$ [i.e., $\dim E(\lambda) = k$]. Let $\mathbf{v}_1, \mathbf{v}_2, \ldots, \mathbf{v}_k$ be a basis for $E(\lambda)$. Extend this basis to a basis $\mathbf{v}_1, \mathbf{v}_2, \ldots, \mathbf{v}_k, \mathbf{v}_{k+1}, \ldots, \mathbf{v}_n$ of $R^n$. If

$$M = \begin{bmatrix} | & & | & | & & | \\ \mathbf{v}_1 & \cdots & \mathbf{v}_k & \mathbf{v}_{k+1} & \cdots & \mathbf{v}_n \\ | & & | & | & & | \end{bmatrix},$$

then

$$AM = \begin{bmatrix} | & & | & | & & | \\ \lambda\mathbf{v}_1 & \cdots & \lambda\mathbf{v}_k & A\mathbf{v}_{k+1} & \cdots & A\mathbf{v}_n \\ | & & | & | & & | \end{bmatrix}.$$

Using the fact that $M$ is invertible, we define

$$B = \begin{bmatrix} | & & | & | & & | \\ \lambda\mathbf{e}_1 & \cdots & \lambda\mathbf{e}_k & M^{-1}A\mathbf{v}_{k+1} & \cdots & M^{-1}A\mathbf{v}_n \\ | & & | & | & & | \end{bmatrix}.$$

Then $AM = MB$ so that

$$B = M^{-1}AM.$$

Since the matrices $A$ and $B$ are similar, they have the same characteristic equation and hence the same eigenvalues with the same multiplicities. From the cofactor expansion of the determinant of $B - \mu I$ and from the form of $B$ it is clear that $\det(B - \mu I) = \alpha(\lambda - \mu)^k$, where $\alpha$ is a polynomial in $\mu$. Therefore, $\lambda$ is a root with multiplicity at least $k$ of the characteristic equation of $B$, and hence of the characteristic equation of $A$. This completes the proof.

With this result we can prove the following theorem, which gives the best result obtainable in the case of repeated eigenvalues.

***Theorem 2*** *Let $A$ be an $n \times n$ matrix and let $\lambda_1, \ldots, \lambda_r$ be the distinct eigenvalues of $A$. Then $A$ is diagonalizable if and only if the geometric multiplicity of each eigenvalue is equal to its algebraic multiplicity.*

**Proof** Let $d_1, \ldots, d_r$ be the algebraic multiplicities of $\lambda_1, \ldots, \lambda_r$, respectively. Then $d_1 + \cdots + d_r = n$ since the degree of the characteristic polynomial is $n$. If $A$ is diagonalizable, then it has $n$ independent eigenvectors. Each of these eigenvectors is associated with a unique eigenvalue and by Theorem 1, at most $d_1$ are associated with $\lambda_1$, $d_2$ with $\lambda_2$, and so on. Therefore, exactly $d_i$ of the eigenvectors are associated with the eigenvalue $\lambda_i$. Hence the geometric multiplicity of each eigenvalue is equal to its algebraic multiplicity. Conversely, if this condition holds, then reversing the steps of this argument shows that $A$ has $n$ independent eigenvectors and hence is diagonalizable.

The proof of Theorem 2 amounts to showing that the matrix $A$ has $n$ independent eigenvectors if and only if the dimension of each eigenspace $E(\lambda)$ is equal to the algebraic multiplicity of $\lambda$. But in order to find the dimension of the eigenspace $E(\lambda)$, we must find the number of independent eigenvectors associated with the eigenvalue $\lambda$. Thus to establish the condition mentioned in the theorem we must in effect find $n$ independent eigenvectors, and we already know (Theorem 1 of Section 8.3) that this is a necessary and sufficient condition for $A$ to be diagonalizable. However, Theorem 2 is useful for showing that a matrix is not diagonalizable. For if we are able to find an eigenvalue $\lambda$ of a matrix $A$ whose geometric multiplicity is less than its algebraic multiplicity, then the matrix $A$ is not diagonalizable.

## EXAMPLE 1

The eigenvalue 5 of matrix $A$ in Example 3 of Section 8.2 has algebraic multiplicity 2 and geometric multiplicity 1. Therefore, $A$ is not diagonalizable.

If a matrix is not diagonalizable, then the best we can hope to accomplish is to find a nonsingular matrix $M$ such that $J = M^{-1}AM$ is as diagonal as possible. The best possible matrix $J$ is called the **Jordan normal form** of $A$. It is a matrix that has the following properties:

1. The eigenvalues of $A$ are on the main diagonal.
2. The diagonal above the main diagonal contains all zeros and ones.
3. All other entries are zero.

The proof of this theorem and its applications are beyond the scope of this book.

## PROBLEMS 8.5

*1. Let

$$A = \begin{bmatrix} 2 & 1 & 0 & 0 \\ 0 & 2 & 0 & 0 \\ 0 & 0 & 2 & 0 \\ 0 & 0 & 0 & 1 \end{bmatrix}.$$

(a) Show that $\lambda = 2$ is an eigenvalue of $A$ with algebraic multiplicity 3 and geometric multiplicity 2.
(b) What is the other eigenvalue of $A$?

2. Let

$$A = \begin{bmatrix} 5 & 1 & 0 & 0 \\ 0 & 5 & 1 & 0 \\ 0 & 0 & 5 & 0 \\ 0 & 0 & 0 & 5 \end{bmatrix} \quad \text{and}$$

$$B = \begin{bmatrix} 5 & 1 & 0 & 0 \\ 0 & 5 & 0 & 0 \\ 0 & 0 & 5 & 1 \\ 0 & 0 & 0 & 5 \end{bmatrix}.$$

Show that $\lambda = 5$ is an eigenvalue of $A$ and $B$ with algebraic multiplicity 4 and geometric multiplicity 2.

*3. Let

$$A = \begin{bmatrix} 4 & 1 & 0 & 0 & 0 \\ 0 & 4 & 0 & 0 & 0 \\ 0 & 0 & -1 & 1 & 0 \\ 0 & 0 & 0 & -1 & 1 \\ 0 & 0 & 0 & 0 & -1 \end{bmatrix}.$$

(a) Show that $\lambda = 4$ is an eigenvalue of $A$ with algebraic multiplicity 2 and geometric multiplicity 1.
(b) Show that $\lambda = -1$ is an eigenvalue of $A$ with algebraic multiplicity 3 and geometric multiplicity 1.

4. Let

$$D = \begin{bmatrix} 1 & & \\ & 1 & \\ & & 2 \end{bmatrix}.$$

Show that $\lambda = 1$ is an eigenvalue of $D$ with algebraic and geometric multiplicity 2.

*5. Let

$$A = \begin{bmatrix} 0 & -1 & 0 \\ 0 & 0 & 1 \\ -1 & -3 & 3 \end{bmatrix}.$$

(a) Show that $\lambda = 1$ is the only eigenvalue of $A$.
(b) Find the eigenvectors for $\lambda = 1$.
(c) Is $A$ diagonalizable?

6. Let $D$ be a diagonal matrix. Show that any eigenvalue of $D$ with algebraic multiplicity $r$ also has geometric multiplicity $r$.

7. Show that if $\lambda$ has algebraic multiplicity 1, then it also has geometric multiplicity 1. Use this result and Theorem 2 to prove Theorem 2 of Section 8.3.

8. Let $A$ be an $n \times n$ matrix and let $\mathbf{v}_1, \mathbf{v}_2, \ldots, \mathbf{v}_n$ be $n$ independent eigenvectors of $A$. Show that
*(a) If $\mathbf{v} = \mathbf{v}_1 + \mathbf{v}_2 + \cdots + \mathbf{v}_n$ and $A\mathbf{v} = \mu\mathbf{v}$, show that $A = \mu I$.
(b) If $\mathbf{v} = \mathbf{v}_2 + \mathbf{v}_3 + \cdots + \mathbf{v}_n$ and $A\mathbf{v} = \mu\mathbf{v}$, show that $\mu$ is an eigenvalue with algebraic multiplicity at least $n - 1$.
(c) Generalize parts (a) and (b).

9. Let $A$ be an $n \times n$ matrix. Explain why the nullity of $A$ is at most equal to the multiplicity of 0 as a root of the characteristic equation of $A$.

*10. The characteristic polynomial of a $15 \times 15$ matrix is $(x-1)^{10}(x+1)^5$. What is the minimum number of linearly independent eigenvectors the matrix can have? Explain.

# 8.6 Symmetric Matrices

In Section 8.5 we investigated the relationship between the eigenvalues of a matrix and the diagonalizability of the matrix. We turn now to the question of whether there are any reasonable conditions on the matrix itself which will guarantee that it is diagonalizable.

An $n \times n$ matrix $A$ is called **symmetric** if $A = A^T$, that is, if $a_{ij} = a_{ji}$ for all $i$ and $j$ (see Problem 17 in Section 1.6). There are two surprising and important facts about symmetric matrices: All of their eigenvalues are real, and they are diagonalizable. The proofs of these statements are given in Chapter 10 (see Theorems 1 and 3 in Section 10.3).

***Theorem 1*** *Let A be a symmetric matrix.*

(*a*) *All of the eigenvalues of A are real.*
(*b*) *A is diagonalizable.*

Still more is true about symmetric matrices. Suppose that $A$ is a symmetric matrix and that $\mathbf{v}_1$ and $\mathbf{v}_2$ are eigenvectors of $A$ associated with different eigenvalues $\lambda_1$ and $\lambda_2$. We have

$$A\mathbf{v}_1 = \lambda_1\mathbf{v}_1, \qquad A\mathbf{v}_2 = \lambda_2\mathbf{v}_2, \qquad \text{and} \qquad \lambda_1 \neq \lambda_2.$$

Moreover, since $A$ is symmetric, $\langle A\mathbf{v}_1, \mathbf{v}_2\rangle = \langle \mathbf{v}_1, A\mathbf{v}_2\rangle$, since

$$\begin{aligned}\langle A\mathbf{v}_1, \mathbf{v}_2\rangle &= (A\mathbf{v}_1)^T\mathbf{v}_2 = \mathbf{v}_1^T A^T \mathbf{v}_2 = \mathbf{v}_1^T A\mathbf{v}_2 \\ &= \langle \mathbf{v}_1, A\mathbf{v}_2\rangle.\end{aligned}$$

Hence

$$\begin{aligned}\lambda_1\langle \mathbf{v}_1, \mathbf{v}_2\rangle &= \langle \lambda_1\mathbf{v}_1, \mathbf{v}_2\rangle = \langle A\mathbf{v}_1, \mathbf{v}_2\rangle = \langle \mathbf{v}_1, A\mathbf{v}_2\rangle = \langle \mathbf{v}_1, \lambda_2\mathbf{v}_2\rangle \\ &= \lambda_2\langle \mathbf{v}_1, \mathbf{v}_2\rangle.\end{aligned}$$

Since $\lambda_1 \neq \lambda_2$ it follows that $\langle \mathbf{v}_1, \mathbf{v}_2\rangle = 0$. This proves the following result.

**Result 1** Let $A$ be a symmetric matrix. Eigenvectors of $A$ associated with different eigenvalues are orthogonal.

The student should compare this result with Theorem 3 in Section 8.3. We know that for any matrix the eigenvectors associated with different eigenvalues are independent. For symmetric matrices, they are not only independent, they are also orthogonal.

### EXAMPLE 1

Let us find the eigenvalues and eigenvectors of the matrix

$$A = \begin{bmatrix} 1 & -2 & 2 \\ -2 & 1 & 2 \\ 2 & 2 & 1 \end{bmatrix}$$

and verify that the eigenvectors associated with different eigenvalues are orthogonal. The characteristic equation of $A$ is

$$\det(A - \lambda I) = \det \begin{bmatrix} 1-\lambda & -2 & 2 \\ -2 & 1-\lambda & 2 \\ 2 & 2 & 1-\lambda \end{bmatrix}$$
$$= -(\lambda - 3)^2(\lambda + 3) = 0.$$

Thus the eigenvalues of $A$ are 3, 3, and $-3$. The reduced echelon form of $A - 3I$ and $A + 3I$ is, respectively,

$$\begin{bmatrix} 1 & 1 & -1 \\ 0 & 0 & 0 \\ 0 & 0 & 0 \end{bmatrix} \quad \text{and} \quad \begin{bmatrix} 1 & 0 & 1 \\ 0 & 1 & 1 \\ 0 & 0 & 0 \end{bmatrix}.$$

Thus a basis for the eigenspace $E(3)$ is $(1, 0, 1)$ and $(-1, 1, 0)$ and a basis for the eigenspace $E(-3)$ is $(-1, -1, 1)$. Therefore, the eigenvectors associated with the eigenvalue 3 are the vectors of the form

$$\alpha(1, 0, 1) + \beta(-1, 1, 0) = (\alpha - \beta, \beta, \alpha)$$

with not both $\alpha$ and $\beta$ equal to zero. Similarly, the eigenvectors associated with the eigenvalue $-3$ are the vectors of the form $(-\gamma, -\gamma, \gamma)$, $\gamma \neq 0$. Since

$$\langle(\alpha - \beta, \beta, \alpha), (-\gamma, -\gamma, \gamma)\rangle = -\gamma(\alpha - \beta) - \beta\gamma + \alpha\gamma = 0,$$

these vectors are indeed orthogonal.

There is one more important result about symmetric matrices which we will prove.

***Theorem 2*** *Let $A$ be an $n \times n$ symmetric matrix. Then $A$ has $n$ orthonormal eigenvectors.*

**Proof** Let $\lambda_1, \ldots, \lambda_r$ be the distinct eigenvalues of $A$, and let $d_1, \ldots, d_r$ be the algebraic multiplicities of $\lambda_1, \ldots, \lambda_r$, respectively. Since $A$ is diagonalizable, the dimension of each of the eigenspaces $E(\lambda_i)$ is $d_i$, $i = 1, \ldots, r$. Using the Gram–Schmidt process, we can find an orthogonal basis for each of the eigenspaces. But since eigenvectors associated with different eigenvalues are orthogonal, the collection of all of these eigenvectors is a collection of $n$ orthogonal vectors. Thus $A$ has $n$ orthogonal eigenvectors. Dividing each of these vectors by its length, we obtain $n$ orthonormal eigenvectors.

The proof of this theorem provides the following algorithm for finding the $n$ orthonormal eigenvectors of an $n \times n$ symmetric matrix.

1. Find the eigenvalues of $A$. (They are all real.)
2. Find a basis for each of the eigenspaces in the usual way.
3. Apply the Gram–Schmidt process to obtain an orthonormal basis for each of the eigenspaces.

**EXAMPLE 2**

In Example 1 we found that $(1,0,1)$ and $(-1,1,0)$ formed a basis for the eigenspace $E(3)$ and $(-1,-1,1)$ formed a basis for the eigenspace $E(-3)$. Applying the Gram–Schmidt process to the vectors $(1,0,1)$ and $(-1,1,0)$, we obtain the orthonormal vectors

$$\frac{1}{\sqrt{2}}(1,0,1) \qquad \text{and} \qquad \frac{1}{\sqrt{6}}(-1,2,1).$$

Thus

$$\frac{1}{\sqrt{2}}(1,0,1), \qquad \frac{1}{\sqrt{6}}(-1,2,1), \qquad \text{and} \qquad \frac{1}{\sqrt{3}}(-1,-1,1)$$

are three orthonormal eigenvectors of $A$.

Given an $n \times n$ symmetric matrix $A$, we know that it has $n$ orthonormal eigenvectors. Let $M$ be the matrix whose columns are these orthonormal eigenvectors. Then $\Lambda = M^{-1}AM$ is a diagonal matrix whose diagonal entries are the eigenvalues of $A$. But because $M$ has orthonormal columns, it has the additional property that

$$M^TM = I = MM^T. \tag{1}$$

Thus $M^{-1} = M^T$ and hence $\Lambda = M^TAM$. Matrices that satisfy (1) are called **orthogonal** matrices. Matrices that can be diagonalized by an orthogonal matrix are called **orthogonally diagonalizable**. Using this terminology we can state Theorem 2 as follows.

***Theorem 3*** *Symmetric matrices are orthogonally diagonalizable.*

**EXAMPLE 3**

In Example 2 we found that

$$\frac{1}{\sqrt{2}}(1,0,1), \qquad \frac{1}{\sqrt{6}}(-1,2,1), \qquad \text{and} \qquad \frac{1}{\sqrt{3}}(-1,-1,1)$$

are orthonormal eigenvectors of the matrix $A$ given in Example 1. Letting

$$M = \begin{bmatrix} 1/\sqrt{2} & -1/\sqrt{6} & -1/\sqrt{3} \\ 0 & 2/\sqrt{6} & -1/\sqrt{3} \\ 1/\sqrt{2} & 1/\sqrt{6} & 1/\sqrt{3} \end{bmatrix},$$

it is easy to verify that $M^TM = I$ and that

$$\begin{bmatrix} 3 & & \\ & 3 & \\ & & -3 \end{bmatrix} = M^TAM.$$

If $A$ is an $n \times n$ symmetric matrix and $\mathbf{v}_1, \ldots, \mathbf{v}_n$ are $n$ orthonormal eigenvectors of $A$ with corresponding eigenvalues $\lambda_1, \ldots, \lambda_n$, then

$$\begin{aligned} A\mathbf{x} &= M\Lambda M^{-1}\mathbf{x} \\ &= \begin{bmatrix} | & & | \\ \mathbf{v}_1 & \cdots & \mathbf{v}_n \\ | & & | \end{bmatrix} \begin{bmatrix} \lambda_1 & & \\ & \ddots & \\ & & \lambda_n \end{bmatrix} \begin{bmatrix} -\mathbf{v}_1^T- \\ \vdots \\ -\mathbf{v}_n^T- \end{bmatrix} \begin{bmatrix} | \\ \mathbf{x} \\ | \end{bmatrix} \\ &= \begin{bmatrix} | & & | \\ \lambda_1\mathbf{v}_1 & \cdots & \lambda_n\mathbf{v}_n \\ | & & | \end{bmatrix} \begin{bmatrix} \langle \mathbf{v}_1, \mathbf{x} \rangle \\ \vdots \\ \langle \mathbf{v}_n, \mathbf{x} \rangle \end{bmatrix} \\ &= \lambda_1\mathbf{v}_1\langle \mathbf{v}_1, \mathbf{x} \rangle + \cdots + \lambda_n\mathbf{v}_n\langle \mathbf{v}_n, \mathbf{x} \rangle \\ &= \lambda_1\mathbf{v}_1\mathbf{v}_1^T\mathbf{x} + \cdots + \lambda_n\mathbf{v}_n\mathbf{v}_n^T\mathbf{x} \\ &= (\lambda_1\mathbf{v}_1\mathbf{v}_1^T + \cdots + \lambda_n\mathbf{v}_n\mathbf{v}_n^T)\mathbf{x}. \end{aligned}$$

From our work in Chapter 3 we recognize that $\mathbf{v}_i\mathbf{v}_i^T$ is a projection matrix that projects a vector $\mathbf{x}$ onto the vector $\mathbf{v}_i$. Thus $A\mathbf{x}$ is a linear combination of the projections of $\mathbf{x}$ onto the orthonormal eigenvectors $\mathbf{v}_1, \ldots, \mathbf{v}_n$. This decomposition is called the **spectral decomposition** for symmetric matrices.

***Theorem 4*** *(Spectral Theorem for Symmetric Matrices) Let $A$ be an $n \times n$ symmetric matrix. Then $A$ can be decomposed into*

$$A = \lambda_1\mathbf{v}_1\mathbf{v}_1^T + \cdots + \lambda_n\mathbf{v}_n\mathbf{v}_n^T, \tag{2}$$

*where $\mathbf{v}_1, \ldots, \mathbf{v}_n$ are orthonormal eigenvectors and $\lambda_1, \ldots, \lambda_n$ are associated eigenvalues.*

## EXAMPLE 4

Consider the matrix

$$A = \begin{bmatrix} 1 & -2 & 2 \\ -2 & 1 & 2 \\ 2 & 2 & 1 \end{bmatrix}$$

of Example 1. In Example 2 we found that

$$\frac{1}{\sqrt{2}}(1,0,1), \qquad \frac{1}{\sqrt{6}}(-1,2,1), \qquad \text{and} \qquad \frac{1}{\sqrt{3}}(-1,-1,1)$$

are orthonormal eigenvectors corresponding to the eigenvalues $\lambda_1 = 3$, $\lambda_2 = 3$, $\lambda_3 = -3$ of $A$. By Theorem 4 the spectral decomposition of $A$ is

$$\begin{aligned} A &= \lambda_1 \mathbf{v}_1 \mathbf{v}_1^T + \lambda_2 \mathbf{v}_2 \mathbf{v}_2^T + \lambda_3 \mathbf{v}_3 \mathbf{v}_3^T \\ &= 3\begin{bmatrix} \frac{1}{2} & 0 & \frac{1}{2} \\ 0 & 0 & 0 \\ \frac{1}{2} & 0 & \frac{1}{2} \end{bmatrix} + 3\begin{bmatrix} \frac{1}{6} & -\frac{1}{3} & -\frac{1}{6} \\ -\frac{1}{3} & \frac{2}{3} & \frac{1}{3} \\ -\frac{1}{6} & \frac{1}{3} & \frac{1}{6} \end{bmatrix} - 3\begin{bmatrix} \frac{1}{3} & \frac{1}{3} & -\frac{1}{3} \\ \frac{1}{3} & \frac{1}{3} & -\frac{1}{3} \\ -\frac{1}{3} & -\frac{1}{3} & \frac{1}{3} \end{bmatrix}. \end{aligned}$$

## PROBLEMS 8.6

**1.** Find an orthogonal matrix that orthogonally diagonalizes each of the following symmetric matrices.

*(a) $\begin{bmatrix} 1 & 2 \\ 2 & 1 \end{bmatrix}$

(b) $\begin{bmatrix} 1 & 0 \\ 0 & -1 \end{bmatrix}$

(c) $\begin{bmatrix} 3 & 1 \\ 1 & 2 \end{bmatrix}$

*(d) $\begin{bmatrix} -1 & 1 & 1 \\ 1 & -1 & 1 \\ 1 & 1 & -1 \end{bmatrix}$

(e) $\begin{bmatrix} 5 & 2 & 2 \\ 2 & 5 & 2 \\ 2 & 2 & 5 \end{bmatrix}$

*(f) $\begin{bmatrix} 1 & 2 & 0 \\ 2 & 1 & 0 \\ 0 & 0 & -1 \end{bmatrix}$

(g) $\begin{bmatrix} 3 & 0 & -1 \\ 0 & 3 & 0 \\ -1 & 0 & 3 \end{bmatrix}$

(h) $\begin{bmatrix} 3 & 0 & -1 \\ 0 & 2 & 0 \\ -1 & 0 & 3 \end{bmatrix}$

(i) $\begin{bmatrix} 1 & 1 & 0 \\ 1 & 0 & -1 \\ 0 & -1 & 1 \end{bmatrix}$

(j) $\begin{bmatrix} 3 & 0 & 1 \\ 0 & 2 & 0 \\ 1 & 0 & 3 \end{bmatrix}$

(k) $\begin{bmatrix} 1 & 1 & 1 & 1 \\ 1 & 1 & 1 & 1 \\ 1 & 1 & 1 & 1 \\ 1 & 1 & 1 & 1 \end{bmatrix}$

*(l) $\begin{bmatrix} 1 & 0 & 1 & 0 \\ 0 & 1 & 0 & 1 \\ 1 & 0 & 1 & 0 \\ 0 & 1 & 0 & 1 \end{bmatrix}$

**2.** Find the spectral decomposition for each of the matrices in Problem 1.

**3.** Find a symmetric matrix $A$ with:

(a) Eigenvalues 1, 3 and corresponding eigenvectors $(1, 1)$ and $(-2, 2)$.

*(b) Eigenvalues 1, 1, 2 and corresponding eigenvectors $(-1, 2, 2)$, $(2, -1, 2)$, $(2, 2, -1)$.

(c) Eigenvalues $-1$, 0, 1 and corresponding eigenvectors $(1, 1, 1)$, $(1, -1, 0)$, $(1, 1, -2)$.

*4. Prove the converse of Theorem 3 by showing that if $A$ is orthogonally diagonalizable, then $A$ is a symmetric matrix.

5. Show that if $A$ is a symmetric matrix, then $A^2, A^3, \ldots$ are also symmetric matrices.

*6. Let $A$ be a symmetric matrix with one eigenvalue $\lambda$. Show that $A = \lambda I$.

7. Suppose that the eigenspaces of a diagonalizable matrix $A$ are orthogonal. (See Problem 16 in Section 3.4 for the definition of orthogonal subspaces.) Prove that the matrix $A$ is symmetric.

8. Explain why Theorem 1 of Section 3.7 is a special case of Theorem 4.

*9. Show that if $A$ is an $n \times n$ symmetric matrix with spectral decomposition (2), then

$$A^k = \lambda_1^k \mathbf{v}_1 \mathbf{v}_1^T + \cdots + \lambda_n^k \mathbf{v}_n \mathbf{v}_n^T.$$

10. Let $A$ be a projection matrix. Prove the following.
   (a) $A$ is diagonalizable.
   (b) If $A \neq 0$ and $A \neq I$, then $\lambda = 0$ and $\lambda = 1$ are eigenvalues of $A$.
   (c) Show that the eigenspace $E(0)$ is the orthogonal complement of the eigenspace $E(1)$.

*11. Two matrices $A$ and $B$ are called **orthogonally similar** if $B = M^T A M$ where $M$ is an orthogonal matrix.
   (a) Show that if $A$ and $B$ are orthogonally similar, then $A$ is symmetric if and only if $B$ is symmetric.
   (b) Show that $A$ is orthogonal if and only if $B$ is orthogonal.

12. Let $A$ be a symmetric matrix. If $A^2 + bA + cI$ is singular, show that the polynomial $\lambda^2 + b\lambda + c$ has two real roots. [*Hint:* Let $\mathbf{x}$ be a nonzero vector in $N(A^2 + bA + cI)$ and show that since

$$\begin{aligned} 0 &= \langle \mathbf{x}, (A^2 + bA + cI)\mathbf{x} \rangle \\ &= \mathbf{x}^T (A^2 + bA + cI)\mathbf{x}, \end{aligned}$$

$$\|(A + (b/2)I)\mathbf{x}\|^2 = (b^2/4 - c)\|\mathbf{x}\|^2.]$$

*13. (a) Let $A$ be a symmetric matrix. Show that $A$ is orthogonal if and only if its eigenvalues are $\pm 1$.
   (b) Construct a $3 \times 3$ symmetric orthogonal matrix (other than $\pm I$).

14. Show that the eigenvalues of a symmetric matrix $A$ are all zero if and only if $A = 0$. More generally, show that all eigenvalues of a symmetric matrix $A$ are equal if and only if $A = \lambda I$ for some scalar $\lambda$. What happens when $A$ is not symmetric?

15. Let $A$ be a symmetric matrix. Show that $A$ is a projection matrix if and only if the only possible eigenvalues of $A$ are 0 and 1.

16. Let $A$ be a symmetric matrix. If $A^k = I$ for some positive integer $k$, show that $A^2 = I$.

# 8.7 Quadratic Forms

In Example 7 of Section 2.5 we considered the equation

$$5x^2 - 6xy + 5y^2 = 8. \tag{1}$$

We showed that a rotation by $\pi/4$ of the coordinate axes determined by the standard basis vectors $\mathbf{e}_1$ and $\mathbf{e}_2$ transformed this equation to

$$\frac{x'^2}{4} + y'^2 = 1$$

in the new $x'y'$-coordinate system determined by the new basis vectors

$$\mathbf{v}_1 = \frac{1}{\sqrt{2}}(1, 1), \qquad \mathbf{v}_2 = \frac{1}{\sqrt{2}}(-1, 1)$$

(see Figure 8.2). How did we know that a rotation by $\pi/4$ would produce this result? We are now in a position to reveal the answer.

We begin by observing that the equation

$$5x^2 - 6xy + 5y^2 = 8$$

can be written as

$$\mathbf{x}^T A\mathbf{x} = [x \quad y]\begin{bmatrix} 5 & -3 \\ -3 & 5 \end{bmatrix}\begin{bmatrix} x \\ y \end{bmatrix} = 8.$$

Notice that the $2 \times 2$ matrix $A$ appearing in this equation is symmetric. Hence it is orthogonally diagonalizable. Since

$$\det(A - \lambda I) = (\lambda - 2)(\lambda - 8)$$

the eigenvalues of $A$ are $\lambda = 2$ and $\lambda = 8$. The vectors

$$\mathbf{v}_1 = \frac{1}{\sqrt{2}}(1, 1) \qquad \text{and} \qquad \mathbf{v}_2 = \frac{1}{\sqrt{2}}(-1, 1)$$

are corresponding unit eigenvectors. Thus if $M$ and $\Lambda$ are the matrices

$$M = \frac{1}{\sqrt{2}}\begin{bmatrix} 1 & -1 \\ 1 & 1 \end{bmatrix}, \qquad \Lambda = \begin{bmatrix} 2 & \\ & 8 \end{bmatrix},$$

then

$$\Lambda = M^T A M \qquad \text{or} \qquad A = M\Lambda M^T.$$

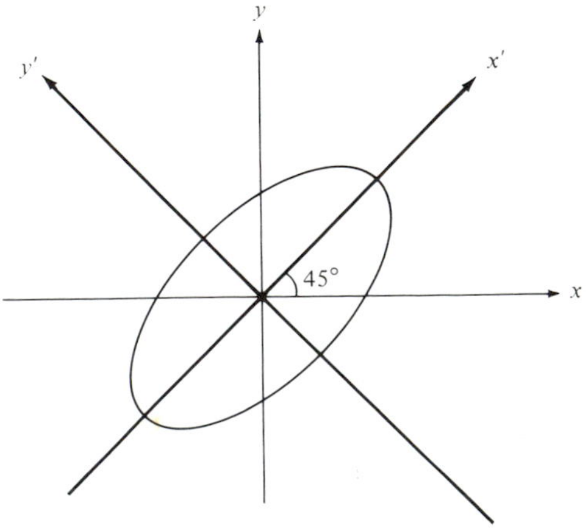


Figure 8.2

Substituting this expression for $A$ into the expression $\mathbf{x}^T A\mathbf{x}$ yields

$$\begin{aligned}\mathbf{x}^T A\mathbf{x} &= \mathbf{x}^T(M\Lambda M^T)\mathbf{x}\\ &= (\mathbf{x}^T M)\Lambda(M^T\mathbf{x})\\ &= (M^T\mathbf{x})^T\Lambda(M^T\mathbf{x}).\end{aligned}$$

Thus if we set

$$\mathbf{x}' = M^T\mathbf{x},$$

then

$$\begin{aligned}\mathbf{x}^T A\mathbf{x} &= \mathbf{x}'^T\Lambda\mathbf{x}'\\ &= [x' \quad y']\begin{bmatrix}2 & \\ & 8\end{bmatrix}\begin{bmatrix}x'\\ y'\end{bmatrix}\\ &= 2x'^2 + 8y'^2.\end{aligned}$$

We have transformed our original equation

$$\mathbf{x}^T A\mathbf{x} = 5x^2 - 6xy + 5y^2 = 8$$

to

$$\mathbf{x}'^T\Lambda\mathbf{x}' = 2x'^2 + 8y'^2 = 8$$

or

$$\frac{x'^2}{4} + y'^2 = 1.$$

The eigenvectors $\mathbf{v}_1$ and $\mathbf{v}_2$ determine an orthonormal basis for $R^2$, which in turn determines the $x'y'$-coordinate system. The orthogonal matrix $M$ having the vectors $\mathbf{v}_1$ and $\mathbf{v}_2$ as its columns is the change of coordinates matrix; it changes the standard coordinates to the $\mathbf{v}$-coordinates. Since $\mathbf{v}_1$ and $\mathbf{v}_2$ make an angle of $\pi/4$ with $\mathbf{e}_1$ and $\mathbf{e}_2$, $M$ rotates the coordinate axes by $\pi/4$.

Almost all has been revealed. What remains is to note that we have some choice in how we choose the diagonal matrix $\Lambda$ and the change of coordinates matrix $M$. For instance, suppose that we choose

$$M_1 = \frac{1}{\sqrt{2}}\begin{bmatrix}-1 & 1\\ 1 & 1\end{bmatrix} \qquad \text{and} \qquad \Lambda = \begin{bmatrix}8 & \\ & 2\end{bmatrix}.$$

Then

$$\Lambda_1 = M_1^T A M_1 \qquad \text{or} \qquad A = M_1\Lambda_1 M_1^T$$

and hence

$$\mathbf{x}^T A\mathbf{x} = (M_1^T\mathbf{x})^T\Lambda_1(M_1^T\mathbf{x}).$$

Setting $\mathbf{x}'' = M_1^T\mathbf{x}$, we obtain

$$\mathbf{x}^T A\mathbf{x} = \mathbf{x}''^T\Lambda_1\mathbf{x}'' = 8x''^2 + 2y''^2.$$

Thus our original equation (1) becomes

$$8x''^2 + 2y''^2 = 8 \qquad \text{or} \qquad x''^2 + \frac{y''^2}{4} = 1.$$

The columns of $M_1$ are $\mathbf{v}_2$ and $\mathbf{v}_1$ (in that order) and these vectors determine the $x''y''$-coordinate system. The graph in this coordinate system is of course the same as the graph in the $xy$-coordinate system or the $x'y'$-coordinate system. It is simply a question of how we label the axes. However, $M_1$ is not a rotation; it is a reflection in the line $y = x$ followed by a rotation by $\pi/4$.

These are two other possible choices for $M$:

$$M_2 = \frac{1}{\sqrt{2}}\begin{bmatrix} 1 & 1 \\ 1 & -1 \end{bmatrix} \qquad \text{and} \qquad M_3 = \frac{1}{\sqrt{2}}\begin{bmatrix} 1 & 1 \\ -1 & 1 \end{bmatrix}.$$

The corresponding diagonal matrices are

$$\Lambda_2 = \begin{bmatrix} 2 & \\ & 8 \end{bmatrix} \qquad \text{and} \qquad \Lambda_3 = \begin{bmatrix} 8 & \\ & 2 \end{bmatrix}.$$

$M_2$ represents a reflection in the line $y = x$ followed by a rotation by $-\pi/4$ and $M_3$ represents a rotation by $-\pi/4$.

***Definition*** Let $A$ be an $n \times n$ symmetric matrix. The function $Q\colon R^n \to R$ defined by

$$Q(\mathbf{x}) = \mathbf{x}^T A \mathbf{x}$$

is called a **quadratic form** in $x_1, x_2, \ldots, x_n$.

For example, a quadratic form in $x_1$ and $x_2$ is an expression

$$Q(\mathbf{x}) = [x_1 \quad x_2]\begin{bmatrix} a & b \\ b & c \end{bmatrix}\begin{bmatrix} x_1 \\ x_2 \end{bmatrix} = ax_1^2 + 2bx_1x_2 + cx_2^2,$$

and a quadratic form in $x_1, x_2, x_3$ is

$$\begin{aligned} Q(\mathbf{x}) = [x_1 \quad x_2 \quad x_3]&\begin{bmatrix} a & b & c \\ b & d & e \\ c & e & f \end{bmatrix}\begin{bmatrix} x_1 \\ x_2 \\ x_3 \end{bmatrix} \\ &= ax_1^2 + 2bx_1x_2 + 2cx_1x_3 + dx_2^2 + 2ex_2x_3 + fx_3^2. \end{aligned}$$

When $\Lambda$ is an $n \times n$ diagonal matrix, then the quadratic form determined by $\Lambda$ has a particularly nice form; it is a sum of squares. That is,

$$\mathbf{x}^T \Lambda \mathbf{x} = \lambda_1 x_1^2 + \lambda_2 x_2^2 + \cdots + \lambda_n x_n^2,$$

where $\lambda_1, \lambda_2, \ldots, \lambda_n$ are the diagonal entries in $\Lambda$.

Now let's investigate what happens to a quadratic form $\mathbf{x}^T A \mathbf{x}$ when we orthogonally diagonalize the symmetric matrix $A$. If $M$ is an orthogonal matrix which diagonalizes $A$, then

$$\Lambda = M^T A M \qquad \text{or} \qquad A = M \Lambda M^T.$$

The columns $\mathbf{v}_1, \mathbf{v}_2, \ldots, \mathbf{v}_n$ of $M$ are eigenvectors of $A$ and the diagonal entries $\lambda_1, \lambda_2, \ldots, \lambda_n$ are the corresponding eigenvalues. Moreover,

$$\begin{aligned}\mathbf{x}^T A \mathbf{x} &= \mathbf{x}^T (M \Lambda M^T) \mathbf{x} = (M^T \mathbf{x})^T \Lambda (M^T \mathbf{x}) \\ &= \mathbf{y}^T \Lambda \mathbf{y} \qquad (\mathbf{y} = M^T \mathbf{x}) \\ &= \lambda_1 y_1^2 + \lambda_2 y_2^2 + \cdots + \lambda_n y_n^2.\end{aligned}$$

Thus orthogonally diagonalizing the matrix $A$ allows us to express the quadratic form as a sum of squares in the coordinate system determined by the diagonalizing matrix $M$. The eigenvectors of $A$ determine the coordinate system. The eigenvalues are the coefficients in the expression of $\mathbf{x}^T A \mathbf{x}$ as a sum of squares.

**EXAMPLE 1**

Consider a quadratic equation in two variables $x$ and $y$:

$$ax^2 + bxy + cy^2 = d, \tag{2}$$

or in matrix form

$$\mathbf{x}^T A \mathbf{x} = [x \quad y] \begin{bmatrix} a & b/2 \\ b/2 & c \end{bmatrix} \begin{bmatrix} x \\ y \end{bmatrix} = d.$$

Let $\lambda_1$ and $\lambda_2$ be the eigenvalues of $A$ and $\mathbf{v}_1$ and $\mathbf{v}_2$ be corresponding unit eigenvectors. Then if $\Lambda$ is the diagonal matrix having $\lambda_1$ and $\lambda_2$ as its columns and $M$ is the orthogonal matrix having $\mathbf{v}_1$ and $\mathbf{v}_2$ as its columns, then

$$A = M \Lambda M^T.$$

If $\mathbf{x}' = M^T \mathbf{x}$, then

$$\begin{aligned}\mathbf{x}^T A \mathbf{x} &= (M^T \mathbf{x})^T \Lambda (M^T \mathbf{x}) \\ &= \lambda_1 x'^2 + \lambda_2 y'^2.\end{aligned}$$

Thus the quadratic equation (2) becomes

$$\lambda_1 x'^2 + \lambda_2 y'^2 = d \tag{3}$$

in the $x'y'$-coordinate system. Consequently, the quadratic equation (2) represents a (possibly degenerate) conic. Specifically, from (3) we have:

1. If $\lambda_1 \lambda_2 > 0$, then $\lambda_1$ and $\lambda_2$ are either both positive or both negative. Thus (3) is the equation of an ellipse or a circle if $d$ has the same sign as $\lambda_1$ and $\lambda_2$. If $d = 0$, then $(0, 0)$ is the only solution and if the sign of $d$ is opposite that of $\lambda_1$ and $\lambda_2$, there are no solutions.
2. If $\lambda_1 \lambda_2 < 0$, then $\lambda_1$ and $\lambda_2$ are nonzero numbers with opposite sign. If $d \neq 0$, then (3) is the equation of a hyperbola. If $d = 0$, then it is the equation of two lines intersecting at the origin.

3. If $\lambda_1\lambda_2 = 0$, then at least one of $\lambda_1$ and $\lambda_2$ is zero. If one of $\lambda_1$ and $\lambda_2$ is not zero, say $\lambda_1$, then (3) is the equation of two parallel lines if $d$ has the same sign as $\lambda_1$; if the sign of $d$ is opposite that of $\lambda_1$, then there are no solutions of the equation; if $d = 0$, its graph is the $y'$-axis.

When (3) does not represent a circle, an ellipse, or a hyperbola, it is said to represent a **degenerate conic section.**

### EXAMPLE 2

Consider the quadratic equation in the three variables $x$, $y$, $z$:

$$\mathbf{x}^T A\mathbf{x} = [x \quad y \quad z]\begin{bmatrix} 5 & -1 & -1 \\ -1 & 3 & 1 \\ -1 & 1 & 3 \end{bmatrix}\begin{bmatrix} x \\ y \\ z \end{bmatrix} = 1. \tag{4}$$

The eigenvalues of $A$ are $\lambda = 2$, $\lambda = 3$, and $\lambda = 6$ and the corresponding unit eigenvectors are

$$\frac{1}{\sqrt{2}}(0, 1, -1), \qquad \frac{1}{\sqrt{3}}(1, 1, 1), \qquad \frac{1}{\sqrt{6}}(-2, 1, 1).$$

If $M$ is the matrix having these vectors as its columns and $\Lambda$ is the diagonal matrix having 2, 3, 6 as its diagonal entries, then the substitution

$$A = M\Lambda M^T$$

reduces (4) to

$$(M^T\mathbf{x})^T\Lambda(M^T\mathbf{x}) = 2x'^2 + 3y'^2 + 6z'^2 = 1.$$

The graph of this equation is an ellipsoid (see Figure 8.3).

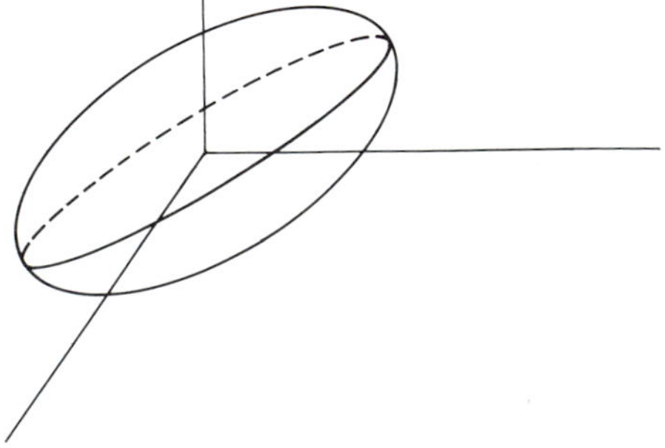

Figure 8.3

A Cartesian equation of the form

$$ax^2 + bxy + cy^2 + dx + ey = f \tag{5}$$

represents a conic section (i.e., a circle, ellipse, hyperbola, parabola) or a degenerate conic section (i.e., the empty set, a single point, a straight line, two straight lines). Moreover, the type of conic is determined by the quadratic form

$$ax^2 + bxy + cy^2.$$

To prove this we begin by writing (5) as a matrix equation:

$$[x \quad y]\begin{bmatrix} a & b/2 \\ b/2 & c \end{bmatrix}\begin{bmatrix} x \\ y \end{bmatrix} + [d \quad e]\begin{bmatrix} x \\ y \end{bmatrix} = f. \tag{6}$$

If we diagonalize $A$ with an orthogonal matrix $M$ we obtain

$$\begin{aligned} \mathbf{x}^T A\mathbf{x} &= (M^T\mathbf{x})^T\Lambda(M^T\mathbf{x}) \\ &= \mathbf{x}'^T\Lambda\mathbf{x}', \qquad \mathbf{x}' = M^T\mathbf{x}. \end{aligned}$$

Thus $\mathbf{x} = M\mathbf{x}'$ and (6) reduces to

$$\lambda_1 x'^2 + \lambda_2 y'^2 + [d \quad e]M\begin{bmatrix} x' \\ y' \end{bmatrix} = f.$$

If we let

$$[d' \quad e'] = [d \quad e]M,$$

then this equation reduces to

$$\lambda_1 x'^2 + \lambda_2 y'^2 + d'x' + e'y' = f. \tag{7}$$

If $\lambda_1$ and $\lambda_2$ are not zero, then by completing the square in $x'$ and $y'$ we obtain an equation of the form

$$\lambda_1(x' - x'_0)^2 + \lambda_2(y' - y'_0)^2 = \gamma.$$

This equation specifies the same conic (ellipse or hyperbola if the conic is not degenerate) as

$$\lambda_1 x'^2 + \lambda_2 y'^2 = \gamma$$

but with center at $(x'_0, y'_0)$. If $\lambda_1$ or $\lambda_2$ is zero, then (7) can be written as

$$\lambda_1(x' - x'_0)^2 + e'y' = \gamma \qquad (\lambda_2 = 0)$$

or

$$\lambda_2(y' - y'_0)^2 + d'x' = \gamma \qquad (\lambda_1 = 0).$$

These equations represents the same conics (a parabola if the conic is not degenerate) as

$$\lambda_1 x'^2 + e'y' = \gamma$$

or

$$\lambda_2 y'^2 + d'x' = \gamma.$$

***Theorem 1*** *Suppose that*

$$ax^2 + bxy + cy^2 + dx + ey = f \tag{8}$$

*specifies a nondegenerate conic. If $\lambda_1, \lambda_2$ are the eigenvalues of*

$$A = \begin{bmatrix} a & b/2 \\ b/2 & c \end{bmatrix},$$

*then* (8) *represents:*

(*a*) *an ellipse if $\lambda_1$ and $\lambda_2$ have the same sign;*
(*b*) *a hyperbola if $\lambda_1$ and $\lambda_2$ have opposite sign;*
(*c*) *a parabola if either $\lambda_1$ or $\lambda_2$ is zero.*

### EXAMPLE 3

Consider the equation

$$5x^2 - 6xy + 5y^2 + 20\sqrt{2}x + 4\sqrt{2}y = -64. \tag{9}$$

We have shown that if

$$A = \begin{bmatrix} 5 & -3 \\ -3 & 5 \end{bmatrix}, \qquad M = \frac{1}{\sqrt{2}}\begin{bmatrix} 1 & -1 \\ 1 & 1 \end{bmatrix}, \qquad \text{and} \qquad \Lambda = \begin{bmatrix} 2 & \\ & 8 \end{bmatrix},$$

then $A = M\Lambda M^T$ and

$$5x^2 - 6xy + 5y^2 = 2x'^2 + 8y'^2,$$

where $\mathbf{x}' = M^T\mathbf{x}$. Thus $\mathbf{x} = M\mathbf{x}'$ or

$$\begin{bmatrix} x \\ y \end{bmatrix} = \frac{1}{\sqrt{2}}\begin{bmatrix} 1 & -1 \\ 1 & 1 \end{bmatrix}\begin{bmatrix} x' \\ y' \end{bmatrix} = \frac{1}{\sqrt{2}}\begin{bmatrix} x' - y' \\ x' + y' \end{bmatrix}$$

and substituting

$$x = \frac{1}{\sqrt{2}}(x' - y'), \qquad y = \frac{1}{\sqrt{2}}(x' + y')$$

into (9), we obtain

$$\begin{aligned} 2x'^2 + 8y'^2 + 20(x' - y') + 4(x' + y') \\ = 2x'^2 + 8y'^2 + 24x' - 16y' = -64. \end{aligned}$$

After completing the square we have

$$2(x' + 6)^2 + 8(y' - 1)^2 = 16$$

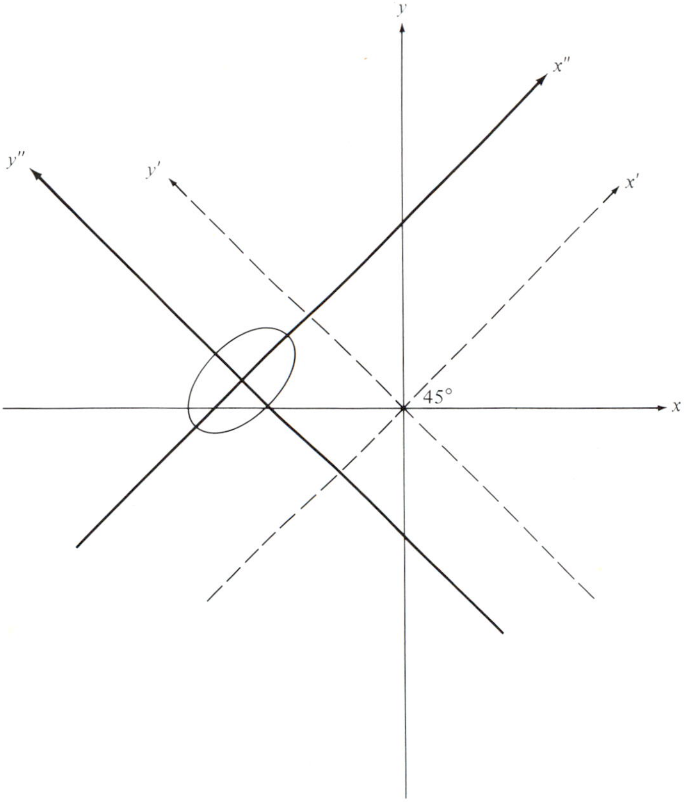


Figure 8.4

or

$$\frac{(x'+6)^2}{8} + \frac{(y'-1)^2}{2} = 1.$$

Setting $x'' = x' + 6$ and $y'' = y' - 1$, this equation further reduces to

$$\frac{x''^2}{8} + \frac{y''^2}{2} = 1$$

(see Figure 8.4). Geometrically, the last substitution is the same as introducing yet another coordinate system, the $x''y''$-coordinate system. The $x''y''$-coordinate axes are parallel to the $x'y'$-coordinate axes but with the new origin at $(-6, 1)$, the center of the ellipse.

So far we have concentrated on quadratic equations

$$\mathbf{x}^T A \mathbf{x} = d,$$

where $d$ is a constant. We now turn our attention to the quadratic form itself. The answers to certain questions in the calculus depend on whether a certain quadratic form is always positive, always negative, or both positive and negative. [For instance, it follows from Taylor's theorem that the nature of a critical point $(x_0, y_0)$ of a function $z = f(x, y)$ depends on whether the quadratic form

$$x^2 \frac{\partial^2 f}{\partial x^2}(x_0, y_0) + 2xy \frac{\partial^2 f}{\partial x \partial y}(x_0, y_0) + y^2 \frac{\partial^2 f}{\partial y^2}(x_0, y_0)$$

is always positive, always negative, or both positive and negative. The matrix of this quadratic form is called the **Hessian matrix**.]

***Definition*** Let $A$ be an $n \times n$ symmetric matrix. The quadratic form $Q(\mathbf{x}) = \mathbf{x}^T A\mathbf{x}$ and the matrix $A$ are called:

(a) **positive definite** if $Q(\mathbf{x}) > 0$ for all $\mathbf{x} \neq \mathbf{0}$;
(b) **positive semidefinite** if $Q(\mathbf{x}) \geq 0$ for all $\mathbf{x}$;
(c) **indefinite** if $Q(\mathbf{x}) > 0$ for some $\mathbf{x}$ and $Q(\mathbf{y}) < 0$ for some $\mathbf{y}$;
(d) **negative definite** if $Q(\mathbf{x}) < 0$ for all $\mathbf{x} \neq \mathbf{0}$;
(e) **negative semidefinite** if $Q(\mathbf{x}) \leq 0$ for all $\mathbf{x}$.

Clearly, a quadratic form $Q(\mathbf{x}) = \mathbf{x}^T A\mathbf{x}$ is negative definite if and only if $-Q(\mathbf{x}) = -\mathbf{x}^T A\mathbf{x}$ is positive definite, and negative semidefinite if and only if $-Q(\mathbf{x}) = -\mathbf{x}^T A\mathbf{x}$ is positive semidefinite.

***Theorem 2*** *Let $A$ be an $n \times n$ symmetric matrix, and let $Q(\mathbf{x}) = \mathbf{x}^T A\mathbf{x}$ be a quadratic form.*

*(a) $Q$ is positive definite if and only if all eigenvalues of $A$ are positive.*
*(b) $Q$ is positive semidefinite if and only if all eigenvalues of $A$ are nonnegative.*
*(c) $Q$ is indefinite if and only if $A$ has both positive and negative eigenvalues.*

**Proof** Let $\lambda_1, \lambda_2, \ldots, \lambda_n$ be the eigenvalues of $A$ and $\mathbf{v}_1, \mathbf{v}_2, \ldots, \mathbf{v}_n$ be corresponding unit eigenvectors. If $M$ is the matrix whose columns are $\mathbf{v}_1, \mathbf{v}_2, \ldots, \mathbf{v}_n$ and $\Lambda$ is the matrix whose diagonal entries are $\lambda_1, \lambda_2, \ldots, \lambda_n$, then

$$A = M\Lambda M^T.$$

Thus

$$Q(\mathbf{x}) = \mathbf{y}^T \Lambda \mathbf{y}, \qquad \mathbf{y} = M^T\mathbf{x}.$$

To prove (a), first suppose that $Q$ is positive definite. Then $Q(\mathbf{x}) > 0$ for all $\mathbf{x} \neq \mathbf{0}$. Since $M^T$ is nonsingular, if $\mathbf{x}$ is the solution of the equation $\mathbf{e}_k = M^T\mathbf{x}$, then $\mathbf{x} \neq \mathbf{0}$ and

$$0 < Q(\mathbf{x}) = \lambda_k.$$

Therefore, the eigenvalues of $A$ are positive. Conversely, suppose that all eigenvalues of $A$ are positive. Then

$$\mathbf{y}^T \Lambda \mathbf{y} > 0 \qquad \text{for all } \mathbf{y} \neq \mathbf{0}. \tag{10}$$

Since $\mathbf{x} = M\mathbf{y}$ and $M$ is nonsingular, $\mathbf{y} \neq \mathbf{0}$ if and only if $\mathbf{x} \neq \mathbf{0}$. Therefore, we can conclude from (10) that $Q(\mathbf{x}) > 0$ for all $\mathbf{x} \neq \mathbf{0}$. This proves part (a). The proofs of parts (b) and (c) are entirely analogous.

## PROBLEMS 8.7

**1.** For each of the following quadratic forms $Q$, do the following.
(i) Write it as a sum of squares.
(ii) Is it positive definite?
(iii) Graph $Q(\mathbf{x}) = 1$.
*(a) $3x^2 - 2\sqrt{2}xy + 2y^2$
(b) $4x^2 + 4xy + y^2$
*(c) $xy$
(d) $x^2 + 2xy + y^2$
*(e) $x^2 + 8xy + 7y^2$
(f) $2x^2 + 4\sqrt{3}xy - 2y^2$
*(g) $5x^2 - 6xy + 5y^2$
(h) $5x^2 + 14xy + 5y^2$
*(i) $4x^2 + 2xy + 4y^2$
(j) $6x^2 + 4xy + 6y^2$
*(k) $3x^2 + 2\sqrt{2}xy + 4y^2$

**2.** Repeat Problem 1 for each of the following quadratic forms in three variables.
*(a) $3x^2 + 2xy + 2xz + 4yz$
(b) $3x^2 + 2xz + 2y^2 + 3z^2$
*(c) $2x^2 + 2xy + 2xz + 2y^2 + 2yz + 2z^2$
(d) $x^2 + 2xy + 2xz + y^2 + 2yz + z^2$
*(e) $5x^2 + 4xy + 4xz + 5y^2 + 4yz + 5z^2$
(f) $x^2 + 2xy - 2yz + z^2$
*(g) $2x^2 - 2xy + 2y^2 - 2yz + 2z^2$

**3.** Sketch the graph of each of the following quadratic equations.
*(a) $16x^2 + 24xy + 9y^2 - 215x - 130y = 0$
(b) $x^2 - 2xy + y^2 + 16x + 16y = 0$
*(c) $4x^2 + 4xy + y^2 - x = 0$
(d) $2x^2 - 8xy - 4y^2 - 12\sqrt{5}x - 6\sqrt{5}y = 9$
*(e) $x^2 + y^2 + z^2 - 2x - 4y - 6z = 12$
(f) $2x^2 + 5y^2 + 2z^2 - 4xy - 2xz + 4yz + \sqrt{3}x + \sqrt{3}z = 1/2$
*(g) $3x^2 - 2xy + 3y^2 - 4z^2 + 8z = 8$
(h) $xy - z = 0$

**4.** Find a symmetric matrix $A$ such that

$$\mathbf{x}^TA\mathbf{x} = (a_1x_1 + a_2x_2 + a_3x_3)^2.$$

Generalize.

***5.** Show that the quadratic form

$$Q(\mathbf{x}) = x_1^2 + (x_1 + x_2)^2 + (x_1 + x_2 + x_3)^2$$

is positive definite. Is the quadratic form

$$Q(\mathbf{x}) = x_1^2 + (x_1 + x_2)^2 + (x_1 + x_2 + x_3)^2 + (x_1 + x_2 + x_3 + x_4)^2$$

positive definite? Generalize.

**6.** Let $A$ and $B$ be symmetric matrices. Show that $\mathbf{x}^TA\mathbf{x} = \mathbf{x}^TB\mathbf{x}$ for all $\mathbf{x}$ if and only if $A = B$.

***7.** Let $A$ be a symmetric matrix. Show that $A$ is positive definite if and only if $A$ is nonsingular and $A^{-1}$ is symmetric and positive definite.

**8.** Show that a matrix is positive definite and symmetric if and only if there is a nonsingular matrix $B$ such that $A = B^TB$. (*Hint:* When $A = B^TB$, compute $\mathbf{x}^TA\mathbf{x}$. When $A$ is positive definite and symmetric, let $B = \Lambda_1 M$ where $M$ is an orthogonal matrix that diagonalizes $A$ and $\Lambda_1$ is a diagonal matrix having diagonal entries $\sqrt{\lambda_1}, \sqrt{\lambda_2}, \ldots, \sqrt{\lambda_n}$.)

**9.** Let $A$ be a positive definite matrix. Prove each of the following.
(a) $A$ is nonsingular.
(b) $A^k$ is positive definite for all integers $k$.
(c) $M^{-1}AM$ is positive definite for any nonsingular matrix $M$.
(d) $M^TAM$ is positive definite for any nonsingular matrix $M$.

(e) If $B$ is also positive definite, then $A + B$ is positive definite.

*10. Let $A$ be an $n \times n$ symmetric matrix and let $Q(\mathbf{x}) = \mathbf{x}^T A\mathbf{x}$. Suppose that $Q$ is positive definite.
(a) Is $A$ nonsingular?
(b) Is the quadratic form $Q_1(\mathbf{x}) = \mathbf{x}^T A^2 \mathbf{x}$ positive definite?

11. Let $A$ and $B$ be $n \times n$ symmetric matrices. Show that if $A$ and $B$ are positive definite, then $AB$ is positive definite. (*Hint:* Use Theorem 2.)

*12. Show that if $A$ is nonsingular, then $A^T A$ is positive definite.

13. What condition on a matrix $W$ will guarantee that the quadratic form $\mathbf{x}^T(W^T W)\mathbf{x}$ is positive definite?

14. Show that $ax^2 + bxy + cy^2 = d$, $d \neq 0$, is the equation of:
(a) A circle, an ellipse, or a degenerate conic if $\det A = ac - b^2/4 > 0$.
(b) A hyperbola if $\det A = ac - b^2/4 < 0$.
(c) A degenerate conic section if $\det A = ac - b^2/4 = 0$.

15. Prove parts (b) and (c) of Theorem 2.

16. Let $A$ be an $n \times n$ symmetric matrix.
*(a) Show that if $a_{ii} = 0$ for some $i$, then $A$ is not positive definite nor negative definite. Give an example to show, however, that such a matrix can be positive semidefinite or negative semidefinite.
(b) Suppose that $a_{ii} = a_{jj} = 0$ for $i \neq j$. Show that if $a_{ij} \neq 0$, then $A$ is indefinite. (*Hint:* Consider $\mathbf{x}^T A\mathbf{x}$ when $\mathbf{x} = \mathbf{e}_i \pm \mathbf{e}_j$.)
(c) Show that if $a_{ii} = 0$ for all $i$ and $A \neq 0$, then $A$ is indefinite.
(d) Show that if $A$ has both positive and negative diagonal entries, then $A$ is indefinite.

# 8.8 Diagonal Representation of a Linear Transformation

In this section we bring together our results on linear transformations and matrices. We have just proved that an $m \times m$ matrix $A$ is diagonalizable (i.e., similar to a diagonal matrix) if and only if there is a basis for $R^m$ consisting of eigenvectors of $A$. In Section 5.5 we proved that a linear transformation $T\colon R^m \to R^m$ can be represented by a diagonal matrix if and only if there is a basis for $R^m$ consisting of eigenvectors of $T$. In Section 6.3 we observed that this last fact remains true when $T\colon V \to V$ is a linear transformation from any $m$-dimensional vector space $V$ into itself. If $v_1, v_2, \ldots, v_m$ is a basis for $V$, then any linear transformation $T\colon V \to V$ is represented by an $m \times m$ matrix $A$ with respect to this basis. The question that is begging to be asked is: What is the relationship (if any) between $T$ being represented by a diagonal matrix with respect to some basis for $V$ and $A$ being diagonalizable? In view of what we have just said, we will be able to answer the question if we can figure out the relationship between the eigenvectors of $T$ and the eigenvectors of $A$.

We already know this relationship in one important special case. When $T\colon R^m \to R^m$, then the eigenvectors of $T$ are exactly the eigenvectors of its standard matrix. This proves

***Theorem 1*** *Let $T: R^m \to R^m$ be a linear transformation, and let $A$ be its standard matrix. Then $T$ can be represented by a diagonal matrix if and only if $A$ is diagonalizable.*

Now let's move to the general situation. Let $V$ be an $m$-dimensional vector space and let $T: V \to V$ be a linear transformation. Choose a basis $v_1, v_2, \ldots, v_m$ for $V$ and let $A$ be the matrix of $T$ with respect to this basis. A nonzero vector $x$ in $V$ is an eigenvector of $T$ if there is a scalar $\lambda$ such that

$$T(x) = \lambda x.$$

The matrix $A$ represents $T$ by sending the $v$-coordinates of $x$ to the $v$-coordinates of $T(x)$:

$$\mathbf{c}_v(T(x)) = A\mathbf{c}_v(x).$$

Therefore, if $T(x) = \lambda x$, then since the coordinate map $\mathbf{c}_v: V \to R^m$ is linear and one-to-one we have that $\mathbf{c}_v(x) \neq \mathbf{0}$ and

$$\lambda\mathbf{c}_v(x) = \mathbf{c}_v(\lambda x) = \mathbf{c}_v(T(x)) = A\mathbf{c}_v(x).$$

Thus the vector $\mathbf{c}_v(x)$ is an eigenvector of $A$ with eigenvalue $\lambda$. Conversely, let $\mathbf{u}$ be an eigenvector of $A$ with eigenvalue $\mu$. Then $\mathbf{u} = (u_1, u_2, \ldots, u_m) \neq \mathbf{0}$ and the vector $x = u_1v_1 + u_2v_2 + \cdots + u_mv_m$ is the unique nonzero vector in $V$ such that $\mathbf{c}_v(x) = \mathbf{u}$. Then

$$A\mathbf{u} = \mu\mathbf{u}$$

implies that

$$\mathbf{c}_v(T(x)) = A\mathbf{c}_v(x) = A\mathbf{u} = \mu\mathbf{u} = \mu\mathbf{c}_v(x) = \mathbf{c}_v(\mu x).$$

Since the coordinate map $\mathbf{c}_v$ is one-to-one,

$$T(x) = \mu x.$$

Thus $x$ is an eigenvector of $T$ with eigenvalue $\mu$. Thus the coordinate map sets up a one-to-one correspondence between the eigenvectors of $T$ and the eigenvectors of $A$.

***Theorem 2*** *Suppose that $V$ is an m-dimensional vector space and $v_1, v_2, \ldots, v_m$ is a basis for $V$. Let $T: V \to V$ be a linear transformation and $A$ be the matrix of $T$ with respect to the basis $v_1, v_2, \ldots, v_m$ for $V$. Then:*

(*a*) *$x$ is an eigenvector of $T$ with eigenvalue $\lambda$ if and only if $\mathbf{c}_v(x)$ is an eigenvector of $A$ with eigenvalue $\lambda$.*
(*b*) *$\lambda$ is an eigenvalue of $T$ if and only if $\lambda$ is an eigenvalue of $A$.*

**Proof** We have already proved part (a). Part (b) follows from (a). For if $\lambda$ is an eigenvalue of $T$, then there is a nonzero vector $x$ in $V$ such that $T(x) = \lambda x$. Then as we have shown above $A\mathbf{c}_v(x) = \lambda\mathbf{c}_v(x)$ and so $\lambda$ is an eigenvalue of $A$.

Conversely, if $\mu$ is an eigenvalue of $A$, then there is a nonzero vector $\mathbf{u}$ in $R^m$ such that $A\mathbf{u} = \mu\mathbf{u}$. If $x$ is the unique nonzero vector in $V$ such that $\mathbf{c}_v(x) = \mathbf{u}$, then as we have shown above, $Tx = \mu x$. This completes the proof.

Theorem 2 provides us with a computational procedure for finding the eigenvalues and eigenvectors of a linear transformation $T: V \to V$ when $V$ is an $m$-dimensional vector space. Simply choose a basis for $V$ and let $A$ be the matrix of $T$ with respect to this basis. The eigenvalues of $A$ are the eigenvalues of $T$ and the eigenvectors of $A$ are the coordinate vectors of the eigenvectors of $T$.

### EXAMPLE 1

Let $T: P_2(R) \to P_2(R)$ be the linear transformation defined by

$$T(p)(x) = (1 + x^2)p(x) + (2x^2 - x^3)p'(x) + \left(\frac{1}{2} - 2x^3 + \frac{x^4}{2}\right)p''(x).$$

[Check that $T$ does, in fact, map $P_2(R)$ into $P_2(R)$.] Since

$$\begin{aligned} T(1) &= 1 + 0\cdot x + x^2 \\ T(x) &= 0 + x + 2x^2 \\ T(x^2) &= 1 + 0\cdot x + x^2, \end{aligned}$$

the matrix of $T$ with respect to the basis $1, x, x^2$ is

$$A = \begin{bmatrix} 1 & 0 & 1 \\ 0 & 1 & 0 \\ 1 & 2 & 1 \end{bmatrix}.$$

In Example 1 of Section 8.2 we showed that the eigenvalues of $A$ are 0, 1, and 2 and the corresponding eigenvectors are

$$(-1, 0, 1), \quad (-2, 1, 0), \quad (1, 0, 1).$$

Thus these vectors are the coordinate vectors with respect to the basis $1, x, x^2$ of the eigenvectors of $T$. Therefore, the eigenvectors of $T$ are

$$-1 + x^2, \quad -2 + x, \quad 1 + x^2.$$

Now that we have worked out the relationship between the eigenvectors and eigenvalues of a linear transformation and the eigenvectors and eigenvalues of a matrix representing the linear transformation, let us return to our original question. Suppose that $V$ is an $m$-dimensional vector space and $v_1, v_2, \ldots, v_m$ is a basis for $V$. Let $T: V \to V$ be a linear transformation and let $A$ be the matrix of $T$ with respect to the basis $v_1, v_2, \ldots, v_m$. What is the relationship between $T$ being representable by a diagonal matrix with respect to some basis and $A$ being diagonalizable? Our work so far clearly suggests that both conditions are equivalent to there being a basis for $V$ (or equivalently for $R^m$) consisting of eigenvectors of $T$ (or equivalently of eigenvectors of $A$). Let's see if it is right. $T: V \to V$ can be represented by a diagonal matrix if and only if there is a basis for $V$ consisting of eigenvectors of $T$. Since the coordinate map

$$\mathbf{c}_v\colon V \to R^m$$

is an isomorphism that maps eigenvectors of $T$ to eigenvectors of $A$, there is a basis for $V$ consisting of eigenvectors of $T$ if and only if there is a basis for $R^m$ consisting of eigenvectors of $A$. This last condition is in turn equivalent to $A$ being diagonalizable.

***Theorem 3*** *Suppose that $V$ is a vector space of dimension $m$ and $v_1, v_2, \ldots, v_m$ is a basis for $V$. Let $T\colon V \to V$ be a linear transformation and let $A$ be the matrix of $T$ with respect to this basis. Then $T$ can be represented by a diagonal matrix if and only if $A$ is diagonalizable.*

The following theorem is an immediate consequence of this theorem (see Problem 8).

***Theorem 4*** *Let $V$ be an $m$-dimensional vector space, and let $T\colon V \to V$ be a linear transformation. $T$ can be represented by a diagonal matrix if and only if any matrix that represents $T$ is diagonalizable.*

## EXAMPLE 2

Let $V = P_2(R)$ be the collection of all polynomials of degree $\leq 2$, and let $T\colon V \to V$ be the linear transformation defined by

$$T(p)(x) = (x + 1)^2 p''(x) + 3(x + 1)p'(x) + 2p(x).$$

We will show that $T$ can be represented by a diagonal matrix. We begin by choosing the basis $1, x, x^2$ for $V$. Since

$$\begin{aligned} T(1) &= 2 \\ T(x) &= 3(x + 1) + 2x = 3 + 5x \\ T(x^2) &= 2(x + 1)^2 + 6(x + 1)x + 2x^2 = 2 + 10x + 10x^2, \end{aligned}$$

the matrix of $T$ with respect to this basis is

$$A = \begin{bmatrix} 2 & 3 & 2 \\ 0 & 5 & 10 \\ 0 & 0 & 10 \end{bmatrix}.$$

Since $A$ has three distinct eigenvalues, $A$ is diagonalizable. Hence $T$ can be represented by a diagonal matrix. In fact, since the eigenvalues of $T$ are the eigenvalues of $A$, $T$ can be represented by the diagonal matrix

$$\Lambda = \begin{bmatrix} 2 & & \\ & 5 & \\ & & 10 \end{bmatrix}.$$

Let $v_1, v_2, \ldots, v_m$ be a basis for $V$ and let $A$ be the matrix of a linear transformation $T\colon V \to V$ with respect to this basis. If $T$ can be represented by a diagonal matrix $\Lambda$, what is the relationship between $\Lambda$ and $A$? The diagonal matrix $\Lambda$ represents $T$ with respect to a basis $x_1, x_2, \ldots, x_m$ for $V$ consisting of eigenvectors of $T$, and the diagonal entries of $\Lambda$ are the corresponding eigenvalues. By Theorem 3 of Section 6.3,

$$\Lambda = M^{-1}AM, \tag{1}$$

where $M$ is the change of coordinates matrix which changes $x$-coordinates to $v$-coordinates:

$$M\mathbf{c}_x(y) = \mathbf{c}_v(y).$$

The columns of $M$ are the vectors

$$\mathbf{c}_v(x_1), \mathbf{c}_v(x_2), \ldots, \mathbf{c}_v(x_m).$$

By Theorem 2, these vectors are the eigenvectors of $A$, and the diagonal entries of $\Lambda$ are the corresponding eigenvalues of $A$. This is in exact agreement with Theorem 1 of Section 8.3! Equation (1) says that $A$ is diagonalizable and that the columns of $M$ are the eigenvectors of $A$ and the diagonal entries of $\Lambda$ are the corresponding eigenvalues. Everything is as it ought to be! When $T$ is represented by a diagonal matrix, then $A$ is diagonalizable. The change of coordinates matrix which determines the relationship between the two matrices $A$ and $\Lambda$ representing $T$ is exactly the diagonalizing matrix.

Now let's work through the reasoning beginning with the assumption that $A$ diagonalizable. Then by Theorem 1 of Section 8.3,

$$\Lambda = M^{-1}AM,$$

where $M$ is the matrix whose columns are the eigenvectors $\mathbf{v}_1, \mathbf{v}_2, \ldots, \mathbf{v}_m$ of $A$ and $\Lambda$ is the diagonal matrix whose diagonal entries are the corresponding eigenvalues. By Theorem 2 the vectors $x_1, x_2, \ldots, x_m$ in $V$ such that $\mathbf{c}_v(x_1) = \mathbf{v}_1, \mathbf{c}_v(x_2) = \mathbf{v}_2, \ldots, \mathbf{c}_v(x_m) = \mathbf{v}_m$ are eigenvectors of $T$. Thus the matrix of $T$ with respect to the basis $x_1, x_2, \ldots, x_m$ is the diagonal matrix $\Lambda$. Since the columns of the matrix $M$ are $\mathbf{c}_v(x_1), \mathbf{c}_v(x_2), \ldots, \mathbf{c}_v(x_m)$, $M$ is the change of coordinates matrix which changes $x$-coordinates to $\mathbf{v}$-coordinates. This is in exact agreement with Theorem 3 of Section 6.3.

## EXAMPLE 3

Let $V = P_2(R)$. In Example 2 we showed that the linear transformation $T\colon V \to V$ defined by

$$T(p)(x) = (x+1)^2 p''(x) + 3(x+1)p'(x) + 2p(x)$$

can be represented by a diagonal matrix

$$\Lambda = \begin{bmatrix} 2 & & \\ & 5 & \\ & & 10 \end{bmatrix}.$$

We concluded this from the fact that the matrix of $T$ with respect to the basis $1, x, x^2$ is

$$A = \begin{bmatrix} 2 & 3 & 2 \\ 0 & 5 & 10 \\ 0 & 0 & 10 \end{bmatrix}.$$

Now let's use $A$ to determine the basis of eigenvectors of $T$. The eigenvectors of $A$ corresponding to the eigenvalues 2, 5, 10 are easily found to be

$$(1, 0, 0), \quad (1, 1, 0), \quad (1, 2, 1).$$

By Theorem 2, these eigenvectors are the coordinates of the eigenvectors of $T$ with respect to the basis $1, x, x^2$. Therefore, the eigenvectors of $T$ are

$$\begin{aligned} 1 &= 1 \cdot 1 + 0 \cdot x + 0 \cdot x^2 \\ 1 + x &= 1 \cdot 1 + 1 \cdot x + 0 \cdot x^2 \\ (1 + x)^2 &= 1 \cdot 1 + 2 \cdot x + 1 \cdot x^2. \end{aligned}$$

## EXAMPLE 4

In Example 5 of Section 6.3 we showed that if $T\colon P_3(R) \to P_3(R)$ is the linear transformation defined by

$$T(p)(x) = xp'(x)$$

is represented by the diagonal matrix

$$\Lambda = \begin{bmatrix} 0 & & & \\ & 1 & & \\ & & 2 & \\ & & & 3 \end{bmatrix}$$

with respect to the basis $1, x, x^2, x^3$. Thus its matrix with respect to any other basis must be diagonalizable. For instance, choose the basis

$$1 + x, \quad 1 - x, \quad 2 + x^2, \quad 1 + x + x^3.$$

Since

$$\begin{aligned} T(1 + x) &= x &&= (\tfrac{1}{2})(1 + x) - (\tfrac{1}{2})(1 - x) \\ T(1 - x) &= -x &&= (-\tfrac{1}{2})(1 + x) + (\tfrac{1}{2})(1 - x) \\ T(2 + x^2) &= 2x^2 &&= -2(1 + x) - 2(1 - x) + 2(2 + x^2) \\ T(1 + x + x^3) &= x + 3x^3 &&= (-\tfrac{5}{2})(1 + x) - (\tfrac{1}{2})(1 - x) + 3(1 + x + x^3), \end{aligned}$$

the matrix of $T$ with respect to this basis is

$$A = \begin{bmatrix} \frac{1}{2} & -\frac{1}{2} & -2 & -\frac{5}{2} \\ -\frac{1}{2} & \frac{1}{2} & -2 & -\frac{1}{2} \\ 0 & 0 & 2 & 0 \\ 0 & 0 & 0 & 3 \end{bmatrix}.$$

By Theorem 3 of Section 6.3,

$$\Lambda = M^{-1}AM,$$

where $M$ is the change of coordinate matrix whose columns are the coordinates of the vectors $1, x, x^2, x^3$ with respect to the basis $1 + x$, $1 - x$, $2 + x^2$, $1 + x + x^3$. Since

$$\begin{aligned} 1 &= (\tfrac{1}{2})(1 + x) + (\tfrac{1}{2})(1 - x) \\ x &= (\tfrac{1}{2})(1 + x) - (\tfrac{1}{2})(1 - x) \\ x^2 &= -1(1 + x) - 1(1 - x) + 1(2 + x^2) \\ x^3 &= -1(1 + x) + 1(1 + x + x^3), \end{aligned}$$

$M$ is the matrix

$$M = \begin{bmatrix} \frac{1}{2} & \frac{1}{2} & -1 & -1 \\ \frac{1}{2} & -\frac{1}{2} & -1 & 0 \\ 0 & 0 & 1 & 0 \\ 0 & 0 & 0 & 1 \end{bmatrix}.$$

Check that $\Lambda = M^{-1}AM$. The columns of $M$ are the eigenvectors of $A$ and the diagonal entries in $\Lambda$ are the corresponding eigenvalues.

## PROBLEMS 8.8

**1.** Determine which of the following linear transformations $T$ can be represented by a diagonal matrix. If the linear transformation can be represented by a diagonal matrix, find a diagonal matrix that represents $T$ and find the basis that gives rise to this matrix.

*(a) $T(x_1, x_2) = (2x_1 + 2x_2, 2x_1 + 5x_2)$

(b) $T(x_1, x_2) = (x_1 + x_2, x_2)$

*(c) $T(x_1, x_2, x_3) = (2x_1 + x_2 + x_3, 2x_1 + 3x_2 + 2x_3, 3x_1 + 3x_2 + 4x_3)$

(d) $T(x_1, x_2, x_3) = (2x_1, 3x_1 + 2x_2, 5x_1 + 2x_2 - x_3)$

**2.** Let $V$ be the vector space of all $2 \times 2$ matrices.

*(a) Show that $T: V \to V$ defined by $T(A) = A + A^T$ is a linear transformation.

*(b) Show that $T$ can be represented by a diagonal matrix, and find the corresponding basis for $V$.

(c) Repeat parts (a) and (b) for $T(A) = \alpha A + \beta A^T$, where $\alpha$ and $\beta$ are fixed scalars.

***3.** Let $V$ be a three-dimensional vector space, let $v_1, v_2, v_3$ be a basis for $V$, and let $T: V \to V$ be a linear transformation. Suppose that the matrix of $T$ with respect to the basis $v_1, v_2, v_3$ is

$$\begin{bmatrix} 2 & 2 & 1 \\ 1 & 3 & 1 \\ 1 & 2 & 2 \end{bmatrix}.$$

(a) Find the eigenvalues of $T$.

(b) Express the eigenvectors of $T$ in terms of the basis $v_1, v_2, v_3$.

(c) If possible, represent $T$ by a diagonal matrix.

**4.** Repeat Problem 3 when the matrix of $T$ with respect to a given basis $v_1, v_2, v_3$ is

$$\begin{bmatrix} 1 & -1 & 0 \\ -1 & 2 & -1 \\ 0 & -1 & 1 \end{bmatrix}.$$

***5.** Let $T: P_2(R) \to P_2(R)$ be the linear transformation defined by

$$T(p)(x) = 2p(x) + (-x^2 + x - 5)p'(x) + \left(x^3 - \frac{x^2}{2} + \frac{9}{2}x + \frac{5}{2}\right)p''(x).$$

Show that $T$ can be represented by a diagonal matrix, and find the corresponding basis for $P_2(R)$.

***6.** Let $T: P_2(R) \to P_2(R)$ be the linear transformation defined by

$$T(p)(x) = (4x^2 - 1)p(x) + (1 + 6x - 2x^2 - 4x^3)p'(x) + (2x^4 + 2x^3 - 3x^2 - x)p''(x).$$

Can $T$ be represented by a diagonal matrix?

**7.** Let $T: P_2(R) \to P_2(R)$ be the linear transformation defined by

$$T(p)(x) = p(x) - 2(1 + x)p'(x) + (1 + 3x + 2x^2)p''(x).$$

Can $T$ be represented by a diagonal matrix? What are its eigenvalues and eigenvectors?

**8.** Prove Theorem 4.

**9.** Interpret the following results about similar matrices in terms of linear transformations.

(a) Similar matrices have the same eigenvalues.

(b) If $A$ is similar to $B$ and $B$ is diagonalizable, then $A$ is diagonalizable.

**10.** Let $V$ be an $m$-dimensional inner product space and let $W$ be an $n$-dimensional subspace of $V$. Define the projection of $V$ onto $W$ and show that it can be represented by a diagonal matrix.

**11.** Let $V$ be an $n$-dimensional vector space and let $T: V \to V$ be a linear transformation. Suppose that $T$ can be represented by a diagonal matrix. Let $\lambda_1, \lambda_2, \ldots, \lambda_k$ be the distinct eigenvalues of $T$ and let $E(\lambda_1), E(\lambda_2), \ldots, E(\lambda_k)$ be the corresponding eigenspaces.

(a) Show that any vector $v$ in $V$ can be expressed uniquely as

$$v = v_1 + v_2 + \cdots + v_k,$$

where $v_i$ is in $E(\lambda_i)$, $i = 1, \ldots, k$.

(b) Show that for each $i = 1, \ldots, k$ there is a unique linear transformation $S_i: V \to V$ such that

$$S_i(v) = v_i$$

where $v = v_1 + v_2 + \cdots + v_k$ as in part (a).

(c) Show that $S_i^2 = S_i$.

(d) Show that $S_i \circ S_j = 0$ if $i \neq j$.

(e) Show that $S_1 + \cdots + S_k = I$, where $I: V \to V$ is the identity mapping.

(f) Show that the image of $V$ under $S_i$ is $E(\lambda_i)$.

(g) Show that $T = \lambda_1 S_1 + \cdots + \lambda_k S_k$.

**12.** Let $V$ be an $n$-dimensional vector space and let $T: V \to V$ be a linear transformation. Suppose that there are $k$ distinct scalars $\lambda_1, \lambda_2, \ldots, \lambda_k$ and $k$ nonzero linear transformations $S_1, S_2, \ldots, S_k$ mapping $V$ into itself such that

(a) $S_i \circ S_j = 0$ if $i \neq j$ and $S_i^2 = S_i$.

(b) $S_1 + S_2 + \cdots + S_k = I$

(c) $T = \lambda_1 S_1 + \lambda_2 S_2 + \cdots + \lambda_k S_k$.

Show that $T$ is diagonalizable and $\lambda_1, \lambda_2, \ldots, \lambda_k$ are the eigenvalues of $T$.

## 8.9 Richardson's Model for an Arms Race

In this section we study Richardson's model for an arms race between two nations. The formulation of this model leads naturally to a system of differential equations that will be studied in detail in the next four sections.

Suppose that there are two nations and that each nation is worried about an attempted takeover by the other. In such a situation each nation keeps a watchful eye on the armament expenditures of the other nation. The reason, of course, is obvious. If one nation increases its armament expenditures, then it increases its war potential and hence it becomes a greater threat to the security of the other nation. In an attempt to defend itself, the other nation may think it necessary to increase its own war potential and hence it may increase its armament expenditures. What effect will this have on the first nation? The increase in the expenditures of the second nation may lead the first nation to increase its expenditures even more. The process is self-perpetuating. However, since there are many other demands on the nations' resources, the cost of armaments will clearly have a restraining effect on this process.

Let $y = y(t)$ and $z = z(t)$ denote the armament expenditures of these two nations, and let $t$ stand for time in years. Our assumption about the foreign policies of these two nations is that the more one nation spends on armaments, the more the other nation is inclined to spend on armaments. There are many possible ways to interpret this assumption. We will interpret it to mean that each nation increases its armaments expenditures at a rate that is proportional to the existing armaments expenditures of the other nation. Similarly, we will assume that the cost of armaments creates a restraining effect that is proportional to the current armament expenditures. These assumptions lead to the following differential equations:

$$\frac{dy}{dt} = -ay + bz, \qquad \frac{dz}{dt} = cy - dz \qquad (a, b, c, d > 0).$$

This is a *system* of two differential equations. Our objective is to find the functions $y = y(t)$ and $z = z(t)$ that solve this system and hence describe the arms race between the two nations. We now turn our attention to this problem.

First, we rewrite the system in matrix notation. If we let $\mathbf{x} = (y, z)$, then

$$\frac{d\mathbf{x}}{dt} = \left(\frac{dy}{dt}, \frac{dz}{dt}\right)$$

and the system becomes

$$\frac{d\mathbf{x}}{dt} = A\mathbf{x}, \qquad \text{where} \qquad A = \begin{bmatrix} -a & b \\ c & -d \end{bmatrix}. \tag{1}$$

If we ignore momentarily the fact that this is a matrix equation, it looks exactly like the differential equation in Example 7 of Section 4.2. We found there that any solution to $dx/dt = \alpha x$ has the form $x(t) = ce^{\alpha t}$, where $c$ is a constant. This suggests in the present situation that any solution to $d\mathbf{x}/dt = A\mathbf{x}$ should be of the form $\mathbf{x}(t) = \mathbf{c}e^{At}$, where $\mathbf{c}$ is a vector. The only problem with this is that we do not know how to raise $e$ to a matrix power.

We must try something else. Perhaps our next best guess is a solution of the form $\mathbf{x}(t) = e^{\lambda t}\mathbf{v}$, where $\lambda$ is a scalar and $\mathbf{v}$ is a constant vector. If $\mathbf{v}$ is the zero vector, then $\mathbf{x}(t) = e^{\lambda t}\mathbf{v} = \mathbf{0}$ is a solution to (1). The interpretation of this solution is that there will be no arms race provided that the initial armament expenditure level of each nation was zero [for $\mathbf{x}(0) = (y(0), z(0))$ gives the initial armament levels of each nation]. If either nation has armaments at the beginning of our study, we must find a solution $\mathbf{x}(t)$ such that $\mathbf{x}(0) \neq \mathbf{0}$ and hence we must suppose that $\mathbf{v} \neq \mathbf{0}$. Let $y_0$ and $z_0$ denote the initial armament expenditures of each nation. The formulation of our model leads to the *initial value problem*

$$\frac{d\mathbf{x}}{dt} = A\mathbf{x}, \qquad \mathbf{x}(0) = \begin{bmatrix} y_0 \\ z_0 \end{bmatrix}. \tag{2}$$

If $\mathbf{x}(t) = e^{\lambda t}\mathbf{v}$, then

$$\frac{d\mathbf{x}}{dt} = \frac{d}{dt}\begin{bmatrix} e^{\lambda t}v_1 \\ e^{\lambda t}v_2 \end{bmatrix} = \begin{bmatrix} \lambda e^{\lambda t}v_1 \\ \lambda e^{\lambda t}v_2 \end{bmatrix} = \lambda e^{\lambda t}\mathbf{v},$$
$$A\mathbf{x} = Ae^{\lambda t}\mathbf{v} = e^{\lambda t}A\mathbf{v},$$

and our differential equation $d\mathbf{x}/dt = A\mathbf{x}$ becomes $\lambda e^{\lambda t}\mathbf{v} = e^{\lambda t}A\mathbf{v}$. Since $e^{\lambda t}$ is never zero, we may divide both sides of this equation by $e^{\lambda t}$ and obtain

$$A\mathbf{v} = \lambda\mathbf{v}.$$

We have obtained partial success; $\mathbf{x}(t) = e^{\lambda t}\mathbf{v}$ is a solution to the differential equation (1) if and only if $\lambda$ is an eigenvalue of $A$ with eigenvector $\mathbf{v}$.

Let us suppose we have determined that the matrix $A$ in (2) is

$$A = \begin{bmatrix} -2 & 4 \\ 1 & -2 \end{bmatrix}.$$

Since

$$\det(A - \lambda I) = \det\begin{bmatrix} -2-\lambda & 4 \\ 1 & -2-\lambda \end{bmatrix} = \lambda(\lambda + 4),$$

the eigenvalues of $A$ are $\lambda = 0$ and $\lambda = -4$. The eigenspaces $E(0)$ and $E(-4)$ are spanned by the eigenvectors $(2, 1)$ and $(-2, 1)$, respectively.

Thus we have arrived at two solutions to our differential equation: namely,

$$\mathbf{x}(t) = e^{0t}(2, 1) = (2, 1) \qquad \text{and} \qquad \mathbf{x}(t) = e^{-4t}(-2, 1).$$

Furthermore, it is clear that any linear combination

$$\mathbf{x}(t) = c_1(2, 1) + c_2e^{-4t}(-2, 1)$$

of these two solutions is also a solution. (Verify this.)

This last observation is important for the following reason. If we suppose that $y_0 = 2$ and $z_0 = 2$, we find that it is impossible to choose a constant $c$ such that

$$\mathbf{x}(t) = c(2, 1) \qquad \text{or} \qquad \mathbf{x}(t) = ce^{-4t}(-2, 1)$$

will satisfy $\mathbf{x}(0) = (2, 2)$. However, it is possible to choose two constants $c_1$ and $c_2$ such that

$$\mathbf{x}(t) = c_1(2, 1) + c_2e^{-4t}(-2, 1)$$

satisfies $\mathbf{x}(0) = (2, 2)$. For

$$c_1(2, 1) + c_2(-2, 1) = (2, 2)$$

yields the following system of equations:

$$\begin{bmatrix} 2 & -2 \\ 1 & 1 \end{bmatrix}\begin{bmatrix} c_1 \\ c_2 \end{bmatrix} = \begin{bmatrix} 2 \\ 2 \end{bmatrix}.$$

The solution of this system is $c_1 = \frac{3}{2}$ and $c_2 = \frac{1}{2}$, and hence the solution to the initial value problem (2) is

$$\mathbf{x}(t) = \frac{3}{2}(2, 1) + \frac{1}{2}e^{-4t}(-2, 1)$$

or

$$y(t) = 3 - e^{-4t}$$
$$z(t) = \frac{3}{2} + \frac{1}{2}e^{-4t}.$$

This is the mathematical solution to our model. Since

$$\lim_{t\to\infty} y(t) = 3 \qquad \text{and} \qquad \lim_{t\to\infty} z(t) = \frac{3}{2},$$

it predicts that the amount that each nation will spend on armaments will approach a limit. The first nation will tend to increase its expenditures by 50% and the second nation will decrease its expenditures by 25%.

## PROBLEMS 8.9

*1. Formulate a model for an arms race between three nations.

2. What changes in the assumptions of our model would have to be made in order to replace (1) by

$$\frac{d\mathbf{x}}{dt} = A\mathbf{x}, \qquad \text{where}$$
$$A = \begin{bmatrix} a & b \\ c & d \end{bmatrix}, \qquad a, b, c, d > 0.$$

3. Let $\lambda_1$ and $\lambda_2$ be distinct eigenvalues of the matrix

$$A = \begin{bmatrix} -a & b \\ c & -d \end{bmatrix},$$

where $a, b, c, d > 0$. Show that Richardson's model makes the following predictions.
   (a) If $\lambda_1 > 0$ and $\lambda_2$ is arbitrary, then there will be a runaway arms race.
   (b) If $\lambda_1 < 0$ and $\lambda_2 \leq 0$, then stability occurs.
   (c) If $\lambda_1 < 0$ and $\lambda_2 < 0$, then each nation disarms.

4. Let $x_1(t)$ and $x_2(t)$ be the populations of two species at time $t$. Discuss the assumptions that would lead to the equation $d\mathbf{x}/dt = A\mathbf{x}$ with

   (a) $A = \begin{bmatrix} a & b \\ c & d \end{bmatrix}$

   (b) $A = \begin{bmatrix} a & -b \\ -c & d \end{bmatrix}$

   (c) $A = \begin{bmatrix} a & -b \\ c & -d \end{bmatrix}$

Here $a, b, c, d > 0$.

# 8.10 Systems of Differential Equations

Richardson's model led to a system of differential equations of the form

$$\frac{dx_1}{dt} = a_{11}x_1 + a_{12}x_2$$
$$\frac{dx_2}{dt} = a_{21}x_1 + a_{22}x_2.$$

We saw that this system is represented in matrix notation by

$$\frac{d\mathbf{x}}{dt} = A\mathbf{x},$$

where

$$A = \begin{bmatrix} a_{11} & a_{12} \\ a_{21} & a_{22} \end{bmatrix}, \qquad \mathbf{x} = \begin{bmatrix} x_1 \\ x_2 \end{bmatrix}, \qquad \text{and} \qquad \frac{d\mathbf{x}}{dt} = \begin{bmatrix} dx_1/dt \\ dx_2/dt \end{bmatrix}.$$

In general, *a system of n differential equations* is a system of the form

$$\begin{aligned} \frac{dx_1}{dt} &= a_{11}x_1 + \cdots + a_{1n}x_n \\ &\vdots \\ \frac{dx_n}{dt} &= a_{n1}x_1 + \cdots + a_{nn}x_n. \end{aligned} \tag{1}$$

Using matrix notation, this system can be expressed by the single matrix equation

$$\frac{d\mathbf{x}}{dt} = A\mathbf{x}, \tag{1'}$$

where

$$A = \begin{bmatrix} a_{11} & \cdots & a_{1n} \\ \vdots & & \vdots \\ a_{n1} & \cdots & a_{nn} \end{bmatrix}, \qquad \mathbf{x} = \begin{bmatrix} x_1 \\ \vdots \\ x_n \end{bmatrix}, \qquad \text{and} \qquad \frac{d\mathbf{x}}{dt} = \begin{bmatrix} dx_1/dt \\ \vdots \\ dx_n/dt \end{bmatrix}.$$

A *solution* of (1) is a set of $n$ functions $x_1 = x_1(t), \ldots, x_n = x_n(t)$ which satisfy the system (1). A *solution* of (1′) is a vector-valued function $\mathbf{x}(t) = (x_1(t), \ldots, x_n(t))$ which satisfies (1′).

**EXAMPLE 1**

Consider the system of two differential equations

$$\frac{dx_1}{dt} = 2x_1 + 2x_2$$
$$\frac{dx_2}{dt} = x_1 + 3x_2.$$

In matrix notation this system becomes

$$\frac{d\mathbf{x}}{dt} = A\mathbf{x}, \qquad \text{where} \qquad A = \begin{bmatrix} 2 & 2 \\ 1 & 3 \end{bmatrix} \quad \text{and} \quad \mathbf{x} = \begin{bmatrix} x_1 \\ x_2 \end{bmatrix}.$$

If $\mathbf{x}(t) = c_1 e^t \begin{bmatrix} -2 \\ 1 \end{bmatrix} + c_2 e^{4t} \begin{bmatrix} 1 \\ 1 \end{bmatrix}$, then

$$\frac{d\mathbf{x}}{dt} = c_1 e^t \begin{bmatrix} -2 \\ 1 \end{bmatrix} + 4c_2 e^{4t} \begin{bmatrix} 1 \\ 1 \end{bmatrix}$$

and

$$\begin{aligned} A\mathbf{x} &= c_1 e^t \begin{bmatrix} 2 & 2 \\ 1 & 3 \end{bmatrix} \begin{bmatrix} -2 \\ 1 \end{bmatrix} + c_2 e^{4t} \begin{bmatrix} 2 & 2 \\ 1 & 3 \end{bmatrix} \begin{bmatrix} 1 \\ 1 \end{bmatrix} \\ &= c_1 e^t \begin{bmatrix} -2 \\ 1 \end{bmatrix} + c_2 e^{4t} \begin{bmatrix} 4 \\ 4 \end{bmatrix} \\ &= \frac{d\mathbf{x}}{dt}. \end{aligned}$$

Thus $\mathbf{x}(t)$ is a solution.

As an exercise verify that $\mathbf{x}(t) = (c_1 e^{2t} + c_2 e^{4t}, -c_1 e^{2t} + c_2 e^{4t})$ is a solution to the system of differential equations

$$\begin{aligned} \frac{dx_1}{dt} &= 3x_1 + x_2 \\ \frac{dx_2}{dt} &= x_1 + 3x_2. \end{aligned}$$

Then write the system of differential equations in matrix notation and verify that $\mathbf{x}(t) = c_1 e^{2t} \begin{bmatrix} 1 \\ -1 \end{bmatrix} + c_2 e^{4t} \begin{bmatrix} 1 \\ 1 \end{bmatrix}$ is a solution.

As we found in the study of Richardson's model, it is sometimes necessary to impose **initial conditions** on the system (1) of the form

$$x_1(t_0) = x_1^{(0)}, \ldots, x_n(t_0) = x_n^{(0)}, \tag{2}$$

where $t_0$ is a specified value and $x_1^{(0)}, \ldots, x_n^{(0)}$ are specified numbers. The system (1) together with the initial conditions (2) is called an **initial value problem**. A solution to an initial value problem is a solution of (1) that satisfies the initial conditions (2).

### EXAMPLE 2

The function $\mathbf{x}(t) = (-2e^t + e^{4t}, e^t + e^{4t})$ is a solution of the initial value problem

$$\frac{d\mathbf{x}}{dt} = \begin{bmatrix} 2 & 2 \\ 1 & 3 \end{bmatrix} \mathbf{x}, \qquad \mathbf{x}(0) = \begin{bmatrix} -1 \\ 2 \end{bmatrix}.$$

In Example 1 we verified that it is a solution to the differential equation. Since $\mathbf{x}(0) = (-1, 2)$, it also satisfies the initial condition. As an exercise, show that $\mathbf{x}(t) = (e^{2t} + 2e^{4t}, -e^{2t} + 2e^{4t})$ is a solution to the initial value problem $d\mathbf{x}/dt = A\mathbf{x}$, $\mathbf{x}(0) = (3, 1)$, where $d\mathbf{x}/dt = A\mathbf{x}$ is the system given by

$$A = \begin{bmatrix} 3 & 1 \\ 1 & 3 \end{bmatrix}.$$

Given a system of differential equations, the goal is of course to find a solution. This immediately raises two questions: (1) Does the system of differential equations always have a solution? (2) If so, does there exist a general method for finding all the solutions?

The answer to the first question as posed is obvious. The zero function [$\mathbf{x}(t) = \mathbf{0}$ for all $t$] is clearly a solution; it is called the **trivial solution**. Hence the real question that we should ask is whether the system of differential equations has a **nontrivial** solution (i.e., a solution that is different from the trivial solution).

The following important result settles this question; nontrivial solutions always exist. We answer the second question in Section 8.12.

***Theorem 1*** *(The Existence–Uniqueness Theorem)*

(*a*) *The differential equation* $d\mathbf{x}/dt = A\mathbf{x}$ *always has a nontrivial solution.*
(*b*) *The initial value problem*

$$\frac{d\mathbf{x}}{dt} = A\mathbf{x}, \qquad \mathbf{x}(t_0) = \mathbf{x}^{(0)} = (x_1^{(0)}, \ldots, x_n^{(0)}),$$

*has one and only one solution.*

The proof is difficult and will be omitted. However, we ask the reader to think about why the first part of the theorem follows immediately from the second part.

We conclude this section with a simple but subtle application of the existence-uniqueness theorem. The theorem states that the initial value problem

$$\frac{d\mathbf{x}}{dt} = A\mathbf{x}, \qquad \mathbf{x}(t_0) = (x_1^{(0)}, \ldots, x_n^{(0)})$$

has one and only one solution. Let $\mathbf{y}(t)$ be this unique solution and suppose that for some $t_1, \mathbf{y}(t_1) = \mathbf{0}$. Now consider the initial value problem

$$\frac{d\mathbf{x}}{dt} = A\mathbf{x}, \qquad \mathbf{x}(t_1) = \mathbf{0}.$$

The solution $\mathbf{y}(t)$ to the first initial value problem is also a solution to this initial value problem. But so is the trivial solution $\mathbf{x}(t) = \mathbf{0}$ for all $t$. Since there can be only one solution to the second initial value problem, $\mathbf{y}(t)$ must be the trivial solution $\mathbf{y}(t) = \mathbf{0}$ for all $t$. This proves the following result.

**Result 1** Let $\mathbf{y}(t)$ be a solution of $d\mathbf{x}/dt = A\mathbf{x}$. Then either $\mathbf{y}(t)$ is the trivial solution or $\mathbf{y}(t) \neq \mathbf{0}$ for all $t$.

# PROBLEMS 8.10

1. In each of the following problems, express the system of differential equations in matrix form and verify that the given functions are solutions.

(a) $\dfrac{dx_1}{dt} = x_2,$

$\dfrac{dx_2}{dt} = 2x_1 - x_2,$

$$x(t) = e^{t}\begin{bmatrix}1\\1\end{bmatrix}$$

$$x(t) = e^{-2t}\begin{bmatrix}1\\-2\end{bmatrix}$$

(b) $\dfrac{dx_1}{dt} = x_1 + x_3,$

$\dfrac{dx_2}{dt} = x_2,$

$\dfrac{dx_3}{dt} = x_1 + 2x_2 + x_3,$

$$x(t) = 4\begin{bmatrix}-1\\0\\1\end{bmatrix}$$

$$x(t) = 2e^{t}\begin{bmatrix}-2\\1\\0\end{bmatrix}$$

$$x(t) = e^{2t}\begin{bmatrix}1\\0\\1\end{bmatrix}$$

(c) $\dfrac{dx_1}{dt} = 2x_1 - 5x_2 + 5x_3,$

$\dfrac{dx_2}{dt} = -3x_2 - x_3,$

$\dfrac{dx_3}{dt} = -x_2 + 3x_3,$

$$x(t) = e^{2t}\begin{bmatrix}0\\1\\1\end{bmatrix}$$

$$x(t) = e^{2t}\begin{bmatrix}1\\0\\0\end{bmatrix}$$

$$x(t) = e^{4t}\begin{bmatrix}5\\-1\\1\end{bmatrix}$$

2. Consider the $n$th-order homogeneous differential equation

$$\frac{d^n y}{dt^n} + a_{n-1}\frac{d^{n-1}y}{dt^{n-1}} + \cdots + a_1\frac{dy}{dt} + a_0 y = 0.$$

(a) Show that the substitution

$$x_1 = y, x_2 = \frac{dy}{dt}, \ldots, x_n = \frac{d^{n-1}y}{dt^{n-1}}$$

converts this $n$th-order differential equation to the following system of $n$ first-order equations

$$\frac{dx_1}{dt} = x_2$$
$$\frac{dx_2}{dt} = x_3$$
$$\vdots$$
$$\frac{dx_{n-1}}{dt} = x_n$$
$$\frac{dx_n}{dt} = -(a_0x_1 + \cdots + a_{n-1}x_n)$$

(b) Show that $\mathbf{x}(t) = (x_1(t), \ldots, x_n(t))$ is a solution of this system if and only if $y(t) = x_1(t)$ is a solution of the $n$th-order equation.

3. Convert each of the following equations to a system of first-order equations.

*(a) $y'' - y = 0$
(b) $y'' + 2y' - 3y = 0$
*(c) $y''' + y'' - 2y' = 0$
(d) $y''' + 4y'' - 7y' - 10y = 0$
*(e) $y'''' - 2y''' - y'' + 2y = 0$

# 8.11 Applications of Linear Algebra to Systems of Differential Equations

This section is devoted to the study of the algebraic properties of the set of solutions to a system of differential equations. We begin by introducing some notation. When dealing with a single solution $\mathbf{x}(t)$ to a system, we denote the components of $\mathbf{x}(t)$ by $x_1(t), \ldots, x_n(t)$. That is,

$$\mathbf{x}(t) = (x_1(t), \ldots, x_n(t)).$$

When dealing with several solutions of a system, we will use superscripts in parentheses to denote different solutions and subscripts to denote the components of these solutions. Thus our notation for the $j$th solution will be

$$\mathbf{x}^{(j)}(t) = (x_1^{(j)}(t), \ldots, x_n^{(j)}(t)).$$

***Theorem 1*** *The set of solutions of* $\dfrac{d\mathbf{x}}{dt} = A\mathbf{x}$ *is a function space.*

**Proof** Let $\mathbf{x}^{(1)}(t)$ and $\mathbf{x}^{(2)}(t)$ be two solutions and let $\mathbf{x}(t) = c_1\mathbf{x}^{(1)}(t) + c_2\mathbf{x}^{(2)}(t)$, where $c_1$ and $c_2$ are scalars. Then

$$\begin{aligned}\frac{d\mathbf{x}}{dt} &= c_1\frac{d\mathbf{x}^{(1)}}{dt} + c_2\frac{d\mathbf{x}^{(2)}}{dt} = c_1 A\mathbf{x}^{(1)} + c_2 A\mathbf{x}^{(2)} \\ &= A(c_1\mathbf{x}^{(1)} + c_2\mathbf{x}^{(2)}) = A\mathbf{x}.\end{aligned}$$

Thus $\mathbf{x}(t)$ is a solution and the set of solutions is closed under addition and scalar multiplication. Finally, $\mathbf{x}(t) = \mathbf{0}$ for all $t$ is also a solution.

In view of this result, we call the set of solutions to $d\mathbf{x}/dt = A\mathbf{x}$ the **solution space**. The fact that the solution space of $d\mathbf{x}/dt = A\mathbf{x}$ is a function space is important. For if we are given a basis for the solution space, we can express any solution to the differential equation as a linear combination of the solutions in the basis. Thus the problem of finding all solutions is reduced to the problem of finding a basis for the solution space. If $\mathbf{x}^{(1)}(t), \ldots, \mathbf{x}^{(n)}(t)$ form a basis for the solution space of $d\mathbf{x}/dt = A\mathbf{x}$, then any solution of this equation can be written in the form

$$\mathbf{x}(t) = c_1\mathbf{x}^{(1)}(t) + \cdots + c_n\mathbf{x}^{(n)}(t).$$

This last expression is called the **general solution**.

### EXAMPLE 1

We will show in the next section that $\mathbf{x}^{(1)}(t) = (-2e^t, e^t)$ and $\mathbf{x}^{(2)}(t) = (e^{4t}, e^{4t})$ form a basis for the solution space of $d\mathbf{x}/dt = A\mathbf{x}$, where

$$A = \begin{bmatrix} 2 & 2 \\ 1 & 3 \end{bmatrix}.$$

Hence the general solution to the equation is

$$\begin{aligned}\mathbf{x}(t) &= c_1(-2e^t, e^t) + c_2(e^{4t}, e^{4t}) \\ &= (-2c_1e^t + c_2e^{4t}, c_1e^t + c_2e^{4t}).\end{aligned}$$

A basis for a function space is a set of independent functions that span the space. It is in general difficult to recognize when a set of functions in a function space is independent. However, when the function space is the solution space of a system of differential equations, there is an elegant and remarkably simple method for determining whether or not a set of solutions is independent.

**Result 1** (Test for Independence of Solutions) Let $\mathbf{x}^{(1)}(t), \ldots, \mathbf{x}^{(m)}(t)$ be solutions of $d\mathbf{x}/dt = A\mathbf{x}$, and let $t_0$ be any number. Then $\mathbf{x}^{(1)}(t), \ldots, \mathbf{x}^{(m)}(t)$ are linearly independent solutions if and only if $\mathbf{x}^{(1)}(t_0), \ldots, \mathbf{x}^{(m)}(t_0)$ are linearly independent vectors.

**Proof** Let $\mathbf{x}^{(1)}(t), \ldots, \mathbf{x}^{(m)}(t)$ be linearly independent solutions and let $t_0$ be a number. Suppose that $c_1, \ldots, c_m$ are scalars such that

$$c_1\mathbf{x}^{(1)}(t_0) + \cdots + c_m\mathbf{x}^{(m)}(t_0) = \mathbf{0}.$$

Let

$$\mathbf{x}(t) = c_1\mathbf{x}^{(1)}(t) + \cdots + c_m\mathbf{x}^{(m)}(t).$$

Since $\mathbf{x}(t)$ is a linear combination of solutions, it is a solution. But $\mathbf{x}(t_0) = \mathbf{0}$, so Result 1 of Section 8.10 implies that $\mathbf{x}(t)$ is the trivial solution; that is, $\mathbf{x}(t) = \mathbf{0}$ for all $t$. Thus

$$c_1\mathbf{x}^{(1)}(t) + \cdots + c_m\mathbf{x}^{(m)}(t) = \mathbf{0} \qquad \text{for all } t.$$

Since $\mathbf{x}^{(1)}(t), \ldots, \mathbf{x}^{(m)}(t)$ are independent solutions, this equation implies that $c_1 = \cdots = c_m = 0$. Thus $\mathbf{x}^{(1)}(t_0), \ldots, \mathbf{x}^{(m)}(t_0)$ are independent.

Conversely, suppose that for some number $t_0$ the vectors $\mathbf{x}^{(1)}(t_0), \ldots, \mathbf{x}^{(m)}(t_0)$ are independent. If $c_1, \ldots, c_m$ are scalars such that

$$c_1\mathbf{x}^{(1)}(t) + \cdots + c_m\mathbf{x}^{(m)}(t) = \mathbf{0} \qquad \text{for all } t,$$

then letting $t = t_0$, we obtain

$$c_1\mathbf{x}^{(1)}(t_0) + \cdots + c_m\mathbf{x}^{(m)}(t_0) = \mathbf{0}.$$

Since $\mathbf{x}^{(1)}(t_0), \ldots, \mathbf{x}^{(m)}(t_0)$ are independent, it follows that $c_1 = \cdots = c_m = 0$. Thus $\mathbf{x}^{(1)}(t), \ldots, \mathbf{x}^{(m)}(t)$ are independent.

### EXAMPLE 2

Consider the differential equation $d\mathbf{x}/dt = A\mathbf{x}$, where

$$A = \begin{bmatrix} 0 & 1 \\ 2 & -1 \end{bmatrix}.$$

You can easily verify that $\mathbf{x}^{(1)}(t) = (e^t, e^t)$ and $\mathbf{x}^{(2)}(t) = (e^{-2t}, -2e^{-2t})$ are solutions of this equation. Since $\mathbf{x}^{(1)}(0) = (1, 1)$ and $\mathbf{x}^{(2)}(0) = (1, -2)$ are independent vectors in $R^2$, it follows from Result 1 that $\mathbf{x}^{(1)}(t)$ and $\mathbf{x}^{(2)}(t)$ are independent solutions.

Our next result gives the dimension of the solution space in terms of the size of the square matrix $A$.

***Theorem 2*** *Let $A$ be an $n \times n$ matrix. The dimension of the solution space of $d\mathbf{x}/dt = A\mathbf{x}$ is $n$.*

**Proof** We prove the result in two steps. First we show that there exist $n$ independent solutions. Then we show that any set consisting of more than $n$ solutions is dependent.

1. *There exist $n$ independent solutions.* Consider the $n$ different initial value problems

$$\frac{d\mathbf{x}}{dt} = A\mathbf{x}, \qquad \mathbf{x}(0) = \mathbf{e}_j, \qquad j = 1, \ldots, n.$$

By the existence-uniqueness theorem, each of these initial value problems has a unique solution which we denote by $\mathbf{x}^{(j)}(t)$, $j = 1, \ldots, n$. By the previous result these $n$ solutions are linearly independent, since $\mathbf{x}^{(j)}(0) = \mathbf{e}_j$, $j = 1, \ldots, n$, and the vectors $\mathbf{e}_1, \ldots, \mathbf{e}_n$ are independent in $R^n$.

2. *Any set with more than $n$ solutions is dependent.* If $\mathbf{x}^{(1)}(t), \ldots, \mathbf{x}^{(m)}(t)$, $m > n$, are solutions of $d\mathbf{x}/dt = A\mathbf{x}$, then $\mathbf{x}^{(1)}(0), \ldots, \mathbf{x}^{(m)}(0)$ are $m$ vectors in $R^n$. Since $m > n$, they are dependent. Thus, by the previous result, $\mathbf{x}^{(1)}(t), \ldots, \mathbf{x}^{(m)}(t)$ are dependent.

### EXAMPLE 3

In Example 2 we found two independent solutions to the system $d\mathbf{x}/dt = A\mathbf{x}$, where $A$ was a $2 \times 2$ matrix. In view of this theorem, these solutions form a basis for the solution space.

## PROBLEMS 8.11

**1.** In Problem 1 of Section 8.10 we gave solutions to certain differential equations. For each of these differential equations, show that the given solutions form a basis for the solution space.

***2.** In each of the following, determine whether the given solutions of $d\mathbf{x}/dt = A\mathbf{x}$ form a basis for the solution space.

(a) $A = \begin{bmatrix} 1 & 6 \\ 0 & -1 \end{bmatrix}$, $\mathbf{x}^{(1)}(t) = e^t \begin{bmatrix} 1 \\ 0 \end{bmatrix}$,

$\mathbf{x}^{(2)}(t) = e^{-t} \begin{bmatrix} -3 \\ 1 \end{bmatrix}$

(b) $A = \begin{bmatrix} 1 & 1 & 1 \\ 0 & 0 & 1 \\ 0 & 0 & -1 \end{bmatrix}$,

$\mathbf{x}^{(1)}(t) = \begin{bmatrix} 2 \\ -2 \\ 0 \end{bmatrix}$, $\mathbf{x}^{(2)}(t) = e^{t}\begin{bmatrix} 4 \\ 0 \\ 0 \end{bmatrix}$,

$\mathbf{x}^{(3)}(t) = e^{-t}\begin{bmatrix} 0 \\ -3 \\ 3 \end{bmatrix}$

(c) $A = \begin{bmatrix} 2 & 0 & 0 \\ 0 & 3 & -1 \\ 0 & -1 & 3 \end{bmatrix}$,

$\mathbf{x}^{(1)}(t) = e^{2t}\begin{bmatrix} -1 \\ 0 \\ 0 \end{bmatrix}$,

$\mathbf{x}^{(2)}(t) = e^{2t}\begin{bmatrix} 0 \\ 1 \\ 1 \end{bmatrix}$,

$\mathbf{x}^{(3)}(t) = e^{4t}\begin{bmatrix} 0 \\ 1 \\ -1 \end{bmatrix}$

**3.** Show that the conclusion of Result 1 fails for the functions

$$\mathbf{f}(t) = (t, t) \qquad \text{and} \qquad \mathbf{g}(t) = (t^2, t^2).$$

What can you conclude from this?

**4.** Consider the $n$th-order homogeneous differential equation

$$\frac{d^n y}{dt^n} + a_{n-1}\frac{d^{n-1}y}{dt^{n-1}} + \cdots + a_0 y = 0.$$

(a) Show that the set of solutions to this equation forms a subspace of $C^n(R)$.
(b) Show that the dimension of this subspace is $n$.

**5.** Let $A$ be an $n \times n$ matrix, let $\mathbf{b}$ be an $n$-vector, and let $\mathbf{x}^{(1)}(t), \ldots, \mathbf{x}^{(n)}(t)$ be $n$ linearly independent solutions of the differential equation $d\mathbf{x}/dt = A\mathbf{x}$. If $\mathbf{x}^{(p)}(t)$ is a solution to the nonhomogeneous equation $d\mathbf{x}/dt = A\mathbf{x} + \mathbf{b}$, show that any solution to this differential equation can be written as

$$\mathbf{x}(t) = \mathbf{x}^{(p)}(t) + c_1\mathbf{x}^{(1)}(t) + \cdots + c_n\mathbf{x}^{(n)}(t)$$

where $c_1, \ldots, c_n$ are scalars.

# 8.12 Eigenvalues and Differential Equations

Let us now turn our attention to the problem of finding a basis for the solution space. It follows from Theorem 2 of Section 8.11 that a basis for the solution space of $d\mathbf{x}/dt = A\mathbf{x}$, $A$ an $n \times n$ matrix, consists of $n$ independent solutions. Once we have found solutions, we can use Result 1 of Section 8.11 to determine whether or not they are independent. What remains to be done, therefore, is to develop a method for finding solutions.

In Section 8.9 we explained that it is reasonable to try to find a solution of the form $\mathbf{x}(t) = e^{\lambda t}\mathbf{v}$, where $\lambda$ is a scalar and $\mathbf{v}$ is a nonzero constant $n$-vector. Since

$$\frac{d}{dt}(e^{\lambda t}\mathbf{v}) = \lambda e^{\lambda t}\mathbf{v} \qquad \text{and} \qquad A(e^{\lambda t}\mathbf{v}) = e^{\lambda t}A\mathbf{v},$$

it follows that $\mathbf{x}(t) = e^{\lambda t}\mathbf{v}$ is a solution of $d\mathbf{x}/dt = A\mathbf{x}$ if and only if $\lambda e^{\lambda t}\mathbf{v} = e^{\lambda t}A\mathbf{v}$.

Since $e^{\lambda t}$ is never zero, we may divide both sides of this equation by $e^{\lambda t}$ to obtain $A\mathbf{v} = \lambda\mathbf{v}$. This proves the following result.

**Result 1** Let $A$ be an $n \times n$ matrix, $\mathbf{v}$ a nonzero $n$-vector, and $\lambda$ a scalar. The function $\mathbf{x}(t) = e^{\lambda t}\mathbf{v}$ is a solution of $d\mathbf{x}/dt = A\mathbf{x}$ if and only if $\lambda$ is an eigenvalue of $A$ and $\mathbf{v}$ is an eigenvector associated with $\lambda$.

Thus each eigenvalue $\lambda$ of $A$ determines an infinite number of solutions to $d\mathbf{x}/dt = A\mathbf{x}$, namely, $\mathbf{x}(t) = e^{\lambda t}\mathbf{v}$, where $\mathbf{v}$ is any eigenvector associated with $\lambda$. The number of independent solutions of this form depends on the number of independent eigenvectors associated with $\lambda$. This follows immediately from Result 1 of Section 8.11 since the solutions

$$\mathbf{x}^{(1)}(t) = e^{\lambda t}\mathbf{v}_1, \ldots, \mathbf{x}^{(k)}(t) = e^{\lambda t}\mathbf{v}_k$$

are independent if and only if

$$\mathbf{x}^{(1)}(0) = \mathbf{v}_1, \ldots, \mathbf{x}^{(k)}(0) = \mathbf{v}_k$$

are independent.

**Result 2** Let $A$ be an $n \times n$ matrix that has $n$ independent eigenvectors $\mathbf{v}_1, \ldots, \mathbf{v}_n$. Let $\lambda_1, \ldots, \lambda_n$ be the corresponding eigenvalues. Then the solutions

$$\mathbf{x}^{(1)}(t) = e^{\lambda_1 t}\mathbf{v}_1, \ldots, \mathbf{x}^{(n)}(t) = e^{\lambda_n t}\mathbf{v}_n$$

form a basis for the solution space of $d\mathbf{x}/dt = A\mathbf{x}$. Therefore, the general form of the solution is

$$\mathbf{x}(t) = c_1 e^{\lambda_1 t}\mathbf{v}_1 + \cdots + c_n e^{\lambda_n t}\mathbf{v}_n.$$

**Proof** Since $\mathbf{x}^{(1)}(0) = \mathbf{v}_1, \ldots, \mathbf{x}^{(n)}(0) = \mathbf{v}_n$ and $\mathbf{v}_1, \ldots, \mathbf{v}_n$ are independent vectors, the result follows from Result 1 and Theorem 2 of Section 8.11.

## EXAMPLE 1

Consider the differential equation

$$\frac{d\mathbf{x}}{dt} = A\mathbf{x} \qquad \text{where} \qquad A = \begin{bmatrix} 1 & 0 & 1 \\ 0 & 1 & 0 \\ 1 & 2 & 1 \end{bmatrix}.$$

In Example 1 of Section 8.2 we showed that the eigenvalues of $A$ are 0, 1, and 2 and that the eigenspaces $E(0)$, $E(1)$, and $E(2)$ are spanned by $(-1, 0, 1)$, $(-2, 1, 0)$, and $(1, 0, 1)$, respectively. These eigenvectors are independent. Since $A$ is a $3 \times 3$ matrix with three independent eigenvectors, the general solution of $d\mathbf{x}/dt = A\mathbf{x}$ is

$$\begin{aligned} \mathbf{x}(t) &= c_1 e^{0t}(-1, 0, 1) + c_2 e^{t}(-2, 1, 0) + c_3 e^{2t}(1, 0, 1) \\ &= (-c_1 - 2c_2 e^{t} + c_3 e^{2t}, c_2 e^{t}, c_1 + c_3 e^{2t}). \end{aligned}$$

## EXAMPLE 2

Consider the differential equation

$$\frac{d\mathbf{x}}{dt} = A\mathbf{x} \qquad \text{where} \qquad A = \begin{bmatrix} 2 & -5 & 5 \\ 0 & 3 & -1 \\ 0 & -1 & 3 \end{bmatrix}.$$

From Example 2 of Section 8.2 we know that $A$ has three independent eigenvectors $(0, 1, 1)$, $(1, 0, 0)$, and $(5, -1, 1)$ with corresponding eigenvalues 2, 2, and 4, respectively. By the previous result, the general solution of $d\mathbf{x}/dt = A\mathbf{x}$ is

$$\begin{aligned}\mathbf{x}(t) &= c_1e^{2t}(0, 1, 1) + c_2e^{2t}(1, 0, 0) + c_3e^{4t}(5, -1, 1) \\ &= (c_2e^{2t} + 5c_3e^{4t}, c_1e^{2t} - c_3e^{4t}, c_1e^{2t} + c_3e^{4t}).\end{aligned}$$

## PROBLEMS 8.12

1. Find the general solution to the differential equation $d\mathbf{x}/dt = A\mathbf{x}$ for each diagonalizable matrix in Problem 1 of Section 8.2.

2. Suppose that $A$ is a diagonalizable matrix, $\Lambda = P^{-1}AP$, where $\Lambda$ is a diagonal matrix.
   (a) Show that the substitution $\mathbf{x} = P\mathbf{y}$ reduces the equation $d\mathbf{x}/dt = A\mathbf{x}$ to $d\mathbf{y}/dt = \Lambda\mathbf{y}$.
   (b) Find the general solution to $d\mathbf{y}/dt = \Lambda\mathbf{y}$.
   (c) Show that if $\mathbf{y}$ is the general solution to $d\mathbf{y}/dt = \Lambda\mathbf{y}$, then $\mathbf{x} = P\mathbf{y}$ is the general solution to $d\mathbf{x}/dt = A\mathbf{x}$.
   (d) Show that the above gives a different proof of Result 2.

3. Apply the results of Problem 2 to each of the diagonalizable matrices in Problem 1 of Section 8.2.

4. Suppose that $A$ is a diagonalizable matrix, $A = P\Lambda P^{-1}$, with $\lambda_1, \ldots, \lambda_n$ being the diagonal entries of $\Lambda$. Show that the solution of the initial value problem

$$\frac{d\mathbf{x}}{dt} = A\mathbf{x}, \quad \mathbf{x}(0) = \mathbf{x}_0$$

is

$$\mathbf{x}(t) = P\begin{bmatrix} e^{\lambda_1 t} & & \\ & \ddots & \\ & & e^{\lambda_n t} \end{bmatrix}P^{-1}\mathbf{x}_0.$$

5. We define

$$e^{At} = I + At + \frac{1}{2!}(At)^2 + \frac{1}{3!}(At)^3 + \cdots.$$

It can be shown that this series converges for any matrix $A$. Note that $e^{At}$ is an $n \times n$ matrix.
   (a) Show that

$$e^{\Lambda t} = \begin{bmatrix} e^{\lambda_1 t} & & \\ & \ddots & \\ & & e^{\lambda_n t} \end{bmatrix} \quad \text{if}$$

$$\Lambda = \begin{bmatrix} \lambda_1 & & \\ & \ddots & \\ & & \lambda_n \end{bmatrix}.$$

   (b) Show that if $A$ is a diagonalizable matrix, $A = P\Lambda P^{-1}$, where $\Lambda$ is a diagonal matrix, then

$$e^{At} = Pe^{\Lambda t}P^{-1}.$$

From this and Problem 4 it follows

that the solution to the initial value problem $d\mathbf{x}/dt = A\mathbf{x}$, $\mathbf{x}(0) = \mathbf{x}_0$, is

$$\mathbf{x}(t) = e^{At}\mathbf{x}_0.$$

(c) Find $e^{At}$ for each of the diagonalizable matrices in Problem 1 of Section 8.2.

(d) Show that:

(i) $e^{-At}$ is the inverse of $e^{At}$.

(ii) If $\lambda$ is an eigenvalue of $A$ with associated eigenvector $\mathbf{v}$, then $e^{\lambda t}$ is an eigenvalue of $e^{At}$ with associated eigenvector $\mathbf{v}$. Hence zero is not an eigenvalue of $e^{At}$.

It follows from either (i) or (ii) that $e^{At}$ is nonsingular for any value of $t$.

## SUPPLEMENTARY PROBLEMS

**1.** (a) Complete the following definitions:

(i) $\lambda$ is an eigenvalue of an $n \times n$ matrix $A$ if....

(ii) An $n \times n$ matrix $A$ is diagonalizable if....

Let

$$A = \begin{bmatrix} 5 & -3 & 0 \\ 6 & -4 & 0 \\ 6 & -3 & -1 \end{bmatrix}.$$

*(b) Find the characteristic polynomial of $A$.

*(c) Find all eigenvalues of $A$ and their algebraic multiplicities.

*(d) Find the geometric multiplicities of the eigenvalues of $A$.

*(e) Determine whether $A$ is diagonalizable. If so, find the corresponding diagonal matrix and the diagonalizing matrix.

**2.** For each of the following matrices, do the following.

(i) Compute the eigenvalues and a basis for the corresponding eigenspaces.

(ii) Determine whether the matrix is diagonalizable. If so, find a matrix $M$ and a diagonal matrix $\Lambda$ such that $\Lambda = M^{-1}AM$.

*(a) $\begin{bmatrix} 1 & 2 & 3 \\ 4 & -1 & 5 \\ 0 & 0 & 3 \end{bmatrix}$

(b) $\begin{bmatrix} 2 & 1 & 1 \\ 0 & 1 & 0 \\ 0 & 1 & 1 \end{bmatrix}$

*(c) $\begin{bmatrix} 0 & 2 & 2 \\ 2 & 0 & 2 \\ 2 & 2 & 0 \end{bmatrix}$

(d) $\begin{bmatrix} 0 & 1 & 1 \\ 0 & 0 & 2 \\ 0 & 0 & 1 \end{bmatrix}$

*(e) $\begin{bmatrix} 0 & 0 & 2 \\ 0 & 0 & 0 \\ 2 & 0 & 0 \end{bmatrix}$

(f) $\begin{bmatrix} 0 & 0 & 0 \\ 0 & 1 & 0 \\ 1 & 0 & 1 \end{bmatrix}$

*(g) $\begin{bmatrix} 2 & 0 & 1 \\ 0 & 1 & 2 \\ 0 & 0 & 1 \end{bmatrix}$

(h) $\begin{bmatrix} 1 & 1 & 1 \\ 0 & 1 & 0 \\ 0 & 1 & 2 \end{bmatrix}$

*(i) $\begin{bmatrix} 5 & -3 & 2 \\ 6 & -4 & 4 \\ 4 & -4 & 5 \end{bmatrix}$

(j) $\begin{bmatrix} 3 & 1 & 1 & 1 \\ 0 & 2 & 2 & 1 \\ 0 & 0 & 5 & 4 \\ 0 & 0 & 0 & 7 \end{bmatrix}$

*(k) $\begin{bmatrix} 0 & 1 & 0 \\ -4 & 4 & 0 \\ -2 & 1 & 2 \end{bmatrix}$

(l) $\begin{bmatrix} -1 & 0 & 0 \\ 0 & 14 & 4 \\ 0 & -25 & -6 \end{bmatrix}$

***3.** Is there a nonsingular matrix $M$ such that

$$\begin{bmatrix} 1 & & & \\ & 4 & & \\ & & 3 & \\ & & & 2 \end{bmatrix} = M^{-1} \begin{bmatrix} 1 & & & \\ & 2 & & \\ & & 3 & \\ & & & 4 \end{bmatrix} M \quad ?$$

If so, find $M$. If not, why not?

**4.** Let $A$ be a $2 \times 2$ matrix. Find a necessary and sufficient condition on $a_{11}$, $a_{12}, a_{21}, a_{22}$ that $A$ be diagonalizable.

**5.** Find two $2 \times 2$ matrices $A$ and $B$ such that:

(a) $A$ and $B$ have only positive eigenvalues.

(b) $AB$ has only negative eigenvalues.

(*Hint:* Let $A$ be an upper triangular matrix and let $B$ be a lower triangular matrix.)

**6.** Let $A$ be an $m \times m$ matrix and let $\lambda_1, \ldots, \lambda_k$ be the distinct eigenvalues of $A$. Suppose that $\mathbf{w}$ is a vector in $R^m$ and the projection of $\mathbf{w}$ onto each of the eigenspaces $E(\lambda_1), E(\lambda_2), \ldots, E(\lambda_k)$ is not zero. Show that if $A\mathbf{w} = \mu\mathbf{w}$, then $A = \mu I$.

**7.** Why are the eigenvalues of $A^2$ always nonnegative?

**8.** Let

$$A = \begin{bmatrix} 0 & 2 & 2 \\ 2 & 0 & 2 \\ 2 & 2 & 0 \end{bmatrix}.$$

*(a) Find the eigenvalues and eigenvectors of $A$.

*(b) Find an orthogonal matrix $M$ and diagonal matrix $\Lambda$ such that $\Lambda = M^{-1}AM$.

(c) Find the spectral decomposition of $A$.

**9.** Show that two idempotent matrices are similar if and only if they have the same rank. (*Hint:* Use Problem 10 in Section 8.3.)

# 9

# Numerical Methods

In previous chapters we have ignored virtually all questions concerning the speed and accuracy of computation. In addition, the matrices in our examples, exercises, and problems have been somewhat unrealistic. Both the size of the matrices and the numbers used in them have been chosen so that solutions could be reasonably calculated by hand. These restrictions are both unnecessary and unwise in the age of computers. Computers enable us to solve problems involving large matrices very quickly.

In this chapter we discuss how to maximize the speed and increase the accuracy with which we solve systems, perform matrix operations, and compute eigenvalues and eigenvectors. We also discuss some new methods for solving these problems, methods that are particularly appropriate for use on computers.

## 9.1 Operation Counts

This section is concerned with estimating the time required to complete certain computations. Such estimates have two important uses. First, we can determine how a change in the size of the problem affects the time required for its solution. This knowledge helps us determine what size problems can be solved in a reasonable amount of time. Second, we can compare the times required by two different procedures. This helps us decide which of the procedures is more efficient.

The procedures we have developed in this book involve three types of operations: multiplication-type operations (multiplications and divisions), addition-type operations (additions and subtractions), and movement-type operations (interchanging numbers). Since multiplications and divisions are much more time consuming than the other operations, it is appropriate to use the number of multiplication-type operations as a rough measure of the time necessary to complete a procedure. Of course, if the other operations are many times more numerous than the multiplication-type operations, this measure may be in error. However, in the cases that we consider, no error of this kind will arise.

Let us begin our analysis by examining the Gaussian elimination procedure. We must count the total number of operations required to solve an $n \times n$ system $A\mathbf{x} = \mathbf{b}$ which has a unique solution. The system is solved by performing row operations on the augmented matrix $[A : \mathbf{b}]$, a matrix with $n$ rows and $n + 1$ columns. In the interest of simplicity we will assume that the system can be solved without any pivoting. This assumption has no effect on our results.

The first step in forward elimination is to eliminate the first entry from each of the last $n - 1$ rows. Each of these eliminations involves computing a multiplier for the initial first row (one division), multiplying every entry in the initial row (except the first) by this multiplier ($n$ multiplications), and subtracting each of these $n$ products from the appropriate entry in some row ($n$ subtractions). Of course, we do not count or perform operations which are not needed. We already know that the new first entries in each row will be zero, so it is pointless to compute them. Each elimination therefore involves $n + 1$ multiplication type operations and the $n - 1$ different eliminations involve $(n - 1)(n + 1) = n^2 - 1$ multiplication-type operations.

The next step in the forward elimination process is to eliminate the second entry from each of the last $n - 2$ rows. Each of these eliminations requires one division, $n - 1$ multiplications, and $n - 1$ substractions. Thus the second step of the forward elimination process takes

$$n(n - 2) = [(n - 1) + 1][(n - 1) - 1] = (n - 1)^2 - 1$$

multiplication-type operations. The following table presents a summary of the operations required for the complete forward elimination process.

The total number of operations required for forward elimination can be computed from the table using the following formulas.

$$\begin{aligned} 1 + 2 + 3 + \cdots + (k - 1) + k &= \frac{k(k + 1)}{2} \\ 1^2 + 2^2 + 3^2 + \cdots + (k - 1)^2 + k^2 &= \frac{k(k + 1)(2k + 1)}{6} \end{aligned} \tag{1}$$

**Forward Elimination Operation Count**

| | Each Elimination Requires | | | | Total Operation Count | |
|---|---|---|---|---|---|---|
| **Step** | $\div s$ | $\times s$ | $-s$ | **Elimination** | **Addition Type** | **Multiplication Type** |
| 1 | 1 | $n$ | $n$ | $n - 1$ | $n(n - 1)$ | $(n + 1)(n - 1) = n^2 - 1$ |
| 2 | 1 | $n - 1$ | $n - 1$ | $n - 2$ | $(n - 1)(n - 2)$ | $n(n - 2) = (n - 1)^2 - 1$ |
| 3 | 1 | $n - 2$ | $n - 2$ | $n - 3$ | $(n - 2)(n - 3)$ | $(n - 1)(n - 3) = (n - 2)^2 - 1$ |
| $\vdots$ | | | | | | |
| $n - 2$ | 1 | 3 | 3 | 2 | $3 \cdot 2$ | $4 \cdot 2 = 3^2 - 1$ |
| $n - 1$ | 1 | 2 | 2 | 1 | $2 \cdot 1$ | $3 \cdot 1 = 2^2 - 1$ |
| $n$ | — | — | — | 0 | 0 | $0 = 1^2 - 1$ |

First we compute the number of multiplication-type operations. The last column of the table gives this as

$$(n^2 - 1) + ([n - 1]^2 - 1) + \cdots + (2^2 - 1) + (1^2 - 1)$$

which equals

$$1^2 + 2^2 + 3^2 + \cdots + (n - 1)^2 + n^2 - n.$$

Using the second formula in (1) this expression reduces to

$$\frac{2n^3 + 3n^2 - 5n}{6}.$$

To count the number of addition-type operations, we first note that $k(k-1) = k^2 - k$. Since the terms we must add up are all of this form, we see that the total number of addition-type operations is

$$(n^2 - n) + ([n - 1]^2 - [n - 1]) + \cdots + (2^2 - 2) + (1^2 - 1).$$

Using the formulas in (1) this expression reduces to

$$\frac{n(n + 1)(2n + 1)}{6} - \frac{n(n + 1)}{2} = \frac{n^3 - n}{3}.$$

The complete solution of the system is now obtained by back substitution. Solving for $x_n$ requires one division. Solving for $x_{n-1}$ requires one multiplication, one subtraction, and one division. Solving for $x_{n-2}$ requires two multiplications, two subtractions, and one division. The last step is to solve for $x_1$, which takes $n - 1$ multiplications, $n - 1$ subtractions, and one division. (The reader should of course verify these statements by examining the equations that are being solved.) The total number of multiplication-type operations needed for back substitution is thus

$$1 + 2 + 3 + \cdots + (n - 1) \text{ multiplications} + n \text{ divisions}$$

which equals $(n^2 + n)/2$. It is easy to show that the number of addition-type operations needed for back substitution is

$$1 + 2 + 3 + \cdots + (n - 1) = \frac{n^2 - n}{2}.$$

Since the total number of operations required by Gaussian elimination is the number of operations required for forward elimination plus the number of operations required for back substitution, we have the following result.

**Result 1** Gaussian elimination requires $(n^3 + 3n^2 - n)/3$ multiplication-type operations and $(2n^3 + 3n^2 - 5n)/6$ addition-type operations to solve an $n \times n$ system which has a nonsingular coefficient matrix.

We can use this result to estimate the time required to solve a system of a given size. The following table gives the number of operations required for systems of various sizes.

| Size of system (number of equations and variables) | Operation count (multiplication-type operations) |
|---|---|
| 5 | 65 |
| 10 | 430 |
| 20 | 3,060 |
| 30 | 9,890 |
| 40 | 22,920 |
| 50 | 44,150 |
| 100 | 343,300 |
| 200 | 2,706,600 |
| 1000 | 334,333,000 |

If each multiplication requires 10 seconds to do by hand, we can deduce from this table that a $20 \times 20$ system will require more than 8 hours to solve by hand. (This estimate is high if the numbers are all one-digit integers but low if we have to perform multiplications such as 2.435 times 8.732.) Even a $10 \times 10$ problem takes at least 1 hour to complete. This explains why square systems with 10 or more variables are conspicuously absent from the problems in this book. Problems with 10 or more variables are not reasonable problems to solve by hand calculation.

If we perform our calculations using a computer, the picture changes considerably. A conservative estimate of the speed of a computer is 100,000 multiplications per second. Under this assumption a $20 \times 20$ system can be solved in slightly more than 0.03 second and a $100 \times 100$ system takes less than 4 seconds. Faster machines (millions of multiplications per second!) will solve these systems in correspondingly less time.

In the discussion above we have ignored the time taken by additions, subtractions, and data movement. Addition-type operations can be done 10 to 100 times faster than multiplications. Since the number of addition-type operations is about the same as the number of multiplication-type operations, we would have to increase our time estimates by no more than 10%. Thus the time taken by the addition-type operations can be safely ignored. Pivoting takes hardly any time and can also be ignored.

Let us now compute the operation count for Gauss-Jordan elimination. This procedure also requires us to perform row operations on the augmented matrix. At each step of the Gauss-Jordan process we must first make the pivot equal to one and then eliminate $n - 1$ entries since we must eliminate every entry in the pivot column except the pivot itself. The first step involves dividing $n$ entries by the pivot and then performing $n - 1$ eliminations each requiring $n$ multiplications. This gives a total of $n^2$ multiplication-type operations. The next step involves dividing $n - 1$ entries by the pivot and then performing $n - 1$ eliminations each requiring $n - 1$ multiplications. This gives $n(n - 1)$ multiplication-type operations. The total number of multiplication-type operations for the whole process is

$$n^2 + n(n - 1) + n(n - 2) + \cdots + n \cdot 1.$$

Using equations (1) to simplify this sum, we obtain the following result.

**Result 2** Gauss-Jordan elimination requires $(n^3 + n^2)/2$ multiplication-type operations to solve an $n \times n$ system with a nonsingular coefficient matrix.

We now use Results 1 and 2 to compare Gaussian elimination with Gauss–Jordan elimination so that we may determine the more efficient procedure. These two results give precise operation counts for the two procedures. But for the comparison we have in mind it is easier to use approximate values for the operation counts. When $n$ is moderately large (say 10), $n^3$ is quite a bit larger than both $n^2$ and $n$. This says that the $n^3$ term in the results is the important (or dominant) term. This means that Gaussian elimination requires about $n^3/3$ multiplication-type operations. Gauss–Jordan elimination, on the other hand, requires about $n^3/2$ multiplication-type operations. Thus Gauss–Jordan elimination requires 50% more operations than Gaussian elimination. This justifies our remark in Chapter 1 that Gaussian elimination is the more efficient procedure.

We conclude this section with a discussion of the inversion process. One way to measure the work required to invert an $n \times n$ matrix by Gaussian elimination is to count the operations required to solve $n$ different systems with identical coefficient matrices (see Problem 4). But when we invert a matrix, the $n$ columns on the right side of the augmented matrix are very special. If we take advantage of this and omit multiplications by 0 and 1, we can easily compute the number of operations required. The following two tables give the operation counts.

**Forward Elimination**

| Step | Each elimination: Division | Each elimination: Multiplications | Number of eliminations | Total |
|---|---|---|---|---|
| 1 | 1 | $n-1$ | $n-1$ | $(n-1)n$ |
| 2 | 1 | $n-1$ | $n-2$ | $(n-2)n$ |
| ⋮ | | | | |
| $n-1$ | 1 | $n-1$ | 1 | $1n$ |
| $n$ | — | — | 0 | 0 |

Grand total $(n^3 - n^2)/2$

**Back Substitution**

| Step | To make pivot 1: Divisions | Each elimination: Multiplications | Number of eliminations | Total |
|---|---|---|---|---|
| 1 | $n$ | $n$ | $n-1$ | $n(n-1)+n$ |
| 2 | $n$ | $n$ | $n-2$ | $n(n-2)+n$ |
| 3 | $n$ | $n$ | $n-3$ | $n(n-3)+n$ |
| ⋮ | | | | |
| $n-1$ | $n$ | $n$ | 1 | $n(1)+n$ |
| $n$ | $n$ | — | 0 | $0+n$ |

Grand total $(n^3 + n^2)/2$

**Result 3** Gaussian elimination requires $n^3$ multiplication-type operations to invert an $n \times n$ nonsingular matrix.

It is important to compare Results 1 and 3. Solving $A\mathbf{x} = \mathbf{b}$ by Gaussian elimination requires approximately $n^3/3$ multiplication-type operations. Just computing $A^{-1}$ requires three times as many operations. Thus solving the system $A\mathbf{x}=\mathbf{b}$ by computing $A^{-1}$ and then finding $A^{-1}\mathbf{b}$ is much less efficient than solving by the Gaussian elimination procedure. A related question is discussed in Problem 6.

## PROBLEMS 9.1

***1.** Let $\mathbf{v}$ be an $n$-vector, $A$ and $B$ be $n \times n$ matrices. Determine the number of multiplications and the number of additions required to compute the following.
(a) $\langle \mathbf{v}, \mathbf{v} \rangle$ (b) $A\mathbf{v}$ (c) $AB$
(d) $A^2$ (e) $A^3$ (f) $(A^2)^2$
(g) $A(A(AA))$

**2.** Show that the most efficient way to compute $A^7$ is to compute $(A^2)^2A^2A$.

**3.** Show that Gauss-Jordan elimination requires $(n^3 - n)/2$ addition-type operations to solve an $n \times n$ system with a nonsingular coefficient matrix.

***4.** The solution of $k$ different systems $A\mathbf{x} = \mathbf{b}_1$, $A\mathbf{x} = \mathbf{b}_2, \ldots, A\mathbf{x} = \mathbf{b}_k$ with identical $n \times n$ nonsingular coefficient matrices can be accomplished by using row operations on the augmented matrix $[A\colon \mathbf{b}_1 \;\; \mathbf{b}_2 \; \ldots \; \mathbf{b}_k]$.
(a) Determine how many multiplication-type operations are required when using Gaussian elimination.
(b) Determine how many multiplication-type operations are required when using Gauss-Jordan elimination.

**5.** Determine the number of multiplications and divisions required to invert an $n \times n$ nonsingular matrix using Gauss–Jordan elimination.

***6.** (a) One method for solving the $k$ different systems, $A\mathbf{x} = \mathbf{b}_1$, $A\mathbf{x} = \mathbf{b}_2, \ldots,$ $A\mathbf{x} = \mathbf{b}_k$, is to determine $A^{-1}$ and then compute the $k$ products $A^{-1}\mathbf{b}_1, A^{-1}\mathbf{b}_2, \ldots, A^{-1}\mathbf{b}_k$. Combine Result 3 and the answer to Problem 1(b) to find the number of multiplication-type operations required by this method.
(b) Compare the answer to part (a) with the number given in Problem 4(a). Under what circumstances will Gaussian elimination (applied to the augmented matrix with $k$ different right-hand columns) be a more efficient procedure than the invert-and-multiply method of part (a)?

# 9.2 Pivoting Strategies

In this section we discuss how numbers are represented in a computer, the nature of the errors involved in computer computations, and some strategies that are used to improve the accuracy of computer solutions.

## Round-off Error

Given an integer $b > 1$ called the **base**, any nonzero number can be uniquely represented in the form

$$s \cdot b^e \cdot m,$$

where $s = \pm 1$ is the **sign**, $e$ is the **exponent**, and $m$, $b^{-1} \le m < +1$, is the **mantissa**. In this representation the first digit of $m$ after the decimal point is never equal to zero. Representing a number by storing its sign, its exponent, and its mantissa is called **floating-point representation**. This is the usual way numbers are represented and stored in computers.

### EXAMPLE 1

The following table gives some examples of floating-point representation.

| Number | Sign | Base | Exponent | Mantissa |
|---|---|---|---|---|
| 35.7 | +1 | 10 | 2 | 0.357 |
| −0.005 | −1 | 10 | −2 | 0.5 |
| 15 | +1 | 10 | 2 | 0.15 |
| 4 | +1 | 2 | 3 | 0.1 |
| 15 | +1 | 2 | 4 | 0.1111 |

Notice that a different choice of base leads to a different representation for the number.

There is an unavoidable error involved in the floating-point representation of numbers. It arises from the finite number of digits (in whatever base is being used) available to represent the mantissa. The fixed length of the mantissa means that many numbers, indeed all numbers in an actual interval, will all have the same representation. For example, if the base is 10 and four digits are used to represent the mantissa, then 9.87649 and 9.8762 will both have 0.9876 as their mantissa. If mantissas are *rounded* to four digits, then 9.8756 also has the same mantissa. *Truncation* of all digits after the fourth means that 9.8769 would also have 0.9876 as its mantissa. All of the errors that result from the floating-point representation of numbers are called **round-off errors.**

Round-off errors are not necessarily small. Representing 278,342,000,000 in base 10 with a four-digit mantissa involves an error of 42 million. This example shows that the **absolute error** (the actual value of the error) is not a good measure of the significance of the error. An error of 42 million in $10^{100}$ is trivial; an error of 42 million in 100 million is disastrous. We use **relative error** (the ratio of the error to the number itself) to measure the error because it gives a clearer indication of the actual importance of the error.

One subtle form of round-off error results from the use of different bases to represent numbers. Computers usually use 2 for a base instead of 10. A number that can be represented exactly in base 10 may not be represented exactly in base 2. For example, 0.1 in base 10 has a repeating representation in base 2 (0.0001100110011...). Thus it can never be represented exactly in base 2, no matter how many digits are available for the mantissa.

Round-off error can also occur when two numbers are added. (In the remainder of this chapter we will assume that the base is 10 and that four digits are available to represent the mantissa; that is, we will work to four significant decimal digits.) For example,

$$0.1234 \cdot 10^1 + 0.5123 \cdot 10^{-2} = 1.234 + 0.005123 = 1.239123,$$

which will be represented as $0.1239 \cdot 10^1$. Notice that all the information in the last three digits of the second number $0.5123 \cdot 10^{-2}$ has been lost.

A much more serious form of round-off error occurs when two nearly equal number are subtracted.

$$0.9876 - 0.9875 = 0.0001 = 0.1000 \cdot 10^{-4}.$$

Because of the inaccuracy of the representation of the original number, the true answer is somewhere between 0 and 0.0002. When we subtracted we lost more than three significant digits. The relative error is 100%.

## Ill-conditioned Systems

Certain systems are very sensitive to small changes in the constant vector. This makes them very difficult to solve. As an example, let us consider the matrix $A = \begin{bmatrix} 1.000 & 1.001 \\ 1.000 & 1.000 \end{bmatrix}$. We wish to study how the solution of $A\mathbf{x} = \mathbf{b}$ is affected by small changes in $\mathbf{b}$. The following table gives the solution for several different choices of the vector $\mathbf{b}$.

| **b** | **Solution of $A\mathbf{x} = \mathbf{b}$** |
|---|---|
| (1.002, 1.001) | (0.001, 1.000) |
| (1.001, 1.001) | (1.001, 1.000) |
| (1.001, 1.000) | (0.000, 1.000) |
| (1.000, 1.000) | (1.000, 0.000) |
| (1.000, 1.001) | (2.001, −1.000) |

This table gives striking evidence that the system $A\mathbf{x} = \mathbf{b}$ is extremely sensitive to small changes in $\mathbf{b}$. Small changes in $\mathbf{b}$, changes that are barely detectable when our numbers are represented with four-digit mantissas, cause changes in the first significant digits of the answer. This behavior is caused by the coefficient matrix of the system. A matrix that gives rise to such an unstable system is called **ill-conditioned**.

There is no cure for the numerical instability caused by ill-conditioned matrices. Whenever the coefficient matrix is ill-conditioned the number of significant digits in the answer will be much less than the number of significant digits in the given constant vector. But even though we cannot cure this problem, we should at least be able to recognize it.

At first glance one might suppose that the instability in this problem has to do with the determinant being close to 0 ($\det A = -0.001$). But the conditioning of a matrix cannot be measured by the determinant because the matrix $1000A$ has a determinant that is quite far from 0 ($\det 1000A = -1000$) and this matrix is just as ill-conditioned as $A$ is. You can verify that $1000A$ is just as ill-conditioned as $A$ is by solving $(1000A)\mathbf{x} = \mathbf{b}$ for the following vectors.

(a) (1002, 1001) (b) (1001, 1001) (c) (1001, 1000)

Then compare with the table of solutions of $A\mathbf{x} = \mathbf{b}$ given above.

The proper way to measure the conditioning of a matrix $A$ is to look at what $A$ (considered as a linear transformation) does to the set of all unit vectors (the unit sphere). Every linear transformation takes the unit sphere into an "ellipsoid" and if $A$ produces a distorted ellipsoid, then it is ill-conditioned. A useful measure of the distortion of the ellipsoid is the ratio of the longest axis to the shortest axis. This number is called the **condition number** of $A$, denoted cond $A$. If $\lambda_{\text{max}}$ is the eigenvalue with the largest absolute value and $\lambda_{\text{min}}$ is the eigenvalue with the smallest absolute value, then $\text{cond } A = |(\lambda_{\text{max}})/(\lambda_{\text{min}})|$. When cond $A$ is large, $A$ is ill-conditioned. Indeed, the size of the condition number of $A$ gives a very good indication of how much accuracy is lost when solving the system $A\mathbf{x} = \mathbf{b}$.

### EXAMPLE 2

The eigenvalues of the matrix $A$ are the roots of $x^2 - 2x - 0.001 = 0$. Correct to four significant figures, cond $A = 4002$.

## Pivoting Strategies

There is always the possibility that the round-off errors that occur while solving a system may "build up" to such magnitudes as to produce significant errors in the answer. We must find some way to minimize the total effect of the round-off errors.

Let us illustrate this buildup of round-off errors in the following example. Let

$$A\mathbf{x} = \begin{bmatrix} 0.1000 \cdot 10^{-5} & 0.1000 \cdot 10^{1} \\ 0.1000 \cdot 10^{1} & 0.1000 \cdot 10^{1} \end{bmatrix} \begin{bmatrix} x_1 \\ x_2 \end{bmatrix} = \begin{bmatrix} 0.1000 \cdot 10^{1} \\ 0.2000 \cdot 10^{1} \end{bmatrix} = \mathbf{b}. \tag{1}$$

We apply Gaussian elimination to solve the system, using four-digit mantissas as we proceed. Forward elimination gives

$$\begin{bmatrix} 0.1000 \cdot 10^{-5} & 0.1000 \cdot 10^{1} & : & 0.1000 \cdot 10^{1} \\ 0 & -0.1000 \cdot 10^{7} & : & -0.1000 \cdot 10^{7} \end{bmatrix}$$

where the nonzero entries in the bottom row have been rounded to four significant figures. Back substitution yields $x_2 = 0.1000 \cdot 10^1$, $x_1 = 0$ as the solution.

The best that can be said for this answer is that it is wrong. $A\begin{bmatrix}0\\1\end{bmatrix}=\begin{bmatrix}1\\1\end{bmatrix}\neq\begin{bmatrix}1\\2\end{bmatrix}$. The problem is that the small error in the computation of $x_2$ has led to a large error in the computation of $x_1$. The (small) initial error has been magnified by a subsequent computation.

If we interchange the equations, this problem does not arise. We start with the equivalent system whose augmented matrix is

$$\begin{bmatrix}0.1000\cdot 10^1 & 0.1000\cdot 10^1 & : & 0.2000\cdot 10^1\\ 0.1000\cdot 10^{-5} & 0.1000\cdot 10^1 & : & 0.1000\cdot 10^1\end{bmatrix}. \qquad (1^*)$$

Forward elimination (using four-significant-digit arithmetic) gives us

$$\begin{bmatrix}0.1000\cdot 10^1 & 0.1000\cdot 10^1 & : & 0.2000\cdot 10^1\\ 0 & 0.1000\cdot 10^1 & : & 0.1000\cdot 10^1\end{bmatrix}.$$

Back substitution then yields the answer $x_1 = x_2 = 0.1000\cdot 10^1$. This answer *is* correct to four significant digits. (The answer correct to seven significant digits is $x_1 = 0.1000001\cdot 10^1$, $x_2 = 0.9999990\cdot 10^0$.)

Clearly, if changing the order of the equations can change the effects of round-off error, then some pivoting strategy must be adopted. *A common strategy is to choose the pivot equation to be that equation which has the largest (in absolute value) coefficient of the pivot variable.* This process is called **partial pivoting**.

For example, partial pivoting applied to system (1) leads to system (1*). This system was solved above and gave us an answer correct to four significant digits.

Partial pivoting uses the largest coefficient (in absolute value) of the "next" variable to determine the pivot equation. Another pivoting strategy called *total pivoting* involves choosing the pivot variable as well as the pivot equation. This strategy requires a search for the largest coefficient (in absolute value) in the remaining equations. The location of this coefficient determines both the pivot variable and the pivot equation. When using total pivoting it is essential that we remember which variables have been used as pivot variables as well as the order in which these variables have been used.

## Scaling

Partial pivoting does not resolve all of the difficulties that can arise. The system (1) is equivalent to

$$\begin{bmatrix}0.1000\cdot 10^1 & 0.1000\cdot 10^7\\ 0.5000\cdot 10^0 & 0.5000\cdot 10^0\end{bmatrix}\begin{bmatrix}x_1\\x_2\end{bmatrix}=\begin{bmatrix}0.1000\cdot 10^7\\ 0.1000\cdot 10^1\end{bmatrix}. \qquad (2)$$

Applying partial pivoting to this system gets us into the same problems as before. The problem here is that the number we pivot on, the one in the first column, is small in comparison to the other entries in that row. To see this solve system (2) using partial pivoting and four-significant-digit arithmetic.

To avoid this problem we must make sure that the numbers in our problem are in some sense about the same size. Procedures that accomplish this are called **scaling** techniques. One simple (although not inevitably successful) method to scale a system properly is to first multiply each equation by a constant so that the largest coefficient (in absolute value) in each equation is 1. If this is done to (2), then partial pivoting will succeed in finding a good solution.

## PROBLEMS 9.2

***1.** Determine the mantissa and the exponent if the base is 10 and four digits are available for the mantissa.
(a) 0.78764 (b) 787.64 (c) $\frac{5}{8}$
(d) $\frac{1}{3}$ (e) $\frac{1}{7}$ (f) $\frac{1}{70}$
(g) 7000 (h) $\frac{1}{7000}$

***2.** What is the largest relative error that can result from representation of numbers in base 10 with four-digit mantissas?

***3.** Let $A = \begin{bmatrix} 4{,}999 & 15{,}000 \\ 1{,}664 & 4{,}993 \end{bmatrix}$.
(a) Compute $A\mathbf{x}$ for each of the following vectors using four-significant-digit arithmetic.
$\mathbf{x} = (300, 400)$, $\mathbf{x} = (303, 399)$, $\mathbf{x} = (330, 390)$, and $\mathbf{x} = (600.1, 300)$.
(b) Estimate cond $A$. How does this explain the results in part (a)?

**4.** For each of the following systems, describe the scaling and pivoting that should be done before eliminating the first variable. Do not solve the systems.
(a)
$$\begin{aligned} 0.01x_1 + 10x_2 &= 55 \\ 0.1x_1 - \phantom{10}x_2 &= 26 \end{aligned}$$
(b)
$$\begin{aligned} 10x_1 + 3x_2 + 10x_3 &= 3 \\ 10x_1 + 7x_2 - 11x_3 &= -4 \\ 7x_1 - 2x_2 + 15x_3 &= 21 \end{aligned}$$

# 9.3 Iterative Solutions

There are times when Gaussian elimination is an inappropriate method to solve a linear system. Situations arise in some applications where the number of variables may be too large for the system to be solved by elimination. There are also other systems that arise in applications where most of the entries in the coefficient matrix are small or zero. In this section we study two methods for solving systems which are at times more efficient or more practical than Gaussian elimination.

## Jacobi Iteration

Given an $n \times n$ system

$$\begin{aligned} a_{11}x_1 + a_{12}x_2 + \cdots + a_{1n}x_n &= b_1 \\ a_{21}x_1 + a_{22}x_2 + \cdots + a_{2n}x_n &= b_2 \\ \vdots \qquad\quad \vdots \qquad\qquad\quad \vdots \quad &\quad \vdots \\ a_{n1}x_1 + a_{n2}x_2 + \cdots + a_{nn}x_n &= b_n \end{aligned} \tag{1}$$

with all diagonal entries nonzero, then for each $i$ we can solve the $i$th equation for $x_i$.

We obtain the following equations.

$$\begin{aligned} x_1 &= \frac{b_1 - a_{12}x_2 - a_{13}x_3 - \cdots - a_{1n}x_n}{a_{11}} \\ x_2 &= \frac{b_2 - a_{21}x_1 - a_{23}x_3 - \cdots - a_{2n}x_n}{a_{22}} \\ &\vdots \\ x_n &= \frac{b_n - a_{n1}x_1 - a_{n2}x_2 - \cdots - a_{n,n-1}x_{n-1}}{a_{nn}} \end{aligned} \tag{2}$$

If we have some way to make a guess at an approximate solution to (1) we can use equations (2) to refine our solution. We substitute the values of $x_1, x_2, \ldots, x_n$ (the guesses) in the right-hand sides of (2) and get new values for $x_1, x_2, \ldots, x_n$. In certain circumstances, the answers we obtain will be closer to the true solution than the original guesses. The process is then repeated (*iterated*) to get another approximation to the true solution. When these successive approximations get closer to the true answer, they are said to *converge*. Of course, there is no guarantee that this will happen. The successive approximations could *diverge*; that is, they could fail to get closer to the true answer. We will shortly give a sufficient condition for convergence.

## EXAMPLE 1

Consider the system

$$\begin{aligned} 21x_1 + x_2 - x_3 &= 16 \\ x_1 - 12x_2 + 2x_3 &= 31 \\ -2x_1 + x_2 + 20x_3 &= 56. \end{aligned} \tag{3}$$

Rewriting this system in the form of (2), we obtain

$$\begin{aligned} x_1 &= \frac{16 - x_2 + x_3}{21} \\ x_2 &= \frac{-31 + x_1 + 2x_3}{12} \\ x_3 &= \frac{56 + 2x_1 - x_2}{20}. \end{aligned} \tag{4}$$

Let us take $x_1 = x_2 = x_3 = 0$ as our initial guess. We now use equations (4) to obtain new values for $x_1$, $x_2$, and $x_3$: $x_1 = 0.7619$, $x_2 = -2.583$, and $x_3 = 2.800$.

We now iterate. We take these new values for the $x_i$'s and substitute them in the right-hand side of equations (4) and determine new values for $x_1$, $x_2$ and $x_3$.

$$\begin{aligned} x_1 &= \frac{16 + 2.583 + 2.800}{21} = 1.018 \\ x_2 &= \frac{-31 + 0.7619 + 2\cdot 2.800}{12} = -2.053 \\ x_3 &= \frac{56 + 2\cdot 0.7619 + 2.583}{20} = 3.005 \end{aligned}$$

The following table gives the results of several iterations. Four significant digits were used throughout the computation.

| Iteration | $x_1$ | $x_2$ | $x_3$ |
|---|---|---|---|
| 0 | 0 | 0 | 0 |
| 1 | 0.7619 | −2.583 | 2.800 |
| 2 | 1.018 | −2.053 | 3.005 |
| 3 | 1.003 | −1.998 | 3.004 |
| 4 | 1.000 | −1.999 | 3.000 |
| 5 | 1.000 | −2.000 | 3.000 |

After only five iterations we have found the answer correct to four significant digits. (Actually, this answer is exact.) This iterative method to solve systems is called **Jacobi iteration**.

In this example the approximations converge to the true answer because the diagonal terms are so much larger than the other terms.

***Definition*** A matrix is called **diagonally dominant** if for each row the absolute value of the diagonal entry is greater than the sum of the absolute values of all other entries in that row.

The coefficient matrix for system (3) is diagonally dominant because $|21| > |1| + |-1|$, $|-12| > |1| + |2|$, and $|20| > |-2| + |1|$. The following result, which we present without proof, explains why Jacobi iteration works for system (3).

**Result 1** If a system has a diagonally dominant coefficient matrix, then the approximations obtained by Jacobi iteration will converge to the solution of the system.

It should be noted that although this condition ensures convergence, it is not a necessary condition. Convergence can occur when the coefficient matrix is not diagonally dominant.

## Gauss–Seidel Iteration

There is a simple modification of the Jacobi iteration process that sometimes converges more rapidly to the solution. The Jacobi process uses equations (2) to find new values for the $x_i$'s simultaneously. All the old $x$'s are used at one time to find all the new $x$'s. But if the new value of $x_1$ is "better" than the old one, why not use it when we compute $x_2$? There is no reason why we cannot use a new value for $x_i$ which has already been computed whenever it is needed in a subsequent equation. This method of solving systems is called **Gauss–Seidel iteration**.

## EXAMPLE 2

Let us solve system (3) by Gauss–Seidel iteration. We take $x_1 = x_2 = x_3 = 0$ as our initial approximation. The first iteration gives

$$x_1 = \frac{16 - 0 + 0}{21} = 0.7619,$$

$$x_2 = \frac{-31 + 0.7619 + 2\cdot 0}{12} = -2.520,$$

$$x_3 = \frac{56 + 2\cdot 0.7619 + 2.520}{20} = 3.002.$$

The second iteration gives

$$x_1 = \frac{16 + 2.520 + 3.002}{21} = 1.025,$$

$$x_2 = \frac{-31 + 1.025 + 2\cdot 3.002}{12} = -1.998,$$

$$x_3 = \frac{56 + 2\cdot 1.025 + 1.998}{20} = 3.002.$$

The following table summarizes the first three iterations.

| **Iteration** | $x_1$ | $x_2$ | $x_3$ |
|---|---|---|---|
| 0 | 0 | 0 | 0 |
| 1 | 0.7619 | −2.520 | 3.002 |
| 2 | 1.025 | −1.998 | 3.002 |
| 3 | 1.000 | −2.000 | 3.000 |

This example shows that Gauss–Seidel iteration can be more rapid than Jacobi iteration. However, the reader should be aware that Jacobi iteration sometimes converges more rapidly than Gauss–Seidel iteration. Even more amazing is the existence of systems where one method converges and the other one does not. There are systems where Jacobi iteration converges while Gauss–Seidel diverges and other systems where the exact opposite occurs. A more thorough analysis of these matters will be found in books on numerical analysis. Here we only mention that Result 1 holds for Gauss–Seidel iteration as well. If the coefficient matrix is diagonally dominant, then Gauss–Seidel iteration will converge.

Before ending this section it is appropriate to say a few words about operation counts when using iterative methods. Each iteration (Jacobi or Gauss–Seidel) requires $n^2$ multiplication-type operations. However, computing an operation count is impossible because we have no idea of the number of iterations required. Even though this problem cannot be overcome, we should note that the number of iterations does not depend on the number of variables in the problem but on the nature of the equations (2). This means that $k$ iterations require $kn^2$ operations, whereas Gaussian elimination requires $n^3/3$. If convergence occurs before $n/3$ iterations have been made (i.e., if $k < n/3$), then iteration is a more efficient method. When $n$ is 100 and convergence occurs within 33 iterations, then iteration is more efficient than elimination.

## PROBLEMS 9.3

*1. Solve each system by Jacobi iteration.

(a) $$\begin{aligned} 100x_1 + x_2 + x_3 &= 72 \\ x_1 + 200x_2 - x_3 &= 301 \\ x_1 - x_2 + 10x_3 &= 42 \end{aligned}$$

(b) $$\begin{aligned} 1000x_1 + 900x_2 + x_3 &= 10 \\ x_1 + 10x_2 + x_3 &= 43 \\ -10x_1 + 45x_2 - 370x_3 &= 27 \end{aligned}$$

(c) $$\begin{aligned} 370x_1 - 45x_2 + 10x_3 &= 35 \\ x_1 + 10x_2 + x_3 &= 47 \\ 10x_1 + 900x_2 - 1000x_3 &= 75 \end{aligned}$$

2. Solve the systems in Problem 1 by Gauss–Seidel iteration.

3. Transform each of the following systems into a diagonally dominant system.

(a) $$\begin{aligned} 2x_1 + 50x_2 - x_3 &= 5 \\ 50x_1 - 10x_2 + x_3 &= 4 \\ x_1 + x_2 + 10x_3 &= 5 \end{aligned}$$

(b) $$\begin{aligned} x_1 + 2x_2 + 30x_3 &= 2 \\ 150x_1 + 100x_2 + x_3 &= 40 \\ 10x_1 + 80x_2 - 3x_3 &= 15 \end{aligned}$$

4. Modify the definition of a diagonally dominant system so that the systems in Problem 3 are included in the definition.

5. In Problem 14 of Section 3.8 the $QR$ decomposition of a square matrix $A$ was used to reduce the normal equations $A^TA\mathbf{x} = A^T\mathbf{b}$ to the equivalent system $R\mathbf{x} = Q^T\mathbf{b}$.

(a) Show that the system $R\mathbf{x} = Q^T\mathbf{b}$ is a triangular system and hence can be solved by back substitution.

(b) Compare the two methods for solving the normal equations (Gaussian elimination versus $QR$ decomposition) for numerical stability.

# 9.4 Eigenvalues and Eigenvectors

In this section we present a method for approximating the largest (in absolute value) eigenvalue of a matrix. We could, of course, compute the eigenvalues by determining the characteristic polynomial and then finding its roots. Unfortunately, this method is so lengthy and so numerically unstable that it is virtually useless for large matrices. The method we present here (like most numerical methods for computing eigenvalues) takes advantage of special characteristics of the matrix and will not work in all cases.

Let $A$ be diagonalizable with eigenvalues $\lambda_1, \lambda_2, \ldots, \lambda_n$. Let $\mathbf{v}_1, \mathbf{v}_2, \ldots, \mathbf{v}_n$ be eigenvectors associated with each of these eigenvalues. Suppose in addition that

$$|\lambda_1| > |\lambda_2| \geq |\lambda_3| \geq \cdots \geq |\lambda_n|.$$

Since $\mathbf{v}_1, \mathbf{v}_2, \ldots, \mathbf{v}_n$ form a basis for $R^n$, we can express any vector $\mathbf{u}$ in $R^n$ in the form

$$\mathbf{u} = \alpha_1\mathbf{v}_1 + \alpha_2\mathbf{v}_2 + \cdots + \alpha_n\mathbf{v}_n.$$

Since the v's are eigenvectors, we have

$$A\mathbf{u} = \alpha_1 A\mathbf{v}_1 + \alpha_2 A\mathbf{v}_2 + \cdots + \alpha_n A\mathbf{v}_n = \alpha_1\lambda_1\mathbf{v}_1 + \alpha_2\lambda_2\mathbf{v}_2 + \cdots + \alpha_n\lambda_n\mathbf{v}_n.$$

It is not hard to verify that for any positive integer $k$

$$A^k\mathbf{u} = \alpha_1\lambda_1^k\mathbf{v}_1 + \alpha_2\lambda_2^k\mathbf{v}_2 + \cdots + \alpha_n\lambda_n^k\mathbf{v}_n.$$

Now $|\lambda_1|$ is larger than $|\lambda_i|$, $i = 2, \ldots, n$, so the first term of this expression for $A^k\mathbf{u}$ will eventually dominate the others. This means that when $k$ is large enough, $A^k\mathbf{u}$ is approximately $\alpha_1\lambda_1^k\mathbf{v}_1$, that is, $A^k\mathbf{u}$ is approximately an eigenvector associated with the eigenvalue $\lambda_1$. (We are of course assuming here that $\alpha_1 \neq 0$.)

If the vector **u** is close to an eigenvector **v** associated with the eigenvalue $\lambda_1$, then

$$\frac{\langle \mathbf{u}, A\mathbf{u}\rangle}{\langle \mathbf{u}, \mathbf{u}\rangle} \quad \text{is approximately} \quad \frac{\langle \mathbf{v}, A\mathbf{v}\rangle}{\langle \mathbf{v}, \mathbf{v}\rangle} = \lambda_1.$$

This quotient

$$\frac{\langle \mathbf{u}, A\mathbf{u}\rangle}{\langle \mathbf{u}, \mathbf{u}\rangle}$$

is called the **Rayleigh quotient**.

We can use the discussion above to approximate $\lambda_1$. The **power method** for approximating the dominant eigenvalue consists of computing the Rayleigh quotient for the vectors

$$\mathbf{u}_0 = \mathbf{u}, \quad \mathbf{u}_1 = A\mathbf{u}_0, \quad \mathbf{u}_2 = A^2\mathbf{u} = A\mathbf{u}_1, \quad \mathbf{u}_3 = A^3\mathbf{u} = A\mathbf{u}_2, \ldots, \mathbf{u}_n = A\mathbf{u}_{n-1}$$

until the Rayleigh quotients get close enough to $\lambda_1$.

## EXAMPLE 1

Let $A = \begin{bmatrix} 9 & -4 \\ 4 & -1 \end{bmatrix}$ and $\mathbf{u}_0 = (1, 1)$. The following table gives the results of the power method.

| Iteration ($n$) | $\mathbf{u}_n$ | $\langle \mathbf{u}_{n-1}, \mathbf{u}_n\rangle / \langle \mathbf{u}_{n-1}, \mathbf{u}_{n-1}\rangle$ |
|---|---|---|
| 0 | (1, 1) | — |
| 1 | (5, 3) | 4 |
| 2 | (33, 17) | 6.353 |
| 3 | (229, 115) | 6.903 |
| 4 | (1,601, 801) | 6.986 |
| 5 | (11,210, 5,603) | 6.998 |
| 6 | (78,480, 39,240) | 7.001 |
| 7 | (549,400, 274,700) | 7.000 |
| 8 | (3,846,000, 1,923,000) | 7.000 |

It is clear that the eigenvalue is approximately 7 and that any of the last three $\mathbf{u}_i$'s would be a good approximation to an eigenvector associated with this eigenvalue. [The eigenvectors associated with 7 are the nonzero multiples of (2, 1).]

Example 1 illustrates the unpleasantly large numbers that are frequently produced by the power method. These numbers can be avoided. We merely multiply each **u** by a scalar after computing the Rayleigh quotient and use this new vector in the iteration. A common procedure is to choose the scalar so that the absolute value of every coordinate in **u** is $\leq 1$.

### EXAMPLE 2

Let us repeat Example 1 using the idea above. The following table summarizes the procedure.

| **u** | $A\mathbf{u}$ | **Rayleigh quotient** | **New** $u$ |
|---|---|---|---|
| (1, 1) | (5.000, 3.000) | 4.000 | (1, 0.6000) |
| (1, 0.6000) | (6.600, 3.400) | 6.353 | (1, 0.5152) |
| (1, 0.5152) | (6.939, 3.485) | 6.902 | (1, 0.5022) |
| (1, 0.5022) | (6.991, 3.498) | 6.986 | (1, 0.5003) |
| (1, 0.5003) | (6.999, 3.500) | 6.998 | (1, 0.5000) |
| (1, 0.5000) | (7.000, 3.500) | 7.000 | (1, 0.5000) |

The power method finds a single eigenvalue. Fortunately, there are many problems where we are interested only in determining this one eigenvalue. Procedures to find the remaining eigenvalues are treated in books on numerical analysis.

## PROBLEMS 9.4

*1. Approximate the dominant eigenvalue and an associated eigenvector by the power method.

(a) $\begin{bmatrix} 1 & 0 & 1 \\ 0 & 1 & 0 \\ 1 & 2 & 1 \end{bmatrix}$ (b) $\begin{bmatrix} 1 & 2 & 0 \\ 2 & 2 & 1 \\ 0 & 1 & 3 \end{bmatrix}$

(c) $\begin{bmatrix} 1 & 1 & 0 \\ 1 & 5 & -2 \\ 1 & 3 & -1 \end{bmatrix}$

**2.** Let

$$A = \begin{bmatrix} 1 & -1 & 0 \\ -1 & 2 & -1 \\ 0 & -1 & 1 \end{bmatrix}.$$

(a) Approximate the dominant eigenvalue of $A$ by the power method using $(1, 1, 1)$ as $\mathbf{u}_0$.

(b) Approximate the dominant eigenvalue of $A$ by the power method using $(1, 2, 3)$ as $\mathbf{u}_0$.

(c) Explain the differences in the results in the preceding parts.

**3.** Let

$$A = \begin{bmatrix} 344 & 1029 \\ -98 & -293 \end{bmatrix}$$

and

$$B = \begin{bmatrix} 9 & 21 \\ -2 & -4 \end{bmatrix}.$$

*(a) Find the eigenvalues and eigenvectors of both $A$ and $B$ by solving the appropriate equations given in Chapter 8.

(b) Use the power method to find the dominant eigenvalue of $A$ and an associated eigenvector. How many iterations are necessary?

(c) Use the power method to find the dominant eigenvalue of $B$ and an associated eigenvector. How many iterations are necessary?

*(d) Can you explain the difference in the rates of convergence in parts (b) and (c)?

# 10

# The Vector Space $C^m$

There is rich soil to be cultivated when we move from $m$-tuples of real numbers to $m$-tuples of complex numbers. The Gaussian elimination procedure and all results and procedures in Chapter 2 remain valid. The soil begins to get richer when we turn to inner product spaces. We obtain several generalizations of concepts studied in Chapter 3. Then we turn to the study of eigenvalues and eigenvectors of a square matrix with complex entries. Here the soil is really fertile. We will not cultivate it extensively but head straight for the spectral theorem for Hermitian matrices.

## 10.1
## Complex Numbers

Undoubtedly, you have already been introduced to complex numbers. They are the numbers of the form

$$z = a + bi,$$

where $a$ and $b$ are real numbers and $i$ is that special number that satisfies the equation $i^2 = -1$. The number $a$ is called the **real part** of $z$ and $b$ is called the **imaginary part** of $z$. Two complex numbers are equal if their real and imaginary parts are equal. Thus $a + bi = c + di$ means that $a = c$ and $b = d$. Addition and multiplication are defined as follows:

$$(a + bi) + (c + di) = (a + c) + (b + d)i$$
$$(a + bi)\cdot(c + di) = (ac - bd) + (ad + bc)i.$$

There is no need to memorize the formula for multiplication. Simply multiply in the obvious way and use the fact that $i^2 = -1$.

One of the simplest and most useful operations on complex numbers is conjugation. The **complex conjugate** of $z = a + bi$ is $\bar{z} = a - bi$. $z$ is a real number if and only if $z = \bar{z}$. If the real part of a complex number $z$ is zero, then $\bar{z} = -z$.

Complex numbers have a beautiful and useful geometric interpretation. The complex number $z = a + bi$ corresponds to the ordered pair $(a, b)$ (see Figure 10.1). The conjugate $\bar{z}$ of $z$ is the reflection of $z$ in the real axis. The number $z\bar{z} = a^2 + b^2$ is the square of the distance of $z$ from the origin. Its square root is called the **modulus** of $z$ and is denoted by $|z|$.

If $\theta$ is the angle between $z = a + bi$ and the real axis, then

$$a = |z| \cos\theta \qquad \text{and} \qquad b = |z| \sin\theta.$$

Hence

$$z = a + bi = |z|(\cos\theta + i \sin\theta).$$

We define

$$e^{i\theta} = \cos\theta + i \sin\theta \qquad \text{(Euler's identity)}.$$

Then

$$z = |z| e^{i\theta};$$

this is called the **polar representation** of $z$.

Using complex conjugates it is easy to compute the quotient of two complex numbers. If $w \neq 0$, then

$$\frac{z}{w} = \frac{z}{w} \cdot \frac{\bar{w}}{\bar{w}} = \frac{z\bar{w}}{|w|^2}.$$

For instance,

$$\frac{1 + i}{2 + 3i} = \frac{(1 + i)(2 - 3i)}{13} = \frac{5}{13} - \frac{1}{13} i.$$

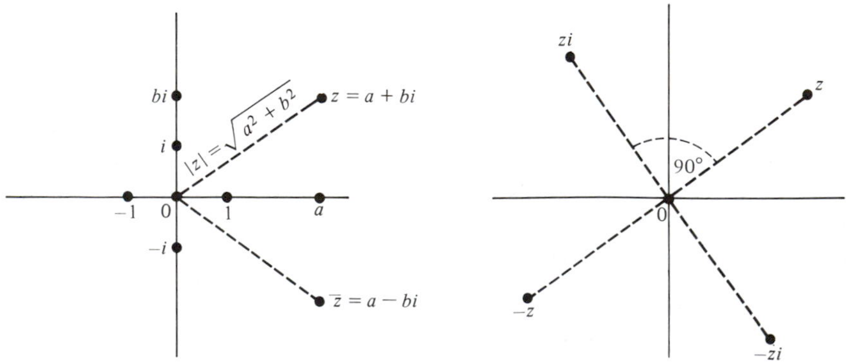


Figure 10.1

Let $C^m$ denote the collection of all $m$-tuples each of whose components is a complex number. Addition and scalar multiplication are defined componentwise, but this time the scalars are complex numbers. It is a routine exercise for you to check that $C^m$ is a vector space with respect to these two operations. All results and procedures in Chapter 2 are valid in $C^m$.

### EXAMPLE 1

Consider the system

$$\begin{bmatrix} i & 1 \\ 1-i & 2-i \end{bmatrix}\begin{bmatrix} x_1 \\ x_2 \end{bmatrix} = \begin{bmatrix} 1 \\ 5-4i \end{bmatrix}.$$

Gaussian elimination reduces the augmented matrix

$$\begin{bmatrix} i & 1 & : & 1 \\ 1-i & 2-i & : & 5-4i \end{bmatrix} \quad \text{to} \quad \begin{bmatrix} 1 & 0 & : & 1+i \\ 0 & 1 & : & 2-i \end{bmatrix}.$$

Thus the solution is $x_1 = 1 + i$, $x_2 = 2 - i$.

### EXAMPLE 2

The three vectors $(1, 2, 1 - i)$, $(1, i, 1 + i)$, and $(1, 1 + i, i)$ are linearly independent. We can show this by row-reducing

$$\begin{bmatrix} 1 & 1 & 1 \\ 2 & i & 1+i \\ 1-i & 1+i & i \end{bmatrix} \quad \text{to} \quad \begin{bmatrix} 1 & 1 & 1 \\ 0 & 1 & (3-i)/5 \\ 0 & 0 & 1 \end{bmatrix}.$$

The row-reduced matrix is obviously of rank 3.

### EXAMPLE 3

The inverses of nonsingular matrices with complex entries can be computed by the method of Chapter 1.

$$\begin{bmatrix} i & 1 & : & 1 & 0 \\ 1+i & 0 & : & 0 & 1 \end{bmatrix} \quad \text{becomes} \quad \begin{bmatrix} 1 & -i & : & -i & 0 \\ 0 & -1+i & : & -1+i & 1 \end{bmatrix},$$

which reduces to

$$\begin{bmatrix} 1 & 0 & : & 0 & (1-i)/2 \\ 0 & 1 & : & 1 & (-1-i)/2 \end{bmatrix}.$$

This shows that the inverse of

$$\begin{bmatrix} i & 1 \\ 1+i & 0 \end{bmatrix} \quad \text{is} \quad \begin{bmatrix} 0 & (1-i)/2 \\ 1 & (-1-i)/2 \end{bmatrix}.$$

# PROBLEMS 10.1

*1. Perform the indicated operations.
   (a) $(3+4i)\cdot(2-4i)$
   (b) $(1+3i)+(2-6i)$
   (c) $(1-4i)/(3-4i)$
   (d) $(1+i)\cdot(1-i)$
   (e) $(1+i)\cdot(1+i)$
   (f) $(1+i)^n/(1-i)^n$

*2. Find the indicated quantities.
   (a) $\operatorname{Re}(-3+4i)$
   (b) $\operatorname{Im}(-3+4i)$
   (c) $|-3+4i|$
   (d) $|i|$
   (e) $\operatorname{Re}(-3-4i)$
   (f) $\operatorname{Im}(-3-4i)$
   (g) $|-3-4i|$
   (h) $|-i|$

3.*(a) Find two square roots of $i$ by solving $(x+yi)^2 = i$ for $x$ and $y$.
   *(b) Find two square roots of $-i$ by solving $(x+yi)^2 = -i$ for $x$ and $y$.
   (c) The answers in parts (a) and (b) are the four fourth roots of $-1$. Plot these four roots in the complex plane.

4. The quadratic formula gives the roots of $x^2+x+1$ as

$$w_1 = -\frac{1}{2} + \frac{\sqrt{3}}{2} i$$

and

$$w_2 = -\frac{1}{2} - \frac{\sqrt{3}}{2} i.$$

Compute $w_1^2$, $w_1^3$, $w_2^2$, $w_2^3$, and $w_1 w_2$.

*5. (a) Find a formula for the reciprocal of a complex number $z = a+bi$.
   (b) Is the reciprocal of the conjugate equal to the conjugate of the reciprocal? In symbols, is the formula

$$\frac{1}{\bar{z}} = \overline{\left(\frac{1}{z}\right)}$$

true?

6. Verify that the following rules are true.
   (a) $\overline{z_1+z_2} = \overline{z_1} + \overline{z_2}$
   (b) $\overline{z_1 z_2} = \overline{z_1}\cdot\overline{z_2}$
   (c) $\overline{(z_1/z_2)} = \overline{z_1}/\overline{z_2}$
   (d) $|z_1 z_2| = |z_1||z_2|$
   (e) $|\bar{z}| = |z|$

7. (a) Compute $i^3$, $i^4$, $i^5$, $i^6$, and $i^7$.
   (b) Give a formula for $i^n$, where $n$ is a positive integer.

8. Show that the polynomial $x^4+4$ has $1+i$, $1-i$, $-1+i$, and $-1-i$ as roots.

9. Show that the system

$$\begin{aligned} cx - dy &= a \\ dx + cy &= b \end{aligned}$$

has a unique solution unless $c = d = 0$.

10. (a) Solve $\bar{z} = z^2$. (b) Solve $\bar{z} = z^3$.

11. Show that $|z_1+z_2|^2 + |z_1-z_2|^2 = 2(|z_1|^2 + |z_2|^2)$. Interpret this identity geometrically.

12. We defined $e^{i\theta}$ by the formula

$$e^{i\theta} = \cos\theta + i\sin\theta.$$

   (a) Show that $e^{i\theta}e^{i\phi} = e^{i(\theta+\phi)}$.
   (b) Show that if $z = re^{i\theta}$, then $z^n = r^n e^{in\theta}$.
   (c) Let $z = re^{i\theta}$. Show that the solutions of the equation

$$w^n = z$$

are

$$w = r^{1/n} e^{i(\theta+2k\pi)/n}, \quad k = 0, 1, \ldots, n-1.$$

   (d) Use part (c) to show that the cube roots of 1 are

$$1, \quad -1/2 + i\sqrt{3}/2, \quad -\sqrt{3}/2 - i/2.$$

*13. Show that the vectors $(i, 1)$ and $(1, -i)$ are linearly dependent.

14. Show that $(i, 1)$ and $(-1, -i)$ form a basis for $C^2$.

**15.** Find the dimension of the subspace spanned by the vectors $(1, 1+i, -i)$, $(i, -1+i, 1)$, and $(1-i, 2, -1-i)$?

***16.** Solve the following systems.

(a) $$\begin{aligned}(1+i)x_1 + (1-i)x_2 &= 1-i\\ 2x_1 + \qquad ix_2 &= 4i\end{aligned}$$

(b) $$\begin{aligned}(2+i)x_1 + (3-i)x_2 &= 15\\ x_1 + \qquad x_2 &= 5\end{aligned}$$

(c) $$\begin{aligned}(1+i)x_1 + (2+i)x_2 &= -1+i\\ ix_1 - (1+i)x_2 &= 2+5i\end{aligned}$$

(d) $$\begin{aligned}x_1 \qquad + \qquad ix_3 &= 0\\ ix_1 + x_2 + (-1+i)x_3 &= -1+2i\\ (1+i)x_1 + x_2 + \qquad 2ix_3 &= 3i\end{aligned}$$

***17.** Find the inverses of the following matrices.

(a) $\begin{bmatrix} i & 1 \\ 1-i & 2-i \end{bmatrix}$

(b) $\begin{bmatrix} 1 & 1 \\ 2+i & 3-i \end{bmatrix}$

(c) $\begin{bmatrix} 1 & 0 & i \\ i & 1 & -1+i \\ 1+i & 1 & 2i \end{bmatrix}$

(d) $\begin{bmatrix} -i & 1 & 0 \\ 1+i & 1 & 0 \\ 2i & -1 & 1 \end{bmatrix}$

**18.** Show that $\mathbf{x}_1, \mathbf{x}_2, \ldots, \mathbf{x}_k$ are independent vectors in $C^m$ if and only if $\bar{\mathbf{x}}_1, \bar{\mathbf{x}}_2, \ldots, \bar{\mathbf{x}}_k$ are independent.

# 10.2
## Inner Products on $C^m$

Here life changes. The usual inner product on $R^m$ is defined to be the sum of the products of corresponding components:

$$\langle \mathbf{x}, \mathbf{y} \rangle = x_1 y_1 + \cdots + x_m y_m. \tag{1}$$

When $\mathbf{x} = \mathbf{y}$ we obtain

$$\|\mathbf{x}\|^2 = \langle \mathbf{x}, \mathbf{x} \rangle = x_1^2 + \cdots + x_m^2.$$

If we use this formula when $\mathbf{x} = (1, i)$, we obtain

$$\|\mathbf{x}\|^2 = 1 - 1 = 0.$$

We hope you agree that that is a problem. When $\mathbf{x}$ is in $R^m$ we can rewrite (1) as

$$\|\mathbf{x}\|^2 = |x_1|^2 + \cdots + |x_m|^2$$

since $|x_i|^2 = x_i^2$. If $\mathbf{x}$ is now in $C^m$, $|x_i|$ is the modulus of $x_i$ and we have a cure. The modulus of a nonzero complex number is a positive real number. Since

$$|z|^2 = z\bar{z} = \bar{z}z,$$

we have two obvious ways to define an inner product on $C^m$ in such a way that $\langle \mathbf{x}, \mathbf{x} \rangle = |x_1|^2 + \cdots + |x_m|^2$:

$$\langle \mathbf{x}, \mathbf{y} \rangle = x_1 \bar{y}_1 + \cdots + x_m \bar{y}_m \tag{2}$$

and

$$\langle \mathbf{x}, \mathbf{y} \rangle = \bar{x}_1 y_1 + \cdots + \bar{x}_m y_m. \tag{3}$$

They are equivalent definitions. We prefer the latter formula because it makes some other formulas nicer.

***Definition*** If $\mathbf{x} = (x_1, \ldots, x_m)$ and $\mathbf{y} = (y_1, \ldots, y_m)$ are vectors in $C^m$, then the **inner product** of $\mathbf{x}$ and $\mathbf{y}$ is defined by

$$\langle \mathbf{x}, \mathbf{y} \rangle = \bar{x}_1 y_1 + \cdots + \bar{x}_m y_m.$$

$C^m$ together with this inner product is called a **complex inner product space.**

The crucial properties of this inner product are these (compare with Theorem 1 of Section 3.2).

(a) $\langle \mathbf{x}, \mathbf{y} \rangle = \overline{\langle \mathbf{y}, \mathbf{x} \rangle}$.
(b) $\langle \mathbf{x}, \mathbf{y} + \mathbf{z} \rangle = \langle \mathbf{x}, \mathbf{y} \rangle + \langle \mathbf{x}, \mathbf{z} \rangle$.
(c) $\langle \alpha\mathbf{x}, \mathbf{y} \rangle = \bar{\alpha}\langle \mathbf{x}, \mathbf{y} \rangle$ and $\langle \mathbf{x}, \alpha\mathbf{y} \rangle = \alpha\langle \mathbf{x}, \mathbf{y} \rangle$.
(d) $\langle \mathbf{x}, \mathbf{x} \rangle \geq 0$ and $\langle \mathbf{x}, \mathbf{x} \rangle = 0$ if and only if $\mathbf{x} = \mathbf{0}$.

Note that when we restrict to $R^m$ our definition of an inner product on $C^m$ reduces to the usual inner product of $R^m$ and these properties reduce to the properties given in Theorem 1 of Section 3.2.

We defined our inner product on $C^m$ in such a way that the norm of a vector is the square root of its inner product with itself. Since the Cauchy–Schwarz inequality is valid (see Problem 3) we can define the angle between two vectors in $C^m$ with the same formula we used for $R^m$:

$$\cos\theta = \frac{\langle \mathbf{x}, \mathbf{y} \rangle}{\|\mathbf{x}\| \cdot \|\mathbf{y}\|}.$$

Thus $\mathbf{x}$ and $\mathbf{y}$ are **orthogonal** if and only if $\langle \mathbf{x}, \mathbf{y} \rangle = 0$.

Let $A$ be an $m \times n$ matrix with complex entries. We define $\bar{A} = [\bar{a}_{ij}]$ to be the **conjugate** of $A$ and $A^* = (\bar{A})^T$ to be the conjugate transpose of $A$. The conjugate transpose will play the role for matrices with complex entries analogous to the role played by the transpose for real matrices. For instance, if

$$A = \begin{bmatrix} i & 1+i \\ 2 & 3+2i \\ 0 & 1 \end{bmatrix},$$

then

$$\bar{A} = \begin{bmatrix} -i & 1-i \\ 2 & 3-2i \\ 0 & 1 \end{bmatrix} \quad \text{and} \quad A^* = \begin{bmatrix} -i & 2 & 0 \\ 1-i & 3-2i & 1 \end{bmatrix}.$$

Similarly, if

$$\mathbf{x} = \begin{bmatrix} -i \\ -1+i \\ 2 \end{bmatrix}, \quad \text{then} \quad \bar{\mathbf{x}} = \begin{bmatrix} i \\ -1-i \\ 2 \end{bmatrix} \quad \text{and} \quad \mathbf{x}^* = [i \quad -1-i \quad 2].$$

Three of the four properties of the transpose given in Chapter 1 remain valid for the conjugate transpose.

(1) $A^{**} = A$.
(2) $(\alpha A)^* = \bar{\alpha} A^*$.
(3) $(A + B)^* = A^* + B^*$.
(4) $(AB)^* = B^* A^*$.

Note the change in property (2). The inner product can be expressed in terms of the conjugate transpose as

$$\langle \mathbf{x}, \mathbf{y} \rangle = \mathbf{x}^* \mathbf{y}.$$

[Our choice between the two equivalent definitions of an inner product given by (2) and (3) was made so that this formula would hold. Choosing (2) as the definition leads to the formula $\langle \mathbf{x}, \mathbf{y} \rangle = \overline{\mathbf{x}^* \mathbf{y}}$.] Using this fact together with properties (1) and (4), it follows that

$$\langle \mathbf{x}, A\mathbf{y} \rangle = \langle A^* \mathbf{x}, \mathbf{y} \rangle.$$

When a matrix has complex entries, the generalization of a symmetric matrix is called a Hermitian matrix.

***Definition*** A square matrix $A$ is called **Hermitian** if $A^* = A$.

For instance,

$$A = \begin{bmatrix} -1 & 2 - i \\ 2 + i & 4 \end{bmatrix} = A^*.$$

Clearly, a matrix $A$ is Hermitian if and only if $a_{ij} = \bar{a}_{ji}$ for all $i$ and $j$. In particular, the diagonal entries of a Hermitian matrix $A$ must be real because $a_{ii} = \bar{a}_{ii}$.

Note that if $A$ has real entries, then $A^* = A^T$ and hence a Hermitian matrix with real entries is a symmetric matrix. Also note that if $A$ is Hermitian, then

$$\langle \mathbf{x}, A\mathbf{y} \rangle = \langle A\mathbf{x}, \mathbf{y} \rangle.$$

For real matrices, we defined a square matrix to be orthogonal if its columns are orthonormal. We use the same definition for matrices with complex entries, but change the name.

***Definition*** A square matrix $U$ is called **unitary** if its columns are orthonormal.

In $R^m$, the equation $\mathbf{x}^T \mathbf{y} = \langle \mathbf{x}, \mathbf{y} \rangle$ gave us that $A$ is symmetric if and only if $A^T A = I$. In $C^m$, we have $\mathbf{x}^* \mathbf{y} = \langle \mathbf{x}, \mathbf{y} \rangle$ and hence $U$ is unitary if and only if $U^* U = I$.

# PROBLEMS 10.2

**1.** Verify properties (a)–(d) of the inner product on $C^m$.

**2.** (a) Show that the properties given in Problem 1 reduce to the properties given in Theorem 1 of Section 3.2 in the event that all numbers involved are real.

(b) Why is

$$\langle \mathbf{x}, \mathbf{y} \rangle + \langle \mathbf{y}, \mathbf{x} \rangle = 2\,\mathrm{Re}(\langle \mathbf{x}, \mathbf{y} \rangle)?$$

**3.** (a) Show that the Cauchy–Schwarz inequality holds in $C^m$. (*Hint:* Use the proof in Section 3.3 with an appropriate $\alpha$.)

(b) Show that

$$|\langle \mathbf{x}, \mathbf{y} \rangle + \langle \mathbf{y}, \mathbf{x} \rangle| \leq 2\|\mathbf{x}\| \cdot \|\mathbf{y}\|.$$

(c) Show that the triangle inequality holds in $C^m$.

**4.** Show that $\|\alpha\mathbf{x}\| = |\alpha|\,\|\mathbf{x}\|$.

**5.** Let $A = [a_{ij}]$ be an $m \times n$ matrix with complex entries. Verify the following properties.

(a) If $A$ has real entries, $A^* = A^T$.

(b) $\overline{A\mathbf{x}} = \bar{A}\bar{\mathbf{x}}$.

(c) $A + \bar{A}$ is a real matrix.

**6.** Prove properties (1)–(4) of the conjugate transpose.

**7.** Recall that an $m \times m$ matrix $H$ is called Hermitian if $H^* = H$.

(a) Show that a real Hermitian matrix is symmetric.

(b) Show that $\langle \mathbf{x}, H\mathbf{y} \rangle = \langle H\mathbf{x}, \mathbf{y} \rangle$ for all $\mathbf{x}, \mathbf{y}$ in $C^m$ if and only if $H$ is Hermitian.

(c) Show that if $H_1$ and $H_2$ are Hermitian, then $H_1 + H_2$ is Hermitian.

(d) Are all diagonal matrices Hermitian?

(e) If $A$ is any matrix, show that $A^*A$ is Hermitian.

**8.** Recall that an $m \times m$ matrix $U$ is called unitary if its columns form an orthonormal set of vectors in $C^m$. Prove each of the following.

(a) $U$ is unitary if and only if $U^*U = I = UU^*$.

(b) $U$ is unitary if and only if $U^{-1} = U^*$.

(c) Use the fact that $\det \bar{A} = \overline{\det A}$ to prove that if $U$ is unitary, then $|\det U| = 1$.

(d) $U$ is unitary if and only if $\bar{U}$ is unitary.

(e) If $U$ is unitary, then $U^k$ is unitary for every integer $k$.

(f) Any product of unitary matrices is a unitary matrix.

(g) If $U$ is unitary, then $U^2 = I$ if and only if $U$ is Hermitian.

(h) If $H$ is Hermitian, then $H^2 = I$ if and only if $H$ is unitary.

(i) $U$ is unitary if and only if $\langle U\mathbf{x}, U\mathbf{y} \rangle = \langle \mathbf{x}, \mathbf{y} \rangle$ for all $\mathbf{x}, \mathbf{y}$ in $C^m$.

(j) Let $\mathbf{v}_1, \mathbf{v}_2, \ldots, \mathbf{v}_m$ be an orthonormal basis for $C^m$. $U$ is unitary if and only if $\langle U\mathbf{v}_i, U\mathbf{v}_j \rangle = \langle \mathbf{v}_i, \mathbf{v}_j \rangle$ for all $i$ and $j$.

(k) $U$ is unitary if and only if $\|U\mathbf{x}\| = \|\mathbf{x}\|$ for all $\mathbf{x}$ in $C^m$.

(l) If $H$ is Hermitian, then $I + iH$ is nonsingular and $(I + iH)^{-1}(I - iH)$ is unitary.

**9.** Let $N$ be an $m \times m$ matrix. $N$ is called **normal** if $NN^* = N^*N$. Prove the following.

*(a) Hermitian matrices are normal.

(b) Unitary matrices are normal.

(c) Define skew-Hermitian and show that a skew-Hermitian matrix is normal.

(d) Any diagonal matrix is normal.

**10.** How would you define an inner product on the collection of all complex-valued functions that are continuous on the interval $[0, 1]$?

# 10.3
# Eigenvalues and Eigenvectors

An eigenvalue of an $m \times m$ complex matrix $A$ is a root of the polynomial $\det(A - \lambda I)$. This polynomial has exactly $m$ complex roots (counting multiplicities). Hence every $m \times m$ matrix with complex entries has $m$ eigenvalues.

In Chapter 8 we considered real matrices all of whose eigenvalues were real numbers. In general, the polynomial $\det(A - \lambda I)$ may have complex roots even when $A$ is a real matrix. It is impossible to find an eigenvector in $R^m$ for such complex eigenvalues. For example, the eigenvalues of the matrix $\begin{bmatrix} 0 & -1 \\ 1 & 0 \end{bmatrix}$ are $i$ and $-i$. The eigenvectors associated with the eigenvalue $i$ are the nonzero multiples of $(i, 1)$ and the eigenvectors for $-i$ are the nonzero multiples of $(-i, 1)$. These eigenvectors are not in $R^2$; they are in $C^2$.

The following example shows that the complex eigenvalues and complex eigenvectors of a real matrix are important in the applications.

### EXAMPLE 1

Consider a fish population that has been divided into three age groups. Suppose that the Leslie matrix (see Section 8.4) for this population is

$$A = \begin{bmatrix} 0 & 60 & 100 \\ \frac{1}{64} & 0 & 0 \\ 0 & \frac{1}{2} & 0 \end{bmatrix}.$$

Then the eigenvalues of $A$ are the roots of the equation $-\lambda^3 + (60/64)\lambda + (50/64) = 0$, namely $\lambda_1 = 5/4$, $\lambda_2 = (-5 + i\sqrt{15})/8$, and $\lambda_3 = \bar{\lambda}_2$. (Note that the two nonreal complex eigenvalues are complex conjugates of each other. This is always the case when the matrix $A$ is a real matrix. In this case the characteristic polynomial is a polynomial with real coefficients. If $\lambda$ is a root of a polynomial with real coefficients then $\bar{\lambda}$ is also a root.) Since these eigenvalues are distinct, we can find three linearly independent eigenvectors in $C^3$. An eigenvector for $\lambda_1$ is $\mathbf{v}_1 = (400, 5, 2)$. The eigenvector $\mathbf{v}_2$ for $\lambda_2$ and the eigenvector $\mathbf{v}_3$ for $\lambda_3$ can be chosen so that $\mathbf{v}_3 = \bar{\mathbf{v}}_2$, because $A\bar{\mathbf{v}}_2 = \bar{A}\bar{\mathbf{v}}_2 = \overline{A\mathbf{v}_2} = \overline{\lambda_2\mathbf{v}_2} = \bar{\lambda}_2\bar{\mathbf{v}}_2$.

Using the notation of Example 2 of Section 8.4, we have $\mathbf{x}_k = A^k\mathbf{x}_0 = (P\Lambda P^{-1})^k\mathbf{x}_0 = P\Lambda^k P^{-1}\mathbf{x}_0$, where

$$P = \begin{bmatrix} | & | & | \\ \mathbf{v}_1 & \mathbf{v}_2 & \bar{\mathbf{v}}_2 \\ | & | & | \end{bmatrix} \quad \text{and} \quad \Lambda = \begin{bmatrix} \lambda_1 & & \\ & \lambda_2 & \\ & & \bar{\lambda}_2 \end{bmatrix}.$$

Letting

$$P^{-1}\mathbf{x}_0 = \begin{bmatrix} \alpha \\ \beta \\ \gamma \end{bmatrix},$$

we have $\mathbf{x}_k = (5/4)^k\alpha\mathbf{v}_1 + (\lambda_2)^k\beta\mathbf{v}_2 + (\bar{\lambda}_2)^k\gamma\bar{\mathbf{v}}_2$. We have expressed the (real) vector $\mathbf{x}_k$ as a linear combination of the (complex) eigenvectors.

This last expression enables us to show how the population grows. Consider $\|(\lambda_2)^k\beta\mathbf{v}_2\| = |\lambda_2|^k|\beta|\cdot\|\mathbf{v}_2\|$. Since $|\lambda_2| < 1$, the number $|\lambda_2|^k$ is very close to 0 when $k$ is large and we can conclude that the vector $(\lambda_2)^k\beta\mathbf{v}_2$ is very small (i.e., has a very small norm). Similarly, the vector $(\lambda_2)^k\gamma\bar{\mathbf{v}}_2$ has a small norm when $k$ is large. This says that when $k$ is large, $\mathbf{x}_k$ is approximately equal to $(5/4)^k\alpha\mathbf{v}_1$. Since $\alpha$ and $\mathbf{v}_1$ are fixed, we can conclude that this population will eventually increase at about 25% per year and that the age distribution will tend toward the distribution given by $\mathbf{v}_1 = (400, 5, 2)$.

Similar arguments can be used to investigate the behavior for large $k$ of the solution $\mathbf{x}_k = A^k\mathbf{x}_0$ of the difference equation $\mathbf{x}_k = A\mathbf{x}_{k-1}$. When the real matrix $A$ is diagonalizable over the complex numbers and has a real eigenvalue $\lambda$ whose absolute value is larger than the absolute value of every other eigenvalue, then for large $k$, the vector $\mathbf{x}_k = A^k\mathbf{x}_0$ is approximately $\lambda^k\alpha\mathbf{v}$ for some real number $\alpha$, where $\mathbf{v}$ is an eigenvector for $\lambda$. In the case of Leslie matrices when there is a real eigenvalue $\lambda$ which is larger in absolute value than all other eigenvalues, then the distribution of the population will usually tend toward the distribution given by an eigenvector of $\lambda$.

We now generalize the spectral theorem for symmetric matrices to Hermitian matrices. We begin by proving that all eigenvalues of a Hermitian (and hence symmetric) matrix are real, and that the eigenspaces of different eigenvalues are orthogonal.

***Theorem 1*** *If A is Hermitian, then:*

*(a) All eigenvalues of A are real.*
*(b) Eigenvectors associated with distinct eigenvalues are orthogonal.*

**Proof of (a)** If $\lambda$ is an eigenvalue of $A$ and $\mathbf{v}$ is a corresponding eigenvector, then

$$\begin{aligned}\lambda\langle\mathbf{v},\mathbf{v}\rangle &= \langle\mathbf{v},\lambda\mathbf{v}\rangle = \langle\mathbf{v},A\mathbf{v}\rangle = \langle A^*\mathbf{v},\mathbf{v}\rangle \\ &= \langle A\mathbf{v},\mathbf{v}\rangle = \langle\lambda\mathbf{v},\mathbf{v}\rangle = \bar{\lambda}\langle\mathbf{v},\mathbf{v}\rangle.\end{aligned}$$

Since $\langle\mathbf{v},\mathbf{v}\rangle \neq 0$, we conclude that $\lambda = \bar{\lambda}$.

**Proof of (b)** Suppose that $\lambda_1$ and $\lambda_2$ are distinct eigenvalues and let $\mathbf{v}_1$ and $\mathbf{v}_2$ be corresponding eigenvectors. Then since $\lambda_1$ is real,

$$\begin{aligned}\lambda_1\langle\mathbf{v}_1,\mathbf{v}_2\rangle &= \bar{\lambda}_1\langle\mathbf{v}_1,\mathbf{v}_2\rangle = \langle\lambda_1\mathbf{v}_1,\mathbf{v}_2\rangle = \langle A\mathbf{v}_1,\mathbf{v}_2\rangle = \langle\mathbf{v}_1,A^*\mathbf{v}_2\rangle \\ &= \langle\mathbf{v}_1,A\mathbf{v}_2\rangle = \langle\mathbf{v}_1,\lambda_2\mathbf{v}_2\rangle = \lambda_2\langle\mathbf{v}_1,\mathbf{v}_2\rangle.\end{aligned}$$

Since $\lambda_1 \neq \lambda_2$, we conclude that $\langle \mathbf{v}_1, \mathbf{v}_2 \rangle = 0$.

Our next step is to prove that any matrix is **unitarily similar** to a triangular matrix. That is, given any matrix $A$ there is a unitary matrix $U$ such that $U^*AU$ is a triangular matrix.

***Theorem 2*** *(The Schur Theorem)* *Let $A$ be a square matrix. There is a unitary matrix $U$ such that $B = U^*AU$ is an upper triangular matrix.*

**Proof** First suppose that $A$ is a $2 \times 2$ matrix. Let $\lambda_1$ be an eigenvalue of $A$ and $\mathbf{v}_1$ an associated eigenvector normalized so that $\langle \mathbf{v}_1, \mathbf{v}_1 \rangle = 1$. Let $\mathbf{u}_2$ be a unit vector in $C^2$ orthogonal to $\mathbf{v}_1$. Then $\mathbf{v}_1$ and $\mathbf{u}_2$ form an orthonormal basis for $C^2$. Now let $U$ be the matrix whose columns are $\mathbf{v}_1, \mathbf{u}_2$. $U$ is unitary and

$$U^*AU = U^*A\begin{bmatrix} | & | \\ \mathbf{v}_1 & \mathbf{u}_2 \\ | & | \end{bmatrix}$$

$$= U^*\begin{bmatrix} | & | \\ \lambda_1\mathbf{v}_1 & A\mathbf{u}_2 \\ | & | \end{bmatrix}$$

$$= \begin{bmatrix} | & | \\ \lambda_1\mathbf{e}_1 & U^*A\mathbf{u}_2 \\ | & | \end{bmatrix},$$

which is a triangular matrix.

Now suppose that $A$ is a $3 \times 3$ matrix. As in the $2 \times 2$ case, let $\lambda_1$ be an eigenvalue of $A$ and let $\mathbf{v}_1$ be a unit eigenvector associated with $\lambda_1$. Extend to an orthonormal basis for $C^3$ by choosing an orthonormal basis $\mathbf{u}_2, \mathbf{u}_3$ for the orthogonal complement of the subspace spanned by $\mathbf{v}_1$. Let $U_1$ be the matrix whose columns are $\mathbf{v}_1, \mathbf{u}_2, \mathbf{u}_3$. Then

$$U_1^*AU_1 = U_1^*A\begin{bmatrix} | & | & | \\ \mathbf{v}_1 & \mathbf{u}_2 & \mathbf{u}_3 \\ | & | & | \end{bmatrix}$$

$$= \begin{bmatrix} | & | & | \\ \lambda_1\mathbf{e}_1 & U_1^*A\mathbf{u}_2 & U_1^*A\mathbf{u}_3 \\ | & | & | \end{bmatrix}$$

$$= \begin{bmatrix} \lambda_1 & * & * \\ 0 & * & * \\ 0 & * & * \end{bmatrix}.$$

The first column is right. Now consider the $2 \times 2$ matrix in the lower right; call it $M$. We have just shown that there is a $2 \times 2$ unitary matrix $W$ such that

$W^*MW$ is upper triangular. Let $U_2$ be the matrix

$$U_2 = \begin{bmatrix} 1 & 0 & 0 \\ 0 & & \\ 0 & \multicolumn{2}{c}{W} \end{bmatrix}.$$

It is easy to see that $U_2$ is a unitary matrix. Then

$$U_2^*(U_1^*AU_1)U_2 = U_2^* \begin{bmatrix} \lambda_1 & * & * \\ 0 & & \\ 0 & \multicolumn{2}{c}{M} \end{bmatrix} U_2$$

$$= \begin{bmatrix} 1 & 0 & 0 \\ 0 & & \\ 0 & \multicolumn{2}{c}{W^*} \end{bmatrix} \begin{bmatrix} \lambda_1 & * & * \\ 0 & & \\ 0 & \multicolumn{2}{c}{M} \end{bmatrix} \begin{bmatrix} 1 & 0 & 0 \\ 0 & & \\ 0 & \multicolumn{2}{c}{W} \end{bmatrix}$$

$$= \begin{bmatrix} \lambda_1 & * & * \\ 0 & & \\ 0 & \multicolumn{2}{c}{W^*MW} \end{bmatrix}$$

is a triangular matrix. Since $U = U_1 U_2$ is a product of unitary matrices, it is unitary [see Problem 8(f) in Section 10.2] and the proof for $3 \times 3$ matrices is complete. Clearly, this process can be continued. Having proved the theorem for $(n-1) \times (n-1)$ matrices we prove it for $n \times n$ matrices as follows. Let $A$ be an $n \times n$ matrix, let $\lambda_1$ be an eigenvalue of $A$, and let $\mathbf{v}_1$ be a unit eigenvector associated with $\lambda_1$. Extend $\mathbf{v}_1$ to an orthonormal basis $\mathbf{v}_1, \mathbf{u}_2, \ldots, \mathbf{u}_n$ of $C^n$ and let $U_1$ be the unitary matrix having these vectors as its columns. Then

$$U_1^*AU_1 = \begin{bmatrix} \lambda_1 & | & & | \\ 0 & U_1^*A\mathbf{u}_2 & \cdots & U_1^*A\mathbf{u}_n \\ \vdots & & & \\ 0 & | & & | \end{bmatrix}$$

$$= \begin{bmatrix} \lambda_1 & * & \cdots & * \\ 0 & & & \\ \vdots & & M & \\ 0 & & & \end{bmatrix},$$

where $M$ denotes the $(n-1) \times (n-1)$ matrix in the lower right corner. If the result has been established for $(n-1) \times (n-1)$ matrices, then there is an $(n-1) \times (n-1)$ unitary matrix $W$ such that $W^*MW$ is triangular. Let

$$U_2 = \begin{bmatrix} 1 & 0 & \cdots & 0 \\ 0 & & & \\ \vdots & & W & \\ 0 & & & \end{bmatrix}$$

and $U = U_1U_2$. Then $U$ is unitary and $U^*AU = U_2^*(U_1^*AU_1)U_2$ is upper triangular. This completes the proof.

## EXAMPLE 2

We illustrate the proof of Theorem 2 with the matrix

$$A = \begin{bmatrix} 0 & -2 \\ 8 & 0 \end{bmatrix}.$$

Since $\det(A - \lambda I) = \lambda^2 + 16 = (\lambda - 4i)(\lambda + 4i)$, the eigenvalues of $A$ are $\lambda = \pm 4i$. A unit eigenvector associated with the eigenvalue $\lambda = 4i$ is

$$\mathbf{v}_1 = \frac{1}{\sqrt{5}}(i, 2)$$

and a unit vector orthogonal to $\mathbf{v}_1$ is

$$\mathbf{u}_2 = \frac{1}{\sqrt{5}}(2, i).$$

Let

$$U = \frac{1}{\sqrt{5}}\begin{bmatrix} i & 2 \\ 2 & i \end{bmatrix}.$$

$U$ is unitary and

$$\begin{aligned} U^*AU &= \frac{1}{\sqrt{5}}\begin{bmatrix} -i & 2 \\ 2 & -i \end{bmatrix}\begin{bmatrix} 0 & -2 \\ 8 & 0 \end{bmatrix}\frac{1}{\sqrt{5}}\begin{bmatrix} i & 2 \\ 2 & i \end{bmatrix} \\ &= \frac{1}{5}\begin{bmatrix} -i & 2 \\ 2 & -i \end{bmatrix}\begin{bmatrix} -4 & -2i \\ 8i & 16 \end{bmatrix} \\ &= \begin{bmatrix} 4i & 6 \\ 0 & -4i \end{bmatrix}. \end{aligned}$$

## EXAMPLE 3

In this example we illustrate the proof of Theorem 2 for the $3 \times 3$ matrix

$$\begin{bmatrix} 1 & 3\sqrt{2}i/2 & 1 \\ \sqrt{2}i/2 & -1 & -\sqrt{2}i/2 \\ 0 & -\sqrt{2}i/2 & 2 \end{bmatrix}.$$

After some computations we find that $\lambda = 2$ is an eigenvalue and $\mathbf{v}_1 = i/\sqrt{2}(1, 0, 1)$ is a corresponding unit eigenvector. An orthonormal basis for $C^3$ containing this vector is

$$\mathbf{v}_1 = \frac{i}{\sqrt{2}}(1, 0, 1), \qquad \mathbf{u}_2 = (0, 1, 0), \qquad \mathbf{u}_3 = \frac{1}{\sqrt{2}}(1, 0, -1).$$

Letting $U_1$ be the matrix with these vectors as its columns, we obtain (after some computations)

$$U_1^* A U_1 = \begin{bmatrix} 2 & 1 & i \\ 0 & -1 & i \\ 0 & 2i & 1 \end{bmatrix}.$$

Let $M$ be the $2 \times 2$ matrix on the lower right:

$$M = \begin{bmatrix} -1 & i \\ 2i & 1 \end{bmatrix}.$$

$\lambda = i$ is an eigenvalue of this matrix and $\mathbf{v}_1 = 1/\sqrt{6}(1 + i, 2)$ is a unit eigenvector. An orthonormal basis for $C^2$ containing this vector is

$$\mathbf{v}_2 = \frac{1}{\sqrt{6}}(1 + i, 2), \qquad \mathbf{u}_2' = \frac{1}{\sqrt{3}}(1 + i, -1).$$

Letting $W$ be the matrix with these vectors as its columns, we obtain (after some computations)

$$W^* M W = \begin{bmatrix} i & (-3 + i)/\sqrt{2} \\ 0 & -i \end{bmatrix}.$$

Let

$$U_2 = \begin{bmatrix} 1 & 0 & 0 \\ 0 & (1 + i)/\sqrt{6} & (1 + i)/\sqrt{3} \\ 0 & 2/\sqrt{6} & -1/\sqrt{3} \end{bmatrix}.$$

Then

$$\begin{aligned} U_2^*(U_1^* A U_1)U_2 &= \begin{bmatrix} 1 & 0 & 0 \\ 0 & (1 - i)/\sqrt{6} & 2/\sqrt{6} \\ 0 & (1 - i)/\sqrt{3} & -1/\sqrt{3} \end{bmatrix} \begin{bmatrix} 2 & 1 & i \\ 0 & -1 & i \\ 0 & 2i & 1 \end{bmatrix} U_2 \\ &= \begin{bmatrix} 2 & 1 & i \\ 0 & (-1 + 5i)/\sqrt{6} & (3 + i)/\sqrt{6} \\ 0 & (-1 - i)/\sqrt{3} & i/\sqrt{3} \end{bmatrix} \\ &\quad \times \begin{bmatrix} 1 & 0 & 0 \\ 0 & (1 + i)/\sqrt{6} & (1 + i)/\sqrt{3} \\ 0 & 2/\sqrt{6} & -1/\sqrt{3} \end{bmatrix} \\ &= \begin{bmatrix} 2 & (1 + 3i)/\sqrt{6} & 1/\sqrt{3} \\ 0 & i & (-3 + i)/\sqrt{2} \\ 0 & 0 & -i \end{bmatrix}. \end{aligned}$$

Thus if $A$ is any square matrix, then there is a unitary matrix $U$ such that

$$U^*AU = B,$$

where $B$ is a triangular matrix. If $A$ is Hermitian, then $A^* = A$ and hence

$$B^* = (U^*AU)^* = U^*AU = B.$$

Therefore, $B$ is triangular *and* Hermitian and hence must be a diagonal matrix. We have proved:

***Theorem 3*** *Any Hermitian matrix can be diagonalized by a unitary matrix. Therefore, any symmetric matrix can be diagonalized by an orthogonal matrix.*

If $A$ is Hermitian and $U$ is an unitary matrix such that $U^*AU = \Lambda$, a diagonal matrix, then the columns $\mathbf{v}_1, \mathbf{v}_2, \ldots, \mathbf{v}_m$ of $U$ are orthonormal eigenvectors of $A$ and the diagonal entries $\lambda_1, \lambda_2, \ldots, \lambda_m$ of $\Lambda$ are the corresponding eigenvalues. Since

$$\begin{aligned} A &= U \Lambda U^* \\ &= \lambda_1 \mathbf{v}_1 \mathbf{v}_1^* + \cdots + \lambda_m \mathbf{v}_m \mathbf{v}_m^*, \end{aligned}$$

we have the spectral theorem for Hermitian matrices.

***Theorem 4*** (*Spectral Theorem for Hermitian Matrices*) *Let $A$ be an $m \times m$ Hermitian matrix. Then $A$ can be decomposed into*

$$A = \lambda_1 \mathbf{v}_1 \mathbf{v}_1^* + \cdots + \lambda_m \mathbf{v}_m \mathbf{v}_m^*,$$

*where $\mathbf{v}_1, \ldots, \mathbf{v}_m$ are orthonormal eigenvectors of $A$ and $\lambda_1, \ldots, \lambda_m$ are corresponding eigenvalues.*

## EXAMPLE 4

The matrix

$$A = \begin{bmatrix} 0 & -i & 0 \\ i & 0 & 0 \\ 0 & 0 & 1 \end{bmatrix}$$

is Hermitian. Its eigenvalues are $\lambda = 1$, $\lambda = 1$, $\lambda = -1$ and

$$\mathbf{v}_1 = (0, 0, 1), \qquad \mathbf{v}_2 = \frac{1}{\sqrt{2}}(-i, 1, 0), \qquad \mathbf{v}_3 = \frac{1}{\sqrt{2}}(i, 1, 0)$$

are corresponding orthonormal eigenvectors. If $U$ is the unitary matrix having these vectors as its columns, then

$$U^*AU = \begin{bmatrix} 1 & & \\ & 1 & \\ & & -1 \end{bmatrix}.$$

The spectral decomposition of $A$ is

$$A = \mathbf{v}_1\mathbf{v}_1^* + \mathbf{v}_2\mathbf{v}_2^* - \mathbf{v}_3\mathbf{v}_3^*$$
$$= \begin{bmatrix} 0 & 0 & 0 \\ 0 & 0 & 0 \\ 0 & 0 & 1 \end{bmatrix} + \begin{bmatrix} -1/2 & -i/2 & 0 \\ -i/2 & 1/2 & 0 \\ 0 & 0 & 0 \end{bmatrix} - \begin{bmatrix} -1/2 & i/2 & 0 \\ i/2 & 1/2 & 0 \\ 0 & 0 & 0 \end{bmatrix}.$$

## PROBLEMS 10.3

*1. The matrix

$$A = \begin{bmatrix} 0 & 0 & 110 \\ 0.05 & 0 & 0 \\ 0 & 0.2 & 0 \end{bmatrix}$$

was introduced in Problem 7 of Section 8.4 as the Leslie matrix of a certain population.
(a) Show that $A$ has three distinct eigenvalues, only one of which is real.
(b) Show that the absolute values of the three eigenvalues are equal.
(c) Discuss the distribution of this population over extended periods of time.

2. (a) Show that the real matrix $A = \begin{bmatrix} 0 & -2 \\ 8 & 0 \end{bmatrix}$ cannot be diagonalized by a real matrix.
(b) Show that $A$ has two complex eigenvalues and find complex eigenvectors associated with the eigenvalues.
(c) Show that $A$ can be diagonalized by a complex matrix.

*3. Find the (complex) eigenvalues and the associated complex eigenvectors for the following complex matrices.

(a) $\begin{bmatrix} 1 & 1-i \\ 1 & -2i \end{bmatrix}$

(b) $\begin{bmatrix} 2-7i & 10+2i \\ 5+i & -1+8i \end{bmatrix}$

4. Show that each of the following matrices is similar to a diagonal matrix.

(a) $\begin{bmatrix} -3 & 2 \\ -1 & -1 \end{bmatrix}$

(b) $\begin{bmatrix} 1 & 0 & 0 \\ 0 & 1 & -1 \\ 0 & 1 & 1 \end{bmatrix}$

5. Diagonalize each of the following Hermitian matrices.

*(a) $\begin{bmatrix} 3\sqrt{2} & -1+i \\ -1-i & 3\sqrt{2} \end{bmatrix}$

(b) $\begin{bmatrix} 0 & 2\sqrt{2}(1-i) \\ 2\sqrt{2}(1+i) & 0 \end{bmatrix}$

*(c) $\begin{bmatrix} 1 & -2i \\ 2i & 1 \end{bmatrix}$

(d) $\begin{bmatrix} -3 & i \\ -i & -3 \end{bmatrix}$

(e) $\frac{1}{5}\begin{bmatrix} 6 & -2i & 0 \\ 2i & 9 & 0 \\ 0 & 0 & 10 \end{bmatrix}$

(f) $\frac{1}{6}\begin{bmatrix} 3 & -3i & -6i \\ 3i & 3 & -6 \\ 6i & -6 & 6 \end{bmatrix}$

6. Construct a Hermitian matrix with the following properties:
*(a) Eigenvalues $9, -1$; eigenvectors $(-2i, 1), (1, -2i)$.
(b) Eigenvalues $3, -1$; eigenvectors $(i, 1), (-i, 1)$.
(c) Eigenvalues $7, 7, 3$ and eigenvectors $(-i, 0, 1), (0, 1, 0), (i, 0, 1)$.

**7.** Find a unitary matrix $U$ such that $U^*AU$ is triangular for each of the following matrices.

*(a) $\begin{bmatrix} 0 & -4 \\ 4 & 0 \end{bmatrix}$ (b) $\begin{bmatrix} 1 & 2 \\ -1 & -1 \end{bmatrix}$

*(c) $\begin{bmatrix} 1 & 1 \\ -1 & 1 \end{bmatrix}$ (d) $\begin{bmatrix} 0 & 0 & -4 \\ 0 & 8 & 0 \\ 1 & 0 & 0 \end{bmatrix}$

*(e) $\begin{bmatrix} 0 & 1 & 2 \\ 0 & -1 & 1 \\ 0 & -1 & -1 \end{bmatrix}$

(f) $\begin{bmatrix} 4 & -5 & 0 \\ 2 & -3 & 0 \\ 0 & 0 & 0 \end{bmatrix}$

*(g) $\frac{1}{9}\begin{bmatrix} 22 & 2 & -4 \\ 8 & 13 & 1 \\ 8 & 13 & 1 \end{bmatrix}$

*(h) $\frac{1}{2}\begin{bmatrix} i & \sqrt{2} & -2-i \\ 4\sqrt{2} & 2 & -4\sqrt{2} \\ i-2 & -\sqrt{2} & -i \end{bmatrix}$

**8.** The first step in the proof of Schur's theorem reduces a square matrix $A$ to the form

$$U^*AU = \begin{bmatrix} \lambda_1 & * & \cdots * \\ 0 & & \\ \vdots & & M \\ 0 & & \end{bmatrix}.$$

Prove that the eigenvalues of $M$ are eigenvalues of $A$.

**9.** Use Schur's theorem to prove that if a square matrix has only 0 as an eigenvalue, then $A$ is nilpotent.

**10.** Prove that if $A$ is nilpotent, then $A$ is similar to a triangular matrix with diagonal entries 0.

**11.** Let $A$ be a square matrix.

*(a) Show that if $A^*A\mathbf{v} = \mathbf{0}$, then $A\mathbf{v} = \mathbf{0}$.

(b) Show that $A^*A$ is positive semidefinite.

(c) Show that if $A$ is nonsingular, then $A^*A$ is positive definite.

**12.** Let $U$ be a unitary matrix.

(a) If $\lambda$ is an eigenvalue of $U$, show that $|\lambda| = 1$. (*Hint:* Consider $\langle U\mathbf{x}, U\mathbf{x}\rangle$, where $\mathbf{x}$ is an eigenvector for $\lambda$.)

(b) Show that eigenvectors of $U$ belonging to distinct eigenvalues are orthogonal.

(c) Show that $U$ is Hermitian if and only if the eigenvalues of $U$ are $\pm 1$.

(d) If $U^*AU$ and $U^*BU$ are both diagonal matrices, then $AB = BA$.

**13.** Let $N$ be a normal matrix, i.e., $NN^* = N^*N$.

*(a) Show that if $N\mathbf{v} = \mathbf{0}$, then $N^*\mathbf{v} = \mathbf{0}$.

(b) Show that for any scalar $\alpha$, $(N - \alpha I)$ is normal.

(c) Show that the eigenvalues of $N^*$ are the conjugates of the eigenvalues of $N$. How are the eigenvectors related?

**14.** Prove the following.

*(a) Let $U$ be unitary. Prove that $N$ is normal if and only if $U^*NU$ is normal.

(b) If $U$ is unitary and $U^*NU$ is a diagonal matrix, then $N$ is normal.

(c) Conversely, if $N$ is normal, then there is a unitary matrix $U$ such that $U^*NU$ is a diagonal matrix. [*Hint:* Use Schur's theorem to find a unitary $U$ such that $U^*NU = T$, a triangular matrix. Use part (a) to deduce $T$ is normal. Finally, show that every entry above the diagonal is zero using $T^*T = TT^*$.]

(d) A matrix is unitarily diagonalizable if and only if it is normal.

(e) State and prove the spectral theorem for normal matrices.

**15.** Prove that if $A$ is a unitary matrix, then there is a unitary matrix $U$ such that

$$U^*AU$$

is diagonal.

**16.** Prove that if $H$ is Hermitian, then there

is a unitary matrix $U$ such that

$$U^*HU$$

is diagonal.

**17.** Show that if $A$ is normal and has real eigenvalues, then $A$ is Hermitian.

**18.** Show that if $N$ is normal and all eigenvalues of $N$ have absolute value 1, then $N$ is unitary.

**19.** Suppose that $A$ and $B$ are $n \times n$ matrices and that all eigenvalues of $A$ and $B$ are positive.

(a) Can $AB$ have only negative eigenvalues?

(b) If $A$ and $B$ are both Hermitian, show that $AB$ has only positive eigenvalues.

**20.** We have studied six special types of matrices: Hermitian, skew-Hermitian, symmetric, skew-symmetric, orthogonal, and unitary. Suppose you have a matrix $A$ with characteristic polynomial $(x + 1)(x^2 + 1)$. $A$ might possibly represent one or more of the foregoing types of matrices. For each type explain why $A$ could or could not represent that type of matrix.

# Solutions

## CHAPTER 1

### SECTION 1.1

**3.** (a), (b), (f).

**5.** (b), (c), (d).

**7.** (a), (b), (e).

**10.** Any value except $b = 3$.

**12.** For example, $3x_1 + 2x_3 = 5$ or $2x_1 + x_2 - x_3 = 7$.

**16.** Each equation of the system represents a line. Two lines meeting in more than one point must coincide. Thus if the system has two distinct solutions, each equation of the system represents the same line, and every point on this line is a solution of the system.

**18.** (a) Algebraic argument: $x_1 = 0$, $x_2 = 0$ is a solution of the system. Geometric argument: The equations represent two lines passing through the origin.

### SECTION 1.2

**1.** (a)
$$\begin{aligned} 2t + 3u &= 37 \\ 10t + u &= 10u + t + 9. \end{aligned}$$

(b)
$$\begin{aligned} x_1 + x_2 + x_3 &= 51 \\ x_1 &= 2x_2 + 5 \\ x_2 &= 3x_3 + 2. \end{aligned}$$

(d)
$$\begin{aligned} 3A - 5B &= 16 \\ 6B - 2A &= 24. \end{aligned}$$

(f)
$$\begin{aligned} \tfrac{5}{8}A + \tfrac{2}{8}B + \tfrac{3}{8}C &= 3 \\ \tfrac{2}{8}A + \tfrac{5}{8}B + \tfrac{1}{8}C &= 3 \\ \tfrac{1}{8}A + \tfrac{1}{8}B + \tfrac{4}{8}C &= 3. \end{aligned}$$

**2.** (a)
$$\begin{aligned} I_1 - I_2 - I_3 - I_4 &= 0 \\ -I_1 + I_2 + I_3 + I_4 &= 0. \end{aligned}$$

(c)
$$\begin{aligned} I_1 - I_2 - I_3 \qquad\qquad\qquad &= 0 \\ I_3 + I_4 \qquad - I_6 &= 0 \\ -I_1 \qquad\qquad + I_5 + I_6 &= 0 \\ I_2 \qquad - I_4 - I_5 \qquad &= 0. \end{aligned}$$

(e)
$$\begin{aligned} I_1 - I_2 - I_3 \qquad\qquad\qquad &= 0 \\ I_3 \qquad + I_5 - I_6 &= 0 \\ - I_4 - I_5 + I_6 &= 0 \\ - I_1 + I_2 \qquad + I_4 \qquad\qquad &= 0. \end{aligned}$$

**3.** Part (a)

(a)
$$\begin{aligned} I_1 + I_2 \qquad\qquad - 1 &= 0 \\ I_1 \qquad + I_3 \qquad - 1 &= 0 \\ I_1 \qquad\qquad + I_4 - 1 &= 0 \\ I_2 - I_3 \qquad\qquad &= 0 \\ I_2 \qquad - I_4 \qquad &= 0 \\ I_3 - I_4 \qquad &= 0. \end{aligned}$$

(c)
$$\begin{aligned} I_1+I_2 \qquad\qquad +I_5 \quad -1&=0\\ I_1+I_2 \qquad +I_4 \qquad +I_6-1&=0\\ I_1 \qquad +I_3 \qquad\qquad +I_6-1&=0\\ I_2-I_3+I_4 \qquad\qquad &=0\\ I_2-I_3 \qquad +I_5-I_6 &=0\\ I_4-I_5+I_6 &=0. \end{aligned}$$

Part (b)

(a)
$$\begin{aligned} I_1+2I_2 \qquad\qquad -1&=0\\ I_1 \qquad +3I_3 \qquad -1&=0\\ I_1 \qquad\qquad +4I_4-1&=0\\ 2I_2-3I_3 \qquad &=0\\ 2I_2 \qquad -4I_4 &=0\\ 3I_3-4I_4 &=0. \end{aligned}$$

(c)
$$\begin{aligned} I_1+2I_2 \qquad\qquad +5I_5 \qquad -1&=0\\ I_1+2I_2 \qquad +4I_4 \qquad +6I_6-1&=0\\ I_1 \qquad +3I_3 \qquad\qquad +6I_6-1&=0\\ 2I_2-3I_3+4I_4 \qquad\qquad &=0\\ 2I_2-3I_3 \qquad +5I_5-6I_6 &=0\\ 4I_4-5I_5+6I_6 &=0. \end{aligned}$$

**4.** (a) $\frac{1}{2}, \frac{1}{3}, \frac{1}{6}$.

**6.**
$$\begin{aligned} \tfrac{1}{3}p_1+\tfrac{1}{5}p_2+\tfrac{1}{6}p_3+\tfrac{1}{8}p_4&=p_1\\ \tfrac{1}{6}p_1+\tfrac{2}{5}p_2+\tfrac{1}{6}p_3+\tfrac{1}{4}p_4&=p_2\\ \tfrac{1}{3}p_1+\tfrac{1}{5}p_2+\tfrac{1}{2}p_3+\tfrac{1}{8}p_4&=p_3\\ \tfrac{1}{6}p_1+\tfrac{1}{5}p_2+\tfrac{1}{6}p_3+\tfrac{1}{2}p_4&=p_4. \end{aligned}$$

**10.** $a_{1j}+a_{2j}+a_{3j}+\cdots+a_{nj}=1$.

**12.**
$$\begin{aligned} c&=-2\\ a+b+c&=6\\ 4a-2b+c&=0\\ 16a+4b+c&=66. \end{aligned}$$

**13.**
$$\begin{aligned} m+b&=-1\\ 5m+b&=7\\ -2m+b&=-7\\ b&=-3. \end{aligned}$$

## SECTION 1.3

**3.** $\mathbf{u}+\mathbf{v}=(5,4)$, $8\mathbf{u}=(24,-8)$, $-3\mathbf{w}=(0,-12)$, $6\mathbf{u}+2\mathbf{v}-\mathbf{w}=(22,0)$.

**5.** (a) $\frac{1}{3}(4,7,-1,16)$.

**7.** Add $-\mathbf{v}$ to both sides of the equation.

**12.** Let $\mathbf{u}=(u_1,u_2,\ldots,u_n)$. Then $\alpha\mathbf{u}=(\alpha u_1,\alpha u_2,\ldots,\alpha u_n)=(0,0,\ldots,0)$ so that $\alpha u_1=0$, $\alpha u_2=0,\ldots,\alpha u_n=0$. If $\alpha\neq 0$, then these equations imply that $u_1=0$, $u_2=0,\ldots,u_n=0$. Another solution: If $\alpha\neq 0$, then

$$\mathbf{u}=[(1/\alpha)\alpha]\mathbf{u}=(1/\alpha)\alpha\mathbf{u}=(1/\alpha)\mathbf{0}=\mathbf{0}.$$

## SECTION 1.4

**2.** (a) $\mathbf{x}=\alpha(1,2,3,1)$.
(b) $\mathbf{x}=(2,1,0,5)+\alpha(1,2,3,1)$.

**3.** (a) and (c) lie on the line.

**5.** (a) The three vectors lie on a line.
(c) The three vectors do not lie on a line.

**6.** The line through $(1,0,1)$ and $(2,1,2)$ is parallel to the line through $(3,2,2)$ and $(4,3,3)$ and the line through $(1,0,1)$ and $(3,2,2)$ is parallel to the line through $(2,1,2)$ and $(4,3,3)$.

**8.** $x_3=2$.

**9.** $\mathbf{x}=\alpha(1,2,3)+\beta(-1,1,-1)$.

**11.** $\mathbf{x}=(1,0,0)+\alpha(-1,1,0)+\beta(-1,0,1)$.

**13.** $\mathbf{x}=(5,5,0)+\alpha(-1,1,0)+\beta(-1,0,1)$.

**15.** $\mathbf{x}=(1,5,4)+\alpha(1,0,0)+\beta(0,1,0)$.

**17.** $\mathbf{x}=(1,0,0)+\alpha(1,-1,0)+\beta(0,0,1)$.

**19.** If the line is not parallel to the plane, then they intersect in a point. If the line is parallel to the plane but does not lie in the plane, then they do not intersect. Finally, if the line is parallel to the plane and intersects the plane, then the line is contained in the plane and the intersection is the line.

**21.** Both planes contain the origin.

**23.** Let $\mathbf{v}$ be any vector that is not collinear with $(1,2,3)$. Then $\mathbf{x}=(1,1,1)+$

$\alpha(1,2,3) + \beta\mathbf{v}$ is such a plane. Every direction determined by a vector that is not collinear with $(1,2,3)$ determines such a plane. Thus there are infinitely many such planes.

**25.** Any two distinct points on the line together with the point not on the line uniquely determine the plane.

**27.** If the line and the plane intersect, then there are scalars $\alpha$, $\beta$, and $\gamma$ such that $\mathbf{a} + \gamma\mathbf{u} = \alpha\mathbf{u} + \beta\mathbf{v}$. Thus $\mathbf{a} = (\alpha - \gamma)\mathbf{u} + \beta\mathbf{v}$ and we conclude that $\mathbf{a}$ is on the plane.

**30.** (a) An equation of the line is $\mathbf{x} = \mathbf{u} + \alpha(\mathbf{v} - \mathbf{u})$, which can be written in the form $\mathbf{x} = (1 - \alpha)\mathbf{u} + \alpha\mathbf{v}$.

(b) If $0 \le \alpha \le 1$, $\alpha(\mathbf{v} - \mathbf{u})$ is collinear with $\mathbf{v} - \mathbf{u}$ and between the origin and $\mathbf{v} - \mathbf{u}$. As $\alpha$ increases from 0 to 1, $\alpha(\mathbf{v} - \mathbf{u})$ moves from $\mathbf{0}$ to $\mathbf{v} - \mathbf{u}$. Therefore, as $\alpha$ increases from 0 to 1, $\mathbf{u} + \alpha(\mathbf{v} - \mathbf{u})$ moves along the line from $\mathbf{u}$ to $\mathbf{v}$ in the direction of $\mathbf{v} - \mathbf{u}$.

**31.** (a) An equation of the plane is $\mathbf{x} = \mathbf{u} + \alpha'(\mathbf{v} - \mathbf{u}) + \beta'(\mathbf{w} - \mathbf{u})$, which can be written in the form $\mathbf{x} = (1 - \alpha' - \beta')\mathbf{u} + \alpha'\mathbf{v} + \beta'\mathbf{w}$.

**34.** $\mathbf{a}$ and $\mathbf{u}$ are collinear.

**35.** Yes, if $\mathbf{a}$ lies on the plane through the origin, $\mathbf{u}$, and $\mathbf{v}$.

**36.** (a) When one pair of vectors lies on the plane determined by the other pair.

## SECTION 1.5

**2.** $A + B = \begin{bmatrix} -2 & 0 & 3 & 5 \\ 9 & 9 & -4 & 4 \end{bmatrix}$,

$-2A = \begin{bmatrix} 6 & -2 & -4 & 0 \\ -18 & -12 & 8 & -4 \end{bmatrix}$,

$3A - B = \begin{bmatrix} -10 & 4 & 5 & -5 \\ 27 & 15 & -12 & 4 \end{bmatrix}$,

$2A + 3B = \begin{bmatrix} -3 & -1 & 7 & 15 \\ 18 & 21 & -8 & 10 \end{bmatrix}$.

**4.** $A\mathbf{u} = \begin{bmatrix} 0 \\ 18 \end{bmatrix}$, $A\mathbf{v} = \begin{bmatrix} -5 \\ 31 \end{bmatrix}$,

$A(\mathbf{u} + \mathbf{v}) = \begin{bmatrix} -5 \\ 49 \end{bmatrix}$, $B\mathbf{u} = \begin{bmatrix} 2 \\ 8 \end{bmatrix}$,

$B(\mathbf{u} + \mathbf{v}) = \begin{bmatrix} -6 \\ 10 \end{bmatrix}$, $A(3\mathbf{v}) = \begin{bmatrix} -15 \\ 93 \end{bmatrix}$,

$G\mathbf{v} = \begin{bmatrix} 9 \\ 4 \\ -7 \\ -2 \end{bmatrix}$.

**8.** (a) $\begin{bmatrix} -1 & 1 & 2 \\ 2 & -3 & -1 \\ -2 & 1 & 7 \end{bmatrix} \begin{bmatrix} x_1 \\ x_2 \\ x_3 \end{bmatrix} = \begin{bmatrix} 1 \\ 5 \\ 9 \end{bmatrix}$.

(c) $\begin{bmatrix} 3 & -1 & 1 \\ 3 & 1 & 0 \\ 0 & 1 & -3 \\ 6 & 1 & -1 \end{bmatrix} \begin{bmatrix} x_1 \\ x_2 \\ x_3 \end{bmatrix} = \begin{bmatrix} 1 \\ 0 \\ -1 \\ 2 \end{bmatrix}$.

(e) $\begin{bmatrix} 1 & 1 & 1 & 1 & 1 \\ 0 & 0 & 2 & 0 & -1 \\ 2 & 2 & 1 & -1 & -1 \\ 1 & 1 & -1 & -1 & 1 \end{bmatrix} \begin{bmatrix} x_1 \\ x_2 \\ x_3 \\ x_4 \end{bmatrix} = \begin{bmatrix} 1 \\ 1 \\ 1 \\ 1 \end{bmatrix}$.

(g) $[2 \quad 3] \begin{bmatrix} x_1 \\ x_2 \end{bmatrix} = [1]$.

**9.** (a) 0. (c) 10. (e) 22.

**10.** Yes. If $A\mathbf{x} = \mathbf{0}$ for all $\mathbf{x}$ in $R^n$, then for $i = 1, 2, \ldots, n$, the $i$th column of $A = A\mathbf{e}_i = \mathbf{0}$. Since each column of $A$ is all zeros, $A = 0$.

**12.** Let $A = [a_{ij}]$, $B = [b_{ij}]$, and $C = [c_{ij}]$. Then

$$\begin{aligned}(A + B) + C &= ([a_{ij}] + [b_{ij}]) + [c_{ij}]\\ &= [a_{ij} + b_{ij}] + [c_{ij}]\\ &= [(a_{ij} + b_{ij}) + c_{ij}]\\ &= [a_{ij} + (b_{ij} + c_{ij})]\\ &= [a_{ij}] + [b_{ij} + c_{ij}]\\ &= [a_{ij}] + ([b_{ij}] + [c_{ij}])\\ &= A + (B + C).\end{aligned}$$

**15.** Let $A = [a_{ij}]$. Then

$$\begin{aligned}(\alpha\beta)A &= (\alpha\beta)[a_{ij}] = [(\alpha\beta)a_{ij}]\\ &= [\alpha(\beta a_{ij})] = \alpha[\beta a_{ij}] = \alpha(\beta A).\end{aligned}$$

## SECTION 1.6

**2.** $AB = \begin{bmatrix} 7 & -1 \\ 10 & 9 \end{bmatrix}$, $(AB)\mathbf{v} = \begin{bmatrix} 5 \\ 28 \end{bmatrix}$,

$B\mathbf{v} = \begin{bmatrix} 8 \\ -2 \\ -3 \end{bmatrix}$, $A(B\mathbf{v}) = (AB)\mathbf{v}$.

**4.** $AB = [1]$, $BA = \begin{bmatrix} 2 & 4 & -2 \\ 0 & 0 & 0 \\ 1 & 2 & -1 \end{bmatrix}$.

**5.** (a) $AB = \begin{bmatrix} 10 & 3 & 17 \\ -6 & 7 & -8 \end{bmatrix}$,

$AC = \begin{bmatrix} 9 & 7 \\ -1 & -13 \end{bmatrix}$,

$BD = \begin{bmatrix} 9 & 5 & 16 \\ 0 & 1 & 12 \end{bmatrix}$, $BF = \begin{bmatrix} -2 \\ 13 \end{bmatrix}$,

$EF = [4]$, $FE = \begin{bmatrix} -2 & -1 & -3 \\ 6 & 3 & 9 \\ 2 & 1 & 3 \end{bmatrix}$.

**10.** (a) Here are a couple of examples:

$A = \begin{bmatrix} -1 & 1 \\ 2 & -2 \end{bmatrix}$, $B = \begin{bmatrix} 1 & 2 \\ 1 & 2 \end{bmatrix}$; and

$A = \begin{bmatrix} 2 & -1 \\ 1 & -\frac{1}{2} \\ -3 & \frac{3}{2} \end{bmatrix}$, $B = \begin{bmatrix} 2 & -1 \\ 4 & -2 \end{bmatrix}$.

**14.** (a) Its first component is multiplied by 3, its second component by 1, and its third component by $-1$.

**15.** (a) $I$, $-I$, and $\begin{bmatrix} a & b \\ c & -a \end{bmatrix}$, where $a^2 + bc = 1$.

**16.** (c)

$$\begin{aligned}\operatorname{tr}(AB) = {} & (a_{11}b_{11} + a_{12}b_{21} + \cdots + a_{1n}b_{n1})\\ & + (a_{21}b_{12} + a_{22}b_{22} + \cdots\\ & + a_{2n}b_{n2}) + \cdots + (a_{n1}b_{1n}\\ & + a_{n2}b_{2n} + \cdots + a_{nn}b_{nn}).\\ \operatorname{tr}(BA) = {} & (b_{11}a_{11} + b_{12}a_{21} + \cdots + b_{1n}a_{n1})\\ & + (b_{21}a_{12} + b_{22}a_{22} + \cdots\\ & + b_{2n}a_{n2}) + \cdots + (b_{n1}a_{1n}\\ & + b_{n2}a_{2n} + \cdots + b_{nn}a_{nn}).\end{aligned}$$

The first term in the parentheses defining $\operatorname{tr}(BA)$, namely,

$$(b_{11}a_{11} + b_{12}a_{21} + \cdots + b_{1n}a_{n1}),$$

is the sum of the first term in each of the parenthesized terms in the sum defining $\operatorname{tr}(AB)$; similarly for each of the other terms in the sum defining $\operatorname{tr}(BA)$.

(d) No. Compute the trace of both sides of the equation.

**17.** (d) $(A^TA)^T = A^TA^{TT} = A^TA$.

**26.** (a) Both $A$ and $B$ must be $n \times n$.
(c) $A$ $m \times n$, $B$ and $C$ both $n \times n$.

**28.** $\langle A\mathbf{x}, \mathbf{y}\rangle = (A\mathbf{x})^T\mathbf{y} = (\mathbf{x}^TA^T)\mathbf{y} = \mathbf{x}^T(A^T\mathbf{y}) = \langle \mathbf{x}, A^T\mathbf{y}\rangle$.

## SECTION 1.7

**2.** (a) $\begin{bmatrix} 1 & 1 \\ 1 & -1 \end{bmatrix}\begin{bmatrix} x_1 \\ x_2 \end{bmatrix} = \begin{bmatrix} 2 \\ 4 \end{bmatrix}$;

solution: $\mathbf{x} = (3, -1)$.

(c) $\begin{bmatrix} 1 & 4 & -1 \\ 1 & 1 & 1 \\ 2 & 0 & 3 \end{bmatrix}\begin{bmatrix} x_1 \\ x_2 \\ x_3 \end{bmatrix} = \begin{bmatrix} 1 \\ 0 \\ 0 \end{bmatrix}$;

solution: $\mathbf{x} = (3, -1, -2)$.

**3.** (a) $\mathbf{x} = (4, -1)$. (c) $\mathbf{x} = (-1, -1, 0)$. (e) $\mathbf{x} = (0, 0, 0)$. (g) $\mathbf{x} = (1, -1, -2, 3)$.

**5.** $\mathbf{x} = (\frac{2}{5}, -\frac{3}{5}, \frac{7}{5})$.

**7.** $a = 2, b = -3, c = 1$.

**9.** (a) They intersect at $(3, 5, 7)$.

**10.** (a) $(0, 1, -4)$.

## SECTION 1.8

**2.** (a) Pivots 2, 3. (c) Pivots 1, 1, 1. (d) Pivots 1, 2. (f) Pivots 1, 1. (h) Pivots 1, 1. (i) Pivot $-1$.

**3.** (c), (f), (h) are in reduced echelon form.

**5.** (a) $\mathbf{x} = (2, 2)$.
(d) $\mathbf{x} = (6, 0, 1) + x_2(-2, 1, 0)$.
(f) $\mathbf{x} = (0, 3, 1)$.
(h) $\mathbf{x} = (2, 0, 4, 2, 0) + x_2(-\frac{3}{2}, 1, 0, 0, 0) + x_5(\frac{1}{2}, 0, -1, 1, 1)$.
(l) $\mathbf{x} = (-\frac{1}{2}, 4, -1, 0, 4, 0) + x_4(-\frac{1}{2}, 0, 0, 1, 0, 0) + x_6(\frac{1}{2}, -1, 0, 0, -1, 1)$.
(p) $\mathbf{x} = (17, 0, 0, 0, 0, 0) + x_2(-2, 1, 0, 0, 0, 0) + x_3(-3, 0, 1, 0, 0, 0) + x_4(-4, 0, 0, 1, 0, 0) + x_5(-5, 0, 0, 0, 1, 0) + x_6(-6, 0, 0, 0, 0, 1)$.

**6.** (a) $[I : \mathbf{d}]$, where $\mathbf{d} = (-1, 3, 0)$.
(c) For example,

$$[1 \quad -3 \quad -1 \quad : \quad 0]$$

or

$$\begin{bmatrix} 1 & -3 & -1 & : & 0 \\ 0 & 0 & 0 & : & 0 \\ 0 & 0 & 0 & : & 0 \end{bmatrix}.$$

**7.** (a) $R$ has a zero row that is not matched by a zero entry in $\mathbf{d}$.

**8.** (a) If the $i$th row of $E$ is a zero row, then $E\mathbf{x} = \mathbf{c}$ is inconsistent whenever $\mathbf{c}$ is a vector in $R^m$ whose $i$th entry is not zero. Conversely, if $E$ has no zero rows, then $E\mathbf{x} = \mathbf{c}$ can always be solved by back substitution. Therefore, $E\mathbf{x} = \mathbf{c}$ is consistent for every vector $\mathbf{c}$ in $R^m$.

**9.** (a) $E$ has a zero on its diagonal if and only if $E$ has a nonpivot column. Since $E$ is square, this holds if and only if $E$ has a zero row.

## SECTION 1.9

**2.** (a) (i) Rank $= 2$.
(ii) $\mathbf{x} = \mathbf{0}$.
(iii) $\mathbf{x} = (0, 2)$.
(iv) $\mathbf{b} = 2$(column 2).

(c) (i) Rank $= 1$.
(ii) $\mathbf{x} = x_2(-3, 1)$.
(iii) $A\mathbf{x} = \mathbf{b}$ is inconsistent.

(e) (i) Rank $= 2$.
(ii) $\mathbf{x} = x_3(-\frac{11}{19}, \frac{2}{19}, 1)$.
(iii) $\mathbf{x} = (\frac{3}{19}, -\frac{4}{19}, 0) + x_3(-\frac{11}{19}, \frac{2}{19}, 1)$.
(iv) $\mathbf{b} = \frac{3}{19}$(column 1) $- \frac{4}{19}$(column 2).

More generally,

$\mathbf{b} = (\frac{3}{19} - \frac{11}{19}x_3)$(column 1)
$+(-\frac{4}{19} + \frac{2}{19}x_3)$(column 2)
$+x_3$(column 3).

In the remaining answers to this question we will write $\mathbf{b}$ as a linear combination of the columns of $A$ with all free variables set to 0.

(g) (i) Rank = 3.
(ii) $\mathbf{x} = \mathbf{0}$.
(iii) $\mathbf{x} = (2, -1, 1)$.
(iv) $\mathbf{b} = 2$(column 1) − (column 2) + (column 3).

(i) (i) Rank = 3.
(ii) $\mathbf{x} = \mathbf{0}$.
(iii) $\mathbf{x} = (-\frac{1}{3}, -\frac{2}{3}, 1)$.
(iv) $\mathbf{b} = -\frac{1}{3}$(column 1) $-\frac{2}{3}$(column 2) + (column 3).

(k) (i) Rank = 3.
(ii) $\mathbf{x} = \mathbf{0}$.
(iii) $\mathbf{x} = (1, \frac{2}{3}, -\frac{1}{3})$.
(iv) $\mathbf{b} =$ (column 1) + $\frac{2}{3}$(column 2) $-\frac{1}{3}$(column 3).

(m) (i) Rank = 2.
(ii) $\mathbf{x} = x_3(-\frac{1}{2}, 0, 1, 0) + x_4(\frac{1}{2}, 0, 0, 1)$.
(iii) $A\mathbf{x} = \mathbf{b}$ is inconsistent.

(o) (i) Rank = 3.
(ii) $\mathbf{x} = \mathbf{0}$.
(iii) $\mathbf{x} = (-\frac{52}{5}, 3, -\frac{14}{5})$.
(iv) $\mathbf{b} = -\frac{52}{5}$(column 1) + 3(column 2) $-\frac{14}{5}$(column 3).

(q) (i) Rank = 3.
(ii) $\mathbf{x} = x_4(0, 0, -1, 1, 0) + x_5(-1, 1, -1, 0, 1)$.
(iii) $\mathbf{x} = (4, 1, -7, 0, 0) + x_4(0, 0, -1, 1, 0) + x_5(-1, 1, -1, 0, 1)$.
(iv) $\mathbf{b} = 4$(column 1) + (column 2) − 7(column 3).

(s) (i) Rank = 4.
(ii) $\mathbf{x} = \mathbf{0}$.
(iii) $\mathbf{x} = (\frac{123}{29}, \frac{27}{29}, -\frac{50}{29}, \frac{34}{29})$.
(iv) $\mathbf{b} = \frac{123}{29}$(column 1) + $\frac{27}{29}$(column 2) $-\frac{50}{29}$(column 3) + $\frac{34}{29}$(column 4).

(u) (i) Rank = 1.
(ii) $\mathbf{x} = x_2(-2, 1, 0, 0, 0, 0, 0) + x_3(1, 0, 1, 0, 0, 0, 0) + x_4(2, 0, 0, 1, 0, 0, 0) + x_5(-3, 0, 0, 0, 1, 0, 0)$.
(iii) $\mathbf{x} = (1, 0, 0, 0, 0, 0, 0)$ + the general solution of the homogeneous equation.
(iv) $b = 1\cdot 1 + 0\cdot 2 + 0\cdot(-1) + 0\cdot(-2) + 0\cdot 3$.

**3.** (a) $\mathbf{x} = (\frac{5}{9}, -\frac{2}{9}, \frac{3}{9})$, $\mathbf{x} = (-\frac{4}{9}, -\frac{11}{9}, \frac{12}{9})$, $\mathbf{x} = (\frac{5}{3}, -\frac{2}{3}, 1)$, $\mathbf{x} = (-\frac{1}{3}, -\frac{5}{3}, 1)$, $\mathbf{x} = (\frac{5}{9}, \frac{7}{9}, \frac{12}{9})$.

**4.** (a) $\alpha = 7$.

**5.** (a) When $\alpha = \pm 1$, no solution; otherwise, the system has a unique solution, namely

$$\mathbf{x} = \frac{1}{(1 - \alpha^2)}(1 - 4\alpha, 4 - \alpha).$$

(c) When $\alpha = 2$, no solution; when $\alpha = 0$, infinitely many solutions $\mathbf{x} = (-1, 0, 2) + x_2(-1, 1, 0)$; otherwise, the system has the unique solution $\mathbf{x} = \frac{1}{(\alpha - 2)}(1, 1, \alpha - 4)$.

**6.** (a) $-x_1 + x_2 + x_3 = 0$.
(c) $-x_2 + x_3 = 1$.
(e) $7x_1 + 2x_2 - 5x_3 = 2$.

**7.** (a) Planes intersect in the line $\mathbf{x} = (0, 1, 0) + \delta(1, -2, 1)$.
(b) Planes do not intersect.
(c) Equations define the same plane; $\alpha = -2 + \gamma + 2\delta$, $\beta = 4 - 3\gamma + \delta$.

**8.** Yes, the parabola $y = x^2 + 2x + 5$. There is no cubic of the given form that passes through these points.

**9.** (b) $\mathbf{p} = p_3(\frac{36}{55}, \frac{48}{55}, 1)$.

**10.** There are infinitely many such matrices. All such matrices are of the form

$$\begin{bmatrix} 14 - 13\alpha & 2 - 13\beta & 26 - 13\gamma \\ -4 + 5\alpha & -1 + 5\beta & -9 + 5\gamma \\ \alpha & \beta & \gamma \end{bmatrix}$$

where $\alpha$, $\beta$, and $\gamma$ are scalars.

**14.** Since $A\mathbf{x} = \mathbf{0}$ has no nontrivial solutions, there are no free variables. Thus rank $A = 4$.

## SECTION 1.10

**1.** (a) $b_1 + b_2 - 2b_3 = 0.$
(c) $7b_1 - 3b_2 + b_3 = 0.$
(e)
$$\begin{aligned} b_1 - 2b_2 + b_3 \quad &= 0 \\ 2b_1 - 3b_2 + \quad b_4 &= 0. \end{aligned}$$
(g) No constraint equations.
(i)
$$\begin{aligned} -3b_1 + b_2 + b_3 \qquad &= 0 \\ -2b_2 + \quad b_4 \quad &= 0 \\ -b_1 + b_2 + \qquad b_5 &= 0. \end{aligned}$$

**2.** (a) The solution set is empty. Each plane is parallel to the line of intersection of the other two.
(c) The planes intersect in the point $(\frac{1}{2}, -2, \frac{5}{2})$.
(e) The solution set is empty. The first two equations represent planes that are parallel and distinct, and the first and third equations represent the same plane.

**3.** (a) $\mathbf{x} = \alpha(-\frac{1}{2}, 1, 0) + \beta(\frac{1}{2}, 0, 1).$
(c) $\mathbf{x} = \alpha(-\frac{3}{2}, 1, 0) + \beta(-\frac{1}{2}, 0, 1).$
(e) $\mathbf{x} = \alpha(1, 1, 0) + \beta(-2, 0, 1).$

**4.** (a) $A\mathbf{x} = \mathbf{b}$, where
$$A = \begin{bmatrix} 5 & -3 & 1 & 0 & 0 \\ 1 & -1 & 0 & 1 & 0 \\ 8 & -4 & 0 & 0 & 1 \end{bmatrix}$$
and
$$\mathbf{b} = \begin{bmatrix} 2 \\ 1 \\ 1 \end{bmatrix}.$$

**5.** The system must have at least one free variable.

**10.** (a) Possible. Let $A$ be a $4 \times 3$ matrix with a zero row.
(b) Possible. Whenever $A$ has rank 3 and the system is consistent, the solution is unique. For example, let $A$ be the matrix
$$\begin{bmatrix} 1 & 0 & 0 \\ 0 & 1 & 0 \\ 0 & 0 & 1 \\ 1 & 0 & 1 \end{bmatrix}.$$
(c) Not possible. If the system has more than one solution, then it has a free variable and hence it has infinitely many solutions.
(d) Possible. Whenever rank $A < 3$ and the system is consistent, the system has infinitely many solutions. For example, let $A$ be the matrix
$$\begin{bmatrix} 1 & 0 & 0 \\ 0 & 1 & 0 \\ 1 & 1 & 0 \\ 2 & 1 & 0 \end{bmatrix}.$$

**12.** Rank $A < n$.

**18.** Let $\mathbf{c}_1, \mathbf{c}_2, \ldots, \mathbf{c}_p$ denote the columns of $C$. Since $A$ has rank $m$, the system $A\mathbf{x} = \mathbf{c}_i$ has a solution for $i = 1, 2, \ldots, p$ (Theorem 2). The solutions form the columns of the matrix $B$.

**20.** (a) $\mathbf{x} = (1, 1, \ldots, 1)$ is a nontrivial solution of the homogeneous equation $A\mathbf{x} = \mathbf{0}$. Thus by Theorem 4, rank $A < n$.

**22.** No. Consider the matrices
$$\begin{bmatrix} 1 & 0 \\ 0 & 1 \end{bmatrix} \quad \text{and} \quad \begin{bmatrix} 1 & 0 & 1 \\ 0 & 1 & 1 \end{bmatrix}.$$
More generally, let $A$ be an $m \times n$ matrix and $B$ be an $m \times p$ matrix both with rank $m$. Then the set of vectors $\mathbf{b}$ such that $A\mathbf{x} = \mathbf{b}$ is consistent is $R^m$, as is the set of vectors $\mathbf{c}$ such that $B\mathbf{x} = \mathbf{c}$ is consistent. However, if $n \neq p$, the corresponding homogeneous systems do not have the same solution sets.

## SECTION 1.11

**2.** (a) $\frac{1}{12}\begin{bmatrix} 6 & -3 \\ 2 & 1 \end{bmatrix}$.

(c) Singular.

(e) $\frac{1}{4}\begin{bmatrix} -3 & -1 & 3 \\ -5 & 1 & 1 \\ 2 & 2 & -2 \end{bmatrix}$.

(g) $\frac{1}{3}\begin{bmatrix} 5 & -6 & 4 \\ -6 & 9 & -6 \\ 4 & -6 & 5 \end{bmatrix}$.

(i) $\begin{bmatrix} 0 & 1 & -1 \\ 2 & -2 & -1 \\ -1 & 1 & 1 \end{bmatrix}$.

(k) $\frac{1}{2}\begin{bmatrix} 1 & 0 & 0 & 1 \\ 0 & 0 & 1 & -1 \\ 0 & 1 & -1 & 0 \\ 1 & -1 & 0 & 0 \end{bmatrix}$.

(m) $\begin{bmatrix} 1 & -2 & 1 & 0 \\ 1 & -2 & 2 & -3 \\ 0 & 1 & -1 & 1 \\ -2 & 3 & -2 & 3 \end{bmatrix}$.

**3.** (a) $\frac{1}{8}\begin{bmatrix} 11 & -3 & -1 \\ -9 & 1 & 3 \\ 6 & 2 & -2 \end{bmatrix}$;

$\mathbf{x} = A^{-1}\mathbf{b} = \frac{1}{8}\begin{bmatrix} 11 \\ -1 \\ -2 \end{bmatrix}$.

$\mathbf{b} = (1, -1, 3) = \frac{11}{8}(1, 0, 3) - \frac{1}{8}(1, 2, 5) - \frac{1}{4}(1, 3, 2)$.

(c) $\frac{1}{8}\begin{bmatrix} 13 & -4 & -1 \\ -15 & 4 & 3 \\ 10 & 0 & -2 \end{bmatrix}$;

$\mathbf{x} = A^{-1}\mathbf{b} = \begin{bmatrix} 12 \\ -14 \\ 10 \end{bmatrix}$;

$\mathbf{b} = (8, 2, 0) = 12(1, 0, 5) - 14(1, 2, 5) + 10(1, 3, 1)$.

**4.** (a) $\begin{bmatrix} 1 & -a & ac-b \\ 0 & 1 & -c \\ 0 & 0 & 1 \end{bmatrix}$.

**5.** (a) $B$ can be any matrix of the form

$$\begin{bmatrix} 5-\alpha & -2-\alpha \\ -2-\alpha & 1-\alpha \\ \alpha & \alpha \end{bmatrix}.$$

(b) There is no such matrix $B$.

**6.** $\begin{bmatrix} 2 & 1 \\ -2 & 3 \end{bmatrix}$.

**7.** If the rank of $A$ is equal to $m$, then the system $A\mathbf{x} = \mathbf{b}$ has a solution for every $\mathbf{b}$ in $R^m$. In particular, $A\mathbf{x} = \mathbf{e}_i$ has a solution for $i = 1, 2, \ldots, m$. The solutions form the columns of an $n \times m$ matrix $B$ such that $AB = I$.

**9.** If $A^T = A, (A^{-1})^T = (A^T)^{-1} = A^{-1}$. Yes.

**11.** No. Simply subtract any nonsingular matrix from itself.

**12.** (a) Since $A^2 - I = 0$, $A^2 = I$ and it follows from Theorem 4 that $A$ is nonsingular and $A^{-1} = A$.

(b) No. For example, let $A$ be the matrix $\begin{bmatrix} 1 & 0 \\ 0 & -1 \end{bmatrix}$.

**17.** (a) Let $A$ be an $n \times n$ upper triangular matrix. From the last equation of the system $A\mathbf{x} = \mathbf{e}_1$ we conclude that $x_n = 0$. Now conclude from the $(n-1)$ st equation that $x_{n-1} = 0$. Continue this reasoning to conclude that $x_{n-2}, x_{n-3}, \ldots, x_2 = 0$. Thus the solution is a multiple of $\mathbf{e}_1$.

**18.** (a) Simply multiply out the product on the right.

**23.** Let $\mathbf{b}_1, \mathbf{b}_2, \ldots, \mathbf{b}_n$ be the columns of $B$, and $\mathbf{c}_1, \mathbf{c}_2, \ldots, \mathbf{c}_n$ be the columns of $C$. $[A:B]$ reduces to $[I:C]$ if and only if $A\mathbf{c}_i = \mathbf{b}_i$ for all $i$, and this holds if and only if $\mathbf{c}_i = A^{-1}\mathbf{b}_i$ for all $i$.

**SUPPLEMENTARY PROBLEMS**

**1.** (a) True. Since rank $A < n$, the system has at least one free variable.

(c) False. The system may be inconsistent. Every consistent system has infinitely many solutions.

(e) False. If $r < n$, then every consistent system has infinitely many solutions. If $r = n$, the statement is true.

(g) False. The system may be inconsistent. Every consistent system has a unique solution.

(i) True. Existence of solutions for every $\mathbf{b}$ in $R^m$ implies that $r = m$ and uniqueness of solutions implies that $r = n$.

(k) True. Existence of solutions for every $\mathbf{b}$ in $R^m$ implies that $r = m$.

(m) False. If $r < m$, it is not the case that the system is consistent for every $\mathbf{b}$ in $R^m$.

(o) False. $E$ has $r$ nonzero rows and $m - r$ zero rows.

(q) False. $r$ is the number of nonzero rows in any echelon form of $A$.

(s) False. $r$ is at most equal to the smaller of $m$ and $n$.

(u) True. $A$ is nonsingular if and only if $r = n$, which holds if and only if the diagonal entries of $A$ are nonzero.

**2.** (a) (i) Rank $A = 2$.

(ii) $\mathbf{x} = x_2(-2, 1, 0, 0) + x_4(-2, 0, 4, 1)$.

(iii) $\mathbf{x} = (2, 0, -1, 0) + x_2(-2, 1, 0, 0) + x_4(-2, 0, 4, 1)$.

(c) (i) Rank $A = 3$.

(ii) $\mathbf{x} = x_4(0, 1, -2, 1, 0) + x_5(-10, 7, -3, 0, 1)$.

(iii) $\mathbf{x} = (-1, -7, 7, 0, 0) + x_4(0, 1, -2, 1, 0) + x_5(-10, 7, -3, 0, 1)$.

**3.** (a) $A^{-1} = \frac{1}{8}\begin{bmatrix} 4 & -4 & 0 \\ 1 & -7 & 2 \\ -3 & 5 & 2 \end{bmatrix}$,

$\mathbf{x} = \frac{1}{8}\begin{bmatrix} -4 \\ -13 \\ 7 \end{bmatrix}$.

(c) $A^{-1} = \begin{bmatrix} 1 & 0 & -1 \\ 0 & 1 & -1 \\ 0 & 0 & 1 \end{bmatrix}$, $\mathbf{x} = \begin{bmatrix} -1 \\ -1 \\ 2 \end{bmatrix}$.

**5.** (a), (c), (e), (f).

**7.** (a) The line in $R^2$ that passes through $(\frac{3}{2}, 0)$ with slope $-2$.

(b) The line in $R^3$ that passes through $(2, 0, 0)$ and is parallel to the line spanned by $(-2, 1, 1)$.

(c) The plane in $R^3$ that contains $(9, 0, 0)$ and is parallel to the plane spanned by $(2, 1, 0)$ and $(-4, 0, 1)$.

**9.** Yes. In both cases, the solution set consists of all vectors $\mathbf{x}$ in $R^4$ that satisfy $x_1 + 2x_2 - x_4 = 4$ and $x_3 - 2x_4 = 1$.

**12.** (b) $\mathbf{x} = (-3, -1, 1)$.

**13.** (a) $I = AA^{-1} = A^2A^{-1} = AAA^{-1} = AI = A$.

**14.** If $A$ is a diagonal matrix that is nonsingular, then the diagonal entries of $A$ are nonzero and $A^{-1}$ is the diagonal matrix whose diagonal entries are the reciprocals of the corresponding diagonal entries of $A$.

# CHAPTER 2

## SECTION 2.1

**2.** (a) Subspace.

(c) Subspace.

(e) Subspace.

(g) Not a subspace; it is not closed under vector addition.

(i) Subspace.

(k) Not a subspace; it does not contain the zero vector and it is not closed under addition or scalar multiplication.
(m) Not a subspace; it is not closed under scalar multiplication.
(o) Not a subspace; it is not closed under addition.
(q) Not a subspace; it does not contain the zero vector and it is not closed under addition or scalar multiplication.
(s) Subspace.

**4.** Certainly $\mathbf{0}$ is in the collection. If $\mathbf{x}$ and $\mathbf{y}$ are in the collection, then $A\mathbf{x} = B\mathbf{x}$ and $A\mathbf{y} = B\mathbf{y}$. Thus $A(\mathbf{x} + \mathbf{y}) = A\mathbf{x} + A\mathbf{y} = B\mathbf{x} + B\mathbf{y} = B(\mathbf{x} + \mathbf{y})$, and if $\alpha$ is any scalar, then $A(\alpha\mathbf{x}) = \alpha A\mathbf{x} = \alpha B\mathbf{x} = B(\alpha\mathbf{x})$. Therefore, the collection is closed under addition and scalar multiplication.

**5.** When $\alpha = \beta = -1$, the sum is $\mathbf{0}$. Thus the plane passes through the origin and hence is a subspace. To verify directly that the collection is closed under addition and scalar multiplication, let $\mathbf{x} = (1,2,1) + \alpha(1,1,0) + \beta(0,1,1)$ and $\mathbf{y} = (1,2,1) + \gamma(1,1,0) + \delta(0,1,1)$. Then

$$\begin{aligned}\mathbf{x} + \mathbf{y} &= (2,4,2) + (\alpha + \gamma)(1,1,0) \\ &\quad + (\beta + \delta)(0,1,1) \\ &= (1,2,1) + (\alpha + \gamma + 1)(1,1,0) \\ &\quad + (\beta + \delta + 1)(0,1,1), \\ \sigma\mathbf{x} &= \sigma(1,2,1) + \sigma\alpha(1,1,0) \\ &\quad + \sigma\beta(0,1,1) \\ &= (1,2,1) + (\sigma\alpha + \sigma - 1)(1,1,0) \\ &\quad + (\sigma\beta + \sigma - 1)(0,1,1).\end{aligned}$$

**7.** (a) No. It does not contain the origin, it is not closed under addition, and it is not closed under scalar multiplication.

**13.** $\{\mathbf{0}\}$, lines through the origin, and $R^2$.

**16.** If $\mathbf{x}$ is in $V$, then $0\mathbf{x} = \mathbf{0}$ is in $V$ because $V$ is closed under scalar multiplication.

**18.** If $U$ is a subspace of $V$(or vice versa), then $U \cup V = V$(or $U$). Conversely, suppose that neither $U$ nor $V$ is a subspace of the other. Then there are vectors $\mathbf{x}$ and $\mathbf{y}$ in $R^n$ such that $\mathbf{x}$ is in $U$ and not in $V$ and $\mathbf{y}$ is in $V$ and not in $U$. Then $\mathbf{x}$ and $\mathbf{y}$ are both in $U \cup V$ but $\mathbf{x} + \mathbf{y}$ is not in $U \cup V$. Why? Because if it were, then it must be in $U$ or $V$. If $\mathbf{x} + \mathbf{y}$ is in $U$, then $(\mathbf{x} + \mathbf{y}) - \mathbf{x} = \mathbf{y}$ is also in $U$ because $U$ is a subspace. Similarly, if $\mathbf{x} + \mathbf{y}$ is in $V$, then it follows that $\mathbf{x}$ is in $V$. For a concrete example, consider two distinct lines through the origin in $R^2$.

**19.** (a) $\mathbf{0}$ is in $U + V$ because $\mathbf{0}$ is in both $U$ and $V$ and $\mathbf{0} = \mathbf{0} + \mathbf{0}$. Suppose that $\mathbf{w}_1$ and $\mathbf{w}_2$ are two vectors in $U + V$ and $\alpha$ is a scalar. Then there are vectors $\mathbf{u}_1, \mathbf{u}_2, \mathbf{v}_1, \mathbf{v}_2$ such that

$$\mathbf{w}_1 = \mathbf{u}_1 + \mathbf{v}_1 \quad \text{and} \quad \mathbf{w}_2 = \mathbf{u}_2 + \mathbf{v}_2.$$

Thus

$$\begin{aligned}\mathbf{w}_1 + \mathbf{w}_2 &= (\mathbf{u}_1 + \mathbf{v}_1) + (\mathbf{u}_2 + \mathbf{v}_2) \\ &= (\mathbf{u}_1 + \mathbf{u}_2) + (\mathbf{v}_1 + \mathbf{v}_2)\end{aligned}$$

is in $U + V$ because $\mathbf{u}_1 + \mathbf{u}_2$ is in $U$ and $\mathbf{v}_1 + \mathbf{v}_2$ is in $V$. Similarly, $\alpha\mathbf{w}_1 = \alpha(\mathbf{u}_1 + \mathbf{v}_1) = \alpha\mathbf{u}_1 + \alpha\mathbf{v}_1$ is in $U + V$.

**21.** (a) If $U$ is a line not passing through the origin, then there are noncollinear vectors $\mathbf{a}$ and $\mathbf{v}$ such that $\mathbf{x}$ is in $U$ if and only if $\mathbf{x} = \mathbf{a} + \alpha\mathbf{v}$. Let $V$ be the subspace consisting of all possible scalar multiples of $\mathbf{v}$. Then $U = \mathbf{a} + V$.

## SECTION 2.2

**2.** (a) Yes. (c) No. (e) Yes.

**3.** (a) Vectors span; the $3 \times 3$ matrix having these vectors as its columns has rank 3.
(c) Vectors do not span; the vector $(0,0,1)$ is not a linear combination of the given vectors.
(e) Vectors span; the $3 \times 3$ matrix

having these vectors as its columns has rank 3.

**4.** (a) Vectors do not span; (0, 0, 0, 1) is not a linear combination of the given vectors.
(c) Vectors span; the $4 \times 4$ matrix having these vectors as its columns has rank 4.
(e) Vectors do not span; (1, 0, 0, 0) is not a linear combination of the given vectors.

**5.** (a) No constraints, the vectors span $R^3$.
(c) $20b_1 + 9b_2 - 8b_3 + b_4 = 0$.
(e) $b_1 + b_2 + b_3 = 0$, $-2b_1 + b_4 = 0$, $57b_1 - 17b_2 + 11b_5 = 0$.

**6.** (a) $-3b_1 + b_2 = 0$ and $-2b_1 + b_3 = 0$.
(c) No constraints; the column space is $R^3$.
(e) No constraints; the column space is $R^3$.
(g) $-4b_1 - b_2 - b_3 + b_4 = 0$.
(i) $-2b_1 + b_2 = 0$, $-3b_1 + b_3 = 0$, $5b_1 + b_4 = 0$, $9b_1 + b_5 = 0$.

**7.** (a) $(\frac{3}{2}, 1)$.
(c) $(-2, 1, 0, 0, 0)$, $(-1, 0, 1, 1, 0)$, $(-1, 0, 2, 0, 1)$.
(e) The nullspace is $R^4$; any set that spans $R^4$ will span $N(A)$ (e.g., $\mathbf{e}_1, \mathbf{e}_2, \mathbf{e}_3$, and $\mathbf{e}_4$).
(g) $(1, -2, 1, 0, 0)$, $(2, -3, 0, 1, 0)$, $(3, -4, 0, 0, 1)$.
(i) The nullspace is the trivial subspace of $R^1$.

**8.** (a) $\begin{bmatrix} -3 & 1 & 0 \\ -2 & 0 & 1 \end{bmatrix}$.
(c) For example, $[0 \quad 0 \quad 0]$.
(e) For example, $[0 \quad 0 \quad 0]$.
(g) $[-4 \quad -1 \quad -1 \quad 1]$.
(i) $\begin{bmatrix} -2 & 1 & 0 & 0 & 0 \\ -3 & 0 & 1 & 0 & 0 \\ 5 & 0 & 0 & 1 & 0 \\ 9 & 0 & 0 & 0 & 1 \end{bmatrix}$.

**9.** In each part, let $B$ be the matrix whose columns are the vectors that span the nullspace of $A$.

**10.** (a) Same subspace, namely, the collection of all vectors $\mathbf{b}$ in $R^3$ whose components satisfy $b_1 - b_2 + b_3 = 0$.
(c) Different subspaces. The constraint equation for the first subspace is $2b_2 - b_3 - b_4 = 0$; the constraint equation for the second subspace is $b_1 + 3b_2 - b_3 + 2b_4 = 0$.

**11.** (a) The columns of $A^T$ are the rows of $A$.

**13.** The zero matrix.

**17.** Suppose that $\mathbf{u}_3 = \beta_1\mathbf{u}_1 + \beta_2\mathbf{u}_2$. If

$$\mathbf{v} = \alpha_1\mathbf{u}_1 + \alpha_2\mathbf{u}_2 + \alpha_3\mathbf{u}_3,$$

then $\mathbf{v} = (\alpha_1 + \alpha_3\beta_1)\mathbf{u}_1 + (\alpha_2 + \alpha_3\beta_2)\mathbf{u}_2$. Thus any linear combination of $\mathbf{u}_1$, $\mathbf{u}_2$, and $\mathbf{u}_3$ is also a linear combination of $\mathbf{u}_1$ and $\mathbf{u}_2$. Conversely, any linear combination of $\mathbf{u}_1$ and $\mathbf{u}_2$ is also a linear combination of $\mathbf{u}_1$, $\mathbf{u}_2$, and $\mathbf{u}_3$.

**20.** We have $\mathbf{v}_1 = \mathbf{u}_1$, $\mathbf{v}_2 = \mathbf{u}_2 - \mathbf{u}_1$, and $\mathbf{v}_3 = \mathbf{u}_3 - \mathbf{u}_2$. If $\mathbf{v}$ is in $V$, then $\mathbf{v} = \alpha_1\mathbf{v}_1 + \alpha_2\mathbf{v}_2 + \alpha_3\mathbf{v}_3$ and substituting we obtain

$$\mathbf{v} = (\alpha_1 - \alpha_2)\mathbf{u}_1 + (\alpha_2 - \alpha_3)\mathbf{u}_2 + \alpha_3\mathbf{u}_3.$$

**22.** If $W$ is a subspace that contains the vectors $\mathbf{v}_1, \mathbf{v}_2, \ldots, \mathbf{v}_n$, then $W$ contains any linear combination of these vectors. Since any vector in $V$ is a linear combination of $\mathbf{v}_1, \mathbf{v}_2, \ldots, \mathbf{v}_n$, $W$ contains $V$.

**24.** (a) If $\mathbf{x}$ is in $N(B)$, then $B\mathbf{x} = \mathbf{0}$ and hence $AB\mathbf{x} = A\mathbf{0} = \mathbf{0}$. Thus $\mathbf{x}$ is in $N(AB)$.
(b) Suppose that $A$ is nonsingular. If $\mathbf{x}$ is in $N(AB)$, then $AB\mathbf{x} = \mathbf{0}$ and hence $B\mathbf{x} = A^{-1}(AB\mathbf{x}) = A^{-1}\mathbf{0} = \mathbf{0}$. Thus if $\mathbf{x}$ is in $N(AB)$, then $\mathbf{x}$ is in $N(B)$. Combining this result with part (a) proves that $N(B) = N(AB)$.

**27.** (a) Suppose that $R(A) = R(B)$. By Theorem 7, the row space of a matrix is the same as the row space of its reduced echelon form. Thus $A$ and $B$ have the *same* reduced echelon form because different reduced echelon matrices have different row spaces. It follows that $A$ and $B$ have the same nullspaces.

(c) Suppose that $N(A) = N(B)$. Row operations do not change the solutions of *homogeneous* equations; that is, if $E$ is any echelon form of $A$, then $A\mathbf{x} = \mathbf{0}$ and $E\mathbf{x} = \mathbf{0}$ have the same solutions. Hence $N(A) = N(E)$. Moreover, different reduced echelon matrices have different nullspaces. Thus if $N(A) = N(B)$, then $A$ and $B$ have the same reduced echelon form and hence the same row space (Theorem 7).

(e) False. Consider the matrices

$$\begin{bmatrix} 1 & 0 \\ 0 & 0 \end{bmatrix} \quad \text{and} \quad \begin{bmatrix} 0 & 1 \\ 0 & 0 \end{bmatrix}.$$

**31.** (a) $\begin{bmatrix} -1 \\ -3 \\ -2 \end{bmatrix} = -\frac{1}{3}\begin{bmatrix} 3 \\ 9 \\ 6 \end{bmatrix}$.

(c) All columns are pivot columns.

(e) All columns are pivot columns.

(g) $\begin{bmatrix} 1 \\ 1 \\ 1 \\ 6 \end{bmatrix} = 5\begin{bmatrix} 1 \\ -1 \\ -1 \\ 2 \end{bmatrix} + 0\begin{bmatrix} -1 \\ 2 \\ 3 \\ 1 \end{bmatrix} - 2\begin{bmatrix} 2 \\ -3 \\ -3 \\ 2 \end{bmatrix}$.

(i) The only column is a pivot column.

**33.** (a) If $\mathbf{v}$ is in $C(A)$, then there is a vector $\mathbf{x}$ such that $A\mathbf{x} = \mathbf{v}$. Thus $A\mathbf{v} = A(A\mathbf{x}) = A^2\mathbf{x} = A\mathbf{x} = \mathbf{v}$. Conversely, if $A\mathbf{v} = \mathbf{v}$, then by definition $\mathbf{v}$ is in $C(A)$.

## SECTION 2.3

**2.** (a) Independent. (c) Dependent.
(e) Independent. (g) Dependent.
(i) Dependent.

**5.** *Note:* In the answers, the independent vectors that span the row space of A are those from the *reduced* echelon form of $A$.

(a) $C(A)$: $(1, 0)$, $(0, 1)$.
$N(A)$: $(1, -3, 1, 0)$, $(2, -4, 0, 1)$.
$R(A)$: $(1, 0, -1, -2)$, $(0, 1, 3, 4)$.
Rank $(A) = 2$.

(c) $C(A)$: $(1, 0)$, $(0, 1)$
$N(A)$: $(-1, 1, 0, 0, 0)$, $(0, 0, -1, 1, 0)$, $(-1, 0, 0, 0, 1)$.
$R(A)$: $(1, 1, 0, 0, 1)$, $(0, 0, 1, 1, 0)$.
Rank $(A) = 2$.

(e) $C(A)$: $(1, 2, 3)$, $(-2, -3, -4)$.
$N(A)$: $(-2, 1, 0, 0)$, $(1, 0, 2, 1)$.
$R(A)$: $(1, 2, 0, -1)$, $(0, 0, 1, -2)$.
Rank $(A) = 2$.

(g) $C(A)$: $(1, 0, 1, -1)$, $(1, 0, -1, 1)$, $(1, 1, 2, -3)$.
$N(A)$: $(1, 0, -1, 1, 0)$, $(1, 0, -1, 0, 1)$.
$R(A)$: $(1, 0, 0, -1, -1)$, $(0, 1, 0, 0, 0)$, $(0, 0, 1, 1, 1)$. Rank $A = 3$.

(i) $C(A)$: $(1, 2, -1, 3)$, $(0, 1, 2, 4)$, $(1, 0, 2, 1)$.
$N(A)$: $(-3, 1, 0, 0, 0)$, $(-3, 0, 2, 1, 0)$.
$R(A)$: $(1, 3, 0, 3, 0)$, $(0, 0, 1, -2, 0)$, $(0, 0, 0, 0, 1)$. Rank $A = 3$.

**7.** Suppose that $\alpha_1\mathbf{u}_1 + \alpha_2\mathbf{u}_2 + \alpha_3\mathbf{u}_3 = \mathbf{0}$. Substituting, we obtain

$$\alpha_1(2\mathbf{v}_1) + \alpha_2(\mathbf{v}_1 + \mathbf{v}_2) + \alpha_3(-\mathbf{v}_1 + \mathbf{v}_3) = (2\alpha_1 + \alpha_2 - \alpha_3)\mathbf{v}_1 + \alpha_2\mathbf{v}_2 + \alpha_3\mathbf{v}_3 = \mathbf{0}$$

Since the $\mathbf{v}_i$'s are independent, it follows that $2\alpha_1 + \alpha_2 - \alpha_3 = \alpha_2 = \alpha_3 = 0$. Hence $\alpha_1 = \alpha_2 = \alpha_3 = 0$.

**8.** (a) Suppose that

$$\alpha_1(1, 0, 0) + \alpha_2(y_1, 1, 0) + \alpha_3(z_1, z_2, 1) = (0, 0, 0).$$

Equating components, it follows immediately that $\alpha_1 = \alpha_2 = \alpha_3 = 0$. One can also prove this by forming the matrix whose columns are the given vectors and observing that the matrix has rank 3 since it is a triangular matrix with no zeros on its diagonal.

**10.** $\lambda = 4$.

**14.** If $\mathbf{v}$ is in the subspace spanned by $\mathbf{v}_1, \mathbf{v}_2, \ldots, \mathbf{v}_n$, then by Theorem 5 the vectors $\mathbf{v}_1, \mathbf{v}_2, \ldots, \mathbf{v}_n, \mathbf{v}$ are dependent.

**17.** (a) Suppose that $\mathbf{u}_1$, $\mathbf{u}_2$, and $\mathbf{u}_3$ are nonzero dependent vectors. Then there are scalars $\alpha_1$, $\alpha_2$, and $\alpha_3$ not all zero such that $\alpha_1\mathbf{u}_1+\alpha_2\mathbf{u}_2+\alpha_2\mathbf{u}_3=\mathbf{0}$. In fact, at least two of the $\alpha$'s must be nonzero since none of the $\mathbf{u}$'s are zero. If $\alpha_3 \neq 0$, then $\mathbf{u}_3 = (-\alpha_1/\alpha_3)\mathbf{u}_1+(-\alpha_2/\alpha_3)\mathbf{u}_2$. If $\alpha_3=0$, then $\mathbf{u}_2=(-\alpha_1/\alpha_2)\mathbf{u}_1$ since $\alpha_1\mathbf{u}_1+\alpha_2\mathbf{u}_2=\mathbf{0}$ and $\alpha_2\neq 0$.

**19.** By Theorem 3 of Section 2.2, the rank of $A$ is $n$. Thus the columns of $A$ are independent (Theorem 3).

**22.** (a) The columns of $A$ are independent if and only if rank $A=m$. The rows of $A$ are independent if and only if the columns of $A^T$ are independent which holds if and only if rank $A^T=m$. It follows easily from Theorems 1 and 2 of Section 1.11 that for an $m\times m$ matrix, rank $A=m$ if and only if rank $A^T=m$. Hence both conditions are equivalent to the condition that rank $A=m$.

**23.** If $\alpha_1A\mathbf{v}_1+\alpha_2A\mathbf{v}_2+\cdots+\alpha_nA\mathbf{v}_n=\mathbf{0}$, then $A(\alpha_1\mathbf{v}_1+\alpha_2\mathbf{v}_2+\cdots+\alpha_n\mathbf{v}_n)=\mathbf{0}$. Since the columns of the matrix $A$ are independent, $N(A)=\{\mathbf{0}\}$ (Theorem 2). Thus $\alpha_1\mathbf{v}_1+\alpha_2\mathbf{v}_2+\cdots+\alpha_n\mathbf{v}_n=\mathbf{0}$ and since the $\mathbf{v}_i$'s are independent, the $\alpha_i$'s are all 0.

## SECTION 2.4

**2.** (a) Yes.
(c) No. Three vectors in $R^2$ cannot be independent.
(e) No. Any list of vectors containing the zero vector is dependent.

**3.** (a) Yes. (c) No. (e) Yes.

**4.** (a) Yes.
(c) No. $R^4$ cannot be spanned by fewer than four vectors.
(e) No.

**5.** (a) $V$ is a plane through the origin. Basis: $(2,1,0), (-1,0,1)$.
(c) $V=N(A)$, where

$$A=[1 \quad 2 \quad -1 \quad 0].$$

Basis: $(-2,1,0,0), (1,0,1,0), (0,0,0,1)$.
(e) $V$ is the column space of the matrix whose columns are the given vectors. Basis: $(1,0,1,2,1)$, $(1,0,1,2,2)$, $(2,1,0,1,2)$.
(g) An equation of the plane is

$$\mathbf{x}=(0,1,2)+\alpha(-2,4,3)+\beta(-4,1,-8).$$

Since the plane passes through the origin ($\alpha=-\frac{2}{7}, \beta=\frac{1}{7}$), $V$ is a subspace. A Cartesian equation of the plane is $5x_1+4x_2-2x_3=0$. Basis: $(\frac{2}{5},0,1), (-\frac{4}{5},1,0)$.
(i) A Cartesian equation of the first plane is $x_1+x_2-3x_3=0$. Therefore, $V$ is the set of solutions of the homogeneous system

$$\begin{aligned} x_1+\ x_2-3x_3&=0\\ 2x_1+3x_2-\ x_3&=0.\end{aligned}$$

Basis: $(8,-5,1)$.
(k) $V=N(A-B)$. Basis: $(-2,1,0,0,0)$, $(-9,0,7,1,0)$, $(14,0,-11,0,1)$.

**6.** (a) $(1,2,3), (1,0,0), (0,1,0)$.
(c) $(1,1,2), (1,0,0), (0,1,0)$.
(e) $(1,1,0,0), (1,2,3,1), (1,0,0,0), (0,0,1,0)$.

**9.** In the answers, we give the nonzero rows of the *reduced* echelon matrix as a basis of the row space.
(a) Basis for $C(A)$: $(1,2,3), (2,3,4)$.
Basis for $N(A)$: $(2,-3,1,0)$, $(3,-4,0,1)$.
Basis for $R(A)$: $(1,0,-2,-3)$, $(0,1,3,4)$.
Rank $A=2$.
$\mathbf{x}=(-5,10,0,0)+x_3(2,-3,1,0)+x_4(3,-4,0,1)$.
(c) Basis for $C(A)$: $(3,-6,2), (-1,2,5)$.
Basis for $N(A)$: $N(A)=\{\mathbf{0}\}$.
Basis for $R(A)$: $(1,0), (0,1)$.
Rank $A=2$.
$\mathbf{x}=(1,2)$.

(e) Basis for $C(A)$: $(1, -2, 3)$.
Basis for $N(A)$: $(1, 1, 0, 0)$, $(0, 0, 1, 0)$, $(-2, 0, 0, 1)$.
Basis for $R(A)$: $(1, -1, 0, 2)$.
Rank $A = 1$.
The system is inconsistent.

(g) Basis for $C(A)$: $(1, 0, 0, 0)$, $(6, -2, 3, 2)$, $(0, 1, 4, 5)$.
Basis for $N(A)$: $(-1, -1, 1, 1)$.
Basis for $R(A)$: $(1, 0, 0, 1)$, $(0, 1, 0, 1)$, $(0, 0, 1, -1)$.
Rank $A = 3$.
$\mathbf{x} = (3, 3, -2, 0) + x_4(-1, -1, 1, 1)$.

(i) Basis for $C(A)$: $(1, 1, 2, 2)$, $(-1, -1, -2, 0)$, $(2, 2, 5, -1)$.
Basis for $N(A)$: $(-2, 1, 0, 0, 0)$, $(-\frac{1}{2}, 0, \frac{1}{2}, 1, 0)$.
Basis for $R(A)$: $(1, 2, 0, \frac{1}{2}, 0)$, $(0, 0, 1, -\frac{1}{2}, 0)$, $(0, 0, 0, 0, 1)$.
Rank $A = 3$.
$\mathbf{x} = (1, 0, 1, 0, 1) + x_2(-2, 1, 0, 0, 0) + x_4(-\frac{1}{2}, 0, \frac{1}{2}, 1, 0)$.

(k) Basis for $C(A)$: $(1, 2, -1, 2, 3)$, $(-2, 0, 0, 0, 0)$, $(0, 2, 2, -2, 2)$, $(1, 3, 1, 3, 2)$.
Basis for $N(A)$: $(\frac{3}{5}, \frac{1}{10}, 1, -\frac{12}{5}, 1)$.
Basis for $R(A)$: $(1, 0, 0, 0, -\frac{3}{5})$, $(0, 1, 0, 0, -\frac{1}{10})$, $(0, 0, 1, 0, -1)$, $(0, 0, 0, 1, \frac{12}{5})$.
Rank $A = 4$.
$\mathbf{x} = (1, 0, 1, 1, 0) + x_5(\frac{3}{5}, \frac{1}{10}, 1, -\frac{12}{5}, 1)$.

(m) Basis for $C(A)$: $(0, 9, 3)$.
Basis for $N(A)$: Empty set.
Basis for $R(A)$: $(9)$
Rank $A = 1$.
$x = -\frac{1}{3}$.

**12.** Suppose that $\alpha_1\mathbf{w}_1 + \alpha_2\mathbf{w}_2 + \alpha_3\mathbf{w}_3 = \mathbf{0}$. Substituting, we obtain $(\alpha_1 + 3\alpha_2 + 4\alpha_3)\mathbf{v}_1 + (2\alpha_1 + 2\alpha_2)\mathbf{v}_2 + \alpha_2\mathbf{v}_3 = \mathbf{0}$. Since the $\mathbf{v}_i$'s are independent, $\alpha_1 + 3\alpha_2 + 4\alpha_3 = 2\alpha_1 + 2\alpha_2 = \alpha_2 = 0$ and hence $\alpha_1 = \alpha_2 = \alpha_3 = 0$. Thus the $\mathbf{w}_i$'s are independent. To show that they span, we must show that any $\mathbf{v}$ in $V$ can be written as a linear combination of the $\mathbf{w}_i$'s. Let $\mathbf{v}$ be in $V$. Since the $\mathbf{v}_i$'s span, $\mathbf{v} = \alpha_1\mathbf{v}_1 + \alpha_2\mathbf{v}_2 + \alpha_3\mathbf{v}_3$ and substituting, we obtain $\mathbf{v} = (\alpha_2/2 - \alpha_3)\mathbf{w}_1 + \alpha_3\mathbf{w}_2 + (\alpha_1/4 - \alpha_2/8 - \alpha_3/2)\mathbf{w}_3$. Thus the $\mathbf{w}_i$'s span $V$.

**17.** The only way to express the zero vector as a linear combination of these vectors is with all coefficients equal to zero. Thus the vectors are independent. Since the vectors also span, they form a basis.

**19.** (a) If the vectors $\mathbf{v}_1, \mathbf{v}_2, \ldots, \mathbf{v}_m$ do not span $R^m$, then there is a vector $\mathbf{v}$ in $R^m$ that is not a linear combination of these vectors. Thus $\mathbf{v}_1, \mathbf{v}_2, \ldots, \mathbf{v}_m, \mathbf{v}$ are independent (Theorem 4). This contradicts the fact that $R^m$ cannot contain more than $m$ independent vectors (Theorem 4 of Section 2.3). Here is another solution. Form the $m \times m$ matrix having the vectors $\mathbf{v}_1, \mathbf{v}_2, \ldots, \mathbf{v}_m$ as its columns. By Theorem 3 of Section 2.3, the rank of $A$ is $m$. Therefore, by Theorem 3, the vectors form a basis for $R^m$.

**22.** No. If $\alpha_{11} = 0$ or $\alpha_{22} = 0$ or $\alpha_{33} = 0$, the vectors are dependent. When all of the $\alpha_{ij} \neq 0$, the vectors are independent. In general, the vectors are independent if and only if $\alpha_{ii} \neq 0$ for $i = 1, 2, 3$. Try your hand at proving that.

## SECTION 2.5

**2.** (a) The given vectors are independent and hence form a basis for $V$. Thus $\dim V = 3$ and $V = R^3$.
(c) The first, second, and fourth vectors form a basis for $V$. Thus $\dim V = 3$.

**3.** (a) The first two vectors form a basis for $V$; $\dim V = 2$.
(c) The first and third vectors form a basis for $V$; $\dim V = 2$.

**4.** (a) $\dim C(A) = 2, \dim N(A) = 2,$ $\dim R(A) = 2.$
(c) $\dim C(A) = 2, \dim N(A) = 0,$ $\dim R(A) = 2.$
(e) $\dim C(A) = 1, \dim N(A) = 3,$ $\dim R(A) = 1.$
(g) $\dim C(A) = 3, \dim N(A) = 1,$ $\dim R(A) = 3.$
(i) $\dim C(A) = 3, \dim N(A) = 2,$ $\dim R(A) = 3.$
(k) $\dim C(A) = 4, \dim N(A) = 1,$ $\dim R(A) = 4.$
(m) $\dim C(A) = 1, \dim N(A) = 0,$ $\dim R(A) = 1.$

**5.** (a) $\dim C(A) = 2, \dim N(A) = 2,$ $\dim R(A) = 2.$
(c) $\dim C(A) = 2, \dim N(A) = 3,$ $\dim R(A) = 2.$
(e) $\dim C(A) = 2, \dim N(A) = 2,$ $\dim R(A) = 2.$
(g) $\dim C(A) = 3, \dim N(A) = 2,$ $\dim R(A) = 3.$
(i) $\dim C(A) = 3, \dim N(A) = 2,$ $\dim R(A) = 3.$

**6.** (a) (i) $\dim C(A) = 2, \dim N(A) = 4,$ $\dim R(A) = 2.$
(ii) No constraints.
(iii) $B = 0.$

(iv) $$\begin{bmatrix} -2 & 1 & \frac{3}{2} & -2 \\ 1 & 0 & 0 & 0 \\ 0 & 1 & 0 & 0 \\ 0 & 0 & -\frac{1}{2} & 1 \\ 0 & 0 & 1 & 0 \\ 0 & 0 & 0 & 1 \end{bmatrix}.$$

(c) (i) $\dim C(A) = 2, \dim N(A) = 0,$ $\dim R(A) = 2.$
(ii) $-b_1 + b_2 + b_3 = 0.$
(iii) $B = [-1 \quad 1 \quad 1]$
(iv) $B = 0.$

(e) (i) $\dim C(A) = 2, \dim N(A) = 1,$ $\dim R(A) = 2.$
(ii) $3b_1 - 4b_2 + b_3 = 0,$
$6b_1 - 7b_2 + b_4 = 0,$
$8b_1 - 9b_2 + b_5 = 0.$

(iii) $$B = \begin{bmatrix} 3 & -4 & 1 & 0 & 0 \\ 6 & -7 & 0 & 1 & 0 \\ 8 & -9 & 0 & 0 & 1 \end{bmatrix}.$$

(iv) $$B = \begin{bmatrix} 1 \\ -2 \\ 1 \end{bmatrix}.$$

**8.** In each part, let $A$ be the matrix whose columns are the given vectors.
(a) (i) Rank $A = 3$.
(ii) $(2, -3, 4)$.
(iii) $M = A^{-1}$

$$= \frac{1}{46}\begin{bmatrix} -4 & -12 & 7 \\ 2 & 6 & 8 \\ 16 & 2 & -5 \end{bmatrix}.$$

(c) (i) Rank $A = 3$.
(ii) $(3, 1, -1)$.

(iii) $$M = A^{-1} = \begin{bmatrix} -4 & 2 & 1 \\ 1 & 0 & -\frac{1}{3} \\ 2 & -1 & -\frac{1}{3} \end{bmatrix}.$$

**9.** A basis for $V$ is $(1, -1, 0, 2)$, $(3, 1, 0, 1)$, $\dim V = 2$. The coordinate vectors are $(1, 0)$, $(-2, 0)$, $(0, 1)$, $(-3, 2)$, $(-5, 2)$.

**18.** When $\alpha \neq 1, -2$, the dimension is 3; when $\alpha = 1$, the dimension is 1; when $\alpha = -2$, the dimension is 2.

**20.** Let $\dim V = k$. Then $V$ has a basis consisting of $k$ vectors. These $k$ vectors are independent vectors in $R^m$, and since $R^m$ cannot contain more than $m$ independent vectors, $k \leq m$. If $k = m$, then these vectors form a basis for $R^m$ (Theorem 8). Finally, if $V = R^m$, then $\dim V = \dim R^m = m$.

**29.** $\mathbf{p} = p_3(\frac{1}{2}, 0, 1, 0) + p_4(0, 1, 0, 1)$.

## SUPPLEMENTARY PROBLEMS

**1.** (a) False. $V$ may be the trivial subspace.
(c) True. Make all the coefficients zero.
(e) True. See Theorem 7 of Section 2.5.
(g) True. If $\alpha_2\mathbf{u}_2 + \alpha_3\mathbf{u}_3 + \cdots + \alpha_n\mathbf{u}_n = \mathbf{0}$

and not all $\alpha$'s are zero, then $0\mathbf{u}_1 + \alpha_2\mathbf{u}_2 + \alpha_3\mathbf{u}_3 + \cdots + \alpha_n\mathbf{u}_n = \mathbf{0}$ and not all $\alpha$'s are zero.

(i) False. $R^6$ cannot be spanned by fewer than six vectors.

(k) True. Geometrically, two planes through the origin in $R^3$ must contain more than the origin in common. Algebraically, let $\mathbf{v}_1, \mathbf{v}_2$ and $\mathbf{w}_1, \mathbf{w}_2$ be bases for the two subspaces. Since any four vectors in $R^3$ are dependent, there are scalars $\alpha_1$, $\alpha_2$, $\beta_1$, $\beta_2$ not all zero such that

$$\alpha_1\mathbf{v}_1 + \alpha_2\mathbf{v}_2 + \beta_1\mathbf{w}_1 + \beta_2\mathbf{w}_2 = \mathbf{0}.$$

Thus

$$\alpha_1\mathbf{v}_1 + \alpha_2\mathbf{v}_2 = -(\beta_1\mathbf{w}_1 + \beta_2\mathbf{w}_2) \neq \mathbf{0},$$

and the two subspaces share this nonzero vector.

**2.** (a) True. The rank of $A$ is at most $n$ and hence there are at most $n$ independent rows.

(c) True. $A$ is an $n \times n$ matrix with rank $n$.

(e) False. $\dim N(A) = n - r \geq n - m$.

(g) False. $r$ may be less than $m$ and hence the system may be inconsistent.

(i) False. $r$ may be less than $m$.

(k) False. This is reflected in the fact that row operations transform $A\mathbf{x} = \mathbf{b}$ into $E\mathbf{x} = \mathbf{c}$, *not* $E\mathbf{x} = \mathbf{b}$.

(m) True. See Theorem 5 of Section 2.5.

(o) True. $r$ is equal to the number of pivot columns of $E$, and the pivot columns of $E$ form a basis for $C(E)$.

(q) True. $C(A) = R^m$.

(s) True. See Theorem 2 of Section 2.3.

(u) False. If $A\mathbf{x} = A\mathbf{y}$, then $\mathbf{x} - \mathbf{y}$ is in $N(A)$ and $N(A) \neq \{\mathbf{0}\}$ when $r < n$.

(w) True. By pivoting, $A$ can by transformed to the identity.

(y) False. Let $A = \begin{bmatrix} 0 & 1 \\ 0 & 0 \end{bmatrix}$, $B = \begin{bmatrix} 1 \\ 0 \end{bmatrix}$.

**3.** (c) Independent.

**4.** (c) Vectors form a basis.

**5.** (d) Let $U$ be the hyperplane determined by $u_1 + u_2 + u_3 = 0$ and let $V$ be the hyperplane determined by $2v_1 + v_3 = 0$. Then the collection of all vectors $\mathbf{x} = \mathbf{u} + \mathbf{v}$, where $\mathbf{u}$ is in $U$ and $\mathbf{v}$ is in $V$, is $W = U + V$. A basis for the intersection of the two subspaces $U$ and $V$ is

$$\mathbf{w}_1 = (-1, -1, 2, 0), \quad \mathbf{w}_2 = (0, 0, 0, 1).$$

Extend this basis to a basis $\mathbf{w}_1, \mathbf{w}_2, \mathbf{u}_3$ for $U$ and then extend it again to a basis $\mathbf{w}_1$, $\mathbf{w}_2$, $\mathbf{v}_4$ for $V$. Then show that $\mathbf{w}_1, \mathbf{w}_2, \mathbf{u}_3, \mathbf{v}_4$ form a basis for $W$. Thus $W = R^4$.

**6.** No. The vectors span a two-dimensional subspace of $R^3$, namely, the subspace determined by $b_1 - 2b_2 + b_3 = 0$. Any vector that does not satisfy this constraint equation is not in the subspace [e.g., $(1, 0, 0)$].

**11.** (a) $\dim C(A) = 3$, $\dim N(A) = 2$, $\dim R(A) = 3$.

(b) Yes, $\alpha = 3$.

**14.** (a) The subspace is $N(A - 2I)$.

(b) Basis: $(1, 0)$.

**22.** (a) A basis for the subspace is $(1, 1, 1)$ and $(2, 2, 5)$. It is a two-dimensional subspace of $R^3$ (i.e., a plane).

(c) Since the system is inconsistent, the set is empty. Therefore, it is not a subspace.

**23.** (a) Both collections span $R^2$.

(b) The two collections of vectors span two different planes.

(c) The two collections of vectors span the same plane.

**24.** (a) The plane contains the line since $(5, -2, 3)$ is in the plane.

(c) The plane does not contain the line since $(7, 3, 9, 5)$ is on the line but not on the plane.

**29.** (a) Always.
(b) True. If $C$ is nonsingular, then $I = (AB)C^{-1} = A(BC^{-1})$, so that $A$ has an inverse.
(c) Sometimes.

**32.** Begin with a $4 \times 3$ matrix whose first two *rows* are the given vectors in the row space and whose last two rows are the zero vectors. Then use row operations to construct the matrix

$$A = \begin{bmatrix} 2 & 3 & 2 \\ 2 & 5 & 10 \\ 0 & 1 & 4 \\ 1 & 3 & 7 \end{bmatrix}.$$

**33.** By Problems 14, 15, and 16 of Section 2.5,
rank $C \leq$ rank $A$ and rank $C \leq$ rank $B$,
rank $C =$ rank $A$ if $B$ is nonsingular,
rank $C =$ rank $B$ if $A$ is nonsingular.

All combinations not listed in the table are impossible because of these facts (verify). Let

$$X = \begin{bmatrix} 1 & 0 & 0 \\ 0 & 0 & 0 \\ 0 & 0 & 0 \end{bmatrix}, \quad Y = \begin{bmatrix} 0 & 0 & 0 \\ 0 & 1 & 0 \\ 0 & 0 & 0 \end{bmatrix},$$

$$Z = \begin{bmatrix} 1 & 0 & 0 \\ 0 & 1 & 0 \\ 0 & 0 & 0 \end{bmatrix}, \quad W = \begin{bmatrix} 0 & 0 & 0 \\ 0 & 1 & 0 \\ 0 & 0 & 1 \end{bmatrix}.$$

To show that rank $A =$ rank $B = 2$ and rank $C = 0$ is impossible, we proceed as follows. Suppose that $A$ has rank 2 and $C$ has rank 0. Then $C = 0$, so $AB = 0$. Show that this implies that rank $B = 1$. *Hint:* $N(A)$ is one-dimensional and each column of $B$ is in $N(A)$.

| Rank $A$ | Rank $B$ | Rank $C$ | Example |
|---|---|---|---|
| 3 | 3 | 3 | $A = B = C = I$ |
| 3 | 2 | 2 | Example given in problem. |
| 3 | 1 | 1 | $A = I$, $B = C =$ any rank 1 matrix. |
| 3 | 0 | 0 | $A = I$, $B = C = 0$ |
| 2 | 3 | 2 | $B = I$, $A = C =$ any rank 2 matrix. |
| 2 | 2 | 2 | $A = B = C = Z$. |
| 2 | 2 | 1 | $A = Z$, $B = W$, $C = Y$. |
| 2 | 2 | 0 | Impossible. See above. |
| 2 | 1 | 1 | $A = Z$, $B = Y$, $C = Y$. |
| 2 | 1 | 0 | $A = W$, $B = X$, $C = 0$. |
| 2 | 0 | 0 | $A = Z$, $B = C = 0$. |
| 1 | 3 | 1 | $B = I$, $A = C =$ any rank 1 matrix. |
| 1 | 2 | 1 | $A = C = X$, $B = Z$. |
| 1 | 2 | 0 | $A = X$, $B = W$, $C = 0$. |
| 1 | 1 | 1 | $A = B = C = X$. |
| 1 | 1 | 0 | $A = X$, $B = Y$, $C = 0$. |
| 1 | 0 | 0 | $B = C = 0$, $A$ any rank 1 matrix. |
| 0 | 3 | 0 | $A = C = 0$, $B = I$. |
| 0 | 2 | 0 | $A = C = 0$, $B$ any rank 2 matrix. |
| 0 | 1 | 0 | $A = C = 0$, $B$ any rank 1 matrix. |
| 0 | 0 | 0 | $A = B = C = 0$. |

# CHAPTER 3

## SECTION 3.1

**1.** $$\begin{bmatrix} 1 & 1 \\ 1 & 2 \\ 1 & 3 \\ 1 & 4 \\ 1 & 5 \end{bmatrix} \begin{bmatrix} d_0 \\ v \end{bmatrix} = \begin{bmatrix} 3 \\ 5 \\ 9 \\ 11 \\ 12 \end{bmatrix}.$$

**2.** $$\begin{bmatrix} 1 & -1 & 1 \\ 1 & 0 & 0 \\ 1 & 1 & 1 \\ 1 & 2 & 4 \end{bmatrix} \begin{bmatrix} a \\ b \\ c \end{bmatrix} = \begin{bmatrix} -2 \\ -1 \\ 0 \\ 3 \end{bmatrix}.$$

## SECTION 3.2

**1.** (a) $\sqrt{6}$. (b) $\sqrt{37/12}$.
(e) $\sqrt{1+\pi^2}$. (f) $\sqrt{1+a^2b^2}/|a|$.

**2.** (a) $\sqrt{3}$. (c) $\sqrt{74}/6$.

**3.** (b), (c), and (d).

**6.** (a), (b).

**7.** (a) For example, $(-1, 2, 1)$.

**8.** (a) For example, $(1, 1, -2)$, $(-1, 1, 0)$.

**9.** (a) For example, $(-2, 1, 0, 0)$, $(-1, -2, 5, 0)$.

**12.** (a) $\|\mathbf{x} \pm \mathbf{y}\|^2 = \langle \mathbf{x} \pm \mathbf{y}, \mathbf{x} \pm \mathbf{y} \rangle = \|\mathbf{x}\|^2 \pm 2\langle \mathbf{x}, \mathbf{y} \rangle + \|\mathbf{y}\|^2$. Hence $\|\mathbf{x} + \mathbf{y}\|^2 = \|\mathbf{x} - \mathbf{y}\|^2$ if and only if $2\langle \mathbf{x}, \mathbf{y} \rangle = -2\langle \mathbf{x}, \mathbf{y} \rangle$, (i.e., $\langle \mathbf{x}, \mathbf{y} \rangle = 0$).
(c) $\|\mathbf{x} - \mathbf{y}\|^2 = \langle \mathbf{x} - \mathbf{y}, \mathbf{x} - \mathbf{y} \rangle = \|\mathbf{x}\|^2 - 2\langle \mathbf{x}, \mathbf{y} \rangle + \|\mathbf{y}\|^2 = 2$.
(h) $\|\mathbf{x} + \mathbf{y}\|^2 - \|\mathbf{x} - \mathbf{y}\|^2 = \|\mathbf{x}\|^2 + 2\langle \mathbf{x}, \mathbf{y} \rangle + \|\mathbf{y}\|^2 - (\|\mathbf{x}\|^2 - 2\langle \mathbf{x}, \mathbf{y} \rangle + \|\mathbf{y}\|^2) = 4\langle \mathbf{x}, \mathbf{y} \rangle$.

**13.** $A^T = (\mathbf{x}\mathbf{x}^T)^T = \mathbf{x}^{TT}\mathbf{x}^T = \mathbf{x}\mathbf{x}^T = A$. $A^2 = (\mathbf{x}\mathbf{x}^T)(\mathbf{x}\mathbf{x}^T) = \mathbf{x}(\mathbf{x}^T\mathbf{x})\mathbf{x}^T = \mathbf{x}\mathbf{x}^T = A$, since $\mathbf{x}^T\mathbf{x} = \|\mathbf{x}\|^2 = 1$.

**15.** Expand $\|\mathbf{x} + \mathbf{y}\|^2 + \|\mathbf{x} - \mathbf{y}\|^2$ as in Problem 12.

**16.** (a) If $\langle \mathbf{u}, \mathbf{x} \rangle = \langle \mathbf{v}, \mathbf{x} \rangle$ for all $\mathbf{x}$, then $\langle \mathbf{u} - \mathbf{v}, \mathbf{x} \rangle = 0$ for all $\mathbf{x}$. Thus $\mathbf{u} - \mathbf{v} = \mathbf{0}$ by Result 1.

**17.** (a) Let $\mathbf{c}_1, \mathbf{c}_2, \ldots, \mathbf{c}_n$ be the columns of $A$. $A$ is orthogonal if and only if $\langle \mathbf{c}_i, \mathbf{c}_j \rangle = 0$ whenever $i \neq j$ and $\langle \mathbf{c}_i, \mathbf{c}_i \rangle = 1$ for $i = 1, 2, \ldots, n$, and this holds if and only if $A^TA = I$.
(b) Let $\mathbf{r}_1, \mathbf{r}_2, \ldots, \mathbf{r}_n$ be the rows of $A$. The rows of $A$ form an orthonormal set if and only if $\langle \mathbf{r}_i, \mathbf{r}_j \rangle = 0$ whenever $i \neq j$ and $\langle \mathbf{r}_i, \mathbf{r}_i \rangle = 1$ for $i = 1, 2, \ldots, n$, and this holds if and only if $AA^T = I$. Finally, $AA^T = I$ if and only if $A^{-1} = A^T$, which holds if and only if $A$ is orthogonal.
(e) $\langle A\mathbf{x}, A\mathbf{y} \rangle = \langle A^TA\mathbf{x}, \mathbf{y} \rangle = \langle \mathbf{x}, \mathbf{y} \rangle$.
(h) $(I - 2\mathbf{x}\mathbf{x}^T)^T = I^T - 2(\mathbf{x}\mathbf{x}^T)^T = I - 2\mathbf{x}\mathbf{x}^T$.
Now use part (a).
(k) Diagonal matrices whose diagonal entries are $\pm 1$.

## SECTION 3.3

**1.** (a) $\theta = \arccos(3/\sqrt{10})$.
(c) $\theta = \pi/6$.
(e) $\theta = \arccos(2/\sqrt{6})$.

**2.** (a) $(\frac{1}{2}, \pm\sqrt{3}/2)$.

**3.** (a) $(1/\sqrt{2}, \pm 1/\sqrt{2})$.

**4.** (a) $(\frac{1}{2}, \frac{1}{2}, \pm 1/\sqrt{2})$.

**6.** Since $1 = \langle \mathbf{x}, \mathbf{y} \rangle = \|\mathbf{x}\| \cdot \|\mathbf{y}\| \cos\theta = \cos\theta$, $\theta = 0$ and hence $\mathbf{x} = \mathbf{y}$. When $\langle \mathbf{x}, \mathbf{y} \rangle = -1$ we conclude similarly that $\cos\theta = -1$. Hence $\theta = \pi$ and $\mathbf{x} = -\mathbf{y}$.

**10.** $\cos\alpha = \dfrac{\langle (1,2),(1,3) \rangle}{\sqrt{5} \cdot \sqrt{10}} = \dfrac{7}{5\sqrt{2}} = \dfrac{\langle (1,1),(3,4) \rangle}{\sqrt{2} \cdot \sqrt{25}} = \cos\beta$.

**11.** The angle whose vertex is at $(7,7)$ is obtuse. For the triangle in $R^4$ all angles are acute.

## SECTION 3.4

**1.** (a) $(1, -1)$.
(c) $(-1, 1, 0)$.
(e) $(5, -5, 0, 2), (2, -2, 1, 0)$.
(g) $(-1, 0, 0, 0, 1), (1, 0, 0, 1, 0), (-1, 0, 1, 0, 0), (1, 1, 0, 0, 0)$.
(i) $(4, 3, -2, 1, 0), (-3, 1, 2, 0, 2)$.
(k) $(1, 0, 0, 0), (0, 0, -1, 1)$.
(m) $(-4, 2, 5, 0), (-3, -1, 0, 5)$.

**3.** (a) The plane spanned by $(3, 1, 0)$ and $(-4, 0, 1)$.
(c) The line spanned by $(-9, 3, 1)$.
(e) The line spanned by $(3, 2, -5)$.

**5.** (a) $(1, 1, 0, 0, 0), (0, 1, 1, 0, 0), (0, 0, 1, 1, 1)$.
(b) $(-1, 1, -1, 1, 0), (-1, 1, -1, 0, 1)$.
(c)
$$\begin{aligned} -x_1 + x_2 - x_3 + x_4 \quad &= 0 \\ -x_1 + x_2 - x_3 + \quad x_5 &= 0. \end{aligned}$$
(d)
$$\begin{aligned} x_1 + x_2 \qquad\qquad &= 0 \\ x_2 + x_3 \qquad\quad &= 0 \\ x_3 + x_4 + x_5 &= 0. \end{aligned}$$

**10.** (a) Let $\mathbf{x}$ and $\mathbf{y}$ be in $S^{\perp}$ and $\alpha$ be a scalar. Then $\langle \mathbf{x} + \mathbf{y}, \mathbf{s} \rangle = \langle \mathbf{x}, \mathbf{s} \rangle + \langle \mathbf{y}, \mathbf{s} \rangle = 0 + 0 = 0$ and $\langle \alpha\mathbf{x}, \mathbf{s} \rangle = \alpha \langle \mathbf{x}, \mathbf{s} \rangle = \alpha \cdot 0 = 0$ for all $\mathbf{s}$ in $S$. Therefore, $\mathbf{x} + \mathbf{y}$ and $\alpha\mathbf{x}$ are in $S^{\perp}$.
(c) If $\mathbf{s}$ is in $S$, then $\langle \mathbf{s}, \mathbf{t} \rangle = 0$ for all $\mathbf{t}$ in $S^{\perp}$. Hence $\mathbf{s}$ is in $S^{\perp\perp}$.

**13.** (a) Basis for $C(A)^{\perp}$: $(2, 1)$; basis for $R(A)^{\perp}$: $(3, 1)$.
(b) Basis for $C(A)^{\perp}$: $(-4, -3, 2)$; $R(A)^{\perp} = \{\mathbf{0}\}$.

**16.** (b) No. Dim $V^{\perp} = 3$ and dim $W = 2$.
(d) Since $W$ is orthogonal to $V$, $W$ is a subspace of $V^{\perp}$. To show that $W = V^{\perp}$, let $\mathbf{u}$ be in $V^{\perp}$. By assumption, $\mathbf{u} = \mathbf{v} + \mathbf{w}$ for some $\mathbf{v}$ in $V$ and $\mathbf{w}$ in $W$. Hence $0 = \langle \mathbf{u}, \mathbf{v} \rangle = \langle \mathbf{v}, \mathbf{v} \rangle + \langle \mathbf{w}, \mathbf{v} \rangle = \langle \mathbf{v}, \mathbf{v} \rangle$ and $\mathbf{v} = \mathbf{0}$. Therefore, $\mathbf{u} = \mathbf{w}$, so $\mathbf{u}$ is in $W$.

## SECTION 3.5

**1.** (a) $(0, 1)$. (d) $\frac{10}{3}(1, -1, 0, 1)$.

**3.** (a) Projection onto $V$: $(\frac{5}{3}, \frac{3}{2}, \frac{5}{3}, \frac{3}{2}, \frac{5}{3})$; projection onto $V^{\perp}$: $(-\frac{2}{3}, \frac{1}{2}, \frac{4}{3}, -\frac{1}{2}, -\frac{2}{3})$.
(b) $\sqrt{114}/6$.
(c) The projection of $\mathbf{u}$ onto $V$.

**5.** (a) $\mathbf{p} = \frac{1}{17}(11, 8, 23)$.
(b) $\sqrt{153}/17$.
(c) $\mathbf{p}$.

**8.** (a) $4/\sqrt{13}, \frac{1}{13}(5, 1)$.
(c) $5/\sqrt{2}, \frac{1}{2}(9, 3)$.
(e) $2\sqrt{21}/7, \frac{1}{7}(-1, 5, 11)$.
(g) $2\sqrt{6}/3, \frac{1}{3}(3, 1, 4, 1)$.

**10.** (a) $3/\sqrt{38}, \frac{1}{38}(61, 44, -47)$.
(c) $\frac{3}{7}, \frac{1}{49}(43, -187, -165)$.
(e) $\sqrt{3}, (1, 2, -2)$.
(g) $18/\sqrt{19}, \frac{1}{19}(35, 19, 37, 35)$.
(i) $3/\sqrt{7}, \frac{1}{7}(22, 24, 11, 7, 0, -10)$.

**11.** (a)
$$\frac{1}{42}\begin{bmatrix} 41 & 2 & 1 & 6 \\ 2 & 38 & -2 & -12 \\ 1 & -2 & 41 & -6 \\ 6 & -12 & -6 & 6 \end{bmatrix}.$$
(b) $\frac{1}{42}(41, 2, 1, 6)$ and $\frac{1}{42}(43, 40, -1, -6)$.
(c) $V^{\perp}$ = the line through the origin spanned by $(-1, 2, 1, 6)$.

(d) $C(P) = V$.

12. (a) $\frac{1}{3}\begin{bmatrix} 2 & -1 & 1 \\ -1 & 2 & 1 \\ 1 & 1 & 2 \end{bmatrix}$.

(c) $\frac{1}{4}\begin{bmatrix} 3 & -1 & -1 & -1 \\ -1 & 3 & -1 & -1 \\ -1 & -1 & 3 & -1 \\ -1 & -1 & -1 & 3 \end{bmatrix}$.

19. (a) $P^2 = [A(A^TA)^{-1}A^T][A(A^TA)^{-1}A^T]$
$= A(A^TA)^{-1}(A^TA)(A^TA)^{-1}A^T$
$= A(A^TA)^{-1}A^T = P$.

25. Suppose that $V$ is the set of all vectors $\mathbf{x}$ such that $A\mathbf{x} = \mathbf{x}$ and that $V = N(A)^\perp$. We must show that for any $\mathbf{x}$ in $R^n$, $A\mathbf{x}$ is in $V$ and $\mathbf{x} - A\mathbf{x}$ is in $V^\perp$. Since $V = N(A)^\perp$, $V^\perp = N(A)$. Given any $\mathbf{x}$ in $R^n$, $\mathbf{x} = \mathbf{v} + \mathbf{v}^\perp$ where $\mathbf{v}$ is in $V$ and $\mathbf{v}^\perp$ is in $V^\perp$ (Theorem 1). Therefore, $A\mathbf{x} = A(\mathbf{v} + \mathbf{v}^\perp) = A\mathbf{v} + A\mathbf{v}^\perp = A\mathbf{v} = \mathbf{v}$ and $\mathbf{x} - A\mathbf{x} = (\mathbf{v} + \mathbf{v}^\perp) - \mathbf{v} = \mathbf{v}^\perp$.

## SECTION 3.6

1. (a) $\bar{\mathbf{x}} = (\frac{5}{6}, \frac{1}{2})$.
(b) $\bar{\mathbf{x}} = x_3(-4, 2, 1) + (\frac{3}{17}, \frac{3}{17}, 0)$.

2. (a) $y = 0.8 + 2.4t$.
(b) $y = -\frac{13}{10} + \frac{11}{10}t + \frac{1}{2}t^2$.
(c) $y = \frac{1}{6}t - t^2 + \frac{5}{6}t^3$.
(d) $y = -\frac{5}{3} + \frac{10}{9}t^2$.

4. (a) $y = (8 + x)/3$.

5. (a) $y = 1 + \frac{3}{2}x + \frac{3}{2}x^2$.

6. (a) $y = -\frac{4}{5} - \frac{2}{5}x_1 - \frac{37}{15}x_2$.
(b) $y = -\frac{2}{3}x_1 + \frac{5}{3}x_2$.

13. $\log\alpha = \frac{7}{6}$, $\log\beta = \frac{11}{2}$.

## SECTION 3.7

1. (a) $\bar{\mathbf{x}} = (\frac{4}{3}, \frac{1}{2})$.
(c) $\bar{\mathbf{x}} = (\frac{1}{3}, 0)$.
(e) $\bar{\mathbf{x}} = (\frac{3}{25}, -\frac{6}{25})$.
(g) $\bar{\mathbf{x}} = (-\frac{2}{25}, -\frac{2}{25}, \frac{1}{25})$.
(i) $\bar{\mathbf{x}} = (1, \frac{1}{3}, \frac{13}{12})$.

2. $(6, 5, 0, -7) = (1, 1, 1, 1) + 2(1, -1, 1, -1) + 3(1, 2, -1, -2)$.

5. $(1, 1, 0, 0)$ is not in the subspace because the system

$$\begin{bmatrix} 1 & 1 & 1 & : & 1 \\ 1 & -1 & 2 & : & 1 \\ 1 & 1 & -1 & : & 0 \\ 1 & -1 & -2 & : & 0 \end{bmatrix}$$

is inconsistent. The projection is $\mathbf{p} = \frac{1}{2}(1, 1, 1, 1) + \frac{3}{10}(1, 2, -1, -2)$.

7. $y = 1 - t$.

10. (a) Use Theorem 2 to expand $\mathbf{u}$ and $\mathbf{v}$ in terms of the orthonormal basis.

## SECTION 3.8

1. (a) $(1, 1, -1), (-1, 2, 1), (1, 0, 1)$.
(c) $(1, 2, 3), (-2, 1, 0), (-\frac{3}{5}, -\frac{6}{5}, 1)$.

3. (a) $(1, 1, 0, 0), (-\frac{1}{2}, \frac{1}{2}, 1, 0), (-\frac{2}{3}, \frac{2}{3}, -\frac{2}{3}, 1)$.

4. (b) $(1, 1, 1), (-2, 1, 1), (0, -1, 1)$.
(d) $(-1, 2, 0, 2), (0, 0, 1, 0), (0, 1, 0, -1)$.

6. (a) $1/\sqrt{2}(1, 1, 0, 0)$, $(0, 0, 1, 0)$, $(0, 0, 0, 1)$.

(b) $P = \begin{bmatrix} \frac{1}{2} & \frac{1}{2} & 0 & 0 \\ \frac{1}{2} & \frac{1}{2} & 0 & 0 \\ 0 & 0 & 1 & 0 \\ 0 & 0 & 0 & 1 \end{bmatrix}$.

(c) $\mathbf{p} = (\frac{1}{2}, \frac{1}{2}, 0, 0)$.

**15.** (b) $Q = \begin{bmatrix} 1/\sqrt{3} & -2/\sqrt{6} & 0 \\ 1/\sqrt{3} & 1/\sqrt{6} & -1/\sqrt{2} \\ 1/\sqrt{3} & 1/\sqrt{6} & 1/\sqrt{2} \end{bmatrix}$,

$R = \begin{bmatrix} \sqrt{3} & 2/\sqrt{3} & 1/\sqrt{3} \\ 0 & \sqrt{2}/\sqrt{3} & 1/\sqrt{6} \\ 0 & 0 & 1/\sqrt{2} \end{bmatrix}$.

(d) $Q = \begin{bmatrix} -\frac{1}{3} & 0 & 0 \\ \frac{2}{3} & 0 & 1/\sqrt{2} \\ 0 & 1 & 0 \\ \frac{2}{3} & 0 & -1/\sqrt{2} \end{bmatrix}$,

$R = \begin{bmatrix} 3 & -6 & 3 \\ 0 & 1 & 1 \\ 0 & 0 & \sqrt{2} \end{bmatrix}$.

## SUPPLEMENTARY PROBLEMS

**1.** (a) True. $C(P) = V$, and rank $P = \dim C(P) = \dim V$.
(c) False. $\dim V + \dim V^{\perp} = m$.
(e) True. The columns of $P$ span $V$.
(g) False. It is true if and only if $\mathbf{b}$ is in $V^{\perp}$, in which case $P\mathbf{b} = \mathbf{0}$.
(i) True. Every vector is orthogonal to $\mathbf{0}$.

**2.** (a) True. If $V = C(A)$ and $\mathbf{b}$ is any vector in $R^m$, then Theorem 1 of Section 3.5 guarantees that the projection $\mathbf{p}$ of $\mathbf{b}$ onto $V$ exists. Now apply Theorem 3 of Section 3.5.
(c) True. $P$ is the projection matrix for the projection onto $C(A)$. Therefore, $C(P) = C(A)$ and rank $P = \dim C(P) = \dim C(A) = \text{rank } A$.
(e) False. $R(A)$ and $N(A)$ are orthogonal subspaces.
(g) True. Since $A^T$ is an $n \times m$ matrix with rank $m$, $N(A^T) = \{\mathbf{0}\}$. If $A^TA\mathbf{x} = A^T\mathbf{b}$, then $A^T(A\mathbf{x} - \mathbf{b}) = \mathbf{0}$ and hence $A\mathbf{x} - \mathbf{b} = \mathbf{0}$.
(i) False. If $A$ is an $m \times n$ matrix, then $A^TA$ is nonsingular if and only if rank $A = n$. An $m \times n$ matrix of rank $n$ is nonsingular if and only if $m = n$. Therefore, the statement is true if and only if $m = n$.

**3.** (a) (i) $\mathbf{p} = \frac{1}{6}(-1, 8, 17)$.
(ii) $\mathbf{p}$.
(iii) $7/\sqrt{6}$.

(c) (i) $\mathbf{p} = \frac{1}{3}(1, 1, 2)$.
(ii) $\mathbf{p}$.
(iii) $2/\sqrt{3}$.

(e) (i) $\mathbf{p} = \frac{1}{9}(4, -1, -1)$.
(ii) $\mathbf{p}$.
(iii) $5/3$.

(g) (i) $\mathbf{p} = (1, 2, 0)$.
(ii) $\mathbf{p}$.
(iii) 3.

(i) (i) $\mathbf{p} = (5, -2, -2)$.
(ii) $\mathbf{p}$.
(iii) $5/\sqrt{2}$.

(k) (i) $\mathbf{p} = \frac{1}{21}(25, 13, 17, -3)$.
(ii) $\mathbf{p}$.
(iii) $4\sqrt{2}/\sqrt{21}$.

(m) (i) $\mathbf{p} = \frac{2}{3}(2, 1, 0, 1)$.
(ii) $\mathbf{p}$.
(iii) $2/\sqrt{3}$.

**5.** $\bar{\mathbf{x}} = \frac{52}{29}$.

**16.** (a) $C(P) = U$.
(b) $N(P) = U^{\perp}$.
(c) Rank $P = 2$.

**21.** $8/\sqrt{30}$.

**23.** $2x - y + 2z - 8 = 0$.

**25.** 1.

**28.** $19/\sqrt{14}$.

# CHAPTER 4

## SECTION 4.1

**1.** (a) Subspace.
(b) Not a subspace; it does not contain the zero function and it is not closed under addition or scalar multiplication.
(c) Subspace.
(d) Subspace.
(e) Subspace.
(f) Subspace.
(g) Not a subspace; it is not closed under scalar multiplication.

**2.** (a) Not a subspace; it is not closed under scalar multiplication.
(b) Subspace.
(c) Not a subspace; it does not contain the zero function and is not closed under addition or scalar multiplication.
(d) Not a subspace; it does not contain the zero function and is not closed under addition or scalar multiplication.
(e) Subspace.
(f) Subspace.

**4.** No; it is not closed under addition.

**8.** (a) Subspace.
(b) Subspace.
(c) Not a subspace; it does not contain the zero function and is not closed under addition or scalar multiplication.

**9.** (a) Subspace.
(b) Subspace.
(c) Subspace.
(d) Not a subspace; it does not contain the zero function and is not closed under addition or scalar multiplication.

## SECTION 4.2

**3.** (b) and (d) are independent.

**4.** (a), (b), and (c) are independent.

**6.** (a) through (f) are independent.

**10.** (a) Basis: $1, 1-t$. Dim $V = 2$.
(c) Basis: $\sin x, \cos x, \sin 2x$. Dim $V = 3$.
(e) Basis: $e^x, \sin x$. Dim $V = 2$.

**11.** (a) Basis: $x,\ 3x^2 - 1,\ x^3$. Dim $V = 3$.
(c) Basis: $1, x$. Dim $V = 2$.
(e) Basis: $1 + x, x^2, x^3$. Dim $V = 3$.

**12.** (a) If $p$ is in $V$, then $p(x) = (x-2)(x-5)(ax^2+bx+c)$. Therefore, a basis for $V$ is $(x-2)(x-5)$, $x(x-2)(x-5)$, and $x^2(x-2)(x-5)$. Dim $V = 3$.
(c) Basis: $1, x$. Dim $V = 2$.
(e) Basis: $3x-2,\ 2x^2-1,\ 5x^3-2,\ 3x^4-1$. Dim $= 4$.

**17.** $e^x$ and $e^{-x}$ are independent and they are solutions of the differential equation. Since the dimension of the solution space is 2, they form a basis for the solution space.

## SECTION 4.3

**1.** (a) 1. (d) $\sqrt{\pi}/2$.

**2.** (a) $\|e^x\|^2 = (e^2 - e^{-2})/2$.

**3.** (a) $\frac{1}{3}$. (c) $\log\sqrt{2}$.

**4.** (a) For example: $3x - 2, 2x^2 - 1$.
(c) For example: $4x - 3, 5x^2 - 3$.

**7.** $6x^2 - 6x + 1$.

**9.** (a) For example: $1, x^2, -3x + 5x^3$.

(c) For example: $-1+3x^2$, $-3x+5x^3$.

(e) For example: $-3+4x+5x^2$, $-3x+5x^3$.

**10.** (a) $x^3 = \frac{3}{5}x + \frac{1}{5}(-3x+5x^3)$,
$1+x^2 = 1+x^2$,
$1+x^3 = 1+\frac{3}{5}x+\frac{1}{5}(-3x+5x^3)$.

(c) $x^3 = \frac{3}{5}x + \frac{1}{5}(-3x+5x^3)$,
$1+x^2 = \frac{4}{3} + (\frac{1}{3})(-1+3x^2)$,
$1+x^3 = 1+(\frac{3}{5})x+(\frac{1}{5})(-3x+5x^3)$.

(e) $x^3 = \frac{1}{35}[9(1+x) - 15x^2 + 3(-3+4x+5x^2) + 7(-3x+5x^3)]$.
$1+x^2 = \frac{1}{35}[20(1+x)+60x^2 - 5(-3+4x+5x^2)]$.
$1+x^3 = \frac{1}{35}[29(1+x)+10x^2 - 2(-3+4x+5x^2) + 7(-3x+5x^3)]$.

## SECTION 4.4

**1.** (b) $h(x) = \frac{7}{4}x$.

**2.** $h(x) = 2\sin x$.

**3.** $h(x) = x - \frac{1}{6}$.

**4.** $4\pi^2/3 - 4\pi\sin x + 4\cos x, \frac{1}{2}, 0$.

**5.** $\pi - \sin 2x, 4\pi^2/3 - 2\pi\sin 2x, 0$.

**6.** (a) $1, x - \frac{1}{2}, x^2 - x + \frac{1}{6}$.
(c) $x, e^x - 3x$.
(e) $x, x^2, x^3 - \frac{3}{5}x$.

**10.** $h(x) = \dfrac{3}{\pi}x + \dfrac{175}{8}\dfrac{4\pi^2-60}{5\pi^3}(x^2 - \frac{3}{5}x)$.

**11.** $h(x) = \pi$, $h(x) = 0$.

**12.** (a) $h(x) = 3x/\pi^2$.
(c) $h(x) = \frac{3}{5}x$.
(e) $h(x) = (15/2\pi^2) - (45/2\pi^4)x^2$.

**14.** $h(x) = (e - e^{-1})/2 + 3e^{-1}x$.

**16.** $h(x) = -\frac{1}{2} + e^x/(e-1)$.

**18.** $h(x) = 2/\pi$.

**22.** $(\bar{a}, \bar{b}, \bar{c}) = (0, 3/\pi^2, 0)$.

# CHAPTER 5

## SECTION 5.1

**2.** (a) $\begin{bmatrix} 0 & 1 & 0 \\ 1 & 0 & 0 \end{bmatrix}$. (c) Not linear.
(e) Not linear. (g) Not linear.

**3.** (a) $\begin{bmatrix} 2 & 1 & 0 \\ 0 & 1 & -1 \end{bmatrix}$.

(c) $\begin{bmatrix} 0 & 0 \\ 0 & 1 \\ 1 & 0 \\ 1 & 1 \end{bmatrix}$.

(e) $\begin{bmatrix} 1 & 1 & 0 \\ 0 & 1 & 1 \\ 1 & 0 & 1 \end{bmatrix}$.

**4.** (a) $(\sqrt{3}/2 - 1, \frac{1}{2} + \sqrt{3})$.
(c) $(-\sqrt{2}/2, 3\sqrt{2}/2)$.

**6.** (a) $P = \begin{bmatrix} 1 & 0 & 0 \\ 0 & \frac{1}{2} & \frac{1}{2} \\ 0 & \frac{1}{2} & \frac{1}{2} \end{bmatrix}$.

(b) $\begin{bmatrix} 1 \\ \frac{3}{2} \\ \frac{3}{2} \end{bmatrix}$.

(c) The plane spanned by $(1,0,0)$, $(0,1,1)$.

(d) $P^2$, $P^3$, $P^n$.

**9.** $\dfrac{1}{9}\begin{bmatrix} 8 & -2 & 2 \\ -2 & 5 & 4 \\ 2 & 4 & 5 \end{bmatrix}$.

**13.** The sum is $S(\mathbf{x}) = (2x_1, x_1 + x_2 - x_3)$. Its standard matrix is

$$\begin{bmatrix} 2 & 0 & 0 \\ 1 & 1 & -1 \end{bmatrix}.$$

The standard matrices for 3(a) and 3(f) are

$$\begin{bmatrix} 2 & 1 & 0 \\ 0 & 1 & -1 \end{bmatrix}$$

and

$$\begin{bmatrix} 0 & -1 & 0 \\ 1 & 0 & 0 \end{bmatrix}.$$

**15.** The composition is

$$S(\mathbf{x}) = (0, x_2 - x_3, 2x_1 + x_2, 2x_1 + 2x_2 - x_3).$$

Its standard matrix is

$$\begin{bmatrix} 0 & 0 & 0 \\ 0 & 1 & -1 \\ 2 & 1 & 0 \\ 2 & 2 & -1 \end{bmatrix}.$$

The standard matrices for 3(c) and 3(a) are

$$\begin{bmatrix} 0 & 0 \\ 0 & 1 \\ 1 & 0 \\ 1 & 1 \end{bmatrix} \quad \text{and} \quad \begin{bmatrix} 2 & 1 & 0 \\ 0 & 1 & -1 \end{bmatrix}.$$

**16.** (a) Composition:

$$S(\mathbf{x}) = (3x_1 + 3x_2 + 2x_3, x_2).$$

Standard matrix:

$$\begin{bmatrix} 3 & 3 & 2 \\ 0 & 1 & 0 \end{bmatrix}.$$

The standard matrices for 3(a) and 3(d) are

$$\begin{bmatrix} 2 & 1 & 0 \\ 0 & 1 & -1 \end{bmatrix} \quad \text{and} \quad \begin{bmatrix} 1 & 1 & 1 \\ 1 & 1 & 0 \\ 1 & 0 & 0 \end{bmatrix}.$$

(c) Composition:

$$S(\mathbf{x}) = (2x_1 + 2x_2 + x_3, 2x_1 + x_2, 2x_1 + x_2 + x_3).$$

Standard matrix:

$$\begin{bmatrix} 2 & 2 & 1 \\ 2 & 1 & 0 \\ 2 & 1 & 1 \end{bmatrix}.$$

The standard matrices for 3(e) and 3(d) are

$$\begin{bmatrix} 1 & 1 & 0 \\ 0 & 1 & 1 \\ 1 & 0 & 1 \end{bmatrix} \quad \text{and} \quad \begin{bmatrix} 1 & 1 & 1 \\ 1 & 1 & 0 \\ 1 & 0 & 0 \end{bmatrix}.$$

(e) Composition:

$$S(\mathbf{x}) = (-x_1 - x_2, x_1 + x_2 + x_3).$$

Standard matrix:

$$\begin{bmatrix} -1 & -1 & 0 \\ 1 & 1 & 1 \end{bmatrix}.$$

The standard matrices for 3(f) and 3(d) are

$$\begin{bmatrix} 0 & -1 & 0 \\ 1 & 0 & 0 \end{bmatrix} \quad \text{and} \quad \begin{bmatrix} 1 & 1 & 1 \\ 1 & 1 & 0 \\ 1 & 0 & 0 \end{bmatrix}.$$

(g) Composition:

$$S(\mathbf{x}) = (0, x_1, -x_2, x_1 - x_2).$$

Standard matrix:

$$\begin{bmatrix} 0 & 0 & 0 \\ 1 & 0 & 0 \\ 0 & -1 & 0 \\ 1 & -1 & 0 \end{bmatrix}.$$

The standard matrices for 3(c) and 3(f) are

$$\begin{bmatrix} 0 & 0 \\ 0 & 1 \\ 1 & 0 \\ 1 & 1 \end{bmatrix}$$

and

$$\begin{bmatrix} 0 & -1 & 0 \\ 1 & 0 & 0 \end{bmatrix}.$$

(i) Composition:

$$S(\mathbf{x}) = (3x_1 + 2x_2 + x_3, \\ 2x_1 + 2x_2 + x_3, x_1 + x_2 + x_3).$$

Standard matrix:

$$\begin{bmatrix} 3 & 2 & 1 \\ 2 & 2 & 1 \\ 1 & 1 & 1 \end{bmatrix}.$$

The standard matrix for 3(d) is

$$\begin{bmatrix} 1 & 1 & 1 \\ 1 & 1 & 0 \\ 1 & 0 & 0 \end{bmatrix}.$$

**17.** $T(\mathbf{x}) = \frac{1}{2}(x_1 - x_2, x_1 - x_2)$.

(*b*) $T = P \circ R$, where $R(\mathbf{x}) = (x_1, -x_2)$ and $P(\mathbf{x}) = \frac{1}{2}(x_1 + x_2, x_1 + x_2)$. The standard matrices of $P$, $R$, and $T$ are, respectively,

$$\begin{bmatrix} \frac{1}{2} & \frac{1}{2} \\ \frac{1}{2} & \frac{1}{2} \end{bmatrix}, \quad \begin{bmatrix} 1 & 0 \\ 0 & -1 \end{bmatrix},$$
$$\begin{bmatrix} \frac{1}{2} & -\frac{1}{2} \\ \frac{1}{2} & -\frac{1}{2} \end{bmatrix}.$$

(c) $(-\frac{3}{2}, -\frac{3}{2})$.

(d) The line spanned by $(1, 1)$.

(e) $(0, 0)$.

(f) The vectors on the line spanned by $(1, 1)$.

**19.** $T(\mathbf{x}) = \frac{1}{2}(x_2 - x_1, x_2 - x_1)$.

(b) $T = R \circ P$ where $R(\mathbf{x}) = (x_2, -x_1)$ and $P(\mathbf{x}) = \frac{1}{2}(x_1 - x_2, -x_1 + x_2)$. The standard matrices of $R$, $P$, and $T$ are, respectively,

$$\begin{bmatrix} 0 & 1 \\ -1 & 0 \end{bmatrix}, \quad \begin{bmatrix} \frac{1}{2} & -\frac{1}{2} \\ -\frac{1}{2} & \frac{1}{2} \end{bmatrix},$$
$$\begin{bmatrix} -\frac{1}{2} & \frac{1}{2} \\ -\frac{1}{2} & \frac{1}{2} \end{bmatrix}.$$

(c) $(\frac{3}{2}, \frac{3}{2})$.

(d) The line spanned by $(1, 1)$.

(e) $(0, 0)$.

(f) The vectors on the line spanned by $(1, 1)$.

**20.** $T(\mathbf{x}) = \frac{1}{2}(x_1 + x_2, -x_1 - x_2)$.

(b) $T = P \circ R$, where $R(\mathbf{x}) = (x_2, -x_1)$ and $P(\mathbf{x}) = \frac{1}{2}(x_1 - x_2, -x_1 + x_2)$. The standard matrices of $P$, $R$, and $T$ are, respectively,

$$\begin{bmatrix} \frac{1}{2} & -\frac{1}{2} \\ -\frac{1}{2} & \frac{1}{2} \end{bmatrix}, \quad \begin{bmatrix} 0 & 1 \\ -1 & 0 \end{bmatrix},$$
$$\begin{bmatrix} \frac{1}{2} & \frac{1}{2} \\ -\frac{1}{2} & -\frac{1}{2} \end{bmatrix}.$$

(c) $(\frac{13}{2}, -\frac{13}{2})$.

(d) The line spanned by $(1, -1)$.

(e) $(0, 0)$.

(f) The vectors on the line spanned by $(1, -1)$.

**22.** $T(\mathbf{x}) = \frac{1}{10}((2\sqrt{3} - 1)x_1 - (2 + \sqrt{3})x_2, \\ (4\sqrt{3} - 2)x_1 - (4 + 2\sqrt{3})x_2)$.

(b) $T = P \circ R$, where

$$R(\mathbf{x}) = \tfrac{1}{2}(-x_1 - \sqrt{3}x_2, \sqrt{3}x_1 - x_2)$$

and

$$P(\mathbf{x}) = \tfrac{1}{5}(x_1 + 2x_2, 2x_1 + 4x_2).$$

The standard matrices of $P$, $R$, and $T$ are, respectively,

$$\begin{bmatrix} \frac{1}{5} & \frac{2}{5} \\ \frac{2}{5} & \frac{4}{5} \end{bmatrix}, \quad \begin{bmatrix} -\frac{1}{2} & -\sqrt{3}/2 \\ \sqrt{3}/2 & -\frac{1}{2} \end{bmatrix},$$
$$\frac{1}{10}\begin{bmatrix} -1 + 2\sqrt{3} & -2 - \sqrt{3} \\ -2 + 4\sqrt{3} & -4 - 2\sqrt{3} \end{bmatrix}.$$

(c) $(2\sqrt{3} - 21, 4\sqrt{3} - 42)$.

(d) The line spanned by $(1, 2)$.

(e) $(0, 0)$.

(f) The vectors on the line spanned by $(8 + 5\sqrt{3}, 11)$.

**25.** The standard matrices of $S$ and $T$ are:

$$\frac{1}{3}\begin{bmatrix} 2 & -1 & -1 \\ -1 & 2 & -1 \\ -1 & -1 & 2 \end{bmatrix}$$

and

$$\frac{1}{3}\begin{bmatrix} 2 & 1 & -1 \\ 1 & 2 & 1 \\ -1 & 1 & 2 \end{bmatrix}.$$

The standard matrices of $S \circ T$ and $T \circ S$ are

$$\frac{1}{9}\begin{bmatrix} 4 & -1 & -5 \\ 1 & 2 & 1 \\ -5 & -1 & 4 \end{bmatrix}$$

and

$$\frac{1}{9}\begin{bmatrix} 4 & 1 & -5 \\ -1 & 2 & -1 \\ -5 & 1 & 4 \end{bmatrix}.$$

**31.** (a) $T(x_1, x_2, x_3) = (x_1 \cos\theta - x_2 \sin\theta, x_1 \sin\theta + x_2 \cos\theta, x_3)$. Thus the $x_1$ and $x_2$ coordinates are rotated through an angle $\theta$ and the $x_3$ coordinate remains fixed.

**32.** (a) The standard matrices for $S$ and $T$ are

$$\begin{bmatrix} \frac{1}{2} & \frac{1}{2} & 0 \\ \frac{1}{2} & \frac{1}{2} & 0 \\ 0 & 0 & 1 \end{bmatrix}$$

and

$$\begin{bmatrix} 1 & 0 & 0 \\ 0 & \cos\theta & -\sin\theta \\ 0 & \sin\theta & \cos\theta \end{bmatrix}.$$

The standard matrices of $S \circ T$ and $T \circ S$ are

$$\frac{1}{2}\begin{bmatrix} 1 & \cos\theta & -\sin\theta \\ 1 & \cos\theta & -\sin\theta \\ 0 & 2\sin\theta & 2\cos\theta \end{bmatrix}$$

and

$$\frac{1}{2}\begin{bmatrix} 1 & 1 & 0 \\ \cos\theta & \cos\theta & -2\sin\theta \\ \sin\theta & \sin\theta & 2\cos\theta \end{bmatrix}.$$

**38.** (b) The columns of the standard matrix are $T(\mathbf{e}_1), T(\mathbf{e}_2), \ldots, T(\mathbf{e}_n)$. Since $\langle T(\mathbf{e}_i), T(\mathbf{e}_j)\rangle = \langle \mathbf{e}_i, \mathbf{e}_j\rangle$, it follows that these vectors are orthonormal. Finally, since there are $n$ of them, they form a basis for $R^n$.

**41.** If $P$ is the projection matrix for $V$, then $A = 2P - I$, $P^T = P$, and $P^2 = P$ (see Problem 19 in Section 3.5).

(a) $A$ is symmetric because

$$\begin{aligned} A^T &= (2P - I)^T = 2P^T - I^T \\ &= 2P - I = A. \end{aligned}$$

$A$ is orthogonal because

$$\begin{aligned} A^T A &= A^2 = (2P - I)^2 \\ &= 4P^2 - 4P + I \\ &= 4P - 4P + I = I. \end{aligned}$$

(b)
$$\begin{aligned} \langle T(\mathbf{x}), T(\mathbf{y})\rangle &= \langle 2P\mathbf{x} - \mathbf{x}, 2P\mathbf{y} - \mathbf{y}\rangle \\ &= 4\langle P\mathbf{x}, P\mathbf{y}\rangle \\ &\quad - 2\langle P\mathbf{x}, \mathbf{y}\rangle \\ &\quad - 2\langle \mathbf{x}, P\mathbf{y}\rangle + \langle \mathbf{x}, \mathbf{y}\rangle \\ &= 4\langle P^T P\mathbf{x}, \mathbf{y}\rangle \\ &\quad - 2\langle P\mathbf{x}, \mathbf{y}\rangle \\ &\quad - 2\langle P^T\mathbf{x}, \mathbf{y}\rangle + \langle \mathbf{x}, \mathbf{y}\rangle \\ &= 4\langle P^2\mathbf{x}, \mathbf{y}\rangle - 4\langle P\mathbf{x}, \mathbf{y}\rangle \\ &\quad + \langle \mathbf{x}, \mathbf{y}\rangle \\ &= 4\langle P\mathbf{x}, \mathbf{y}\rangle - 4\langle P\mathbf{x}, \mathbf{y}\rangle \\ &\quad + \langle \mathbf{x}, \mathbf{y}\rangle \\ &= \langle \mathbf{x}, \mathbf{y}\rangle. \end{aligned}$$

(c) $A^2 = I$.

**44.** (c) Let $B$ be the standard matrix of $T^{-1}$. Then the standard matrix of $T \circ T^{-1}$ is $AB$ and that of $T^{-1} \circ T$ is $BA$. Since $T \circ T^{-1} = T^{-1} \circ T = I$, $AB = BA = I$. Therefore, $B = A^{-1}$.

**46.** (a) $S$ is affine if and only if there is a linear transformation $T$ and a vector $\mathbf{b}$ such that $S = T + \mathbf{b}I$, and this holds if and only if $T = S - \mathbf{b}I$.

## SECTION 5.2

**2.** (a) (1, 2, 0), (1, 3, 1). (c) (1, 4, 2), (1, 3, 1).

**3.** (a) (1, 0), (0, 1). (c) (1, 2), (1, 1).

**4.** (b) and (d).

**6.** (a) Rank $T = 3$, nullity $T = 1$.
(b) (1, 1, 1), (0, 1, 0).

**8.** (a) $(14, -13, 5, 0), (3, -1, 0, 5)$.
(b) (3, 2), (4, 1).
(c) $\mathbf{x} = (-1, 1, 0, 0) + x_3(14, -13, 5, 0) + x_4(3, -1, 0, 5)$.

**11.** (a) (i) Ker $T$ is the line spanned by (1, 1). $T(R^2)$ is also the line spanned by (1, 1).
(ii) If $\mathbf{v} \neq \mathbf{0}$ is on $L$, then an equation of $L$ is $\mathbf{x} = \alpha\mathbf{v}$. Thus $T(L)$ consists of all vectors of the form $\mathbf{y} = \alpha T(\mathbf{v})$. If $T(\mathbf{v}) \neq \mathbf{0}$, then $T(L)$ is a line and hence is equal to $T(R^2)$. If $T(\mathbf{v}) = \mathbf{0}$, then $T(L) = \{\mathbf{0}\}$.
(c) (i) Ker $T$ is the line spanned by $(1, -1)$. $T(R^2)$ is also the line spanned by $(1, -1)$.
(ii) If $\mathbf{v} \neq \mathbf{0}$ is on $L$, then an equation of $L$ is $\mathbf{x} = \alpha\mathbf{v}$. Thus $T(L)$ consists of all vectors of the form $\mathbf{y} = \alpha T(\mathbf{v})$. If $T(\mathbf{v}) \neq \mathbf{0}$, then $T(L)$ is a line and hence is equal to $T(R^2)$. If $T(\mathbf{v}) = \mathbf{0}$, then $T(L) = \{\mathbf{0}\}$.

**13.** (a) (i) $(-1, -1, 1, 1)$.
(ii) (1, 0, 0, 0), (1, 1, 0, 0), (1, 1, 1, 1).
(iii) $x_3 - x_4 = 0$.
(c) (i) $(-2, 1, 0, 0, 0, 0, 0)$, $(3, 0, -9, 1, 4, 0, 0)$, $(-5, 0, -5, -7, 0, 8, 0)$, $(15, 0, -9, -11, 0, 0, 8)$.
(ii) (1, 2, 1, 1), (3, 3, 1, 6), (4, 1, 2, 3).
(iii) $2x_1 + x_2 - 3x_3 - x_4 = 0$.

**14.** (d) Let $L_1$ and $L_2$ be two parallel lines represented by $\mathbf{x} = \mathbf{a} + \alpha\mathbf{u}$ and $\mathbf{x} = \mathbf{b} + \beta\mathbf{u}$, $\mathbf{u} \neq \mathbf{0}$, respectively. Then $T(L_1)$ and $T(L_2)$ consists of all vectors of the form $T(\mathbf{a}) + \alpha T(\mathbf{u})$ and $T(\mathbf{b}) + \beta T(\mathbf{u})$, respectively. Since $T$ is one-to-one, $T(\mathbf{u}) \neq \mathbf{0}$ because $\mathbf{u} \neq \mathbf{0}$ and $T(\mathbf{a}) \neq T(\mathbf{b})$ if $\mathbf{a} \neq \mathbf{b}$.

**15.** (a) $T(\mathbf{x}) = (x_1 - 2x_3 - 4x_4, 3x_1 - 2x_2 - 2x_4, x_2 - 3x_3 - 5x_4)$.

**16.** $\begin{bmatrix} -36 & 21 \\ -70 & 41 \end{bmatrix}$.

**18.** $\begin{bmatrix} 1 & 0 \\ 1 & 1 \\ 1 & 0 \end{bmatrix}$.

**20.** Since $A$ is nonsingular, ker $T = N(A) = \{\mathbf{0}\}$ and $T(R^n) = C(A) = R^n$.

**22.** If $U$ is an affine subspace of $R^n$, then $U = \mathbf{a} + V$ for some vector $\mathbf{a}$ and some subspace $V$ of $R^n$. Therefore, $S(U) = S(\mathbf{a}) + S(V)$ and since $V$ is a subspace, $S(V)$ is a subspace. Thus $S(U)$ is an affine subspace.

**25.** (a) If $\mathbf{v}$ is any vector in $R^n$, then $\mathbf{v} = \alpha_1\mathbf{v}_1 + \alpha_2\mathbf{v}_2 + \cdots + \alpha_n\mathbf{v}_n$. Thus

$$\begin{aligned} S(\mathbf{v}) &= \alpha_1 S(\mathbf{v}_1) + \alpha_2 S(\mathbf{v}_2) \\ &\quad + \cdots + \alpha_n S(\mathbf{v}_n) \\ &= \alpha_1 T(\mathbf{v}_1) + \alpha_2 T(\mathbf{v}_2) \\ &\quad + \cdots + \alpha_n T(\mathbf{v}_n) = T(\mathbf{v}). \end{aligned}$$

(b) Of course. Simply let $S$ and $T$ be given by two $m \times n$ matrices having the same column space but different nullspaces, for example,

$$\begin{bmatrix} 1 & 0 & 1 \\ 0 & 1 & 0 \end{bmatrix} \quad \text{and} \quad \begin{bmatrix} 1 & 0 & 2 \\ 0 & 1 & 1 \end{bmatrix}.$$

**31.** (a) $\begin{bmatrix} 1 & \alpha \\ 0 & 1 \end{bmatrix}$.

(b) Any horizontal line L can be represented by an equation of the form $\mathbf{x} = \mathbf{a} + \beta\mathbf{e}_1$. Since $T(\mathbf{e}_1) = \mathbf{e}_1$, $T(\mathbf{x}) = T(\mathbf{a}) + \beta T(\mathbf{e}_1) = T(\mathbf{a}) + \beta\mathbf{e}_1$. Thus the image of $L$ is the horizontal line through $T(\mathbf{a})$. Since $T(\mathbf{a}) = T(a_1, a_2) = (a_1 + \alpha a_2, a_2)$, $T(\mathbf{a})$ lies on $L$ and hence $T(L) = L$.

(c) $\|T(\mathbf{x}) - \mathbf{x}\| = |\alpha| \cdot |x_2|$

(d) If $1 + \alpha m = 0$, the image is a vertical line. If $1 + \alpha m \neq 0$, then the image is $x_2 = (mx_1 + b)/(1 + \alpha m)$.

(e) The parallelogram with vertices $(0,0)$, $(1,0)$, $(1 + \alpha, 1)$, $(\alpha, 1)$.

(f) Multiples of $(1,0)$.

## SECTION 5.3

**2.** (a) $(1,-1,2)$. (b) $(2,-1,2)$.

**3.** $\begin{bmatrix} -1 & 0 & -\frac{12}{8} \\ 1 & \frac{14}{8} & \frac{7}{8} \end{bmatrix}$.

**4.** $\begin{bmatrix} 2 & 3 & 6 \\ -2 & -1 & -1 \end{bmatrix}$.

**8.** (a) Basis for ker $T$: $(-4,6,3)$.

(b) $\begin{bmatrix} -5 & -7 & -5 \\ 8 & 14 & 7 \end{bmatrix}$.

(c) Basis for $N(A)$: $(-21,5,14)$.

(d) The vectors in $N(A)$ are the coordinate vectors with respect to the given basis of $R^3$ of the vectors in ker $T$: For if $\mathbf{v}$ is in ker $T$, then

$$\mathbf{v} = \alpha(-4,6,3) = \frac{\alpha}{4}[-21(1,0,1) + 5(1,2,1) + 14(0,1,2)],$$

so that $\alpha/4(-21,5,14)$ is the coordinate vector of $\mathbf{v}$ with respect to the given basis.

**9.** $T(\mathbf{x}) = (3x_1 + 2x_2 + x_3, x_2 - x_3)$; standard matrix: $\begin{bmatrix} 3 & 2 & 1 \\ 0 & 1 & -1 \end{bmatrix}$.

**11.** (a) $\begin{bmatrix} 1 & 2 & 1 \\ 1 & 0 & 1 \\ 0 & 1 & 1 \end{bmatrix}$.

**14.** (a) First suppose that $T$ is one-to-one. If $A\mathbf{u} = \mathbf{0}$, then letting $\mathbf{x}$ be the unique vector such that $\mathbf{c}_{\mathbf{v}}(\mathbf{x}) = \mathbf{u}$, we have $\mathbf{0} = A\mathbf{u} = A\mathbf{c}_{\mathbf{v}}(\mathbf{x}) = T(\mathbf{x})$. Since $T$ is one-to-one, $\mathbf{x} = \mathbf{0}$ and hence $\mathbf{u} = \mathbf{0}$. Therefore, $N(A) = \{\mathbf{0}\}$ and rank $A = n$. Conversely, suppose that rank $A = n$. If $T(\mathbf{x}) = \mathbf{0}$, then $A\mathbf{c}_{\mathbf{v}}(\mathbf{x}) = \mathbf{0}$. Since $N(A) = \{\mathbf{0}\}$, it follows that $\mathbf{c}_{\mathbf{v}}(\mathbf{x}) = \mathbf{0}$ and hence that $\mathbf{x} = \mathbf{0}$. Thus $T$ is one-to-one.

## SECTION 5.4

**2.** (a) $\begin{bmatrix} \sqrt{2}/2 & -\sqrt{2}/2 \\ \sqrt{2}/2 & \sqrt{2}/2 \end{bmatrix}$.

(b) Same matrix as part (a).

**3.** $A = M^{-1}BM$, where

$$A = \begin{bmatrix} 1 & 1 & 1 \\ 1 & -1 & 3 \\ 0 & 1 & 0 \end{bmatrix}, \quad B = \begin{bmatrix} 2 & 1 & 0 \\ 0 & 0 & 1 \\ 0 & 1 & -2 \end{bmatrix}, \quad M = \frac{1}{2}\begin{bmatrix} 1 & 0 & -1 \\ -1 & 0 & 3 \\ -1 & 2 & -1 \end{bmatrix}.$$

**4.** $A = M^{-1}BM$, where

$$A = \begin{bmatrix} 1 & 1 & 1 \\ 1 & 1 & 0 \\ 0 & 0 & 1 \end{bmatrix}, \quad B = \begin{bmatrix} 2 & 1 & 1 \\ 0 & 1 & 1 \\ 0 & 0 & 0 \end{bmatrix}, \quad M = \frac{1}{2}\begin{bmatrix} 1 & 1 & -1 \\ 1 & -1 & 1 \\ -1 & 1 & 1 \end{bmatrix}.$$

**7.** $(1,2)$, $(-2,1)$.

**10.** (a) Basis: (0, 1, 0), (1, 0, 0), (0, 0, 1).

$$M = \begin{bmatrix} 0 & 1 & 0 \\ 1 & 0 & 0 \\ 0 & 0 & 1 \end{bmatrix}.$$

**11.** $A = \begin{bmatrix} 7 & -8 & -4 \\ 10 & -11 & -5 \end{bmatrix}$,

$B = \begin{bmatrix} 1 & 1 & 2 \\ 1 & -1 & -1 \end{bmatrix}$,

$M_1 = \begin{bmatrix} -3 & 2 \\ 2 & -1 \end{bmatrix}$,

$M_2 = \begin{bmatrix} 2 & -2 & -1 \\ -1 & 2 & 1 \\ -1 & 1 & 1 \end{bmatrix}$.

**13.** Because the $\mathbf{w}_j$'s form an orthonormal basis,

$$T(\mathbf{v}_i) = \langle T(\mathbf{v}_i), \mathbf{w}_1 \rangle \mathbf{w}_1 + \langle T(\mathbf{v}_i), \mathbf{w}_2 \rangle \mathbf{w}_2 + \cdots + \langle T(\mathbf{v}_i), \mathbf{w}_n \rangle \mathbf{w}_n.$$

Therefore, the $ij$th entry of the matrix is $\langle T(\mathbf{v}_j), \mathbf{w}_i \rangle$.

**14.** $\begin{bmatrix} 0 & 0 & 0 & \cdots & 0 & a_1 \\ 1 & 0 & 0 & \cdots & 0 & a_2 \\ 0 & 1 & 0 & \cdots & 0 & a_3 \\ \vdots & \vdots & \vdots & \vdots & \vdots & \vdots \\ 0 & 0 & 0 & \cdots & 1 & a_n \end{bmatrix}$.

**18.** (a) If $B$ is any matrix representing $T$, then $T(R^n) = C(B)$. Therefore, rank $T = \dim T(R^n) = \dim C(B) =$ rank $B$.

**19.** (g) Suppose that $A = M^{-1}BM$. If $A$ is idempotent, then $A^2 = A$ and hence $(M^{-1}BM)^2 = M^{-1}BM$. Since $(M^{-1}BM)^2 = M^{-1}B^2M$, it follows that $B^2 = B$. On the other hand, if $B$ is idempotent, then $B^2 = B$ and hence

$$A^2 = (M^{-1}BM)^2 = M^{-1}B^2M = M^{-1}BM = A.$$

**20.** The matrix of $T$ with respect to this basis will be an $m \times m$ matrix whose $(n+1)$st through $m$th rows have their first $n$ entries equal to zero.

**21.** (a) Suppose that $T$ is nilpotent with index of nilpotency $k$. If $T(\mathbf{v}) = \alpha\mathbf{v}$, then $\mathbf{0} = T^k(\mathbf{v}) = \alpha^k\mathbf{v}$. Therefore, $\alpha = 0$ or $\mathbf{v} = \mathbf{0}$.

## SECTION 5.5

**1.** (a) (i) $A = \begin{bmatrix} 7 & 6 \\ -9 & -8 \end{bmatrix}$.

(ii) $T(1, -1) = (1, -1) = 1 \cdot (1, -1)$.
$T(-2, 3) = (4, -6) = -2 \cdot (-2, 3)$.

(iii) Basis: $(1, -1), (-2, 3)$;
matrix: $\begin{bmatrix} 1 & 0 \\ 0 & -2 \end{bmatrix}$.

(iv) $M = \begin{bmatrix} 1 & -2 \\ -1 & 3 \end{bmatrix}$.

(c) (i) $A = \begin{bmatrix} 3 & 1 \\ 0 & 4 \end{bmatrix}$.

(ii) $T(1, 0) = (3, 0) = 3 \cdot (1, 0)$.
$T(1, 1) = (4, 4) = 4 \cdot (1, 1)$.

(iii) Basis: $(1, 0), (1, 1)$;
matrix: $\begin{bmatrix} 3 & 0 \\ 0 & 4 \end{bmatrix}$.

(iv) $M = \begin{bmatrix} 1 & 1 \\ 0 & 1 \end{bmatrix}$.

(e) (i) $A = \begin{bmatrix} 3 & -1 & 0 \\ -1 & 2 & -1 \\ 0 & -1 & 3 \end{bmatrix}$.

(ii) $T(1, 2, 1) = (1, 2, 1) = 1 \cdot (1, 2, 1)$.
$T(1, 0, -1) = (3, 0, -3) = 3 \cdot (1, 0, -1)$.

$T(1,-1,1)=(4,-4,4)$
$= 4\cdot(1,-1,1).$

(iii) Basis: $(1,2,1)$, $(1,0,-1)$,

$(1,-1,1)$; matrix: $\begin{bmatrix}1&0&0\\0&3&0\\0&0&4\end{bmatrix}$.

(iv) $M=\begin{bmatrix}1&1&1\\2&0&-1\\1&-1&1\end{bmatrix}$.

(g) (i) $A=\begin{bmatrix}1&-2&2\\0&-1&2\\0&0&1\end{bmatrix}$.

(ii) $T(1,1,0)=(-1,-1,0)$
$= -1\cdot(1,1,0).$
$T(1,0,0)=(1,0,0)$
$= 1\cdot(1,0,0).$
$T(0,1,1)=(0,1,1)$
$= 1\cdot(0,1,1).$

(iii) Basis: $(1,1,0)$, $(1,0,0)$, $(0,1,1)$;

matrix: $\begin{bmatrix}-1&0&0\\0&1&0\\0&0&1\end{bmatrix}$.

(iv) $M=\begin{bmatrix}1&1&0\\1&0&1\\0&0&1\end{bmatrix}$.

**2.** (a) Eigenvectors: $(1,1,1)$, $(2,0,1)$, $(1,0,0)$. Eigenvalues: 0, 2.

(i) $A=\begin{bmatrix}0&-2&4\\0&2&0\\0&0&2\end{bmatrix}$.

(ii) $T(1,1,1)=(2,2,2)=2\cdot(1,1,1)$
$T(2,0,1)=(4,0,2)=2\cdot(2,0,1)$
$T(1,0,0)=(0,0,0)=0\cdot(1,0,0)$

(iii) Basis: $(1,1,1)$, $(2,0,1)$, $(1,0,0)$;

matrix: $\begin{bmatrix}2&0&0\\0&2&0\\0&0&0\end{bmatrix}$.

(iv) $M=\begin{bmatrix}1&2&1\\1&0&0\\1&1&0\end{bmatrix}$.

**3.** If $A$ is the standard matrix of $T$, $M$ is the matrix having the given eigenvectors as its columns, and $\Lambda$ is the diagonal matrix whose $i$th diagonal entry is the eigenvalue associated with the $i$th column of $M$, then $\Lambda = M^{-1}AM$. Therefore, $A = M\Lambda M^{-1}$.

(a) $T(x,y,z)=(x+z, y+z, 2z)$.

(i) $A=\begin{bmatrix}1&0&1\\0&1&1\\0&0&2\end{bmatrix}$.

(ii) $T(1,0,0)=(1,0,0)=1\cdot(1,0,0).$
$T(1,1,0)=(1,1,0)=1\cdot(1,1,0).$
$T(1,1,1)=(2,2,2)=2\cdot(1,1,1).$

(iii) Basis: $(1,0,0)$, $(1,1,0)$, $(1,1,1)$;

matrix: $\begin{bmatrix}1&0&0\\0&1&0\\0&0&2\end{bmatrix}$.

(iv) $M=\begin{bmatrix}1&1&1\\0&1&1\\0&0&1\end{bmatrix}$.

(c) $T(x,y,z)=(-18x-y+5z, -39x-8y+15z, -61x-13y+24z)$.

(i) $A=\begin{bmatrix}-18&-1&5\\-39&-8&15\\-61&-13&24\end{bmatrix}$.

(ii) $T(1,2,3)=(-5,-10,-15)$
$= -5\cdot(1,2,3).$
$T(1,3,4)=(-1,-3,-4)$
$= -1\cdot(1,3,4).$
$T(1,3,5)=(4,12,20)$
$= 4\cdot(1,3,5).$

(iii) Basis: $(1,2,3)$, $(1,3,4)$, $(1,3,5)$;

matrix: $\begin{bmatrix}-5&0&0\\0&-1&0\\0&0&4\end{bmatrix}$.

(iv) $M=\begin{bmatrix}1&1&1\\2&3&3\\3&4&5\end{bmatrix}$.

**5.** Show that $T(x,y)=\lambda(x,y)$ has a nontrivial solution if and only if $\lambda = 2$ or $\lambda = 4$. Basis: $(1, 3), (1, 1)$; matrix: $\begin{bmatrix} 2 & 0 \\ 0 & 4 \end{bmatrix}$.

**6.** (a) A rotation about an axis leaves every vector on the axis fixed. Hence $T(\mathbf{v}) = \mathbf{v}$, so $\mathbf{v}$ is an eigenvector of $T$ and $\lambda = 1$ is the corresponding eigenvalue.

**7.** (a) With respect to any orthonormal basis of the form $(0, 0, 1)$, $\mathbf{v}_2$, $\mathbf{v}_3$, the matrix of $T$ is $\begin{bmatrix} 1 & 0 & 0 \\ 0 & 0 & -1 \\ 0 & 1 & 0 \end{bmatrix}$.

(c) With respect to any orthonormal basis of the form $(1, 0, 0)$, $\mathbf{v}_2$, $\mathbf{v}_3$, the matrix of $T$ is $\begin{bmatrix} 1 & 0 & 0 \\ 0 & -1 & 0 \\ 0 & 0 & -1 \end{bmatrix}$.

(e) With respect to any orthonormal basis of the form $(\frac{2}{7}, \frac{3}{7}, -\frac{6}{7})$, $\mathbf{v}_2, \mathbf{v}_3$, the matrix of $T$ is

$$\begin{bmatrix} 1 & 0 & 0 \\ 0 & \sqrt{2}/2 & -\sqrt{2}/2 \\ 0 & \sqrt{2}/2 & \sqrt{2}/2 \end{bmatrix}.$$

**8.** (a) (i) $T(1, 0, 1) = 1 \cdot (1, 0, 1)$.

(ii) $\mathbf{v}_1 = 1/\sqrt{2}(1, 0, 1)$.

(iii) $\mathbf{v}_1 = 1/\sqrt{2}(1, 0, 1)$,
$\mathbf{v}_2 = (0, 1, 0)$,
$\mathbf{v}_3 = 1/\sqrt{2}(-1, 0, 1)$.

$$\begin{bmatrix} 1 & 0 & 0 \\ 0 & -\frac{1}{3} & -2\sqrt{2}/3 \\ 0 & 2\sqrt{2}/3 & -\frac{1}{3} \end{bmatrix}.$$

(iv) Axis: $(1, 0, 1)$, $\theta = \arccos(-\frac{1}{3})$.

**11.** Choose an orthonormal basis $\mathbf{v}_1, \mathbf{v}_2, \mathbf{v}_3$ for $R^3$ with $\mathbf{v}_1$ spanning the axis of the rotation. The matrix $A$ of the rotation $T$ with respect to this basis is the orthogonal matrix given in Problem 6(c). Moreover, the change of coordinates matrix $M$ that changes $\mathbf{v}$-coordinates to standard coordinates is an orthogonal matrix because its columns are the orthonormal vectors $\mathbf{v}_1, \mathbf{v}_2, \mathbf{v}_3$. Since $M$ is orthogonal, $M^{-1} = M^T$ and

$$\begin{aligned} \langle \mathbf{c}_\mathbf{v}(\mathbf{x}), \mathbf{c}_\mathbf{v}(\mathbf{y}) \rangle &= \langle M^T\mathbf{x}, M^T\mathbf{y} \rangle \\ &= \langle MM^T\mathbf{x}, \mathbf{y} \rangle = \langle \mathbf{x}, \mathbf{y} \rangle. \end{aligned}$$

Thus

$$\begin{aligned} \|T(\mathbf{x})\|^2 &= \langle T(\mathbf{x}), T(\mathbf{x}) \rangle \\ &= \langle \mathbf{c}_\mathbf{v}(T(\mathbf{x})), \mathbf{c}_\mathbf{v}(T(\mathbf{x})) \rangle \\ &= \langle A\mathbf{c}_\mathbf{v}(\mathbf{x}), A\mathbf{c}_\mathbf{v}(\mathbf{x}) \rangle \\ &= \langle A^TA\mathbf{c}_\mathbf{v}(\mathbf{x}), \mathbf{c}_\mathbf{v}(\mathbf{x}) \rangle \\ &= \langle \mathbf{c}_\mathbf{v}(\mathbf{x}), \mathbf{c}_\mathbf{v}(\mathbf{x}) \rangle \\ &= \langle \mathbf{x}, \mathbf{x} \rangle = \|\mathbf{x}\|^2. \end{aligned}$$

**14.** (a) Suppose that $\alpha_1\mathbf{v}_1 + \alpha_2\mathbf{v}_2 = \mathbf{0}$. Then

$$\begin{aligned} T(\alpha_1\mathbf{v}_1 + \alpha_2\mathbf{v}_2) &= \alpha_1 T(\mathbf{v}_1) + \alpha_2 T(\mathbf{v}_2) \\ &= \alpha_1\lambda_1\mathbf{v}_1 + \alpha_2\lambda_2\mathbf{v}_2 \\ &= \mathbf{0}. \end{aligned}$$

Multiplying the first equation by $-\lambda_1$ and adding it to the second, we obtain $\alpha_2(\lambda_2 - \lambda_1)\mathbf{v}_2 = \mathbf{0}$. Something in this equation must be zero. $\mathbf{v}_2 \neq \mathbf{0}$ since it is an eigenvector. $\lambda_2 - \lambda_1 \neq 0$ by assumption. Therefore, $\alpha_2 = 0$. Thus the first equation reduces to $\alpha_1\mathbf{v}_1 = \mathbf{0}$ and we conclude that $\alpha_1 = 0$ because $\mathbf{v}_1 \neq \mathbf{0}$.

**19.** $T$ is invertible if and only if $\ker T = \{\mathbf{0}\}$. $\lambda = 0$ is an eigenvalue of $T$ if and only if $\ker T \neq \{\mathbf{0}\}$.

**22.** (a) Verify that the sum of two eigenvectors and a scalar multiple of an eigenvector is an eigenvector. Thus the collection of all eigenvectors fails to be a subspace simply because it fails to contain the zero vector. We can also prove the proposition by simply observing that $E(\lambda) = \ker(T - \lambda I)$.

**24.** Suppose that $\lambda$ is an eigenvalue of $T$. Then there is a nonzero vector $\mathbf{v}$ such that $T(\mathbf{v}) = \lambda\mathbf{v}$. Therefore, $\lambda\mathbf{v} = T(\mathbf{v}) = T^2(\mathbf{v}) = \lambda^2\mathbf{v}$, so $(\lambda - \lambda^2)\mathbf{v} = \mathbf{0}$. Since $\mathbf{v} \neq \mathbf{0}$, $\lambda - \lambda^2 = 0$.

# CHAPTER 6

## SECTION 6.1

**1.** Yes. **2.** No.

**9.** (b) Let $A_{ij}$ be the $m \times n$ matrix whose entry in the $i$th row and $j$th column is 1 and which has all other entries equal to zero. Then the $mn$ matrices $A_{ij}$, $i = 1, \ldots, m$, $j = 1, \ldots, n$, are a basis for $V$. Hence $\dim V = mn$.

(c) No.

**10.** (a) The matrices $A_{ij}$ [see the solution of Problem 9(b)] with $i \leq j$ are a basis for $W$. Hence $\dim W = 1 + 2 + 3 + \cdots + n = n(n+1)/2$.

(b) The matrices $A_{ii}$, $i = 1, \ldots, n$, are a basis for $U$. Hence $\dim U = n$.

(c) The dimension is $1 + 2 + 3 + \cdots + n = n(n+1)/2$.

(d) The dimension is $1 + 2 + 3 + \cdots + (n-1) = n(n-1)/2$.

**11.** (a), (d). **16.** No.

## SECTION 6.2

**2.** Let $u = (\sqrt{3} - 2, 1)$.

**3.** For example,

$$\begin{aligned}\langle \mathbf{u}, \mathbf{v} + \mathbf{w} \rangle &= u_1(v_1 + w_1) \\ &\quad + u_2(v_1 + w_1) \\ &\quad + u_1(v_2 + w_2) \\ &\quad + 2u_2(v_2 + w_2) \\ &= \langle \mathbf{u}, \mathbf{v} \rangle + \langle \mathbf{u}, \mathbf{w} \rangle.\end{aligned}$$

Also, $\langle \mathbf{u}, \mathbf{u} \rangle = u_1^2 + 2u_1u_2 + 2u_2^2$ and since the roots of the equation $\langle \mathbf{u}, \mathbf{u} \rangle = 0$ are given by $u_1 = u_2(-1 \pm \sqrt{-1})$, it follows that $\langle \mathbf{u}, \mathbf{u} \rangle \geq 0$ and $\langle \mathbf{u}, \mathbf{u} \rangle = 0$ if and only if $\mathbf{u} = \mathbf{0}$.

**7.** (a) $\pi/2$.

(b) Acute. Since $\langle (0,1), (-1,1) \rangle = 5 > 0$, $\cos\theta > 0$.

**9.** (a) The vectors on the line spanned by $(1, -9)$.

**10.** (b) Any scalar multiple of $p(x) = x - (e^2 + 1)/4$.

## SECTION 6.3

**1.** (a), (b), and (d).

**2.** (b) $\operatorname{Ker} T$ is the line in $C^\infty$ spanned by $e^x$,

(c) $T$ is not one-to-one since $\ker T \neq \{0\}$.

**4.** (a) $\begin{bmatrix} 1 & 1 & 2 & 0 \\ 0 & 2 & 6 & 6 \\ 0 & 0 & 5 & 15 \\ 0 & 0 & 0 & 10 \end{bmatrix}$.

(b) $\begin{bmatrix} 1 & & & \\ & 2 & & \\ & & 5 & \\ & & & 10 \end{bmatrix}$.

(c) $\begin{bmatrix} 1 & -1 & 1 & -1 \\ 0 & 1 & -2 & 3 \\ 0 & 0 & 1 & -3 \\ 0 & 0 & 0 & 1 \end{bmatrix}$.

**5.** (a) $\begin{bmatrix} 0 & -1 \\ 1 & 0 \end{bmatrix}$.

(c) $\begin{bmatrix} 0 & -1 & 1 & 0 \\ 1 & 0 & 0 & 1 \\ 0 & 0 & 0 & -1 \\ 0 & 0 & 1 & 0 \end{bmatrix}$.

**8.** (d) Antiderivative.

**9.** $(x-2)(x+5)$, $x(x-2)(x+5)$.

**10.** (a) $T(R^3)$ is the line spanned by $(1, 2)$;

dim $T(R^3) = 1$. Ker $T$ is the plane spanned by $(-2, 1, 0)$ and $(-5, 0, 1)$; dim ker $T = 2$. $T$ is not one-to-one or onto.

(c) $T(P_1) = V$ because any vector $\mathbf{v} = \alpha(-3, 2, 1) + \beta(3, -1, 1)$ in the plane is the image under $T$ of the polynomial $p(x) = (\alpha - 2\beta) + \beta x$. Therefore, dim $T(P_1) = 2$ and $T$ is onto. On the other hand, since $(-3, 2, 1)$ and $(3, -1, 1)$ are independent, a polynomial $p$ is in ker $T$ if and only if $p(2) = p'(5) = 0$. The only polynomial satisfying these conditions is the zero polynomial. Therefore, ker $T = \{0\}$ and $T$ is one-to-one.

**13.** (a) $\begin{bmatrix} 2 & 3 & 0 & 0 \\ 2 & 6 & -1 & 0 \\ 0 & -1 & 3 & 3 \\ 0 & 0 & 2 & 5 \end{bmatrix}$.

(b) $\begin{bmatrix} 0 & \frac{1}{2} & 1 & -\frac{3}{2} \\ 0 & -\frac{1}{2} & 1 & \frac{3}{2} \\ 0 & \frac{1}{2} & -1 & \frac{3}{2} \end{bmatrix}$.

**15.** $T(cx^n) = x(cnx^{n-1}) = n(cx^n)$.

**18.** Basis for $R^3$: $(7, 2, 0), (-3, 0, 2), (2, -7, 3)$. Basis for $V$: $(7, 2, 0), (-3, 0, 2)$.

**21.** (b) $\begin{bmatrix} 0 & 0 & 2 & 0 & 0 \\ 0 & 0 & 0 & 0 & 0 \\ 0 & 0 & 0 & 0 & 0 \\ 0 & 0 & 0 & -1 & 0 \\ 0 & 0 & 0 & 0 & -1 \end{bmatrix}$.

(e) $\alpha + \beta x - \sin x$.

**22.** (b) Since $e^x$ and $e^{-x}$ are independent, they form a basis for ker $L$.

**27.** $$\|T(f)\|^2 = \int_{-1}^{1} f^2(-x)\,dx = \int_{-1}^{1} f^2(u)\,du = \|f\|^2.$$

**30.** (d) $\begin{bmatrix} 1 & 0 & 1 \\ 0 & 0 & 0 \\ 1 & 0 & 1 \end{bmatrix}$, $\begin{bmatrix} 0 & 1 & 0 \\ 1 & 0 & 1 \\ 0 & 1 & 0 \end{bmatrix}$, $\begin{bmatrix} 0 & 0 & 0 \\ 0 & 1 & 0 \\ 0 & 0 & 0 \end{bmatrix}$.

# CHAPTER 7

## SECTION 7.1

**1.** (a) $-17$. (b) 12. (c) 4. (d) 0. (e) 0. (f) $-\frac{19}{120}$. (g) 37. (h) $-18$. (i) 113.

**4.** (a) $-ab$. (b) $-abd$. (c) $abdg$.

**5.** The determinant of the matrix is $(1 - x^3)^2$. Therefore, the matrix is singular if and only if $x = 1$.

**8.** (a) $\frac{1}{2}$. (b) $-45$. (c) 0. (d) $-\frac{6}{5}$.

**10.** $$\det(AB) = (\det A)(\det B) = (\det B)(\det A) = \det(BA).$$

**15.** If $n$ is odd, then (see Problem 9)
$$\det A = \det A^T = \det(-A) = (-1)^n \det A = -\det A,$$
hence $2 \det A = 0$, that is, $\det A = 0$.
$$A = \begin{bmatrix} 0 & 1 \\ -1 & 0 \end{bmatrix}$$
is skew-symmetric, but $\det A = 1$.

**18.** Let $r$ be the remainder obtained by dividing $n$ by 4. Then
$$\det A = (-1)^k a_{1n} a_{2,n-1} \cdots a_{n-1,2} a_{n1},$$
where $k = 0$ if $r = 0$, 1 and $k = 1$ if $r = 2, 3$.

## SECTION 7.2

**1.** (a) 206. (b) $-6$. (c) $\frac{5}{9}$.
(d) $-3$. (e) 4. (f) 60.

**3.** $\det\begin{bmatrix} 3-\lambda & 1 \\ -6 & -4-\lambda \end{bmatrix} = \lambda^2+\lambda-6 = (\lambda-2)(\lambda+3)$; $\lambda=2$, $\lambda=-3$.

**4.** (a) No real number $\lambda$. (b) $\lambda = 1$.

## SECTION 7.3

**1.** (a) $(-3,-3,5)$, area $= \sqrt{43}$.
(b) $(1,17,-11)$, area $= \sqrt{411}$.
(c) $(13,-5,-4)$, area $= \sqrt{210}$.

**2.** (a) 9. (b) 7.

**3.** (a) 5. (b) 34.

**4.** (a) 5. (b) $\sqrt{59}$.

## SECTION 7.4

**1.** (a) $\frac{1}{43}\begin{bmatrix} -3 & 10 & 2 \\ 11 & -8 & 7 \\ 9 & 13 & -6 \end{bmatrix}$.

(b) $\begin{bmatrix} 3 & 2 & 1 \\ 5 & 4 & 2 \\ 10 & 7 & 4 \end{bmatrix}$.

(c) $\frac{1}{10}\begin{bmatrix} 9 & 2 & -6 \\ -15 & 10 & 0 \\ 21 & -2 & 6 \end{bmatrix}$.

**2.** $\begin{bmatrix} \frac{1}{a} & -\frac{b}{ad} & \frac{be}{adf} - \frac{c}{af} \\ 0 & \frac{1}{d} & -\frac{e}{df} \\ 0 & 0 & \frac{1}{f} \end{bmatrix}$.

4. (a) $(\frac{23}{37}, -\frac{13}{37})$. (b) $(\frac{3}{2}, -2, -\frac{1}{2})$.

# CHAPTER 8

## SECTION 8.1

**3.** Eigenvalues: $2, -3$.

**4.** (a) Eigenvalues: $-3, 3, 6$.
(b) Eigenvalues: $-1, -1, 2$.

**6.** (a) Eigenvalues: $-1, -2, -3, -4$.
(b) Eigenvalues: $1, 1, 2, 2$.

**7.** (a) Basis for $E(0)$: $(-3,2)$;
basis for $E(8)$: $(1,2)$.
(c) Basis for $E(1)$: $(1,0,0)$;
basis for $E(2)$: $(1,0,1)$.

**9.** $A(A\mathbf{v}) = A(\lambda\mathbf{v}) = \lambda(A\mathbf{v})$.

**11.** If $\Lambda$ is a diagonal matrix with diagonal entries $\lambda_1, \lambda_2, \ldots, \lambda_n$, then $\Lambda\mathbf{e}_i = i$th column of $\Lambda = \lambda_i\mathbf{e}_i$. The eigenvalues of $\Lambda$ are $\lambda_1, \lambda_2, \ldots, \lambda_n$.

**13.** By Problem 12, $A$ is invertible if and only if 0 is not an eigenvalue of $A$. Therefore, $\lambda$ is an eigenvalue of $A$ if and only if there is a nonzero vector $\mathbf{v}$ such that $A\mathbf{v} = \lambda\mathbf{v}$, and this holds if and only if $A^{-1}\mathbf{v} = \lambda^{-1}\mathbf{v}$. Thus $\lambda$ is an eigenvalue of $A$ if and only if $\lambda^{-1}$ is an eigenvalue of $A^{-1}$ and $A$ and $A^{-1}$ have the same eigenvectors.

**17.** (a) Suppose that $\lambda$ is an eigenvalue of $A$. Then there is a nonzero vector $\mathbf{v}$ such that $A\mathbf{v} = \lambda\mathbf{v}$. Thus

$$\lambda\mathbf{v} = A\mathbf{v} = A^2\mathbf{v} = A(A\mathbf{v}) = A(\lambda\mathbf{v}) = \lambda A\mathbf{v} = \lambda^2\mathbf{v}.$$

Since $\lambda\mathbf{v} = \lambda^2\mathbf{v}$, $(\lambda - \lambda^2)\mathbf{v} = \mathbf{0}$ and since $\mathbf{v} \neq \mathbf{0}$ we conclude that $\lambda - \lambda^2 = 0$. Thus $\lambda = 0$ or 1.

(b) Since $A \neq 0$, there is a nonzero vector $\mathbf{v}$ such that $A\mathbf{v} \neq \mathbf{0}$. It follows that $A\mathbf{v}$ is an eigenvector with eigenvalue 1 because $A(A\mathbf{v}) = A^2\mathbf{v} = A\mathbf{v}$.

## SECTION 8.2

**1.** (a) $\lambda = 2$, basis $(1, 1)$; $\lambda = 3$, basis $(0, 1)$.

(c) $\lambda = -1$, basis $(1, 1)$; $\lambda = 3$, basis $(-1, 3)$.

(e) $\lambda = 0$, basis $(-1, 1, 2)$; $\lambda = 1$, basis $(2, 0, 1)$; $\lambda = 4$, basis $(1, 3, 2)$.

(g) $\lambda = 0$, basis $(-1, 0, 1)$; $\lambda = 1$, basis $(0, 1, 0)$, $(0, 0, 1)$.

(i) $\lambda = 0$, basis $(1, 0, 0)$; $\lambda = 1$, basis $(3, 2, 1)$.

(k) $\lambda = 0$, basis $(1, 1, 1)$; $\lambda = 1$, basis $(-1, 0, 1)$; $\lambda = 3$, basis $(1, -2, 1)$.

(m) $\lambda = 1$, basis $(1, -2, 2)$; $\lambda = 3$, basis $(1, 0, 0)$.

(o) $\lambda = 1$, basis $(-1, 4, 1)$; $\lambda = 3$, basis $(1, 2, 1)$; $\lambda = -2$, basis $(-1, 1, 1)$.

(q) $\lambda = 1$, basis $(1, 0, 0, 0)$; $\lambda = -2$, basis $(0, 0, 1, 0)$, $(0, 0, 0, 1)$.

**5.** Since $\lambda^2 + \lambda - 6 = (\lambda - 2)(\lambda + 3)$, the eigenvalues of $A$ are 2 and $-3$. By Problem 14 of Section 8.1 the eigenvalues of $A^2$ are 4 and 9. Therefore, the characteristic polynomial of $A^2$ is $\pm(\lambda - 4)(\lambda - 9)$.

**7.** If $A = M^{-1}BM$, then

$$\begin{aligned}\det(A - \lambda I) &= \det(M^{-1}BM - \lambda I)\\ &= \det(M^{-1}BM - \lambda M^{-1}IM)\\ &= \det[M^{-1}(B - \lambda I)M]\\ &= \det(M^{-1})\det(B - \lambda I)\det(M)\\ &= \det(B - \lambda I).\end{aligned}$$

If $\mathbf{v}$ is an eigenvector of $A$, then $M\mathbf{v}$ is an eigenvector of $B$.

## SECTION 8.3

**1.** (a) $P = \begin{bmatrix} 1 & 0 \\ 1 & 1 \end{bmatrix}$, $\Lambda = \begin{bmatrix} 2 & 0 \\ 0 & 3 \end{bmatrix}$.

(c) $P = \begin{bmatrix} 1 & -1 \\ 1 & 3 \end{bmatrix}$, $\Lambda = \begin{bmatrix} -1 & 0 \\ 0 & 3 \end{bmatrix}$.

(e) $P = \begin{bmatrix} -1 & 2 & 1 \\ 1 & 0 & 3 \\ 2 & 1 & 2 \end{bmatrix}$, $\Lambda = \begin{bmatrix} 0 & 0 & 0 \\ 0 & 1 & 0 \\ 0 & 0 & 4 \end{bmatrix}$.

(g) $P = \begin{bmatrix} -1 & 0 & 0 \\ 0 & 1 & 0 \\ 1 & 0 & 1 \end{bmatrix}$, $\Lambda = \begin{bmatrix} 0 & 0 & 0 \\ 0 & 1 & 0 \\ 0 & 0 & 1 \end{bmatrix}$.

(i) Not diagonalizable.

(k) $P = \begin{bmatrix} 1 & -1 & 1 \\ 1 & 0 & -2 \\ 1 & 1 & 1 \end{bmatrix}$, $\Lambda = \begin{bmatrix} 0 & 0 & 0 \\ 0 & 1 & 0 \\ 0 & 0 & 3 \end{bmatrix}$.

(m) Not diagonalizable.

(o) $P = \begin{bmatrix} -1 & 1 & -1 \\ 4 & 2 & 1 \\ 1 & 1 & 1 \end{bmatrix}$, $\Lambda = \begin{bmatrix} 1 & 0 & 0 \\ 0 & 3 & 0 \\ 0 & 0 & -2 \end{bmatrix}$.

(q) Not diagonalizable.

**2.** $A = M\Lambda M^{-1}$

$$= \begin{bmatrix} 1 & 1 \\ 3 & 4 \end{bmatrix}\begin{bmatrix} 1 & 0 \\ 0 & -1 \end{bmatrix}\begin{bmatrix} 4 & -1 \\ -3 & 1 \end{bmatrix}$$

$$= \begin{bmatrix} 7 & -2 \\ 24 & -7 \end{bmatrix}.$$

**5.** No. Think about the identity matrix.

**7.** No. The $i$th column of $M$ may be any eigenvector associated with the eigenvalue in the $i$th diagonal entry of $\Lambda$, as long as eigenvectors associated with repeated eigenvalues are chosen to be independent.

**8.** No. The eigenvalues can be listed in any order along the diagonal of $\Lambda$. Moreover, repeated eigenvalues do not have to be listed together.

**9.** If $A$ is diagonalizable, then there is a nonsingular matrix $M$ such that $M^{-1}AM = \lambda I$. Hence $A = \lambda I$.

**15.** $A = \lambda I$.

**16.** $\begin{bmatrix} -5\alpha + 6\beta & 3\alpha - 3\beta \\ -10\alpha + 10\beta & 6\alpha - 5\beta \end{bmatrix}$.

**17.** No. Consider the $2 \times 2$ identity $I$ and the matrix $\begin{bmatrix} 1 & 1 \\ 0 & 1 \end{bmatrix}$.

**18.** Yes.

**19.** Yes. The eigenvalues of both matrices are 1, 5, and $-2$. Call these matrices $A$ and $B$, respectively. The corresponding eigenvectors of the first matrix $A$ are $\mathbf{e}_1$, $\mathbf{e}_2$, and $\mathbf{e}_3$. Since $A$ is diagonalizable, if $M$ is the matrix whose columns are $\mathbf{e}_2$, $\mathbf{e}_1$, and $\mathbf{e}_3$ in that order, then the columns of $M$ are eigenvectors of $A$ and the diagonal entries of $B$ are the corresponding eigenvalues. Thus $B = M^{-1}AM$.

**22.** Just the identity matrix.

**23.** Suppose that $A = M^{-1}BM$. If $\mathbf{v}$ is an eigenvector of $A$, then $M\mathbf{v}$ is an eigenvector of $B$. If $\mathbf{w}$ is an eigenvector of $B$, then $M^{-1}\mathbf{w}$ is an eigenvector of $A$.

## SECTION 8.4

**1.** $x^{(k)} = (1 + r/1200)^k x^{(0)}$.

**3.** (a) Eigenvalues of $I + A$: 0.6 and 1. Corresponding eigenvectors: (1, 1) and (1, 2). The ratio $p_1^{(k)}/p_2^{(k)}$ tends to $\frac{1}{2}$.

**5.** (c) $\lambda$. (d) $\lambda > 0$.

**6.** $A = \begin{bmatrix} 0 & 0 & 0 & 1100 \\ 0.01 & 0 & 0 & 0 \\ 0 & 0.2 & 0 & 0 \\ 0 & 0 & 0.5 & 0 \end{bmatrix}$,

$A^4 = (1.1)I$.

The population will increase by 10% every four years.

**8.** (a) $9.96N_0$. (b) $0.47N_0$. (c) $15.18N_0$.

**9.** Let $\mathbf{v}_1 = (160, 5, 1)$. If $\mathbf{x}_0 = \alpha\mathbf{v}_1$, then $A\mathbf{x}_0 = A\alpha\mathbf{v}_1 = \alpha A\mathbf{v}_1 = \alpha\mathbf{v}_1 = \mathbf{x}_0$.

**10.** (a) Eigenvalues: 1, $-\frac{1}{4}$, $-\frac{3}{4}$. Corresponding eigenvectors: (16, 8, 1), (1, −2, 1), (9, −6, 1)

$$\mathbf{x}_k = \alpha(16, 8, 1) + (-\tfrac{1}{4})^k\beta(1, -2, 1) + (-\tfrac{3}{4})^k\gamma(9, -6, 1).$$

$$\mathbf{x}_\infty = \alpha(16, 8, 1), \quad N_\infty = 25\alpha,$$

where $\alpha = (a_0 + 2b_0 + 3c_0)/35$.

**15.** (a) Eigenvalues: 1, $-\frac{3}{8}$. Corresponding eigenvectors: (5, 6), (−1, 1).

$$\mathbf{x}_k = \alpha(5, 6) + \beta(-\tfrac{3}{8})^k(-1, 1),$$

where $\alpha = \frac{5}{11}$, $\beta = -\frac{8}{11}$. $\mathbf{x}_\infty = \frac{5}{11}(5, 6)$.

(c) Eigenvalues: 1, $\frac{1}{6}$. Corresponding eigenvectors: (1, 4), (−1, 1).

$$\mathbf{x}_k = \alpha(1, 4) + \beta(\tfrac{1}{6})^k(-1, 1),$$

where $\alpha = (a_0 + b_0)/5$, $\beta = (-4a_0 + b_0)/5$. $\mathbf{x}_\infty = \alpha(1, 4)$.

(e) Eigenvalues: 1, −1, −2.

Corresponding eigenvectors: $(1,1,0)$, $(0,2,1)$, $(0,1,1)$.

$$\mathbf{x}_k = \alpha(1,1,0) + \beta(-1)^k(0,2,1) + \gamma(-2)^k(0,1,1)$$

where $\alpha = a_0$, $\beta = b_0 - a_0 - c_0$, $\gamma = a_0 - b_0 + 2c_0$.

**16.** $(a_k, b_k) = (a_0 + b_0)/3(1,2) + (-2a_0 + b_0)/3(1/2)^k(-1,1)$. $(a_\infty, b_\infty) = (a_0 + b_0)/3(1,2)$.

**18.** $(a_k, b_k) = \frac{1}{3}(1,2) + (0.7/3)(0.7)^k(-1,1)$. $(a_\infty, b_\infty) = \frac{1}{3}(1,2)$ and is independent of $a_0$ and $b_0$.

**20.** (a) $\mathbf{x}_{k+1} = A\mathbf{x}_k$, where

$$A = \begin{bmatrix} 1 & \frac{1}{4} & 0 \\ 0 & \frac{1}{2} & 0 \\ 0 & \frac{1}{4} & 1 \end{bmatrix}.$$

**23.** Since the columns of a transition matrix add to 1, the columns of $A - I$ add to 0. Thus $A - I$ is singular and that means that 1 is an eigenvalue.

## SECTION 8.5

**1.** (a) $\det(A - \lambda I) = (2-\lambda)^3(1-\lambda)$. $E(2)$ has basis $(1,0,0,0)$, $(0,0,1,0)$, hence $\dim E(2) = 2$.
(b) $\lambda = 1$.

**3.** $\det(A - \lambda I) = (4-\lambda)^2(-1-\lambda)^3$. $E(4)$ has basis $(1,0,0,0,0)$, hence $\dim E(4) = 1$. $E(-1)$ has basis $(0,0,1,0,0)$, hence $\dim E(-1) = 1$.

**5.** (a) $\det(A - \lambda I) = (1-\lambda)^3$.
(b) Basis for $E(1)$: $(-1,1,1)$.
(c) No.

**8.** (a) Let $\lambda_1, \lambda_2, \ldots, \lambda_n$ be eigenvalues associated with the eigenvectors $\mathbf{v}_1, \mathbf{v}_2, \ldots, \mathbf{v}_n$, respectively. If $\mathbf{v} = \mathbf{v}_1 + \mathbf{v}_2 + \cdots + \mathbf{v}_n$ and $A\mathbf{v} = \mu\mathbf{v}$, then

$$\begin{aligned} A\mathbf{v} &= \lambda_1\mathbf{v}_1 + \lambda_2\mathbf{v}_2 + \cdots + \lambda_n\mathbf{v}_n \\ &= \mu\mathbf{v} = \mu\mathbf{v}_1 + \mu\mathbf{v}_2 + \cdots + \mu\mathbf{v}_n. \end{aligned}$$

Since the $\mathbf{v}_i$'s are independent, $\mu = \lambda_1 = \lambda_2 = \cdots = \lambda_n$.

**10.** Since $\lambda = \pm 1$ are eigenvalues, the matrix has two distinct eigenvalues and hence at least two independent eigenvectors.

## SECTION 8.6

**1.** (a) $\dfrac{1}{\sqrt{2}}\begin{bmatrix} 1 & 1 \\ -1 & 1 \end{bmatrix}.$

(d) $\dfrac{1}{\sqrt{6}}\begin{bmatrix} -\sqrt{3} & -1 & \sqrt{2} \\ 0 & 2 & \sqrt{2} \\ \sqrt{3} & -1 & \sqrt{2} \end{bmatrix}.$

(f) $\dfrac{1}{\sqrt{2}}\begin{bmatrix} 0 & -1 & 1 \\ 0 & 1 & 1 \\ \sqrt{2} & 0 & 0 \end{bmatrix}.$

(l) $\dfrac{1}{\sqrt{2}}\begin{bmatrix} 0 & -1 & 0 & 1 \\ -1 & 0 & 1 & 0 \\ 0 & 1 & 0 & 1 \\ 1 & 0 & 1 & 0 \end{bmatrix}.$

**3.** (b) $\dfrac{1}{9}\begin{bmatrix} 13 & 4 & -2 \\ 4 & 13 & -2 \\ -2 & -2 & 10 \end{bmatrix}.$

**4.** $A^T = (M\Lambda M^T)^T = M^{TT}\Lambda^T M^T = M\Lambda M^T = A.$

**6.** $A = M(\lambda I)M^T = \lambda MM^T = \lambda I$, if $M^{-1} = M^T$.

**9.** If $A$ is given by (2), then

$$\begin{aligned} A^k &= A^{k-1}A = A^{k-1}(\lambda_1\mathbf{v}_1\mathbf{v}_1^T + \lambda_2\mathbf{v}_2\mathbf{v}_2^T + \cdots + \lambda_n\mathbf{v}_n\mathbf{v}_n^T) \\ &= \lambda_1A^{k-1}\mathbf{v}_1\mathbf{v}_1^T + \lambda_2A^{k-1}\mathbf{v}_2\mathbf{v}_2^T + \cdots + \lambda_nA^{k-1}\mathbf{v}_n\mathbf{v}_n^T \\ &= \lambda_1^k\mathbf{v}_1\mathbf{v}_1^T + \lambda_2^k\mathbf{v}_2\mathbf{v}_2^T + \cdots + \lambda_n^k\mathbf{v}_n\mathbf{v}_n^T. \end{aligned}$$

**11.** (a) If $A^T = A$, then $B^T = (M^TAM)^T = M^TA^TM^{TT} = M^TAM = B$. Similarly, if $B$ is symmetric, so is $A$.
(b) If $A^T = A^{-1}$ and $B = M^TAM$, then

$$B^T = (M^TAM)^T = M^TA^TM^{TT} = M^{-1}A^{-1}(M^T)^{-1} = B^{-1}.$$

Similarly, if $B$ is orthogonal, so is $A$.

**13.** (a) Suppose that $A = A^T$. If $A$ is orthogonal, then $A^{-1} = A^T = A$, so that $A^2 = I$. If $\lambda$ is an eigenvalue of $A$, then there is a nonzero vector $\mathbf{v}$ such that $A\mathbf{v} = \lambda\mathbf{v}$. Thus $\mathbf{v} = A^2\mathbf{v} = \lambda^2\mathbf{v}$. Since $\mathbf{v} \neq \mathbf{0}$, $\lambda^2 = 1$ so that $\lambda = \pm 1$. Now suppose that the eigenvalues of $A$ are $\pm 1$. Since $A$ is symmetric, $A$ is orthogonally similar to a diagonal matrix $\Lambda$ whose diagonal entries are $\pm 1$. Thus $\Lambda^{-1} = \Lambda$ and it follows that $A^{-1} = A$.
(b) Let $\Lambda$ be any diagonal matrix having both 1 and $-1$ on its diagonal, let $M$ be any orthogonal matrix, and form the product $M^T \Lambda M$.

## SECTION 8.7

**1.** (a) $A = \begin{bmatrix} 3 & -\sqrt{2} \\ -\sqrt{2} & 2 \end{bmatrix}$.
Eigenvalues: 1, 4.
Eigenvectors: $1/\sqrt{3}(1, \sqrt{2})$,
$1/\sqrt{3}(-\sqrt{2}, 1)$.
(i) $Q(\mathbf{x}') = x'^2 + 4y'^2$, where

$$\mathbf{x}' = M^T\mathbf{x}$$

and

$$M = \frac{1}{\sqrt{3}}\begin{bmatrix} 1 & -\sqrt{2} \\ \sqrt{2} & 1 \end{bmatrix}.$$

(ii) Yes: both eigenvalues are positive.
(iii) Ellipse.

(c) $A = \begin{bmatrix} 0 & \frac{1}{2} \\ \frac{1}{2} & 0 \end{bmatrix}$.
Eigenvalues: $\frac{1}{2}$, $-\frac{1}{2}$.
Eigenvectors: $1/\sqrt{2}(1, 1)$,
$1/\sqrt{2}(-1, 1)$.
(i) $Q(\mathbf{x}') = 1/2(x'^2 - y'^2)$, where

$$\mathbf{x}' = M^T\mathbf{x}$$

and

$$M = \frac{1}{\sqrt{2}}\begin{bmatrix} 1 & -1 \\ 1 & 1 \end{bmatrix}.$$

(ii) No: one eigenvalue is negative.
(iii) Hyperbola.

(e) $A = \begin{bmatrix} 1 & 4 \\ 4 & 7 \end{bmatrix}$.
Eigenvalues: 9, $-1$.
Eigenvectors: $1/\sqrt{5}(1, 2)$,
$1/\sqrt{5}(-2, 1)$.
(i) $Q(\mathbf{x}') = 9x'^2 - y'^2$, where

$$\mathbf{x}' = M^T\mathbf{x}$$

and

$$M = \frac{1}{\sqrt{5}}\begin{bmatrix} 1 & -2 \\ 2 & 1 \end{bmatrix}.$$

(ii) No: one eigenvalue is negative.
(iii) Hyperbola.

(g) $A = \begin{bmatrix} 5 & -3 \\ -3 & 5 \end{bmatrix}$.
Eigenvalues: 2, 8.
Eigenvectors: $1/\sqrt{2}(1, 1)$,
$1/\sqrt{2}(-1, 1)$.
(i) $Q(\mathbf{x}') = 2x'^2 + 8y'^2$, where

$$\mathbf{x}' = M^T\mathbf{x}$$

and

$$M = \frac{1}{\sqrt{2}}\begin{bmatrix} 1 & -1 \\ 1 & 1 \end{bmatrix}.$$

(ii) Yes: both eigenvalues are positive.
(iii) Ellipse.

(i) $A = \begin{bmatrix} 4 & 1 \\ 1 & 4 \end{bmatrix}$.
Eigenvalues: 3, 5.

Eigenvectors: $1/\sqrt{2}(1, -1)$,
$1/\sqrt{2}(1, 1)$.

(i) $Q(\mathbf{x}') = 3x'^2 + 5y'^2$, where

$$\mathbf{x}' = M^T\mathbf{x}$$

and

$$M = \frac{1}{\sqrt{2}}\begin{bmatrix} 1 & 1 \\ -1 & 1 \end{bmatrix}.$$

(ii) Yes: both eigenvalues are positive.

(iii) Ellipse.

(k) $A = \begin{bmatrix} 3 & \sqrt{2} \\ \sqrt{2} & 4 \end{bmatrix}$.

Eigenvalues: 2, 5.

Eigenvectors: $1/\sqrt{3}(\sqrt{2}, -1)$,
$1/\sqrt{3}(1, \sqrt{2})$.

(i) $Q(\mathbf{x}') = 2x'^2 + 5y'^2$, where

$$\mathbf{x}' = M^T\mathbf{x}$$

and

$$M = \frac{1}{\sqrt{3}}\begin{bmatrix} \sqrt{2} & 1 \\ -1 & \sqrt{2} \end{bmatrix}.$$

(ii) Yes: both eigenvalues are positive.

(iii) Ellipse.

**2.** (a) $A = \begin{bmatrix} 3 & 1 & 1 \\ 1 & 0 & 2 \\ 1 & 2 & 0 \end{bmatrix}$.

Eigenvalues: $-2, 1, 4$.

Eigenvectors: $1/\sqrt{2}(0, -1, 1)$,
$1/\sqrt{3}(-1, 1, 1)$,
$1/\sqrt{6}(2, 1, 1)$.

(i) $Q(\mathbf{x}') = -2x'^2 + y'^2 + 4z'^2$.

(ii) No: one eigenvalue is negative.

(iii) Hyperboloid of one sheet.

(c) $A = \begin{bmatrix} 2 & 1 & 1 \\ 1 & 2 & 1 \\ 1 & 1 & 2 \end{bmatrix}$.

Eigenvalues: 1, 1, 4.

Eigenvectors: $1/\sqrt{2}(-1, 1, 0)$,
$1/\sqrt{2}(-1, 0, 1)$,
$1/\sqrt{3}(1, 1, 1)$.

(i) $Q(\mathbf{x}') = x'^2 + y'^2 + 4z'^2$.

(ii) Yes: all eigenvalues are positive.

(iii) Ellipsoid.

(e) $A = \begin{bmatrix} 5 & 2 & 2 \\ 2 & 5 & 2 \\ 2 & 2 & 5 \end{bmatrix}$.

Eigenvalues: 3, 3, 9.

Eigenvectors: $1/\sqrt{2}(-1, 1, 0)$,
$1/\sqrt{2}(-1, 0, 1)$,
$1/\sqrt{3}(1, 1, 1)$.

(i) $Q(\mathbf{x}') = 3x'^2 + 3y'^2 + 9z'^2$.

(ii) Yes: all eigenvalues are positive.

(iii) Ellipsoid.

(g) $A = \begin{bmatrix} 2 & -1 & 0 \\ -1 & 2 & -1 \\ 0 & -1 & 2 \end{bmatrix}$.

Eigenvalues: $2, 2 + \sqrt{2}, 2 - \sqrt{2}$.

Eigenvectors: $1/\sqrt{2}(-1, 0, 1)$,
$1/2(1, -\sqrt{2}, 1)$,
$1/2(1, \sqrt{2}, 1)$.

(i) $Q(\mathbf{x}') = 2x'^2 + (2 + \sqrt{2})y'^2 + (2 - \sqrt{2})z'^2$.

(ii) Yes: all eigenvalues are positive.

(iii) Ellipsoid.

**3.** (a) $A = \begin{bmatrix} 16 & 12 \\ 12 & 9 \end{bmatrix}$.

Eigenvalues: 0, 25.

Eigenvectors: $\frac{1}{5}(-3, 4)$, $\frac{1}{5}(4, 3)$.

Graph is the parabola

$$x' + (y' - 5)^2 = 25.$$

(c) $A = \begin{bmatrix} 4 & 2 \\ 2 & 1 \end{bmatrix}$.

Eigenvalues: 0, 5.

Eigenvectors: $1/\sqrt{5}(-1, 2)$,
$1/\sqrt{5}(2, 1)$.

Graph is the parabola

$$\frac{1}{5\sqrt{5}}x' + \left[y' - \frac{1}{5\sqrt{5}}\right]^2 = \frac{1}{125}.$$

(e) $A = I$.
Eigenvalues: 1, 1, 1.
Eigenvectors: $(1,0,0)$, $(0,1,0)$, $(0,0,1)$.
Graph is the sphere

$$(x-1)^2+(y-2)^2+(z-3)^2=26.$$

(g) $A = \begin{bmatrix} 3 & -1 & 0 \\ -1 & 3 & 0 \\ 0 & 0 & -4 \end{bmatrix}$.
Eigenvalues: 2, 4, $-4$.
Eigenvectors: $\frac{1}{\sqrt{2}}(1,1,0)$, $\frac{1}{\sqrt{2}}(-1,1,0)$, $(0,0,1)$.
Graph is the hyperbolic paraboloid

$$2x'^2+4y'^2-4(z'-1)^2=4.$$

**5.** Clearly, $Q(\mathbf{x}) \geq 0$ for all $\mathbf{x}$. If $Q(\mathbf{x})=0$, then

$$x_1^2=0,\ (x_1+x_2)^2=0,\ (x_1+x_2+x_3)^2=0.$$

Hence $\mathbf{x}=\mathbf{0}$.

**7.** Let $A$ be a symmetric matrix. If $A$ is positive definite, then all of the eigenvalues of $A$ are positive (Theorem 2). In particular, 0 is not an eigenvalue of $A$ and hence $A$ is nonsingular. Since the eigenvalues of $A^{-1}$ are the reciprocals of the eigenvalues of $A$, they are all positive. Therefore, $A^{-1}$ is positive definite (Theorem 2). $A^{-1}$ is symmetric because $(A^{-1})^T=(A^T)^{-1}=A^{-1}$. The converse follows easily from Theorem 2.

**10.** (a) Yes: zero is not an eigenvalue of $A$.
(b) Yes: the eigenvalues of $A^2$ are the squares of those of $A$ and hence are positive.

**12.** Since $A$ is nonsingular, $A\mathbf{x} \neq \mathbf{0}$ for all $\mathbf{x} \neq \mathbf{0}$. Thus $\mathbf{x}^T(A^TA)\mathbf{x} = \langle A\mathbf{x}, A\mathbf{x}\rangle > 0$ for all $\mathbf{x} \neq \mathbf{0}$.

**16.** (a) Consider $\mathbf{e}_i^T A\mathbf{e}_i$. Example:

$$A=\begin{bmatrix} 0 & 0 \\ 0 & \pm 1 \end{bmatrix}.$$

## SECTION 8.8

**1.** (a) Matrix $\begin{bmatrix} 1 & 0 \\ 0 & 6 \end{bmatrix}$, basis $(-2,1)$, $(1,2)$.

(c) Matrix $\begin{bmatrix} 1 & 0 & 0 \\ 0 & 1 & 0 \\ 0 & 0 & 7 \end{bmatrix}$, basis $(-1,1,0)$, $(-1,0,1)$, $(1,2,3)$.

**2.** (a)
$$\begin{aligned} T(A+B) &= (A+B)+(A+B)^T \\ &= A+A^T+B+B^T \\ &= T(A)+T(B). \\ T(\alpha A) &= (\alpha A)+(\alpha A)^T \\ &= \alpha(A+A^T) \\ &= \alpha T(A). \end{aligned}$$

(b) Eigenvalues of $T$ are: 0, 2.
Basis for the eigenspace $E(0)$:

$$\begin{bmatrix} 0 & 1 \\ -1 & 0 \end{bmatrix}.$$

Basis for the eigenspace $E(2)$:

$$\begin{bmatrix} 1 & 0 \\ 0 & 0 \end{bmatrix},\quad \begin{bmatrix} 0 & 0 \\ 0 & 1 \end{bmatrix},\quad \begin{bmatrix} 0 & 1 \\ 1 & 0 \end{bmatrix}.$$

$T$ can be represented by a diagonal matrix whose diagonal entries are 0, 2, 2, 2.

**3.** Eigenvalues: 1, 1, 5. Eigenvectors of the matrix: For $\lambda=1$, $(-2,1,0)$ and $(-1,0,1)$; for $\lambda=5$, $(1,1,1)$. These vectors are the $v$-coordinate vectors of the eigenvectors of $T$. $T$ can be represented by a diagonal matrix whose diagonal entries are 1, 1, 5.

**5.** The matrix of $T$ with respect to the basis $1$, $x$, $x^2$ is the matrix in Example 2 of Section 8.3. Since that matrix is diagonalizable, $T$ can be represented by a diagonal matrix. The matrix of $T$ with respect to the basis $1, x+x^2, 5-x+x^2$ is the diagonal matrix whose diagonal entries are 2, 2, 4.

**6.** No. The matrix of $T$ with respect to the basis $1$, $x$, $x^2$ is the matrix in Example 3 of Section 8.3. Since this matrix is not diagonalizable, $T$ cannot be represented by a diagonal matrix.

## SECTION 8.9

**1.** $dx/dt = -ax + by + cz$
$dy/dt = dx - ey + fz$
$dz/dt = gx + hy - iz$ where $a, b, c, d, e, f, g, h, i \geq 0$.

## SECTION 8.10

**3.** (a) $d\mathbf{x}/dt = A\mathbf{x}$, where $\mathbf{x} = \begin{bmatrix} y \\ y' \end{bmatrix}$ and

$$A = \begin{bmatrix} 0 & 1 \\ 1 & 0 \end{bmatrix}.$$

(c) $d\mathbf{x}/dt = A\mathbf{x}$, where $\mathbf{x} = \begin{bmatrix} y \\ y' \\ y'' \end{bmatrix}$ and

$$A = \begin{bmatrix} 0 & 1 & 0 \\ 0 & 0 & 1 \\ 0 & 2 & -1 \end{bmatrix}.$$

(e) $d\mathbf{x}/dt = A\mathbf{x}$, where $\mathbf{x} = \begin{bmatrix} y \\ y' \\ y'' \\ y''' \end{bmatrix}$ and

$$A = \begin{bmatrix} 0 & 1 & 0 & 0 \\ 0 & 0 & 1 & 0 \\ 0 & 0 & 0 & 1 \\ -2 & 0 & 1 & 2 \end{bmatrix}.$$

## SECTION 8.11

**2.** In parts (a), (b), and (c) the given solutions are a basis for the solution space.

## SUPPLEMENTARY PROBLEMS

**1.** (b) $\operatorname{Det}(A - \lambda I) = -(\lambda + 1)^2(\lambda - 2)$.
(c) Eigenvalues: $-1, 2$; algebraic multiplicities: 2, 1.
(d) Geometric multiplicities: 2, 1.
(e) $\Lambda = \begin{bmatrix} -1 & & \\ & -1 & \\ & & 2 \end{bmatrix}$,

$$M = \begin{bmatrix} 1 & 0 & 1 \\ 2 & 0 & 1 \\ 0 & 1 & 1 \end{bmatrix}.$$

**2.** (a) (i) Eigenvalues: $-3, 3, 3$. Basis for $E(-3)$: $(-1, 2, 0)$. Basis for $E(3)$: $(1, 1, 0)$.
(ii) Not diagonalizable.

(c) (i) Eigenvalues: $-2, -2, 4$. Basis for $E(-2)$: $(-1, 1, 0)$, $(-1, 0, 1)$. Basis for $E(4)$: $(1, 1, 1)$.
(ii) $\Lambda = \begin{bmatrix} -2 & & \\ & -2 & \\ & & 4 \end{bmatrix}$,

$$M = \begin{bmatrix} -1 & -1 & 1 \\ 1 & 0 & 1 \\ 0 & 1 & 1 \end{bmatrix}.$$

(e) (i) Eigenvalues: $-2, 0, 2$. Basis for $E(-2)$: $(-1, 0, 1)$. Basis for $E(0)$: $(0, 1, 0)$. Basis for $E(2)$: $(1, 0, 1)$.
(ii) $\Lambda = \begin{bmatrix} -2 & & \\ & 0 & \\ & & 2 \end{bmatrix}$,

$$M = \begin{bmatrix} -1 & 0 & 1 \\ 0 & 1 & 0 \\ 1 & 0 & 1 \end{bmatrix}.$$

(g) (i) Eigenvalues: 1, 1, 2. Basis for

$E(1)$: $(0,1,0)$. Basis for $E(2)$: $(1,0,0)$.

(ii) Not diagonalizable.

(i) (i) Eigenvalues: $\lambda = 1,\ 2,\ 3$. Basis for $E(1)$: $(1,2,1)$. Basis for $E(2)$: $(1,1,0)$. Basis for $E(3)$: $(1,2,2)$.

(ii) $\Lambda = \begin{bmatrix} 1 & & \\ & 2 & \\ & & 3 \end{bmatrix}$,

$$M = \begin{bmatrix} 1 & 1 & 1 \\ 2 & 1 & 2 \\ 1 & 0 & 2 \end{bmatrix}.$$

(k) (i) Eigenvalues: $\lambda = 2, 2, 2$. Basis for $E(2)$: $(1,2,0)$, $(0,0,1)$.

(ii) Not diagonalizable.

**3.** Yes. The columns of $M$ are $\mathbf{e}_1, \mathbf{e}_4, \mathbf{e}_3, \mathbf{e}_2$.

**8.** (a) Eigenvalues: $-2, -2, 4$. Orthonormal basis for $E(-2)$: $1/\sqrt{2}(-1,1,0)$, $1/\sqrt{6}(1,1,-2)$. Orthonormal basis for $E(4)$: $1/\sqrt{3}(1,1,1)$.

(b) $\Lambda = \begin{bmatrix} -2 & & \\ & -2 & \\ & & 4 \end{bmatrix}$,

$$M = \begin{bmatrix} -1/\sqrt{2} & 1/\sqrt{6} & 1/\sqrt{3} \\ 1/\sqrt{2} & 1/\sqrt{6} & 1/\sqrt{3} \\ 0 & -2/\sqrt{6} & 1/\sqrt{3} \end{bmatrix}.$$

# CHAPTER 9

## SECTION 9.1

**1.**

| | Multiplications | Additions |
|---|---|---|
| (a) | $n$ | $n-1$ |
| (b) | $n^2$ | $n^2 - n$ |
| (c) | $n^3$ | $n^3 - n^2$ |
| (d) | $n^3$ | $n^3 - n^2$ |
| (e) | $2n^3$ | $2n^3 - 2n^2$ |
| (f) | $2n^3$ | $2n^3 - 2n^2$ |
| (g) | $3n^3$ | $3n^3 - 3n^2$ |

**4.** (a) $(n^3 - n)/3 + kn^2$.
(b) $(n^3 - n^2)/2 + kn^2$.

**6.** (a) $n^3 + kn^2$. (b) Always.

## SECTION 9.2

**1.**

| | Mantissa | Exponent |
|---|---|---|
| (a) | 0.7876 | 0 |
| (b) | 0.7876 | 3 |
| (c) | 0.6250 | 0 |
| (d) | 0.3333 | 0 |
| (e) | 0.1428 | 0 |
| (f) | 0.1428 | $-1$ |
| (g) | 0.7000 | 4 |
| (h) | 0.1428 | $-3$ |

**2.** 0.0005.

**3.** (a) For each $\mathbf{x}$, $A\mathbf{x} = (0.7500 \cdot 10^7, 0.2496 \cdot 10^7)$.
(b) $0.1426 \cdot 10^8$.

## SECTION 9.3

**1.** (a) $x_1 = 0.6619$, $x_2 = 1.523$, $x_3 = 4.286$.
(b) $x_1 = -4.182$, $x_2 = 4.658$, $x_3 = 0.6065$.
(c) $x_1 = 0.5121$, $x_2 = 4.271$, $x_3 = 3.774$.

### SECTION 9.4

1. (a) 2, (1, 0, 1).
   (b) 4.14, (0.64, 1, 0.86).
   (c) 4, (0.33, 1, 0.67).
   (d) The smaller the value of $\lambda_1/\lambda_2$ the more rapid will be the convergence of the power method.

3. (a) For $A$: 50, (1, −0.286); 1, (−0.333, 1).
   For $B$: 3, (1, −0.286); 2, (−0.333, 1).

# CHAPTER 10

### SECTION 10.1

1. (a) $22 - 4i$.
   (b) $3 - 3i$.
   (c) $\frac{19}{25} - \frac{8}{25}i$.
   (d) 2.
   (e) $2i$.
   (f) $(-2)^n$.

2. (a) −3.
   (b) 4.
   (c) 5.
   (d) 1.
   (e) −3.
   (f) −4.
   (g) 5.
   (h) 1.

3. (a) $\sqrt{2}/2 + (\sqrt{2}/2)i$ and $-\sqrt{2}/2 - (\sqrt{2}/2)i$.
   (b) $-\sqrt{2}/2 + (\sqrt{2}/2)i$ and $\sqrt{2}/2 - (\sqrt{2}/2)i$.

5. (a) $a/(a^2 + b^2) - (b/(a^2 + b^2))i = \bar{z}/|z|$.
   (b) Yes.

13. $(i, 1) = i(1, -i)$.

15. 1.

16. (a) $\mathbf{x} = (i, 2)$. (b) $\mathbf{x} = (2 - i, 3 + i)$.
    (c) $\mathbf{x} = (3, -2)$. (d) $\mathbf{x} = (1 - i, i, 1 + i)$.

17. (a) $\begin{bmatrix} (-1 - 2i)/3 & i/3 \\ (1 + i)/3 & i/3 \end{bmatrix}$.

    (b) $\begin{bmatrix} 1 + i & (-1 - 2i)/5 \\ -i & (1 + 2i)/5 \end{bmatrix}$.

    (c) $\begin{bmatrix} 1 & 1 & 0 \\ 1 + i & i & 0 \\ 1 - i & -i & 1 \end{bmatrix}$.

    (d) $\begin{bmatrix} 1 + i & i & -i \\ 0 & 1 + i & -i \\ -1 & -1 & 1 \end{bmatrix}$.

### SECTION 10.2

9. (a) $HH^* = HH = H^*H$.

### SECTION 10.3

1. (a) The eigenvalues are the roots of $\lambda^3 - 1.1 = 0$. They are $\sqrt[3]{1.1}$ and $\sqrt[3]{1.1}(-1 \pm \sqrt{3}\,i)/2$.
   (b) The absolute value of each eigenvalue equals $\sqrt[3]{1.1}$.
   (c) The size of the population will increase by 10% every three years. After three years, the distribution of the population will return to its original distribution.

3. (a) Eigenvalues: $-i$, $1 - i$.
   Corresponding eigenvectors: $(i, 1)$, $(1 - i, -i)$.
   (b) Eigenvalues: $3i$, $1 - 2i$.
   Corresponding eigenvectors: $(1, i)$, $(2i, -1)$.

**5.** (a) Eigenvalues: $2\sqrt{2}, 4\sqrt{2}$.
Eigenvectors: $\frac{1}{2}(1-i, \sqrt{2})$, $\frac{1}{2}(1-i, -\sqrt{2})$.

(c) Eigenvalues: $3, -1$.
Eigenvectors: $1/\sqrt{2}(i, -1)$, $1/\sqrt{2}(-1, i)$.

**6.** (a) $A = U\Lambda U^* = \begin{bmatrix} 7 & -4i \\ 4i & 1 \end{bmatrix}$, where

$$U = \frac{1}{\sqrt{5}}\begin{bmatrix} -2i & 1 \\ 1 & -2i \end{bmatrix}$$

and $\Lambda = \begin{bmatrix} 9 & \\ & -1 \end{bmatrix}$.

**7.** (a) $U = \frac{1}{\sqrt{2}}\begin{bmatrix} i & 1 \\ 1 & i \end{bmatrix}$.

(c) $U = \frac{1}{\sqrt{2}}\begin{bmatrix} -i & 1 \\ 1 & -i \end{bmatrix}$.

(e) $U = \frac{1}{\sqrt{2}}\begin{bmatrix} \sqrt{2} & 0 & 0 \\ 0 & -i & 1 \\ 0 & 1 & -i \end{bmatrix}$.

(g) $U_1 = \frac{1}{3}\begin{bmatrix} 1 & 2 & 2 \\ 2 & 1 & -2 \\ 2 & -2 & 1 \end{bmatrix}$,

$U_2 = \frac{1}{\sqrt{2}}\begin{bmatrix} \sqrt{2} & 0 & 0 \\ 0 & -1 & 1 \\ 0 & 1 & 1 \end{bmatrix}$,

$U = U_1 U_2$.

(h) $U_1 = \frac{1}{\sqrt{2}}\begin{bmatrix} 1 & 0 & 1 \\ 0 & \sqrt{2} & 0 \\ 1 & 0 & -1 \end{bmatrix}$,

$U_2 = \frac{1}{\sqrt{5}}\begin{bmatrix} \sqrt{5} & 0 & 0 \\ 0 & -2 & 1 \\ 0 & 1 & 2 \end{bmatrix}$,

$U = U_1 U_2$.

**11.** (a) If $A^*A\mathbf{v} = \mathbf{0}$, then $\langle \mathbf{v}, A^*A\mathbf{v}\rangle = \langle A\mathbf{v}, A\mathbf{v}\rangle = 0$. Thus $A\mathbf{v} = \mathbf{0}$.

**13.** (a) Show that $\langle N^*\mathbf{v}, N^*\mathbf{v}\rangle = 0$.

**14.** (a) Suppose that $U$ is unitary (i.e., $U^*U = I$). If $N$ is normal, then

$$\begin{aligned}(U^*NU)(U^*NU)^* &= U^*NUU^*N^*U \\ &= U^*NN^*U \\ &= U^*N^*NU \\ &= U^*N^*UU^*NU \\ &= (U^*NU)^*(U^*NU).\end{aligned}$$

Conversely, suppose that $U^*NU$ is normal. Then

$$\begin{aligned}(U^*NU)(U^*NU)^* &= U^*NUU^*N^*U \\ &= U^*NN^*U, \\ (U^*NU)^*(U^*NU) &= U^*N^*UU^*NU \\ &= U^*N^*NU.\end{aligned}$$

Thus $N$ is normal.

# Index

## A



## B



*Numbers in parentheses refer to problems.

## C

## G



## H



## I



## J

## K



## L



## M

## N



## O



## P

## Q



## R



## S

## W



## Z